Carbon Nanotubes

Carbon Nanotubes

Edited by

Jose Mauricio Marulanda

Carbon Nanotubes,
Edited by Jose Mauricio Marulanda

Republished by InTech

Janeza Trdine 9, 51000 Rijeka, Croatia

Edition 2014

Carbon Nanotubes,
Edited by Jose Mauricio Marulanda
p. cm.
ISBN 978-953-307-054-4

Preface

Carbon nanotubes (CNT) discovered in 1991 have been a subject of intensive research for a wide range of applications. These one-dimensional (1D) graphene sheets rolled into a tubular form have been the target of many researchers around the world due to their exceptional electrical and mechanical properties. This book provides a unique insight on the phenomena encountered in the physical electronics when operating with carbon nanotubes. This book also presents the reader with useful information on the fabrication and applications of these outstanding materials in current technology.

In the past decades, carbon nanotubes have undergone massive research from countless researchers around the world. Nevertheless, considering the success of Silicon, it has been difficult to appreciate the potential influence of carbon nanotubes in current technology. The main objective of this book is to give in-depth understanding of the physical electronics and mechanical properties of carbon nanotubes. In other words, this book is intended to provide a profound knowledge of these electrical and mechanical characteristics allowing the reader to further appreciate the potential and applications of carbon nanotubes.

Readers of this book should have a strong background on physical electronics and semiconductor device physics. Philanthropists and readers with strong background in quantum transport physics and semiconductors materials could definitely benefit from the results presented in the chapters of this book. Especially, those with research interests in the areas of nanoparticles and nanotechnology.

This book has been outlined as follows: A review on the literature and increasing research interests in the field of carbon nanotubes. Fabrication techniques followed by an analysis on the physical properties of carbon nanotubes. The device physics of implemented carbon nanotubes applications along with proposed models in an effort to describe their behavior in circuits and interconnects. And ultimately, the book pursues a significant amount of work in applications of carbon nanotubes in sensors, nanoparticles and nanostructures, and biotechnology.

A list of the chapters is given below. It is highly recommended for the reader to go over the following descriptions, as they provide excellent insights on the contents and results of the book's chapters.

Chapter 1. Carbon Nanotubes - A Scientometric study

The book starts with an outline on the time dependent and overall number of carbon nanotube's articles written in the past decades. This chapter reveals the still strongly increasing research activity in this field.

Chapter 2. Chemical Vapour Deposition of CNTs Using Structural Nanoparticle Catalysts

The role of the catalyst in the selective growth of single walled carbon nanotubes (SWNTs) by chemical vapour deposition (CVD) is studied. In addition, SWNT growth was shown to be possible from ceramic, noble metal, and semiconducting catalysts.

Chapter 3. Properties of Carbon Nanotubes under External Factors

The electronic and magnetic carbon nanotube properties under external factors are discussed. Radial deformation and external electric fields are investigated. The combined effects of radial deformation and external electric field are given in terms of band gap modulations and electronic band structure changes, due to the strength of the applied electric field and electric field orientation.

Chapter 4. Electronic Structure of Fluorinated Carbon Nanotubes

The study of the electronic structure of fluorinated single-walled (F-SWCNTs) and multiwalled (F-MWCNT) carbon nanotubes is conducted. High-energy resolution X-ray Carbon 1s and Fluor 1s absorption and photoelectron spectra of F-SWCNTs with different fluorine contents (35 and 40%) and F-MWCNTs with fluorine contents (10 to 55%) are measured.

Chapter 5. Microwave Dielectric Properties of Carbon Nanotube Composites

The complex permittivity of single-walled carbon nanotubes (SWCNTs), double-walled carbon nanotubes (DWCNTs), and multi-walled carbon nanotubes (MWCNTs) at microwave band is obtained experimentally using the impedance method.

Chapter 6. Environmental effects on photoluminescence of single-walled carbon nanotubes

The environmental effects on the optical properties of single-walled carbon nanotubes (SWNTs) are discussed. It is shown that a detail understanding of the environmental effects contributes into making, a spectroscopic technique under a standardized methodology, to evaluate the chirality abundance and the quality of SWNTs.

Chapter 7. Charge Transport in Carbon Nanotube Films and Fibers

Electrical and magneto transport properties of various types of carbon nanotubes arrays are reported. It is experimentally shown that the electrical and magneto transport properties of the investigated carbon nanotubes arrays are defined not only by the types of pristine carbon nanotubes (multi- or single-walled carbon nanotubes with different chiralities), but also by the quality of inter-tube coupling between separate nanotubes from which these assemblies consist of.

Chapter 8. Doped Carbon Nanotube Properties

The effect of impurities from the atoms of Aluminum (Al), Boron (B), Nitrogen (N) and Phosphorous (P) on models of carbon nanotubes with chiral vectors (10,0) and (6,5), consisting of 200 atoms is reviewed in terms of the performances of their energy-band structure. The magnification of number of atoms of carbon of the doped nanostructure feebly influences its energy distribution, results shows a diminution of the forbidden gap of the nanotube by approximately, 0.1 eV.

Chapter 9. Fundamental Physical Aspects of Carbon Nanotube Transistors

Two simulation methodologies are reliable for modeling carbon nanotube field effect transistors (CNT-FETs): The non equilibrium green's function (NEGF) transport equation self-consistently with Poisson's equation and the semi-classical Wentzel-Kramers-Brillouin (WKB) method for calculating the tunneling current through the Schottky barriers (SB) in the SB-CNT-FETs. Using the simulation methods discussed, trade-offs are studied for the different parameters of a particular CNT-FET design.

Chapter 10. Gate controlled Particle-Wave duality in a single walled carbon nanotube hole-transistor

After successfully fabricating and demonstrating a multi-functional quantum transistor using the particle nature and wave nature of holes in single-walled carbon nanotubes (SWNTs); a transistor that is able to operate in the wave nature mode as an RTT (resonant tunneling transistor) and in the particle nature mode as an SHT (single hole transistor) is accomplished and described.

Chapter 11. Numerical Modeling of the I-V Characteristics of Carbon Nanotube Field Effect Transistors

A numerically based model for the carbon nanotube potential and current transport of carbon nanotube field effect transistors (CNT-FETs) is presented. The tight binding model approximation for the electronic band structure of a carbon nanotube is used to find the carrier concentration. Poisson's equation is solved inside a CNT-FET and used to characterize the charge transport, resulting in the I-V characteristics. Results provide designers with useful mathematical relations describing the properties in conductivity of carbon nanotubes and their response in circuit applications.

Chapter 12. Compact Modeling of Carbon Nanotube Transistor and Interconnects

A detailed procedure for developing compact models for carbon nanotube transistors and interconnects is presented. The developed model does not use any iteration-based calculations; therefore, it is scalable with process and design parameters while keeping a high accuracy.

Chapter 13. Measurement of High-Frequency Characteristics of CNTFETs and Equivalent Circuit Model Analysis

A method for accurately measuring and modeling the high-frequency characteristics of CNTFETs is described. In addition, a high-density multiple-channel CNTFET structure with focus on accurately measuring the S-parameters is developed. Consequently, a cut-off frequency (fT) of 10.3 GHz and a maximum oscillation frequency (fmax) of 3.5 GHz are achieved.

Chapter 14. Carbon Nanotubes Interconnects for Nanoelectronics Circuits

The use of carbon nanotubes as possible innovative materials to fabricate high-speed interconnects at various hierarchical levels of the future nanoelectronics technology is studied. The electromagnetic model presented is used to describe the electrical propagation along the carbon nanotube interconnects and it is derived from a semi-classical solution of Boltzmann's transport equation.

Chapter 15. Interconnect Challenges and Carbon Nanotube as Interconnect in Nano VLSI Circuits

The study of the interconnect challenges and the behaviour of carbon nanotubes (CNTs) as interconnects in VLSI circuits is presented. In this chapter, the two main structures of carbon nanotubes, including carbon nanotube bundles and multi-walled carbon nanotubes (MWCNTs) are reviewed. Results show extremely good performance using parallel singlewalled carbon nanotubes (SWCNTs) in a bundle and using the parallel shells of MWCNTs.

Chapter 16. Affinity of CNT for metal - Its importance to application: Molecular dynamics approach

A molecular dynamics (MD) simulation approach for carbon nanotubes (CNT) analysis is explained. A simulation is conducted for carbon nanotubes using physical vapour deposition (PVD) and using coating; the simulations presented show good agreement with the experimental results from the fabricated CNTs.

Chapter 17. Carbon nanotube field emitters

The field emission from carbon nanotube field emitters in a diode configuration between a flat anode and a cathode is theoretically investigated. It is concluded that carbon nanotubes are excellent field emitters and have good potential in the field emission display market.

Chapter 18. Sensing mechanisms of carbon nanotube based NH3 gas detectors

Sensing mechanisms using carbon nanotube (CNT) based NH3 detection are studied. Selective Si3N4 passivation clearly shows that the Schottky barrier (SB) modulation at the CNT/metal contacts dominates the sensing performance at room temperature, and the sensor exhibits high sensitivity and good tuning ability under the appropriate gate voltages.

Chapter 19. Gas sensors based on decorated carbon nanotubes

It is shown that the study of gas sensors based on short multi-walled carbon nanotubes (MWCNTs), in a chemical resistor configuration (where nanotubes can be heated considerably by current), reveals two main effects. These effects occur under the exposure to simple low-pressure gases like Ar, N2 and O2, and show significantly different results for supported (over SiO2) and suspended nanotubes.

Chapter 20. Suspended Carbon Nanotubes: Applications in Physical Sensors and Actuators

An overview of the applications of suspended or freestanding carbon nanotubes (CNTs) as sensors and actuators is provided. A scheme for using 5-10 mm long, diffusively contacted single-walled carbon nanotubes (SWNTs) as thermal conductivity-based pressure sensors is presented. The actuator applications of suspended CNTs are also presented. The use of direct current plasma-enhanced chemical vapour deposition (dc PECVD) for forming isolated vertically oriented tubes with manufacturability techniques, where such tubes have applications in three-dimensional (3D) electronics is also provided.

Chapter 21. Study of Carbon NanoTube Field Effect Transistors for NEMS

Given the chemical and physical structure, low mass, and exceptional stiffness of carbon nanotubes (CNTs); CNTs are presented as ideal candidates for nanoelectromechanical systems (NEMS). It is shown that even with pessimistic assumptions, CNT-FET nanoelectronics can achieve significantly greater performance than Silicon technology.

Chapter 22. Solid phase (micro) extraction tools based on carbon nanotubes and related

Nanostructures Carbon nanoparticles (nanotubes, fullerenes, and nanocones) have been proved to positively contribute to the development of analytical methods further enhancing the properties of carbon nanotubes. It is shown that the present research on carbon nanoparticles is also capable of reducing the negative aspect related to the aggregation tendency of the carbon nanotubes.

Chapter 23. Broadband Terahertz Source based on Photomixing in Laser-Assisted Field Emission

with Clusters of Carbon Nanotubes It is demonstrated that photo mixing in laser-assisted field emission shows considerable promises as the means to generate microwave or THz radiation with an extremely large tuneable bandwidth. The purpose of this chapter is to obtain fundamental research and understanding of the process of quantum tunneling for applications in the THz range.

Chapter 24. Liquid crystal dispersions of carbon nanotubes: dielectric, electro-optical and structural peculiarities

A class of unique composites with fascinating electrical, optical, electro-optical, nonlinear optical, and structural properties is presented after the combination of carbon nanotubes and liquid crystals (LCs). Carbon nanotubes (CNTs) shunt double electric layers in the LC cells and in this way; they change essentially a spatial distribution of the electric field applied in these cells.

Chapter 25. Functionalization of carbon nanotubes with luminescent silicon nanocrystals upon nanosecond laser processing in liquid media

Scalable and low cost approaches introducing the silicon nanocrystals (Si-ncs) within carbon nanotubes (CNTs) and lipid nanotubes (LNTs) are discussed. It is shown that after opening the CNT ends and freeze-draying the LNTs; it is possible to introduce Si-ncs inside the emptied cavity of both type of nanotubes.

Chapter 26. Microstructured Optical Fibers filled with Carbon Nanotubes: Photonic Bandgap Modification and Sensing Applications

Research activities devoted to the integration of microstructured optical fibers (MOFs) with single-walled carbon nanotubes (SWCNTs) in order to develop new in-fiber active and passive optoelectronic devices are reviewed. On the basis of the retrieved results, a novel opto-chemical sensor for VOCs (Volatile Organic Compounds) detection is developed and its sensing capability is proved by the exposure to VOC traces in a proper test chamber.

Chapter 27. Carbon Nanotubes as a New Solid Phase Extraction Sorbent for Analysis of Environmental Pollutants

Several comparative studies present results that show carbon nanotubes (CNTs) to be more effective than other commonly used adsorbents such as the Carbon 18 bonded silica, activated Carbon, or macro porous resins. It is reported that CNTs may be re-used more than 100 times before the need for a proper cleaning and reconditioning.

Chapter 28. Hydrogen storage by carbon materials

An overview of the experimental work on systems dealing with storing hydrogenin is provided. The experimental work together with an outline of the theoretical studies that have been under taken to estimate the practical limits of the amount of hydrogen that can be stored per unit weight are also presented.

Chapter 29. Carbon Nanotube Supercapacitors

Carbon nanotubes (CNTs) are explored as a new type of electrode materials for supercapacitors. Due to their unique properties and large surface area, both randomly entangled and highly aligned CNTs are investigated for high capacitance and high rate composites applications.

Chapter 30. Carbon Nanotube Membrane Solar Sails A Challenge for Extremely Fast Space Flight

A better understanding and appreciation of the recent technological advancements in the sail material through a proper development of carbon nanotube membranes is presented. This allows the design of a solar sail featuring a cruising speed on the order of thousands the speed of light. This shows a speed much higher than the speed attainable by any other propulsion method.

Chapter 31. Carbon Nanotube-Nanoparticle Hybrid Structures

The electrostatic force directed assembly (ESFDA) technique is used to efficiently coat randomly dispersed and vertically-aligned carbon nanotubes (CNTs) with various aerosol nanoparticles (NPs). The final NP size distribution and areal density can be controlled through flow residence time/electric field or assembly time, respectively. For verticallyaligned CNTs, the electric field distribution near the CNT surface is computed using a simple model for the CNT as a cylinder with a hemispherical cap. It is shown that the field enhancement factor increases from the root to the tip of the CNT for the two cases studied.

Chapter 32. SWC of Al/CNT Two Phase Systems

Micro-nanosystem research on preparing materials and using shock wave compaction (SWC) for future developments is presented. Al-MWCNT (Aluminum multi-walled carbon nanotube) aggregate 2-phase composites of 2 and 5% volume fraction are fabricated and presented using a single tube shockwave consolidation process.

Chapter 33. Superconductivity in carbon nanotubes

Superconductivity is reported at 12 K in arrays of multi-walled carbon nanotube (MWNTs) and Boron-doped single-walled carbon nanotube (SWNTs) at 19 K. The possibility of quantum spin entangler and quantum bit utilizing a hybrid system of carbon nanotubes (CNTs)/superconductor is presented. This allows the feasibility of CNTs as onedimensional (1D) superconductors and the understanding of 1D electron correlation for quantum devices.

Chapter 34. Nucleic acid interaction and interfaces with single-walled carbon nanotubes

An experimental and theoretical study of the DNA-interface coupling is performed to achieve better understanding of the properties of carbon nanotubes based biosystems, as well as the study of the novel phenomena caused by the interaction of carbon nanotubes with biomolecules.

Chapter 35. DNA-Wrapped CNTs From Synthesis to Application

It is shown that working with the chemical functionality of carbon nanotubes (CNTs) is the most efficient method to increase their solubility. DNA-CNT hybrids can take advantage of the unique properties of CNTs and the outstanding recognition properties of DNA. The non-covalent interaction between CNTs and DNA wrapping facilitates separations applications and allows the functionality of CNTs without altering their unique properties.

Chapter 36. The Unlikely Surfactant: DNA as a Ligand for Single-Walled Carbon Nanotubes

Oligonucleotide-SWCNTs (single-walled carbon nanotubes) complexes are discussed. Results show an emphasis on the fact that DNA has surfactant properties that make it useful for solubilizing carbon nanotubes. It is shown that after sonication and separation processes, a stable, largely mono-disperse suspension can be obtained, such as density gradient ultracentrifugation.

Acknowledgments

I would like to thank the authors of the chapters of this book for their excellent contributions in their areas of expertise and for the efforts place in the publication of their work. I am certainly sure that the material published in this book will be of great help and genuinely appreciated by students, professors, and researchers all around the world.

Jose Mauricio Marulanda
Department of Electrical and Computer Engineering
Louisiana State University
Baton Rouge, LA 70803
USA

Contents

1

Carbon Nanotubes – A scientometric study

Werner Marx and Andreas Barth
Max Planck Institute for Solid State Research, D-70569 Stuttgart (Germany)
FIZ Karlsruhe, D-76344 Eggenstein-Leopoldshafen (Germany)

1. Introduction

In contrast to our previous study (Barth & Marx, 2008) dealing with a currently decreasing research field (high-temperature superconductors) we analyzed here a topic which has raised a strongly increasing interest among researchers: research activities around carbon nanotubes (CNTs or NTs). Carbon nanotubes (often named only nanotubes) are graphite sheets rolled up into cylinders with diameters of the order of a few nanometers and up to some millimeters in length with at least one end capped with a hemisphere of the fullerene structure. There are two main types of nanotubes: the single-walled nanotubes (SWCNTs or SWNTs) and the multi-walled nanotubes (MWCNTs or MWNTs), in particular the double-walled nanotubes (DWCNTs or DWNTs). MWCNTs consist of a single sheet of graphite rolled in around itself (like a rolled up newspaper) or consist of multiple layers of graphite arranged in concentric cylinders (like a Russian Doll).

Nanotubes exhibit some remarkable properties: They feature extraordinary strength, show efficient conductivity of heat, and unique electrical properties (metallic conductivity and semiconductivity). These properties make them potentially useful in a wide range of applications like in materials science, electronics, and nanotechnology. The one-atom thick single graphite layers building up the nanotube cylinders are named graphene, the newest member of this structural family. This species was presumed not to exist in the free state before it was discovered in the year 2004.

The large number of articles with respect to nanotubes has brought about that scientists being active in this research field have increasingly problems to overview their discipline. On the other hand, modern information systems offer databases and analysis tools providing remedy. However, due to lack of access and experience, many scientists do not take advantage of them. In this analysis we demonstrate the potential of such tools with respect to different kinds of meta-information. The data presented here are not expected to reveal surprising insights for experts working in this research field. However, they provide a quantification of (1) the productivity of the active players and (2) of the impact of their works. Moreover, the data could also be interesting for scientists working in neighboring research fields.

2. Methodology and Information Sources

The data presented in this study are based on the Science Citation Index (SCI) including the Conference Proceedings Citation Index, Science (CPCI-S) under the Web of Science (WoS). The SCI under the WoS stretches back to 1900 and the CPCI-S covers conference proceedings published since 1992. The WoS is accessible under the Web of Knowledge (WoK), the search platform provided by Thomson Reuters (Thomson Reuters, 2009), the former Institute for Scientific Information (ISI). The research field analyzed here stretches throughout most natural sciences disciplines being covered by the multi-disciplinary SCI and CPCI-S. However, the WoS source journals selected by the Thomson Reuters staff as contributing to the progress of science do not cover all publications being relevant here, in particular with respect to application and technology.

Therefore, the literature file of the Chemical Abstracts Service (CAS), a division of the American Chemical Society (ACS), has been consulted as an alternative information source. The CAS literature file is available via the online service STN International (STN International, 2009). The literature file CAplus is seen as the most extensive source of substance related publications (either articles or patents) in the fields of chemistry, materials science, and physics. Specific functions of the STN search system for carrying out statistical investigations have made it possible to perform extensive scientometric studies. Additional information is accessible via STN AnaVist, an analysis tool developed by STN International. However, the competent use of such databases and search systems requires some experience and awareness of the possibilities and pitfalls: e.g. about the coverage of the research disciplines by the various databases, the appropriate search and analyze functions available under the different search systems or the significance and the limitations of citation analysis (bibliometry).

3. Overall Productivity: Nanotubes vs Fullerenes and Graphene

The WoS offers two search modes: The General Search and the Cited Reference Search (the latter is not relevant for this study). The General Search mode reveals publications which appeared in WoS source journals (in particular articles, reviews, and meeting abstracts - no books, no popular scientific publications, no conference proceedings unless they appear in source journals or in the CPCI-S). The number of articles published in the WoS source journals has become a standard measure for scientific productivity (output in terms of the number of publications). The number of publications per year can easily be plotted as a function of the publication years using the WoS analyze function.

At the date of search (01-07-2009) the SCI including the CPCI-S revealed altogether 57128 publications related to nanotubes. The terms "nanotub*" or "nano tub*" (* = wildcard allowing to include the plural or synonyms like tubulus) were searched in the title and the abstract search fields. The search has not been restricted to WoS specific document types. Additionally searching the relevant abbreviations in common use to distinguish single walled, double walled, and multi walled nanotubes (SWCNT, SWNT, DWCNT, DWNT, MWCNT, MWNT) increased the total number of papers only marginally (57208). Due to the fact that the abbreviations hardly appear without additionally mentioning the full term and the potential ambiguity, the abbreviations were not taken into consideration here. Figure 1 shows the time curve of the articles of the entire nanotubes research field. The time evolution of the related fullerene and graphene literature are shown for comparison. The total number of articles covered by the SCI is included as a measure for the growth of the overall scientific literature.

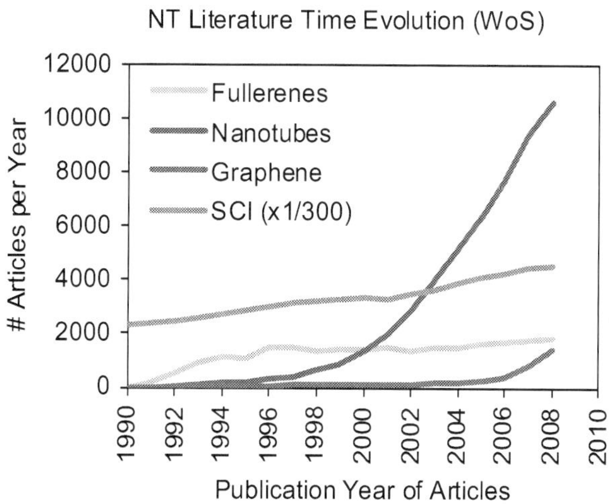

Fig. 1. Time dependent number of articles dealing with fullerenes, nanotubes, and graphene. The total number of articles covered by the SCI is shown as a rough measure for the growth of scientific literature. Source: SCI and CPCI-S under WoS.

According to Figure 1, the productivity (total number of articles per year) of the research activities dealing with nanotubes steadily increased, reaching about 11000 papers published in the year 2008 (compared to "only" 2000 fullerene papers published in the same year). The output increased by a factor of ten since 2000, which is far above the growth of the overall scientific literature in the same time period (about a factor of 1.25 with respect to the literature appearing in the source journals covered by the SCI).

In contrast to the nanotubes productivity, the time evolution of the fullerene literature shows a distinct saturation since about five years after discovery (Braun, 1992). Obviously, the nanotubes are one of the hottest research topics within the last decades with an undamped evolution. Since 2000 they have supplanted the (firstly) more promising fullerenes. The graphene related articles (meanwhile about 4500) show a rather similar increase, obviously starting a follow-up boom beside the ongoing fullerenes and nanotubes research activities.

Google and Google Scholar have become powerful search engines for web resources. Searching the world wide web for "nanotube(s)" with Google results in the large number of 3.1 million entries, while Google Scholar results in 0.27 million hits. The Google searches were carried out without any limitations concerning format, language, or time.

4. Productivity: Authors and Research Organizations

The almost 60,000 articles dealing with nanotubes and selected under the WoS were analyzed using the WoS analyze function. The most productive authors and research organizations, the countries of authors, and the leading journals were determined and are given in the Tables 1-4 below (date of search: 01-07-2009).

Rank	Author	Country	# Articles
1	Iijima, S	Japan	333
2	Bando, Y	Japan	307
3	Ajayan, PM	USA	294
4	Dresselhaus, MS	USA	291
5	Chen, Y	PR China	288
6	Roth, S	Germany	269
7	Lee, YH	South Korea	262
8	Chen, J	PR China	259
9	Wang, J	USA	258
10	Wang, Y	PR China	258
11	Zhang, Y	PR China	258
12	Zhang, J	PR China	248
13	Kataura, H	Japan	246
14	Goldberg, D	Japan	239
15	Li, Y	PR China	238
16	Terrones M	England	232
17	Wang, X	PR China	230
18	Liu, J	USA	225
19	LI, J	PR China	212
20	Smalley, RE	USA	209
21	Liu, Y	PR China	208
22	Zhang, L	PR China	208

Table 1. Top authors with at least 200 nanotubes articles based on the SCI and the CPCI-S under WoS.

Among the authors is a clear dominance of researchers from the East Asian countries and the US. Only a single author from a European country (S. Roth, Germany) is found in the group of the top ten authors in Table 1. Please note: Except for reprint authors, the SCI author addresses (and countries) are not allocated to the corresponding author names. In addition, some authors changed their affiliation. Asian names with only one forename initial comprise namesakes. Hence, the countries of authors given in Table 1 are not fully definite.

The top research organizations with respect to the number of articles dealing with nanotubes are shown in Table 2. Among the top positions are many Chinese research organizations and universities. The "weight" of the Chinese nanotubes research is confirmed further with Peoples Republic of China on rank two of the top countries of authors given in Table 3 further below.

Rank	Research Organization	# Articles	% Articles
1	Chinese Acad Sci	2840	5.0
2	Tsing Hua Univ	903	1.6
3	Russian Acad Sci	881	1.5
4	Peking Univ	684	1.2
5	Tohoku Univ	630	1.1
6	Rice Univ	628	1.1
7	Univ Sci & Technol China	616	1.1
8	Univ Cambridge	609	1.1
9	MIT	607	1.1
10	Univ Tokyo	607	1.1
11	Osaka Univ	573	1.0
12	Nanjing Univ	556	1.0
13	Zhejiang Univ	551	1.0
14	Natl Univ Singapore	514	0.9
15	NASA	513	0.9
16	Univ Illinois	497	0.9
17	CNRS	496	0.9
18	Seoul Natl Univ	479	0.8
19	Univ Calif Berkeley	476	0.8
20	Penn State Univ	460	0.8
21	Natl Inst Mat Sci	453	0.8
22	Natl Inst Adv Ind Sci & Technol	449	0.8
23	Natl Tsing Hua Univ	445	0.8
24	Sungkyunkwan Univ	427	0.7
25	Rensselaer Polytech Inst	414	0.7
26	Nanyang Technol Univ	406	0.7
27	Georgia Inst Technol	401	0.7

Table 2. Top research organizations with at least 400 nanotubes articles based on the SCI and the CPCI-S under WoS.

Please note: (1) In contrast to author names and journal titles, author addresses are not fully standardized in literature databases. Hence, the data offer only a rough picture of the leading research organizations and do not provide an exact ranking. (2) Many publications have been assigned to more than one country of author resulting in a substantial overlap.

5. Productivity: Countries and Continents

The ranking of the countries of authors having published articles dealing with nanotubes is given in Table 3.

Rank	Country of Author	# Articles	% Articles
1	USA	15845	27.7
2	Peoples R China	13386	23.4
3	Japan	6683	11.7
4	South Korea	3660	6.4
5	Germany	3423	6.0
6	France	2469	4.3
7	England	2391	4.2
8	Taiwan	1743	3.1
9	Russia	1691	3.0
10	Italy	1394	2.4
11	India	1335	2.3
12	Spain	1177	2.1
13	Canada	1094	1.0
14	Australia	994	1.7
15	Singapore	965	1.7
16	Switzerland	756	1.3
17	Belgium	694	1.2
18	Brazil	662	1.2
19	Poland	599	1.0
20	Israel	589	1.0
21	Sweden	523	0.9
22	Mexico	508	0.9

Table 3. Top countries of authors with at least 500 nanotubes articles based on the SCI and the CPCI-S under WoS.

The countries of authors ranking given in Table 3 is based on articles dealing with any type of nanotubes. A ranking based on articles dealing only with multi-walled nanotubes reveals that the top position is taken up by the Peoples Republic of China. A broader view is given in Figure 2, showing the share of the continents of the authors publishing nanotubes related articles being covered by the WoS source journals.

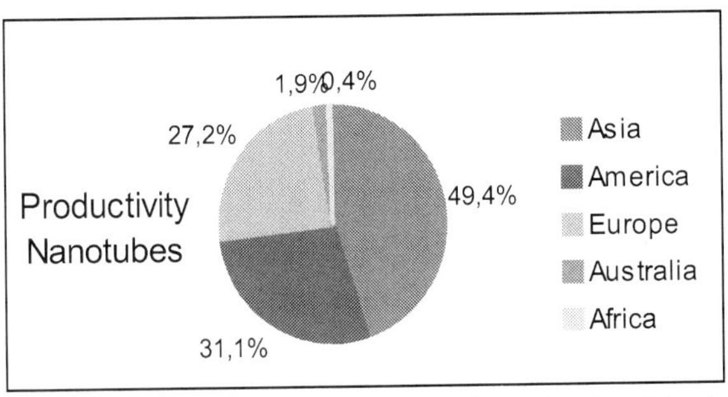

Fig. 2. Share of the continents of the authors of nanotubes related articles (with Europe incl. Russia and Australia incl. New Zealand).

6. Research Output: Journal Articles and Patents

The scientific literature dealing with nanotubes has been published in about 500 different WoS source journals. Almost half of this literature appeared in only 25 key-journals (according to Bradford's law of scattering). The leading journals each with their number of nanotubes articles are given in Table 4. Among these journals about one fourth are journals around nanotechnology or material science. For scientists, such information could be valuable with respect to the selection of the appropriate journals for publishing their own results.

Rank	SCI Source Journal	# Articles	% Articles
1	Physical Review B	2428	4.3
2	Applied Physics Letters	2242	3.9
3	Nanotechnology	1671	2.9
4	Carbon	1652	2.9
5	Chemical Physics Letters	1295	2.3
6	Nano Letters	1286	2.3
7	Journal of Physical Chemistry B	1164	2.0
8	Journal of Physical Chemistry C	1106	1.9
9	Journal of the American Chemical Society	919	1.6
10	AIP Conference Proceedings	915	1.6
11	Physical Review Letters	880	1.5
12	Journal of Nanoscience and Nanotechnology	865	1.5
13	Advanced Materials	825	1.4
14	Abstracts of Papers of the American Chemical Society	801	1.4
15	Journal of Applied Physics	779	1.4
15	Proceedings of the Society of Photo-Optical Instrumentation Engineers (SPIE)	717	1.3
17	Chemistry of Materials	618	1.1
18	Diamond and Related Materials	610	1.1
19	Langmuir	536	0.9

Table 4. Top WoS source journals with more than 500 nanotubes articles based on the SCI and the CPCI-S under WoS.

A search in the CAS literature file CAplus (based on the corresponding WoS search query) revealed 53772 publications and 13184 patents. The lower number of publications in CAplus compared to the multidisciplinary WoS (57128) results from the narrower field coverage of CAS mainly focusing on chemistry. The publications searched in CAplus have been taken for establishing a nanotubes research landscape (see Figure 4 further below) - this is not possible with the WoS records. At first we had a closer look at the patents and show in Figure 3 the time evolution of the articles and patents as covered by CAS. The patents have been further analyzed with respect to technical applications in broad areas of activity. The country specific number of patents are presented in Table 5 as a heat map.

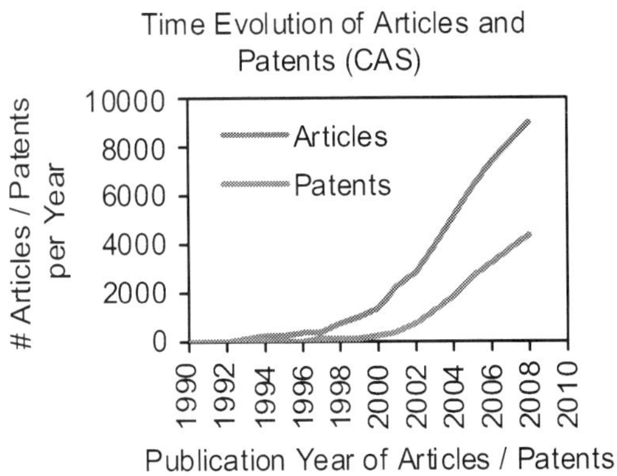

Fig. 3. Time dependent number of articles and patents dealing with nanotubes and covered by the CAS literature file. Source: CAplus under STN International.

	Health Care	Sensors	Electronics & Logic	Mass Data Storage	Displays & Field Emiss.	Energy Storage	Materials & Reactions	Total Sum
Japan	66	368	722	52	919	704	1206	4037
USA	251	552	825	90	627	388	865	3598
S. Korea	19	196	342	44	588	276	385	1850
P. R. China	26	216	235	24	410	202	564	1677
Taiwan	3	66	82	7	182	72	125	537
Germany	17	72	112	20	62	37	107	427
France	12	23	46	4	44	46	67	242
United Kingdom	6	15	28	2	21	11	30	113
Canada	2	12	9	0	13	12	22	70
Switzerland	2	6	14	0	11	13	14	60
Netherlands	1	8	15	3	12	3	10	52
Singapore	2	8	6	0	11	8	16	51
Russia	3	6	8	0	9	5	16	47
Austria	1	7	5	0	10	3	17	43
Italy	0	5	12	4	4	2	11	38
Australia	7	8	5	1	5	4	8	38
Israel	1	5	8	0	10	5	6	35
Sweden	1	3	12	0	6	3	9	34
Belgium	0	7	6	0	7	1	10	31
Brazil	2	2	4	1	4	8	4	25

Table 5. Number of patents with respect to nanotubes published by the top 20 countries (country of author) in the order of ranking with respect to the research areas.

The ranking of nations can be depicted from the last column in Table 5 which shows the total number of patents for each country. The data show a clear dominance of East Asian countries with Japan (1), S. Korea (3), P. R. China (4), and Taiwan (5). USA is second in the ranking while most European countries show a lower rank: Germany (6), France (7), and United Kingdom (8). The same tendency can be seen with respect to the research areas analyzed. In Health Care, Sensors, and Electronics & Logic, however, the ranking between Japan and the United States is flipped. The highest number of patents has been published in the area of Materials & Reactions.

7. Content Analysis: Research Landscape

STN International has recently launched a new interactive analysis tool called STN AnaVist (STN AnaVist, 2009) which focuses mainly on patent analysis (Fischer & Lalyre, 2006). This tool can also be applied to journal articles exploring basic research topics. STN AnaVist has been used to refine the nanotubes literature analysis. The selection of the articles to be analyzed has to be done in the CAS literature file. Currently, the STN AnaVist system limits do not allow to process more than 20,000 articles. Therefore, the literature analysis had to be restricted to the time period 2008-2009.

One of the functions offered by STN AnaVist implies the creation of so-called research landscapes: "Significant keywords and concepts are derived from document titles and abstracts. These keywords are used to determine the similarity between documents. An algorithm uses document similarity scores to position each document relative to one another in a two-dimensional space, with each document positioned at one point. This process is repeated until all documents have been clustered and each assigned to a single x, y coordinate pair. A graphical map is generated. The z coordinates, determining the height of each 'peak', are calculated based on the density of the documents in an area." (STN International, 2009). The research landscape of the past-2007 articles dealing with nanotubes is shown in Figure 4.

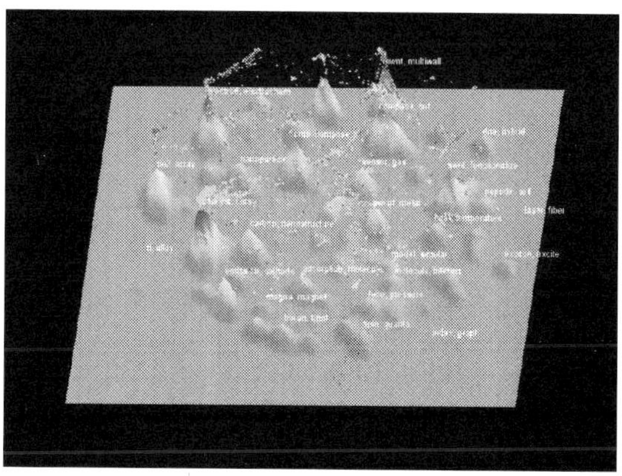

Fig. 4. Research landscape of the nanotubes literature published in the time period 2008-2009 established using STN AnaVist (cnts = carbon nanotubes, swnt = single-walled nanotubes, mwnt = multi-walled nanotubes).

The nanotubes landscape in Figure 4 shows two different layers one upon the other: (1) the mountains with the related keyword pairs based on the text analysis as mentioned and (2) the colored patches above based on the CAS classification categories (technical indicators). The keyword analysis reveals major studies on composites, catalysis, and electrochemistry. The mountains are characterized by the two top clustering words, e. g. "catalyst, cnts" at the left of the landscape or "compose, cnt" top right. The colored dots correspond to documents with different technical indicators (green: vapor deposition process, red: nanocomposites, blue: biosensors, cyan: electric conductivity, and yellow: overlaps). Please note: STN AnaVist is an interactive tool that allows to focus and to zoom according to the specific needs. The research landscape, for example, can be twisted for better analysis.

8. Most Highly-Cited Papers

The number of citations is often taken as a measure of the attention an article, a researcher, an institute or even a country has attracted. Although citation numbers are no ultimate scale of the final importance and quality of articles, they reflect strengths and shortcomings of

Author(s)	Title	Journal	# Citations
Iijima S	Helical microtubules of graphitic carbon	Nature 354, 56-58 (1991)*	8545
Thess A, Lee R, Nikolaev P, et al.	Crystalline ropes of metallic carbon nanotubes	Science 273, 483-487 (1996)	2872
Pan ZW, Dai ZR, Wang ZL	Nanobelts of semiconducting oxides	Science 291, 1947-1949 (2001)	2680
Xia YN, Yang PD, Sun YG, et al.	One-dimensional nanostructures: synthesis, characterization, and applications	Advanced Materials 15, 353-389 (2003)	2621
Tans SJ, Verschueren ARM, Dekker C	Room-temperature transistor based on a single carbon nanotube	Nature 393, 49-52 (1998)	2539
Baughman RH, Zakhidov AA, de Heer WA	Carbon nanotubes - the route toward applications	Science 297, 787-792 (2002)	2299
Morales AM, Lieber CM	A laser ablation method for the synthesis of crystalline semiconductor nanowires	Science 279, 208-211 (1998)	2137
Kong J, Franklin NR, Zhou CW, et al.	Nanotube molecular wires as chemical sensors	Science 287, 622-625 (2000)	2103
Iijima S, Ichihashi T	Single-shell carbon nanotubes of 1-nm diameter	Nature 363, 603-605 (1993)	1975
Kitagawa S, Kitaura R, Noro S	Functional porous coordination polymers	Angewandte Chemie (Int. Edit.) 43, 2334-2375 (2004)	1907

Table 6. The top ten most highly-cited nanotubes papers based on the SCI and the CPCI-S under WoS (date of search: 01.07.09).

research activities and are therefore frequently used for research evaluation. Being cited means that a given article appears as a reference in the article of another author for additional reading. The number of citations of a specific paper is thus a rough measure of the importance or usefulness of the paper within the scientific community. The top ten most highly-cited nanotubes articles until present are given in Table 6. Note that such lists actually imply no real ranking because the various papers accumulated their citations over different time periods.

Please note: The Iijima paper on rank 1 illustrates the problems when selecting literature by using search terms: Primarily, this paper has not been included in the answer set, because the term "nanotube" does not appear within the title or the abstract of this early paper.

The graph displaying the time evolution of the citations of a single article is called its citation history. Each article develops its own life span as it is being cited. With time, the citations per year (citation rate) normally evolve following a similar pattern: The citations generally do not increase substantially until one year after the publication. They reach a summit after about three years, the peak position depending somewhat on the research discipline. Subsequently, as the articles are displaced by newer ones, their impact decreases, accumulating citations at a lower level. Finally, most of the articles are barely cited or forgotten.

Normally, articles receive only a few citations in the first years after the publication. Those which are decisive for research usually garner hundreds or even thousands of citations and keep being cited for a longer time, often for decades. The five most highly-cited nanotubes articles given in Table 6 are no exceptions. Figure 5 shows the citation history of these high-impact nanotubes papers.

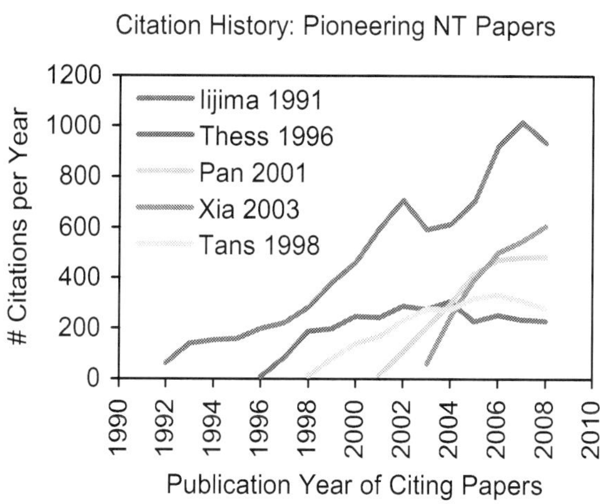

Fig. 5. Time dependent number of citations (citation history) of the five most highly-cited nanotubes papers (see Table 6).

Recently, an article published in Carbon, one of the leading nanotubes journals (see Table 4) discussed the nanotubes history and raised the question, to whom the credit for the discovery should be given (Monthioux & Kuznetsov, 2006). The first evidence for nano-sized carbon tubes is believed to have been published 1952 in the Russian Journal of Physical Chemistry by Radushkevich and Lukyanovich (Radushkevich & Lukyanovich, 1952). The article contains clear images of carbon tubes with about 50 nanometers in diameter based on Transition Electron Microscopy (TEM). However, due to the Russian language and the limited access of Western scientists to Russian publications during the time of the Cold War the discovery was largely unnoticed. Another important article prior to 1991 was published 1976 by Oberlin et al. (Oberlin et al., 1976).

The citation history of these two articles is shown in Figure 6. The Radushkevich paper received altogether 73 citations (many from Russian scientists) and was barely cited until present (19 citations in 2008). The Oberlin article received altogether 600 citations (72 citations prior to 1991) showing a remarkable impact increase (75 citations in 2008). These two articles seem to belong to the few papers in science being nearly ignored for decades and then are "rediscovered" (so-called sleeping beauties).

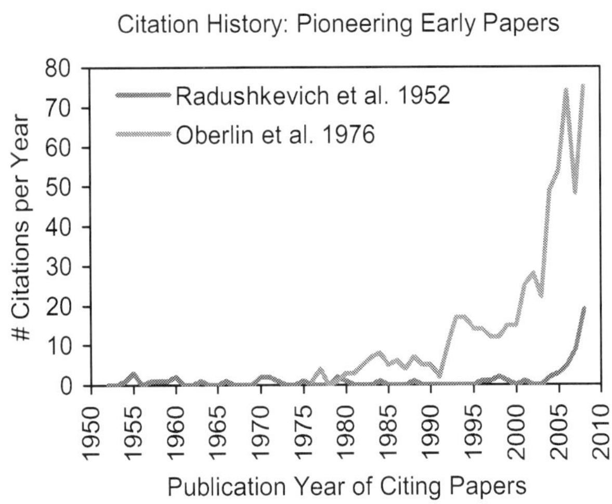

Fig. 6. Time dependent number of citations (citation history) of the Radushkevich & Lukyanovich and the Oberlin paper.

The "weight" of research fields can be indicated by their Hirsch index (h-index, h-number). The h-index is defined as the number of articles in WoS source journals that have had h citations or more. For example, a researcher with an h-index of 40 will have published 40 articles that have received at least 40 citations each. The h-index can be seen as a rough indicator for measuring the output and impact of a researcher in a single number (Hirsch, 2005). This index reflects a researcher´s contribution based on a broad body of publications rather than on a few high-impact papers. The h-index may well be extended to journals (Bornmann et al., 2009) as well as to research fields (Banks, 2006). The h-index for the overall literature related to nanotubes, fullerenes, graphene, and further research fields is given in Table 7. The data can be taken as a rough measure of the relative "weight" of these fields in terms of both publications and citations.

Rank	Research Field	# Articles	h-index
1	Nanotubes	57128	305
2	Fullerenes	24552	204
3	Graphene	4639	112
4	Quantum Dots	38457	245
5	Nanowires	27939	201
6	Amorphous Silicon	18060	145
7	Porous Silicon	8824	134

Table 7. Number of articles related to nanotubes, fullerenes, graphene, and further research fields and their corresponding h-index (date of search: 01-07-2009).

9. Citation Graphs

Using the HistCite® software of Eugene Garfield, we established the citation graph shown in Figure 7, which is based on the most highly-cited nanotubes articles accessible in WoS at the date of search (22.09.09). This graph visualizes the citation network within the high impact nanotubes literature published so far: The nodes represent the papers with at least 1000 citations (31 papers). Lower citation limits increase the number of nodes considerably without changing the node pattern substantially (see Figure 8 further below). The specific papers represented by the circles of the citation graph of Figure 7 are given in Table 8 in short form only. Like STN AnaVist®, also HistCite® is used in an interactive mode and can deliver much more information than given in the limited citation graphs shown here.

The citation graph of Figure 7 shows the Iijima 1991-paper (see Table 6) as the starting point of the nanotubes research field activities at the top. The graph reveals a pronounced cluster of high-impact publications based on nanotubes within the time period 1995-2003. The more recently published articles are increasingly rare because of the time delay of citations. These publications had so far not sufficient time for accumulating enough citations to reach the high limit of 1000 citations specified in Figure 7.

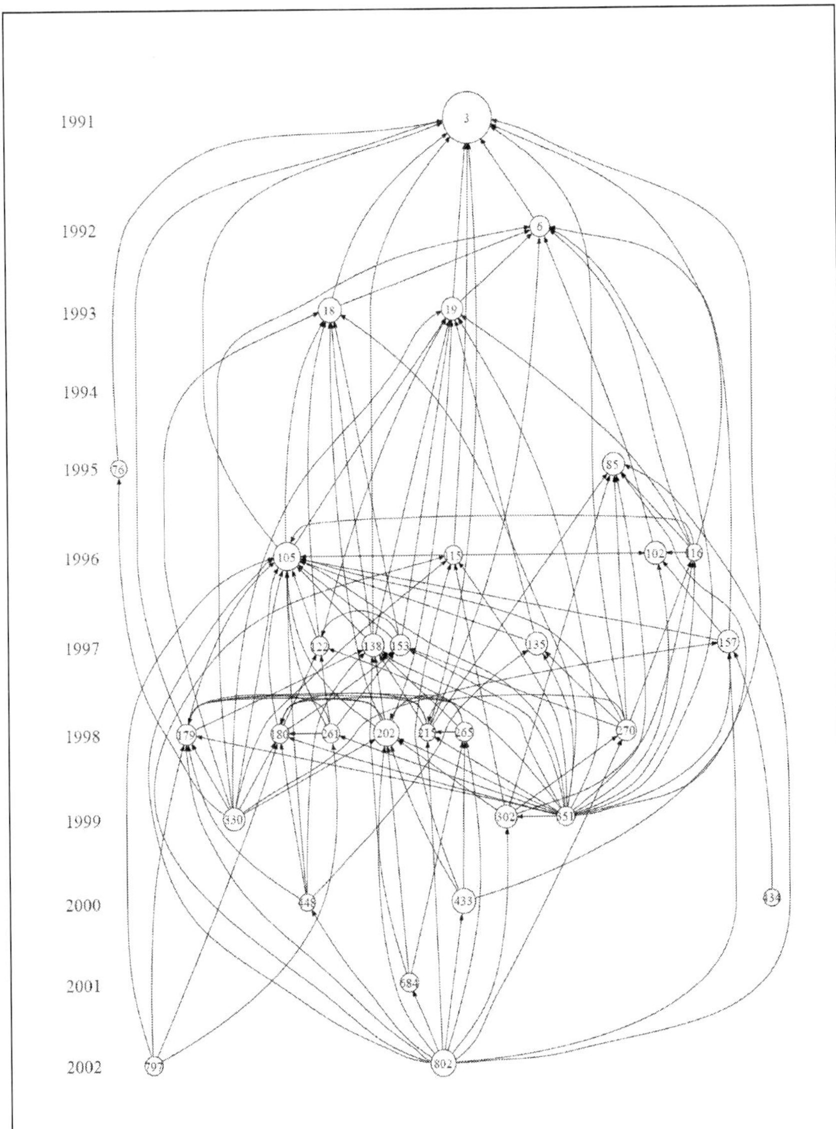

Fig. 7. Citation graph based on the most highly-cited nanotubes articles searched in the WoS. The nodes represent the currently 31 nanotubes papers with at least 1000 citations until the date of search. The circle diameter is proportional to the overall number of citations and the arrows indicate the citation direction. The numbers within the circles originate from the consecutive numbering of the nanotubes papers by the software (see Table 8) and do not represent their citation counts. Source: HistCite® and SCI under WoS.

#	No.	Papers	LCS	GCS
1	3	IIJIMA S, 1991, NATURE, V354, P56	377	8787
2	6	EBBESEN TW, 1992, NATURE, V358, P220	150	1628
3	18	IIJIMA S, 1993, NATURE, V363, P603	161	2005
4	19	BETHUNE DS, 1993, NATURE, V363, P605	125	1686
5	76	CHOPRA NG, 1995, SCIENCE, V269, P966	48	1133
6	85	DEHEER WA, 1995, SCIENCE, V270, P1179	105	1831
7	102	Treacy MMJ, 1996, NATURE, V381, P678	125	1779
8	105	Thess A, 1996, SCIENCE, V273, P483	264	2917
9	115	Dai HJ, 1996, NATURE, V384, P147	68	1245
10	116	Li WZ, 1996, SCIENCE, V274, P1701	75	1066
11	122	Rao AM, 1997, SCIENCE, V275, P187	96	1259
12	135	Dillon AC, 1997, NATURE, V386, P377	75	1854
13	138	Tans SJ, 1997, NATURE, V386, P474	131	1708
14	153	Journet C, 1997, NATURE, V388, P756	111	1519
15	157	Wong EW, 1997, SCIENCE, V277, P1971	102	1756
16	179	Wildoer JWG, 1998, NATURE, V391, P59	141	1497
17	180	Odom TW, 1998, NATURE, V391, P62	121	1194
18	202	Tans SJ, 1998, NATURE, V393, P49	143	2606
19	215	Frank S, 1998, SCIENCE, V280, P1744	57	1143
20	261	Chen J, 1998, SCIENCE, V282, P95	120	1280
21	265	Martel R, 1998, APPL PHYS LETT, V73, P2447	68	1322
22	270	Ren ZF, 1998, SCIENCE, V282, P1105	78	1496
23	302	Fan SS, 1999, SCIENCE, V283, P512	75	1708
24	330	Hu JT, 1999, ACCOUNT CHEM RES, V32, P435	11	1682
25	351	Ajayan PM, 1999, CHEM REV, V99, P1787	45	1153
26	433	Kong J, 2000, SCIENCE, V287, P622	89	2181
27	434	Yu MF, 2000, SCIENCE, V287, P637	35	1036
28	448	Collins PG, 2000, SCIENCE, V287, P1801	63	1060
29	684	Bachtold A, 2001, SCIENCE, V294, P1317	36	1232
30	797	O'Connell MJ, 2002, SCIENCE, V297, P593	60	1243
31	802	Baughman RH, 2002, SCIENCE, V297, P787	48	2432

Table 8. List of the specific papers of the HistCite® based citation graph of Figure 7 represented by the numbered circles. LCS = local citation score: citations collected within the ensemble of the selected nanotubes articles, GCS = global citation score: overall number of citations as given by the WoS under "times cited".

The citation graph of Figure 8 shows more clearly the pattern of the citation network based on the highly-cited papers, in particular within the starting period after 1991. Again, the past-2002 papers are increasingly rare because of the time delay of the citations.

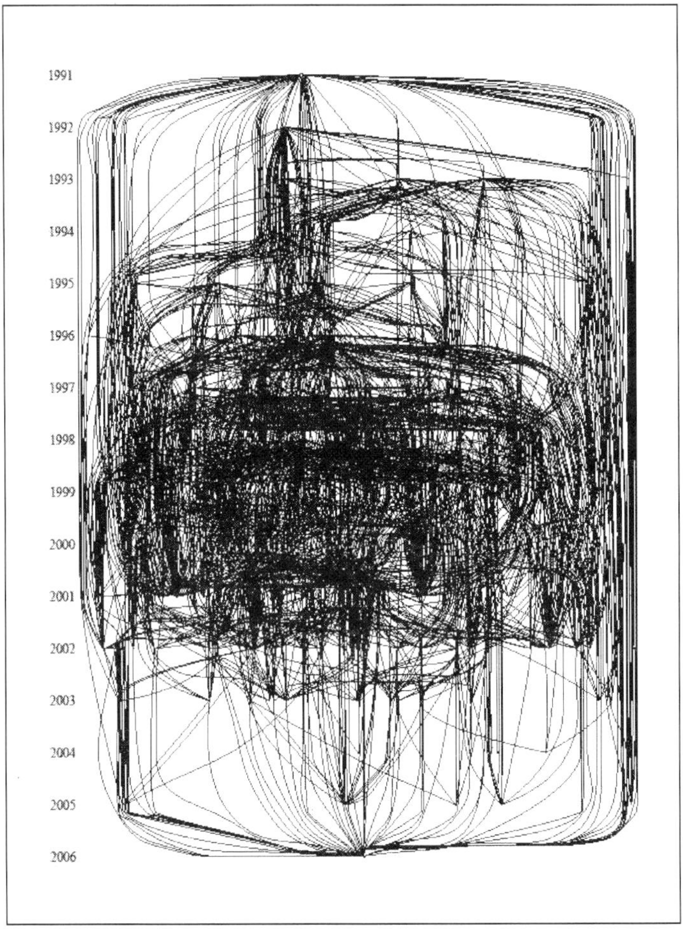

Fig. 8. Citation graph based on the highly-cited nanotubes articles searched in the WoS. The nodes represent the currently 226 nanotubes papers with at least 300 citations until the date of search (rather than only the 31 papers with at least 1000 citations as in Figure 7). Source: HistCite® and SCI under WoS.

Summary

A currently most popular research discipline has been reviewed by scientometric methods. The time dependent overall number of nanotubes articles is given, revealing a still strongly increasing research activity in this field. The nanotubes articles were analyzed with respect to the most productive authors, research organizations, countries of authors, and leading journals. Among the authors is a clear dominance of researchers from the East Asian countries and the US. Among the top positions are many Chinese research organizations and universities. The time evolution of the nanotubes patents shows a similar increase as the evolution of the articles. The number of patents of the various organizations related to broad areas of research and technology were determined and visualized as a heat map. The research landscape of the nanotubes research field was established using the new analysis tool STN AnaVist. The STN AnaVist based landscape shows a pronounced clustering with respect to specific keywords and technical indicators. The time evolution of the most highly-cited nanotubes papers shows an unusually long-lasting impact. Two early papers may be characterized as "sleeping beauties". Finally, a citation graph has been established revealing the historiography of the nanotubes research with regard to the high impact papers.

10. References

Banks, M.G. (2006). An extension of the Hirsch-index: indexing scientific topics and compounds. *Scientometrics* 69, 161-168.

Barth, A. & Marx, W. (2008). Mapping High-Temperature Superconductors - A Scientometric Approach. *Journal of Superconductivity and Novel Magnetism* 21, 113-128. (PDF), URL : http://arxiv.org/abs/cond-mat/0609114

Bornmann, L.; Marx, W. & Schier, H. (2009). Hirsch-type index values for organic chemistry journals: A comparison of new metrics with the Journal Impact Factor. *European Journal of Organic Chemistry* 10, 1471-1476.

Braun, T. (1992). The epidemic spread of fullerene research. *Angew. Chem. Int. Edit.* 31, 588-589.

Fischer, G. & Lalyre, N. (2006). Analysis and visualisation with host-based software - The features of STN®AnaVist™, *World Patent Information.*

Hirsch, J.E. (2005). An index to quantify an individual's scientific research output. *Proceedings of the National Academy of Sciences of the USA* (PNAS) 102, 16569-16572.

Monthioux, M. & Kuznetsov V. L. (2006). Who should be given the credit for the discovery of carbon nanotubes? *Carbon* 44, 1621-1623.

Oberlin, A.; Endo, M. & Koyama, T. (1976). Filamentous growth of carbon through benzene decomposition. *Journal of Crystal Growth* 32, 335-349.

Radushkevich, L.V. & Lukyanovich, V.M. (1952). *Russian Journal of Physical Chemistry* 26, 88-95.

STN International (2009), URL: www.stn-international.de/

Thomson Reuters (2009), URL: www.isinet.com/isi/

STN AnaVist (2009). http://www.stn-international.de/stninterfaces/stnanavist/stn_anavist.html

Corresponding author: Dr. Werner Marx
Max Planck Institute for Solid State Research
Heisenbergstraße 1, D-70569 Stuttgart, Germany
E-mail: w.marx@fkf.mpg.de

E-mail Dr. Andreas Barth: andreas.barth@fiz-karlsruhe.de

2

Chemical Vapour Deposition of CNTs Using Structural Nanoparticle Catalysts

G. N. Ayre, T. Uchino, B. Mazumder, A. L. Hector,
D. C. Smith, P. Ashburn and C. H. de Groot
University of Southampton
United Kingdom

J. L. Hutchison
Oxford University
United Kingdom

1. Introduction

The extraordinary mechanical, electrical and optical properties of CNTs have stimulated extensive research since their discovery by Sumio Iijima of the NEC Corporation in the early 1990s (Iijima, 1991). Possible applications for carbon nanotubes range from nanoelectronics, quantum wire interconnects, sensors and field emitters to composites (Meyyappan et al., 2005). These applications require reliable synthesis techniques capable of generating large quantities of high purity material. In addition, applications in nanoelectronics and photonics may require controlled growth at precise lithographically patterned areas. Understanding how to control the synthesis of SWNTs is vital in order to deterministically integrate such nanostructures into various technologies.

Of the various methodologies developed for carbon nanotube (CNT) synthesis, those based upon catalyst-assisted chemical vapour deposition (CCVD) appear to be best suited to satisfy these requirements. Traditionally, *3d* valence transition metal nanoparticles, such as Fe, Co and Ni, are used to catalyse CNT growth. These catalysts are thought to work in accordance with the model for carbon filament growth, derived by Baker et al. (Baker, 1989; Baker et al., 1972; 1973), from concepts of vapour-solid-liquid theory (Wagner & Ellis, 1964). In this model, hydrocarbons adsorbed on the metal nanoparticle are catalytically decomposed resulting in atomic carbon dissolving into the liquid catalyst particle, and when a supersaturated state is reached, carbon precipitates in a tubular, crystalline form.

Recently, several groups have reported successful growth of CNTs from noble metal (Lee et al., 2005; Takagi et al., 2006; Yoshihara et al., 2008; Yuan et al., 2008; Zhou et al., 2006), ceramic (Liu et al., 2008b; Steiner et al., 2009) and semiconducting nanoparticles (Takagi et al., 2007; Uchino et al., 2009; 2008; 2005b), all of which are regarded as unable to catalyse the dissociation of hydrocarbons. In addition to this, in their bulk form, these materials do not have a catalytic function to produce graphite. This implies that given enough energy, carbon atoms on a nanoparticle are capable of a structural reorganisation into CNTs. This leads to a new in-

terpretation of the role of the catalyst in nanotube growth in which only a nanoscale curvature and carbon adsorption sites are necessary.
This work examines the recent developments in non-traditional CCVD of CNTs with a view to determine the essential role of the catalyst in nanotube growth. Section 2 provides a brief overview of the techniques reliant on the structural reorganization of carbon to form CNTs. An in-depth analysis of CNT synthesis based upon ceramic (Section 3), noble metal (Section 4) and semiconducting nanoparticle catalysts (Section 5) is presented. Various approaches to germanium catalyst preparation are compared in terms of growth density and quality of synthesized nanotubes. Scanning electron microscopy measurements indicate that a technologically relevant density is achievable using non conventional catalysts. Raman measurements have identified the synthesized nanotubes as single walled and, in terms of graphitization and structure, of a high quality. Extensive atomic force microscopy characterisation of the catalyst has been undertaken in order to ascertain the influence of morphology on the ability of the catalyst to yield CNT growth. A model for CNT growth consistent with the experimental results is proposed in Section 6. Finally, a summary of challenges and future directions for investigations is presented in Section 7.

2. Catalyst Free Synthesis of CNTs

Silicon carbide has been the most widely used non metallic catalyst. The synthesis techniques involving this catalyst have produced high densities of carbon nanostructures by annealing either SiC particles (Botti et al., 2004; Takikawa et al., 1998), amorphous SiC films (Botti et al., 2001; Kusonoki et al., 1997) or hexagonal SiC (6H-SiC) (Derycke et al., 2002; Kusonoki et al., 1997) in a vacuum. In these methodologies, the nanotube formation can be explained by the mechanism proposed by Kusonoki et al. (1997). Owing to the low vacuum in the chamber when annealing, the SiC oxidises forming SiO_2. As a consequence, the carbon atoms are free to bond with other atoms. If they bond to neighbouring carbon atoms, graphite fragments are formed containing dangling bonds. Thermodynamics drives the folding of graphitic fragments so that the dangling bonds of opposite edges are saturated. The as-formed nanotube segments act as seeds for the attachment of new carbon atoms, leading to CNT growth. However, these techniques require high temperature annealing at approximately 1650 °C.
An alternate approach involves using carbon nanoparticles as a catalyst. This technique depends on the structural reorganization of carbon aggregates into nanotubes upon annealing. Botti et al. (2002) report a dense array of CNTs grown on silicon by spraying amorphous hydrogenated carbon nanoparticles on a Si substrate. Other similar approaches have been reported in the literature (Koshio et al., 2002; Larciprete et al., 2002; 2003).
Figure 1 shows transmission electron microscope images of carbon nanostructures deposited on a carbon implanted Ge nanoparticle sample. By inspection of Figure 1(a) it can be determined that this is made up of a collection of small multi-walled carbon nanotubes (MWNTs). As there were no catalyst particles detected at either ends of the nanotubes, it is therefore believed that these nanotubes were nucleated by the structural reorganization of amorphous carbon deposits without the influence of the Ge nanoparticles. This hypothesis is further reinforced by Figure 1(b), which shows a double walled nanotube (DWNT) in the early stages of growth. Again, there is no catalyst detected, and it appears that the nanotube was nucleated from the nearby carbon deposits. Both nanostructures demonstrate a good degree of graphitization and furthermore there are few impurities present, such as amorphous carbon, in the materials.

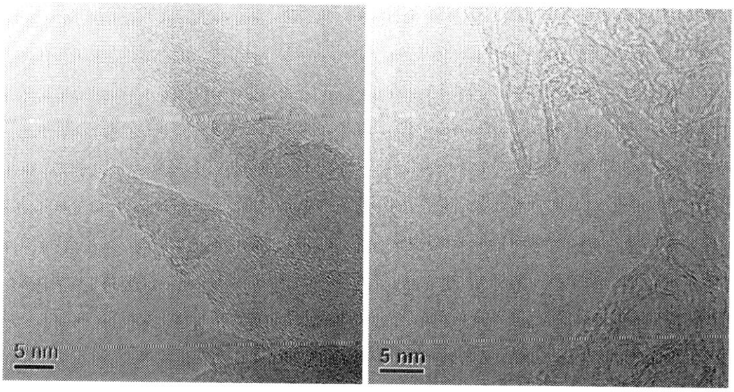

(a) (b)

Fig. 1. Transmission electron micrographs of self-assembled carbon nanostructures deposited on a carbon implanted Ge nanoparticle sample. The TEM sample was prepared by scraping the substrate surface with a surgical blade and collecting the material on a holey-carbon TEM grid. (a) Image of MWNTs and (b) a DWNT in the early stages of growth formed by the structural reorganization of carbon.

These techniques, although not strictly classed as chemical vapour deposition of CNTs, provide some insight into the behaviour of carbon aggregates at elevated temperatures without the influence of an external catalyst with a function to produce graphite. These results also demonstrate that regardless of the catalyst, the formation of CNTs involves two important processes: (i) the diffusion of carbon and (ii) the nucleation of a graphitic cap or fragment followed by the further incorporation of carbon into the growing nanotube. It has been reported that the diffusion process on a nanoparticle surface or across its interior is a rate limiting step (Bartsch et al., 2005; Hofmann et al., 2005), while the chirality of the growing CNT is decided upon the formation of the graphitic cap (Reich et al., 2006; Yazyev & Pasquarello, 2008).

3. Ceramic Nanoparticle Catalysts

Ceramic materials, such as Al_2O_3, have typically been used as a buffer layer to disperse metallic catalyst particles and enhance their catalytic properties in CNT growth (Takagi et al., 2007). However, the simplistic view that the support only plays a catalytically passive role in the formation of carbon nanotubes requires some examination. Rummeli et al. (2007) demonstrated that under typical conditions for CVD growth of CNTs, nanoparticles of difficult-to-reduce metal-oxides are exceedingly good at promoting ordered carbon (graphene) growth. As was expected, there was no observation of ordered carbon formation in bulk/film samples. This difference was attributed to the presence of surface defect sites on the nanoparticle oxides, and it was argued that in the substrate-based CNT synthesis routes, the interface between the catalyst particle and the surface behaves as an annular defect site. These sites would then promote the formation of cylindrical graphene structures, or nanotubes.
A recent study by Liu et al. (2008b) reports the formation of dense CNT layers catalysed by Al_2O_3 nanoparticles. Raman spectra of the synthesized nanotubes indicated that the nan-

otubes synthesized were predominantly single walled and of a good quality. Interestingly, the authors surmise that the mechanism of formation is different from the traditional vapour-liquid-solid mechanism as the nanoparticles are likely to be in the solid state during growth. This finding reinforces the argument of Rummeli et al. (2007), and additionally indicates that the growth of single-walled carbon nanotubes (SWNTs) on flat Al_2O_3 substrates may be possible by nanostructuring their surfaces. This hypothesis was partially confirmed by Liu et al. (2009) using a nanostructured SiO_2 substrate to grow SWNTs.
Another ceramic catalyst reported in the literature is ZrO_2 (Steiner et al., 2009). In this publication, dense growth of either MWNTs or SWNTs was possible, depending on the carbon feedstock used. ZrO_2 was typically deposited on either Al_2O_3 capped SiO_2 supports or Si substrates with an oxynitride support through a chloride salt solution. Samples were pretreated in H_2 prior to the introduction of the carbon feedstock. ZrO_2 is known to not be reduced by H_2 (Mctaggart, 1961) and additionally, carbothermic reduction of ZrO_2 does not yield Zr metal, but results in the formation of ZrC (Berger et al., 1999). In-situ x-ray photoelectron spectroscopy (XPS) revealed that the state of the catalyst after H_2 pretreatment showed two phases; a stoichiometric and an oxygen deficient phase of zirconia. The role of the H_2 pretreatment in this work, while shown not to result in the formation of Zr metal, is thought to introduce surface defect sites into zirconia nanoparticles that aid in enhancing catalytic ability. It should be noted that, as observed by in-situ XPS, CNT growth seems to begin only after the introduction of both the hydrocarbon and hydrogen. It is speculated that the introduction of hydrogen aids in the transformation of the hydrocarbon into other organic precursors which can then be uptaken and catalysed into CNTs.

4. Noble Metal Nanoparticle Catalysts

Nanosized iron-group metals (Fe, Co, Ni) are known for their ability to catalyse SWNT growth in chemical vapour deposition. It has been generally accepted that these metals and their alloys consistently show the highest catalytic activity (Awasthi et al., 2005; Melechko et al., 2005). This is attributed to the solubility of carbon in the metal-solid solution (Deck & Vecchio, 2006). However, noble metals such as Au, Ag or Cu have both negligible carbon solubility and negligible carbide formation, and have recently been identified as catalysts for the growth of CNTs. Takagi et al. (2006) have found that the yield of SWNTs from noble metals is comparable to that of iron-group metals. Moreover, noble metals, in particular Cu, are thought to favour CVD growth of CNT nanotubes at low temperatures with a narrow chirality distribution (Yazyev & Pasquarello, 2008).
Bulk Au is considered a noble metal, as it is highly unreactive and catalytically inactive. Au is the only metal with an endothermic chemisorbtion requirement, and in addition it has d-states so low in energy that the interaction with oxygen $2p$-states is net repulsive. Nevertheless, in its nanoparticle form, Au is capable of catalysing a wide variety of reactions. These include the oxidation of CO (Hvolbaek et al., 2007), the selective hydrogenation of acetylene (Jia et al., 2000), hydrogenation of halogen compounds, reduction of nitrogen oxides and photocatalytic hydrogen production (Haruta, 1997). The origin of this effect is believed to be the increase in the fraction of low-coordinated Au atoms as the size of the Au cluster is reduced. In some cases, the catalytic nature of supported Au clusters can be explained by assuming the Au-support perimeter interface acts as a site for activating at least one of the reactants.
The first demonstration of CNT growth from Au nanoparticle catalysts by Lee et al. (2005), involved the decomposition of acetylene over nanoparticles supported on SiO_2-Al_2O_3. This support showed a good propensity for the decomposition of acetylene and demonstrated

strong interactions between the Au nanoparticles and its surface. The synthesized products were predominantly MWNTs, with average diameters of $\leq$ 20 nm. The first reports of the formation of SWNTs from small Au nanoparticle catalysts were by Takagi et al. (2006) and Bhaviripudi et al. (2007). XPS measurements in both publications showed that CNT growth was only possible from contaminant-free catalyst nanoparticles, once the residual shell of gold oxides or gold chlorides were reduced by H_2. The findings of Liu et al. (2008a) corroborate this finding. Interestingly, neither paper detected any radial breathing modes in the low Raman shift region, corresponding to large diameter nanotubes.

Figure 2(a) shows a scanning electron microscope image of carbon nanotubes synthesized from a Au nanoparticle catalyst. In this experiment, colloidal gold nanoparticles were spin coated on SiO_2 (300 nm) capped Si substrates. Atomic force microscopy (AFM) characterisation of the catalyst revealed that the nanoparticles were approximately 1.4 nm in diameter, with a very narrow particle size distribution, shown in Figure 2(b). The measurements indicated a density of 2500 ± 790 (mean $\pm$ standard deviation) particles/μm^2, which corresponds to an interparticle separation of approximately 20 ± 3 nm. The samples were then pretreated in an H_2 atmosphere for 10 minutes at temperatures ranging from $850 - 1050$ °C, followed by a growth step in a mixture of CH_4 and H_2 at 850 °C. The highest area density was found for samples pretreated at 1000 °C. Raman spectroscopy, with an excitation wavelength of 632.8 nm, was performed in order to evaluate the synthesized nanotubes; a typical spectrum is shown in Figure 2(c). The spectrum exhibits the radial breathing mode feature, indicating that the synthesized nanotubes are predominantly single walled.

From AFM measurements taken to analyse the influence of the pretreatment step on catalyst morphology, we can determine that the initial density of particles is reduced as pretreatment increases. The initial density of 2500 ± 790 particles/μm^2 was reduced to 420 and 290 particles/μm^2 after a pretreatment at 900 °C and 1000 °C, respectively. Broadening of the nanoparticles size distribution is accompanied by a reduction of the modal height as pretreatment temperatures increase. This reduction in density and modal height of the catalyst nanoparticles can be attributed to the evaporation of Au from the substrate in conjuction with the possible diffusion of the Au into the substrate. A similar effect was reported by Bhaviripudi et al. (2007), and may be the cause of the difficulty in synthesizing large diameter nanotubes in other reports. The broadening of the particle size distribution is thought to be owed to particle coalescence by ripening and migration. This change in morphology during pretreatment may even modify the nature of the catalyst surface, thus modifying the catalytic properties of the nanoparticles and consequently the morphology of the carbon products produced (Pisana et al., 2007; Wang et al., 2007).

Typically, the vapour-liquid-solid mechanism is used to explain the mechanism of carbon uptake, supersaturation and precipitation in the catalyst. However, owing to the low solubity of carbon in bulk Au, this must be reviewed for the synthesis of CNTs from Au nanoparticles. By studying the formation of carbon nanowires (CNWs) from Au catalysts, Takagi et al. (2008) inferred that nanosized Au shows some carbon solubility and that Au can form Au-C nanoalloy droplets and produce CNWs by the VLS mechanism. When the catalyst size approaches $\leq$ 5 nm, the carbon solubility and the nanowire nucleation energy increases dramatically, leading to a structural change in the synthesized carbon products from CNW to SWNT. This finding is in agreement with simulations by Yazyev & Pasquarello (2008), who determined that monatomic carbon in Au can diffuse uniformly across the nanoparticle, even at low temperatures. However, it should be noted that no direct evidence of this effect was demonstrated

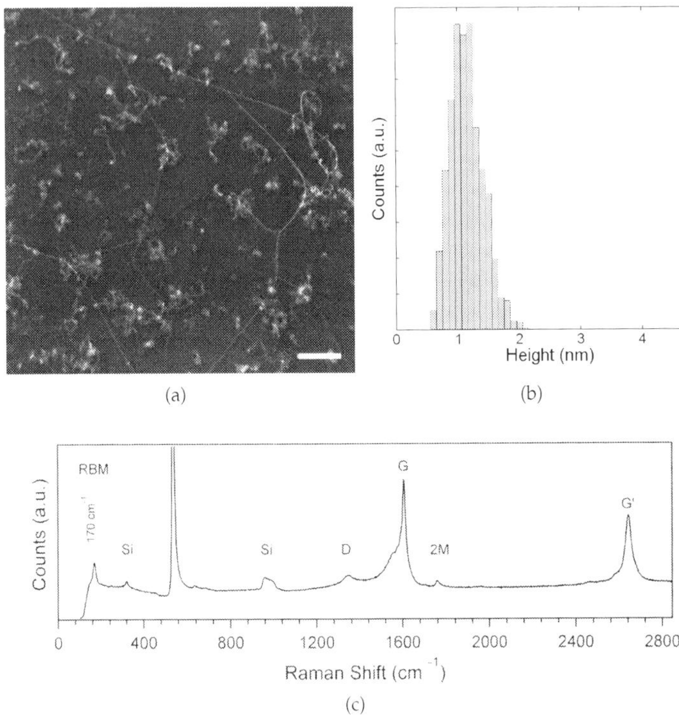

Fig. 2. (a) SEM image of CNTs synthesized from Au nanoparticles pretreated in H_2 at 1000 °C. Scale bar corresponds to 250 nm. (b) Particle size distribution of the as-deposited Au catalyst on a SiO_2 substrate. (c) Typical Raman spectrum from CNTs synthesized from the Au catalyst at the optimum growth condition.

and that similar studies (Yoshihara et al., 2008) could not determine whether carbon atoms were supplied to the nanotube from the Au-C liquid phase or through surface diffusion.

Metallic Cu, long considered to be a contaminant in the growth of SWNTs, has also been reported as an efficient catalyst for SWNT formation in several studies (Takagi et al., 2006; Yuan et al., 2008; Zhou et al., 2006). Figure 3(a) shows a SEM image of CNTs synthesized from a Cu catalyst. In this experiment, Cu nanoparticles were formed by the thermal decomposition of $Cu(NO_3)_2$ in air at 400 °C, deposited from a 1 mM isopropanol solution on a SiO_2 support. The particle size distribution of the catalyst, as determined by AFM, is shown in Figure 3(b). The mean particle size was found to be 1.5 ± 0.4 nm (mean $\pm$ standard deviation), with a particle density 350 ± 50 particles/μm^2. The samples were then pretreated in an H_2 atmosphere for 10 minutes at 900 °C, followed by a growth step in a mixture of CH_4 and H_2 at the same temperature. Raman spectroscopy showed that the synthesized carbon products were predominantly high quality SWNTs (Figure 3(c)).

It has been reported that the carbon solubility in a metallic catalyst should be in the range of 0.5 – 1.5 wt% carbon in order to efficiently form CNTs (Deck & Vecchio, 2006). Therefore,

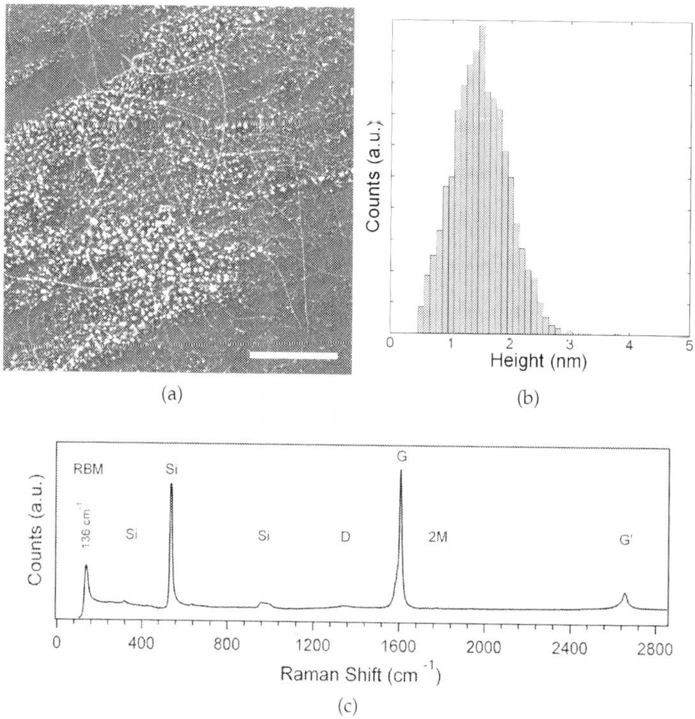

Fig. 3. (a) SEM image of CNTs synthesized from Cu nanoparticles at 900 °C. Cu nanoparticles were fabricated by the thermal decomposition of $Cu(NO_3)_2$, deposited from solution. Scale bar corresponds to 1 μm. (b) Particle size distribution of the decomposed Cu catalyst on a SiO_2 substrate. (c) Typical Raman spectrum from CNTs synthesized from the Cu catalyst at the optimum growth condition.

it is surprising that Cu can be catalytically active, as its carbon solubility is extremely low. However, Zhou et al. (2006) argue that the low solubility of carbon in Cu results in an increased rate of carbon precipitation. Additionally, Cu has a lower catalytic ability for the dissociation of alkanes than traditional catalysts, resulting in a slower supply of carbon in the CVD process. Thus, matching the supply of carbon to the formation rate of nanotubes will result in the production of high quality SWNTs (Lu & Liu, 2006). This argument is supported by Yazyev & Pasquarello (2008), who state that the stability and the diffusion barriers of diatomic carbon on Cu allow one to restrict the diffusion pathways to the nanoparticle surface by choosing an appropriate gas-phase carbon source, resulting in the preferred formation of high quality SWNTs.

Interestingly, Zhou et al. (2006) reported a higher ratio of metallic SWNTs in Cu catalysed samples, determined from Raman analysis. This characteristic was not detected in our experiments. However, only one laser excitation line was used and the sampling size was too small to draw any significant conclusions. Simulations by Yazyev & Pasquarello (2008) also found that the nucleation of graphitic fragments bound to the Cu nanoparticle catalyst favours the formation of metallic nanotubes. In addition, the low melting point and low carbon diffusion barriers suggest that CVD synthesis could take place at much lower temperatures. In these conditions, the chirality preference would be further enhanced.

5. Semiconductor Nanoparticle Catalysts

Results presented in the previous sections demonstrate that hydrocarbon dissociation and graphite formation abilities are not essential in a catalyst to synthesize CNTs. This leads to a new interpretation of the role of the catalyst particle in CNT growth, where only a nanoscale curvature is needed to act as a template for nanotube formation. This assertion is supported by the reports of CNT formation from semiconductor nanoparticles (Takagi et al., 2007; Uchino et al., 2009; 2008; 2005b), from which no catalytic functions were expected.

The first reports of CNT growth from semiconducting catalysts were by Uchino et al. (2005b). In this experiment, carbon-doped SiGe islands, deposited by CVD on Si, form nanoscale clusters through various mechanisms which act as a seed for SWNT growth. These results were supported by the work of Takagi et al. (2007), who showed that CNT growth from Ge, Si and SiC nanoparticles was possible. More recently, there have been various reports of CNT growth from SiO_2 nanoparticles (Huang et al., 2009; Liu et al., 2009), which are thought to be promising catalysts owing to their ability to maintain a narrow size distribution at CNT growth temperatures.

In this section, research on the use of germanium for carbon nanotube growth is reviewed. Four different techniques to synthesize CNTs based upon Ge nanoparticle catalysts are investigated. These are based on SiGe islands, Ge Stranski-Krastanow dots, Ge nanoparticles formed by ion implantation and colloidal Ge nanoparticles. It is shown that in all cases high quality SWNTs can be grown.

5.1 SiGe Islands

A 50 nm thick $Si_{0.7}Ge_{0.3}$ layer was deposited by CVD on Si(001) wafers after the growth of a thin Si buffer layer. To accommodate the stress resulting from the lattice mismatch between Si and Ge, the SiGe layer forms islands on top of a thin wetting layer. The heights of the islands ranged from 20 to 50 nm. Subsequently, the islands were implanted with carbon ions (energy 30 keV, dose of 3×10^{16} cm^{-2}). This heavy ion implantation is thought to induce damage and form an amorphous layer at the surface (Uchino et al., 2005a). The substrates were then dipped in buffered HF solution to remove the native oxide. Chemical oxidation was performed using a 30% hydrogen peroxide (H_2O_2) solution at room temperature. This step was followed by a pretreatment step in a mixture of Ar and H_2 for 10 minutes at 900 °C, followed by the CNT growth step in a mixture of CH_4 and H_2 at 850 °C.

Figure 4(a) shows a SEM image of the as-synthesized products on SiGe islands. In this image, two distinct types of nanostructures are visible. The short and thick nanofibres, approximately 20 nm in diameter and 1 μm in length, are formed during the pretreatment step. These nanostructures were identified as SiO_x nanowires by TEM, Raman and photoluminescence measurements, and are formed by the carbothermic reduction of SiO_2 (Lee et al., 2004; Li et al., 2004). These fibres were easily removed by an HF vapour etch, as shown in Figure 4(b). The second type of nanostructure forms during the growth step and comprises straight and thin fibres of less than 10 nm diameter and approximately 5 μm in length. Raman measurements, shown in Figures 4(c) and 4(d), confirm that these fibres are SWNTs. Despite considerable effort, the disorder induced D-band feature that is normally seen at 1350 cm^{-1} (Dresselhaus

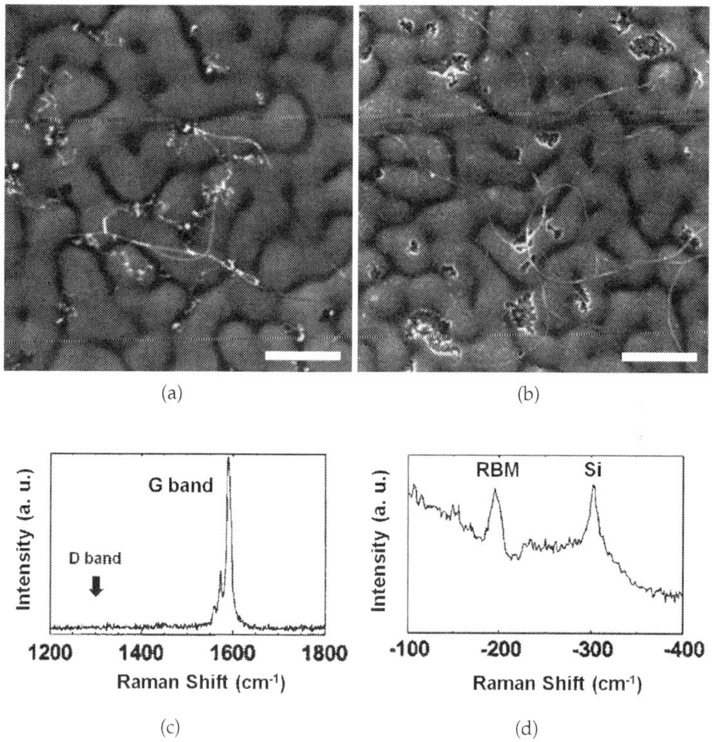

Fig. 4. SEM images of as grown CNTs and SiO_x nanowires synthesized from C implanted SiGe islands (a) before and (b) after HF vapour etching, showing that only carbon nanotubes remain. Scale bar corresponds to 500 nm. Typical Raman spectra of the as-grown CNTs showing (c) G-band characteristic and (d) anti-Stokes spectra showing the radial breathing mode characteristic.

et al., 2005) could not be detected. This indicates that the nanotubes have a low defect density, and thus could be described as high quality.

In this experiment, nanoscale Ge clusters are formed following the chemical oxidation and annealing of the SiGe layers. The oxidation behaviour of SiGe layers has been studied to a great extent (Liou et al., 1991; Paine et al., 1991). Si is known to have a stronger thermodynamic tendency to be oxidised in comparison to Ge. Therefore, the dry oxidation of SiGe alloys, with a low Ge content, results in the formation of SiO_2 and the segregation of Ge clusters from the growing oxide (Sass et al., 2002). It should be noted that this effect is less pronounced with wet oxidation, and the oxide layer typically contains a mixture of Si-O and Ge-O bonds. However, upon annealing in a reducing atmosphere, the Ge-O bonds are preferentially broken owing to a lower stability, resulting in the formation of nanoscale Ge clusters (Paine et al., 1993). These clusters are thought to act as the catalyst for the growth of CNTs in this methodology.

5.2 Ge Stranski-Krastanow Dots

Figure 5(a) shows a TEM image of a bundle of SWNTs grown from Ge Stranki-Krastanow dots. In this experiment, Ge Stranski-Krastanow dots are formed by CVD deposition of Ge atop a thin Si buffer layer. This step forms Ge dots in the form of cones with diameters from 20 to 250 nm and heights between 10 and 25 nm. Subsequently, the islands were implanted with carbon ions (energy 30 keV, dose of 3×10^{16} cm^{-2}). The substrates were then dipped in buffered HF solution to remove the native oxide and subjected to a chemical oxidation using a 30% hydrogen peroxide (H_2O_2) solution at room temperature. This step was followed by a pretreatment step in a mixture of Ar and H_2 for 10 minutes at 900 °C, followed by the CNT growth step in a mixture of CH_4 and H_2 at 850 °C. Raman measurements on the synthesized CNTs (Figure 5(b)) clearly show the radial breathing mode feature and tangential G band mode expected for SWNTs. The radial breathing modes indicate that the diameters of the synthesized CNTs are in the range 1.6 to 2.1 nm, which are slightly larger than those on SiGe islands (Uchino et al., 2005b). Again, the disorder induced D-band could not be detected, indicating that these CNTs are of a high quality.

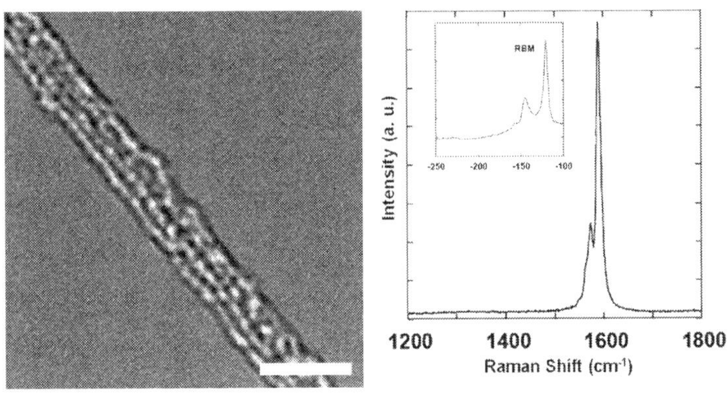

(a) (b)

Fig. 5. (a) TEM image of a bundle of SWNTs synthesized from C implanted Ge Stranski-Krastanow dots. Scale bar corresponds to 10 nm. The TEM sample was prepared by scraping the substrate surface with a surgical blade and collecting the material on a holey-carbon TEM grid. (b) Typical Raman spectra of the as-grown CNTs showing the G-band characteristic. Inset shows anti-Stokes spectra displaying the radial breathing mode characteristic.

It is believed that the mechanism of formation is very similar to that of the CNTs grown from SiGe islands. Upon chemical oxidation of the Ge Stranski-Krastanow dots, a thin layer of SiGe oxide is formed. Following a subsequent anneal in a reducing atmosphere, Ge clusters are nucleated and it is believed that these act as catalysts in this growth technique. Sass et al. (2002) reported that after the oxidation of Ge islands on Si(001), recovery of the original dot structure was not possible. Instead, a reduction of the GeO_2 around the single crystalline core of the non-oxidized Ge dot materials results in only Ge-enriched clouds, surrounded by a matrix of non-reductible material. This assertion is supported by SEM images taken after each stage in the process (not shown), which indicate a definite change of morphology after the chemical oxidation and reduction steps (Uchino et al., 2008). In fact, there have been reports of the formation of ultra-high density Ge nanoparticles, with diameters of about 4 nm, from the oxidation/reduction of Ge/Si surfaces (Nakamura et al., 2004), which further supports this hypothesis.

5.3 Ge Nanoparticles fabricated by ion implantation

In order to further investigate the role of Ge nanoparticles in the growth of carbon nanotubes, Ge nanoparticles were fabricated directly by Ge ion implantation into a layer of thermally oxidised SiO_2, and subsequently annealed at 600 °C. This step was followed by a HF vapour etch to remove the SiO_2 and expose the Ge nanoparticles. AFM characterisation revealed that a uniform layer of nanoparticles had been synthesized with a mean density of 460 ± 30 particles/μm^2 and a modal height of 1.8 nm. The particle size distribution is shown in Figure 6(a). This result shows good agreement with others in the literature, for instance Min et al. (1996) reported the formation of Ge nanocrystals by a similar process, with an average size of 1.9 ± 0.8 nm. Selected samples were then implanted with C. The C implanted samples showed a lower particle density (70 ± 18 particles/μm^2), a lower modal height (0.7 nm), and a narrower size distribution, shown in Figure 6(b). This change in morphology is attributed to a sputtering effect caused by the ion implantation.

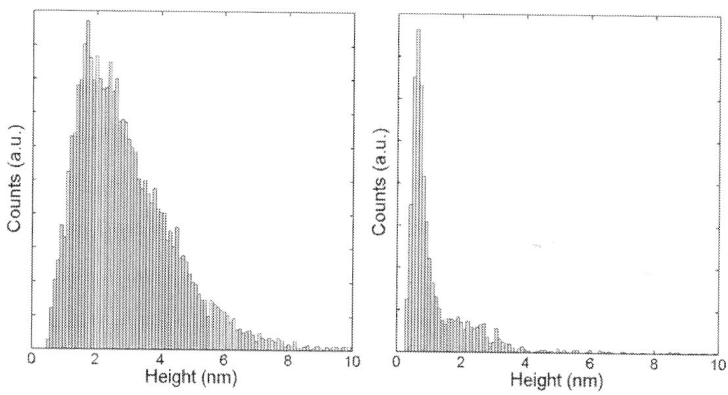

(a) (b)

Fig. 6. (a) AFM particle size distributions of Ge nanoparticles synthesized by Ge ion implantation (20 keV, 5×10^{15} cm^{-2}) into a 30 nm thick SiO_2 layer and annealed at 600 °C for 40 min in N_2 followed by an HF vapour etch to expose the nanoparticles. (b) AFM particle size distribution for samples given a C implant (30 keV, 3×10^{16} cm^{-2})

Figure 7(a) and 7(b) shows a typical SEM image after CNT growth for a sample without and with C implantation, respectively. Both images show that a good density of CNTs are achievable using this methodology. Representative Raman spectra for samples without and with C implantation are shown in Figure 7(c) and 7(d), respectively. All samples (insets) clearly show the radial breathing mode, indicating that single walled nanotubes are present. In the case of CNTs grown without C implantation, a small D-band peak is visible around 1320 cm^{-1}, which can be attributed to disorder in the nanotubes. In contrast, samples grown from the C implanted samples show no D-band peak. This indicates that the SWNTs synthesized from C implanted Ge nanocrystals have a low defect density, and are thus high quality.

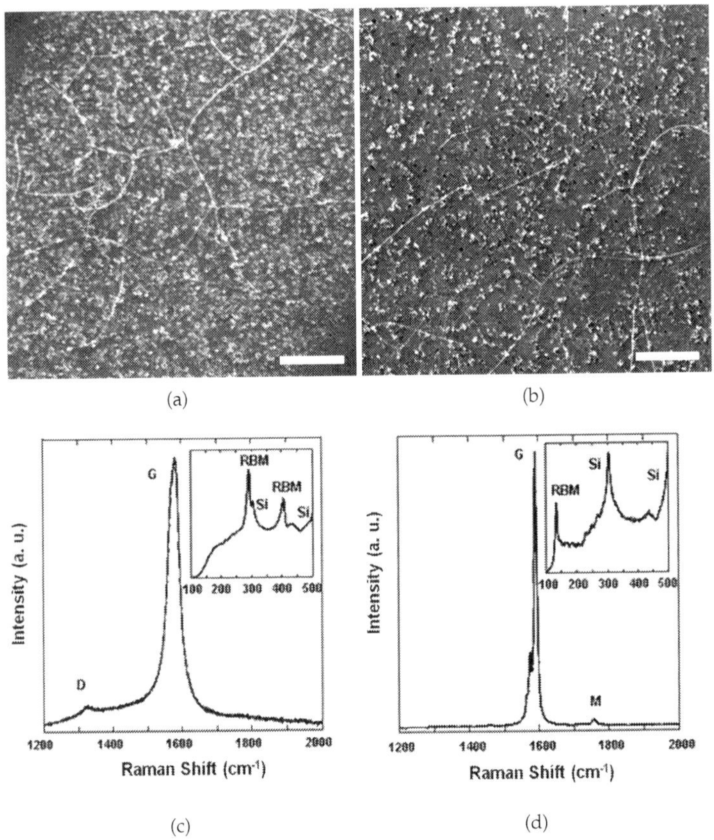

Fig. 7. SEM images of CNTs grown from Ge nanoparticles fabricated by ion implantation from (a) nonimplanted samples and (b) carbon implanted samples. Scale bar corresponds to 500 nm. Typical Raman spectra of the as-grown CNTs from (c) nonimplanted and (d) carbon implanted Ge nanocrystals. Inset shows Stokes spectra displaying the radial breathing mode characteristic.

At the optimum growth condition, there is no statistically significant benefit in terms of area density from the C implant. However, results at other growth conditions show that successful CNT growth can be achieved for a wider range of temperatures when the C implant is formed. AFM measurements taken to analyse the influence of the pretreatment temperature revealed that samples without a C implant show a strong reduction in particle density with increasing pretreatment temperature. In contrast, samples with a C implant show a much smaller decrease in particle density with increasing pretreatment temperature. This suggests that the C implant might increase the Ge melting point through formation of a $Ge_{1-y}C_y$ alloy. This hypothesis is supported by the phase diagram of the Ge-C system presented by Scace & Slack

(1959), which indicates that the presence of a small percentage of C has a strong effect in raising the melting point of Ge. Furthermore, Kanazawa et al. (2001) demonstrate that $Ge_{1-y}C_y$ alloys can be successfully formed by C implantation into Ge.

5.4 Colloidal Germanium Nanoparticles

Figure 8(a) shows an AFM image of Ge nanoparticles deposited from a 1mM colloidal solution on a SiO_2 support by spin coating. The colloidal solution was synthesized by means of an inverse micelle method. The synthesis of a $Ge[N(SiCH_3)_2]_2$ precursor for the formation of Ge nanocrystals was based on the works of Lessing et al. (1977a;b), with some minor modifications. The precursor (50mg) was dissolved in 7 ml of trioctylamine and injected into hot (340 °C), molten hexadecylamine (HDA, 1g). The injection method was inspired from various publications and has been standardised for this purpose (Nair et al., 2002, and references therein). The residue was then dissolved in toluene, re-precipitated with methanol and suspended in tryoctlamine. This process yielded nanoparticles with a narrow size distribution, shown in Figure 8(b). The synthesized nanoparticles had a mean size of 1.5 ± 0.4 nm, and the density of the spin coated layer was 430 ± 60 particles/μm^2.

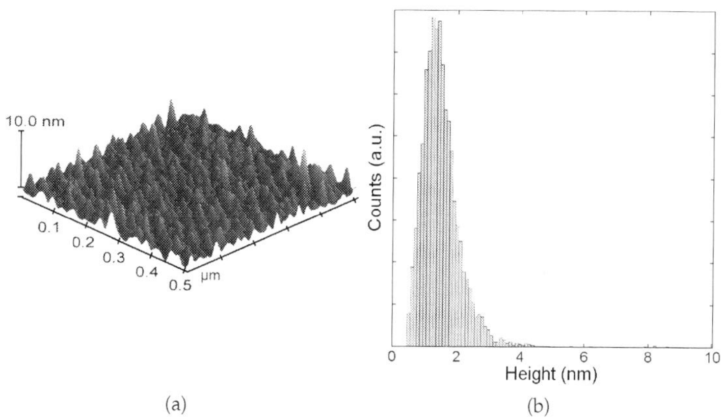

Fig. 8. (a) AFM image and (b) particle size distributions of colloidal Ge nanoparticles, deposited on SiO_2 by spin coating of a 1 mM solution. After deposition, samples were cleaned in a 100W O_2 plasma for 30 minutes.

Samples were then subjected to a 100W O_2 plasma for 30 minutes in order to remove the organic residue left from deposition. Subsequently, samples were pretreated in an H_2 atmosphere for 10 minutes at temperatures ranging from 850 – 1050 °C, followed by a growth step in a mixture of CH_4 and H_2 at 850 °C. The highest area density was found for samples pretreated at 900 °C, shown in Figure 9(a). Raman spectroscopy showed that the synthesized carbon products were predominantly high quality SWNTs (Figure 9(c)). Synthesis on sapphire (Al_2O_3) substrates showed that a slightly higher uniformity and area density of CNTs was achievable on this support medium, shown in Figure 9(b). This is attributed to the ability of the Al_2O_3 support to provide an interface for graphite formation (Rummeli et al., 2007).
In comparison to the non C implanted Ge nanoparticles fabricated by ion implantation, the process window for this methodology was considerably wider, with little change in CNT area

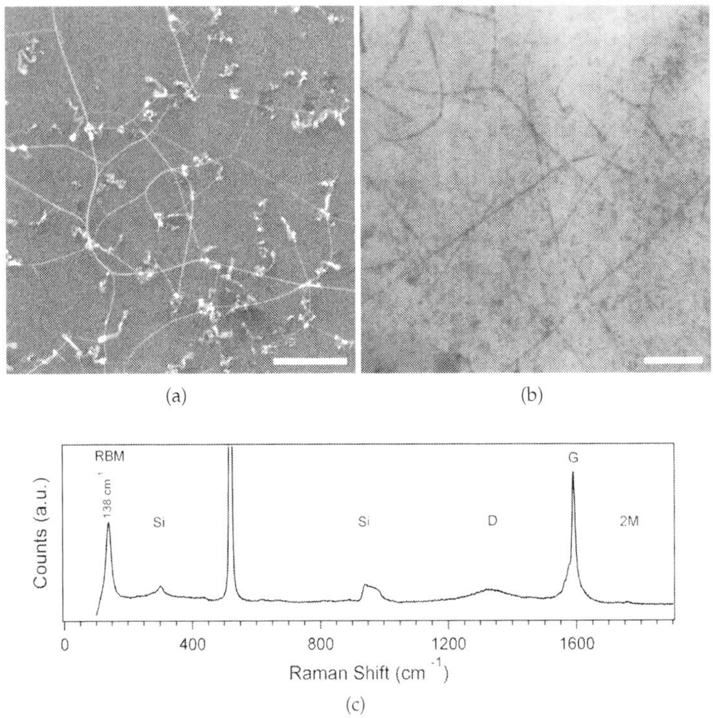

Fig. 9. SEM image of CNTs synthesized from colloidal Ge nanoparticles pretreated at 900 °C in H_2, followed by a CNT growth step at 850 °C in a mixture of CH_4 and H_2 on a (a) SiO_2 support and a (b) sapphire (Al_2O_3) support. Scale bar corresponds to 500 nm in both images. (c) Typical Raman spectrum from CNTs synthesized from the colloidal Ge catalyst at the optimum growth condition.

density in samples pretreated at temperatures from 900 to 1000 °C. AFM studies of the catalyst after H_2 pretreatment revealed that there was no statistically significant reduction in particle density with increasing pretreatment temperature. In addition, rather than a reduction in mean particle size, a very slight increase in size was found with increasing pretreatment temperature. This effect is believed to be due to the organic cap on the nanoparticles acting as a stabilising agent as the furnace ramps up to temperature. It should be noted that at the synthesis temperatures, the organic cap on the nanoparticles is expected to be fully reduced.

6. Growth Mechanism Discussion

Despite enormous strides in the synthesis of carbon nanotubes, the mechanism for growth is still a highly debated issue. As discussed previously, it is generally accepted that the model for carbon filament growth (Baker, 1989; Baker et al., 1972; 1973), derived from concepts of vapour-liquid-solid theory, also applies to carbon nanotube growth. This belief arises from the visual observation (by TEM) of catalyst particles on the ends of nanotubes, as was the case with carbon filaments. In this model, hydrocarbons adsorbed on the metal nanoparticle are catalytically decomposed resulting in atomic carbon dissolving into the liquid catalyst particle, and when a supersaturated state is reached, carbon precipitates in a tubular, crystalline form.

However, the results presented in this work suggest that this belief holds several observational inconsistencies that do not support this mechanism for CVD production. The successful CNT growths from catalyst free, noble metal catalysts and semiconducting catalysts imply that hydrocarbon dissociation ability is not essential in a catalyst. It should be noted that the catalytic behaviour of Cu and Au may be explained by electron donation to the support (Vander Wal et al., 2001), creating *d*-vacancies which may cause hydrocarbon dissocation. However, the ability of catalyst-free and semiconducting catalysts to seed CNT growth cannot be explained by the same mechanism. Reilly & Whitten (2006) argue that a more likely scenario is that a free radical condensate (FRC) provides carbon species through a leaving group, such as hydrogen (or oxygen). FRCs naturally form during hydrocarbon pyrolysis by the breaking of carbon-hydrogen or carbon-carbon bonds with each fragment keeping one electron to form two radicals. The presence of a radical in a hydrocarbon molecule permits rapid rearrangement of carbon bonds. In this case, the catalyst particle's role is to simply provide an interface where carbon rearrangement can occur and act as a template for growth.

Typically, metal catalysts with no *d*-vacancies, such as Cu and Au, do not offer sites to dissolve carbon, such that neither saturation nor precipitation is possible. However, despite the low carbon solubility, these catalysts have demonstrated an ability to catalyse CNT formation. Additionally, catalysts with a high melting point such as Al_2O_3 or ZrO_2 are thought to be solid at CNT synthesis temperatures. However, if nanoparticles of these metals are small enough ($\leq$ 5 nm), the increasing fraction of low-coordinated atoms may lead to surface saturation followed by carbon precipitation, as reported by Takagi et al. (2006). Considering that carbon penetration inside small nanoparticles is unlikely (Raty et al., 2005), the growth of CNTs is most likely a process primarily controlled by surface diffusion (Chadderton & Chen, 1999; Ding et al., 2005). Indeed, it is believed by several groups that the rate-limiting factor in CNT synthesis is the surface diffusion of carbon across the catalyst (Bartsch et al., 2005; Hofmann et al., 2005). Additionally, this factor could explain the influence of the carbon source on the ability of a catalyst to synthesize CNTs. Yazyev & Pasquarello (2008) reported different activation energies for the surface diffusion of C dimers and adatoms on noble metal catalysts, and argued that appropriate choice of a diatomic or monatomic carbon gas-phase source could significantly accelerate diffusion.

The specificity of the growth of nanotubes on nanoparticles with regard to the growth of carbon filaments is their nanometer dimensions. Other mechanisms are therefore required to explain the nucleation of CNTs from nanoparticle catalysts. One such model is the Yarmulke mechanism proposed by Dai et al. (1996). In the Yarmulke mechanism, a graphene cap is assembled on the particle surface with its edges strongly chemisorbed to the catalyst. The graphene cap acts to reduce the high total surface energy of the particle caused by its high curvature, owing to the fact that the basal plane of graphite has an extremely low surface energy. As additional carbon atoms are added, the hemifullerene cap formed on the particle surface lifts off, creating a hollow tube with constant diameter which grows away from the particle (Nikolaev et al., 1999). This model was supported by molecular dynamics simulations by Shibuta & Maruyama (2003). Recent works in high-resolution *in-situ* TEM observation of the catalytic growth of CNTs have verified this mechanism (Helveg et al., 2004; Hofmann et al.,

2007). These studies have also shown that cap stabilisation and nanotube growth involve re-shaping of the catalyst nanoparticle.

7. Conclusions and Future Directions

In this review, the role of the catalyst in the selective growth of SWNTs by CVD has been studied. Evidence of self-assembled carbon nanostructures was presented, indicating that upon annealing, C undergoes a structural reorganization to form graphitic structures. In addition, SWNT growth was shown to be possible from ceramic, noble metal and semiconducting catalysts. This demonstrates that hydrocarbon dissociation and graphitization ability are not essential in a catalyst, and it was argued that carbon species are supplied through a free radical condensate. The results presented show that the commonly utilised model of carbon filament growth is inadequate to describe SWNT growth from non traditional catalysts. A new interpretation of the role of the catalyst was presented where only a nanoscale curvature is necessary to grow CNTs. A mechanism for SWNT growth was suggested where the surface saturation and diffusion of C on the catalyst nanoparticle lead to the formation of a graphitic cap followed by the further incorporation of C into a growing nanotube.

Although there have been tremendous advances in the fabrication of CNTs, the integration of these nanostructures into successful applications and large-scale production processes depend on the understanding of several fundamental issues, which are yet to be addressed. A few of these issues are briefly discussed below.

The role played by the support in the CVD of CNTs is not yet fully understood. The simplistic view that the support only plays a catalytically passive role in the formation of CNTs requires examination. The work of Rummeli et al. (2007) demonstrated that under typical CVD growth conditions, nanoparticles of difficult-to-reduce metal oxides were capable of promoting ordered carbon growth. The authors attributed this to the presence of surface defect sites on the nanoparticle oxides. However, the interface between the catalyst nanoparticle and the support is thought to act as a annular defect site. This would indicate that the nature of the support-nanoparticle interface may be very important to the behaviour of the catalyst.

There appears to be a consensus in the literature concerning the correlation of catalyst size and SWNT diameter. Several groups have observed a direct dependence of the two quantities (Jeong et al., 2005; Wei et al., 2001). However, whether there is a correlation between the atomic structure/facets of the nanoparticle catalyst and the chirality of the synthesized nanotube has not yet been determined. If a good correlation is found between the facets and the chirality of the nanotube, how can the physical structure of the nanoparticles be retained as the sample is heated to the synthesis temperature?

The low-temperature synthesis of CNTs is of significant technological importance. This is essential for synthesis on glass substrates or other applications requiring plastic substrates. However, it is not yet clear whether this is possible and what the effect on the material quality would be. The integration of carbon nanotubes in IC manufacturing would also require large area growth uniformities, however most research is undertaken in small, low-throughput batch reactors, typically on small pieces of wafers. This is expected to change as the market for CNT-based products begins to emerge, creating the need for large-scale commercial reactors.

8. References

Awasthi, K., Srivastava, A. & Srviastava, O. N. (2005). Synthesis of Carbon Nanotubes, *J. Nanosci. Nanotechnol.* **5**: 1616–1636.

Baker, R. T. K. (1989). Catalytic Growth of Carbon Filaments, *Carbon* **27**: 315.

Baker, R. T. K., Barber, M. A., Harris, P. S., Feates, F. S. & White, R. J. (1972). Nucleation and Growth of Carbon Deposites from the Nickel Catalyzed Decomposition of Acetylene, *J. Catal.* **26**: 51.

Baker, R. T. K., Harris, P. S., Thomas, R. B. & Waite, R. J. (1973). Formation of Filamentous Carbon from Iron, Cobalt and Chromium Catalyzed Decomposition of Acetylene, *J. Catal.* **30**: 86.

Bartsch, K., Biedermann, K., Gemming, T. & Leonhardt, A. (2005). On the Diffusion-Controlled Growth of Multiwalled Carbon Nanotubes, *J. Appl. Phys.* **97**: 114301.

Berger, L.-M., Gruner, W., Langholf, E. & Stolle, S. (1999). On the Mechanism of Carbothermal Reduction Processes of TiO_2 and ZrO_2, *Int. J. Refract. Met. Hard Mater.* **17**: 235–243.

Bhaviripudi, S., Mile, E., Steiner, S. A., Zare, A. T., Dresselhaus, M. S., Belcher, A. M. & Kong, J. (2007). CVD Synthesis of Single-Walled Carbon Nanotubes from Gold Nanoparticle Catalysts, *J. Am. Chem. Soc.* **129**: 1516–1517.

Botti, S., Asilyan, C. R. L., Dominicis, L. D., Fabbri, F., Orlanducci, S. & Fiori, A. (2004). Carbon Nanotubes Grown by Laser-annealing of SiC Nano-particles, *Chem. Phys. Lett.* **400**: 264–267.

Botti, S., Asilyan, L. S., Ciardi, R., Fabbri, F., Lortei, S., Santoni, A. & Orlanducci, S. (2001). Catalyst-Free Growth of Carbon Nanotubes by Laser Annealing of Amrphous SiC Films, *Chem. Phys. Lett.* **396**: 1–5.

Botti, S., Ciardi, R., Terranova, M. L., Piccirillo, S., Sessa, V., Rossi, M. & Vittori-Antisari, M. (2002). Self-Assembled Carbon Nanotubes Grown Without Catalyst From Nanosized Carbon Particles Adsorbed on Silicon, *Appl. Phys. Lett.* **80**: 1441.

Chadderton, L. T. & Chen, Y. (1999). Nanotube Growth by Surface Diffusion, *Phys. Lett. A* **263**: 401–405.

Dai, H., Rinzler, A. G., Nikolawv, P., Thess, A., Colbert, D. T. & Smalley, R. E. (1996). Single-wall Nanotubes Produced by Metal-Catalyzed Disproportionation of Carbon Monoxide, *Chem. Phys. Lett.* **260**: 471–475.

Deck, C. P. & Vecchio, K. (2006). Prediction of Carbon Nanotube Growth Success by the Analysis of Carbon-Catalyst Binary Phase Diagrams, *Carbon* **43**: 2654–2663.

Derycke, V., Martel, R., Radosavljevic, M., Ross, F. M. & Avouris, P. (2002). Catalyst Free Growth of Ordered Single Walled Carbon Nanotube Networks, *Nano Lett.* **2**: 1043–1046.

Ding, F., Rosen, A. & Bolton, K. (2005). Dependence of SWNT Growth Mechanism on Temperature and Catalyst Particle Size: Bulk versus Surface Diffusion, *Carbon* **43**: 2215–2217.

Dresselhaus, M. S., Dresselhaus, G., Sait, R. & Jorio, A. (2005). Raman Spectroscopy of Carbon Nanotubes, *Phys. Rep.* **409**: 47.

Haruta, M. (1997). Size- and Support-Dependency in the Catalysis of Gold, *Catal. Today* **36**: 156–166.

Helveg, S., Lopez-Cartes, C., Sehested, J., Hansen, P. L., Clausen, B. S., Rostrup-Nielsen, J. R., Abild-Pedersen, F. & Norskov, J. K. (2004). Atomic-scale Imaging of Carbon Nanofibre Growth, *Nature* **427**: 426–429.

Hofmann, S., Csányi, G., Ferrari, A. C., Payne, M. C. & Robertson, J. (2005). Surface Diffusion: The Low Activation Energy Path for Nanotube Growth, *Phys. Rev. Lett.* **95**: 036101.

Hofmann, S., Sharma, R., Ducati, C., Du, G., Mattevi, C., Cepek, C., Cantoro, M., Pisana, S., Parvez, A., Cervantes-Sodi, F., Ferrari, A. C., Dunin-Borkowski, R., Lizzit, S., Petaccia, L., Goldoni, A. & Robertson, J. (2007). In situ Observations of Catalyst Dynamics during Surface-Bound Carbon Nanotube Nucleation, *Nano Lett.* **7**: 602–608.

Huang, S., Cai, Q., Chen, J., Qian, Y. & Zhang, L. (2009). Metal-Catalyst-Free Growth of Single-Walled Carbon Nanotubes on Substrates, *J. Am. Chem. Soc.* **131**: 2094–2095.

Hvolbaek, B., Janssens, T. V. W., Clausen, B. S., Falsig, H., Christensen, C. H. & Norkov, J. K. (2007). Catalytic Activity of Au Nanoparticles, *Nano Today* **2**: 14–18.

Iijima, S. (1991). Helical Microtubules of Graphitic Carbon, *Nature* **354**: 56.

Jeong, G.-H., Suzuki, S., Kobayashi, Y., Yamazaki, A., Yoshimura, H. & Homma, Y. (2005). Effect of Nanoparticle Density on Narrow Diameter Distribution of Carbon Nanotubes and Particle Evolution During Chemical Vapour Deposition, *J. Appl. Phys.* **98**: 124311.

Jia, J., Haraki, K., Kondo, J. N., Domen, K. & Tamaru, K. (2000). Selective Hydrogenation of Acetylene over Au/Al_2O_3 Catalyst, *J. Phys. Chem. B* **104**: 1153–1156.

Kanazawa, Y., Katayama, K., Nozawa, K., Saitoh, T. & Kubo, M. (2001). Preparation of $Ge_{1-y}C_y$ Alloys by C Implantation into Ge Crystal and Their Raman Spectra, *Jpn. J. Appl. Phys.* **40**: 5880.

Koshio, A., Yudasaka, M. & Iikima, S. (2002). Metal-free Production of High-Quality Multi-Wall Carbon Nanotubes, in which the Innermost Nanotubes Have a Diameter of 0.4 nm, *Chem. Phys. Lett.* **356**: 595–600.

Kusonoki, M., Rokkak, M. & Suzuki, T. (1997). Epitaxial Carbon Nanotube Fil Self Organised by Sublimation Decomposition of Silicon Carbide, *Appl. Phys. Lett.* **71**: 2620.

Larciprete, R., Lizzit, S., Botti, S., Cepek, C. & Goldini, A. (2002). Structural Reorganisation of Carbon Nanoparticles into Single-Walled Nanotubes, *Phys. Rev. B.: Condes. Matter Mater. Phys.* **66**: 121402.

Larciprete, R., Lizzit, S., Cepek, C., Botti, S. & Goldini, A. (2003). Thermal Reactions at the Interface between Si and C Nanoparticles: Nanotube Self-Assembling and Transformation into SiC, *Surf. Sci.* **532**: 886–891.

Lee, K.-H., Yang, H. S., Baik, K. H., Bang, J., Vanfleet, R. R. & Sigmund, W. (2004). Direct Growth of Amorphous Silica Nanowiresby Solid State Transformation of SiO_2 Films, *Chem. Phys. Lett.* **383**: 380.

Lee, S. Y., Yamada, M. & Miyake, M. (2005). Synthesis of Carbon Nanotubes Over Gold Nanoparticle Supported Catalysts, *Carbon* **43**: 2654–2663.

Lessing, J. G. V., Fouche, K. F. & Retief, T. T. (1977a). Redox Extractions from Molten Alkali-Metal Cyanides. Part 1. Mechanism for Extractions with Liquid Zinc Alloys, *Dalton Trans.* **20**: 2020.

Lessing, J. G. V., Fouche, K. F. & Retief, T. T. (1977b). Redox Extractions from Molten Alkali-metal Cyanides. Part 2. Mechanism for Extractions with Liquid Tin Alloys, *Dalton Trans.* **20**: 2004.

Li, S.-H., Zhu, X.-F. & Zhao, Y.-P. (2004). Carbon-Assisted Growth of SiO_x Nanowires, *J. Phys Chem. B* **108**: 17032.

Liou, H. K., Mei, P., Gennser, U. & Yang, E. S. (1991). Effects of Ge Concentration on SiGe Oxidation Behavior, *Appl. Phys. Lett.* **59**: 1200.

Liu, B., Ren, W., Gao, L., Li, S., Pei, S., Liu, C., Jiang, C. & Cheng, H.-M. (2009). Metal-Catalyst-Free Growth of Single-Walled Carbon Nanotubes, *J. Am. Chem. Soc.* **131**: 2082–2083.

Liu, H., Takagi, D., Ohno, H., Chiashi, S., Chokan, T. & Homma, Y. (2008a). Effect of Ambient Gas on the Catalytic Properties of Au in Single Walled Carbon Nanotube Growth, *Jpn. J. Appl. Phys.* **47**: 1966–1970.

Liu, H., Takagi, D., Ohno, H., Chiashi, S., Chokan, T. & Homma, Y. (2008b). Growth of Single Walled Carbon Nanotubes from Ceramic Particles by Alcohol Chemical Vapor Deposition, *Appl. Phys. Express* **1**: 014001.

Lu, C. & Liu, J. (2006). Controlling the Diameter of Carbon Nanotubes in Chemical Vapor Deposition Method by Carbon Feeding, *J. Phys. Chem. B* **110**: 20254.

Mctaggart, F. K. (1961). Reduction of Zirconium and Hafnium Oxides, *Nature* **191**: 1192.

Melechko, A. V., Merlukov, V. I., McKnight, T. E., Guillorn, M. A., Lowndes, D. H. & Simpson, M. L. (2005). Vertically Aligned Carbon Nanofibers and Related Structures: Controlled Synthesis and Directed Assembly, *J. Appl. Phys.* **97**: 041301.

Meyyappan, M., Yamada, T., Sarrazin, P. & Li, J. (2005). *Carbon Nanotubes: Science and Applications*, CRC Press.

Min, K. S., Shcheglov, K. V., Yang, C. M., Atwater, H. A., Brongersma, M. L. & Polman, A. (1996). The Role of Quantum-Confined Excitons vs Defects in the Visible Luminescence of SiO_2 Films Containing Ge Nanocrystals, *Appl. Phys. Lett.* **68**: 2511.

Nair, P. S., Radhakrishnan, T., Revaprasadu, N., Kolawolea, G. & OŠBrien, P. (2002). Cadmium Ethylxanthate: A Novel Single-Source Precursor for the Preparation of CdS Nanoparticles, *J. Mater. Chem.* **12**: 2722.

Nakamura, Y., Nagadomi, Y., Sugie, K., Miyata, N. & Ichikawa, M. (2004). Formation of Ultrahigh Density Ge Nanodots on Oxidized Ge/Si(111) Surfaces, *J. Appl. Phys.* **95**: 5014.

Nikolaev, P., Bronikowski, M. J., Bradley, R. K., Rohmund, F., Colbert, D. T., Smith, K. A. & Smalley, R. E. (1999). Gas-Phase Catalytic Growth of Single-Walled Carbon Nanotubes from Carbon Monoxide, *Chem. Phys. Lett.* **313**: 91–97.

Paine, D. C., Caragianis, C., Kim, T. Y., Shigesato, Y. & Ishahara, T. (1993). Visible Photoluminescence from Nanocrystalline Ge Formed by H_2 Reduction of $Si_{0.6}Ge_{0.4}O_2$, *Appl. Phys. Lett.* **62**: 2842.

Paine, D. C., Caragianis, C. & Schwartzman, A. F. (1991). Oxidation of $Si_{1x}Ge_x$ Alloys at Atmospheric and Elevated Pressure, *J. Appl. Phys.* **70**: 5076.

Pisana, S., Cantoro, M., Parvez, A., Hofmann, S., Ferrari, A. C. & Robertson, J. (2007). The Role of Precursor Gases on the Surface Restructuring of Catalyst Films During Carbon Nanotube Growth, *Physica E* **37**: 1.

Raty, J.-Y., Gygi, F. & Galli, G. (2005). Growth of Carbon Nanotubes on Metal Nanoparticles: A Microscopic Mechanism from *Ab-Initio* Molecular Dynamics , *Phys. Rev. Lett.* **95**: 096103.

Reich, S., Li, L. & Robertson, J. (2006). Control the Chirality of Carbon Nanotubes by Epitaxial Growth, *Chem. Phys. Lett.* **421**: 469.

Reilly, P. T. A. & Whitten, W. B. (2006). The Role of Free Radical Condensates in the Production of Carbon Nanotubes During the Hydrocarbon CVD Process, *Carbon* **44**: 1653–1660.

Rummeli, M. H., Kramberger, C., Gruneis, A., Ayala, P., Gemming, T., Buchner, B. & Pichler, T. (2007). On the Graphitization Nature of Oxides for the Formation of Carbon Nanostructures, *Chemistry of Meterials* **19**: 4105–4107.

Sass, T., Zela, V., Gustafsson, A., Pietzonka, I. & Seifert, W. (2002). Oxidation and Reduction Behavior of Ge-Si Islands, *Appl. Phys. Lett.* **81**: 3455.

Scace, R. I. & Slack, G. A. (1959). Solubility of Carbon in Silicon and Germanium, *J. Chem. Phys.* **30**: 1551.

Shibuta, Y. & Maruyama, S. (2003). Molecular Dynamics Simulation of Formation Process of Single-Walled Carbon Nanotubes by CCVD Method, *Chem. Phys. Lett.* **382**: 381–386.

Steiner, S. A., Baumann, T. F., Bayer, B. C., Blume, R., Worsley, M. A., MoberlyChan, W. J., Shaw, E. L. ., Schlogl, R., Hart, A. J., Hofmann, S. & Wardle, B. L. (2009). Nanoscale Zirconia as a Nonmetallic Catalyst for Graphitization of Carbon and Growth of Single- and Multiwall Carbon Nanotubes, *J. Am. Chem. Soc.* **131**: 12144–12154.

Takagi, D., Hibino, H., Suzuki, S., Kobayashi, Y. & Homma, Y. (2007). Carbon Nanotube Growth from Semiconductor Nanoparticles, *Nano Lett.* **7**: 2272–2275.

Takagi, D., Homma, Y., Hibino, H., Suzuki, S. & Kobayashi, Y. (2006). Single Walled Carbon Nanotube Growth from Highly Activated Metal Nanoparticles, *Nano Lett.* **6**: 2642–2645.

Takagi, D., Kobayashi, Y., Hibino, H., Suzuki, S. & Homma, Y. (2008). Mechanism of Gold-Catalyzed Carbon Material Growth, *Nano Lett.* **8**: 832–835.

Takikawa, H., Miyano, R., Yatsuki, M. & Sakakibara, T. (1998). Carbon Nanotubes on SiC Powder Surface Grown by a Vacuum Heating Process, *Jpn. J. Appl. Phys.*, **2**: L187.

Uchino, T., Ayre, G. N., Smith, D. C., Hutchison, J. L., de Groot, C. H. & Ashburn, P. (2009). Growth of Single-Walled Carbon Nanotubes Using Germanium Nanocrystals Formed by Implantation, *J. Electrochem. Soc.* **156**: K144–K148.

Uchino, T., Bourdakos, K. N., Ayre, G. N., de Groot, C. H., Ashburn, P. & Smith, D. C. (2008). CMOS Compatible Synthesis of Carbon Nanotubes, *Mater. Res. Soc. Symp. Proc.* **1081**: 1081–P01–09.

Uchino, T., Bourdakos, K. N., de Groot, C. H., Ashburn, P., Kiziroglou, M. E., Dilliway, G. D. & Smith, D. C. (2005a). Catalyst Free Low Temperature Direct Growth of Carbon Nanotubes, *Proceedings of 2005 5th IEEE Conference on Nanotechnology* **5**: 1–4.

Uchino, T., Bourdakos, K. N., de Groot, C. H., Ashburn, P., Kiziroglou, M. E., Dilliway, G. D. & Smith, D. C. (2005b). Metal Catalyst-Free Low-Temperature Carbon Nanotube Growth on SiGe Islands, *Appl. Phys. Lett.* **86**: 233110.

Vander Wal, R. L., Tichich, T. M. & Curtis, V. E. (2001). Substrate-support Interactions in Metal-Catalyzed Carbon Nanofiber Growth, *Carbon* **39**: 2277–2289.

Wagner, R. S. & Ellis, W. C. (1964). Vapor-Liquid-Solid Mechanism of Single Crystal Growth, *Appl. Phys. Lett.* **4**: 89–90.

Wang, W. P., Wen, H. C., Jian, S. R., Juang, J. Y., Lai, Y. S., Tsai, C. H., Wi, W. F., Chen, K. T. & Chou, C. P. (2007). The Effects of Hydrogen Plasma Pretreatment on the Formation of Vertically Aligned Carbon Nanotubes, *Appl. Surf. Sci.* **253**: 9248.

Wei, Y. Y., Eres, G., Merlukov, V. I. & Lowndes, D. H. (2001). Effect of Catalyst Film Thickness on Carbon Nanotube Growth by Selective Area Chemical Vapour Deposition, *Appl. Phys. Lett.* **78**: 1394.

Yazyev, O. V. & Pasquarello, A. (2008). Effect of Metal Elements in Catalytic Growth of Carbon Nanotubes, *Phys. Rev. Lett.* **100**: 156102.

Yoshihara, N., Ago, H. & Tsuji, M. (2008). Growth Mechanism of Carbon Nanotubes over Gold-Supported Catalysts, *Jpn. J. Appl. Phys.* **47**: 1944–1948.

Yuan, D., Ding, L., Chu, H., Feng, Y., McNicholas, T. P. & Liu, J. (2008). Horizontally Aligned Single-Walled Carbon Nanotube on Quartz from a Large Variety of Metal Catalysts, *Nano Lett.* **8**: 2576–2579.

Zhou, W., Han, Z., Wang, J., Zhang, Y., Jin, Z., Sun, Z., Zhang, Y., Yan, C. & Li, Y. (2006). Copper Catalyzing Growth of Single-Walled Carbon Nanotubes on Substrates, *Nano Lett.* **6**: 2987–2990.

3

Properties of Carbon Nanotubes under External Factors

Yaroslav V. Shtogun and Lilia M. Woods
University of South Florida
USA

1. Introduction

Carbon nanotubes are quasi-one dimensional nanostructures with many unique properties. These materials are considered as promising building blocks in nanoelectronics, spintronics, and nano-electromechanics. Potential applications of carbon nanotubes have strongly relied on their extraordinary ability of property changes upon external factors, such as mechanical alterations or applied external fields. The success in application of carbon nanotubes has significantly depended on predictions from theoretical calculations and modeling. In particular, first principal simulation methods have proven to be useful techniques in predicting the fundamental nature of changes in carbon nanotube properties and how these properties can be altered under external factors. There is a considerable amount of research on simulations of carbon nanotubes under one type of external factors, such as mechanical defects, mechanical deformations, applied external electric fields, *or* applied external magnetic fields. However, the development and design of current and new carbon nanotube based nanodevices has led to the search of additional ways to modify and tailor their properties. To achieve this flexibility, researchers have begun to study the influence of more than one external factor simultaneously. This is a relatively new area which opens up more possibilities for new devices. Thus, the knowledge of physical property changes in carbon nanotubes under several external factors is crucial for fundamental science and technological prospects.

Our primary focus in this chapter is first principal simulations of electronic and magnetic properties of single wall carbon nanotubes under two external factors. The discussion involves understanding the combined effect of radial deformation *and* mechanical defects, such as a Stone-Wales defect, nitrogen impurity, and mono-vacancy in the carbon network. Furthermore, we will discuss the nanotube property changes due to the application of radial deformation *and* external electric field. To provide the basis for further arguments, we start with a review of important results of carbon nanotubes under one external factor such as mechanical deformations, mechanical defects, or external fields. After we provide the description of the calculation method and model, we arrive at the major issue of radially deformed carbon nanotubes with defects or external electric field. We start with the analysis of the structural changes in defective carbon nanotubes under two types of radial deformations. One is deformation achieved by squeezing of the nanotube between two hard narrow surfaces, which have width smaller than the nanotube cross section. The other one is achieved by squeezing of the nanotube between two hard surfaces with width larger than the nanotube cross section. As a result, we show the characteristic energies of deformation and defect-formation. Further, we show first principal calculations of band gap modulations, band structure evolutions, and changes in magnetic properties of carbon nanotubes in terms of types of defects and degrees of radial deformation. A special consideration is given to the mechanically induced magnetic properties. Finally, we present the combined effect of radial deformation and external electric field on the nanotube electronic structure and band gap modulations. The chapter reveals greater capabilities for carbon nanotube applications in new devices.

2. Carbon Nanotubes under One External Factor

In many earlier theoretical calculations, carbon nanotubes were considered to be with perfect structure, where ideal graphite plane was rolled up into a cylindrical tube. Depending on the particular way of graphene rolling, nanotubes can be metallic or semiconducting. The rolling direction is characterized by a chiral index (n, m). Achiral zigzag $(n,0)$ and armchair (n,n) nanotubes are distinguished from the rest (chiral nanotubes). Armchair tubes are always metallic, while zigzag tubes can be semiconducting or metallic (Reich et al., 2004). It has been revealed that these nanostructures have interesting mechanical and electronic properties strongly depended on chirality, which immediately drew interest to their application in micro-electro-mechanics, spintronics, and nanoelectronics. However, experimental measurements have shown that the nanotube structure is not perfect (Fan et al., 2005; Hashimoto et al., 2004; Suenaga et al., 2007). External factors, such as mechanical deformations, mechanical defects, or applied external fields can induce various changes in nanotube properties In addition, researchers have realized that these external factors can be used in new methods to control and/or tune carbon nanotube properties for practical applications.

2.1 Mechanical Deformation of Carbon Nanotubes

Carbon nanotubes have high elastic modulus, strength, flexibility, and low mass density, which makes them excellent candidates for polymer composite materials and nano-devices (Ashraf et al., 2006; Fennimore et al., 2003; Itkis et al., 2008). They are rigid in the axial direction while they are flexible in the radial direction. In actual experiments, nanotubes undergo structural modifications such as bending, twisting, stretching, and radial deformation (Barboza et al., 2008; Cohen-Karni et al., 2006; Falvo et al., 1997). It has been suggested that the most common mechanical deformation is radial squeezing. Such deformation takes place in bundles of nanotubes, under applied hydrostatic pressure, in interaction of nanotubes with surfaces, nano-indentation experiments, and atom force microscopy (AFM) squeezing of nanotubes (Hertel et al., 1998; Minary-Jolandan & Yu 2008; Tang et al., 2000). In many experiments the carbon nanotubes are manipulated by an AFM tip. This has a major effect not only on the mechanical, but also on the electronic and transport properties of nanotubes. For instance, pressing suspended nanotube over trench by an AFM tip causes the local deformation in the nanotube structure (Fig. 1). Subsequently, the deformed region affects the electron flow between the source and drain electrodes, and as a result, the conductance of carbon nanotubes decreases several times as compared to the theoretically predicted one (Leonard et al., 2005; Maiti et al., 2002).

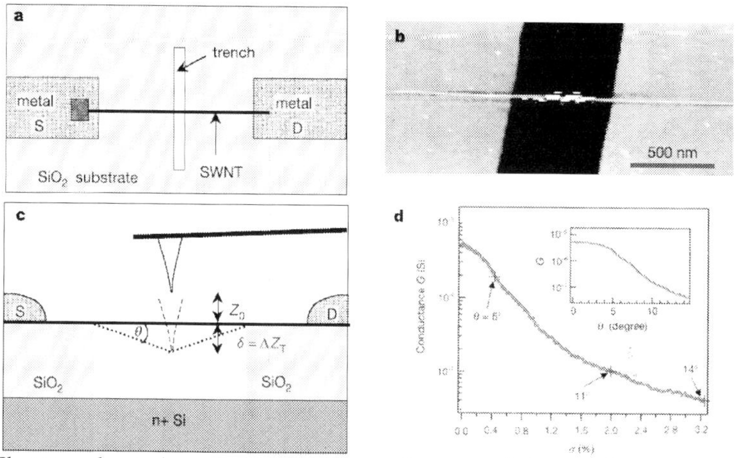

Fig. 1. Electromechanical measurements of a partly suspended nanotube over a trench. a) Device view from above. The substrate has a trench of 500 nm in width and 175 nm in depth. A pair of metal electrodes (S - source and D - drain) is bridged by nanotube suspended over the trench. b) An atom force microscopy image of suspended nanotube. c) Side view of experimental setup when the nanotube is pushed into a trench by atom force microscopy tip. d) Experimentally measured conductance of nanotube as a function of strain. All figures are from (Tombler et al., 2000).

Cross-sectional changes of carbon nanotube geometry under radial deformation are characterized by two applied strains ϵ_{xx} and ϵ_{yy} along x and y axis, respectively, and they are defined as follows:

$$\epsilon_{xx} = \frac{D_0 - a}{D_0} \tag{1}$$

and

$$\epsilon_{yy} = \frac{D_0 - b}{D_0} \tag{2}$$

where D_0 is the diameter of the undeformed nanotube, a and b are the major and minor axes of the ellipse (Fig. 2a-b). For discussion purposes here we use percentage (%) values of the applied strains ϵ_{xx} and ϵ_{yy}. Depending on the degree of radial deformation, the radial cross section of the nanotube can take a different form. When the applied strain is small, the nanotube has an elliptical-like shape. However, if the applied strain continues to increase, then nanotube cross section undergoes Peanut-like or Flat-like forms. The Peanut-like form corresponds to squeezing of the nanotube between two hard surfaces with a cross section smaller than its radial one, while the Flat-like form corresponds to radial squeezing between two hard surfaces with a cross section larger than its radial one (Fig. 2c-d). The Peanut and Flat deformations can also occur during the experimental measurements when the size of AFM tip is smaller or larger than the cross section of the nanotube.

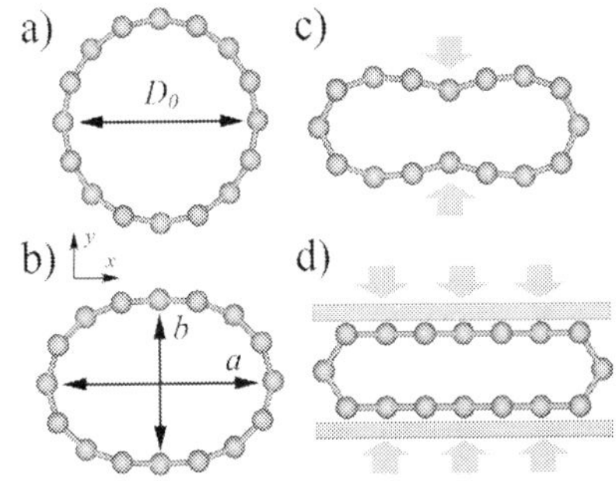

Fig. 2. Cross section view of (8,0) nanotubes under radial deformation. a) Perfect nanotube. b) Nanotube under small radial deformation with elliptical shape of the cross section. c) Peanut-like deformed nanotube. d) Flat-like deformed nanotube.

First principal calculations have shown that the band gap of zigzag semiconductor nanotubes decreases with the increase of radial deformation while for zigzag metallic nanotube it increases and then decreases (Fig. 3). At the same time, if the mirror symmetry of armchair carbon nanotubes is not broken, the radial deformation does not have any effect on the band gap (Gulseren et al., 2002; Park et al., 1999). Therefore, the application of radial deformation causes chirality dependent semiconductor-metal and metal-semiconductor transitions in carbon nanotubes. There are several factors responsible for these transitions, such as σ-π hybridization of the atomic orbitals located on the high curvature regions, interaction between low curvature regions, breaking mirror-symmetry of carbon nanotubes, and geometry deformations (Giusca et al., 2007, Gulseren et al., 2002; Mazzoni & Chacham 2000; Mehrez et al., 2005; Shan et al., 2005; Vitali et al., 2006).

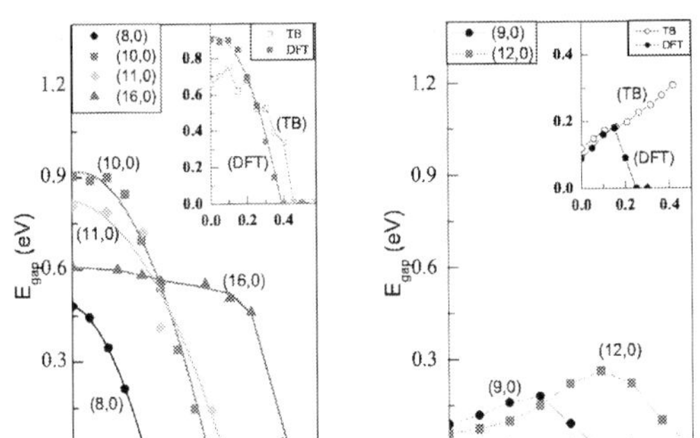

Fig. 3. Density Functional Theory (DFT) calculations of the band gap as a function of the cross sectional flattering for variety of semiconductor (left panel) and metallic (right panel) carbon nanotubes. The insets on both panels compare tight-binding (TB) and DFT results for (10,0) and (9,0) nanotubes, respectively from (Shan et al., 2005).

2.2 Mechanical Defects in Carbon Nanotubes

The presence of mechanical defects in carbon nanotube structure drastically affects all their properties such as transport, mechanical, magnetic, electronic, and optical. Defects can occur during growth, purification, alignment, or device application processes. They can also be introduced intentionally to modify the carbon nanotube properties. Common defects are Stone-Wales defects (Fig.4b), substitutional impurities (Fig. 4c), and vacancies (Fig. 4d). A Stone-Wales defect takes place in a carbon network when one C-C bond is rotated by 90^0 degree (Stone & Wales 1986). Such rotation creates two pairs of a pentagon and a heptagon around the rotated C-C bond. It is also known as a 5-7-7-5 defect (Nardelli et al., 2000). The pentagon/heptagon pairs create a local disturbance which serves as a reactive center for adsorption of various atoms, nanoparticles, and molecules (Bettinger 2005; Picozzi et al., 2004; Yang et al., 2006). Increasing the concentration of Stone-Wales defects leads to semiconductor metal transitions in zigzag nanotubes. It also increases the density of states around the Fermi level for armchair nanotubes due to new defect states, thus, making the tubes more metallic (Crespi & Cohen 1997).

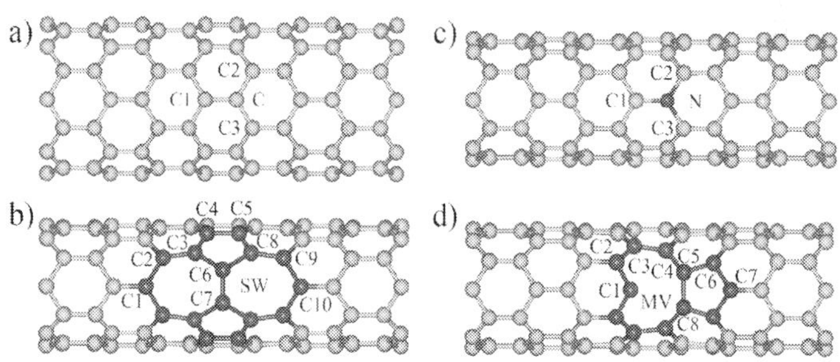

Fig. 4. Side view of (8,0) nanotube structure a) without deformation, and with different defects at ϵ_{yy} = 20% deformation for b) a Stone-Wales defect, c) nitrogen substitutional impurity, and d) mono-vacancy. The atoms of the defective sites are labeled.

A substitutional impurity in nanotube can happen when a carbon atom is substituted by a foreign one (Fig. 4c). It has been considered that the common doping atoms in carbon nanotubes are boron and nitrogen, which can appear during laser-ablation and arc-discharge synthesizing processes or during substitutional reaction methods (Droppa et al., 2002; Gai et al., 2004; Golberg et al., 1999). Since B and N atomic radii have similar size to the C atom, they create a small perturbation in the nanotube structure in comparison to the perfect one. However, other atoms such as Li, K, Br, and Ni, which can create a larger disturbance, are also considered for new nano-device application of nanotubes (Lee et al., 1997; Santos et al., 2008). The B and N serve as acceptors and donors of electrons in the nanotubes, respectively, since B has one electron less and N has one electron more than the carbon atom. Theoretical calculations have revealed that such impurities lead to the shift of the Fermi level, additional impurity states in the band structure, and a break of the nanotube mirror symmetry (Czerw et al., 2001; Moradian & Azadi 2006; Yi & Bernholc 1993). Fig. 5b shows evidence of new impurity levels in the density of states of (10,0) nanotube around the Fermi level under nitrogen and boron doping. The impurity effect on electronic properties becomes more evident if their concentration increases. Nitrogen doping has drawn substantial interest due to the one extra electron provided by a nitrogen atom, which causes magnetic effects in nanotubes, transforms semiconductor nanotube to metallic one, and serves as a reactive center in the interaction of nanotubes with other molecules (Nevidomskyy et al., 2003).

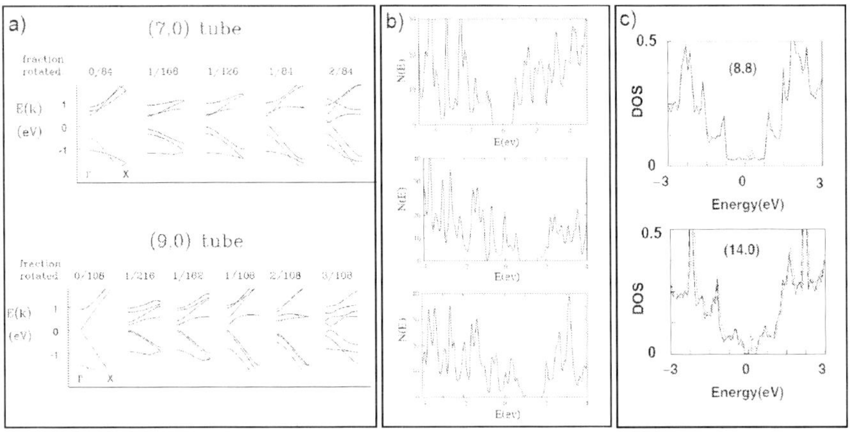

Fig. 5. Band structure and density of states of defective nanotubes. a) Band structure evolutions of (7,0) and (9,0) nanotubes upon increasing the concentration of Stone-Wales defect from (Crespi et al., 1997). b) Density of states (top) of (10,0) nanotube, (middle) where one of carbon atom is substituted by a nitrogen atom, and (bottom) where a carbon atom is substituted by a boron atom in the supercell of 120 carbon atom from (Moradian & Azadi 2006). The nitrogen and boron defect concentration is 0.83%. c) Density of states for (8,8) and (14,0) nanotubes without/with mono-vacancy defect in their structure (solid/dashed line respectively) from (Lu & Pan 2004).

Furthermore, removing a carbon atom leaves a vacancy defect with three dangling carbon bonds in the nanotube network (Rossato et al., 2005). These three dangling bonds are unstable and undergo recombination to make a chemical bond between two of them forming a pentagon ring and one remaining dangling bond – in a nonagon ring (Fig. 4d) (Berber, S. & Oshiyama 2006). The existence of vacancies has been demonstrated experimentally (Hashimoto et al., 2004). It occurs during synthesis of nanotubes or under intentional ion or electron irradiation techniques (Krasheninnikov et al., 2001; Krasheninnikov et al., 2005). The formation of vacancy and its orientation depends on the radius and chirality of nanotubes. Such a disturbance causes significant changes in their transport, magnetic, mechanical, and optical properties (Orellana & Fuentealba 2006; Tien et al., 2008; Yuan & Liew 2009). Fig. 5c shows the density of state of (8,8) and (14,0) carbon nanotubes without a vacancy (solid line) and with vacancy defect (dashed line). The vacancy defect gives the sharp peak in the density of states above the Fermi level, which consists of localized states from the C dangling bond. Such imperfections can reduce the mean-free path and mobility of the free carrier in nanotubes. This leads to the decrease in the nanotube conductivity. On the other hand, the dangling bond can participate in the interaction with other molecules as well as it can be the center of functionalization of nanotubes. Theoretical studies have shown that carbon nanotubes with vacancies demonstrate ferromagnetic ordering due to the dangling bond of the nonagon ring (Orellana & Fuentealba 2006). Moreover, the magnetic moment depends on the defect concentration. It opens up opportunities for the application of carbon nanotubes in spintronics, logical, and memory storage devices.

2.3 Carbon Nanotubes under External Fields

The successful application of carbon nanotubes in nano-devices such as filed-effect-transistors (Tans et al., 1998), rectifiers (Yao et al., 1999), p-n junctions (Lee et al., 2004), or sensors (Ghosh et al., 2003) requires the knowledge of changes in electronic properties of nanotubes under external electric or magnetic fields. However, it is also important to realize that carbon nanotube properties can be tuned by such fields to reach a desirable functionality. The cylindrical structure of nanotubes suggests two high symmetry directions for application of those external fields: parallel and perpendicular to the nanotube axis (Fig. 6). Theoretical and experimental studies have shown that external electric and magnetic fields have a drastic effect on electronic and magnetic properties of carbon nanotubes (Lai, et al., 2008; Li. & Lin 2006; Minot et al., 2004). The general pattern of external electric and magnetic fields is an electronic band structure modulation with semiconductor-metal or metal-semiconductor transition, as well as lifting the degeneracy of sub-bands (Fig. 6).

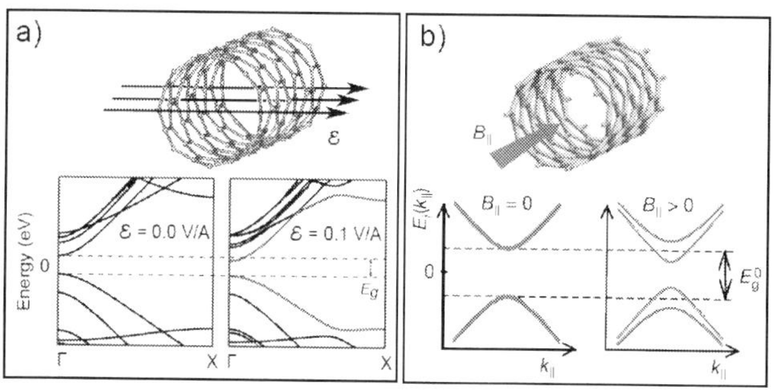

Fig. 6. a) Prospective view of (8,0) nanotube under transverse external electric field (top) and band structure changes under $\mathcal{E}$=0 V/Å and $\mathcal{E}$=0.1 V/Å external electric field (bottom). b) Prospective view of nanotube in the presence of magnetic field along nanotube axis (top) and band structure changes under magnetic field from (Minot et al., 2004). The magnetic and electric fields lift of the degeneracy of the band structure.

As Fig.6 shows, the energy band gap decreases, and the degeneracy of the energy bands around E_F disappears with the increase of the strength of applied fields. Application of the external electric field parallel to the nanotube axis has attracted a lot of attention because of the excellent field emission properties of carbon nanotubes (Jonge et al., 2004). Such properties have effectively been used in prototype devices such as flat panel display (Choi et al., 1999), x-ray tube (Sugie et al., 2001), and scanning x-ray source (Zhang et al., 2005). Nevertheless, the search for new devices is ongoing. Currently, researchers are exploring the

perpendicular direction of field applications as new opportunities of controlling electronic properties of nanotubes.
The application of external magnetic field leads to metal-insulator transition in nanotubes. Moreover, the carbon nanotubes can have diamagnetic or paramagnetic characteristics which depend on their chirality and the orientation of magnetic field (Lu 1995). When the magnetic field is applied along the nanotube axis (Fig.6b), oscillations in the magnetoresistance or in the band gap can be observed due to the Aharonov-Bohm effect, which is a periodic behavior of the wave-function phase factor in the cylindrical geometry (Bachtold et al., 1999; Lee et al., 2000). The Zeeman effect has been found to cause metallization of carbon nanotubes at certain values of the magnetic flux (Jiang et al., 2000). However, perpendicularly applied magnetic field has a different effect on the electronic structure of nanotubes. It opens the band gap at a weak strength and closes that band gap at a high strength of magnetic field for zigzag nanotube. When the magnetic length $l = \sqrt{c\hbar / eH}$ is smaller than the nanotube circumference, the formation of Landau level formations at high magnetic fields closes the band gap, while those levels do not appear in small magnetic fields (Ajiki & Ando 1996). The application of external electric or magnetic fields has also been used in experimental alignment of carbon nanotube samples (Kordas et al., 2007; Kumar et al., 2003).

3. Calculations Method and Model

3.1 Method

First principal simulations are considered to be a state of the art approach for nanoscale design. Such simulations allow investigating a large variety of systems, while these can be very difficult to recreate in actual experiments. They are a powerful tool to explore the functional properties of prospective nanodevices. Among atomistic simulation methods, density functional theory (DFT) is the most successful quantum mechanical technique for calculating the electronic structures of many materials. DFT results are found in a satisfactory agreement with numerous experimental data. Our main method of investigation is based on self-consistent DFT calculations within the local density approximation (LDA) for the exchange-correlation functional implemented within the Vienna Ab Initio Simulation Package (VASP) (Kresse & Furthmuller 1996). The code uses plane-wave basis set and it is a periodic supercell approach. The core electrons of the atoms are treated by ultrasoft Vanderbilt pseudopotentials or by the projector-augmented wave method (Kresse & Joubert 1999). This DFT method is rescalable for large systems, and it requires relatively small plane-wave basis set for each atom in the calculation. It also allows the application of external electric fields in the system (Neugebauer & Scheffler 1992). Over the last decay, DFT-LDA has been a useful model in studying large graphitic structures. It provides an accurate description of the electronic structure along the graphite planes and nanotube axis, where the electronic structure is characterized by sp^2 orbitals of the carbon atoms (Kolmogorov & Crespi 2005; Shtogun et al., 2007; Woods et al., 2007). Based on its past success, we utilize DFT-LDA for our studies. The calculation criteria for relaxation of the atom positions have been 10^{-5} eV for the total energy and 0.005 eV/Å for the total force. Spin-polarization effects are also included in the study for defective carbon nanotubes.

3.2 Model

As a model for investigating the combined effect of radial deformation and defects, we study a zigzag (8,0) single wall carbon nanotube. To avoid the interaction between nearest-neighbor defects we construct a supercell consisting of four unit cells along the z-axis of the nanotube which has the length of 17.03 Å after relaxation. The radial deformation is simulated by applying stress to the cross section of the nanotube along the y-direction from both sides. Consequently, the nanotube is compressed along the y-direction and extended along the x-direction (Fig. 2a-b). The nanotubes are squeezed by two types of strain. The first one is deforming the nanotube between two hard surfaces with a cross section smaller than the nanotube one, which we call "Peanut" deformation (Fig. 2c). The second one represents squashing the nanotube between two hard surfaces with a cross section larger than the nanotube one, which we call "Flat" deformation (Fig. 2d). Such deformations are modeled by fixing the y-coordinates of the top and bottom rows at low curvature in Peanut and Flat structures, while the rest of the atoms are free to relax. We show that the (8,0) nanotubes can undergo large radial deformations - up to ϵ_{yy}=75% and ϵ_{yy}=65% for Peanut and Flat deformation, respectively (Shtogun & Woods 2009a).

After the radial deformation, defects, such as a Stone-Wales defect, nitrogen impurity, or mono-vacancy are introduced on the high curvature of the nanotube. Experimentally this can be achieved, for example, by deforming the tube using an atom force microscope and then intentionally introducing defects using electron or ion irradiation methods. The radially deformed carbon nanotube with or without defects has an elastic response even under high degree of squeezing. When the applied stress is removed, the nanotube comes back to its original circular cross-sectional form regardless of the defect presence.

To study the combined effect of radial deformation and an external electric field we focus on zigzag semiconducting (8,0) and metallic (9,0) single wall carbon nanotubes. The constructed supercell consists of one unit cell along z-axis with length of 4.26 Å for the (8,0) and 4.25 Å for the (9,0) nanotubes after relaxation. The external applied electric field is generated by a saw-tooth electric potential (Neugebauer & Scheffler1992) in three different directions: $\mathcal{E}_x$ and $\mathcal{E}_y$ along x and y–axes, respectively, and $\mathcal{E}_{45^0}$ along 45^0 from the x–axis (Shtogun & Woods 2009b).

4. Carbon Nanotubes under Two External Factors

Many devices associated with different applications are based on the unique properties of carbon nanotubes and the opportunity of controlling them. The search for new methods to modulate nanotube properties has led to the introduction of deformation, mechanical defects, and external fields. Reaching the desirable properties of nanotubes under one external factor, however, can be difficult in device engineering. To overcome such difficulties, researchers now consider how more than one external factors can alter the carbon nanotube properties. Thus, it is important to build fundamental understanding of the combination of more than one external stimulus by starting with applying two stimuli simultaneously.

Studies have shown that the combination of radial deformation and impurity or vacancy, for example, can facilitate tuning electronic properties of nanotubes (Fagan et al., 2003; Fagan et al., 2004). As we have shown above, the radial deformation causes the closing of band gap in semiconductor nanotubes but how fast this band gap is closed can be controlled by various defects such as impurities and vacancies (Fagan et al., 2004). Radial deformation produces two regions with high and low curvatures (Fig.2b). As a result, the atoms on the high curvature tend to have sp^3 configuration, and they are more strained than the ones on the low curvature with sp^2 configuration. It makes it easier to create the defect on the high curvature in comparison to the low one. The flexibility in localization of the defect in radially deformed nanotubes gives more control over nanotube properties.

Band gap modulation can also be achieved by combining external electric fields and defects in nanotubes. This is of particular interest in application of nanotubes in field-effect transistors since the current can be turned off or on by applying external electric field. Such combination provides an opportunity to identify presence of defects in carbon nanotubes. Rotating nanotube in external electric field provides different responses in different directions (Tien et al., 2005). Fig. 7 shows the presence of a vacancy defect in (10,0) nanotubes with an applied external electric field for various directions. We can see that the value of the band gap can be controlled by the presence of vacancy, strength, and direction of electric field. It is interesting that usually semiconducting tubes are used in nanoelectronics. Metallic nanotubes have been discarded because their resistance is not sensitive to gate voltage or transverse electric field (Li et al., 2003). New views on metallic nanotube application are considered due to the combined effect of external electric field and impurities, which allows tuning the resistance by two orders of magnitude. The origin of this change comes from the backscattering of conduction electron by impurities, and it strongly depends on the strength and orientation of the applied electric field (Son et al., 2005).

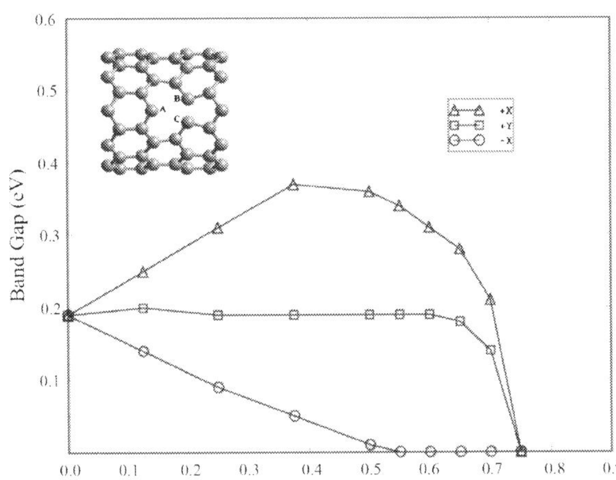

Fig. 7. Energy band gap evolution as a function of external electric field in three directions such as +x-axis, −x-axis, and +y-axis for vacancy defect in (10,0) nanotube with 79 carbon atoms from (Tien et al., 2005). The insert shows the optimized structure of vacancy defect.

Nowadays, most magnetic materials are based on *d* and *f* elements. However, as we mentioned above, sp^2 and sp^3 hybridization due to mechanical defects (vacancy and impurity) can also induce magnetism in carbon nanotubes. This is of great importance for nanotube applications in magnetic and spintronic devices. Magnetism can be further influenced by the combined effect of mechanical deformation and defects. The strong magneto-mechanical coupling has been shown in defective nanotubes under mechanical deformation (Zheng & Zhuang 2008). It can increase or decrease the magnetic moment and cause changes in the spin transport. To further investigate this, we have explored energetic, structural, magneto-mechanical, and electronic changes in carbon nanotubes under combined effect of radial deformation and external electric field along with radial deformation and different defects.

4.1 Structural Changes in Radially Deformed and Defective Carbon Nanotubes

In this section we analyze the changes in the carbon nanotube structure under radial deformation and various defects under two types of deformation - Peanut and Flat. The main focus is on the bonds on the high curvature and the bonds comprising the defects (Fig. 4). When the perfect nanotube is radially deformed, the bonds along the nanotube axis (*C-C1*) decrease while the curvature bonds (C-C2) increase to reduce the strain on the high curvature (Fig. 8a). Under moderate radial deformation (ϵ_{yy}<50%) there is no significant difference between Peanut and Flat deformation. However, at high deformation (ϵ_{yy}>50%) the difference becomes more evident since the Flat deformation applies much more distortion on the high curvature rather than the Peanut does. As a result, the bonds on the high curvature change more in Flat than Peanut deformation, while there are no changes along nanotube axis (Firg. 8a). It is worth mentioning that in the region of (ϵ_{yy}~20÷25%), the (8,0) nanotube shows interesting changes in the structure, which can be seen in the jump of the bond length profile. Later, there will be more discussion of this jump due to the admixture of different orbitals from the C atoms.

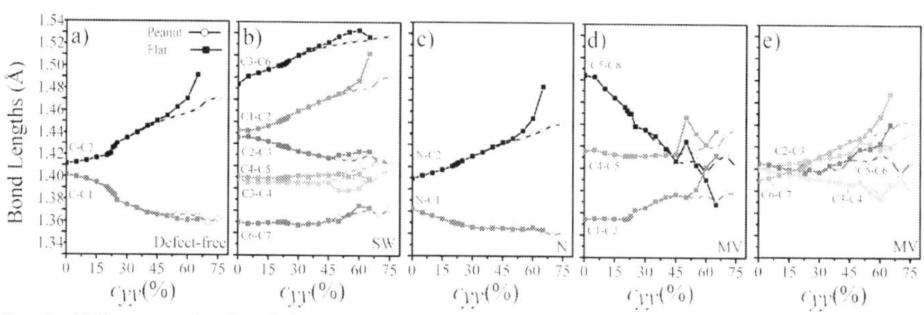

Fig. 8. (8,0) nanotube bond length evolution as a function of applied strain ϵ_{yy} for a) no defect, b) with a Stone-Wales defect, c) with a nitrogen impurity, d) and e) with a mono-vacancy.

Energetically favorable configuration of Stone-Wales defect happens when two heptagon rings align with the nanotube axis (Fig. 4b). Fig. 8b) shows the evolution of various bond changes as a function of ϵ_{yy}. Both, Peanut and Flat, deformations produce almost the same changes in the defect bond lengths with the exception of the *C1-C2* bond. In fact, the bonds in the heptagon rings (*C1-C2, C2-C3,* and *C3-C6*) undergo more changes than the bonds in pentagon ones (*C3-C4, C4-C5,* and *C6-C7*). Such changes are explained by the location of the heptagon rings on the high curvature because the pentagon rings are distant from it.

Moreover, the *C1-C2* functional behavior looks similar to the one for the *C-C2* in the defect-free nanotube.

The presence of a nitrogen impurity on the high curvature of the nanotube does not create much disturbance in its structure (Fig. 4c). Since the atomic size of the nitrogen atom is larger than the carbon one, it leads to the decrease in *N-C1* and *N-C2* bond lengths by 0.01 Å and 0.03 Å as compared to the *C-C1* and *C-C2* in defect-free nanotube for all ϵ_{yy}, respectively (Fig. 8c). Another evidence of small disturbance due to the nitrogen atom can be seen from the functional behavior of the *N-C1* and *N-C2* bonds, which exhibits similar dependence on ϵ_{yy} as the corresponding *C-C* bonds in the defect-free nanotube (Fig. 8a).

Removing a carbon atom from the carbon network creates a vacancy leaving three dangling bonds behind. Such configuration is not stable, and it undergoes reconstruction with the formation pentagon and nonagon along the axis of zigzag nanotube (Fig. 4d). On the contrary to the previous cases (defect-free, Stone-Wales, and nitrogen impurity) where the bond lengths show continuous behavior, vacancy has more dramatic dependence as a function of deformation ϵ_{yy} (Fig. 8d-e). In particular, when the distance between the two low curvatures becomes comparable with graphite interplanar distance, there are some relatively "sudden" jumps and dips in bond lengths for Flat deformation. We again observe that most changes in the bond lengths happen on the high curvature. For instance, the bond lengths of *C1-C2* and *C6-C7* increase by 0.06 Å and 0.08 Å, respectively, while the *C5-C8* decreases by 0.11 Å for the Flat deformation.

As a result, the presence of mechanical defects causes the structural changes in nanotubes where nitrogen impurity creates the lowest perturbation in nanotube network whiles the mono-vacancy - the largest one.

4.2 Characteristic Energies of Deformation and Defect Formation in Carbon Nanotubes

The structural changes described above lead to changes in deformation and defect formation energies. We introduce the characteristic energies of radial deformation and defect formations. The energy required to perform radial deformation is defined as:

$$E^f_{\epsilon_{yy}} = E_{\epsilon_{yy}} - E \tag{3}$$

where $E_{\epsilon_{yy}}$ is the total energy for the deformed nanotube and E is the total energy for the undeformed perfect nanotube. Fig. 9 shows the deformation energy for (8,0) and (9,0) carbon nanotubes as a function of ϵ_{yy}. The deformation energy has non-linear dependence and it rises with the increasing ϵ_{yy} showing that higher deformation requires more energy to overcome strain energy on the high curvature. On the other hand, the deformation energy for (8,0) nanotube is steeper than for (9,0) due to the larger curvature effect in (8,0) nanotube as compared to the (9,0) one.

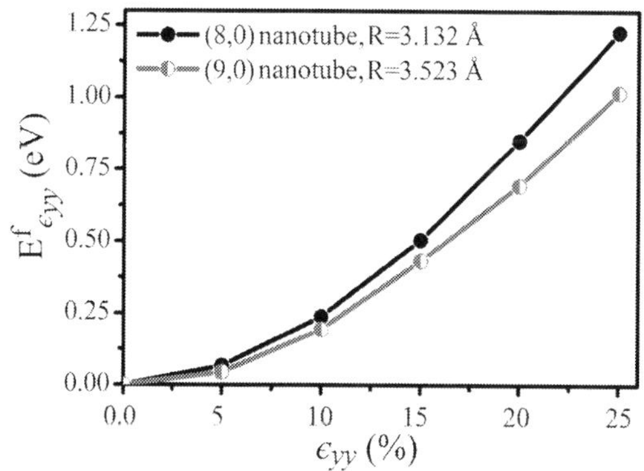

Fig. 9. The deformation energy as a function of radial deformation ϵ_{yy} for one unit cell of (8,0) and (9,0) nanotubes.

The defect formation energies for the Stone-Wales defect, nitrogen impurity, and vacancy at the high curvature of the radially deformed nanotube are calculated, respectively as:

$$E^f_{\epsilon_{yy}/SW} = E_{\epsilon_{yy}/SW} - E_{\epsilon_{yy}} \tag{4}$$

$$E^f_{\epsilon_{yy}/N} = E_{\epsilon_{yy}/N} - E_{\epsilon_{yy}} - \mu_N + \mu_C \tag{5}$$

$$E^f_{\epsilon_{yy}/MV} = E_{\epsilon_{yy}/MV} - E_{\epsilon_{yy}} + \mu \tag{6}$$

where $E_{\epsilon_{yy}/SW,N,MV}$ are the total energies for the deformed nanotube with the appropriate defect (SW - Stone-Wales, N - nitrogen impurity, and MV - mono-vacancy), $\mu_{N,C}$ are the chemical potentials for a free nitrogen and carbon atoms, respectively, and μ is the chemical potential of a carbon atom in the nanotube (energy per C atom in the supercell).

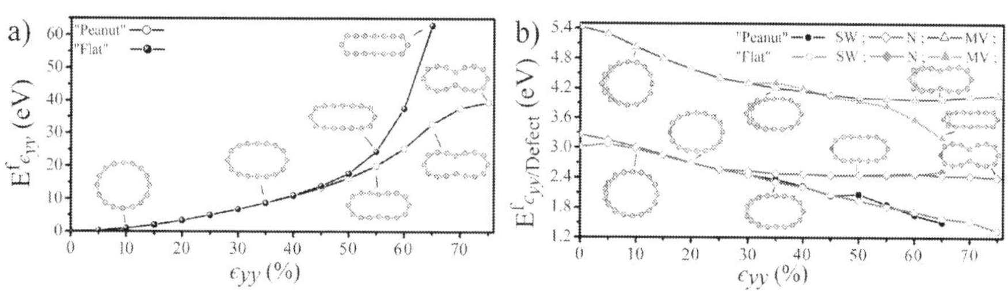

Fig. 10. The energy of a) deformation and b) defect formation as a function of ϵ_{yy} for Peanut and Flat deformations in (8,0) nanotube. Cross-sectional views of the deformed and defective (8,0) nanotubes are also provided.

Applied radial deformation of up to ϵ_{yy}<35% transforms the circular (8,0) nanotube to an elliptical-like one (Fig. 2b). Further increase of ϵ_{yy}>35% results in Peanut or Flat structures.

As it can be seen from Fig. 10a, the deformation energy becomes more distinctive between Peanut and Flat deformations for higher radial stress - ϵ_{yy}>45%. What actually happens for high deformation is that for the Peanut deformation only one row set on the top and the bottom of low curvatures are brought together. Consequently, there are no significant changes on the high curvature and elongation along *a*-axis (Fig. 2c). The Flat deformation, however, has five rows of atoms from top and bottom on the low curvatures brought together (Fig. 2d). As a result, it leads to major changes on high curvature and appropriate elongation along *a*-axis (Fig. 2d). Meanwhile, radial deformation energy shows steeper increase in Flat that in Peanut deformation. In addition, there is a repulsion between two low curvatures of nanotube at ϵ_{yy}>50%, where separation distance becomes smaller than graphite interplanar distance. The repulsion is stronger for Flat rather than Peanut deformation because fewer atoms are brought together in Peanut as compared to Flat deformation.

Fig. 10b also shows variations in the defect formation energy as a function of the deformation. All energies $E^{f}_{\epsilon_{yy}/SW,N,MV}$ decrease for all types of defects as ϵ_{yy} increases. Therefore, the defect formation occurs much easier in radially deformed carbon nanotubes than in the undeformed ones. This has been shown for Si doping of carbon nanotubes (Fagan et al., 2004). Analysis of defect formation energies shows that the lowest amount of energy is required to produce Stone-Wales defect, more energy is needed to make nitrogen impurity and the most energy-expensive is the vacancy formation at high curvature of carbon nanotubes. There is a small difference between Peanut and Flat deformations for Stone-Wales defect and nitrogen impurity, while this difference is significant for mono-vacancy. It relates to the fact that Stone-Wales and nitrogen impurity create small disturbance in nanotube structure for both deformations where the *C-C* bonds in mono-vacancy exhibit more dramatic changes for the Flat that Peanut deformations (Fig. 8d-e). Thus it is easier to break the three bonds to form a mono-vacancy after squashing between two wide surfaces as compared to the case of two narrow ones (Fig. 2c-d). As a result, the higher the deformation ϵ_{yy} is, the easier it is to make defects in carbon nanotube structure.

4.3 Electronic and Magnetic Properties of Carbon Nanotubes under Combined Effect of Radial Deformation and Defects

One of the main questions we consider here is what electronic response carbon nanotubes show under the combined effect of radial deformation and different defects. We address this issue by analyzing the modulation of the band gap E_g of defect-free and defective (8,0) nanotube under radial deformations ϵ_{yy} (Fig. 11). Initially, the defect-free (8,0) nanotube is a semiconductor with E_g=0.55 eV. Applying radial deformation leads to decreasing of E_g until its closure at ϵ_{yy} =23% where semiconductor-metal transition occurs. As we observed earlier, this semiconductor-metal transition is reflected in discontinuous change in the *C-C1* and *C-C2* bonds (Fig. 8a). On the other hand, it corresponds to a strong σ−π hybridization on high curvature which causes structural and semiconductor-metal transformations in (8,0) nanotube. Interestingly, such transition arises before (8,0) nanotube undergoes Peanut or Flat deformations. Further increase of radial deformation does not open second bang gap in the (8,0) nanotube.

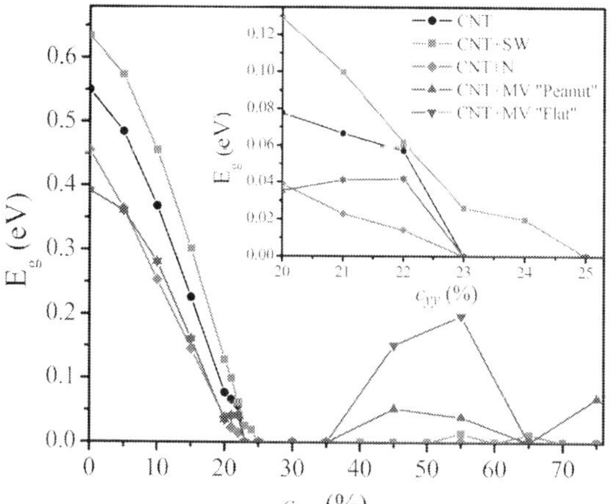

Fig. 11. Energy band gap E_g of perfect and defective (8,0) nanotubes as a function of radial deformation ϵ_{yy}. Insert shows the semiconductor-metal transition in (8,0) nanotube in detail.

The presence of defects at high curvature tunes the value of band gap in nanotubes. For instance, Stone-Wales defect increases E_g, while E_g decreases for nitrogen impurity and mono-vacancy as compared to defect-free nanotube at ϵ_{yy}=0%. Applying radial deformation to the defective nanotube has shown similar trends as in defect-free nanotubes (Fig. 11). Moreover, the semiconductor-metal transition happens at the same degree of deformation ϵ_{yy}=23% for nitrogen impurity and mono-vacancy. However, it occurs at ϵ_{yy}=25% for Stone-Wales defect. Further increasing deformation opens the band gap in nanotube with mono-vacancy and Stone-Wales defect again, while nanotube with nitrogen impurity remains metallic for all deformations. In addition, the band gap can be modulated by Peanut or Flat deformations at high deformation. Still, we can see that strong σ–π hybridization is a major factor in semiconductor-metal transition in defective nanotubes but defect-deformation coupling is significant for further opening of the band gap at high deformations.

To elucidate the mechanism of semiconductor-metal transformation in defect-free and defective nanotubes we carry out comprehensive analysis of the electronic structure changes for various cases. The applied radial deformation removes the degeneracy in sub-bands of defect-free nanotubes (Fig. 12a). Increasing ϵ_{yy} causes the lowest conduction band and the highest valence band to move towards the Fermi level. They cross the Fermi level at the point different than Γ for ϵ_{yy}=23%. The further deformation leads to the shift of crossing point with E_F towards the X point. For sufficiently large ϵ_{yy}>50%, multiple crossings at the Fermi level from higher conduction and lower valence bands occur, which improve metallization of nanotubes and increase the density of states at the Fermi level. Initial analysis reveals that all atoms of carbon network equally contribute to the energy band. However, this input changes with applied deformation. The atoms from high curvature have the main contribution to the energy bands around the Fermi level. For ϵ_{yy}>50% the atoms from the low curvature in the Peanut deformation (top and bottom rows) also contribute to the energy bands at the Fermi level.

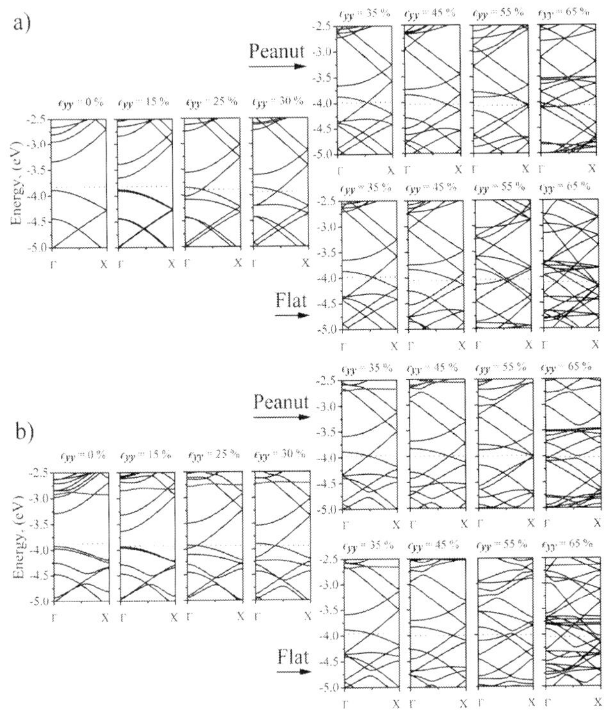

Fig. 12. The energy band structure of a) defect-free (8,0) nanotube, b) (8,0) nanotubes with Stone-Wales defect at high curvature site for several degrees of ϵ_{yy}. The red level is associated with the Stone-Wales defect.

The band structure evolution of the deformed (8,0) nanotube with a Stone-Wales defect shows similar behavior as compared to the defect-free one (Fig. 12b). However, Stone-Wales defect increases initial value of band gap and introduces flat defect level in the conduction band at $E\approx-2.9$ eV (in red). Under applied radial deformations $\epsilon_{yy}>0\%$, this flat level starts to move away from the Fermi level while increasing $\sigma-\pi$ hybridizations brings the lowest conduction and the highest valence bands closer together. Small band gap $E_g\approx0.01$ eV opens up for the Peanut deformation at $\epsilon_{yy}=55\%$ and $\epsilon_{yy}=65\%$. There is no such opening for Flat deformation, and the band gap remains closed. The analysis of conduction and valence bands reveals that the lowest conduction bands consist of the orbital from defect atoms where $\sigma-\pi$ hybridizations is significant.

The applied radial deformation induces nonuniform charge distribution on the nanotube surface (Fig. 13). For a defect-free nanotube, the charge accumulation appears on the high curvature. Even more irregular allocation is caused by Peanut and Flat deformations (Fig. 13a-c). Stone-Wales defect creates local disturbance in the nanotube structure, which reduces charge accumulation on the defect itself, except for the *C6-C7* bond even though it is on the high curvature. Most of the charge is gathered around the atoms surrounding Stone-Wales defect (Fig. 13d-f).

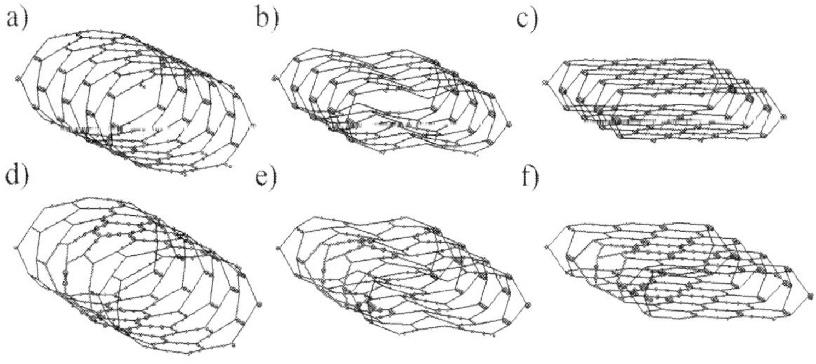

Fig. 13. Total charge density plots with isosurface value is 0.0185 $e/Å^3$ for a) defect-free (8,0) nanotube with ϵ_{yy}=25%, b) Peanut structure with ϵ_{yy}=65%, c) Flat structure with ϵ_{yy}=65%, and d) (8,0) nanotube with Stone-Wales defect with ϵ_{yy}=25%, e) Peanut structure with ϵ_{yy}=65%, f) Flat structure with ϵ_{yy}=65%.

The effects of nitrogen impurity and mono-vacancy in radially deformed nanotubes have been investigated in terms of Density of States (DOS) for different ϵ_{yy}. Since the nitrogen atom supplies one extra electron in nanotube, and mono-vacancy provides dangling bond, the calculations include spin-polarized effect to account for magnetic effect in nanotubes. The possibility of magnetism in nanotube has been predicted by other authors (Lim et al., 2007; Orellana & Fuentealba 2006; Wu & Hagelberg 2009). The density of states of nitrogen doped nanotubes is responsive to the concentration of impurities (Orellana & Fuentealba 2006). Increasing the impurity concentration more than 1% provides a flat level in the band gap. In our calculation we do not observe any flat level because the nitrogen concentration is 0.78% (Fig. 14a). As the radial deformation ϵ_{yy} increases, the band gap decreases, and the conduction and valance bands move towards the Fermi level. For ϵ_{yy}>45%, the DOS for Peanut and Flat deformation raises around the Fermi level which suggests that the nanotube shows more metal-like behavior.

The nitrogen impurity contributes significantly to conduction band in the vicinity of the Fermi level which is partially occupied. The highest valence band consists of contribution from atoms on the high curvature. The extra nitrogen e^- induces spin-polarization effects (magnetism) in the nanotube electronic structure at ϵ_{yy}=0% and sharp features in the DOS for spin "up" around ~ -2.8 eV (Fig. 14a). As ϵ_{yy} increases, the spin-polarized effect is overtaken by σ-π hybridization. Eventually, it completely disappears at ϵ_{yy}>25% except for at ϵ_{yy}=65% for Flat deformation. Thus, applied radial deformation provides the opportunity to control not only electronic but also magnetic properties of nanotubes. To elucidate this effect further, we investigate changes in local magnetic moment m under ϵ_{yy} (Fig. 14b). For small deformations m increases with a maximum at ϵ_{yy}=5% and then it reduces to m=0 at ϵ_{yy}=25% when the nanotube becomes almost a metal (ϵ_{yy}=23%). However, complicated magneto-mechanical coupling produces large magnetic moment m=0.92 μ_B (not shown on Fig. 14b) at ϵ_{yy}=65%. It gives evidence of new methods of tuning nanotube properties and complex interrelation between defects and deformations.

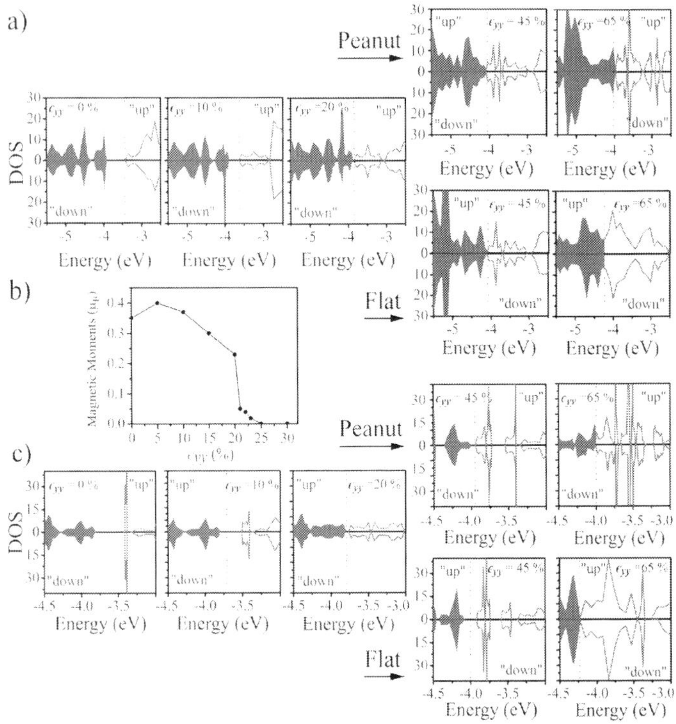

Fig. 14. a) Total Density of States for spin "up" and "down" carriers of (8,0) nanotube with a substitutional nitrogen impurity on its high curvature at different ϵ_{yy} for Peanut and Flat structures. b) The magnetic moment of the (8,0) nanotube with the nitrogen impurity as a function of ϵ_{yy}. c) Total Density of States for spin "up" and "down" carriers of (8,0) nanotube with mono-vacancy defect on its high curvature at different ϵ_{yy} for Peanut and Flat structures.

Finally, we consider a radially deformed (8,0) nanotube with a mono-vacancy (Fig. 14c). The presence of mono-vacancy is manifested in a sharp localized peak around -3.41 eV in the DOS conduction region for the spin "up" and "down" at ϵ_{yy}=0%. This state mainly consists of contributions from the mono-vacancy atoms. As the deformation progresses, the peak intensity decreases and becomes broader, and peak location moves closer to the Fermi level due to σ-π hybridization from the nanotube atoms on high curvature along with the C atoms of mono-vacancy. Such process repeats again for ϵ_{yy}>40%, where sharp peaks around the Fermi level occur and become wider and sharper with increasing ϵ_{yy}. The major contribution to these peaks comes from dangling bond atom - *C1* and the C atoms of the pentagon ring of mono-vacancy (Fig. 4d). DOS for spin "up" and "down" demonstrates that there is no spin-polarized effect in radially deformed nanotubes with mono-vacancy for all ϵ_{yy} except ϵ_{yy}=65% in Peanut deformation where a small magnetic moment m=0.04 μ_B has been found. These results are in partial agreement with previously reported findings that zigzag nanotubes with mono-vacancy do not show magnetic response for given orientation of defect (Berber & Oshiyama 2006). However, the appearance of magnetism at ϵ_{yy}=65% can originate from several reasons such as uncoordinated C atom with a localized unpaired spin, concentration of vacancies, nanotube chirality, sp^3 pyramidization, and related bond length changes, as well as vacancy location with respect to the nanotube structure (Ma et al., 2004; Orellana & Fuentealba 2006; Zheng & Zhuang 2008). Another explanation of the magnetism in radially deformed nanotubes with mono-vacancy is the magnetic flat-band theory together with σ-π hybridization (Fujita et al., 1996; Nakada et al., 1996). The presence of flat bands around the Fermi level leads to increased electron-electron interaction and results in the occurrence of magnetic polarization which is common for edge states in zigzag nanoribbons (Fujita et al., 1996; Nakada et al., 1996). We can consider a mono-vacancy on the high curvature and large deformations such edge states. Our results suggest that there is a complex relation among mechanical, electric, and magnetic coupling in radially deformed nanotubes with the mono-vacancy.

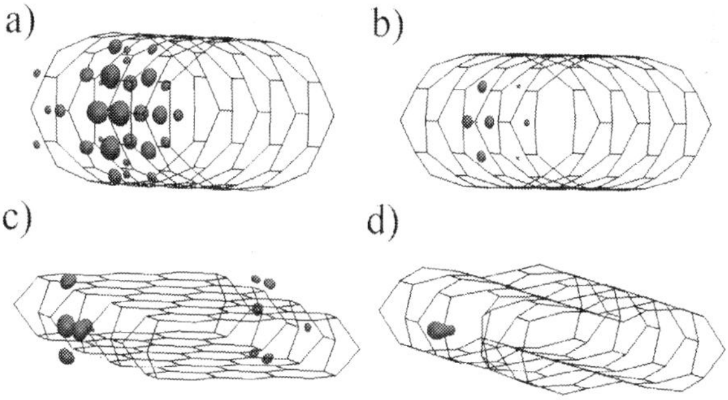

Fig. 15. Spin density isosurface plots for a) (8,0) nanotube with nitrogen impurity for ϵ_{yy}=10% at 0.004 μ_B/Å^3, b) (8,0) nanotube with nitrogen impurity for ϵ_{yy}=20% at 0.004 μ_B/Å^3, c) Flat (8,0) nanotube with nitrogen impurity for ϵ_{yy}=65% at 0.0122 μ_B/Å^3, d) Peanut (8,0) nanotube with mono-vacancy for ϵ_{yy}=65% at 0.018 μ_B/Å^3.

Further insight in the nanotube magnetism can be gained by exploring the spin density distribution for various deformations ϵ_{yy} (Fig. 15). Initially extra nitrogen e^- is localized on the defect site but increasing deformation obliterates this localization. As an example, the unpaired e^- is much more delocalized at ϵ_{yy}=20% in comparison to the ϵ_{yy}=10% case, where the e^- spin density localized around the impurity site is much more intense. Similar behavior has been reported for semiconducting nanotubes, where the nitrogen e^- is localized around the impurity, while for metallic nanotube it is completely delocalized (Nevidomskyy et al., 2003). However, the e^- spin localization around the impurity appears again at ϵ_{yy}=65% in Flat deformation, where the nanotube is a metal, but the magnetic moment has the highest value of - m=0.92 μ_B. Moreover, similar localization of the spin density happens around the dangling bond in the nanotube with mono-vacancy ϵ_{yy}=65%. This suggests that the uncoordinated C atom with unpaired spin is vital in the magnetic properties of defective and deformed nanotubes.

4.4 Electronic Properties of Carbon Nanotubes under Radial Deformation and an External Electric Field

Next, we explore the electronic properties of radially deformed nanotubes in an external radial electric field. We focus on the modulation of the band gap by an external electric field in nanotubes with various degrees of deformation. As we have already seen from earlier discussion, applying radial deformation causes the metal semiconductor transition in semiconductor nanotubes and causes opening and closing of a band gap in metallic nanotubes because of the $\sigma - \pi$ hybridization from the high curvature regions. Fig. 16a-b show these dependences in semiconductor (8,0) and semi-metal (9,0) carbon nanotubes as a function of ϵ_{yy}. The same behavior can be achieved by the application of an external electric field to undeformed (8,0) and (9,0) nanotubes. E_g dependences in Fig. 16c-d and Fig. 16a-b look similar. However, the semiconductor-metal transition happens faster for the (8,0) nanotube over a shorter field strength interval as compared to the (9,0) nanotube. In this case, E_g for (8,0) nanotube is closed at $\mathcal{E}$=0.4 V/Å, while for (9,0) tube the closure occurs at $\mathcal{E}$=0.8 V/Å. Therefore, we can obtain the same electronic characteristics of nanotubes from radial deformation and the applied external electric field. In addition, it provides the flexibility of controlling nanotube properties at different circumstances.

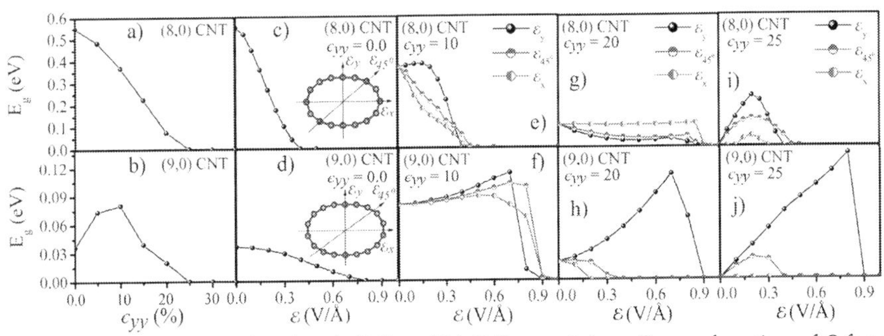

Fig. 16. E_g as a function of ϵ_{yy} for a) (8,0) and b) (9,0) nanotubes; E_g as a function of $\mathcal{E}$ for c) (8,0) at ϵ_{yy}=0%; d) (9,0) at ϵ_{yy}=0%; e) (8,0) at ϵ_{yy}=10%; f) (9,0) at ϵ_{yy}=10%; g) (8,0) at ϵ_{yy}=20%; h) (9,0) at ϵ_{yy}=20%; i) (8,0) at ϵ_{yy}=25%; j) (9,0) at ϵ_{yy}=25%. The insert indicates electric field directions as follows: $\mathcal{E}_x$ is along x-axis, $\mathcal{E}_y$-along y-axis, and $\mathcal{E}_{45^0}$-along 45^0 from x-axis in x-y plane.

The symmetry of the nanotube is decreased under applied radial deformation. Consequently, the x and y directions of the cross section are not equivalent any more (Fig. 2b). It has been shown before that electronic structure of the defective nanotube depends on the orientation of external electric field (Tien et al., 2005). Here, we analyze the E_g evolution of the deformed (8,0) and (9,0) nanotubes under $\mathcal{E}$ along x, y-axes and the 45^0 direction in the x-y plane (insert in Fig. 16c-d).

Radially deformed (8,0) nanotube at ϵ_{yy}=10% still has a relatively large band gap - E_g= 0.37 eV, which decreases under applied electric field and vanishes at $\mathcal{E}$~0.45 V/Å for all three directions. Each E_g dependence has distinct profile for $\mathcal{E}$<0.4 V/Å. E_g initially increases and then starts to decrease for y-direction. However, it sharply decreases for x-direction, and it shows intermediate, almost linear, decline for 45^0-direction (Fig.16e). At ϵ_{yy}=20% the applied electric field has a small effect on E_g over relatively long range of electric field strength for all directions until it closes E_g at $\mathcal{E}$~0.8 V/Å (Fig. 16g). The (8,0) nanotube becomes metallic at ϵ_{yy}=25% but applied electric field opens E_g with its maximum at $\mathcal{E}$~0.2 V/Å for all directions. The largest E_g is for the electric field along y-direction and the lowest - for x-direction (Fig. 16i).

On the other hand, the radially deformed (9,0) nanotube at ϵ_{yy}=10% has the largest E_g=0.08 eV, which continues to increase under applied electric field in all directions until $\mathcal{E}$~0.6÷0.7 V/Å, where E_g is sharply closed (Fig. 16f). This behavior is similar for E_g in the (8,0) nanotube at ϵ_{yy}=20%. More drastic behavior of the band gap is observed at ϵ_{yy}=20% for electric field along y-direction. The original value of the band gap increases by ~ 4 times at $\mathcal{E}$=0.7 V/Å which is followed by its rapid closure at $\mathcal{E}$=0.9 V/Å. At the same time, the weak electric field $\mathcal{E}$ along x- and 45^0- directions closes the band gap. The same dependence has been found at ϵ_{yy}=25%. The band gap reaches maximum for $\mathcal{E}$=0.8 V/Å while there is no effect of applied electric field on the band gap along x-direction.

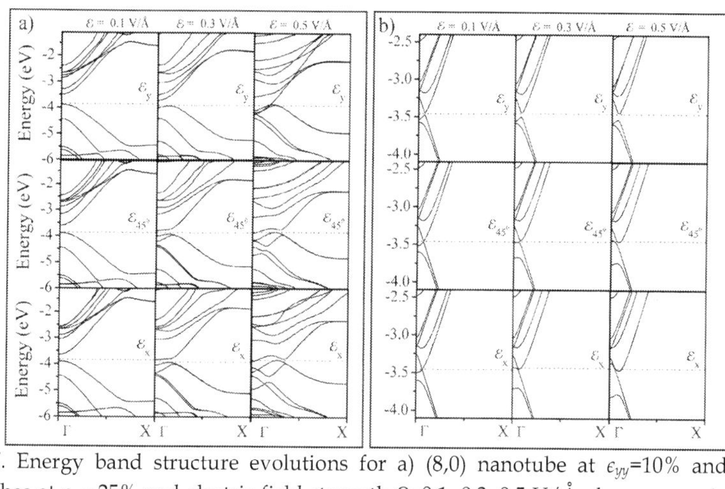

Fig. 17. Energy band structure evolutions for a) (8,0) nanotube at ϵ_{yy}=10% and b) (9,0) nanotubes at ϵ_{yy}=25% and electric field strength $\mathcal{E}$=0.1, 0.3, 0.5 V/Å along x, y-axis, and 45^0 from x-axis in x-y plane, respectively. The Fermi level is shown as a dashed line.

Better understanding of band gap modulations comes from the analysis of the electronic band structure for (8,0) and (9,0) nanotubes at ϵ_{yy}=10% and ϵ_{yy}=25% respectively, for several orientations (Fig. 17). The electronic level of conduction and valance bands are associated with the quantum angular momentum J which defines the rule for admixture of allowed states (Vukovic et al., 2002). It has been shown that electronic states can mix according to the selection rule ΔJ=0, ±2 for radial deformation of nanotubes (Park et al., 1999). This admixture and increasing σ-π hybridization on high curvature bring the lowest conduction and the highest valance bands - closer to the Fermi level. However, the applied electric field mixes the electronic states in accordance with $\Delta J = \pm 1$ and also decreases the values of the band gap to its closure at E_F (Kim & Chang 2001; Li et al., 2006). The combination of radial deformation and external electric field causes the band structure modulation according to ΔJ=0, ±1, ±2 selection rule and σ–π hybridization on the high curvature of nanotubes. As we have seen from the band gap behavior, the strength and orientation of applied electric field with respect to the nanotube cross section are crucial for electronic properties of nanotubes. This has been reflected in the band structures for (8,0) and (9,0) nanotubes (Fig. 17). Most changes happen with the lowest unoccupied single degenerate conduction state for both (8,0) and (9,0) nanotubes. Therefore, application of electric field along x-direction (through the high curvature) has revealed the largest σ–π hybridization on the high curvature. Thus, depending on the interplay between the radial deformation and electric field strength/orientation, the electronic properties of nanotubes can reveal semiconductor-metal or metal semiconductor transformation (Fig. 17).

5. Conclusion

In this chapter we discuss electronic and magnetic carbon nanotube properties under two external factors: radial deformation and various defects along with radial deformation and external electric fields. Our analysis is based on density functional theory calculations. We show that defects, such as Stone-Wales, nitrogen impurity, and mono-vacancy, create a local distortion in the carbon network. This disturbance is lowest for nitrogen, larger for Stone-Wales defect, and largest for mono-vacancy. The analysis of the characteristic energies reveals that it takes more energy to deform a smaller diameter nanotube in comparison to larger one. The larger degree of radial deformation makes it easier to introduce a defect on the high curvature of the nanotube. The Stone-Wales defect requires the lowest formation energy, while it is higher for nitrogen impurity, and it is the largest for mono-vacancy. We find that applied radial deformation causes semiconductor-metal transition in defective nanotubes at approximately the same value of applied strain (ϵ_{yy}=23÷25%) regardless of the type of defect.

The electronic structure calculations show that the main factors responsible for semiconductor-metal transition in defective nanotubes are the σ-π hybridization from the high curvature and the presence of defect states around the Fermi level region. We observe that interplay between radial deformation and type of defects such as Stone-Wales defect and mono-vacancy can open the band gap again, after it has been closed due to the same factors. However, there is not such opening for the nitrogen impurity and defect-free nanotubes regardless of the type of deformation. On the other hand, mutual effect of radial deformation and defects leads to magneto-mechanical coupling in the nanotube structure because of the presence of extra an electron from the nitrogen atom and dangling bond from the mono-vacancy. The magnetic properties of nanotubes can be controlled by the degree of radial deformation. The origin of magnetic properties of nanotubes is also discussed.

The combined effect of radial deformation and external electric field is given in terms of band gap modulations, electronic band structure changes, strength of applied electric field, electric field orientation, and selection rules. We show various responses in semiconductor and metallic carbon nanotubes. The electronic structure is sensitive to the strength and electric field orientation with respect to the cross section of nanotubes. The interplay between selection rules of mixing different states in the radial deformation and applied

electric field along with σ-π hybridization controls and tunes the band gap of nanotubes. Finally, we discuss semiconductor-metal transition in semiconducting nanotubes and opening of the band gap in metallic nanotubes.
This chapter provides a comprehensive study of electronic and magnetic properties of nanotubes under two external factors. It discusses additional opportunities in controlling and tuning carbon nanotube properties for their expanded applications in current and new nano-devices.

Acknowledgments

We acknowledge the donors of the American Chemical Society Petroleum Research Fund for support of this research. In addition, the part of this work was supported by the US Army Medical Research and Material Command under Award No. W81XWH-07-1-0708. We also would like to acknowledge the use of the services provided by the Research Computing Core at the University of South Florida and the TeraGrid Advanced Support Program at the University of Illinois.

6. References

Ajiki, H. & Ando, T. (1996). Energy bands of carbon nanotubes in magnetic fields. *J. Phys. Soc. Jap.*, Vol. 65, No. 2, 505-514, ISSN 00314-9015

Ashraf, B.; Hubert, P. & Vengallatore, S. (2006). Carbon nanotube-reinforced composites as structural materials for microactuators in microelectromechanical systems. *Nanotechnology*, Vol. 17, No. 19, 4895-4903, ISSN 0957-4484

Bachtold, A.; Strunk, C.; Salvetat, J.-P.; Bonard, J.-M.; Forro, L.; Nussbaumer, T. & Schonenberg, C. (1999). Aharonov-Bohm oscillations in carbon nanotubes. *Nature*, Vol. 397, No. 6721, 673-675, ISSN 0028-0836

Barboza, A.; Gomes, A.; Archanjo, B.; Araujo, P.; Jorio, A.; Ferlauto, A.; Mazzoni, M.; Chacham, H. & Neves, B. (2008). Deformation induced semiconductor-metal transition in single wall carbon nanotubes probed by electric force microscopy. *Phys. Rev. Lett.*, Vol. 100, No. 25, 256804-1-4, ISSN 1079-7114

Berber, S. & Oshiyama, A. (2006). Reconstraction of mono-vacancies in carbon nanotubes: atomic relaxation vs. spin polarization. *Physica B*, Vol. 376-377, No. 1, 272-275, ISSN 0921-4526

Bettinger, H. (2005). The reactivity of defects at the sidewalls of single-walled carbon nanotubes: the Stone-Wales defect. *J. Phys. Chem.*, Vol. 109, No. 15, 6922-6924, ISSN 1089-5647

Choi, W.; Chung, D.; Kang, J.; Kim, H.; Jin, Y.; Han, I.; Lee, Y.; Jung, J.; Lee, N.; Park, G. & Kim, J. (1999). Fully sealed, high-brightness carbon-nanotube field-emission display. *Appl. Phys. Lett.*, Vol. 75, No. 20, 3129-3131, ISSN 0003-6951

Chopra, N.; Benedict, L.; Crespi, V.; Cohen, M.; Louie, S. & Zettl, A. (1995). Fully collapsed carbon nanotubes. *Nature*, Vol. 377, No. 6544, 135-138, ISSN 0028-0836

Cohen-Karni, T.; Segev, L.; Srur-Lavi, O.; Cohen, S. & Joselevich, E. (2006). Torsional electromechanical quantum oscillation in carbon nanotubes. *Nature Nanotechnology*, Vol. 1, No. 1, 36-41, ISSN 1748-3387

Crespi, H.; Cohen, L. & Rubio, A. (1997). In situ band gap engineering of carbon nanotubes. *Phys. Rev. Lett.*, Vol. 79, No. 11, 2093-2096, ISSN 1079-7114. Figure 5a is used with permission from American Institute of Physics, Copyright 1997.

Czerw, R.; Terrones, M.; Charlier, J.-C.; Blase, X.; Foley, B.; Kamalakaran, R.; Grobert, N.; Terrones, H.; Tekleab, D.; Ajayan, P.; Blau, W.; Ruhle, M. & Carroll, D. (2001). Identification of electron donor stets in N-doped carbon nanotubes. *Nano Lett.*, Vol. 1, No. 9, 457-460, ISSN 1530-6984

Droppa, R.; Hammer, P.; Carvalho, A.; Santos, M. & Alvarez, F. (2002). Incorporation of nitrogen in carbon nanotubes. *J. Non-Cryst. Sold.*, Vol. 299-302, No. 2, 874-879, ISSN 0022-3093

Fagan, S.; da Silva, L. & Mota, R. (2003). Ab initio study of radial deformation plus vacancy on carbon nanotubes: energetics and electronic properties. *Nano Lett.*, Vol. 3, No. 3, 289-291, ISSN 1530-6984

Fagan, S.; Mota, R.; da Silva, A. & Fazzio, A. (2004). Substitutional Si doping in deformed carbon nanotubes. *Nano Lett.*, Vol. 4, No. 5, 975-977, ISSN 1530-6984

Fan, Y.; Goldsmith, B. & Collins, P. (2005). Identifying and counting point defects in carbon nanotubes. *Nature Materials*, Vol. 4, No. 12, 906-911, ISSN 1476-1122

Favlo, M.; Clary, G.; Taylor II, R.; Chi, V.; Brooks, F.; Washburn, S. & Superfine, R. (1997). Binding and buckling of carbon nanotubes under large strain. *Nature*, Vol. 389, No. 6651, 582-584, ISSN 0028-0836

Fennimore, A.; Yuzvinsky, T.; Han, W.-Q.; Fuhrer, M.; Cumings, J. & Zetti, A. (2003). Rotational actuators based on carbon nanotubes. *Nature*, Vol. 424, No. 6947, 408-410, ISSN 0028-0836

Fujita, M.; Wakabayashi, K.; Nakada, K. & Kusakabe, K. (1996). Peculliar localized states at zig-zag graphite edge. *J. Phys. Soc. Jap.*, Vol. 65, No. 7, 1920-1923, ISSN 0031-9015

Hashimoto, A.; Suenaga, K.; Gloter, A.; Urita, K, & Iijima, S. (2004). Direct evidence for atomic defects in graphene layers. *Nature*, Vol. 430, No. 7002, 870-873, ISSN 0028-0836

Hertel, T.; Walkup, R. & Avouris, P. (1998). Deformation of carbon nanotubes by surface van der Waals forces. *Phys. Rev. B*, Vol. 58, No. 20, 13870-73, ISSN 1098-0121

Gai, P.; Odile, S.; Mcguire, S.; Rao, A.; Dresselhaus, M.; Dresselhaus, G. & Colliex, C. (2004). Structural systematic in boron-doped single wall carbon nanotubes. *J. Mater. Chem.*, Vol. 14, No. 4, 669-675, ISSN 0959-9428

Gao, G.; Cagin, T. & Goddard, W. (1998). Energetics, structure, mechanical and vibrational properties of single-walled carbon nanotubes. *Nanotechnoly*, Vol. 9, No. 3, 184-191, ISSN 0957-4484

Ghosh, A.; Sood, A. & Kumar, N. (2003). Carbon nanotube flow sensors. *Science*, Vol. 299, No. 5609, 1042-1044, ISSN 003-8075

Giusca, C.; Tison, Y. & Silva, S. (2007). Atomic and electronic structure in collapsed carbon nanotubes evidenced by scanning tunneling microscopy. *Phys. Rev. B*, Vol. 76, No. 3, 035429-1-6, ISSN 1098-0121

Golberg, D.; Bando, Y.; Han, W.; Kurashima, K. & Sato, T. (1999). Single-walled B-doped carbon and BN nanotubes synthesized from single walled carbon nanotubes through a substitutional reaction. *Chem. Phys. Lett.*, Vol. 308, No. 3-4, 337-342, ISSN 0009-2614

Gulseren, O.; Yildirim, T.; Ciraci, S. & Kilic, C. (2002). Reversible band-gap engineering in carbon nanotubes by radial deformation. *Phys. Rev. B*, Vol. 65, No. 15, 155410-1-7, ISSN 1098-0121

Itkis, M.; Yu, A. & Haddon, R. (2008). Single-walled carbon nanotube thin film emitter-detector integrated optoelectronic device. *Nano Lett.*, Vol. 8, No. 8, 2224-2228, ISSN 1530-6984

Jiang, J.; Dong, J. & Xing, D. (2000). Zeeman effect on the electronic spectral properties of carbon nanotubes in an axial magnetic field. *Phys. Rev. B*, Vol. 62, No. 19, 13209-13215, ISSN 1098-0121

Jonge, N.; Allioux, M.; Doytcheva, M.; Kaiser, M.; Teo, K.; Lacerda, R. & Milne, W. (2004). Characterization of the field emission properties of individual carbon nanotubes. *Appl. Phys. Lett.*, Vol. 85, No. 9, 1607-1609, ISSN 0003-6951

Kim, Y.-H. & Chang K. (2001). Subband mixing rules in circumferentially perturbed carbon nanotubes: effects of transverse electric fields. *Phys. Rev. B*, Vol. 64, No. 15, 153404-1-4, ISSN 1098-0121

Kolmogorov, A. & Crespi, V. (2005). Registry-dependent interlayer potential for graphitic systems. *Phys. Rev. B*, Vol. 71, No. 23, 235415-1-6, ISSN 1098-0121

Kordas, K.; Mustonen, T.; Toth, G.; Vahakangas, J.; Uusimaki, A.; Jantunen, H.; Gupta, A.; Rao, K.; Vajtai, R. & Ajayan, P. (2007). Magnetic-field induced efficient alignment of carbon nanotubes in aqueous solutions. *Chem. Mater.*, Vol. 19, No. 4, 787-791, ISSN 0897-4756

Krasheninnikov, A.; Nordlund, K.; Sirvio, M.; Salonen, E. & Keinonen, J. (2001). Formation of ion-irradiation-induced atomic-scale defects on walls of carbon nanotubes. *Phys. Rev. B*, Vol. 63, No. 24, 245405-1-6, ISSN 1098-0121

Krasheninnikov, A.; Banhart, F.; Li, J.; Foster, A. & Nieminen, R. (2005). Stability of carbon nanotubes under electron irradiation: role of tube diameter and chirality. *Phys. Rev. B*, Vol. 72, No. 12, 125428-1-6, ISSN 1098-0121

Kresse, G. & Furthmuller, J. (1996). Efficient interactive schemes for ab initio total-energy calculations using a plane-wave basis set. *Phys. Rev. B*, Vol. 54, No. 16, 11169-11186, ISSN 1098-0121

Kresse, G. & Joubert, D. (1999). From ultrasoft pseudopotentials to the projector augmented-wave method. *Phys. Rev. B*, Vol. 59, No. 3, 1758-1775, ISSN 1098-0121

Kumar, M.; Lee, S.; Kim, T.; Kim, T.; Song, S.; Yang, J.; Nahm, K. & Suh, E.-K. (2003). DC electric field assisted alignment of carbon nanotubes on metal electrodes. *Solid State Electron.*, Vol. 47, No. 11, 2075-2080, ISSN 0038-1101

Lai, P.; Chen, S. & Lin, M. (2008). Electronic properties of single-walled carbon nanotubes under electric and magnetic fields. *Physica E*, Vol. 40, No. 6, 2056-2058, ISSN 1386-9477

Lee, J.; Gipp, P. & Heller, C. (2004). Carbon nanotube *p-n* junction diodes. *Apll. Phys. Lett.*, Vol. 85, No. 1, 145-147, ISSN 0003-6951

Lee, J.-O.; Kim, J.-R.; Kim, J.-J.; Kim, J.; Kim, N.; Park, J. & Yoo, K.-H. (2000). Observation of magnetic-field-modulated energy gap in carbon nanotubes. *Solid State Comm.*, Vol. 115, No. 9, 467-471, ISSN 0038-1098

Lee, R.; Kim, H.; Fischer, J.; Thess, A. & Smalley, R. (1997). Conductivity enhancement in single-walled carbon nanotube bundles doped with K and Br. *Nature*, Vol. 388, No. 6639, 255-257, ISSN 0028-0836

Leonard, F.; Jones, F.; Talin, A. & Dentinger P. (2005). Robustness of nanotube electronic transport to conformational deformations. *Appl. Phys. Lett.*, Vol. 86, No. 9, 093112-1-3, ISSN 0003-6951

Li, T. & Lin, M. (2006). Electronic properties of carbon nanotubes under external fields. *Phys. Rev. B*, Vol. 73, No. 7, 075432-1-7, ISNN 1098-0121

Li, Y.; Rotkin, S. & Ravaioli, U. (2003). Electronic response and band structure modulation of carbon nanotubes in a transverse electrical field. *Nano Lett.*, Vol. 3, No. 2, 183-187, ISSN 1530-6984

Lim, S.; Li, R.; Ji, W. & Lin, J. (2007). Effect of nitrogenation on single-walled carbon nanotubes within density functional theory. *Phys. Rev. B*, Vol. 76, No. 19, 195406-1-16, ISSN 1098-0121

Lu, A. & Pan, B. (2004). Nature of single vacancy in achiral carbon nanotubes. *Phys. Rev. Lett.*, Vol. 92, No. 10, 105504-1-4, ISSN 1079-7114. Figure 5c is used with permission from American Institute of Physics, Copyright 2004.

Lu, J. (1995). Novel magnetic properties of carbon nanotubes. *Phys. Rev. Lett.*, Vol. 74, No. 7, 1123-1126, ISSN 1079-7114

Ma, Y.; Lehtinen, P. Foster, A. & Nieminen. (2004). Magnetic properties of vacancies in graphene and single-walled carbon nanotubes. *New J. Phys.*, Vol. 6, No. 1, 68-1-15, ISSN 1367-2630

Maiti, A.; Svizhenko, A. & Anantram, M. (2002). Electronic transport through carbon nanotubes: effects of structural deformation and tube chirality. *Phys. Rev. Lett.*, Vol. 88, No. 12, 126805-1-4, ISSN 1079-7114

Mazzoni, M. & Chacham, H. (2000). Bandgap closure of a flattened semiconductor carbon nanotube: a first principal study. *Appl. Phy. Lett.*, Vol. 76, No. 12, 1561-1563, ISSN 0003-6951

Mehrez, H.; Svizhenko, A.; Anantram, M.; Elstner, M. & Frauenheim, T. (2005). Analysis of band-gap formation in squashed armchair carbon nanotubes. *Phys. Rev. B*, Vol. 71, No. 15, 155421-1-7, ISSN 1098-0121

Minary-Jolandan, M. & Yu, M.-F. (2008). Reversible radial deformation up to the complete flattening of carbon nanotubes in nanoindentation. *J. Appl. Phys.*, Vol. 103. No. 7, 073516-1-5, ISSN 0021-8979

Minot, E.; Yaish, Y.; Sazonova, V & McEuen, P. (2004). Determination of electron orbital magnetic moments in carbon nanotubes. *Nature*, Vol. 428, No. 6982, 536-539, ISSN 0028-0836. Figure 6b is used with permission from Macmillan Publisher Ltd: Nature, Copyright 2004.

Moradian, R. & Azadi, S. (2006). Boron and nitrogen-doped single-walled carbon nanotube. *Physica E*, Vol. 35, No. 1, 157-160, ISSN 1386-9477. Figure 5b is used with permission from Elsevier, Copyright 2005.

Nakada, K.; Fujita, M.; Dresselhaus, G. & Dresselhaus, M. (1996). Edge states in graphene ribbonce: nanometer size effects and edge shape dependence. *Phys. Rev B*, Vol. 54, No. 24, 17954-17961, ISSN 1098-0121

Nardelli, M.; Fattebert, J.-L.; Orlikowski, D.; Roland, C.; Zhao, Q. & Bernholc, J. (2000). Mechanical properties, defects and electronic behavior of carbon nanotubes. *Carbon*, Vol. 38, No. 11-12, 1703-1711, ISSN 0008-6223

Neugebauer, J. & Scheffler, M. (1992). Adsorbate-substrate and adsorbate-adsorbate interactions of Na and K adlayers on Al(111). *Phys. Rev. B*, Vol. 46, No. 24, 16067-106080, ISSN 1098-0121

Nevidomskyy, A.; Csanyi, G. & Payne, M. (2003). Chemically active substitutional nitrogen impurity in carbon nanotubes. *Phys. Rev. Lett.*, Vol. 91, No. 10, 105502-1-4, ISSN 1079-7114

Orellana, W. & Fuentealba, P. (2006). Structural, electronic and magnetic properties of vacancies in single-walled carbon nanotubes. *Surf. Sci.*, Vol. 600, No. 18, 4305-4309, ISSN 0039-6028

Park, C.; Kim, Y. & Chang, K. (1999). Band-gap modification by radial deformation in carbon nanotubes. *Phys. Rev. B*, Vol. 60, No. 15, 10656-10659, ISSN 1098-0121

Picozzi, S.; Santucii, S. & Lozzi, L. (2004). Ozone adsorption on carbon nanotubes: the role of Stone-Wales defects. *J. Chem. Phys.*, Vol. 120, No. 15, 7147-7152, ISSN 0021-9606

Reich, S.; Thomsen, C. & Maultzsch, J. (2004). Carbon Nanotubes. *EILEY-VCH Verlag GmbH & Co. KGaA*, ISBN 3-527-40386-8

Rossato, J.; Baierle, R.; Fazzio, A. & Mota, R. (2005). Vacancy formation process in carbon nanotubes: first-principal approach. *Nano Lett.*, Vol. 5, No. 1, 197-200, ISSN 1530-6984

Santos, E.; Ayuela, A.; Fagan, S.; Filho, J. Azevedo, D.; Souza, A. & Sanchez-Portal, D. (2008). Switching on magnetism in Ni-doped graphene: density functional calculation. *Phys.Rev. B*, Vol. 78, No. 19. 195420-1-5, ISSN 1098-0121

Shan, B.; Lakatos, G.; Peng, S. & Cho, K. (2005). First-principal study of band-gap change in deformed nanotubes. *Appl. Phys. Lett.*, Vol. 87, No. 17, 173109-1-3, ISSN 0003-6951. Figure 3 is used with permission from American Institute of Physics, Copyright 2005.

a Shtogun, Y. & Woods, L. (2009). Electronic and magnetic properties of deformed and defective single wall carbon nanotubes. *Carbon*, Vol. 47, No. 14, 3252-3262, ISSN 0008-6223

b Shtogun, Y. & Woods, L. (2009). Electronic structure modulations of radially deformed single wall carbon nanotubes under transverse external electric fields. *J. Phys. Chem. C*, Vol. 113, No. 12, 4792-4796, ISSN 1932-7447

Shtogun, Y, Woods, L. & Dovbeshko, G. (2007). Adsorption of adenine and thymine and their radicals on single-wall carbon nanotubes. *J. Phys. Chem. C*, Vol. 111, No. 49, 18174-18181, ISSN 1932-7447

Son, Y.; Ihm, J.; Cohen, M.; Louie, S. & Choi H. (2005). Electrical switching in metallic carbon nanotubes. *Phys. Rev. Lett.*, Vol. 95, No. 21, 216602-1-4, ISSN 1079-7114

Stone, A. & Wales, D. (1986). Theoretical studies of icosahedral C_{60} and some related species. *Chem Phys. Lett.*, Vol. 128, No. 5-6, 501-503, ISSN 0009-2614

Suenaga, K.; Wakabayashi, H.; Koshino, M.; Sato, Y.; Urita, K. & Iijima, S. (2007). Imaging active topological defects in carbon nanotubes. *Nature Nanotechnology*, Vol. 2, No. 6, 358-360, ISSN 1748-3387

Sugie, H.; Tanemura, M.; Filip, V.; Iwata, K.; Takahashi, K. & Okuyama, F. (2001). Carbon nanotubes as electron source in an x-ray tube. *Appl. Phys. Lett.*, Vol. 78, No. 17, 2578-2580, ISSN 0003-6951

Tang, J.; Qin, L.-C.; Sasaki, T.; Yudasaka, M.; Matsushita, A. & IIjima, S. (2000). Compressibility and polygonization of single-wall carbon nanotubes under hydrostatic pressure. *Phys. Rev. Lett.*, Vol. 85, No. 9, 1887-1889, ISSN 1079-7114

Tans, S.; Verschueren, A. & Dekker, C. (1998). Room-temperature transistor based on a single carbon nanotube. *Nature*, Vol. 393, No. 6680, 49-52, ISSN 0028-0836

Tien, L.; Tsai, C.; Li, F. & Lee, M. (2005). Band-gap modification of defective carbon nanotubes under a transverse electric field. *Phys. Rev. B*, Vol. 72, No. 24, 245417-1-6, ISSN 1098-0121. Figure 7 is used with permission from American Institute of Physics, Copyright 2005.

Tien, L.-G.; Tsai, C.-H.; Li, F.-Y. & Lee, M.-H. (2008). Influence of vacancy defect on electronic properties of armchair single wall carbon nanotube. *Diam. Relat. Mater.*, Vol. 17, No. 4-5, 563-566, ISSN 0925-9635

Tombler, T.; Zhou, C.; Alexseyev, L.; Kong, J.; Dai, H.; Liu, L.; Jayanthi, C.; Tang, M. & Wu, S. (2000). Reversible electromechanical characteristics of carbon nanotubes under local-probe manipulation. *Nature*, Vol. 405, No. 6788, 769-772, ISSN 0028-0836. Figure 1 is used with permission from Macmillan Publisher Ltd: Nature, Copyright 2000.

Vitali, L.; Burghard, M.; Wahl, P.; Scheider, M. & Kern, K. (2006). Local pressure-induced metallization of a semiconducting carbon nanotube in a crossed junction. *Phys. Rev. Lett.*, Vol. 96, No. 8, 086804-1-4, ISSN 1079-7114

Vukovic, T.; Milosevic, I. & Damjanovic, M. (2002). Carbon nanotubes band assignation, topology, Bloch states, and selection rules. *Phys. Rev. B*, Vol. 65, No. 4, 045418-1-9, ISSN 1098-0121

Wang, C.; Zhou, G.; Liu, H.; Wu, J.; Qiu, Y.; Gu, B.-L. & Duan, W. (2006). Chemical functionalization of carbon nanotubes by carboxyl groups on Stone-Wales defects: a density functional theory study. *J. Phys. Chem. B*, Vol. 110, No. 21, 10266-10271, ISSN 1089-5647

Woods, L.; Badescu, S. & Reinecke, T. (2007). Adsorption of simple benzene derivatives on carbon nanotubes. *Phys. Rev. B*, Vol. 75, No. 15, 155415-1-9, ISSN 1098-0121

Wu, J. & Hagelberg, F. (2009). Magnetism in finite-sized single-walled carbon nanotubes of zizzag type. *Phys. Rev. B*, Vol. 79, No. 11, 115436-1-9, ISSN 1098-0121

Yang, S.; Shin, W. & Kang, J. (2006). Ni adsorption on Stone-Wales defect sites in single-wall carbon nanotubes. *J. Chem. Phys.*, Vol. 125, No. 8, 084705-1-5, ISSN 0021-9606

Yao, Z.; Postma, H.; Balents, L. & Dekker, C. (1999). Carbon nanotube intramolecular junction. *Nature*, Vol. 402, No. 6759, 273-276, ISSN 0028-0836

Yi, J.-Y. & Bernholc, J. (1993). Atomic structure and doping of microtubules. *Phys. Rev. B*, Vol. 47, No. 3, 1708-1711, ISSN 1098-0121

Yuan, J. & Liew, K. (2009). Effects of vacancy defect reconstruction on the elastic properties of carbon nanotubes. *Carbon*, Vol. 47, No. 6, 1526-1533, ISSN 0008-6223

Zhang, J.; Yang, G.; Cheng, Y.; Gao, B.; Qiu, Q. & Lee, Y. (2005). Stationary scanning x-ray source based on carbon nanotube field emitters. *Appl. Phys. Lett.*, Vol. 86, No. 18, 184104-1-3, ISSN 0003-6951

Zheng, G. & Zhuang, H. (2008). Magneto-mechanical coupling behavior of defective single-walled carbon nanotubes. *Nanotechnology*, Vol. 19, No. 32, 325701-1-9, ISSN 0957-6984

4

Electronic Structure of Fluorinated Carbon Nanotubes

Maria Brzhezinskaya and Alexander Vinogradov
ALBA Synchrotron Light Source
Spain
Saint-Petersburg State University
Russia

1. Introduction

Chemical functionalization of carbon nanotubes (CNTs), i.e., attachment of individual atoms/molecules or their aggregates to CNTs, can extend the field of application of these nanosystems in nanoelectronics, sensorics, hydrogen power engineering, bioengineering, medicine, etc. (Dresselhaus & Dresselhaus, 2001; Burghard, 2005). In this respect, the fluorination of CNTs is of special interest because the fluorination results in a considerable decrease in the chemical inertness of the initial systems. Therefore, the fluorination is considered as a promising technological process for the first stage of the CNT chemical functionalization (Mickelson et al., 1998; Mickelson et al., 1999; Khabashesku et al., 2002; Lee, 2007). However, carbon materials react with fluorine over a wide range of external conditions. Consequently, the chemical composition, as well as the atomic and electronic structures of fluorinated carbon nanotubes (F-CNTs), depends substantially on the structure and properties of the initial materials and fluorination conditions, such as the reaction temperature and duration, the presence of catalysts, pressure, and concentration of fluorinating reactants (Touhara & Okino, 2000). All these factors need comprehensive investigations of fluorination products by different experimental methods.

Fitting of fluorination conditions that are necessary to perform the preset controllable restructuring of the CNT electronic structure remains at present the most topical problem of the carbon nanomaterial chemistry. Therefore, their atomic and electronic structures have been studied by using a limited number of experimental techniques. First and foremost, these are different microscopic techniques ensuring their visualization (Mickelson et al., 1998; Mickelson et al., 1999; Khabashesku et al., 2002; Lee, 2007; Touhara & Okino, 2000), Hamwi et al., 1997; Yudanov et. al., 2002; Hayashi et al., 2002; Lee et al., 2003), x-ray diffraction methods characterizing the degree of F-CNT crystallinity (Hamwi et al., 1997; Yudanov et. al., 2002) UV spectroscopy (Mickelson et al., 1999), Raman spectroscopy (Mickelson et al., 1998; Mickelson et al., 1999; Khabashesku et al., 2002; Lee, 2007; Touhara & Okino, 2000; Hamwi et al., 1997; Lee et al., 2003) and C 1*s* and F 1*s* core-level x-ray photoelectron spectroscopy (XPS) (Touhara & Okino, 2000; Hamwi et al., 1997; Lee et al., 2003; An et al, 2002) which provide information on the nature of chemical bonding between carbon C and fluorine F atoms in F-CNTs. The investigations performed made it possible to determine the main conditions for the fluorination of single-walled carbon nanotubes (SWCNTs) and multi-walled carbon nanotubes (MWCNTs) to characterize their atomic structure and to obtain preliminary information on the specific features of their electronic structure due to the chemical bonding between C and F atoms. However, the data obtained in those works even with due regard for the results of the first theoretical calculations (Kudin et al., 2001; Park et al., 2003; Lebedev et al., 2003; Van Lier et al., 2005; Ewels et al., 2006) do not provide clear perception of the CNT fluorination mechanism and properties of fluorination products, which is necessary for expansion of practical use of F-CNTs.

Near-edge x-ray absorption fine-structure spectroscopy (NEXAFS) is at present one of the most efficient experimental methods (Stöhr, 1992; Chen, 1997). NEXAFS spectroscopy uses the relation between the spectral characteristics of the near-edge fine structure of x-ray absorption spectra and the parameters of the local atomic and electronic structures of the material under investigation in the vicinity of absorbing atoms. There are a few works in which NEXAFS spectroscopy has been successfully used to characterize pristine and oxidized CNTs (Kuznetsova et al., 2001; Tang et al., 2002; Schiessling et al., 2003; Banerjee et al., 2004; Banerjee et al., 2005; Hemray-Benny et al., 2006; Zhou et al., 2007). Total electron yield (TEY) mode is the most popular for recording of NEXAFS spectra (Gudat & Kunz, 1972). At the same time, TEY mode with probing depth approximately equal to the average MWCNT diameter (several tens of monolayers) does not allow one to get information about characteristics of investigated fluorinated multi-walled carbon nanotubes (F-MWCNTs) in their near-surface region where fluorination process can differ from that at greater depths. However, these data are of genuine fundamental interest and is indeed necessary for optimization of the MWCNT fluorination techniques. Such information for near-surface region only several graphene layers thick can be obtained, particularly, by XPS providing C 1*s* and F 1*s* core-level spectra of F-MWCNTs under investigation. XPS is an effective surface-sensitive method for investigation of material electronic structure (Huefner, 2003). In this case, the probing depth is adequate to the depth the registered electrons are emitted from, which depends on the electron kinetic energy. Therefore, the probing depth can vary from several tenth of nanometer (two to three atomic monolayers) to nanometers, depending on the energy of excitation photons. Presently, several works with XPS used to characterize fluorinated SWCNTs (F-SWCNTs) (Lee et al., 2007; Lee et al., 2003; An et al., 2002), fluorinated multiwalled nanoparticles (Okotrub et al., 2000), and F- CNTs (Shul'ga et al., 2007) are known.

In this chapter, the results of combined investigation of electronic structure of F-CNTs are presented. The high-resolution NEXAFS and XPS spectroscopies were used. The main goal of the present project is to elucidate the nature of chemical bonding between the carbon and fluorine atoms in the SWCNTs and MWCNTs, to evaluate a chemical state of fluorine and carbon atoms in these CNTs, to analyze modification of electronic structure of CNTs under fluorination, to characterize mechanism of CNT fluorination. The defluorination process of F-CNTs on thermal annealing has been investigated.

2. Experimental Details

MWCNTs were produced by arc-discharge evaporation of graphite rods (extremely pure, spectral purity, 99.9992 wt %) in the helium atmosphere (500 Torr) at current density of 175

A/cm^2 and voltage of 23 V. Material with MWCNTs was withdraw from the cathodic deposit center and had a columnar structure with the column length of up to 10 mm (Kiselev et. al., 1999). The carbon-based material containing nanotubes was only grinded and screened by using a sieve with 0.25 mm mesh without any chemical treatment.

The powdered samples of MWCNTs were fluorinated in a nickel reactor at the temperature T_F=420 °C in a flow of molecular fluorine produced by electrolyzing acidic potassium trifluoride KF·2HF containing up to 3 wt % of HF.

SWCNTs were synthesized by the electroarc method using nickel-yttrium catalyst. Amorphous carbon and metal catalyst were removed from the initial condensation products (containing 15–20 weight % of SWCNTs) by multiple alternating air oxidations at temperatures up to 550 °C and by rinsing with hydrochloric acid (Krestinin et al., 2003a). After purifying, nanotubes in the form of SWCNT powder were obtained, their main substance content being about 80–85 weight %.

Diameters of the purified nanotubes varied within a narrow range around the average value of about 1.5 nm; the nanotubes existed in the SWCNT powder in a highly aggregated state in the form of bundles, micro-crystal films, and carpets with polycrystalline structures (Krestinin et al., 2003b). SWCNTs were isolated from the main impurity by dispersing SWCNT powder in aqueous solution of surfactants. As a result, high-purity SWCNT were obtained in the form of paper (SWCNT paper) where the main substance content was over 98–99 wt %, and also there was a small dope of ultra highly-dispersed particles of graphitized soot less than 20 nm in size.

Direct fluorination was carried out in a stainless steel reactor. The SWCNT sample in an aluminum boat was placed into the reactor, which was evacuated at room temperature until the residual vapor pressure became about 10^{-1} mbar. Next, undiluted fluorine (the amount of impurities in fluorine did not exceed 0.1 volume %) was inserted into the reactor to achieve the pressure of 0.8 bar, and the reactor was heated for one hour to 222 °C (sample 1) and 190 °C (sample 2). Thereafter, the samples were kept at the given temperature for four hours more. Fluorine content was estimated by the increase in the sample weight and appeared to be 35% for sample 1 (SWCNT+F35%) and 40% for sample 2 (SWCNT+F40%).

In this work, the spectra of "white" graphite fluoride (WGF) (C_F=62.4 wt %), a highly oriented pyrolytic graphite (HOPG) crystal (Brzhezinskaya et al., 2008, 2009) and nanodiamond (Brzhezinskaya et al., 2010) were used as the reference spectra.

The C 1*s* and F 1*s* x-ray absorption spectra of the pristine and fluorinated MWCNTs, HOPG, and WGF were measured using monochromatic synchrotron radiation and the facilities of the Russian-German beamline (RGBL) at the BESSY II (Berlin) (Molodtsov et al., 2009). The samples for x-ray absorption and photoelectron measurements were prepared in air. For other details of sample preparation see (Brzhezinskaya et al., 2009).

The NEXAFS spectra were obtained by recording the TEY of the x-ray photoemission (Gudat & Kunz, 1972) in the mode of measurement of the drain current of the sample by varying the energy of incident photons. All x-ray absorption and photoelectron spectra were measured under ultrahigh vacuum (~2x10^{-10} Torr).

The energy resolutions ΔE of the monochromator in the range of the F 1*s* x-ray absorption edge ($h\nu$~680 eV) and the C 1*s* x-ray absorption edge ($h\nu$~285 eV) were equal to ~150 and 70 meV, respectively. The $h\nu$ in the range of the fine structure of the F and C x-ray absorption spectra was calibrated against the energy positions of the first narrow peak in the F 1*s* x-ray absorption spectrum of K_2TiF_6 (683.9 eV) and the C 1*s* x-ray absorption spectrum of HOPG (285.45 eV).

In recording x-ray photoelectron spectra, *hv* varied from 345 to 1130 eV. Energy resolution ΔE of the monochromator with the 200 μm exit slit was 145–750 meV within the above mentioned range of *hv*. Photoelectron spectra of all the samples were measured in the normal photoemission registration mode by using the Phoibos 150 spherical analyzer from Specs whose resolution in recording the spectra remained equal to 200 meV. The analyzer was calibrated against energy based on photoelectron spectra of Au atom $4f7/2,5/2$ electrons. The monochromator was calibrated by recording the basic photoelectron lines of C 1*s* spectra exited by radiation reflected from the diffraction grating in the first and second diffraction orders.

For more experimental details see (Brzhezinskaya et al., 2009; 2010).

3. Experimental Results and Discussion

3.1 NEXAFS spectra

Fig. 1 presents C *K*(1*s*) absorption spectra of initial SWCNTs, MWCNTs, HOPG crystal and nanodiamond measured under the same experimental conditions (Brzhezinskaya et al., 2008, 2009, 2010). All the spectra were normalized to the same level of continuous C 1*s* absorption at photon energy of 315 eV. The number and energy positions of *A–F* fine structure features of the initial MWCNTs, SWCNTs and HOPG C 1*s* spectra are similar. This fact confirms the crucial role a single graphene layer plays in formation of the structure of carbon 1*s* absorption spectra in HOPG, MWCNTs and SWCNTs (Brzhezinskaya et al., 2008, 2009; Krestinin et al., 2009). It also confirms that the SWCNT conduction band structure is slightly affected by nanotube curvature (Louie, 2001). In the scope of this philosophy, we associate the *A* and *B-C* peaks in the HOPG spectrum with dipole-allowed electron transitions from 1*s* carbon states to the unoccupied π and σ symmetry states in the conduction band that are formed from carbon $\pi 2p_z$ and $\sigma 2p_{x,y}$ states oriented perpendicular to and parallel to the carbon layer plane (graphene), respectively. Such states are mostly associated with free electron states of individual carbon atoms. The states are actually quasi-molecular as they are similar to those of benzene molecule C_6H_6 and are localized within one carbon hexagon (Brzhezinskaya et al., 2008; Comelli et al., 1988). The broad *A'* band that determines high level of continuous absorption between the *A* and *B* resonance peaks is associated with electron transitions to unoccupied π type states that are less localized in graphene layer than the state responsible for π resonance peak *A*. The broad *D-F* absorption bands are associated with electron transitions to the unoccupied σ states in the graphite conduction band arising due to interaction between carbon hexagons in graphene layer.

Comprehensive investigation of spectra has revealed slight differences between nanotube and HOPG spectra, as well as between SWCNT and MWCNT spectra. Those are mostly the differences in the low-energy spectral region (*hv*=284-295 eV). All structural features become a little broader in nanotube spectra as compared to HOPG spectrum, the broadening being different for SWCNTs and MWCNTs. This is most clearly seen for π band *A*: its full width at half maximum (FWHM) increases from 1.15 eV (for HOPG spectrum) to 1.3 eV for SWCNT spectra and up to 1.45 eV for MWCNT spectra. Moreover, for SWCNT and MWCNT spectra, extra weak structural features *a* and *a, b, c* arise in the gaps between absorption bands *A* and *B-C*, respectively.

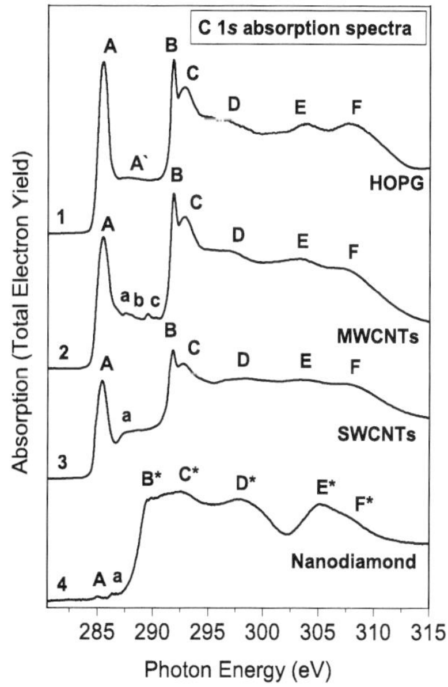

Fig. 1. C 1s absorption spectra of HOPG (1), MWCNTs (2), SWCNTs (3) and nanodiamond (4)

Naturally, the observed broadening of π band *A* in CNTs carbon spectrum as compared to HOPG spectrum can be associated with splitting of the graphite $\pi 2p_z$ conduction sub-band, which is caused by the reduction in local symmetry of nanotubes carbon atoms due to the fact that graphene layers become curved. However, for single-walled nanotubes, one should have expected greater broadening of band *A* since the SWCNT diameters (~1.5 nm) are an order of magnitude less than those of MWCNTs (~10-30 nm). In reality, as shown above, the situation is quite opposite.

In interpreting the differences in the C 1*s* absorption spectra recorded by measuring the total electron yield, it is necessary to take into account that the nanotube spectra under consideration are obtained by averaging a great number of individual SWCNTs in the bundles or in MWCNTs within the area limited by the size of the focused beam spot (0.2 x 0.1 mm^2) on the sample surface and by the probing depth (15-20 nm) (Chen, 1997). In this case, the differences in chemical states of carbon atoms in different tubes or parts of individual tubes result in energy dispersion of C 1*s* electron transitions to π symmetry electron states and, hence, can contribute significantly to broadening of π band *A*. Thus, the greater broadening of π band *A* in MWCNT spectrum may be attributed to greater variations in the carbon atom chemical states in MWCNTs as compared to SWCNTs. Greater variations in diameters and concentration of defects in the MWCNTs as compared to SWCNTs may be the reasons for this. It is also possible that just the higher degree of order of carbon atoms and their electron states in the SWCNT bundles provides high energy of interaction (about 0.5 eV per 1 nm of length) between individual SWCNTs in the bundles.

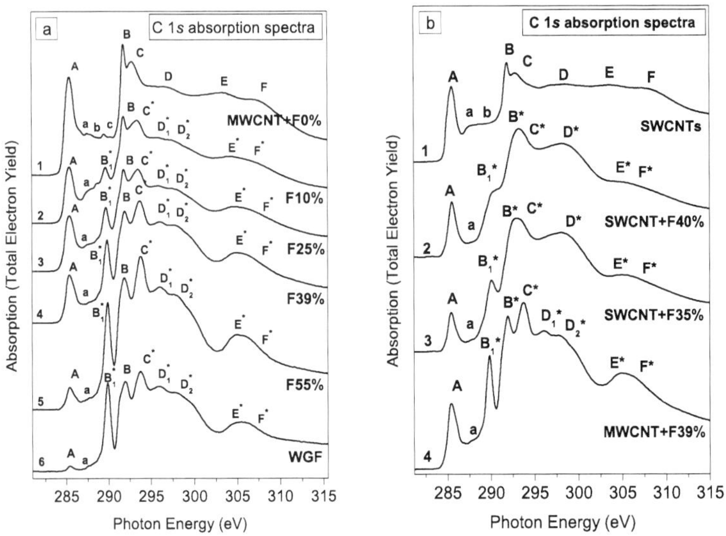

Fig. 2. (a) C 1*s* absorption spectra of (*1*) MWCNTs, (*2*-*5*) F-MWCNTs containing (2) 10%, (3) 25%, (*4*) 39%, (5) 55% and (6) WGF (62.4%); and (b) C 1*s* absorption spectra of SWCNTs (1), F-SWCNTs containing 40% (2) and 35% (3) and MWCNTs+F39% (4)

In its turn, the rise of weak structures *a* and *a, b*, c between π and σ peaks in the nanotubes spectrum is typically attributed to the C 1*s* electrons transitions to free 2*p* states of carbon atoms localized on nanotube surface areas oxidized during synthesis (Banerjee et al., 2004; Kuznetsova et al., 2001). It is quite possible that the weak structures under discussion are partly associated with the 1*s*->2*p* absorption transitions in carbon atoms at the fullerene-like edges of nanotubes.

Comparing the spectra of pristine and fluorinated CNTs (Fig. 2), we can see that their fine structures differ significantly. What is most important is that the C 1*s* NEXAFS spectra of the F-MWCNTs with the highest fluorine content (C_F=55 wt %) (MWCNTs+F55%) and WGF [Fig. 2(a), curves 5 and 6] are very similar and differ substantially from the spectra of HOPG and MWCNTs (Fig. 1, curves 1 and 2). The most significant differences involve a drastic decrease in the intensity of the resonance *A* and the band *A*′ associated with the free π states and the appearance of the high-energy σ band *E*-F**. It is evident that the observed changes are governed by the fluorination of the HOPG and MWCNTs. The resonance *A* is retained in the spectra of the fluorinated samples but has a considerably lower intensity with respect to the other spectral portion associated with the transitions of the 1*s* electrons to the free states of the conduction band. The band *B* is also retained in the spectra, whereas the other specific features in the spectra of HOPG and MWCNTs have no clear analogs in the spectra of the fluorinated samples. New specific features B_1*-*F** in the spectra of the F-MWCNTs and WGF should be unambiguously treated as a result of the transitions of the C 1*s* electrons to the free states of the new phase formed in the MWCNTs and HOPG due to their fluorination. It should be emphasized that in the given case, only one particular phase can be formed. In the presence of several fluorocarbon phases, the C 1s spectra would be virtually structureless and similar in this respect to the spectrum of amorphous carbon (Rosenberg et al., 1986). Hence, the spectra of the F-MWCNTs and WGF are a superposition of the dominant C 1s spectrum of the fluorocarbon phase and the low-intensity spectrum of the pristine sample (MWCNTs and HOPG). In other words, despite the maximum possible fluorine content in the F-MWCNTs and WGF (55.0 and 62.4 wt %, respectively), they are incompletely fluorinated and contain regions composed of the pristine HOPG and MWCNTs. This is most probably explained by the fact that a limiting depth of fluorination exists under these conditions of the synthesis. This fluorination depth is less than the probing depth (d~15 nm) of the sample when the TEY method is employed for recording the NEXAFS spectra (Stohr, 1992; Chen, 1997). Since the sample contains MWCNTs of different sizes (L~0.5-2.0 μm and D~10-30 nm), including nanotubes with a diameter D>15 nm, the measured C 1s spectrum should involve the contributions of the fluorinated and pristine carbon phases.

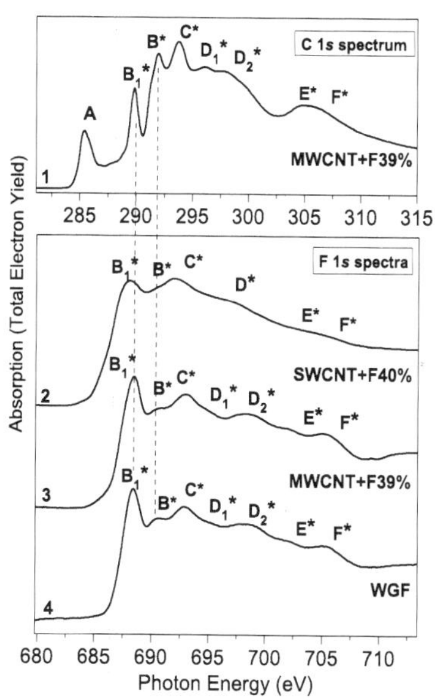

Fig. 3. Comparison of C 1s (1) and F 1s (2) NEXAFS spectra of MWCNTs+F39% and F 1s NEXAFS spectra of SWCNTs+F40% (3) and WGF (4). The F 1s absorption spectra are reduced to the energy scale of the C 1s absorption spectrum with the use of energy spacing ΔE(F 1s–C 1s) = 398.4 eV (measured by x-ray photoemission spectroscopy) between the F 1s and C 1s core levels.

Since the C 1s NEXAFS spectra of the F-MWCNTs and WGF do not contain the intense π resonance A and the π band A', we can make the inference that the C atoms are involved in the chemical bonding upon formation of the fluorocarbon phase. This implies that upon fluorination, the F atoms are attached to the graphene layer perpendicular to it rather than replace the C atoms in the layer. In this case, the F atoms form bonds with the C atoms with the participation of the C $2p_z$ states. As a result, the C atom has a spatial coordination rather than the planar coordination.

One more indirect evidence for the change in the coordination of the C atom in HOPG and MWCNTs during fluorination can be the appearance of a single σ band *E*-F** in the x-ray absorption spectra of the fluorinated samples ($h\nu$~303–310 eV). A similar high-energy band, together with the absence of the π resonance, is considered as a characteristic difference between the C 1*s* NEXAFS spectrum of diamond (Fig. 1, curve 4) with the sp^3 tetrahedral coordination of the C atoms and the spectrum of HOPG with the sp^2 trigonal coordination. This conclusion is in good agreement with the results obtained from the investigation into the mechanism of the formation of graphite fluoride C_xF by Raman spectroscopy (Gupta et al., 2003).

It is interesting to compare the C 1*s* x-ray absorption spectra of the F-MWCNTs with different fluorine contents (0%–55%) [Fig. 2(a)] because they correspond to different stages of the MWCNT fluorination. These spectra demonstrate a monotonic character of variations in the fine structure with an increase in the degree of MWCNT fluorination. As a result, a gradual disappearance of the π structure (the resonance *A* and the band *A'*), as well as the formation of the characteristic resonances and in the region of the σ structure in addition to a single high-energy band *E*-F**, is observed. These changes manifest the uniform character of the interaction between the F and C atoms in all stages of the fluorination process.

In analyzing C 1*s* absorption spectra of fluorinated SWCNTs with the almost equal fluorine content of 35 and 40 wt % [Fig. 2(b), curves 2 and 3] it is necessary to notice, first of all, similarity of their general spectral shape manifesting itself in coincidence of the fine structure features number and energy positions. The only slight difference between those spectra is somewhat better resolution of band B_1* on the spectra of the SWCNTs with lower fluorine content. Obviously, similarity of fine structures of the spectra of two fluorinated SWCNT samples reflect the similarity of their atomic and electronic structures; therefore, this confirms the reproducibility of SWCNTs fluorination results for the fluorination process described above.

Comparing the spectra of pristine and fluorinated SWCNTs, we can see that their fine structures differ significantly. Indeed, after fluorination, the only fine structure feature still remaining in the nanotube spectra is the depressed π resonance of *A*, all other features B_1*-*F** being new. Evidently, the presence of band *A* typical of the pristine nanotube spectrum in the F-SWCNT spectra concurrently with the new absorption bands is the evidence of incomplete fluorination of single-walled nanotube bundles within the probing depth of 15-20 nm. Probably, it is caused by the fact that the SWCNT bundle fluorination depth is much shorter than the probing depth in the absorption spectra under consideration.

The above-described variations in the F-SWCNT carbon absorption spectra fine structures are quite similar but not identical to the variations observed in the structures of F-MWCNT spectrum [Fig. 2(b), curve 4] with respect to the pristine MWCNT spectrum (Fig. 1, curve 2). Indeed, Fig. 3 demonstrates clearly that, having the general form quite similar to that of spectrum of F-MWCNTs with almost the same content of fluorine (39 wt %), the F-SWCNT spectra (F35 wt % and F40 wt %) are less structured and manifest broader absorption bands. Bands *B*-C** and *D** in F-SWCNT spectra were assumed to be the result of superposition of individual absorption bands *B** and *C**, D_1* and D_2* in F-MWCNT spectra.

Getting back to the differences between F-SWCNT and F-MWCNT C 1*s* spectra, we can briefly characterize them as significant broadening and blurring of prominent and well-resolved features of F-MWCNT spectrum as compared with the F-SWCNT spectra, the general spectrum shape character remaining almost unchanged. Assuming that only one fluorine-carbon phase arises in fluorination in both SWCNTs and MWCNTs, we can attribute the observed differences in their spectra to different levels of structural ordering in the phase formed. This could be due to different temperatures of nanotubes fluorination by molecular fluorine: 420 °C for multi-walled (Brzhezinskaya et al., 2008, 2009) and 190 °C and 222 °C for single-walled nanotubes. Indeed, the differences in structural ordering of fluorinated regions of single-walled nanotubes should be accompanied by differences in the strength of chemical carbon-fluorine bonding and, hence, by dispersion of charge (chemical) states of those atoms. In its turn, this should cause noticeable dispersion of energies of identical electron transitions and, hence, broadening and casual blurring of fine structure features of the carbon and fluorine absorption spectra of F-SWCNTs.

However, the nature of differences in C 1*s* absorption spectra of F-SWCNTs and F-MWCNTs is, probably, more complicated. In the case of F-MWCNTs with diameters of approximately 10 to 30 nm, the spectra under consideration are related to the major (bulk) fluorine-carbon phase due to quite large probing depth (15-20 nm) (Brzhezinskaya et al., 2009). This phase is highly ordered because of strong covalent carbon-fluorine bonding; it dominates in the F-MWCNTs over the less structured near-surface phase that coexists with the major phase within 2-3 graphene monolayers (Brzhezinskaya et al., 2009). The two phases revealed in F-MWCNTs are characterized by chemical bonding with different electron transfers between carbon and fluorine atoms, which gives rise to different carbon atomic states in those phases. Relying on these facts, we assume that, in the case of F-SWCNTs (with diameters ~ 1.5 nm), similar phases should coexist in all tubes of the bundles and, consequently, their C 1*s* absorption spectra should manifest structural features of both phases concurrently. So, blurring of fine structure features in the F-SWCNT spectra as compared to F-MWCNT spectrum may be attributed to superposition of the spectral structure features characteristic of the two fluorine-carbon phases whose carbon atoms are in different chemical states. Theoretical calculations (Ewels et al., 2006) also point to the possible presence of two fluorine super-lattices that give rise to fluorine-carbon phases with the C_4F and C_2F stechiometries on the SWCNTs lateral surface in the temperature range of 200-250 °C.

The above conclusion is confirmed by the results of comparison of the F 1*s* x-ray absorption spectra of F-MWCNTs and F-SWCNTs (Fig. 3). All the F 1*s* x-ray absorption spectra of the F-MWCNTs and WGF exhibit a similar fine structure (Brzhezinskaya et al., 2009). The main difference between the spectra is that the measured intensity of the spectra increases monotonically with respect to the background intensity with increasing fluorine content in the F-MWCNTs. This behavior of the F 1*s* x-ray absorption spectra of the F-MWCNTs with an increase in the fluorine content in the samples can be interpreted only as a result of the formation of a fluorocarbon phase, which is identical in all stages of the fluorination process. This conclusion, of course, should correlate with the probing depth (~15 nm) of the nanotube samples during the recording of the x-ray absorption spectra in this work. The F 1*s* x-ray absorption spectra are also characterized by rich and clearly pronounced fine structure. This circumstance indicates a well-defined atomic structural order in the fluorinated samples. In other words, the fluorine atoms, such as the carbon atoms, have an identical coordination, which reflects chemical bonding of the C and F atoms.

The above-made conclusion on the covalent character of fluorine-carbon bonds in F-MWCNTs and retention of laminar character of the carbon atomic structure up to maximal fluorine concentrations ensuring the binary (CF) stoichiometry of fluorination products agrees well with wellknown results of graphite fluorination. It was found out for graphite that when it is maximally fluorinated, fluorinecarbon phase poly(carbon monofluoride) $(CF)_n$ predominates (Gupta et al., 2003; Sato et al., 2003) which is characterized by the retention of laminar graphitelike structure, although the layers are no longer planar but become corrugated because fluorine atoms are attached to them. What is important is that this graphite fluoride is characterized by covalent binding with sp^3 hybridization of carbon-atom valent states. This fact is confirmed also by the similarity of absorption spectra of F-MWCNTs and WGF reference samples. All this confirms the above conclusion that MWCNTs retain their basic structure at high fluorination levels, such as 55 wt %, where almost every carbon atom throughout all CNT layers is in the sp^3 configuration.

F 1s absorption spectra of F-SWCNTs with almost the same fluorine content of 35 and 40 wt % are in general similar: numbers of their spectral features are identical, their energies are close to each other (Brzhezinskaya et al., 2010). The only slight difference between them is somewhat better fine structure resolution of the spectra of tubes with less fluorine content; this is clearly seen when comparing the B_1* bands.

It is very important that both F 1s and C 1s absorption spectra of F-SWCNTs are, in general, very similar to respective F-MWCNT spectra (Fig. 3, curve 4). The main difference is that the fluorine spectra for F-MWCNTs are better structured (as the carbon spectra are). This fact clearly confirms our conclusion made above that the major covalent fluorine/carbon phases arising in fluorination of single- and multi-walled nanotubes are similar to each other. Hence, we can assume that SWCNT fluorination occurs mainly by bonding fluorine atoms to carbon atoms on nanotube sidewalls (similar to MWCNT fluorination); this is accompanied by generation of σ(C-F) bonds by covalent mixing of valence F 2p and C $2p_z\pi$ electronic states without any destruction of the tubular structure of carbon layers (Brzhezinskaya et al., 2008, 2009). Such fluorine bonding transforms carbon triangular arrangement of initial nanotubes into almost tetrahedron one of fluorinated nanotubes, which becomes possible only if sp^2 hybridization of carbon atom valence states in pristine nanotubes transforms into sp^3 hybridization of F-SWCNTs. Such transformation of atomic structure should cause corrugation of nanotube sidewalls.

It is interesting to analyze the fluorinated sample unoccupied electronic state spectrum by using C 1s and F 1s X-ray absorption spectra that reflect energy distribution of free electron states formed from the 2p states of the carbon and fluorine atoms, respectively, because of the dipole nature of absorption transitions. For this purpose, the carbon and fluorine atom spectra for F-MWCNTs with the 35 wt % fluorine content (curves 1 and 3) are given in Fig. 3 on the photon energy scales that are brought into coincidence by using the ΔE (F 1s – C 1s) =398.4 eV energy shift between the fluorine and carbon 1s states in the system. In this work, energy positions of the core levels were measured the by X-ray photoemission technique at the exciting photon energy of 1030 eV.

Comparing the spectra, we can clearly see that both spectra manifest fine structures similar in the number and energy positions of their key features B_1*-F* (the A band in the carbon spectrum is associated with non-fluorinated tubes). Vertical dashed lines in Fig. 3 indicate that energy positions of bands B_1 and B^*-C^* in the carbon and fluorine absorption spectra of F-SWCNTs and F-MWCNTs are close to each other. Such correlation in the F-CNT carbon and fluorine spectra means that carbon and fluorine 1*s* electron transitions to the unoccupied electronic states formed from C 2*p* and F 2*p* electronic states are responsible for the key features of the spectra compared. It indicates the hybrid (covalent) F 2*p* + C 2*p* nature of the free electronic states in the systems discussed, which complies with the conclusion about sp^3 hybridization of the carbon atom valence states, which was made as a result of independent analyses of carbon and fluorine spectra.

3.2 XPS spectra

X-ray photoelectron spectra of SWCNTs and MWCNTs were measured by using exciting photons with energies from 385 to 1030 eV. All the measurements were carried out under the same experimental conditions. For both SWCNTs and MWCNTs, the *A* line in the C 1*s* spectra is a narrow peak at *BE*=284.8 eV which is asymmetrically shaped at the higher binding energy side and whose full-width at half maximum (FWHM) is 0.7 eV.

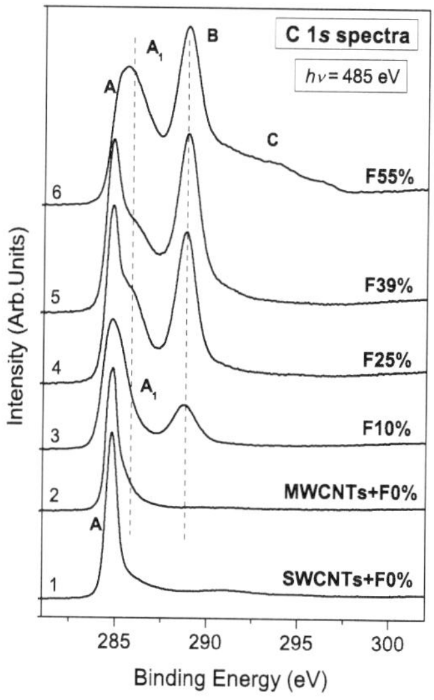

Fig. 4. C 1*s* x-ray photoelectron spectra of (1) SWCNTs, (2) MWCNTs and F-MWCNTs with fluorine concentrations: 10% (3), 25% (4), 39% (5), 55% (6).

Fig. 4 demonstrates that C 1s photoelectron spectrum structure changes radically as pristine MWCNTs are transformed into F-MWCNTs. Shoulder A_1 shows up near the main line *A* at *BE*=285.7 eV, and the second line *B* arises at *BE*=288.8 eV. As the fluorine concentration in F-MWCNTs is growing, the intensities of the spectrum new features increase, while their energy positions do not change. When the concentration reaches C_F=39 wt %, intensity *B* becomes almost equal to the main C 1s line intensity *A*. Spectrum measured at the maximum fluorine concentration (C_F=55 wt %) differs significantly from the other spectra: band *A* almost fully disappears against the background of the intense band into which shoulder *A* is transformed, whereas band *B* becomes the most intense structure feature having an intense, extended and poorly structured tail C on the high-energy side.

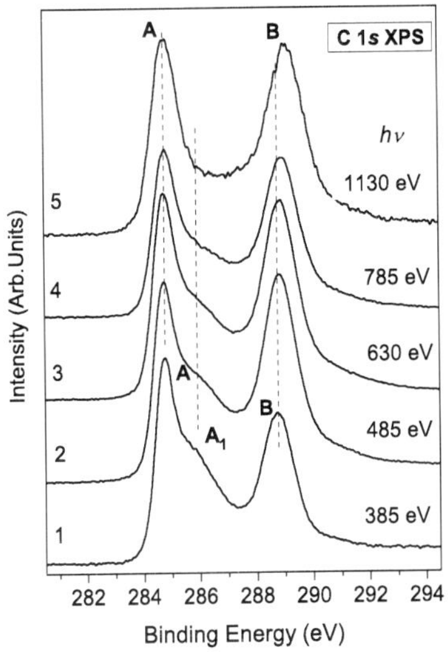

Fig. 5. C 1*s* x-ray photoelectron spectra of MWCNTs+F39% measured at photon energies of 385 eV (1), 485 eV (2), 630 eV (3), 785 eV (4) and 1130 eV (5).

As shown in our work (Brzhezinskaya et al., 2009), appearance of new features in the F-MWCNT C 1*s* spectra structure as compared with those of MWCNTs can be attributed to CNT fluorination; therefore, it can be regarded as a result of carbon-fluorine chemical bonding. This interaction results in formation of fluorine-carbon phases and is accompanied by charge transfer from carbon atoms to fluorine atoms due to the fact that fluorine electronegativity is higher. Charge states (oxidation rates) of carbon atoms in at least two of the fluorine-carbon phases are different and can be characterized by chemical shifts of their C 1*s* photoelectron lines (peaks A_1 and *B*) from the energy position of this line in pristine MWCNT spectrum (peak *A*). Energy positions of bands A_1 and *B* relative to peak *A* are significantly different (0.9 eV and 4.0 eV, respectively). This large difference (>3 eV) indicates considerable differences in the natures of C-F atom interactions in the analyzed fluorine-carbon phases that are formed in F-MWCNTs.
Relative intensities of bands *A* and *B* do not change much with $h\nu$ increasing up to 1030 eV, while the A_1 band intensity decreases, and the corresponding shoulder gets barely visible against band *A* background. It is reasonable to connect those observations with the probing depth increase from ~0.6 nm (1-2 graphene layers) to 2 nm (5-6 layers). The weak relationship between the peaks *A* and *B* relative intensities indicates that the ratio of the initial and fluorinated MWCNTs phases for the samples with C_F=10–39 wt % changes only slightly with depth within several graphene layers. This statement is consistent with the

conclusion that the fluorination process in MWCNTs under investigation remains uniform within the depth limits of ~15 nm (Brzhezinskaya et al., 2008).
A significant decrease in the band A_1 intensity with the probing depth increase from 0.6 to 2 nm shows unambiguously that phase 2 is of the near-surface character and is formed by fluorination of only a few outer graphene layers. This conclusion is consistent with the specific shape of C 1*s* photoelectron spectrum obtained for MWCNTs+F55%.

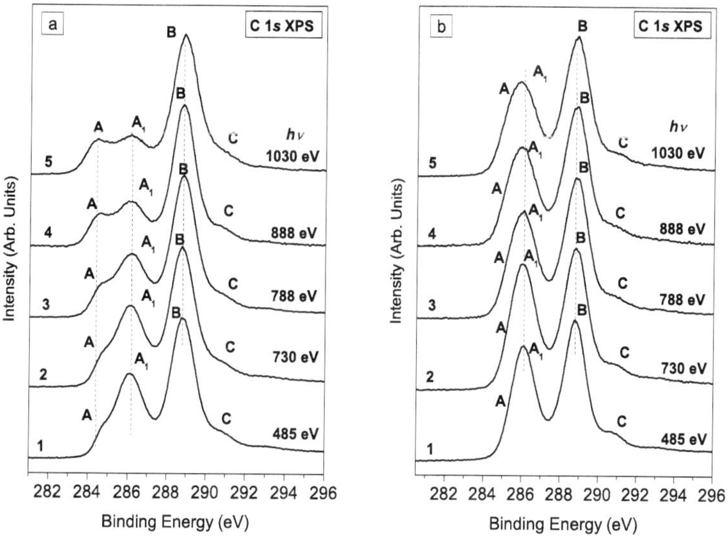

Fig. 6. C 1*s* x-ray photoelectron spectra of (a) SWCNTs+F35% and (b) SWCNTs+F40% measured at photon energies of 385 eV (1), 730 eV (2), 788 eV (3), 888 eV (4), 1030 eV (5).

The energy gap between bands *A* and *B* grows by ~0.5 eV as photon energy changes from *hν*=385 eV to *hν*=1130 eV. It means that chemical shift of the C 1*s* level grows from 4.0 to 4.5 eV with increasing probing depth for spatial fluorocarbon phase. Such an increase in the chemical shift is observed for *hν*>700 eV only. It may be a result of a small change in the chemical state of phase 1 carbon atoms located deeper in the samples. In this case, the chemical shift of 4.5 eV for C 1*s* level is practically equal to the 4.45 eV chemical shift in the C 1*s* absorption spectra (Brzhezinskaya et al., 2008).
The feature A_1 behavior is similar in all F-MWCNTs spectra; Fig. 5 represents MWCNT+F39% to illustrate this fact. The band A_1 intensity gradually decreases with increasing probing depth (*d*=0.4–2 nm and *hν*=345–1130 eV). It is reasonable to connect the gradual band A_1 weakening with formation of fluorocarbon phase in the very near-surface layer of the tubes. This layer is one to two graphene monolayers thick as this is just the depth (0.4–0.8 nm), where from C 1*s* photoelectrons generated by absorption of photon with *hν*=345–630 eV and having kinetic energies of E_k=60–345 eV can be emitted without energy losses (Huefner, 2003; Brown et al., 1978; Martin et al., 1985).

The spectra of F-SWCNTs (Fig. 6) were measured at photon energies from 385 to 1030 eV. According to the universal curve of electron mean free path versus kinetic energy (Brzhezinskaya et al., 2009), the escape depth of C 1*s* photoelectrons excited by photons with energies of 385 to 1030 eV varies from about 0.5 to 1.5 nm; thus, the probing depth lays actually within the subsurface layer about one nanotube diameter wide. As compared with F-MWCNTs (Fig. 4), for F-SWCNTs, C 1*s* photoelectron spectrum structure changes even more radically as initial SWCNTs are being transformed into F-SWCNTs. In the C 1*s* photoelectron spectrum of SWCNTs+F35%, there are 4 peaks, *A* (*BE*=284.6 eV), A_1 (*BE*=286.1 eV), *B* (*BE*=288.8 eV) and *C* (*BE*=290.9 eV). Notice that at small photon energies (<888 eV) band *A* manifests itself as a shoulder next to peak A_1 on the lower binding energy side. In the C 1*s* spectra of SWCNTs+F40% there are only three explicit features: bands A_1 (*BE*=286.1 eV), *B* (*BE*=288.8 eV), and shoulder *C* (*BE*=290.9 eV). The presence of band *A* is indicated by only the large width and slight low energy asymmetry of band A_1. Coincidence of bands A_1 positions and bands *B* positions in the F-SWCNTs and F-MWCNTs indicates their connection with two fluorine-carbon phases that are similar for the single- and multi-walled nanotubes. This fact indicates their relation to two different fluorine-carbon phases that are respectively similar in single- and multi-walled nanotubes. As shown above, a single fluorine-carbon phase uniformly distributed within the sample probing depth is formed in fluorination of both MWCNTs and SWCNTs. Peak *B* in C 1*s* photoelectron spectra of F-MWCNTs is attributed to this phase. Then the same peak *B* should correspond to this phase in the F-SWCNT C 1*s* photoelectron spectra.

Energy positions of bands A_1, *B* and *C* relative to peak *A* are significantly different (1.3 eV, 4.0 eV and 6.1 eV, respectively) for SWCNTs+F35%. For SWCNTs+F40%, energy shifts of peaks A_1, *B* and *C* are the same when measured relative to peak *A* energy position for SWCNTs+F35%. Note that peak *A* cannot be clearly identified in the SWCNTs+F40% spectra [Fig. 6(b)]. This large difference (~3 and 6 eV) indicates considerable differences in the natures of C-F atom interactions in the analyzed fluorine-carbon phases formed in F-SWCNTs.

It is interesting that in the C 1*s* spectra of SWCNTs+F35% [Fig. 6(a)] measuring with photon energy increasing from 485 to 1030 eV and, hence, with probing depth increasing from ~0.5 nm to ~1.5 nm, band A_1 relative intensity becomes more than twice lower than that of band *B*. Probably, this indicates a significant decrease in the fluorination level of inner nanotubes located inside the bundle, namely, all the nanotubes that do not participate in forming the bundle surface layer. This conclusion correlates well with the earlier revealed increase in the SWCNT fluorination level accompanied by improvement of their dispersion, i.e., by the decrease in transverse size of the bundles formed by the nanotubes (Krestinin et al., 2009). However, the decrease in SWCNTs+F40% band A_1 intensity is not so dramatic [Fig. 6(b)]. Generally, shapes of SWCNTs+F40% [Fig. 6(b)] and MWCNTs+F55% (Fig. 4, curve 5) C 1*s* spectra are similar, except for the band *C* shape.

Thus, the first fluorine-carbon phase (phase 1, peak *B*) is apparently formed by fluorine atoms covalent bonding with the graphene layer. Contrary to phase 2 (peak A_1), this phase is uniformly distributed within the sample probing depth (0.5-1.5 nm for XPS). Moreover, this fluorine-carbon phase contributes most to C 1*s* absorption spectra of fluorinated nanotubes at larger probing depths (Brzhezinskaya et al., 2010). In addition, just phase 1 is associated with formation of the C_2F super-lattice on the surfaces of both F-SWCNTs and F-MWCNTs.

This is confirmed by most theoretical calculations (Bettinger et al., 2001; Kudin et al., 2001; Seifert et al., 2000).

Chemical shift in the second phase (phase 2, peak A_1) of the C 1*s* electron binding energy (1.2 eV) is significantly smaller than that in the first phase (4.0 eV). Hence, unlike phase 1, phase 2 formation is accompanied by significantly lower carbon-fluorine electron density transfer resulting in a weaker chemical bond. Phase 2 is most probably associated with formation of C_4F superlattice on the surfaces of both F-SWCNTs and F-MWCNTs. This idea is consistent with the calculation results (Ewels et al., 2006) showing that the most stable phase is (1,2) (C_2F) phase followed by the (1,4) (C_4F) phase. When the probing depth increases from ~0.5 nm to ~1.5 nm, the band A_1 intensity decreases relatively to the band B intensity. For F-SWCNTs, such a transformation of band A_1 is likely to confirm the above conclusion about incomplete SWCNT fluorination inside the bundles. Therefore, phase 2 is developed only on SWCNTs located in the bundle near-surface layers, at least SWCNTs located on the bundle outer periphery having the greatest chance to become completely fluorinated. It is known that individual SWCNTs are arranged in the bundles almost coaxially and practically without tangles, the number of tubes per bundle being about 300. The F-SWCNT defluorination experimental results discussed below also point to a weaker carbon-fluorine bond.

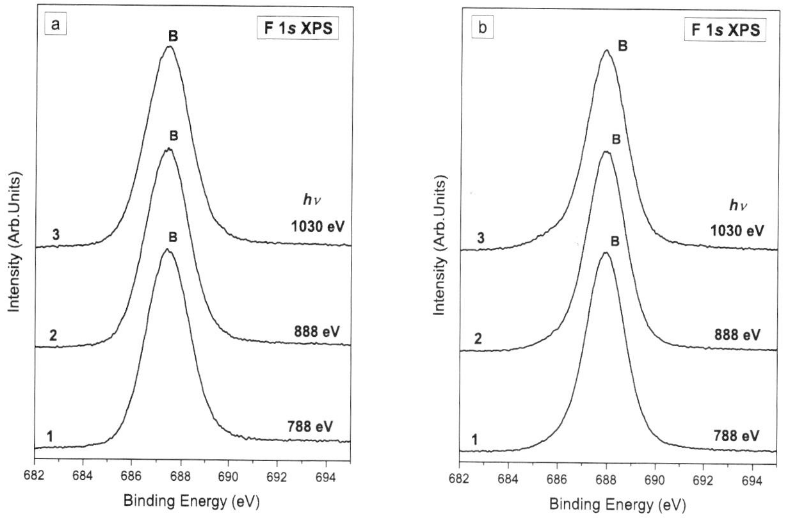

Fig. 7. F 1s x-ray photoelectron spectra of (a) SWCNTs+F40% and (b) MWCNTs+F39% measured at photon energies of 788 eV (1), 888 eV (2), 1030 eV (3).

In addition to new peaks A_1 and B in the F-SWCNT C 1*s* photoelectron spectra there is one more feature C whose intensity is significantly lower than that of peaks A-B. Feature C is approximately 6 eV apart peak A itself in the case of SWCNTs+F35% or from peak A assumed position in the case of SWCNTs+F40%. In this case, chemical shift of feature C

carbon atom 1*s* electron binding energy (~6 eV) is much larger than the shift corresponding to phase 1 (4.0 eV). Hence, this peak is obviously associated with F-SWCNT carbon atoms with higher oxidation levels as compared with carbon atoms of phase 1. Relative intensity of peak *C* is independent of the excited photon energy (Fig. 6). It is possible that, contrary to MWCNTs, an extra fluorine-carbon phase 3 is formed as a result of SWCNT fluorination. As compared with phase 1, formation of phase 3 is accompanied by even higher carbon-fluorine electron density transfer resulting in formation of a stronger chemical bond. Phase 3 is probably formed due to covalent bonding of fluorine atoms to carbon atoms on the SWCNT open edges. Theoretical calculations performed for nanographite and STM measurements (Saito et al., 1999; Maruyama et al., 1994; Kobayashi et al., 2005) also point to such possibility. Notice that destruction of nanotube fullerene-like edges is a particular feature of just the SWCNT purification process (Krestinin et al., 2003). If this is true, fluorine atoms should be able to get attached to SWCNT graphene mesh not only by sharing carbon atom $2\pi p_z$ orbitals but also by using dangling $2\sigma p_{x,y}$ bonds at the SWCNT graphene mesh edges at the points where fullerene-like structures were previously attached. This would result in formation of even stronger carbon-fluorine chemical bond of the CF type due to covalent mixing of the F 2*p* and C $2\sigma p_{x,y}$ electron states.
(Saito et al., 1999) showed, the fluorine atoms get first attached to the edge carbon atoms and only later they begin bonding to carbon atoms located along the nanotube perimeter. It was shown that doped F is more stable at the edge than in the interior region. The dopant´s electron is strongly localized around the F site, and even two F atoms can interact with a single carbon atom, i.e., CF_x (x>1) clusters are formed. There are two ways for fluorine atoms to get attached to the edges of nanotubes: by forming either Klein`s bearded zigzag edge (Klein, 1994) (two fluorine atoms are attached to a single carbon atom at the edge) or Fujita`s normal zigzag edge (Fujita et al., 1996) (one fluorine atom is attached to one carbon atom). This conclusion is confirmed by the peak *C* behavior in the process of SWCNTs+F40% thermal defluorination (see below), which shows that, contrary to peaks *B* and A_1, peak *C* can not be totally reduced. Seifert et al., 2000 calculated that equimolar $(CF)_n$ nanotubes are stable. However, the authors believe that formation of stable $(CF)_n$ nanotubes may be hindered by the height of the energy barrier preventing fluorine atoms penetration into CNTs. But in our case, when the most part of fullerene-like nanotube edges are already destructed, it is easier for fluorine to penetrate into SWCNTs.
This difference in carbon atom chemical states in the fluorine-carbon phases compared should manifest itself in different spatial structures, particularly, in the increase of interatomic spacing when moving from phase 1 to phase 2 and then to phase 3.
F 1*s* spectra with $h\nu$=788–1088 eV were also measured for F-SWCNTs and F-MWCNTs (Fig. 7). Here the photoelectron kinetic energies were about E_k~100-400 eV, which corresponds to the probing depth of about d~0.5÷1.0 nm. F 1*s* spectra of SWCNTs+F35% and SWCNTs+F40% are the same. F 1*s* photoelectron spectra of F-MWCNTs keep their energy positions and, contrary to the C 1*s* spectra, do not exhibit any significant changes in their shapes up to C_F=39 wt %. F 1*s* photoelectron spectra of F-CNTs [(Fig. 7(a)] consist of the main peak *B* only. The binding energy corresponding to the F 1*s* peak of F-SWCNTs is *BE*=687.4 eV; it is shifted by 0.5 eV towards smaller *BE* as compared with MWCNTs+F39% (*BE*=687.9 eV) [Fig. 7(b)] (Brzhezinskaya et al., 2009). At the same time, its FWHM is 2.1 eV, which is 0.2 eV larger than that of F 1*s* peak of F-MWCNTs (1.9 eV).

F 1*s* spectra are less informative as compared to C 1*s* spectra. The effective charge of fluorine atom in F-SWCNTs, as in the most of other compounds, is negative and varies in a limited range of 0 to -1 when the atom´s chemical state changes. A small change in the valence electron density in a fluorine atom is accompanied, due to its participation in chemical bonding with carbon atoms, by a weak change in screening of F 1*s* core level electrons by valence electrons, which manifests itself in small values of chemical shifts of F 1*s* electron bonding energies.

Thus, the fluorine spectrum is a superposition of contributions from all the F 1*s* electrons involved in chemical bonds with carbon atoms. In the case of F-SWCNTs, there is one extra fluorine-carbon phase with much stronger fluorine-carbon bond (as compared with the other two phases observed also in F-MWCNTs) and, hence, with greater carbon-fluorine electron charge transfer. C 1*s* photoelectron spectra showed that the phase 3 contribution is smaller than that from phases 1 and 2, and, hence, the shift of the F-SWCNTs F 1*s* peak as a whole towards the lower binding energies is also insignificant (0.5 eV).

3.3 Results of defluorination experiments

Investigation of CNT defluorination process makes it possible to estimate the type and power of carbon-fluorine bonding and to find out whether the graphene mesh is distorted. Thermal annealing is one of the possible methods of F-CNT defluorination. Our data obtained for F-MWCNT annealing (Brzhezinskaya et al., 2009) showed that annealing started at temperatures almost equal to the fluorination temperature (T_F). As was shown above the results of fluorination process is more complicated for F-SWCNTs (Brzhezinskaya et al., 2010). Therefore, the results of defluorination experiments are presented here for F-SWCNTs only (for F-MWCNTs see Brzhezinskaya et al., 2009). The thermal annealing process was controlled by both NEXAFS and XPS.

The first changes in the C 1*s* absorption spectrum structure of SWCNTs+F40% were observed only after the fifth annealing carried out temperature close to T_{DF} and T_F of F-MWCNTs (T_A=350 °C) [Fig. 8(a) and 8(b), curves 2]. The changes manifested themselves in small reduction of bands B_1* and D* relative intensities and growth of bands A and B intensities [Fig. 8(a), curve 3]. Bands B_1* and D * in the C 1*s* absorption spectrum correspond to the transitions of C 1*s* electrons to the free hybridized states formed from C 2*p* and F 2*p* states. Bands A and B correspond to the spectra of the pristine regions of hexagonal graphene mesh without any fluorine atoms attached. At the same time, no changes are observed in F 1*s* spectrum features [Fig. 8(b), curve 3]. Obviously, such changes in C 1*s* and F 1*s* absorption spectra of F-SWCNTs [Figs. 8(a) and 8(b), curves 4] demonstrate the onset of defluorination process, i.e., partial breakage of C-F bonds, which, apparently, occurs mainly in the outer SWCNTs in SWCNT bundles. After the eighth annealing performed at T_{DF}=420 °C, defluorination becomes even more evident, namely, the F 1*s* spectrum begins changing as intensely as the C 1*s* spectrum does [Figs. 8(a) and 8(b), curves 5]. Therefore, F-SWCNT and F-MWCNT defluorination temperatures are equal to each other and to the F-MWCNT fluorination temperature (T_{DF}=T_F=420 °C). This fact seems to be quite natural. It is generally recognized that the rolling of the graphene sheet induces strain, but for large-diameter CNTs the strain energy per carbon atom E_s is small. It was shown that $E_s \sim D^{-2}$, where D is the tube diameter (Mitmire & White, 1997; Hernandez et al., 1999). That is why the chemistry of wide CNTs should be similar to that of graphite, whereas the narrower tubes are expected to be more reactive due to larger strain of the carbon framework (Bettinger, 2001). Accordingly, fluorination temperature of SWCNTs with 1.4-1.6 nm in diameter (T_F~200 °C) is much lower than fluorination temperature of MWCNTs with 10-30 nm in diameter. We have demonstrated that the absorption and photoelectron spectra of F-SWCNTs and F-MWCNTs and, hence, SWCNT and MWCNT fluorination results are similar, which manifests itself in covalent attachment of fluorine atoms to carbon atom π$2p_z$ orbitals. Thus, at least two of the fluorine-carbon phases formed as a result of fluorination are identical in F-SWCNTs and F-MWCNTs, and, hence, the carbon-fluorine chemical bonds within each phase are equally strong. Consequently, the same amount of energy is needed to break those bonds in either of the tube types. It is due to this that the defluorination temperatures of F-SWCNTs and F-MWCNTs are equal.

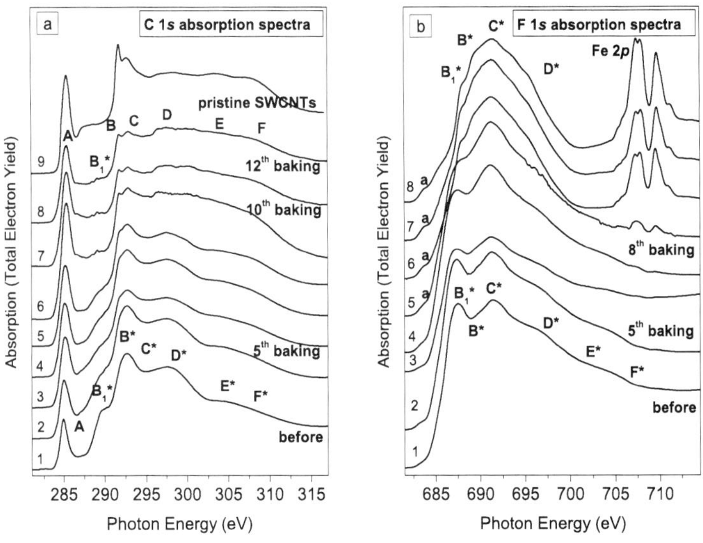

Fig. 8. (a) C 1s and (b) F 1s absorption spectra of SWCNTs+F40% before and during annealing

Notice that even the last 12th annealing did not ensure complete defluorination of F-SWCNTs. It appears from the fact that band B_1*, though dramatically reduced, is still retained on the C 1s absorption spectrum of F-SWCNTs. This fact indirectly confirms the presence of phase 3 with stronger fluorine carbon bonds than in phases 1 and 2.

Contrary to the C 1s spectrum, the F 1s absorption spectrum [Fig. 8(b)] changes stepwise. The intensities of bands B_1*-F* drop instantly, the absorption spectrum as a whole becomes less structured. From the above, it is reasonable to believe that at T_{DF}=T_F (MWCNTs) intense breakage of earlier formed C-F bonds takes place. Therefore, we can assume that the hexagonal carbon mesh destruction is insignificant, though more significant than in this case of F-MWCNTs. This is demonstrated by restoration of C 1s spectrum features typical of the initial MWCNTs and HOPG (π resonance A, σ resonance B, bands D-F). Such restoration of the A-F bands takes place concurrently with the reduction of bands associated with the transitions of C 1s electrons to the hybridized C-F states (bands B_1*-F*). Such a behavior of the main spectrum features does not contradict the interpretation of new absorption bands arising in C 1*s* spectra of F-SWCNTs after fluorination as those reflecting the C 1*s* electrons transition to the hybridized C-F states of the new C-F phase.

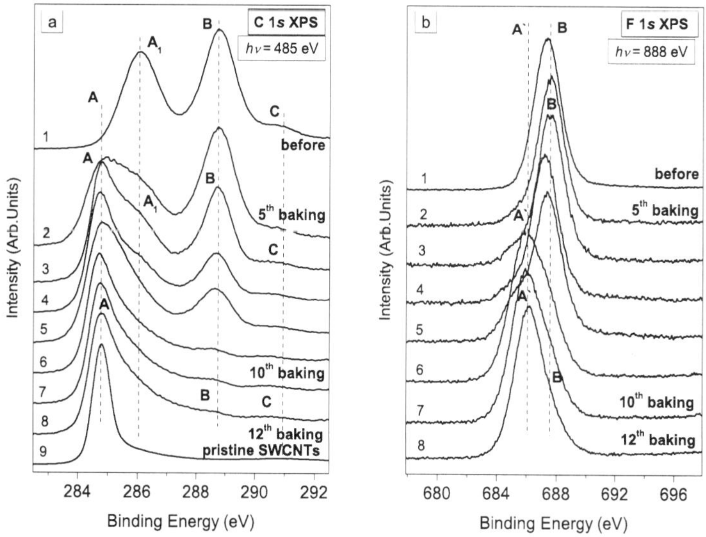

Fig. 9. (a) C 1*s* and (b) F 1*s* x-ray photoelectron spectra of SWCNTs+F40% before and during annealing

Actually, defluorination of F-SWCNTs is accompanied by modification of the *E*-F** bands typical of sp^3 hybridized electron states of carbon atoms into the *E* and *F* bands typical of HOPG with sp^2 hybridization of electron states of carbon atoms. After the further annealing of the sample (T_{DF}=440 °C), the C 1*s* spectrum features reduced after fluorination were restored almost fully [Fig. 8(a), curves 6-8], whereas the F 1*s* spectrum varies only slightly [Fig. 8(b), curves 6-8]. The most important modification of the latter is the emergence of new features at hv=704-713 eV after the 8th annealing. Those new features became well pronounced in Fe 2*p* absorption spectrum [Fig. 8(b), curves 6-8] after the 9th annealing. It should be noticed that, upon the 12th annealing, F 1*s* spectrum does not look like HOPG or CNT spectrum any more. As in the case of the F-MWCNT absorption spectra, we can assume that F 1*s* absorption spectra of F-SWCNTs become similar to the 3*d* transition-metal fluorides spectra. It is notable that in the spectrum there is an extra low-energy structure, namely, feature *a* (hv= 683.6 eV) that is caused by transitions of the F 1*s* electrons to low-lying empty electronic states with the e_g and t_{2g} type symmetry of the transition-metal 3*d* - fluorine 2*p* hybridized character (Vinogradov et al., 2005). In addition, a broad band (~15 eV) at the photon energies of 685-700 eV becomes more and more pronounced in the spectrum structure at the last stages of annealing (after the 9th annealing); this band is caused by core-electron transitions to the unoccupied $4sa_{1g}$ and $4pt_{1u}$ electronic states of the transition-metal atom hybridized with the 2*p* states of the ligand atoms. Notice that the Fe 2*p* spectrum consists of two bands whose intensities grow with heating. Thus, that it is not only the stainless steel sample holder responsible for the signal detected but that fluorine-iron interaction has happened as a result of F-SWCNT defluorination. The presence of two spectral bands under consideration is, most probably, the result of superposition of the iron spectra from the stainless steel and iron fluorides. The substrate signal is detected not only in the absorption spectra (probing depth of 15 nm) but also in overall photoelectron spectra measured at 1030 eV (probing depth of 1.5 nm). This may be the evidence of micro-cracks formation in the sample (the micro-crack number increasing with the annealing time), since no visible cracks were detected in the sample even after the end of heating.

To our point of view, the scanning transmission microscopy (STM) results (Kelly et al., 1999) support this idea. STM images present the fluorinated zones as well-defined closed rings along the nanotube perimeters. At the same time, complete fluorination of C_2F results in radial distortions of CNTs (van Lier et al., 2005). Therefore, we can assume that thermal defluorination of SWCNTs may result in both destruction of bundles and slight distortion of nanotubes themselves. The first of the processes is more probable and preferable as our F-SWCNTs absorption and photoemission spectra show no evidence of any significant amount of amorphous phase in the samples after annealing.

The F 1*s* spectrum shape after the 12th annealing possibly points to formation of iron fluoride FeF_x. These findings (fluorination of metal substrate) testify that separate fluorine atoms are mainly detached during the thermal defluorination of F-SWCNTs, as we found for F-MWCNTs also (Brzhezinskaya et al., 2009). Moreover, formation of FeF_x indirectly proves that in defluorination C-F bonds get broken and fluorine is ejected from the sample not in the form of CF_x clusters but in single atoms as in the case of F-MWCNTs.

At the same time, there is significant difference between F-SWCNT and F-MWCNT defluorination results. For F-MWCNTs, the identity of C 1*s* absorption spectra of defluorinated F-MWCNTs and pristine MWCNTs points to the absence of any remarkable amount of amorphous component in the sample after annealing. As for F-SWCNTs, thermal annealing did not result in complete restoration of resonance *B* associated with C 1*s* electron transitions to the unoccupied σ symmetry states in the conduction band, which arise from $\sigma 2p_{x,y}$ carbon atomic states. This may be the evidence of partial distortion of SWCNT graphene mesh in fluorinated zones provided we assume them to be located around the tube circumference (van Lier et al., 2005). On the other hand, defluorination can lead to destruction of the outer (fully fluorinated) tubes of the bundles and to the increase in the C 1*s* absorption spectrum signal from non-fluorinated tubes from the bundles inner parts. The π resonance is restored almost completely. Therefore, we can conclude that SWCNTs are destructed by defluorination only scarcely, the destruction manifesting itself as a small number of point defects in the graphene mesh. At the same time, defluorination still destructed a negligible number of tubes, which resulted in emergence of a small amount of amorphous component in the sample; the evidence of this fact was some blurring in the high-energy region of the C 1*s* absorption spectra ($h\nu$=295-315 eV) after the 12th annealing [Fig. 8(a), curve 8]. Such blurring may also be the result of incomplete fluorination of SWCNTs in the bundles or of the existence of two different ways of graphene fluorination: the first one is that fluorine atoms are attached to positions (1,2) of the carbon hexagons, the second one is that they get attached to positions (1,4) (Ewels et al., 2006; Khabashesku et al., 2002). These two types of fluorine atoms attachment to carbon correspond to the two fluorine-carbon phases described above (C_2F and C_4F, respectively). As both structures are assumed to be formed on a nanotube wall concurrently (Khabashesku et al., 2002), transition zones with low fluorination degree would appear along the boundaries of the zones with different types of fluorine attachment. On the other hand, defects can emerge in the transition regions in the process of defluorination (Khabashesku et al., 2002).

The photoelectron spectra (Fig. 9) analysis supports our conclusions about the defluorination processes in F SWCNTs. The first changes in the C 1*s* photoelectron spectrum structure were observed only after the firth annealing carried out at the temperature close to T_{DF} and T_F of F-MWCNTs (T_{DF}=350 °C) [Fig. 9(a), curves 2]. The changes manifested themselves in big reduction of peak's *B* and small reduction of peaks' *B* and *C* relative intensities and growth of peak's *A* intensity [Fig. 9(a), curve 3]. Peak *A* corresponds to the spectra of the pristine regions of hexagonal graphene mesh. Fig. 9(a) shows that the band A_1 intensity decreases drastically, and peak *A* increases with respect to band *B* already at the first stages of annealing. Hence, such a behavior of peak A_1 during annealing is unambiguously consistent with its interpretation as a signal from carbon atoms forming a fluorine-carbon phase with fluorine-carbon chemical bond weaker than the bond in phase 1. At the same time, contrary to F 1*s* absorption spectrum, F 1*s* photoelectron spectrum begins to change after the 5th annealing [Fig. 9(b), curve 3]. Shoulder *A`* arises on the low-energy side of peak *B*. These findings become even more evident after the annealing at T_{DF}=400°C. Obviously, such changes in C 1*s* and F 1*s* photoelectron spectra of F-SWCNTs [Figs. 9(a) and 9(b), curves 4] demonstrate the onset of defluorination process, i.e., partial breakage of C-F bonds, which, apparently, occurs mainly in the outer SWCNTs in SWCNT bundles. After the annealing at T_{DF}=420°C defluorination becomes even more evident [Figs. 9(a) and 9(b), curves 5]. Contrary to the absorption spectra, the F 1*s* photoelectron spectrum changes more dynamically. After the 8th annealing, the new structural feature *A`* was still looking like a small bulge on the low-energy side of the main peak *B*; however, immediately after the 9$^{th}$ annealing [Fig. 9(b), curve 6], relative intensities of the two structural features changed dramatically: the *A`* bulge turned into an intense peak having on the high binding energy side a bulge that had been previously peak *B* associated with the new C-F phases. After the 10th annealing this bulge becomes even less intense, and the spectrum exhibits a single peak *A'* with the binding energy of 686 eV and FWHM=2.5 eV [Fig. 9(b), curve 7]; peak *A'* is asymmetric on side of high binding energies. In effect, annealing caused drastic changes in the F 1*s* photoelectron spectrum: peak *B* (687.6 eV, FWHM=2 eV) has disappeared, and new peak *A'* (686 eV, FWHM=2.5 eV) [Fig. 9(b), curve 8] has arisen. Such transformations of the F 1*s* photoelectron spectrum can be easily explained by our hypothesis that the process of F-SWCNT defluorination is accompanied by fluorination of the sample substrate with formation of a new FeF_x phase.

Similarly to the F 1*s* spectrum, C 1*s* photoelectron spectrum undergoes principal changes after the 9th annealing: the features associated with the new C-F phases, namely, A_1, *B* and *C*, are not clearly seen in the spectrum any more. Notice that, contrary to the A_1 and *B* features, feature *C* intensity remains almost the same. This observation supports the idea that band *C* can be best interpreted as the signal from a fluorine-carbon phase with stronger chemical bonds than those in phases 1 and 2. The presence of this phase is also confirmed by the asymmetrical shape of F 1*s* photoelectron spectrum and by the fact that, even after the 12th annealing, initial shape of the C 1*s* spectrum is not restored: C 1*s* peak (FWHM=1.6 eV, *BE*=284.8 eV) becomes evidently broader, which points to surface deformations of the outer tubes in the bundles.

6. Conclusion

The goal of this work was to study electronic structure of fluorinated single-walled and multi-walled carbon nanotubes. High energy resolution X-ray C 1*s* and F 1*s* absorption and photoelectron spectra of F-SWCNTs with different fluorine contents (35 and 40%) and F-MWCNTs with fluorine contents (10-55%) were measured. By comparing fine structures of F-SWCNTs and F-MWCNTs absorption spectra with HOPG, WGF and nanodiamond spectra we have found out that fluorination processes in single- and multi-walled carbon nanotubes are partly similar. Fluorine atoms interact with single-walled and multi-walled carbon nanotubes by getting covalently attached to $\pi 2p_z$ orbitals of the carbon atoms similar to WGF. It was found out that SWCNTs and MWCNTs cannot be completely fluorinated at the given fluorination temperatures (190-222 °C and 420 °C, respectively). The NEXAFS measurements demonstrated that within the probing depth (~15 nm) of F-MWCNTs, the process of fluorination occurs uniformly and does not depend on the fluorine concentration. Our experimental data confirm the previously predicted possibility of existence of two fluorine super-lattices on the F-CNTs lateral surface, which results in developing of fluorine-carbon phases with C_2F and C_4F stoichiometries. Phase 1 (C_2F) content in F-SWCNTs and F-MWCNTs does not change with the increase in probing depth. It was assumed that phase 2 (C_4F) whose electron transfer is smaller than that in phase 1 is formed only on the SWCNTs located in the near-surface layer of the bundles and at the surface of the F-MWCNTs (1-2 monolayers). At the same time, only in the case of F-SWCNTs, phase 3 (CF) exists. It is characterized by greater electron transfer than phase 1. Probably, phase 3 arises due to fluorine atoms attachment to carbon atoms located at the open edges of SWCNTs. We have revealed that at the temperatures of 190 °C and 222 °C the bundles of single-walled nanotubes are not completely fluorinated within the probing depth of 15-20 nm. The process of F-SWCNT and F-MWCNT defluorination by thermal annealing has been investigated. Defluorination temperatures of F-SWCNTs and F-MWCNTs have been found to be equal to each other and to the F-MWCNTs fluorination temperature (T_{DF} = T_F = 420 °C). We have shown that thermal defluorination of F-SWCNTs contrary to F-MWCNTs is impossible without partial deformation of graphene mesh of F-SWCNTs located on the bundle surface.

Acknowledgments

This work was performed within the scope of the "Russian-German Laboratory at BESSY" bilateral Program and supported by the Russian Foundation for Basic Research (Project No. 09-02-01278).

We thank Dr. Muradyan V.E. and Dr. Shul`ga Yu.M. (Institute of Problems of Chemical Physics, the Russian Academy of Sciences, Chernogolovka, Russia) for samples of F-MWCNTs and the Dr. Krestinin A.V., Zvereva G.I. and Dr. Kharitonov A.P. (Institute of Problems of Chemical Physics, the Russian Academy of Sciences, Chernogolovka, Russia) for samples of F-SWCNTs.

7. References

An, K.H.; Heo, J.G.; Jeon, K.G.; Bae, D.J.; Jo, C.; Yang, C.W.; Park, C.-Y.; Lee, Y.H.; Lee, Y.S. & Chung, Y.S. (2002). X-ray photoemission spectroscopy study of fluorinated single-walled carbon nanotubes. *Appl. Phys. Lett*, Vol. 80, No 22 (June 2002) pp. 4235, ISSN 1077-3118

Banerjee, S.; Hemraj-Benny, T.; Balasubramanian, M.; Fisher, D.A.; Misewich, J.A. & Wong, S.S. (2004). Ozonized single-walled carbon nanotubes investigated using NEXAFS spectroscopy. *Chem. Commun.*, No. 7, (2004) pp. 772-773, ISSN 1359-7345

Banerjee, S.; Hemraj-Benny, T.; Sambasivan, S.; Fischer, D.A.; Misewich, J.A. & Wong, S.S. (2005). Near-edge x-ray absorption fine structure investigations of order in carbon nanotube-based systems. *J. Phys. Chem. B*, Vol. 109, No. 17 (February 2005) pp. 8489-8495, ISSN 1520-6106

Bettinger, H.F.; Kudin, K.N. & Scuseria, G.E. (2001). Thermochemistry of fluorinated single wall carbon nanotubes. *J. Am. Chem. Soc.* Vol. 123, No. 51, (December 2001) pp. 12849-12856, ISSN 0002-7863

Brown, F.C.; Bachrach, R.Y. & Bianconi, A. (1978). Fine structure above the carbon K-edge in methane and in the fluoromethanes. *Chem. Phys. Lett*, Vol. 54, No. 3, (March 1978) pp. 425-429, ISSN 0009-2614

Brzhezinskaya, M.M.; Vinogradov, N.A.; Muradyan, V.E.; Shulga, Yu.M.; Polyakova, N.V. & Vinogradov, A.S. (2008). Characterization of fluorinated multiwalled carbon nanotubes by x-ray absorption spectroscopy. *Phys. Solid State*, Vol. 50, No. 3 (March 2008) pp. 587-594, ISSN 1063-7834

Brzhezinskaya, M.M.; Vinogradov, N.A.; Muradyan, V.E.; Shul'ga, Yu.M.; Puettner, R.; Vinogradov, A.S. & Gudat, W. (2009, a). Specific features of the electronic structure of fluorinated multiwalled carbon nanotubes in the near-surface region. *Phys. Solid State*, Vol. 51, No. 9, (September 2009) pp. 1961-1971, ISSN 1063-7834

Brzhezinskaya, M.M.; Muradyan, V.E.; Vinogradov, N.A.; Preobrajenski, A.B.; Gudat, W. & Vinogradov, A.S. Electronic structure of fluorinated multiwalled carbon nanotubes studied using x-ray absorption and photoelectron spectroscopy. *Phys. Rev. B*, Vol. 79, No. 15, (2009, b) pp. 155439(12), ISSN 1098-0121

Brzhezinskaya, M.M.; Vinogradov, A.S.; Krestinin, A.V.; Zvereva, G.I.; Kharitonov, A.P. & Kulakova, I. I. (2010). Characterization of fluorinated singlewalled carbon nanotubes by X-ray absorption spectroscopy. *Phys. Solid State*, Vol. 52, No. 4, (April 2010) (in print), ISSN 1063-7834

Burghard, M. (2005). Electronic and vibrational properties of chemically modified single-wall carbon nanotubes. *Surf. Sci. Rep.*, Vol. 58, Nos. 1-4, (August 2005) pp. 1-109, ISSN 0167-5729

Chen, J.G. (1997). NEXAFS investigations of transition metal oxides, nitrides, carbides, sulfides and other interstitial compounds. *Surf. Sci. Rep.*, Vol. 30, Nos. 1-3, (October 1997) pp. 1-152, ISSN 0167-5729

Dresselhaus, M.S.; Dresselhaus, G. (2001). *Carbon Nanotubes: Synthesis, Structure, Properties and Applications: Topics in Applied Physics*, Springer-Verlag, ISBN 3540410864, Berlin

Ewels, C.P.; Van Lier, G.; Charlier, J.C.; Heggie, M.I. & Briddon, P.R. (2006). Pattern formation on carbon nanotube surfaces. *Phys. Rev. Lett.* Vol. 96, No. 21, (May 2006) pp. 216103(4), ISSN 0031-9007

Fujita, M.; Wakabayashi, K.; Nakada, K. & Kusakabe, K. (1996). Peculiar localized state at zigzag graphite edge. *J. Phys. Soc. Jpn.*, Vol. 65, No. 7, (July 1996) pp. 1920-1923, ISSN 0031-9015

Gudat W. & Kunz, C. (1972). Close Similarity between Photoelectric yield and photoabsorption spectra in the soft-x-ray range. *Phys. Rev. Lett*, Vol. 29, No. 3 (July 1972) pp. 169-172, ISSN 0031-9007

Gupta, V.; Nakajima, T.; Ohzawa, Y. & Zemva, B. 2003. A study on the formation mechanism of graphite fluorides by Raman spectroscopy. *J. Fluorine Chem.*, Vol. 120, No. 2, (April 2003) pp. 143-150, ISSN 0022-1139

Hamwi, A.; Alvergnat, H.; Bonnamy, S. & Béguin, F. (1997). Fluorination of carbon nanotubes. *Carbon*, Vol. 35, No. 6, (May 1997) pp. 723-728, ISSN 0008-6223

Hayashi, T.; Terrones, M.; Scheu, C.; Kim, Y.A.; Rühle, M.; Nakajima, T. & Endo, M. (2002). NanoTeflons: structure and EELS characterization of fluorinated carbon nanotubes and nanofibers. *Nano Lett.*, Vol. 2, No. 5, (May 2002) pp. 491-496, ISSN 1530-6984

Hemraj-Benny, T.; Banerjee, S.; Sambasivan, S.; Fischer, D.A.; Eres, G.; Puretzky, A.A.; Geohegan, D.B.; Lowndes, D.H.; Misewich, J.A. & Wong, S.S. (2006). Imperfect surface order and functionalization in vertical carbon nanotube arrays probed by near edge X-ray absorption fine structure spectroscopy (NEXAFS). Phys. Chem. Chem. Phys., Vol. 8, No. 43, (2006) pp. 5038-5044, ISSN 1463-9076

Hernandez, E.; Goze, C.; Bernier, P. & Rubio, A. (1999). Elastic properties of single-wall nanotubes. *Appl. Phys. A*, Vol. 68, No. 3, (March 1999) pp. 287-292, ISSN 0947-8396

Huefner, S. (2003). *Photoelectron Spectroscopy*, Springer, ISBN 978-3-540-41802-3, Berlin.

Kelly, K.F.; Chiang, I.W.; Mickelson, E.T.; Hauge, R.H.; Margrave, J.L.; Wang, X.; Scuseria, G.E.; Radloff, C. & Halas, N.J. 1999. Insight into the mechanism of sidewall functionalization of single-walled nanotubes: an STM study. *Chem. Phys. Lett*, Vol. 313, Nos 3-4, (November 1999) pp. 445-450, ISSN 0009-2614

Khabashesku, V.N.; Billups, W.E. & Margrave, J.L. (2002). Fluorination of single-wall carbon nanotubes and subsequent derivatization reactions. *Acc. Chem. Res.*, Vol. 35, No. 12, (December 2002) pp. 1087-1095, ISSN 0001-4842

Kiselev, N.A.; Moravsky, A.P.; Ormont, A.B. & Zakharov, D.N. (1999). SEM and HREM study of the internal structure of nanotube rich carbon arc cathodic deposits. *Carbon*, Vol. 37, No. 7, (June 1999) pp. 1093-1103, ISSN 0008-6223

Klein, D.J. 1994. Graphitic polymer strips with edge states. *Chem. Phys. Lett*, Vol. 217, No. 3, (January 1994) pp. 261-265, ISSN 0009-2614

Kobayashi, Y.; Fukui, K.; Enoki, T.; Kusakabe, K. & Kaburagi, Y. 2005. Observation of zigzag and armchair edges of graphite using scanning tunneling microscopy and spectroscopy. *Phys. Rev. B*, Vol. 71, No. 19 (May 2005) pp. 193406(4), ISSN 1098-0121

Krestinin, A.V.; Kiselev, N.A.; Raevskii, A.V.; Ryabenko, A.G.; Zakharov, D.N. & Zvereva, G.I. (2003, a). Perspectives of single-wall carbon nanotube production in the arc discharge process. *Eurasian Chem. Tech. J.*, Vol. 5, No. 1, (January 2003) pp. 7-18, ISSN 1562-3920

Krestinin, A.V.; Raevskii, A.V.; Kiselev, N.A.; Zvereva, G.I.; Zhigalina, O.M. & Kolesova, O.I. (2003, b). Optical activity effect in crystalline structures of purified single-wall carbon nanotubes. *Chem. Phys. Lett*, Vol. 381, Nos. 5-6, (November 2003) pp. 529-534, ISSN 0009-2614

Krestinin, A.V.; Kharitonov, A.P.; Shul'ga, Yu.M.; Gigalina, O.M.; Knerelman, E.I.; Brzhezinskaya, M.M.; Vinogradov, A.S.; Preobrajenski, A.B.; Zvereva, G.I.; Kislov, M.B.; Martinenko, V.M.; Korobov, I.I.; Davidova, G.I.; Gigalina, V.G. & Kiselev, N.A. 2009. The synthesis and characterization of fluorinated single-walled carbon nanotubes. *Nanotechnologies in Russia*, Vol. 4, No. 1-2, (January 2009) pp. 60-78, ISSN 1995-0780

Kudin, K.N.; Bettinger, H.F. & Scuseria, G.E. (2001). Fluorinated single-wall carbon nanotubes. *Phys. Rev. B*, Vol. 63, No. 4 (January 2001) pp. 045413(8), ISSN 1098-0121

Kuznetsova, A.; Popova, I.; Yates, J.T.; Bronikowski, M.J.; Huffman, C.B.; Liu, J.; Smalley, R.E.; Hwu, H.H. & Chen, J.G. (2001). Oxygen-containing functional groups on single-wall carbon nanotubes: NEXAFS and vibrational spectroscopic studies. *J. Am. Chem. Soc.* Vol. 123, No. 43, (October 2001) pp. 10699-10704, ISSN 0002-7863

Lebedev, N.G.; Zaporotskova, I.V. & Chernozatonskii, L.A. (2003). Fluorination of carbon nanotubes within the molecular cluster method. *Microelectron. Eng.*, Vol. 69, Nos. 2-4, (September 2003) p. 511-518, ISSN 0167-9317

Lee, Y.S.; Cho, T.H.; Lee, B.K.; Rho, J.S.; An, K.H. & Lee, Y.H. J. (2003). Surface properties of fluorinated single-walled carbon nanotubes. *J. Fluorine Chem..*, Vol. 120, No. 2, (April 2003) p. 99-104, ISSN 0022-1139

Lee, Y.-S. (2007). Syntheses and properties of fluorinated carbon materials. *J. Fluorine Chem.*, Vol. 128, No. 4, (April 2007) pp. 392-403, ISSN 0022-1139

Louie, S.G. (2001). Electronic properties and defects, In: *Carbon Nanotubes*, Dresselhaus, M.S.; Dresselhaus, G. & Avouris, Ph., (Eds.), pp. 117-152, Springer-Verlag, ISBN 978-3-540-41086-7, Berlin

Martin, C.; Arakawa, E.T.; Callcott, T.A. & Ashley, J.C. (1985). Low energy electron attenuation length studies in thin amorphous carbon films. *J. Electron Spectrosc. Relat. Phenom.* Vol. 35, No. 2, (March 1985) pp. 307-317, ISSN 0368-2048

Maruyama, M.; Kusakabe, K.; Tsuneyuki, S.; Akagi, K.; Yoshimoto, Y. & Yamauchi, J. 2004. Magnetic properties of nanographite with modified zigzag edges. *J. Phys. Chem. Solids*, Vol. 65, Nos. 2-3, pp. 119-122 (March 2004), ISSN 0022-3697

Mickelson, E.T.; Huffman, C.B.; Rinzler, A.G.; Smalley, R.E.; Hauge, R.H. & Margrave, J.L. (1998). Fluorination of single-wall carbon nanotubes. *Chem. Phys. Lett*, Vol. 296, Nos. 1-2, (October 1998) pp. 188-194, ISSN 0009-2614

Mickelson, E.T.; Chiang, I.W.; Zimmerman, J.L.; Boul, P.J.; Lozano, J.; Liu, J.; Smalley, R.F.; Hauge, R.H. & Margrave, J.L. (1999). Solvation of fluorinated single-wall carbon nanotubes in alcohol solvents. *J. Phys. Chem. B.* Vol. 103, No. 21 (May 1999) pp. 4318-4322, ISSN 1520-6106

Mitmire, J.W. & White, C.T. (1997). In: *Carbon nanotubes.* Ebbesen, T.W. (Ed.), pp. 191-224, CRC Press, ISBN 0849396026, Boca Raton, FL

Molodtsov, S.L.; Fedoseenko, S.I.; Vyalikh, D.V.; Iossifov, I.E.; Follath, R.; Gorovikov, S.A.; Brzhezinskaya, M.M.; Dedkov, Yu.S.; Puettner, R.; Schmidt, J.-S.; Adamchuk, V.K.; Gudat, W. & Kaindl, G. (2009). High-resolution Russian-German beamline at BESSY. *Appl. Phys. A*, Vol. 94, No. 3, (March 2009) pp. 501-505, ISSN 0947-8396

Okotrub, A.V.; Yudanov, N.F.; Chuvilin, A.L.; Asanov, I.P.; Shubin, Yu.V.; Bulusheva, L.G.; Gusel'nikov, A.V. & Fyodorov, I.S. (2000). Fluorinated cage multiwall carbon nanoparticles. *Chem. Phys. Lett*, Vol. 322, Nos. 3-4 (May 2000) pp. 231-236, ISSN 0009-2614

Park, K.A.; Choi, Y.S.; Lee, Y.H. & Kim, C. (2003). Atomic and electronic structures of fluorinated single-walled carbon nanotubes. *Phys. Rev. B*, Vol. 68, No. 4 (July 2003) pp. 045429(8), ISSN 1098-0121

Rosenberg, R.A.; Love, P.J. & Rehn, V. 1986. Polarization-dependent C(*K*) near-edge x-ray-absorption fine structure of graphite. *Phys. Rev. B*, Vol. 33, No. 6 (March 1986) pp. 4034-4037, ISSN 1098-0121

Saito, R.; Yagi, M.; Kimura, T.; Dresselhaus, G. & Dresselhaus, M.S. 1999. Electronic structure of fluorine doped graphite nanoclusters. *J. Phys. Chem. Solids*, Vol. 60, No. 6, (June 1999) pp. 715-721, ISSN 0022-3697

Sato, Y.; Hagiwara, R. & Ito, Y. 2003. Refluorination of pyrocarbon prepared from fluorine-GIC. *Solid State Sci.*, Vol. 5, No. 9, (September 2003) pp. 1285-1290, ISSN 1293-2558

Schiessling, J.; Kjeldgaard, L.; Rohmund, F.; Falk, L.K.L.; Campbell, E.E.B.; Nordgren, J. & Brühwiler, P.A. (2003). Synchrotron radiation study of the electronic structure of multiwalled carbon nanotubes. *J. Phys.: Condens. Matter*, Vol. 15, No. 38, (October 2003) pp. 6563-6579, ISSN 0953-8984

Seifert, G.; Kohler, T. & Frauenheim, T. 2000. Molecular wires, solenoids, and capacitors by sidewall functionalization of carbon nanotubes. *Appl. Phys. Lett*, Vol. 77, No 9, (August 2000) pp. 1313(3), ISSN 1077-3118

Stöhr, J. (2003). *NEXAFS Spectroscopy*, Springer, ISBN-13 978-3540544227, Berlin

Shul'ga, Y.M.; Tien, T.-C.; Huang, C.-C.; Lo, S.-C.; Muradyan, V.E.; Polyakova, N.V.; Ling, Y.-C.; Loufty, R.O. & Moravsky, A.P. (2007). XPS study of fluorinated carbon multi-walled nanotubes. *J. Electron Spectrosc. Relat. Phenom.*, Vol. 160, Nos. 1-3, (August 2007) pp. 22-28, ISSN 0368-2048

Tang, Y.H.; Sham, T.K.; Hu, Y.F.; Lee, C.S. & Lee, S.T. (2002). Near-edge X-ray absorption fine structure study of helicity and defects in carbon nanotubes. *Chem. Phys. Lett*, Vol. 366, Nos. 5-6, (December 2002) pp. 636-641, ISSN 0009-2614

Touhara, H. & Okino, F. (2000). Property control of carbon materials by fluorination. *Carbon*, Vol. 38, No. 2, (December 2000) pp. 241-267, ISSN 0008-6223

Van Lier, G. ; Ewels, C.P.; Zuliani, F. ; De Vita, A. & Charlier, J.-C. (2005). Theoretical analysis of fluorine addition to single-walled carbon nanotubes: functionalization routes and addition patterns. *J. Phys. Chem. B*, Vol. 109, No. 13, (March 2005) pp. 6153-6158, ISSN 1520-6106

Vinogradov, A.S.; Fedoseenko, S.I.; Krasnikov, S.A.; Preobrajenski, A.B.; Sivkov, V.N.; Vyalikh, D.V.; Molodtsov, S.L.; Adamchuk, V.K.; Laubschat, C. & Kaindl G. 2005. Low-lying unoccupied electronic states in 3*d* transition-metal fluorides probed by NEXAFS at the F 1*s* threshold. *Phys. Rev. B*, Vol. 71, No. 4 (January 2005) pp. 045127(11), ISSN 1098-0121

Yudanov, N.F.; Okotrub, A.V.; Shubin, Yu.V.; Yudanova, L.I.; Bulusheva, L.G.; Chuvilin, A.L. & Bonard, J.-M. (2002). Fluorination of arc-produced carbon material containing multiwall nanotubes. *Chem. Mater.*, Vol. 14, No. 4, (April 2002) pp. 1472-1476, ISSN 0897-4756

Zhou, J.; Zhou, X.; Sun, X.; Li, R.; Murphy, M.; Ding, Z.; Sun, X. & Sham, T.-K. (2007). Interaction between Pt nanoparticles and carbon nanotubes – An X-ray absorption near edge structures (XANES) study. *Chem. Phys. Lett*, Vol. 437, Nos. 4-6 (April 2007) pp. 229-232, ISSN 0009-2614

5

Microwave Dielectric Properties of Carbon Nanotube Composites

L. Liu[1*], L. B. Kong[1], W. Y. Yin[2], Y. Chen[3] and S. Matitsine[1]
1 Temasek laboratories, National University of Singapore, Singapore, 117508
2 Centre for Optical and EM Research, State Key Lab of MOI, Zhe Jiang University Hangzhou, China, 310058
3 School of Chemical and Biomedical Engineering, Nanyang Technological University, Singapore 637459

1. Introduction

Carbon black (CB) (Donnet *et al.*, 1993) and carbon fiber (CF) (Burchell, 1999) have been used as fillers in radio wave and microwave composites for more than half century. Typical applications include electromagnetic compatibility (EMC) or electromagnetic interference (EMI), microwave absorbing coatings, and anti-electrostatic materials, *etc* due to their good electric/dielectric properties, stability and chemical resistance as compared to composites with metal powders of nano or micron meter in size. CB composites normally have frequency-dependent permittivity which is useful in the design of broadband absorbing or functional materials at microwave band. The frequency dispersion in CB composites is mainly caused by the effect of large agglomerate of CB spherical particles with diameter below 100 nanometers and usually appears in composites with large concentration of fillers. CF composites can have special microwave properties, such as high permittivity at relatively low concentrations of fibers (Lagarkov *et al*, 1998). The dispersive property of CF composites at microwave band is determined by the frequency-dependent response of individual fibers or fiber clusters (Liu, *et al*, 2007a). The dispersion can be observed even at very low concentration of fibers, for example, less than one percent in volume. This feature makes CF composites advantageous for many technical applications, such as lightweight absorbing materials (Neo & Varadan, 2004) and phantom materials for modeling the electromagnetic response of biological issue (Youngs, *et al*, 2002). However, to obtain strong dielectric dispersion at microwave frequencies, CF must contain fibers which are at least millimeters in length to achieve large aspect ratio considering the diameter of commercial CF which is normally close to 10 microns. The large dimension of the fibrous inclusions restricts the practical application of CF composites. Also, the favorite processing methods, such as spray technique, are not applicable for CF composites. The planar CF composites are anisotropic with degraded dielectric response across the plane. Due to the heterogeneity of the composites, a surface layer of millimeters in thickness appears with properties different

* Corresponding author, Email: tslll@nus.edu.sg, Tel: 65-65161007

from those of the bulk composites (Simovski, *et al*, 2000 and Matitsine, *et al*, 2003). This gives rise to difficulty in modeling and designing composites with desired properties.

Carbon nanotubes (CNTs) have been fabricated since the early 1990s (Ijima, 1991). The primary research interests include the synthesis or growth of CNTs, because it was a challenging task to prepare enough amounts of CNTs with desired dimension and purity for measurement purpose during the early stage. With the development of processing method like arc discharge and laser ablation of graphite, as well as more productive chemical vapor deposition (CVD) and plasma enhanced CVD method, high purity CNTs with controllable wall-thickness/length and acceptable price are commercially available now (Meyyappan, 2005). Other research interests include the characterization and theoretical understanding of their basic properties of CNTs. The physical properties of carbon nanotubes were reviewed in literature (Saito, 1998), which have been considered as the classic monograph on CNTs. Due to the unique electronic structure depending on the diameter and chirality, CNTs could be the smallest semiconductor devices likely to be fabricated (Saito, 1998). And the extraordinary mechanical properties of CNTs make them suitable replacement of CF in continuous fiber reinforced composites for defence or aerospace applications. Other potential applications includes single- or multi-walled CNTs sensor for scanning probe microscopy, and biosensor or device applications such as CNT-based diodes, transistors and field emitter, *etc* (Meyyappan, 2005).

Due to their fibrous shape with extremely large in aspect ratio, CNTs may allow low resistivity and high permittivity and frequency dispersion to be obtained in composites with extremely low concentrations. As compared with CF, one significant difference is that CNTs are nanoscaled particles whose diameter is several orders of magnitude smaller than that of CF (typically about 10 μm). It is well known that the nanosized particles usually display distinct properties from microsized particles of the same composition, which is the primary reason for the great attention currently given to the radio and microwave frequency performance of CNTs composites. A number of novel features have been reported on CNTs in the literature, these results demonstrate the possibility to design CNTs composites with electric/dielectric properties which are more diverse than those obtainable with other carbon fillers. That is why in the last few years, main research interest in CNT was shifted from fabrication methods to the novel properties, such as the conductivity at low concentration and dielectric properties at low frequency (DC to 1MHz) and high frequency (above 1MHz, including radio wave, microwave, millimeter wave, Terahertz and Inferred wave frequency). Since the dielectric properties of CNT composites at microwave frequency are the main focus of this chapter, previous research works on the electric properties, dielectric permittivity at high frequency, as well as potential EMC/EMI and absorbing applications which are most relevant to our research objective, will be reviewed sequentially.

After the observation of a conductivity threshold in polymer/CNTs composites (Coleman, et al, 1998), around 200 publications have reported on the electrical percolation threshold of CNTs in different polymer systems by 2009 (W. Bauhofer & J. Z. Kovacs, 2009). Different CNTs, synthesis methods, dispersion treatment and polymer matrix have been employed in the previous research. Since it is established that the percolation theory or power law could explain the electric conductivity of CNTs composite (Martin *et al*, 2004). Minimum percolation ratio and maximum conductivity at low frequency becomes the main research focus. Due to the dispersion method and difference in the holding matrix, percolation varies from less than 0.01 percent to a few percent. Theoretically, typical aspect ratio of 1000 produces percolation value about 0.1% (Balberg *et al*, 1984). Practically, a few order of difference could be found from composite with same CNTs and different polymer matrix (W. Bauhofer & J. Z. Kovacs, 2009). Maximum conductivity of MWCNTs composite could be realized when the weight concentration is larger than 10% (Skakalove *et al*, 2005). And the maximum conductivity seems insensitive to the types of CNTs and treatments. Nevertheless, it is a challenging task to disperse CNTs uniformly at such high concentration. The mechanism of charge transport in CNTs composites could be explained by hopping and tunneling effect among neighboring tubes (Kim, 2006). However, the morphology of CNTs in holding matrix seems difficult to be represented by the classical percolation theory.

Permittivity ($\varepsilon=\varepsilon'-j\,\varepsilon''$) instead of AC conductivity is normally used to describe the dielectric properties of CNTs composites as frequency is above 1MHz. Also, measurement methodologies and potential applications could be different from that at low frequency. Microwave properties are normally measured with impedance or network analyzer through coaxial line or waveguide fixture. A number of special features have been reported on CNT in the literature, such as excitation of localized electronic states resulting in high intrinsic permittivity (Watts *et al*, 2003).The microwave permittivity is found to exhibit smooth frequency-dependent (Browning, *et al*, 1998 & Sandler, *et al*, 2003). These results seem to demonstrate the possibility to obtain CNT composites with microwave properties which are more diverse than those obtainable with other types of carbon fillers, e. g., with various types of dispersive dielectrics, with high dielectric but having relative low dielectric loss (Yang, et al, 2009), and tunable dielectric under small bias voltage (Liu, *et al*, 2008). In addition, the nanoscaled size of CNT in composites may avoid the drawbacks of CF composites, which are limited by the macroscopic size of the fibers for thin structure and component applications.

Since the low resistivity of CNTs composite with low filling factor as compared with CB composite which are commonly employed as filler of rubber, conductivity polymer and shielding coating, the shielding effectiveness (SE) of CNTs composites are mostly investigated at microwave frequency (Huang *et al*, 2007). MWCNTs/Ceramic composites were measured at Ku-band with SE up to 25 to 30dB achieved (Shi & Liang, 2008). Some experimental results showed that absorption is the major shielding mechanism and reflection is the secondary mechanism (M. H. Al-saleh & U. Sundararaj, 2009). SWCNTs thin film demonstrated good shielding of EM wave at terahertz frequency while still maintaining good transparency for visible light (Seo, *et al*, 2008). Measurement results show that CNT sheets are more effective in providing EMI shielding compared with graphite and carbon black sheets over broad frequency range due to better electron transmission (Wang, *et al*, 2009). Radar absorbing structures with MWCNTs and polyurethane foam were designed with optimization method before measured with free space setup (Park, et al, 2006). 10 dB reflection loss at X-band could be achieved by MWCNTs layer of thickness about 3.3mm (Lee *et al*, 2006). Self-sensing and self-actuation response were found out from CNTs composites which could be explained by electrostriction effect (Lee & Shkel, 2007). The tunability of both real and imaginary parts of the permittivity found from CNTs composites has potential applications for smart materials and structures (Liu, *et al*, 2008).

The shielding effectiveness and absorption of EM wave are functions of electric and dielectric properties; the objective of this chapter is to investigate the resistivity and dielectric permittivity of CNTs composites for different wall thicknesses. The organization of this chapter is as follow. Theoretical methods and models concerning dielectric properties of CNTs composites are reviewed briefly in Section 2. The materials, fabrication conditions and measurement method of dielectric properties are addressed in Section 3. The results in Section 4 include electric properties, frequency-dependent permittivity, tunable permittivity under bias voltage and potential applications. Last but not least, the conclusion of this research work are summarized in Section 5.

2. Theory

In CB composites, the microwave behavior is attributed mainly to the large agglomerates of carbon particles. In this case, the properties of the composite are consistent with the percolation theory that accounts for the statistical properties of clusters comprising many inclusions in electrical contact within the composite (Bergman & Imry, 1977). The percolation approach (Satuffer & Aharony, 1991) depends on the empirical percolation threshold p_c that cannot be derived from the percolation theory. The percolation threshold is defined as the least concentration at which a composite with conductive inclusions is capable of conducting direct current. For the permittivity ($\varepsilon=\varepsilon'+i\varepsilon''$) of a composite comprising conductive inclusions and a lossless dielectric host matrix of permittivity ε_h, percolation theory predicts a power dependence on the frequency f (Gefen, et al, 1983):

$$\varepsilon'' = A f^{-x}, \quad \varepsilon' - \varepsilon_{\mathrm{h}} = B f^{-y}, \tag{1}$$

In (1), A and B are proportionality factors, x and y are critical exponents. In the classical percolation theory, the range of concentrations p where scaling law (1) is not expected to be observed in a wide frequency range, but typically very narrow, for example $(p-p_c)/p_c \leq 0.1$ (Gefen, *et al*, 1983). However, in many measured data of CB composites, the scaling frequency dispersion is observed in a wide range of both concentrations and frequencies (Brosseau, 2002). If (1) is valid over the entire frequency range, it follows from the Kramers-Kronig relations that (Jonscher, 1999):

$$x = y, \quad \tan\delta = \frac{A}{B} = \tan\frac{\pi y}{2} \tag{2}$$

where $\tan\delta$ is the dielectric loss tangent, and the second of (2) is valid for $\varepsilon' \gg \varepsilon_h$. Therefore, the critical exponents for ε' and ε'' are the same, while the dielectric loss tangent is frequency-independent. Equation (1) coincides with the "universal relaxation law" established for many types of dielectrics (Jonscher, 1999). In the percolation theory, the critical exponents are considered to be universal, *i.e.* the same for different composites, independent of the type and properties of inclusions comprising the composite. The physical basis is that the behavior of the composite is determined by the statistical properties of the clusters, but not by the parameters of individual inclusions.

The universal values for critical exponents of frequency dependence are usually derived from the exponents for concentration dependence. For this, two different approaches are known, with exponent value of either 0.27 or 0.42 (Song, *et al*, 1986). Many numerical and experimental data are obtained within these limits. For example, numerical studies of random RC networks (Straley, 1977) produce 0.28 for 3-D bond problem, but 0.18 for 3-D site problem. For continuum percolation problem, the value of 0.38 is obtained from the studies of water-oil emulsions (Bordi, *et al*, 1996). Therefore, the critical exponent values are believed to be within the range of 0.2 to 0.4, with the uncertainty attributed to the difference in geometry between the bond, site, and continuum percolation problems that may result in different properties of the clusters. For CNT composites, the critical exponent of 0.34, consistent with percolation theory, is obtained (Kim, *et al*, 2003). In all cases, x is less than 1. Therefore, the AC conductivity of the composite must increase with frequency.

In actual composites, the percolation dispersion governed by (1) appears within a limited range of frequency. The high frequency limit is determined by the frequency at which the conductivity of the composite is equal to the conductivity of the inclusions, and therefore, no longer increases with frequency. The low frequency limit is as considered (Laibowitz & Gefen, 1984). Below this low frequency limit, the real permittivity or conductivity is independent of frequency at concentrations below p_c or above p_c correspondingly. Due to the finite frequency range for percolation dispersion, (2) is not exact for practical materials, with the discrepancy increases as the frequency approaches the limits of the percolation frequency range. If (1) is valid over restricted frequency range, then the dielectric loss tangent increases relatively with respect to the value given by (2), as the high frequency limit of this frequency range is approached. Near the low frequency limit, the loss tangent decreases for concentrations below p_c and increases for concentrations above p_c.

3. Experiment

3.1 Characterization of CNTs

	SWCNT	DWCNT	MWCNT
Purity	95%	90%	95%
Impurities	MWCNT, Ash, Metal	MWCNT, Ash, Metal	Ash, Metal
Length	~30 μm	~50 μm	~50 μm
Diameters	1~2nm	2~4nm	8~15nm
Aspect ratio	~20k	~17k	~2k

Table 1. Specifications of three types of CNTs

Single-, double- and multi-walled carbon nanotubes (SWCNT, DWCNT, and MWCNT) produced by CVD method were obtained from the Timesnano, Chengdu Organic Chemicals Co. Ltd., the Chinese Academy of Sciences. As produced CNTs were purified by the supplier with Oxidation and HCL wash method. The specifications of these CNTs are given in Table 1. Figure 1 shows the image of these CNTs with high resolution field emission SEM (JEOL JSM-6340F). The diameter of SWCNT and DWCNT seems larger than the values given by the supplier. Possible reasons can be the appearance of bundles or small amount of the CNTs may be multi-walled instead of single-walled. The aspect ratio is calculated based on the length over the average diameter of individual CNTs. It can also been observed from Figure 1 that length of CNTs are larger than the scope observed by SEM which is about a few micron meters.

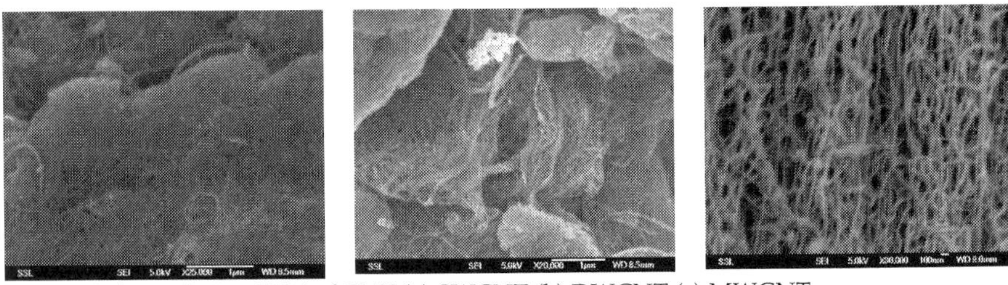

Fig. 1. High-resolution SEM of CNT (a) SWCNT (b) DWCNT (c) MWCNT

The results of thermal gravimetric analysis (TGA) shown in Figure 2 verify that the carbon content in samples is within the ranges provided by the supplier. The residual weights (which is usually contributed by metallic catalyst particles) of the SWCNT and MWCNT samples are almost zero. It suggests that metallic particles can be efficiently removed by the HCl wash. However, about 6 wt% of residues is still remained on the DWCNT sample, shown in Figure 2(b). This may be caused by the different purification method applied by the supplier. Differential Thermal Gravimetry (DTG) results in Figure 2 also demonstrate that most of the carbon materials can be oxidized in a narrow temperature range centered at the peak oxidation temperature (Tg). The narrow oxidation temperature range suggests that carbon species in the samples have similar structures. T_g of the SWCNT sample is at 624 °C, which is higher than the oxidation temperature of purified SWCNTs with diameter of 0.7-1.0 nm (Wei, et al, 2008). The increase of T_g may be contributed by the large diameter tubes or certain amount of impurities, such as MWCNTs. The decrease of T_g for DWCNT samples is caused by the metallic residues remained on the sample which may catalyze the oxidation of carbon and result in the lower oxidation temperature at 604 °C. The DTG result of MWCNT shows a relative broader oxidation peak, as well as a higher T_g. The relative large range of wall thickness might contribute to the observed difference.

Raman spectroscopy was further applied to analyze the three CNT samples. Samples show three typical bands: radial breathing mode (RBM) between 150 and 350 cm^{-1} characteristic for SWCNT or DWCNT, the D band at approximately 1300 cm^{-1} assigned to defective and disordered carbon species, and the peak complex centered around 1600 cm^{-1}, known as the G band, which is characteristic for ordered carbon species such as CNTs and graphite. The sharp RBM peaks from the SWCNT sample in Figure 3(a) and the DWCNT sample in Figure 3(b) verify the high content of SWCNTs and DWCNTs. The very low D band intensities also suggest that little amorphous carbon exists. Figure 3(c) shows the typical Raman spectrum of MWCNTs. However, it is difficult to quantify the purity of different CNT species. Instead, it is confirmed that there are significant amount of CNTs in the three samples.

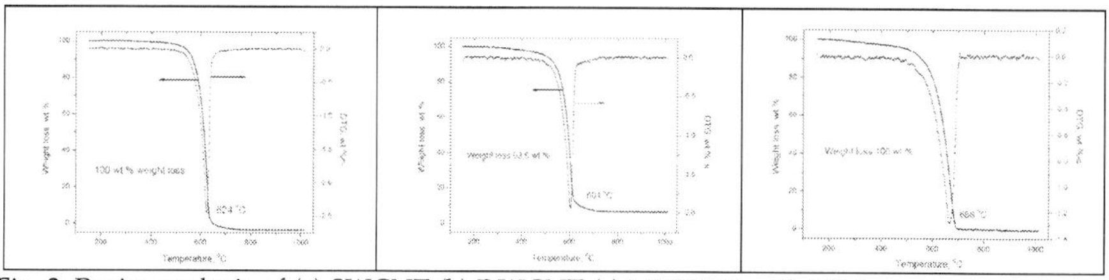

Fig. 2. Purity analysis of (a) SWCNT (b) DWCNT (c) MWCNT with TGA

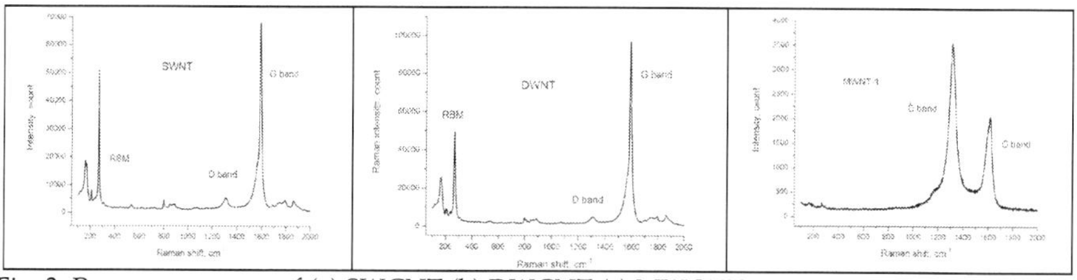

Fig. 3. Raman spectrum of (a) SWCNT (b) DWCNT (c) MWCNT

3.2 Fabrication of CNTs composites

The host polymer to prepare composites was commercial two-component silicone (KE1800T) from Shin Etsu. CNTs and silicone (Component A) were dispersed in Toluene (Methylbenzene) with Ultrasonication processer (Sonics VCX500) for 20 min. The resistance of liquid mixture with 0.01g of SWCNTs, 1g of silicone and 25ml of Toluene was monitored with a parallel-plate sensor connected with insulation multimeter (Megger BMM80). The dependence of resistance on the sonication time plotted in Figure 4 shows that 20 minutes sonication time is probably enough to disperse the mixture to stable stage. A SEM image (shown in Figure 5) of SWCNTs after sonication process shows that big agglomerates in Figure 1(a) were broken and no big bundles (larger than 1μm) could be found.

Mixture of CNTs, silicone and Toluene in a 25ml crucible was heated up with a hotplate at 90°C for 8 hours to remove the solvent. After that, the 10% of liquid hardener (Component B) was added and mixed manually before pouring into mould. The mould was put into convection oven to cure at 70°C for 24 hours. As a control, another group of SWCNT, DWCNT and MWCNT was mixed manually with silicone component A and component B for 15 min before cured under the same conditions. And the control group prepared by manual mixing will be used as the benchmark of the sonication process.

Two types of samples were fabricated with moulding method. Disk shape samples have diameter of 16mm and thickness of 2mm to measure the resistivity and permittivity from 0.001GHz to 1GHz. Toroidal shape samples for permittivity measurement at high frequency (from 0.05GHz to 5GHz) have outer diameter of 14mm, inner diameter of 7mm and thickness of about 2mm.

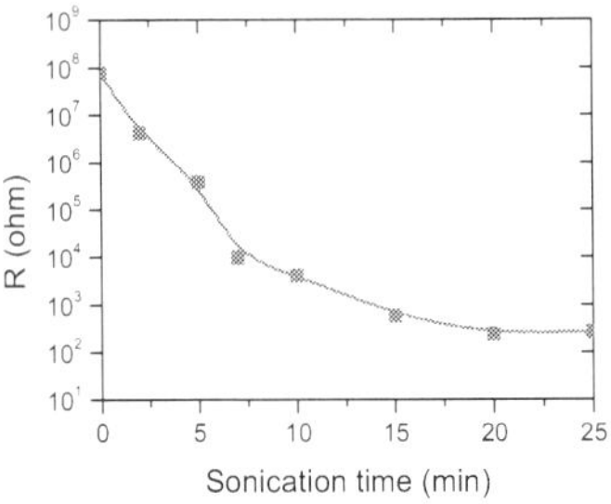

Fig. 4. Resistance vs. sonication time

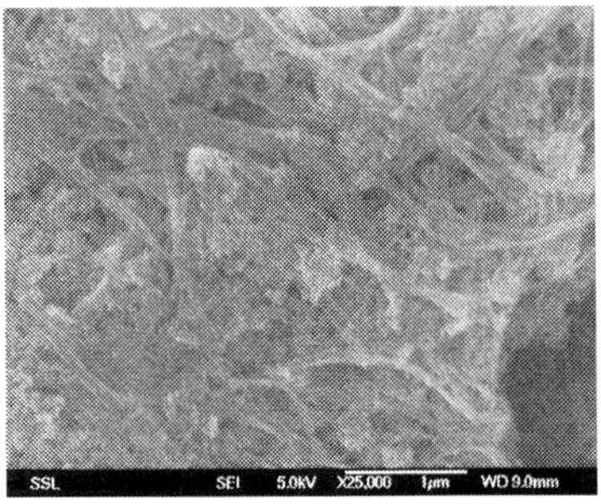

Fig. 5. SWCNT after sonication

3.3 Measurement methods

DC resistance of the samples was measured using a Megger™ BMM80™ insulation multimeter and a parallel plate fixture. Before measurement, the disk-shape samples were cleaned with ethanol to remove the release agent sprayed on the mould prior to casting. This is important since the agent contributed to the measured resistance. Resistivity was calculated from the measured resistance and dimension of the samples. Agilent 4991A RF impedance/materials analyzer was employed for permittivity measurement from 0.001GHz to 1 GHz. Figure 6 shows the test fixture (Agilent 16453A) for measurement of dielectric permittivity. During measurement, a bias voltage up to 40 volts was applied to the sample in between the upper and lower electrodes. The permittivity measured is in the direction of the bias voltage applied. Special care was taken to keep good electrical contact between the electrodes and the sample by applying certain pressure. Since there is no air gap effect, the impedance method can achieve good accuracy for measurement of conductive samples without correction [Liu, *et al*, 2007b, 2008].

To validate the measurement results by impedance analyzer, 14mm coaxial line fixture (as shown in Figure 7) and network analyzer (Agilent N5230A) was used to measure the permittivity of SWCNT composites samples from 0.05 to 5GHz. The permittivity was extracted from the measured S-parameters of the sample placed within the coaxial line. Since the outer diameter of the sample can be slightly smaller than the inner diameter of the test fixture due to the shrinkage of the sample after curing, the air gap in between is the largest contributor of uncertainty, especially when the concentration is high. Therefore, only 4wt% and below measured by the two different methods have been compared. The real and imaginary permittivity of SWCNTs composites of weight concentration from 0.5% to 4% are plotted in Figure 8. It shows rather good agreement between these two techniques over 0.05GHz to 1GHz for concentration up to 3%. Both real and imaginary permittivity measured by 14mm coaxial line fixture at 4% show lower values than these from impedance method, which could be properly explained through air gap effect (Liu *et al*, 2007b and reference therein). After correction of air gap effect, the real permittivity (dash line) shows better agreement with impedance measurement. It is proved that impedance measurement has better accuracy even at high concentration as compared with coaxial line method.

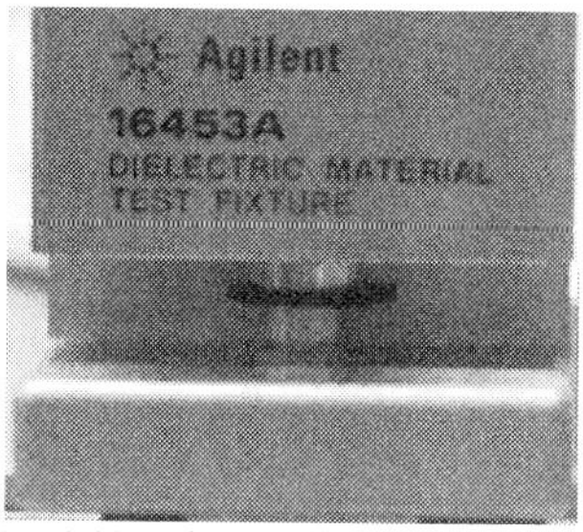

Fig. 6. Test fixture and sample for permittivity from 1MHz to 1GHz

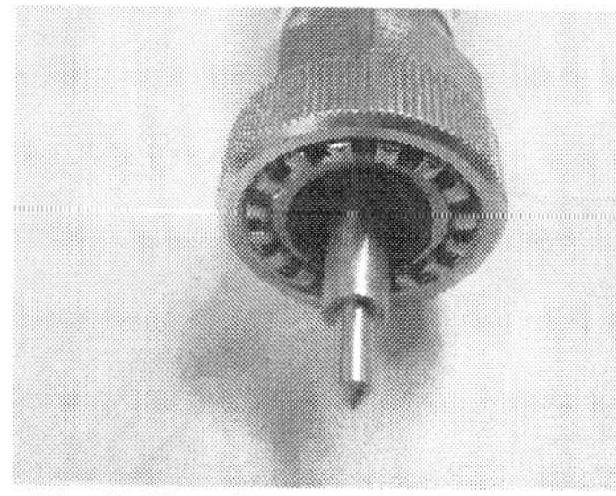

Fig. 7. Test fixture and sample for permittivity from 50MHz to 5GHz

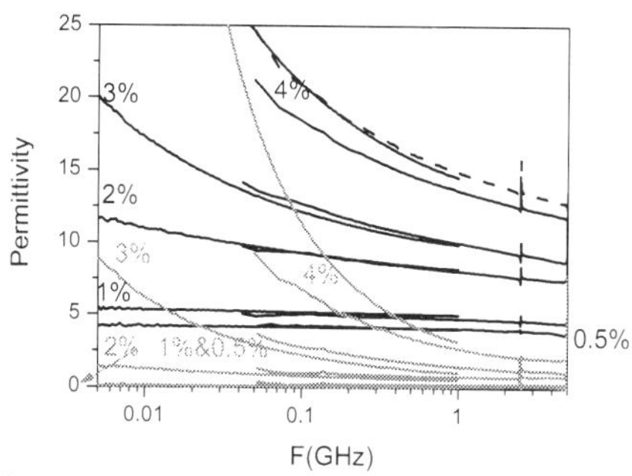

Fig. 8. Permittivity of SWCNT composites measured by Impedance and 14mm coaxial method

4. Results and Discussion

4.1 Electrical properties

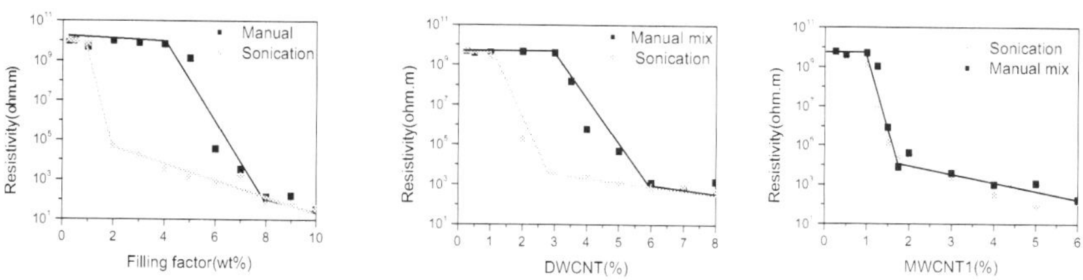

Fig. 9. Resistivity of (a) SWCNT (b)DWCNT (c)MWCNT

The measurement electrical properties of different types of CNTs are plotted in Figure 9. The marks are the measured resistivity of CNTs composites prepared by sonication process (stars) and manual mixing (squares). And the poly-lines are the guide of the dependence of resistivity on the concentration of CNTs instead of fitting curves. Due to the measurement range of the insulation multimeter is up to 200G Ω, the resistibility measured at low concentration might be lower than the real value which could results in larger percolation threshold.

For SWCNT and DWCNT, the percolation threshold is around 1wt% if dispersed by sonication process. The percolation of manually mixed SWCNT and DWCNT composites could be up to 3~4wt%. For MWCNT composites, it seems the resistivity decreases before 1%. The percolation ratio could be between 0.5% and 1%. No obvious difference could be found from MWCNT composites dispersed by two different processing methods. Individual tubes instead of bundles could be observed from SEM image of MWCNT in Figure 1(c), as compared with SWCNT (Figure 1(a)) and DWCNT (Figure 1(b)). Therefore, manual mixing might be enough to disperse the tubes inside polymer matrix. However, sonication mixing could result in more steep percolation transition for SWCNT and DWCNT composites. And the bundle of SWCNT and DWCNT couldn't be broken by the manual mixing process.

4.2 Frequency-dependent permittivity

Figure 10 to Figure 12 show the measured dielectric properties of SWCNT, DWCNT and MWCNT composites processed with sonication method respectively. The percolation for three types of CNTs is around 1%. It is sufficient to investigate concentration up to 5% from percolation phenomena point of view. In all of three figures, solid lines show the measured real (in black) and imaginary (in red) permittivity and dash lines show fitting curved using (1). The permittivity of silicone host (ε_h) is constant value, 3, according to impedance measurement . From the measured dependence of the permittivity, the dispersion curves for three types of CNTs appear to be similar and do not show any absorption peak that is typical of the Debye dispersion law and mixing rules. The frequency dispersion is smooth and resembles the dispersion of the percolation type. Therefore, the fitting of date made with scaling dispersion law instead of the Cole-Davidson law is validated (Liu *et al*, 2007b). The frequency dependence of the real and imaginary permttivity are fitted separatedly, to track deviations from the percolation law given by (2).

The parameters x and y obtained by curve fitting, which depend on the concentration, are shown in Table 2. Both A and B increase steadily with concentration which is understandable due to more conducting channels involved at higher concentrations. Therefore, they are not included in Table 2. Both x and y increase with filling factor. The critical exponent y is smaller than x for most of concentraions for all of three types of CNTs. Exponent x at percolation threshold fall in the the range (0.2 to 0.4) form theoretical calculation, but exponent y is relatively lower (0.1 to 0.2) for 2%. Deviation from the linear dependence (scaling law) can be found from 4% to 5% may explain the higher exponent value than the universal value.

Therefore, the scaling frequency dispersion is observed in the CNTs composites within wide frequency range and concentration around percolation. In the percolation theory, the scaling behavior depends on the geometric spread of the properties of clusters, which is a function of the number of inclusions comprising a cluster. In actual composites, certain reasons lead to an additional spread of the properties of the clusters, such as the dispersion in dimension and in the conductivity of the CNT. Another reason can be the imperfect electrical contacts among the inclusion in the clusters where the conductivity between adjacent inclusions is due to interpaticle tunneling. An additional spread of the dispersion curve may be due to the elongated shape of CNT, because a net of tubes forms clusters of complex shape, which must be described by an assembly of RC circuits rather than by a single circuite. All these phenomana change the distibution of properties of tube clusters and therefore, results different frequency-dispersion from and exponents.

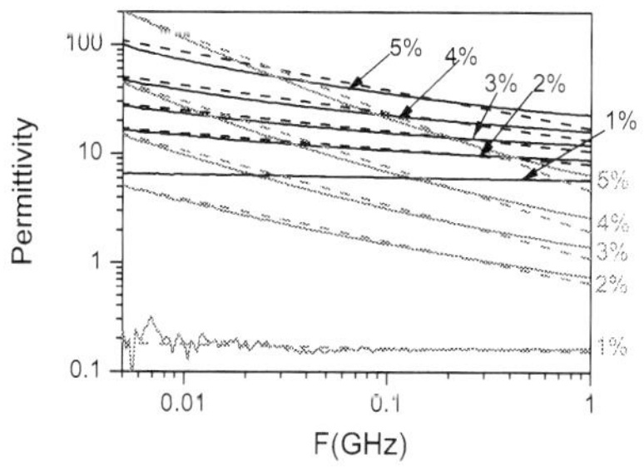

Fig. 10. Permittivity of SWCNT composite

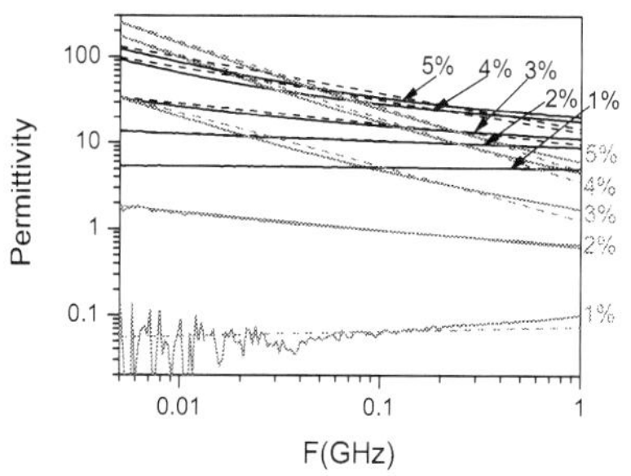

Fig. 11. Permittivity of DWCNT composite

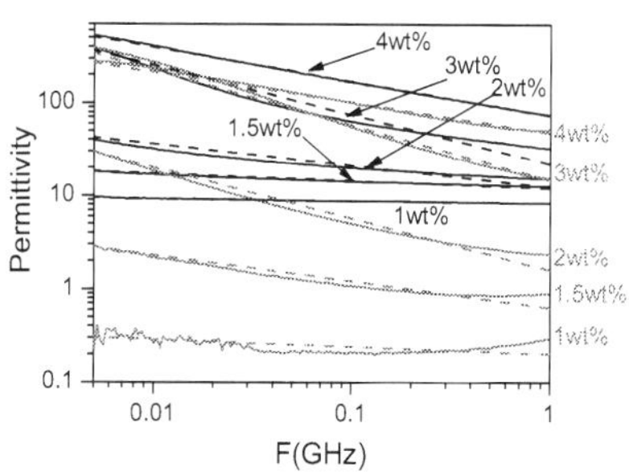

Fig. 12. Permittivity of MWCNT composite

SWCNT			DWCNT			MWCNT		
wt%	y	x	wt%	y	x	wt%	y	x
1	0.04	0.03	1	0.02	-0.05	1	0.03	0.07
2	0.18	0.39	2	0.11	0.21	1.5	0.09	0.28
3	0.217	0.491	3	0.276	0.62	2	0.26	0.55
4	0.275	0.591	4	0.402	0.736	3	0.54	0.59
5	0.368	0.713	5	0.438	0.747	4	0.37	0.32

Table 2. Critical exponents as function of concentration obtained by linear fitting of permittivity

4.3 Tunable permittivity under bias voltage

Figure 13 shows the dependence of normalized real and imaginary permittivity at 1 MHz, of the SWCNT composites at various concentrations, measured under a biasing voltage. The dependence of permittivity on bias voltage can be found at high frequency till a few hundred MHz. Figure 14 plots the tunability of permittivity of SWCNT composite from 1MHz to 100MHz. However, with increasing measurement frequency, the tunability becomes smaller. When the concentration is equal to or smaller than 2%, both real and imaginary permittivities do not change with voltage up to 15 V. It is understandable from the percolation threshold point of view because most of the tubes are out of the conducting network. Therefore, no voltage can be applied to isolate tubes or tube clusters. Once the concentration is higher than 3%, permittivity becomes sensitive to bias voltage. The tunability becomes more pronounced with increasing concentration. Samples with concentration of CNTs higher than 5% were not investigated due to extremely large imaginary permittivity. Obviously, bias voltage can decrease real permittivity but increase the imaginary part as shown in Figure 13. If the whole conducting network is treated as a circuit containing resistors and capacitors, replacing some capacitors with resistors will achieve the effect to reduce the real part and raise the imaginary part of permittivity. This is what was happening when bias voltage was applied to the samples in which junctions are formed by overlapping tubes. Electrons hop among tubes driven by bias voltages. The interaction among tubes becomes more resistive instead of capacitive under bias voltage. The tunability, defined as the change under bias over the original value, is more than 60% for the real permittivity and 80% for the imaginary permittivity for the sample with 5% of tubes. It is important to emphasize that the tunability of the CNT composites was reversible over a few thousand of cycles. Similar effect has also been found out from DWCNT and MWCNT composites when the concentration is large than percolation value within similar frequency range.

The novel tunable permittivity of the CNT composites may have potential applications in electromagnetic smart materials and structures. A smart materials or structures basically possess electrically or magnetically adjustable transmission or reflection coefficient to incident electromagnetic wave. The simplest solution is on-off switch, which was realized by pin diodes. One of the drawbacks of pin diodes is their relatively weak tolerance to high current, which is usually encountered in practical applications. A possible approach is to use a small piece of CNT composite to replace pin diode. Without the application of a bias voltage, the composite is insulating (off), while it becomes conducting under a certain bias voltage (on). Comparing with diodes, CNT composites can withstand higher current, voltage and power. Although the tunability of the SWCNT composite is still not high enough to provide similar on-off effect as a pin diode, we are confident that it can be further improved through proper dispersion or alignment of CNTs within holding matrix.

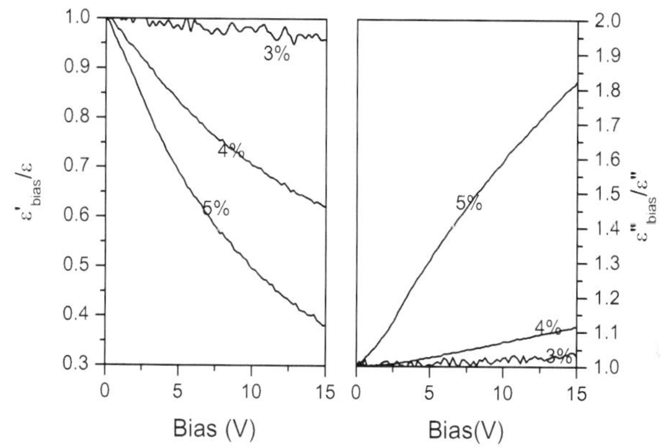

Fig. 13. Tunable permittivity of SWCNT composites (5MHz)

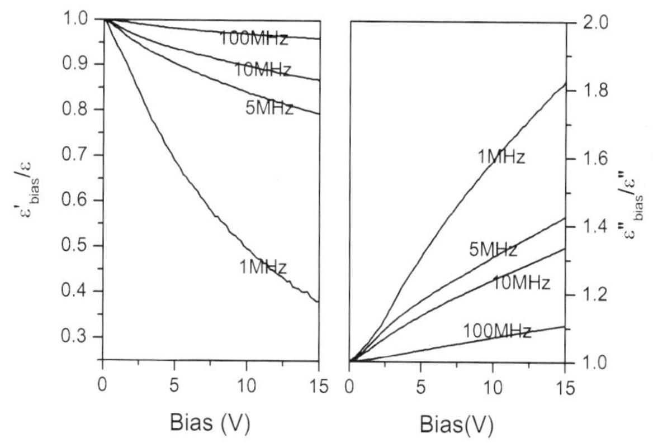

Fig. 14. Tunable permittivity of SWCNT composites (5%)

5. Conclusions

In this chapter, the complex permittivity of SWCNT, DWCNT and MWCNT composites at microwave band is obtained experimentally by the impedance method. The frequency dependence of the permittivity is well described by the scaling law based on the percolation theory. CNT composite is a system comprising clusters and conducting networks, the microwave behavior of which depends on the cluster formation, instead of the properties of individual fibers. Strong frequency-dependent behavior may be observed in measurement due to size effects. The microwave properties of the composites are found to be very sensitive to the preparation method. However, certain deviations are observed in the scaling parameters derived from the percolation theory. Namely the critical exponents are different from conventionally accepted universal values. This may be attributed to the non-perfect electrical contacts within clusters, which typically are not considered in mixing laws.

The CNT composites have large real permittivity at microwaves, adjustable dielectric loss tangent, and noticeable frequency dispersion of permittivity, even at concentrations above the percolation threshold. For these reasons, CNT composites with $p>p_c$ are probably useful as microwave dielectrics and frequency dispersive materials. CNT composites have been found to possess novel tunable effective permittivity under small bias voltages. The tunable properties can be attributed to flowing of electrons in the conducted network formed by tubes and tube clusters. Hence they are only obtainable at concentrations above percolation threshold. The tunability of both real and imaginary permittivity has potential applications in smart material and structure, which is the research subject that is currently on-going in our group.

Acknowledgements

The authors appreciate the fruitful discussion with Dr. K. Rozanov (ITAE, Russia) and Dr. C. R. Deng (DSO, Singapore). The research was supported by Defence Science and Technology Agency (DSTA), Singapore, under project POD0103671. Wen-Yan Yin appreciates the financial support by the National Basic Research Program under Grant 2009CB320204, and the NSFC under Grant 60821062 of CHINA.

6. References

A. N. Lagarkov, S. M. Matitsine, K. N. Roznov, and A. K. Sarychev (1998). Dielectric properties of fiber-filled composites, *J. Appl. Phys.*, **84**, 3806-14

A.K. Jonscher (1999). Dielectric relaxation in solids, J. *Phys. D – Appl. Phys.* **32**, n 14, pp57-70.

B. K. Kim, J. Lee and I. Yu (2003). Electrical properties of single-wall carbon nanotube and epoxy composites, *J. Appl. Phys.*, **94**(10), 6724-8

C. A. Martin, J. K. W. Sandler, M. S. P. Shaffer, M. K. Schwarz, W. Bauhofer, K. Schulte, A. H. Windle (2004), Formation of percolation networks in multi-wall carbon-nanotube-epoxy composites, *Composites science and technologies*, **64**, 2309-16

C. Brosseau (2002), Generalized effective medium theory and dielectric relaxation in particle-filled polymeric resins, *J. Appl. Phys.* **91**(5), 3197-204.

C. P. Neo and V. K. Varadan (2004). Optimization of carbon fiber composite for microwave absorber, *IEEE Trans. Electromagn. Compat.* **46**, 102-6

C. R. Simovsky, S. A. Tretyakov, A. H. Sihvola and M. M. Popov (2000). On the surface effect in thin molecular or composite layers, *Eur. Phys. J. Appl. Phys.* **9**, 195-204

C. Yang, Y. Lin, C. W. Nan, Modified carbon nanotube composites with high dielectric constant, low dielectric loss and large energy density (2009). *Carbon*, **47**, 1096-101

Coleman J. N., Curran S, Dalton AB, Davey AP, Mccarthy B, Blau W, Barklie RC (1998). Percolation-dominated conductivity in a conjugated-polymer-carbon-nanotube composite, *Phys Rev B*, **58**(12), 7492-5

D. Satuffer and A. Aharony (1991), *Introduction to Percolation Theory*, Taylor & Francis, 2nd edition,

D.J. Bergman and Y. Imry (1977). Critical behaviour of the complex dielectric constant near the percolation threshold of a heterogeneous material, *Phys. Rev. Lett.*, **39**, 1222-25

F. Bordi, C. Cametti, J. Rouch, F. Sciortino, and P. Tartaglia (1996). Cluster formation in water-in-oil microemulsions at percolation: evaluation of the electrical properties, *J. Phys.: Condens. Matter*. **8**(25), 19-37.

H. Y. Lee and Y. M. Shkel (2007), Self-sensing and self-actuation response of carbon nanotube composites, *Proc. Of SPIE*, **6526**, 1-10.

I. J. Youngs, A. S. Treen, G. Fixter and S. Holden (2002). Design of solid broadband human tissue simulant materials, *IEE Proc. Sci. Meas. Technol.* **149**, 323-8.

J. B. Donnet, R. C. Bansal, and M. J. Wang (1993). *Carbon Black: Science and Technoloy*, Dekker, New York

J. K. W. Sandler, J. E. Krik, I. A. Kinloch, M. S. P. Shaffer, and A. H. Windle (2003). Ultra-low electrical percolation threshold in carbon-nanotube-epoxy composites, *Polymer*, **44**, 5893-99

J.P. Straley (1977), Critical exponents for the conductivity of random resistor lattices, *Phys, Rev. B*, 15, 5733-7.

K. Park, S. Lee, C. Kim, J. Han (2006), Fabrication and electromagnetic characteristics of electromagnetic wave absorbing sandwich structures, *Composites science and technology*, **66**, 576-84

L. Liu, L. B. Kong and S. Matitsine (2008), Tunable effective permittivity of carbon nanotube composites, *Appl. Phys. Lett.*, 93, 113106

L. Liu, S. Matitsine, Y. B. Gan and K. Rozanov (2007a). Cluster effect in frequency selective composites with randomly distributed long conductive fibres, *J. Phys. D: Appl. Phys.* **40**, 7534–9

L. Liu, S. Matitsine, Y. B. Gan, L. F, Chen and L. B. Kong (2007b), Frequency dependence of effective permittivity of carbon nanotube composites, *J. Appl. Phys.* **101**, 094106.

Li-Li Wang, Beng-kang Tay, Kye-yak See, Zhuo Sun, Lin-Kin Tan, Darren Lua (2009), Electromagnetic interference shielding effectiveness of carbon-based materials prepared by screen printing, *Carbon*, **47**, 1905-10

L. Wei, B. Wang, Q. Wang, L. Li, Y. Yang, and Y. Chen (2008). Effect of Centrifugation on the Purity of Single-Walled Carbon Nanotubes from MCM-41 Containing Cobalt, *J. Phys. Chem.*, **112**, 17567–75

M. A. Seo, J. H. Yim, Y. H. Ahn, F. Rotermund, D. S. Kim, S. Lee, and H. Lim (2008), Terahertz electromagnetic interference shielding using single-walled carbon nanotube flexible films, *Appl. Phys. Lett.*, **93**, 231905

M. Meyyapan (2005). *Carbon Nanotubes: Science and Applications*, CRC Press, LLC

M. H. Al-Saleh, Uttandaraman, Sundararaj (2009), Electromagnetic interference shielding mechanisms of CNT/polymer composites, *Carbon*, **47**, 1738-46.

P. C. P. Watts, D. R. Ponnampalam, W. K. Hsu, A. Barnes and B. Chambers (2003), The complex permittivity of multi-walled carbon nanotube-polystyrene composite films in X-band, *Chem. Phys. Lett.*, **378**, 609-14.

R. B. Laibowitz and Y. Gefen (1984). Dynamic scaling near the percolation threshold in thin Au films, *Phys. Rev. Lett.* , **53**, 380-3

R. Saito, G. Dresselhaus and M. S. Dresselhaus (1998). *Physical Properties of Carbon Nanotubes,* Imperial College Press, London.

S. Iijima, *Nature* (1991). Helical microtubules of graphitic carbon, **354** , 56.

S. L. Browning, J. Jodge, R. R. Price, J. Schelleng, P. E. Schoen and D. Zabetakis (1998). Fabrication and radio frequency characterization of high dielectric loss tubule-based composites near percolation, *J. Appl. Phys.* , **84**, 6109-13.

S. Lee, J. Kang and G. Kim(2006), Fabrication and design of multi-layered radar absorbing structures of MWCNT-filled glass/epoxy plain-weave composites, *Composite structures*, **76**, 397-405

S. Shi and J. Liang (2008), the effect of multi-wall carbon nanotubes on electromagnetic interference shielding of ceramic composites, *Nanotechnology*, 19, 255707

S.M. Matitsine, K.M. Hock, L. Liu, Y.B. Gan, A.N. Lagarkov, K.N. Rozanov (2003), Shift of resonance frequency of long conducting fibers embedded in a composite, *J. Appl. Phys.* **94**, 1146-54.

T. D. Burchell (1999). *Carbon Materials for Advanced Technologies*, Elsevier Science Ltd, Oxford

Y. Gefen, A. Aharony and S. Alexander (1983), Anomalous diffusion on percolating clusters *Phys. Rev. Let.*, **50**(1) 77-80.

Y. Huang, N. Li, Y. Ma, F. Du, F. Li, X. He, X. Lin, H. Gao, Y. Chen (2007), The influence of single-walled carbon nanotube structure on the electromagnetic interference shielding efficiency of its epoxy composites, *Carbon*, 45, 1614-21

Y. Song, T. W. Noh, S. I. Lee and J. R. Gaines (1983), Experimental study of the three-dimensional AC conductivity and dielectric constant of a conductor-insulator composite near the percolation threshold, *Phys. Revs. B*, **33**(2), 904-8

6

Environmental effects on photoluminescence of single-walled carbon nanotubes

Yutaka Ohno, Shigeo Maruyama* and Takashi Mizutani
Department of Quantum Engineering, Nagoya University
**Department of Mechanical Engineering, The University of Tokyo*
Japan

1. Introduction

Spectroscopy is a powerful technique to characterize single-walled carbon nanotubes (SWNTs). In particular, photoluminescence (PL) and excitation spectroscopies have enabled to make great advances in understanding the energy states and transition processes of excitons in SWNTs. Furthermore, PL spectroscopy also has potential for being used to determine the chiral index (n,m) of an SWNT and evaluate the abundance of (n,m) in bulk SWNTs. (Bachilo, Strano et al. 2002; Miyauchi, Chiashi et al. 2004) However, in order to establish PL spectroscopy as a reliable standard technique to characterize SWNTs, it is necessary to understand not only the intrinsic physics of excitons but also the effects of extrinsic factors, namely, *environmental effects*. (Fantini, Jorio et al. 2004) The optical properties of an SWNT are modified sensitively by its environmental conditions because of the excitons existing on the SWNT surface. In this chapter, we focus on the environmental effects on the optical properties of SWNTs. After considering the fundamental physics of the optical transitions in SWNTs and application of PL spectroscopy to characterize SWNTs, various types of environmental effects are discussed with practical examples.

2. Optical transition and photoluminescence

2.1 Density of states and optical transition

In an SWNT, electrons are confined to its surface, and the wave vectors along the circumference are quantized by a periodic boundary condition; however, the carriers are free to move in the direction of the SWNT axis. Therefore, the SWNT has a one-dimensional electron system, where the density of states is characterized by the sharp peaks originating from a van Hove singularity, as shown in Fig. 1(a). (Saito, Dresselhaus et al. 1998)
An optical transition in an SWNT occurs dominantly between two subbands located at van Hove singularities. However, because of the one-dimensional structure, it depends on the polarization of the incident light. If the incident light is polarized linearly with the polarization vector parallel to the SWNT axis, a strong optical transition occurs. In this case, the optical transitions from a subband of a valence band to a subband of the conduction band, and vice versa, with the same quantum number are allowed by the selection rule. For

example, transitions such as those from C1 to V1 and from V2 to C2 are allowed transitions, as shown in Fig. 1(a). On the other hand, if the incident light is polarized perpendicular to the SWNT axis, the optical transition is strongly suppressed by the depolarization effect in which the induced charges cancel the electric field of the light in the SWNT. (Duesberg, Loa et al. 2000)

The energy of an optical transition is given by the difference in the energies of the two subbands, in the case of the single-particle approximation. However, in practice, it is modified by many-body effects, as shown in Fig. 1(b). (Ando 2005) The single-particle bandgap E_{ii}^{gr} is renormalized by the self-energy of an electron due to the repulsive interaction between electrons. In addition, for an optical transition, the exciton effect should be taken into account, that is, the optical transition energy is decreased by the binding energy of the attractive Coulomb interaction between an electron and a hole. In the case of SWNTs, since the exciton binding energy is as large as a few hundred millielectronvolts, the exciton scheme is more suitable than the single-particle scheme. (Wang, Dukovic et al. 2005) Consequently, the optical transition energy E_{ii} between the i-th valence band and the i-th conduction band (i = 1, 2, 3, …) is given by

$$E_{ii} = E_{ii}^{gr} + E_{ii}^{ee} + E_{ii}^{eh}, \quad (1)$$

where E_{ii}^{ee} denotes the energy of repulsive interaction between electrons, and E_{ii}^{eh} (< 0) represents the binding energy of an exciton. Note that every term is perturbed by the environment around the SWNT, as described later.

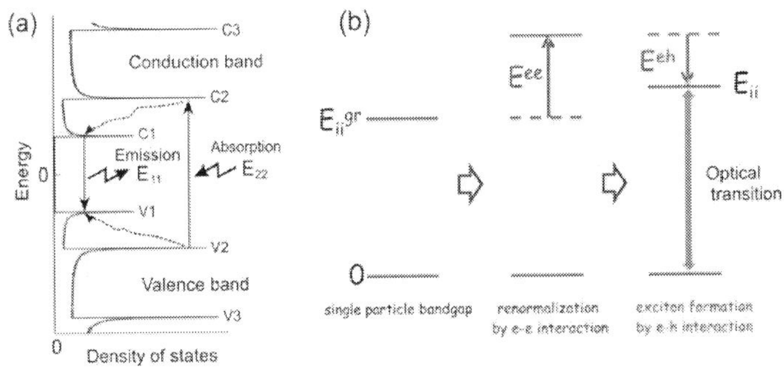

Fig. 1. (a) Density of states of a semiconducting SWNT and optical transition processes in the single particle scheme. (b) Energy diagram of an exciton state, showing optical transition energy in the exciton scheme.

2.2 Photoluminescence spectroscopy

Since the optical transition energy E_{ii} depends on the chiral index (n,m) of an SWNT, (n,m) can be determined by spectroscopic techniques such as Raman scattering spectroscopy (Kataura, Kumazawa et al. 1999) and photoluminescence (PL) spectroscopy. (Bachilo, Strano et al. 2002) In the case of PL spectroscopy of a semiconducting SWNT, the E_{22} state is resonantly excited by incident light so as to generate excitons, as shown in Fig. 1(a). (O'Connell, Bachilo et al. 2002) The excitons relax via phonons to the E_{11} state, where radiative recombination occurs. Figure 2(a) shows the PL spectra of SWNTs measured at various excitation wavelengths. Such measurement is called the PL mapping measurement. Each peak originates from a different (n,m) species. From the relative intensity of each peak, the abundance of each (n,m) species can be evaluated. It should be note that the quantum efficiency of light absorption and emission depends on the diameter and chirality of SWNTs. (Oyama, Saito et al. 2006) Therefore, in order to make accurate evaluation of the (n,m) abundance, a correction of the (n,m) dependence of the quantum efficiency is required. (Okazaki, Saito et al. 2006)

Figure 2(b) shows the E_{22} vs. E_{11} plot for various (n,m) species, which are derived from the PL map of Fig. 2(a). The points connected by a solid line belong to the family of SWNTs that have the same value of $(2n + m)$. (Saito, Dresselhaus et al. 1998) It can be observed that the family patterns are separated into two groups, depending on the value of $(2n + m \bmod 3)$. In this chapter, we define these two groups as Type I if $(2n + m \bmod 3) = 1$, and Type II if $(2n + m \bmod 3) = 2$. When a graphite layer is rolled up to form an SWNT, the quantized energies of the subbands primarily depend on the diameter of the SWNT. In addition, they also depend on the direction to be quantized, namely the chiral angle, because the equi-energy contours of graphite are warped to form a triangular shape around the K point, which is called the trigonal warping effect. (Saito, Dresselhaus et al. 2000) This is the main reason for the appearance of the family pattern in the plot of optical transition energies.

The family patterns are technically important for the analysis of PL spectra. Even though the optical transition energies of SWNTs vary by more than 100 meV, depending on the methods used to prepare SWNTs and on the environment, the identification of the family pattern allows us to easily determine the chirality index (n,m) of the SWNTs.

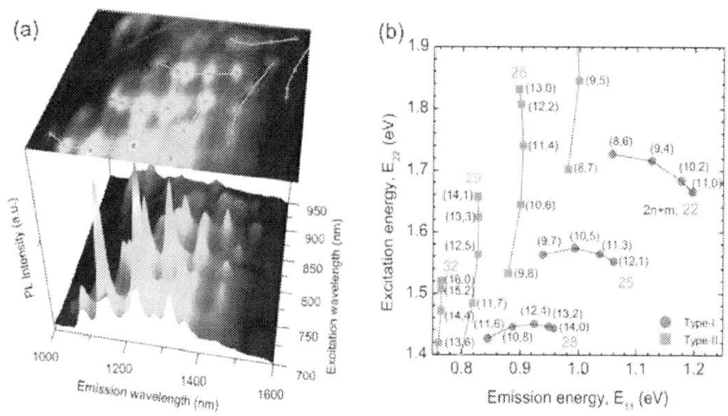

Fig. 2. (a) PL spectra of SWNTs measured for various excitation wavelengths (PL map). (b) E_{22} vs. E_{11} plot. A solid line connects SWNTs belonging to the same family, i.e., SWNTs having the same value of $2n + m$.

2.3 Relaxation process

In PL processes, the excitons excited to the E_{22} state relax to the lowest state E_{11} via phonons in a short time such as ~0.1 ps because the optical phonon energy is as large as 0.2 eV (Ichida, Hamanaka et al. 2002; Ostojic, Zaric et al. 2004) Then, the excitons relax to the ground state via radiative recombination or nonradiative recombination. The radiative recombination lifetime is expected to be about 1 ns. (Spataru, Ismail-Beigi et al. 2005) In contrast, the experimental results of pump-probe measurements and time-resolved PL measurements have shown a much shorter recombination lifetime, of the order of several tens of picoseconds. (Hagen, Moos et al. 2004; Wang, Dukovic et al. 2004; Hirori, Matsuda et al. 2006; Ohno, Kishimoto et al. 2006) Such a short recombination lifetime suggests that the dominant recombination process is nonradiative. The luminescence quantum efficiency has been estimated experimentally to be several percent. (Lefebvre, Austing et al. 2006) Recombination centers formed by defects and interface states may enhance such nonradiative recombination. It has been suggested that nonradiative recombination rate is enhanced if the exciton is localized. (Perebeinos and Avouris 2008)

2.4 Sample preparation for achieving photoluminescence

The soot of SWNTs generally consists of SWNT bundles formed by van der Waals interaction. In a bundle of SWNTs, the energy of photoexcited excitons is transferred to neighboring metallic SWNTs or SWNTs with a smaller bandgap, and then, these excitons relax via phonons. Therefore, in order to obtain luminescence from SWNTs, it is necessary to debundle and individualize SWNTs from a bundle. The PL from individual SWNTs has been observed for the first time by O'Connell et al. by wrapping SWNTs with a surfactant such as sodium-dodecyl-sulfate (SDS) in order to form micelles encasing individual SWNTs. (O'Connell, Bachilo et al. 2002) After dispersing the soot of SWNTs in an aqueous solution of SDS by performing ultrasonication, they obtained individualized SWNTs in the supernatant liquid of ultracentrifugation, as shown in Fig. 3(a).

Another method to obtain individualized SWNTs is to grow SWNTs directly between pillars or on trenches formed on a Si or quartz substrate, as shown in Fig. 3(b). (Lefebvre, Fraser et al. 2004; Ohno, Iwasaki et al. 2006) The free-standing SWNTs bridging such a microstructure provide strong PL, although the PL is suppressed if an SWNT touches the substrate, as will be described later. Further, the PL signal from a free-standing SWNT is sufficiently strong and can hence be observed by a microscopic PL technique.

For the purpose of studying the environmental effects on the PL of SWNTs, free-standing SWNTs grown between pillars or on trenches are suitable because the environmental conditions can be intentionally changed by exposing the SWNTs to a gas (Finnie, Homma et al. 2005; Chiashi, Watanabe et al. 2008) or by immersing the SWNTs in a liquid. (Iwasaki, Ohno et al. 2007; Ohno, Iwasaki et al. 2007)

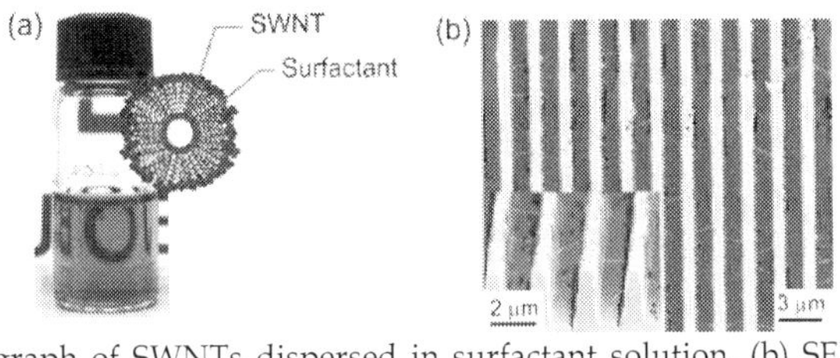

Fig. 3. (a) Photograph of SWNTs dispersed in surfactant solution. (b) SEM image of free-standing SWNTs grown on the trenches formed on a quartz substrate.

3. Environmental effects

The environmental effects on the PL of SWNTs is one of the most important topics as regards the optical properties of SWNTs. (Moore, Strano et al. 2003; Chou, Ribeiro et al. 2004; Fantini, Jorio et al. 2004) It is known that the optical transition energy varies, depending on the type of surfactants used to individualize SWNTs. Lefebvre et al. have also reported that the optical transition energies of SWNTs bridging micropillars show a blueshift as compared to the SDS-wrapped SWNTs. (Lefebvre, Fraser et al. 2004) A later study by Ohno et al. has shown that the energy differences depend on (n,m), particularly on the chiral angle and the type of SWNTs, i.e., whether the SWNTs are of Type I or Type II. (Ohno, Iwasaki et al. 2006) One of the causes of the variation in the optical transition energy is the dielectric screening of the many-body interactions between carriers. (Perebeinos, Tersoff et al. 2004; Ando 2005) In addition to the dielectric screening effect, various environmental factors such as chemical interactions (Finnie, Homma et al. 2005; Hertel, Hagen et al. 2005) and mechanical interactions (Arnold, Lebedkin et al. 2004) should be considered. Comprehensive understanding of the environmental effects is necessary to understand the optical properties of SWNTs. In this section, we discuss the effects of the following factors on the PL of SWNTs: dielectric screening, wrapping SWNTs with surfactants, exposing SWNTs to air, doping SWNTs with carriers, and allowing SWNTs to make contact with a substrate.

3.1 Dielectric screening effect

The variation in the optical transition energy caused by the environmental conditions is primarily due to the dielectric screening effect, whereby the many-body Coulomb interactions between carriers are screened by the surrounding material. As shown in Fig. 4, the electric force lines contributing to the Coulomb interactions pass through not only the inside of an SWNT but also the outside, where the surrounding material screens the electric force lines depending on its dielectric constant.

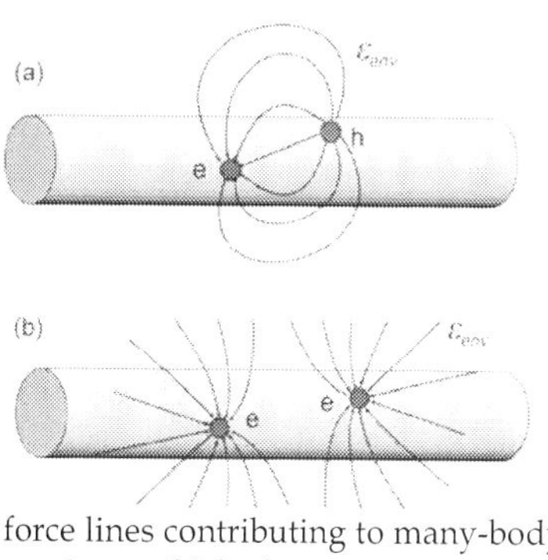

Fig. 4. Schematics of electric force lines contributing to many-body Coulomb interactions: (a) electron-hole attractive interaction, which forms an exciton, and (b) electron-electron repulsive interaction.

Ohno et al. have studied the dielectric screening effect in detail by immersing the SWNTs grown on trenches (Fig. 3(b)) in various organic solvents with dielectric constants from 1.9

to 37. (Iwasaki, Ohno et al. 2007; Ohno, Iwasaki et al. 2007) Figure 5 shows the typical PL spectra measured in the ambient air, hexane (dielectric constant: 1.9), chloroform (4.8), and acetonitrile (37). Note that although in this chapter, the dielectric constants are used considering a dc field, they depend on the frequency of the field. At the frequency corresponding to the exciton lifetime, a dielectric constant is expected to approach unity. Further, although the exact values of ac dielectric constants of these liquids have not been known, they could be related to the dc dielectric constants. The peaks exhibit a redshift and broadening with the increasing dielectric constant of liquids. As indicated by the E_{22} vs. E_{11} plots shown in Fig. 6, both E_{11} and E_{22} exhibit a redshift with the increasing dielectric constant of liquids, with a small (n,m) dependence. The amounts of the redshifts are 20~43 meV for E_{11} and 15~33 meV for E_{22}.

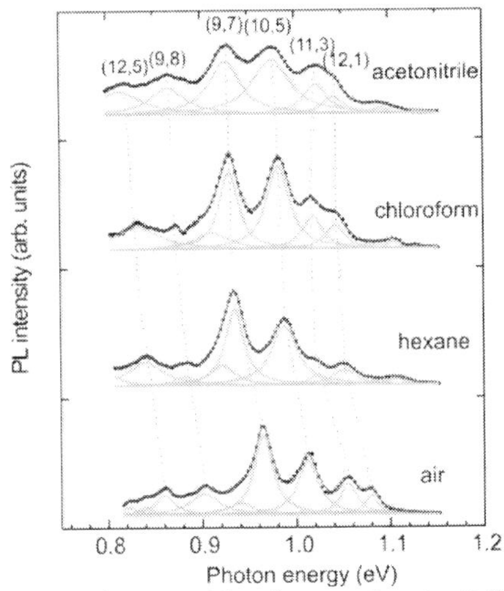

Fig. 5. PL spectra in ambient air, hexane, chloroform, and acetonitrile.

As the dielectric constant increases, the energies of both the electron-electron and electron-hole interactions decrease. The reduction in the exciton binding energy causes a blueshift in the optical transition energy because of the negative energy, whereas the reduction in the electron-electron repulsion energy causes a redshift. The occurrence of redshifts suggests that the variation in the electron-electron repulsion energy is larger than that in the exciton binding energy. This is due to the fact that the repulsion energy is larger in magnitude than the exciton binding energy. (Ando 2005; Spataru, Ismail-Beigi et al. 2005)

Figure 7 shows E_{11} and E_{22} of (8,6)-SWNTs as a function of the dielectric constant of the surrounding liquid. The energy shift is significant in the regime of a low dielectric constant, and exhibits a tendency to saturate at a dielectric constant of ~5. Empirically, the dependence of E_{ii} on the environmental dielectric constant ε_{env} is given by

$$E_{ii} = E_{ii}^{\infty} + A\varepsilon_{env}^{-\alpha}. \quad (2)$$

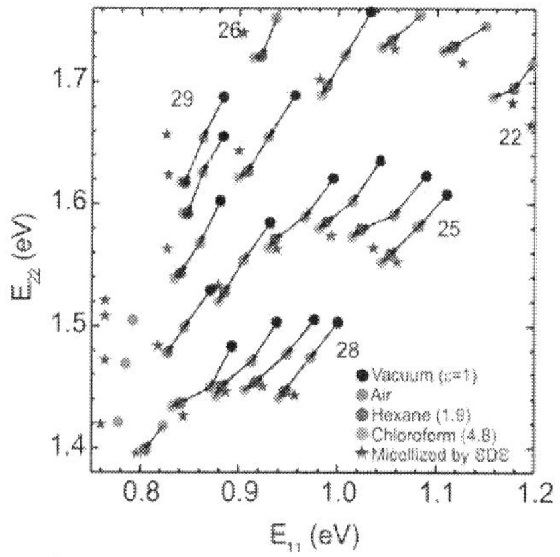

Fig. 6. E_{22} vs. E_{11} plots of SWNTs grown on microtrenches in various environmental conditions: in vacuum (black dots), air (yellow green), hexane (blue), chloroform (orange), and aqueous solution of SDS (red stars).

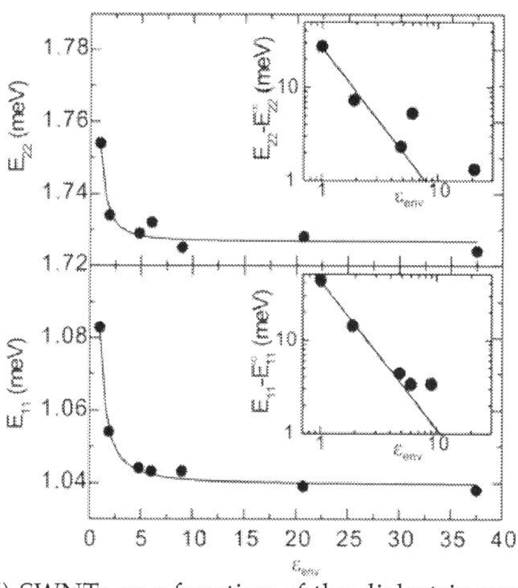

Fig. 7. E_{11} and E_{22} of (8,6)-SWNTs as a function of the dielectric constant of the surrounding liquid. The insets are the log-log plots.

In Eq. (2), the first term on the right-hand side corresponds to the transition energy when ε_{env} is infinity; it consists of the single-particle bandgap and the energies of the many-body interactions, which are determined by the electric force lines traveling only inside the SWNT. The second term shows the screening of Coulomb interactions, including both electron-hole binding and electron-electron repulsive interactions, which are related to the electric force lines traveling outside the SWNT. In the equation, "A" denotes the maximum energy variation. In Fig. 6, the solid curves are fitted to the data by using Eq. (2). Here, A and α are 43 meV and 1.6 for E_{11}, and 27 meV and 1.8 for E_{22}, respectively. Miyauchi et al. have also obtained a power-law-like dependence of the exciton transition energies on the dielectric constant of the material surrounding SWNTs by the tight-binding calculation.(Miyauchi, Saito et al. 2007)

The variation in E_{ii} due to the dielectric screening effect also depends on (n,m), particularly, on the chiral angle, and the type (Type I or Type II). This may be due to the (n,m) dependence of the effective mass of carriers. As descried before, the effective mass depends on the chiral angle because of the trigonal warping effect. (Jorio, Fantini et al. 2005) In addition, the effective mass is one of the scaling parameters for exciton energy. (Perebeinos, Tersoff et al. 2004) Similarly, the repulsive energy of electron-electron interaction could depend on the effective mass.

It should be noted that in addition to exhibiting the redshift, the PL spectrum became broader when the SWNTs were immersed in liquids with a large dielectric constant, as shown in Fig. 5. The full width at half maximum (FWHM) increased from 23 meV in the ambient air to 40 meV in acetonitrile for (9,7)-SWNTs. This linewidth broadening may be attributed to inhomogeneous broadening due to the fluctuation in the local dielectric constant of the environment occurring on the scale of the size of a liquid molecule. It should be mentioned that the dielectric constants used in this chapter are macroscopic and static values. In practice, the size of a molecule of an organic solvent is comparable to that of an exciton; therefore, the local dielectric constant around an exciton could fluctuate depending on the position and orientation of the surrounding organic molecules, which could not follow the dynamic electric field formed by the exciton within the lifetime of the exciton. This results in the dispersion of the energies of many-body interactions.

3.2 Effect of ambient air and intrinsic transition energies

On measuring the PL of free-standing SWNTs in a vacuum, it is found that the transition energies are 20~40 meV higher than those corresponding to the PL measured in the ambient air, as shown by the black dots in Fig. 6. Further, sequential measurements of the PL in the ambient air after the growth of SWNTs exhibit a phenomenon that every E_{ii} shifts to a lower energy with an incubation time ranging from a few hours to several days. The incubation time required for the occurrence of a redshift depends on the diameter of SWNTs, that is, the redshift occurs earlier in the case of SWNTs with a small diameter than that in the case of SWNTs with a large diameter. This variation in E_{ii} is reversible, namely, E_{ii} returns to its former value when the sample is heated up above 50°C in a vacuum. The shifts in E_{ii} can be attributed to the dielectric screening effect because their behavior is similar to that observed in the liquid immersion experiments.

In air, the ambient molecules such as H_2O and C_xH_y are adsorbed on the surface of SWNTs and cause the screening of many-body interactions. Chiashi et al. have investigated the effect of gas adsorption in detail. (Chiashi, Watanabe et al. 2008) They have shown that a redshift occurs at a gas pressure exceeding the transition pressure, which agrees with Langmuir's adsorption model.

Carrier doping due to oxygen molecules should be discussed as one of the effects of the ambient air because the energies of many-body interactions may be modified depending on the carrier density. The effect of carrier doping performed by the field-effect doping technique has been studied by measuring the PL of a free-standing SWNT placed in a field-effect transistor structure. (Ohno, Kishimoto et al. 2006) In addition, the effect of chemical doping with F_4TCNQ (tetrafluorotetracyano-p-quinodimethane), which is a strong p-type dopant for SWNTs (Nosho, Ohno et al. 2007), has also been studied. Even though the carrier densities introduced by these doping techniques were as high as that introduced by the ambient oxygen, the optical transition energies were not significantly modified by doping. This may be because the doping density was probably not sufficiently high to dope carriers in the range of the exciton diameter.

Now, by heating the free-standing SWNTs in a vacuum, the adsorbed molecules can be desorbed, and then the *intrinsic* transition energies of SWNTs can be measured without any environmental effects.

3.3 Effect of wrapping with surfactants

When the free-standing SWNTs grown on trenches were immersed in an aqueous solution of SDS (1 wt%), the E_{22} vs. E_{11} plots shifted to the positions indicated by the red stars in Fig. 6. The plots of these SWNTs are almost identical to those obtained for the SWNTs wrapped with SDS by using O'Connell's method, as mentioned in section 2.4, which implies that the free-standing SWNTs were wrapped with SDS by immersing them in the SDS solution.

Most E_{11} and E_{22} of SDS-wrapped SWNTs exhibit redshifts as compared to the free-standing SWNTs in air, except for E_{22} of near-zigzag Type II SWNTs, which exhibit blueshifts. (Ohno, Iwasaki et al. 2006) This behavior is different from the results obtained for the dependence of E_{11} and E_{22} on the dielectric constant of the surrounding material, as described above, in which all E_{11} and E_{22} exhibited redshifts with an increase in the dielectric constant. The positions of the E_{22} vs. E_{11} plots of the SDS-wrapped SWNTs shift so that the $(2n + m)$-family pattern spreads more widely, in addition to the shift due to the dielectric screening effect. This behavior of the energy shifts is similar to stress-induced shift. (Arnold, Lebedkin et al. 2004) One plausible explanation for the effect of SDS wrapping is that the SWNT is compressed by the surfactant molecules in the radial direction, resulting in the uniaxial strain that modifies the bandgap of the SWNT.

3.4 Effect of substrate

Even though a substrate is one of the indispensable parts for device application of SWNTs and supports the SWNTs and electrodes, the contact of SWNTs with the substrate seriously affects the optical properties of SWNTs. For example, in the case of SWNTs grown on a substrate such as SiO_2, PL is strongly suppressed if the body of the SWNT lies on the substrate. (Ohno, Kishimoto et al. 2006) The quenching of PL suggests that the excitons excited in an SWNT recombine by a nonradiative recombination process, probably via interface states between the SWNT and the substrate. The interface states also cause serious degradation of not only the optical properties but also the electrical properties of SWNTs, for example, the capture and emission of carriers at the interface states may cause noise in SWNT FETs.(Lin, Tsang et al. 2007)

Even though PL spectroscopy is not applicable to such SWNTs directly grown on a substrate, photocurrent spectroscopy is useful to study the optical properties of such SWNTs.(Ohno, Kishimoto et al. 2004) Figure 8 shows the photocurrent spectra of two types of transistors; one transistor is a conventional SWNT transistor with a channel lying on a SiO_2/Si substrate, and the other is a transistor with a channel of a free-standing SWNT, as shown in the inset. The peaks in the spectra are attributed to E_{22} of the semiconducting SWNT. In the case of the free-standing SWNT, the peak is quite sharp and has an FWHM of 30 meV, which is comparable to the excitation spectrum of the PL of an SWNT. On the other hand, in the case of the SWNT lying on SiO_2, the peak structure is not clear, and the FWHM is as broad as 60 meV. The broadening of the spectrum suggests that the energy band of this SWNT is modified inhomogeneously, probably depending on the surface structure of amorphous SiO_2.

Recently, Xie et al. have reported that PL can be observed from SWNTs lying on a SiO_2 surface when the SWNTs are transferred to another substrate after the growth. (Xie, Liu et al. 2007) At room temperature, it is difficult to form interface states between an SWNT and a SiO_2 substrate; this is because both the SWNT and SiO_2 are chemically stable. Nevertheless, during the growth of SWNTs directly on a substrate, chemical bonds may be formed at the interface due to high growth temperature. Such chemical bonds may act as nonradiative recombination centers. Moreover, Xie et al. have also reported that the PL was enhanced when the substrate surface was covered with a self-assembled monolayer. It has been explained on the basis of carrier transfer to the traps present in the SiO_2 layer of the substrate.

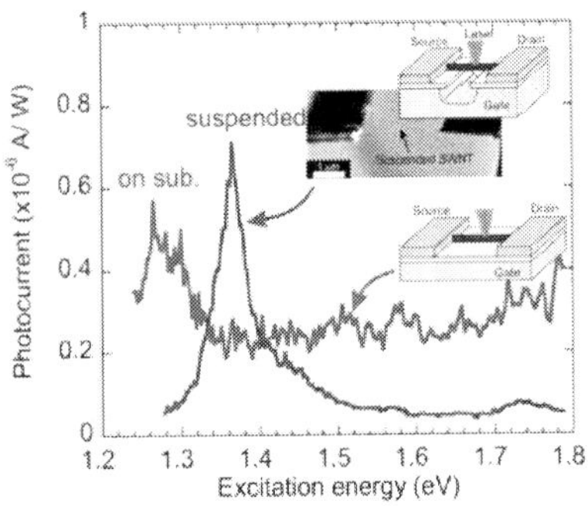

Fig. 8. Photocurrent spectra of CNTFETs with a free-standing SWNT (blue) and a SWNT in contact with a substrate (red).

4. Summary

In this chapter, we discussed environmental effects on the optical properties of SWNTs. Detailed understandings of the environmental effects help us to make a spectroscopic technique a standardized methodology to evaluate the chirality abundance and the quality of SWNTs. It is also interesting to develop sensing tools of nano dimensions for bio and medical applications by utilizing the environmental effects.

Acknowledgments

The authors thank S. Iwasaki, A. Kobayashi, Y. Murakami, Y. Miyauchi, and R. Saito for fruitful collaborations and discussions. This study is partially supported by Grant-in-Aid from MEXT, Japan.

5. References

Ando, T. (2005). "Theory of electronic states and transport in carbon nanotubes." Journal of the Physical Society of Japan **74**(3): 777-817.

Arnold, K., S. Lebedkin, et al. (2004). "Matrix-imposed stress-induced shifts in the photoluminescence of single-walled carbon nanotubes at low temperatures." Nano Letters **4**(12): 2349-2354.

Bachilo, S. M., M. S. Strano, et al. (2002). "Structure-assigned optical spectra of single-walled carbon nanotubes." Science **298**(5602): 2361-2366.

Chiashi, S., S. Watanabe, et al. (2008). "Influence of Gas Adsorption on Optical Transition Energies of Single-Walled Carbon Nanotubes." Nano Letters **8**(10): 3097-3101.

Chou, S. G., H. B. Ribeiro, et al. (2004). "Optical characterization of DNA-wrapped carbon nanotube hybrids." Chemical Physics Letters **397**(4-6): 296-301.

Duesberg, G. S., I. Loa, et al. (2000). "Polarized Raman spectroscopy on isolated single-wall carbon nanotubes." Physical Review Letters **85**(25): 5436-5439.

Fantini, C., A. Jorio, et al. (2004). "Optical transition energies for carbon nanotubes from resonant Raman spectroscopy: Environment and temperature effects." Physical Review Letters **93**(14): 147406.

Finnie, P., Y. Homma, et al. (2005). "Band-gap shift transition in the photoluminescence of single-walled carbon nanotubes." Physical Review Letters **94**(24): 247401.

Hagen, A., G. Moos, et al. (2004). "Electronic structure and dynamics of optically excited single-wall carbon nanotubes." Applied Physics a-Materials Science & Processing **78**(8): 1137-1145.

Hertel, T., A. Hagen, et al. (2005). "Spectroscopy of single- and double-wall carbon nanotubes in different environments." Nano Letters **5**(3): 511-514.

Hirori, H., K. Matsuda, et al. (2006). "Exciton localization of single-walled carbon nanotubes revealed by femtosecond excitation correlation spectroscopy." Physical Review Letters **97**(25): 257401.

Ichida, M., Y. Hamanaka, et al. (2002). "Ultrafast relaxation dynamics of photoexcited states in semiconducting single-walled carbon nanotubes." Physica B-Condensed Matter **323**(1-4): 237-238.

Iwasaki, S., Y. Ohno, et al. (2007). "Environmental dielectric screening effect on exciton transition energies in single-walled carbon nanotubes." arXiv:0704.1018v1 [cond-mat.mtrl-sci].

Jorio, A., C. Fantini, et al. (2005). "Resonance Raman spectroscopy (n,m)-dependent effects in small-diameter single-wall carbon nanotubes." Physical Review B **71**(7): 075401.

Kataura, H., Y. Kumazawa, et al. (1999). "Optical properties of single-wall carbon nanotubes." Synthetic Metals **103**(1-3): 2555-2558.

Lefebvre, J., D. G. Austing, et al. (2006). "Photoluminescence imaging of suspended single-walled carbon nanotubes." Nano Letters **6**(8): 1603-1608.

Lefebvre, J., J. M. Fraser, et al. (2004). "Photoluminescence from single-walled carbon nanotubes: a comparison between suspended and micelle-encapsulated nanotubes." Applied Physics a-Materials Science & Processing **78**(8): 1107-1110.

Lin, Y. M., J. C. Tsang, et al. (2007). "Impact of oxide substrate on electrical and optical properties of carbon nanotube devices." Nanotechnology **18**(29): 295202.

Miyauchi, Y., R. Saito, et al. (2007). "Dependence of exciton transition energy of single-walled carbon nanotubes on surrounding dielectric materials." Chemical Physics Letters **442**(4-6): 394-399.

Miyauchi, Y. H., S. H. Chiashi, et al. (2004). "Fluorescence spectroscopy of single-walled carbon nanotubes synthesized from alcohol." Chemical Physics Letters **387**(1-3): 198-203.

Moore, V. C., M. S. Strano, et al. (2003). "Individually suspended single-walled carbon nanotubes in various surfactants." Nano Letters **3**(10): 1379-1382.

Nosho, Y., Y. Ohno, et al. (2007). "The effects of chemical doping with F(4)TCNQ in carbon nanotube field-effect transistors studied by the transmission-line-model technique." Nanotechnology **18**(41): 415202.

O'Connell, M. J., S. M. Bachilo, et al. (2002). "Band gap fluorescence from individual single-walled carbon nanotubes." Science **297**(5581): 593-596.

Ohno, Y., S. Iwasaki, et al. (2006). "Chirality-dependent environmental effects in photoluminescence of single-walled carbon nanotubes." Physical Review B **73**(23): 235427.

Ohno, Y., S. Iwasaki, et al. (2007). "Excitonic transition energies in single-walled carbon nanotubes: Dependence on environmental dielectric constant." Physica Status Solidi B-Basic Solid State Physics **244**(11): 4002-4005.

Ohno, Y., S. Kishimoto, et al. (2006). Carrier transport property in single-walled carbon nanotubes studied by photoluminescence spectroscopy. Journal of Physics: Conference Series. **38:** 5-8.

Ohno, Y., S. Kishimoto, et al. (2006). "Photoluminescence of single-walled carbon nanotubes in field-effect transistors." Nanotechnology **17**(2): 549-555.

Ohno, Y., S. Kishimoto, et al. (2004). "Chirality assignment of individual single-walled carbon nanotubes in carbon nanotube field-effect transistors by micro-photocurrent spectroscopy." Applied Physics Letters **84**(8): 1368-1370.

Okazaki, T., T. Saito, et al. (2006). "Photoluminescence and population analysis of single-walled carbon nanotubes produced by CVD and pulsed-laser vaporization methods." Chemical Physics Letters **420**(4-6): 286-290.

Ostojic, G. N., S. Zaric, et al. (2004). "Interband recombination dynamics in resonantly excited single-walled carbon nanotubes." Physical Review Letters **92**(11): 117402.

Oyama, Y., R. Saito, et al. (2006). "Photoluminescence intensity of single-wall carbon nanotubes." Carbon **44**(5): 873-879.

Perebeinos, V. and P. Avouris (2008). "Phonon and electronic nonradiative decay mechanisms of excitons in carbon nanotubes." Physical Review Letters **101**(5): 057401.

Perebeinos, V., J. Tersoff, et al. (2004). "Scaling of excitons in carbon nanotubes." Physical Review Letters **92**(25): 257402.

Saito, R., G. Dresselhaus, et al. (1998). Physical Properties of Carbon Nanotubes. Singapore, World Scientific Publishing Co. Pte. Ltd.

Saito, R., G. Dresselhaus, et al. (2000). "Trigonal warping effect of carbon nanotubes." Physical Review B **61**(4): 2981-2990.

Spataru, C. D., S. Ismail-Beigi, et al. (2005). "Theory and ab initio calculation of radiative lifetime of excitons in semiconducting carbon nanotubes." Physical Review Letters **95**(24): 247402.

Wang, F., G. Dukovic, et al. (2004). "Time-resolved fluorescence of carbon nanotubes and its implication for radiative lifetimes." Physical Review Letters **92**(17): 177401.

Wang, F., G. Dukovic, et al. (2005). "The optical resonances in carbon nanotubes arise from excitons." Science **308**(5723): 838-841.

Xie, L. M., C. Liu, et al. (2007). "Photoluminescence recovery from single-walled carbon nanotubes on substrates." Journal of the American Chemical Society **129**(41): 12382.

7

Charge Transport in Carbon Nanotube Films and Fibers

Vitaly Ksenevich[1], Jean Galibert[2] and Vladimir Samuilov[1,3]
[1]*Belarus State University, Minsk*
Belarus
[2]*LNCMI & Universite de Toulouse*
France
[3]*State University of New York at Stony Brook*
USA

1. Introduction

Electrical properties and magnetotransport in carbon nanotubes (CNTs) have attracted much attention due to their importance in verification of existing theories of modern condensed matter physics and a number of possible applications (Robertson, 2007; Dai, 2002). Single-wall carbon nanotube (SWCNT) is a graphene sheet rolled up into a hollow cylinder and show metallic or semiconducting properties dependently upon their diameter and chirality. Due to their unique structure SWCNTs allows to study a large variety of different quantum phenomena like single-electron tunneling (Bockrath et al., 1997), Luttinger liquid behaviour (Bockrath et al., 1999), ballistic transport (Krstic et al., 2000), *etc.* Multi-wall carbon nanotubes (MWCNTs) are more complicated systems. They consist of a several shells of different diameter and chirality. Due to weak coupling between the shells the conductivity in bulk-contacted MWCNTs is defined mostly by the outermost shells. Diffusive transport in majority of experiments for individual MWCNTs was observed. Therefore quantum interference effects inherent for mesoscopic systems (weak localization and universal conductance fluctuation) were reported (Schonenberger et al., 1999). Besides that, MWCNTs with large diameter allows to observe Aahronov-Bohm effect at experimentally available values of magnetic fields (Bachtold et al., 1999; Lassagne et al., 2007). Ballistic transport even at room temperatures was observed by some authors (Frank et al., 1998; Urbina et al., 2003). Possibility to switch between ballistic and diffusive transport regime in the same MWCNTs sample using gate voltage by virtue of an electrostatic change of electron density was reported as well (Strunk et al., 2006; Nanot et al., 2009).

Processing of nanotubes on macroscopic scale and investigation of their synergetic properties is a most important task for realistic application of these materials, especially for fabrication of carbon nanotubes-based gas-, bio- and chemical sensors where signal from the sensor output depends on the conductivity of device (Stetter & Maclay, 2004). Different examples of morphology of the samples of arrays of nanotubes involve definitions of bundles (ropes) (Fischer et al., 1997; Krstic et al., 2000), mats (Fischer et al., 1997; Kaiser et al.,

1998), networks (Kim et al., 2001) and films (Baumgartner et al., 1997). The intertubes barriers and defects, length of the individual nanotubes in the assemblies as well as contact geometry play an essential role in the electrical transport properties of the carbon nanotube arrays. Therefore different charge transport mechanisms can be observed in the arrays of nanotubes: metallic conductivity, variable range hopping (VRH), weak localization (WL), and fluctuation induced tunneling. Combination of various mechanisms is possible as well. Verification of charge transport mechanisms (as well as the ranges of the temperature, electric and magnetic fields etc.) in which these mechanisms play dominating role is very important and non-trivial task for such systems.

In this chapter we show some examples of different CNTs assemblies in which contribution from different charge transport mechanisms can determine electrical properties of the systems. The chapter organized as follows. We briefly describe several methods of preparation of various types of CNTs assemblies (monolayers of MWCNTs and SWCNTs, SWCNTs fibers and SWCNTs coatings of silica fibers) and present experimental results on investigation of their transport properties. Besides traditional experimental technique used for investigations of charge transport (measurements of current-voltage *(I-V)* characteristics and resistance as a function of temperature and magnetic field), effect of the influence of strong microwave field and terahertz radiation on the conductivity of SWCNTs fibers and SWCNTs coatings of silica fibers was studied as well. Mechanisms responsible for the charge transport in CNTs assemblies were identified. Sensitivity of the electrical properties of carbon nanotubes arrays to the types of pristine carbon nanotubes, contacts geometry and to the preparation methods is considered. Special emphasize on the role of quality of intertube contacts in CNTs assemblies is done as a result of experimental data analysis.

2. Methods of preparation of carbon nanotubes arrays

2.1 Assembly of carbon nanotubes monolayers using Langmuir-Blodgett technique

Thin film geometry of CNT arrays is a most promising for their possible commercial applications like antistatic shielding and transparent conducting layers, if the layers are thin enough (Robertson, 2004), interconnections and heat sinks in integrated circuits, nano- and optoelectronic devices. However, assembly of CNTs into thin films or monolayers is hindered by processing and manipulations difficulties. The Langmuir-Blodgett (LB) technique can be used as a method for depositing of monolayers of molecularly ordered ultra-thin films with controlled thickness. However, due to absence of free bonds on the surface of nanotubes, they possess a high chemical stability and are insoluble in the organic solvents. Therefore, to find a possibility of arrangement of CNTs into arrays by means of LB technique, a chemical modification (functionalization) of their surfaces is necessary. Different approaches for CNTs functionalization were proposed, including covalent sidewall functionalization, noncovalent functionalization with surfactants and polymers, defect group functionalization and plasma modification of nanotubes sidewall (Bahr & Tour, 2002; Hirsch, 2002). One of the main advantage of the organic covalent functionalization of CNTs for their further assembly in thin films using LB technique is a possibility to obtain large surfaces covered by CNTs without free space. In contrast, quite low concentration of CNTs in layers was achieved when noncovalent functionalization was used (Krstic et al., 1998). Besides that, applications for chemical and biosensors may require the presence of different functional groups bounded to CNTs.

We used prior organic functionalization of MWCNTs based on the scheme described in (Georgakilas et al., 2002). The MCWNTs produced by chemical vapor deposition (CVD) method with catalyst 5%Fe,Co/CaCO3 with average diameter of about 30 nm and length less than 1 μm were used. The purified MWCNTs were suspended in DMF [N,N-Dimethylformamide HCON(CH3)2] together with an excess of *p*-Anisaldehide (4-methoxybenzaldehyde) CH3OC6H4CHO and 3-methylhippuric acid [m-toluric acid, N-(3-methyl-benzoyl)glycine] CH3C6H4CONHCH2CO2H. The heterogeneous reaction mixture was heated at 130 °C for approx. 120 hours. After the reaction was stopped, the organic phase was separated from unreached material by centrifugation and washing five times with chloroform ($CHCl_3$) and vacuum drying. The obtained dark solid phase was easily soluble in $CHCl_3$ up to a few mg/mL without sonication. The deposition of the layers (arrays) of nanotubes on the surface of the devices with the electrodes was done by using the cell, imitating the LB trough. Once a droplet of the solution of functionalized nanotubes in chloroform was spread on the water surface, a droplet of diblock-copolymer PS-PMMA solution was added in order to create a surface pressure of ~ 9 mN/m. The functionalized nanotubes were self-assembled in a dense arrays (monolayers) covering without free space the whole surfaces of the devices, when the monolayer was picked up from the water surface. The same procedure was done with grown by CVD method SWCNTs nanotubes. Scanning electron microscopy (SEM) image of dense arrays of SWCNTs is shown in Fig. 1.

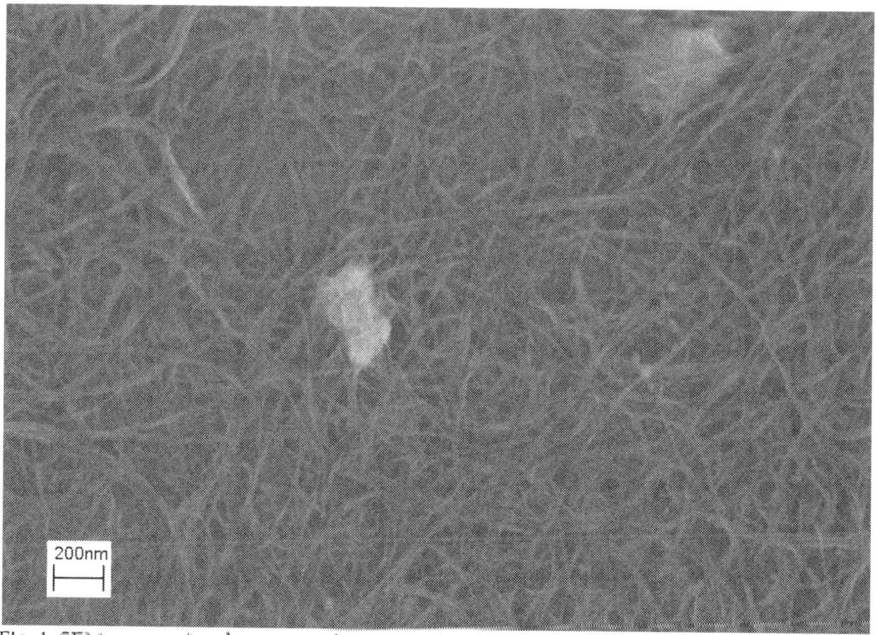

Fig. 1. SEM image of carbon nanotube arrays obtained by LB technique

Substrates with finger-shaped contact geometry with 10 μm distance between contacts were used for investigation of the electrical transport both for MWCNTs and SWCNTs arrays.

2.2 Preparation of SWCNTs fibers using wet-spinning process

Due to the high aspect ratio of the nanotubes and intertube coupling in conductivity of the arrays of carbon nanotubes, enhanced electrical transport properties are expected from the more aligned material. Therefore, different techniques for the alignment of nanotubes in the arrays are proposed: orientation of nanotubes in the films by using high magnetic fields (Smith et al., 2000), fabrication of fibers of aligned nanotubes by using an electrophoretic process (Gommans et al., 2000), synthesis of the fibers with the preferential nanotubes orientation by means of wet spinning process (Vigolo et al., 2000; Vigolo et al., 2002). The last technique can be potentially used for the large scale material production due to their particular simplicity and similarity to the industrial process of polymer fibers fabrication. We used SWCNTs fibers produced by polymer-free spinning method. The principle of the spinning method for the polymer/nanotubes composite fibers was reported in (Vigolo et al., 2000). The presence of non-conductive polymer poly vinyl alcohol (PVA) makes problematic the most applications of these fibers. Polymer-free SWCNTs fibers were produced in the process developed at Rice University (Davis et al., 2004; Zhou et al., 2004). The SWCNTs were dispersed in 102% sulfuric acid (2 wt. % excess SO_3) and then wet-spun into either diethyl ether 5% sulfuric acid, or water. The SWCNT fibers with high electrical and thermal conductivity were obtained using this technique. However, some protonation of material occurred due to long contact of fibers with sulfuric acid. Polymer-free SWCNTs fibers used in our experiments were synthesized without superacids from solutions comprising nanotubes, surfactant and water (Kozlov et al., 2005). SEM image of SWCNTs fiber produced by this method is shown in Fig. 2.

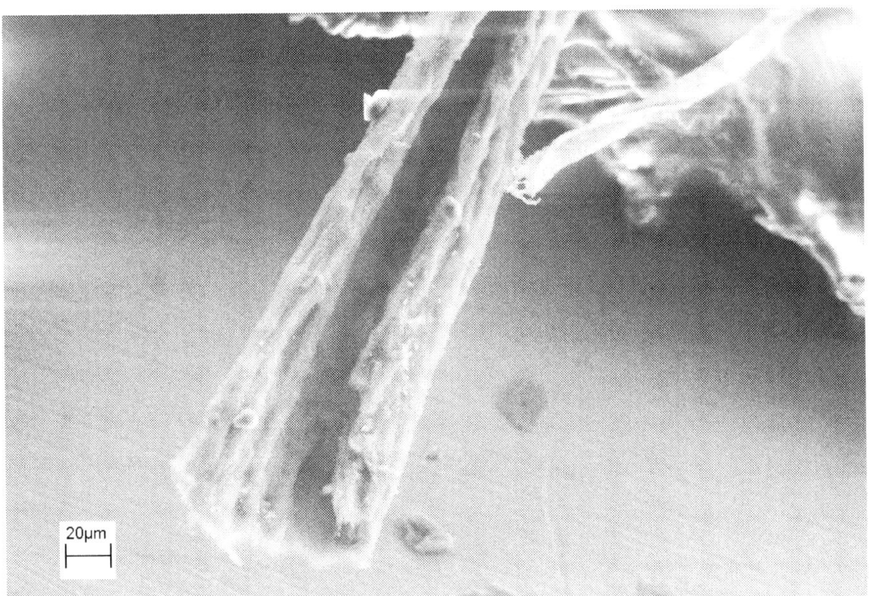

Fig. 2. SEM image of the hollow SWCNTs fiber produced by wet-spinning process

The nanotubes were dispersed using a horn sonicator in an aqueous solution of 1.2 wt.-% lithium-dodecyl-sulfate (LDS) surfactant. A narrow jet of this spinning solution was injected into the flocculation bath containing 37 % hydrochloric acid, which rotated at 33 rpm. The configuration of the rotating bath and spinneret needle was similar to that described for PVA-based coagulation spinning (Vigolo et al., 2000). Flocculation of nanotubes in the spinning solution to form a gel fiber occurred very close to the point of contact of the spinning solution and the acid in the bath. This gel fiber (containing 90 wt % volatilizable liquid, based on gravimetric measurements) was washed in methanol to remove the hydrochloric acid. The fiber, which has very low elasticity both before and after the washing step, was then pulled from the wash bath, stretched over a frame, and dried under tension. The fibers obtained by this method are of 20-100 µm diameter. As was mentioned above, one of the main advantages of this method for fabrication of the SWCNTs macroscopic arrays is the existence of the preferential orientation of individual nanotubes in the fibers due to flow-induced alignment of the nanotubes during processing (Vigolo et al., 2000). This alignment can be seen from the scanning electron microscopy (SEM) image of SWCNTs fiber shown in the Fig. 2.

SWCNTs with different geometry (solid, hollow and ribbon fibers) can be obtained by polymer-free wet spinning process in dependence of injection rate into the rotating bath of solution containing SWCNTs and inner spinneret diameter (Kozlov et al., 2005). HiPco SWCNTs obtained from a carbon monoxide process and SWCNTs prepared by laser ablation purchased from *Carbon Nanotechnologies Inc.* were used as pristine nanotubes for the SWCNTs fibers preparation. Electrical contacts for further measurements of charge transport in the fibers were made by Ag paint.

2.3 Fabrication of SWCNTs coatings of Silica fibers

Due to the high mechanical strength carbon nanotubes are widely used for reinforcement of different types composites materials, replacing conventional reinforcing fillers such as carbon or glass fibers (Shaffer & Sandler, 2007). Manufacturing of hierarchical composites with carbon nanotubes combined with conventional microscopic fibers has attracted growing interest years recently. Carbon nanotubes coatings on the surface of ceramic (Ci et al., 2005), carbon (Thostenson et al., 2002), quartz and aliminium silicate (Zhang et al., 2008) fibers were synthesized. Multi-wall carbon nanotubes coatings on the surface of silica fibers were obtained by chemical bonding (Liu et al., 2009) and using injection CVD method (Qian et al., 2009).

Procedure of preparation of SWCNTs coatings of silica fibers used in our investigations included the following stages: i) dissolving of polyimide claddings of silica fibers by soaking them in toluene for 2 hours, ii) cleaning of the residues of claddings by sonication in saturated KOH/propanol solution for 20 minutes with following sonication during 5 minutes in distilled water (this step was repeated several times), iii) silanization of silica fibers (creation of abundant carbon long alkyl chains on the silica fibers surfaces) in octadecyltrimethoxysilane (OTMS) solution in ethanol at a concentration of 2 mL/g for 2 hours. After that, the silanized silica fibers were soaked in water in which pristine SWCNTs were dispersed. When the silica fiber turns black due to a good coverage of SWCNTs as a result of clinging presumably driven by hydrophobic interactions, they were heated in an oven at 250°C to anneal the bound SWCNTs. SEM image of SWCNTs coating on the surface of silica fiber obtained as a result of the above described procedure is shown in Fig. 3.

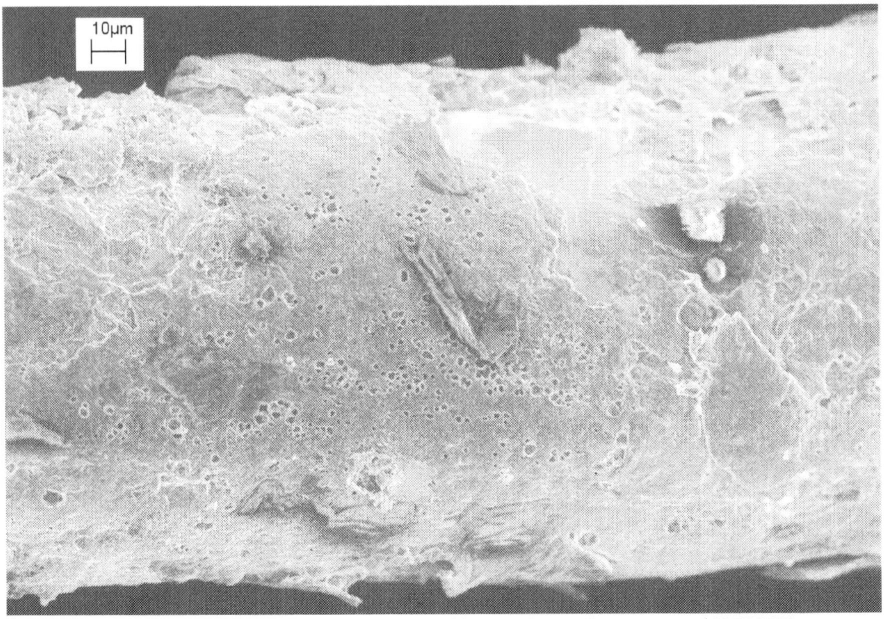

Fig. 3. SEM image of the side view of the silica fiber with good coverage of SWCNTs

For the further electrical and magnetotransport measurements silica fibers with SWCNTs coatings were transferred on solid substrates and electrical contacts were made by Ag paint.

3. Charge transport properties of carbon nanotubes arrays

3.1 Electrical and magnetotransport properties of carbon nanotubes monolayers

Measurements of the electrical and magnetotransport properties is one of the most reliable tool for exploring of different materials and verification of the mechanisms responsible for the charge transport in the system, (for a general revue of electronic and transport properties of carbon nanotubes, see (Charlier et al. 2007)). We used standard four-probe dc- and lock-in technique for the measurements of transport properties of carbon nanotubes assemblies in the temperature range 2-300 K. Magnetoresistance (MR) measurements were carried out in pulsed magnetic fields at the *Laboratoire National des Champs Magnētiques Intenses de Toulouse* (LNCMI) in the temperature range 2-300 K and up to 40 T magnitude fields.

We report and compare here the transport properties of different types of carbon nanotubes arrays. The temperature dependence of the resistance of the layers of short MWCNTs shows the negative temperature coefficient of the resistance ($dR/dT<0$) in the whole investigated temperature range (4.2-300 K) (Ksenevich et al., 2008 a) as one can see in the Fig. 4. Assuming possibility of the VRH conduction in our system due to structural defects and impurities, we fitted $R(T)$ dependence by classical law for variable range hopping (VRH) (Shklovskii & Efros, 1984):

$$R = R_0 \exp(T_0/T)^{1/n} \quad (1)$$

where $T_0=\beta/k_B\xi^3 g(\mu)$, β is a constant, k_B is a Boltzmann constant, ξ denotes the localization length, and $g(\mu)$ is the density of states, $n=d+1=2,3,4$ (d is the dimensionality of the system). In order to find parameter n we used both linearization in the scale $logR$-$T^{-1/n}$ and method proposed by Zabrodskii (Zabrodskii, 1977). We found that $R(T)$ dependence can be approximated by Eq. (1) in the temperature range T=4.2~60 K and the best fitting results were obtained at T below 60 K with parameter n=3, indicating 2D VRH.

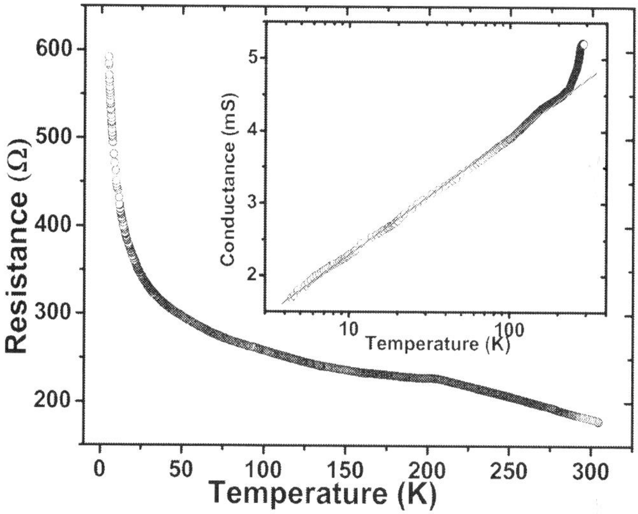

Fig. 4. The temperature dependence of the resistance of the MWCNTs arrays. The inset shows the $R(T)$ dependence in G-$LogT$ scale.

In order to see suitability of the WL theory for our system, we tried to fit $R(T)$ dependence with one-, two- and three-dimensional WL formulas. We found that $R(T)$ dependence follows the 2D weak localization (WL) behavior ($G(T)$~lnT) (Lee & Ramakrishnan, 1985) in the temperature range 4.2~202 K as one can see from the inset to Fig. 4 In the temperature range 202 - 300 K conductivity increases approximately linearly with temperature.

MR data measured at various temperatures are plotted in Fig. 5. The negative MR in the low magnetic fields range was observed. In the low-temperature range upturn of negative MR was observed. The minimum position of the negative MR shifts to the higher fields as the temperature rises. Both low-field negative MR due to changing of phase between alternate hopping paths enclosing a magnetic flux (Nguen et al., 1985; Sivan et al., 1988) and high-field positive MR due to electronic orbit shrinkage (Shklovskii & Efros, 1984) are predicted for systems with hopping conductivity mechanism. However, monotonic decrease of relative value of negative MR with the temperature rising are usually observed. In our samples the opposite situation is clearly seen: relative value of negative MR increased with the temperature rising in the low temperature range where VRH can be responsible for the charge transport mechanism. From the other side negative MR is inherent for the systems where conductivity can be described in the frame of WL theory (Lee & Ramakrishnan, 1985).

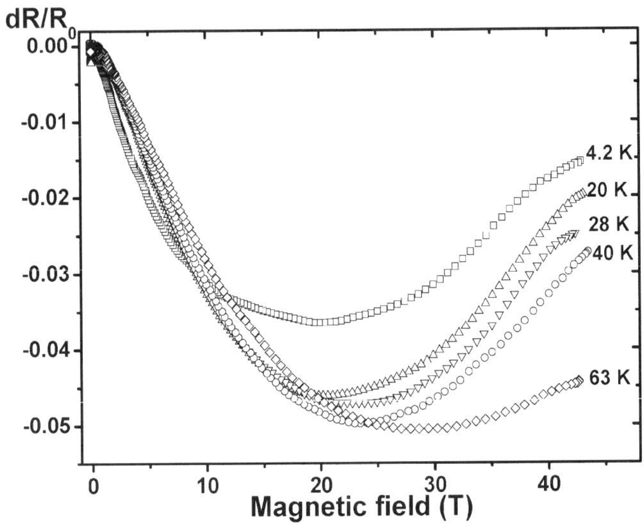

Fig. 5. The dependence of the normalized change in the resistance *dR/R(B=0)* of the MWCNTs arrays on the magnetic field in the temperature range 4.2 - 63 K.

Therefore we made assumption that MR data are the sum of the positive and negative contribution due to MR effects in the VRH and WL regimes, respectively. We found that high-field positive part of magnetoristance can be approximated in frame of Kamimura model for spin-dependent VRH conductivity (Kurobe & Kamimura, 1982):

$$\frac{\Delta G}{G} = -A_{KK}\frac{H^2}{H_{KK}^2 - H^2} \tag{2}$$

where $G=1/R$ is the conductance of the system, H_{KK} is the characteristic field for spin alignment, A_{KK} the saturation value of the magnetoconductance. Using formula (2) and values of parameters H_{KK} and A_{KK} obtained from the approximation data of high field MR data, we calculated positive magnetoresistance for low-field region. Pure negative contribution to MR according our assumption was calculated by subtracting positive MR from the experimental data. These data can be fitted reasonably well by the Eq. (3) for 2D WL:

$$\Delta G = \frac{e^2}{\pi h}[\psi(\frac{1}{2} + x) + \ln(x)] \tag{3}$$

where ψ is the digamma function, $x=L_{Th}^2\, 8\pi EH/hc$, $L_{Th} \sim T^{-p/2}$ is the Thouless length.

The example of the above mentioned procedure for extracting of positive and negative part from the total MR curve for temperature T = 40 K is shown in the Fig. 6.

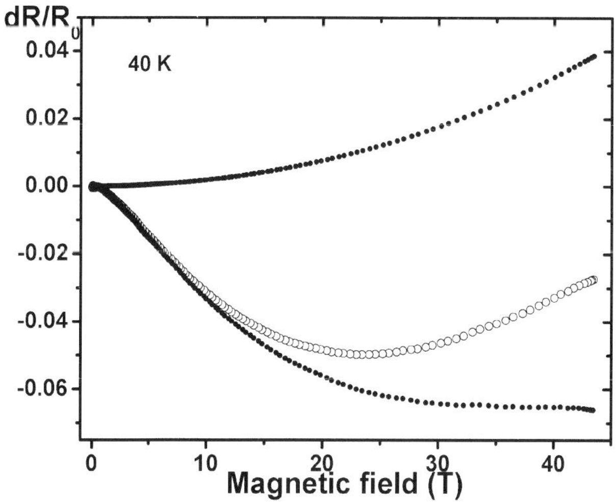

Fig. 6. The dependence of the normalized change in the resistance $dR/R(B=0)$ of the MWCNTs arrays on the magnetic field measured at 40 K (open circle). Calculated value of positive MR obtained by fitting positive part of MR data to Eq. (2) (upper curve). The negative MR (lower curve) obtained by subtracting the calculated positive MR from the experimental MR data.

With this assumption we can explain positive upturn observed on MR curves by adding curves for positive and negative MR. Positive MR may come from the spin-dependent VRH among the defects on the surface of nanotubes. The negative contribution to MR can originate from the WL effects. It should be noted, that similar magnetotransport properties were observed for SWCNTs entangled networks (Kim et al., 1998). Therefore we can assume that similar MR effects in different type of carbon nanotubes arrays are defined not only by the type of pristine nanotubes, but their length and contact geometry as well. For both types of the samples the distance between contacts was much higher than the length of the separate nanotubes in the arrays, thus enhancing the probability to observe the localization effects.

Investigation of transport in SWCNTs monolayers confirms the importance of intertube contacts between separate nanotubes in their assemblies (Ksenevich et al., 2006). The temperature dependence of the resistance $R(T)$ of the SWCNTs arrays is shown in Fig. 7.

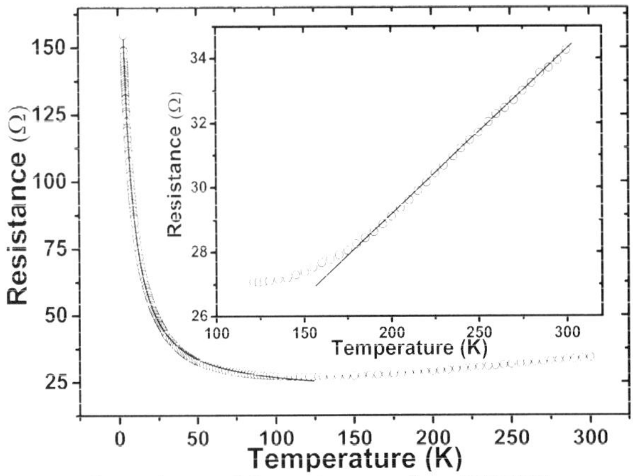

Fig. 7. The temperature dependence of the resistance of the SWCNTs arrays. Solid line is the fitting of the *R(T)* dependence by means of fluctuation-induced tunneling model (second term of Eq. (4)) in the low temperature range. The inset shows the high temperature metallic term of the *R(T)* dependence.

The resistivity minimum on the temperature dependence of the resistance of SWCNTs arrays was observed at T~125K. At T>125 K *R(T)* dependence has a metallic-type behavior ($dR/dT>0$) while in the low temperature range non-metallic sign for *R(T)* dependence ($dR/dT<0$) was observed. In the low-temperature range (T ~ 4.2 - 100 K) the temperature dependence of the resistance is consistent with the exponential law proposed for the fluctuation-induced tunneling model (Sheng, 1980). Thus, temperature dependence of the resistance can be written as:

$$R = \alpha T + R_t \exp(T_c/T+T_s), \qquad (4)$$

where α is the temperature coefficient arising from metallic conductivity, and the second term represents the conductivity through the barriers between the metallic regions in frame of the fluctuation-induced tunnelling model. Such type of the temperature dependence of the resistance with systematic metallic sign and crossover to the non-metallic sign of the *R(T)* curves at lower temperatures was firstly observed by Fischer (Fischer et al., 1997) for SWCNTs ropes and mats. Heterogeneous model for conduction was proposed to explain crossover from metallic to non-metallic sign of the temperature dependence of the resistance as the temperature decreases (Kaiser et al., 1998). This model is based on the simple assumption that conduction process at high temperatures is defined by nanotubes with good metallic conductivity. The presence of small electrical barriers (tangled regions, intertube contacts, structural defects insight nanotubes itself) give rise to low temperatures localization effects. Therefore, the conductivity in the low temperature range is determined by tunneling through these electrical barriers. In our samples in the intermediate temperature range (T ~ 100 - 185 K) the contribution to the total conductivity of the system from the metallic conductivity and fluctuation-induced tunneling are comparable in the absolute value. This assumption explains the behavior of the *R(T)* dependence. Linear rise of the resistance with temperature is clearly seen in the temperature range $T \sim 185 - 300$ K as one can see from the inset to Fig. 7. From the other side, due to a weak temperature dependence of the metallic term and its saturation at low temperatures of the temperature dependence the resultant resistance can be well fitted by fluctuation-induced tunneling law (second term of Eq. (4)).

It should be noted that the crossover temperature T between the metallic and nonmetallic type of $R(T)$ curve depends strongly on the type of SWCNTs arrays and quality of intertube contacts between separate nanotubes and varies from 35 K for a single well-ordered rope to 250 K for a rope with tangled regions (Fischer et al., 1997). Moreover, for SWCNTs with tangled regions the nonmetallic term can be described by proposed for variable range hopping conductivity Mott's law (1), indicating localization of charge carriers and hopping transport instead of fluctuation-induced tunneling (Kaiser et al., 1999). We assume that in our samples the localization effects and the resulting hopping transport can be essential only at very low temperatures where the rapid rise of the resistance is observed as the temperature decrease. This conclusion is confirmed by the shape of the MR curves, shown in Fig. 8.

Positive magnetoresistance with decreasing amplitude inherent for the variable range hopping conductivity was observed while temperature rising in the low temperature range. At temperatures higher than 7 K the MR effect is negligible. Because localization near defects is a possible cause of the barriers for charge carriers in heterogeneous conductivity model, magnetoresistance can exhibit localization effects at low temperatures, inducing positive MR. However absolute value of the positive magnetoresistance can be lower in comparison with uniformly disordered systems as far as the conductivity in our heterogeneous system is enhanced by the well conductive metallic regions (metallic carbon nanotubes).

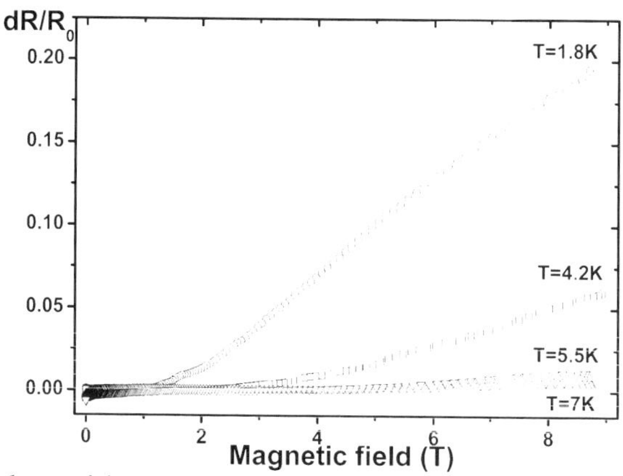

Fig. 8. The dependence of the normalized change in the resistance $dR/R(B=0)$ of the SWCNTs monolayers on the magnetic field in the temperature range 1.8-7 K.

3.2 Charge transport in SWCNTs fibers

Electrical and thermal transport properties of the SWCNT fibers fabricated by means of wet spinning process in correlation with the orientation degree of nanotubes inside fibers are already reported (Zhou et al., 2004; Badaire et al., 2004). For the optimizing of conducting properties of SWCNT fibers for possible applications deeper understanding of the charge transport mechanism in this material is required.

We applied different experimental techniques in order to study transport properties of SWCNTs fibers and to distinguish mechanisms responsible for the charge carrier's transfer in the system (Ksenevich et al., 2008 b; Ksenevich et al., 2008 c). We have measured resistance as a function of temperature and magnetic field, current-voltage (*I-V*) characteristics, and investigated effect of the influence of the strong microwave field and terahertz radiation on the conductivity of SWCNT fibers. The *R*(*T*) dependence of the SWCNT fibers is shown in Fig. 9. In the low-temperature range (4.2 - 80 K) the resistance as a function of temperature exhibits a classical VRH conduction, following the law (1) with exponent $n=4$, assuming 3D character of hopping conductivity. In the temperature range T = 80 - 300 K *R*(*T*) dependence can be approximated by a typical law for fluctuation-induced tunneling conductivity mechanism (Sheng, 1980):

$$R = R_t \exp(T_c/T+T_s), \tag{5}$$

where parameters T_c and T_s are defined by the following equations:

$$T_s = 16\varepsilon_0 \hbar A V_0^{3/2}/(\pi e^2 k_B (2m_e)^{1/2} w^2, \tag{6}$$

$$T_c = 8\varepsilon_0 A V_0^2/(e^2 k_B w), \tag{7}$$

where w is the width of the tunnel junction, A labels its area, and V_0 is the height of the contact potential; k_B stands for the Boltzmann constant, $\hbar$ is a Plank constant, e is the electronic charge, m_e the electron's effective mass, and ε_0 the vacuum permittivity.

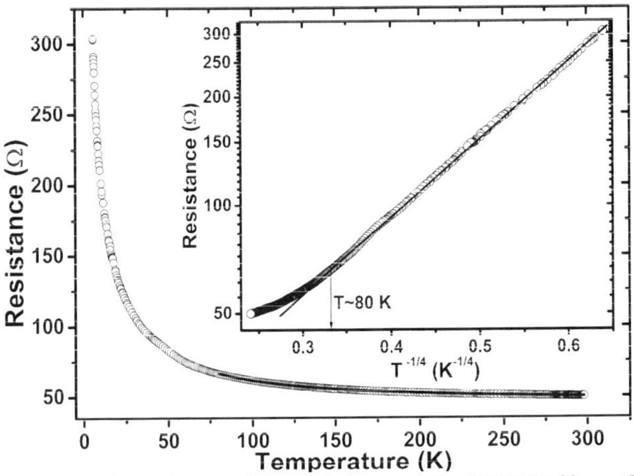

Fig. 9. The temperature dependence of the resistance of the SWCNT fibers. Solid line is the fitting of the *R(T)* dependence by means of fluctuation-induced tunneling model in the temperature range 80 - 300 K. The inset shows the logR-$T^{-1/4}$ plot. (According to ref. (Ksenevich et al., 2008 b)).

Fluctuation-induced tunneling conductivity model was proposed for disordered heterogeneous systems contrary to the systems with hopping charge carrier transport between localized sites. These systems (for example, conductor-insulator composites, granular metals, disordered semiconductors) consist of large conductive segments (where electrons are delocalized and free to move over very large distances as compared to the atomic dimensions) separated by small insulating gaps. Due to small sizes of tunnel junction, thermal voltage fluctuation resulting from the thermal motion of electrons in the conduction region in the vicinity of tunnel barrier can change drastically the electron tunneling probability through the barrier. Different types of carbon nanotubes arrays with existing electrical barriers can be also considered as heterogeneously disordered systems. Therefore this model was used for describing the temperature dependence of conductivity of SWCNTs fibers (Zhou et al., 2004) and networks (Kim et al., 2001). As was mentioned earlier we observed fluctuation induced tunneling conductivity in the low temperature range for SWCNTs layers obtained by LB technique. As one can see from Fig. 9 for our SWCNTs fibers the fitting results also agree well with experimental data thus indicating the importance of intertube contacts between individual carbon nanotubes. Assuming the area A of the tunnel junction being equal to the cross section of a SWCNT with the diameter of about 1.5 nm and using expressions for parameters T_c and T_s, the effective junction width is estimated to be 4.6 nm and the effective barrier height to be 0.172 eV.

I-V curves of SWCNTs fibers measured in the temperature range 6 - 300 K are plotted in Fig. 10. At room temperature, linear *I-V* curves were observed. As the temperature decreases, the non-Ohmic behavior becomes pronounced. Nonlinearity of *I-V* curves is clearly seen at temperatures about 100 K and becomes more significant in the low temperature range. Reproducibility of the *I-V* curves by number of thermal cycling and different sweeping times indicate that the heating effect is negligible. Nonlinear *I-V* characteristics for both charge transport mechanisms (VRH and fluctuation induced tunneling) can be observed.

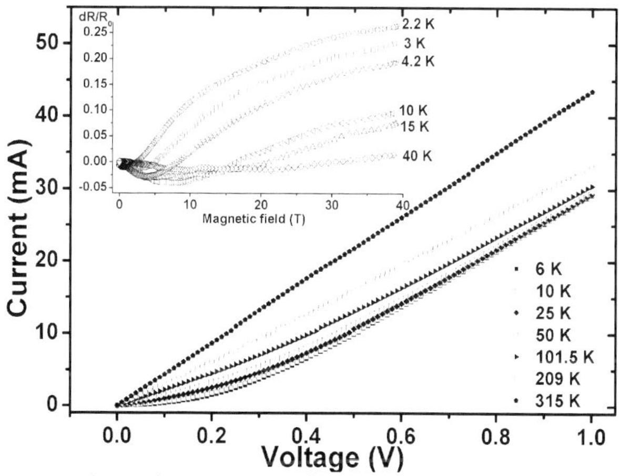

Fig. 10. The current-voltage characteristics of the SWCNT fibers at different temperatures. The inset figure: the dependence of the normalized change in the resistance $dR/R(B=0)$ of the SWCNTs fibers on the magnetic field in the temperature range 2.2-40 K (According to ref. (Ksenevich et al., 2008 b)).

As far as $R(T)$ dependencies in the low temperature range were approximated by Mott's law (1) for VRH conductivity, we assumed that nonlinearity of the *I-V* curves can be explained within the framework of classical theoretical models for hopping conductivity (Pollak & Riess, 1976; Grannan et al., 1992). According to these models, in the low electric fields, the conductivity is expected to be nearly independent of the applied electric field. In this regime hopping of the charge carriers between hopping sites is induced by phonons. In the high electric fields, both temperature and electric fields (with energy scales k_BT and eEl, respectively, where E is the electric field and l is the length parameter related to maximum hopping length between the hopping sites) have similar activation effects on the charge carriers. Low fields are defined as fields $E<Ec$, where $Ec=k_BT/el$. In the region of high electric fields $Ec<E<k_BT/e\xi$ (where ξ is a localization length), non-Ohmic VRH conductivity follows the formula (Pollak & Riess, 1976; Grannan et al., 1992):

$$\sigma(E,T) = \sigma(0,T)\exp(eUl/k_BTd), \tag{8}$$

where U is the voltage measured between voltage probes and d is the distance between them. However, we found rather large discrepancies between experimental data and fitting curves in the region of very high electric field ($E>k_BT/e\xi$) where theoretical models for VRH predict "activationless" conductivity. In this regime charge carriers acquire the energy necessary for hopping between localized sites not from the phonons but from the electric field, and the conductivity becomes temperature independent: $\sigma(E) \sim \exp(-E_0/E)^m$, where the exponent m is usually equal to exponent in formula (1) for the Ohmic regime of hopping conductivity as was found in most experiments. Therefore in order to check the possibility to explain the origin of nonlinearity of *I-V* curves in the frame of the VRH models, we estimated the fitting parameter $(k_Bd/el)T$ at different temperatures from the slope of $ln(U)$ dependencies. From the slope $log(k_Bd/el)T$ versus $logT$ we obtained the temperature dependence of the parameter l (maximum hopping length), $l \sim T^{0.13}$. This means that hopping length slightly increases with the temperature, and this fact contradicts to the classical VRH models. As it was mentioned above, in the low temperature range, the $R(T)$ dependencies of the SWCNT fibers measured in the linear range of *I-V* curves were fitted well by Mott's law (1). According to the Mott's model for 3D VRH, the maximum hopping length can be described as $l \sim T^{-0.25}$. Therefore, the nonlinearity of *I-V* curves cannot be explained by the influence of the high electric fields on the hopping conductivity. We assume that at $T<80$ K, where $R(T)$ dependencies are well fitted by Mott law (1), the VRH conductivity can be considered as the charge transport mechanism in the SWCNT fibers only for low electric fields $E<Ec$. We found that best fitting results of the nonlinear *IV* curves in the whole range of applied electric fields can be obtained by using the extended Sheng's model (Sheng, 1980) proposed by Kaiser and Park (Kaiser & Park, 2005):

$$G = I/V = G_0\exp(V/V_0)/(1+h[\exp(V/V_0 -1)]) \tag{9}$$

where G_0 is the temperature dependent low-field conductance, parameter V_0 is the voltage scale factor that gives an exponential increase in conductance and depends strongly on the barrier energy, and parameter $h=G_0/G_h$, where G_h reflects the large conductance in the absence of barriers. Eq. (9) was obtained by means of numerical calculations of fluctuation induced tunneling and thermal activation (Kaiser & Park, 2005).

We carried out also measurements of the dependence of the resistance on the magnetic field in order to clarify the charge transport picture in SWCNT fibers. The general features of MR at different temperatures behavior can be seen from the insert to Fig. 10.
At low fields, MR is negative. A positive upturn is observed on the MR curves, of which position is shifted to higher magnetic field as the temperature increases. The upturn field of the MR effect was shifted from 1.5 T at 2 K to value of about 20 T at 40 K. The value of positive MR varies as $exp(B^2)$ which changes to $exp(B^{1/3})$ behavior at sufficiently high fields as expected for the VRH transport (Shklovskii & Efros, 1984). The tendency to saturation of the positive MR was observed at low temperatures as one can see from the inset to Fig. 10. However, the negative MR can not be explained in frame of Sivan model (Sivan et al., 1980) proposed for VRH charge transport. According to this model the negative MR should saturate when one flux quantum $\Phi_0=hc/e$ penetrates an area equal $R_M^{3/2}\chi^{1/2}$ (here R_M - the average hopping length and χ - the average distance between the localization centers). The position of the negative MR minimum shifts to the higher fields with increasing temperature due the hopping length decrease and as a consequence, an area bounded by different paths decreases. In addition, the monotonic decrease of the relative value of negative MR with the temperature is usually observed. We found the nonmonotonic behavior of the relative value of the negative MR with the temperature. At T=2 K the relative value of the negative MR was about 0.4 %. The maximum of the relative negative MR value about 4.2 % was observed at T=10 K. At the further temperature increasing the negative MR decreases. Therefore, the contribution of the other mechanism must be considered in order to explain the negative MR behavior. One of the possible explanations of the negative MR can be a contribution of the weak localization effects in total MR (Lee & Ramakrishnan, 1985).
The determined carrier transport features were supported by additional measurements of the change in conductivity in strong 10 GHz microwave fields and measurements of THz radiation induced photocurrent at various lattice temperatures.
Relaxation kinetics of the THz induced photovoltage at different temperatures are shown in Fig. 11. In order to diminish Joule heating of the samples due to absorption of laser radiation pulsed THz radiation with low repetition rate (30 Hz) and duration of pulses of 100 μs (the shortest possible pulse available in this kind of optically-pumped molecular terahertz laser) was used. Two relaxation processes are visible: relatively fast part (with time constant of some hundreds of μs which corresponds to the duration of THz pulses) and relatively slow part (in the range of about ~20 ms) that we associate with the Joule heating. The fast peak we associate with the influence of terahertz radiation on the hopping conductivity. We assume this effect as *terahertz - induced hopping conductivity* (Ksenevich et al., 2008 b), taking into account the following arguments. In the low temperature range where VRH conduction is observed, the conductivity increases with temperature because of activation of charge carriers by thermal energy k_BT on hopping sites located at higher energy levels. Since the terahertz radiation energy is in the order of magnitude of thermal energy (1 THz ~ 4 meV ~ 48 K), that terahertz quanta exceeding the thermal energy k_BT should give rise to additional increase of conductivity. On the other hand, within the temperature rise fluctuation induced tunneling mechanism starts to prevail. As it is evident from the experiments, it becomes negligible at T~30 K. Only slow Thz - induced signal due to heating is observed. We assume that contributions from fluctuation induced tunneling and VRH at this temperature are comparable on the absolute value of magnitude.

The dependencies of the variation of the conductivity of SWCNT fibers on applied microwave field power at the temperatures 77 and 300 K measured using technique described in (Pozela, 1985) are shown in the inset to Fig. 11.

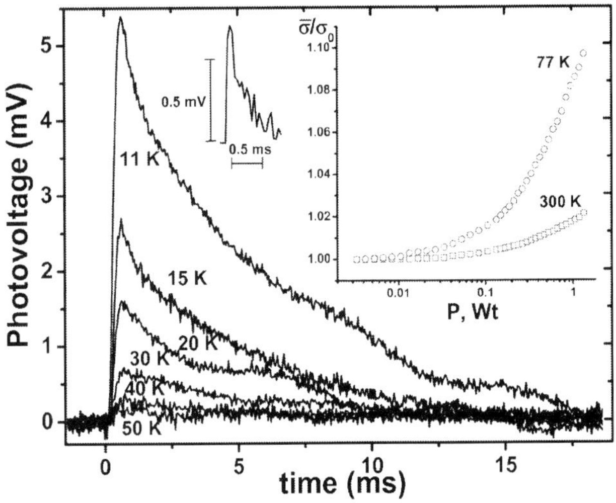

Fig. 11. Relaxation kinetics of THz-induced photovoltage signal at different temperatures. Left inset: terahertz induced photocurrent transient excluding lattice heating. In the right inset the dependence of the conductivity variation on applied microwave power is shown. (According to ref. (Ksenevich et al., 2008 b)).

The increase of the conductivity on microwave power becomes more essential with the temperature decrease as one can see from the right inset to Fig. 11. At the same microwave field power of about 1 W the observed value of relative conductivity increase was about 2 % and 10 % at the temperatures 300 K and 77 K, respectively. It should be noted that the heating of the charge carriers by strong microwave field can give rise to conductivity both in VRH and fluctuation-induced tunneling regimes due to a possible rising of the probability of charge carriers for the hopping between the localized sites and for the tunneling through the barriers, respectively.

3.3 Conductivity mechanisms in SWCNTs coatings of Silica fibers

According to our knowledge the electrical transport in SWCNTs coatings of silica fibers had not been studied earlier.

We used a standard dc 4-probe technique for measurements of the electrical and magnetotransport properties of the SWCNTs-SiO_2 coatings in a close-cycled cryostat in the temperature range 1.8-300 K and in the magnetic fields up to 8 T (Ksenevich et al., 2009). Charge transport in the system was found to have general features similar to the properties of SWCNTs fibers produced by polymer-free wet spinning process. The *R(T)* dependence of the SWCNTs coatings of silica fibers can be well fitted by a typical law (5) for fluctuation-induced tunneling model (Sheng, 1980) almost in the whole range of temperatures

(~8-300 K). The best fit to Eq. (5) was obtained with parameters T_s=13.3 K and T_c=52.8 K. Assuming that the area A of the tunnel junction equals to the cross section of a SWCNT with diameter of about 1.5 nm and using equations (6) and (7) for parameters T_s and T_c, the effective junction width was estimated to be 3.2 nm and the effective barrier height of 68 meV. In the low temperature range (2~8 K) $R(T)$ dependence was found to follows the Mott law (1) for 3D variable range hopping (VRH) conduction (Shklovskii & Efros, 1984). It should be noted that both for SWCNTs fibers and SWCNTs coatings of silica fibers non-metallic sign for $R(T)$ dependence ($dR/dT<0$) was observed in the whole range of temperatures in contrast to SWCNTs monolayers. We believe that possibility to observe metallic sign for $R(T)$ dependence for SWCNTs fibers and SWCNTs coatings of silica fibers is hindered by large distance between contacts (about ~ 1 *mm*) in contrast to SWCNTs monolayers, where length of pristine nanotubes exceeded distance between contacts (10 μm) in finger-shaped structures.
Magnetoresistance data measured at different temperatures are shown in Fig. 12.

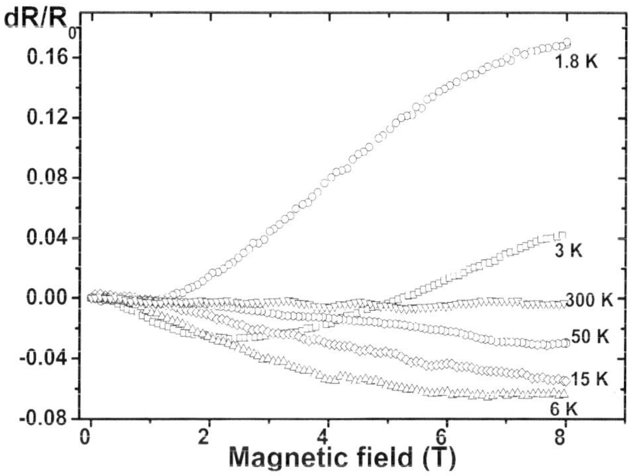

Fig. 12. The dependence of the normalized change in the resistance $dR/R(B=0)$ of the SWCNTs coatings of silica fibers on the magnetic field in the temperature range 1.8 - 300 K.

Both positive and negative MR was observed in the temperature range 1.8 - 6 K. The value of positive MR varies as $exp(B^2)$ as expected for the VRH transport and explained by electronic wave function shrinkage in the magnetic field. Tendency to saturation of positive MR was observed at 1.8 K at the value of magnetic field higher than 6 T. Non-monotonic behavior of relative value of negative MR was observed similar to the experimental data for MR of SWCNTs fibers. The relative value of negative MR increases with temperature in the range of 1.8-6 K. As the temperature exceeds 6 K, only negative MR is observed in the available range of magnetic fields, and the relative value of negative MR then decreases as the temperature rises. One can note, however, that in order to extract the negative contribution to the total magnetoresistance and to clarify the physical origin of this negative MR, a further analysis is needed both for SWCNTs fibers and SWCNTs coatings of SiO_2 fibers.

The experiments in strong microwave field in GHz range and THz radiation-induced photocurrent measurements were carried out as well in order to get an additional view into the electrical transport phenomena in SWCNTs coatings of silica fibers in comparison with transport properties of SWCNTs fibers.

The relaxation kinetics of photocurrent recorded at 1.6 THz frequency measured at different temperatures are shown in Fig. 13. Similar to the THz radiation-induced photocurrent experiments carried out for SWCNTs fibers the relaxation kinetics allows one to distinguish two different processes: relatively fast (with time constant of about of some hundreds of μs, which corresponds to the duration of THz pulses) due to terahertz-induced hopping conductivity, and slow (with time constant in the range of *ms*) due to the Joule heating of the samples. The decreasing of the amplitude of the fast photocurrent signal with the temperature was observed. Quantitative differences between the experimental results on the measurements of the conductivity under THz radiation for SWCNTs coatings of silica fibers and for SWCNT fibers were observed. Firstly, the amplitude of fast photocurrent signal is higher at the same temperatures and at the same frequency of THz radiation for the SWCNTs coatings of silica fibers. Secondly, the time constant for slow part of photocurrent signal is several times lower for SWCNTs-SiO_2 samples than for SWCNT fibers. Thz radiation-induced photocurrent measurements allow us to assume that in the intermediate temperature range both charge transport mechanism (VRH and fluctuation induced tunneling) can coexist. Therefore, the first difference can be explained by the higher degree of alignment of carbon nanotubes in fibers due to specifics of wet spinning method (Kozlov et al., 2005) and as a consequence a higher contribution to the total conductivity of VRH transport in comparison with SWCNTs fibers. A possible reason of lower duration of slow photocurrent signal component in SWCNTs coatings of silica fibers can be the heat exchange between the silica core and SWCNTs coatings.

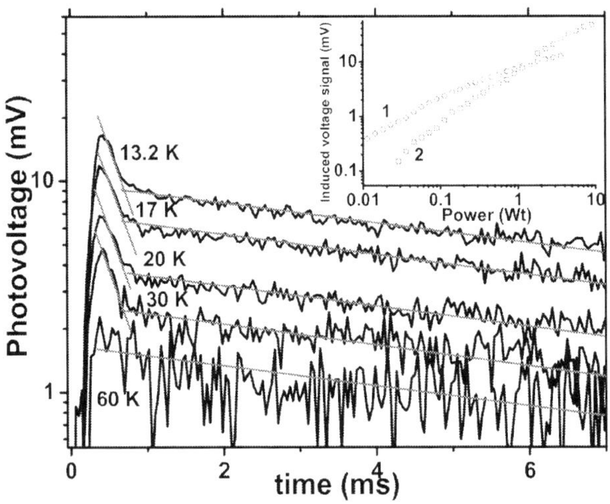

Fig. 13. Terahertz radiation-induced photocurrent at 1.6 THz at different temperatures in SWCNTs coatings of silica fibers. In the inset induced voltage signal proportional to the change of conductivity as function of microwave power at liquid nitrogen temperature for SWCNTs fibers (curve 1) and SWCNTs coatings of silica fibers (curve 2) is shown.

Measurements of the conductivity change of SWCNTs coatings of silica fibers in strong microwave field of 10 GHz at 77 and 300 K have shown similar results with experiments carried out for the SWCNTs fibers produced by wet spinning process. We observed increase in conductivity in microwave fields became more pronounced lowering the temperature. The data measured at 77 K are plotted in the inset to Fig. 13. For comparison, results both for SWCNTs fibers and SWCNTs coatings of silica fibers are given. As one can see, for both types of structures, the dependence is similar, however in SWCNTs-SiO_2 it follows a linear law while for SWCNTs fibers it becomes sublinear with microwave power.

As was mentioned above, heating of charge carriers by strong microwave field can give rise to conductivity both in VRH and fluctuation-induced tunneling regimes. However, according to the *R(T)* dependencies we can assume that the fluctuation-induced tunneling is a prevailing mechanism of charge carrier transport at 77 K for both types of samples. Hence, the difference in transport properties can be only defined by the barrier height. This can be an argument to explain the more rapid increase of conductivity with microwave fields in SWCNT-SiO_2 samples containing lower effective barrier height in comparison to that of SWCNTs fibers (68 meV *vs* 172 meV).

4. Conclusion

Electrical and magnetotransport properties of various types of carbon nanotubes arrays were reported. Different fabrication procedures for assembly of separate carbon nanotubes into monolayers, SWCNTs fibers and coatings of silica fibers were used. Langmuir-Blodgett (LB) technique was applied for self-assembling of SWCNTs and MWCNTs monolayers. SWCNTs fibers were synthesized using wet spinning process from solutions comprising nanotubes, surfactant and water. SWCNTs coatings were produced by means of clinging of pristine SWCNTs dispersed in water to previously silanized silica fibers. We experimentally showed that the electrical and magnetotransport properties of investigated carbon nanotubes arrays are defined not only by the types of pristine carbon nanotubes (multi wall or single wall with different chiralities) but also by the quality of intertube coupling between separate nanotubes from which these assemblies consist of.

Magnetotransport properties of MWCNTs arrays can be interpreted in framework of model assuming existing of two charge transport mechanisms in the system: spin-dependent hopping conductivity and 2D weak localization.

Charge transport properties of 2D arrays of SWCNTs can be interpreted in terms of heterogeneous model for conduction where the presence of structural defects and intertube connections give rise to low temperature localization effects. The conductivity of the system at high temperatures is determined by the charge transport trough metallic nanotubes.

Experimental data for SWCNTs fibers and SWCNTs coatings of silica fibers could be consistent both with the charge transport model for homogeneously disordered materials due to low-temperature localization of charge carriers arising from the defects in the tubes themselves and with the model for heterogeneous materials that emphasizes the role of energy barriers between the nanotubes.

Acknowledgments

We highly appreciate Mikhail Kozlov, David Novitski, Robin Helburn and Qi Lu for sample preparations and Gintaras Valusis, Dalius Seliuta, Irmantas Kasalynas and Zigmas Martunas for providing possibility and kind assistance in microwave and THz experiments and for useful discussions.

5. References

Bachtold, A.; Strunk, C.; Salvetat, J.-P.; Bonard, J.-M.; Forro, L.; Nussbaumer, T. & Schonenberger, C. (1999). Aharonov-Bohm oscillations in carbon nanotubes. *Nature,* Vol. 397, No. 6721, pp. 673-675

Badaire, S.; Pichot, V.; Zakri, C.; Poulin, P.; Launois, P.; Vavro, J.; Guthy, C.; Chen, M. & Fischer, J. E. (2004). Correlation of properties with preferred orientation in coagulated and stretch-aligned single-wall carbon nanotubes. *J. Appl. Phys.*, Vol. 96, No. 12, pp. 7509–7513

Bahr, J.L. & Tour, J.M. (2002). Covalent chemistry of single-wall carbon nanotubes. *J. Mater. Chem.*, Vol. 12, No. 7, pp. 1952–1958

Baumgartner, G.; Carrard, M.; Zuppiroli, L.; Bacsa, W.; de Heer, W. A. & Forro, L. (1997). Hall effect and magnetoresistance of carbon nanotube films. *Phys. Rev. B*, Vol. 55, No. 11, pp. 6704-6707

Bockrath, M.; Cobden, D. H.; McEuen, P. L.; Chopra, N. G.; Zettl, A.; Thess, A. & Smalley, R. E. (1997). Single-Electron Transport in Ropes of Carbon Nanotubes. *Science*, Vol. 275, No. 5308, pp. 1922-1925

Bockrath, M.; Cobden, D. H.; Lu, J.; Rinzler, A. G.; Smalley, R. E.; Balents, L. & McEuen, P. L. (1999). Luttinger-liquid behavior in carbon nanotubes. *Nature*, Vol. 397, No. 6720, pp. 598–601

Charlier, J.-C.; Blase, X. & Roche, S. (2007). Electronic and transport properties of nanotubes *Rev. Mod. Phys.*, Vol. 79, No. 2, pp. 677-732

Ci, L. J.; Zhao, Z. G. & Bai, J. B. (2005). Direct growth of carbon nanotubes on the surface of ceramic fibers. *Carbon*, Vol. 43, No. 4, pp. 883–886

Dai, H. (2002). Carbon nanotubes : opportunities and chalenges. *Surface Science*, Vol. 500, No. 1-3, pp. 218–241

Davis, V. A.; Ericson, L. M.; Parra-Vasques, A. N. G.; Fan, H.; Wang, Y.; Prieto, V.; Longoria, J. A.; Ramesh, S.; Saini, R. K; Kittrell, C; Billups, W. E.; Adams, W. W.; Hauge, R. H; Smalley, R. E. & Pasquali, M. (2004). Phase Behavior and Rheology of SWNTs in Superacids. *Macromolecules,* Vol. 37, No. 1, pp. 154-160

Fischer, J. E.; Dai, H.; Thess, A.; Lee, R.; Hanjani, N. M.; Dehaas, D. L. & Smalley, R. E. (1997). Metallic resistivity in crystalline ropes of single-wall carbon nanotubes. *Phys. Rev. B*, Vol. 55, No. 8, pp. R4921-R4924

Frank, S.; Poncharal, P.; Wang, Z. L. & De Heer, W. (1998). Carbon Nanotube Quantum Resistor. *Science*, Vol. 280, No. 5370, pp. 1744-1746

Georgakilas, V.; Kordatos, K.; Prato, M.; Guldi, D.M.; Holzingeer, M. & A. Hirsch. (2002). Organic Functionalization of Carbon Nanotubes. *J. Am. Chem. Soc.*, Vol. 124, No. 5, pp. 760-761.

Gommans, H. H.; J. Alldredge, W.; Tashiro, H.; Park, J.; Magnuson, J. & Rinzler, A. G. (2000). Fibers of aligned single-walled carbon nanotubes: Polarized Raman spectroscopy. *J. Appl. Phys.*, Vol. 88, No. 5, pp. 2509-2514

Grannan, S. M.; Lange, A. E.; Haller, E. E. & Beeman, J. W. (1992). Non-Ohmic hopping conduction in doped germanium at T<1K. *Phys. Rev. B*, Vol. 45, No. 8, pp. 4516-4519.

Hirsch, A. (2002). Functionalization of Single-Walled Carbon Nanotubes. *Angew. Chem. Int. Ed.*, Vol. 41, No. 11, pp. 1853-1859

Kaiser, A. B.; Dusberg, G. & Roth, S. (1998). Heterogeneous model for conduction in carbon nanotubes. *Phys. Rev. B*, Vol. 57, No. 3, pp. 1418–1421

Kaiser, A. B.; Park, Y. W.; Kim. G. T., Choi, E. S.; Dusberg, G. & Roth, S. (1999). Electronic transport in carbon nanotube ropes and mats. *Synth. Met.*, Vol. 103, No. 1-3, pp. 2547–2550

Kaiser, A. B. & Park, Y. W. (2005). Current-voltage characteristics of conducting polymers and carbon nanotubes. *Synth. Met.*, Vol. 152, No. 1- 3, pp. 181-184

Kim, G. T.; Choi, E. S.; Kim, D. C.; Suh, D. S..; Park, Y. W.; Liu, K.; Dusberg, G. & Roth, S. (1998). Magnetoresistance of an entangled single-wall carbon-nanotube networks. *Phys. Rev. B*, Vol. 58, No. 24, pp. 16064–16069

Kim, G. T.; Jhang, S. H.; Park, J. G.; Park, Y. W. & Roth, S. (2001). Non-ohmic current-voltage characteristics in single-wall carbon nanotube networks. *Synth. Met.*, Vol. 117, No. 1–3, pp. 123–126

Kozlov, M. E.; Capps, R. C.;W. Sampson, M.; Ebron, V. H.; Ferraris, J. P. & Baughman, R. H. (2005). Spinning Solid and Hollow Polymer-Free Carbon Nanotubes Fibers. *Adv. Mater.*, Vol. 17, No. 5, pp. 614-617

Krstic, V.; Duesberg, G.S.; Muster, J.; Burghard, M. & Roth, S. (1998). Langmuir-Blodgett Films of Matrix-Diluted Single-Walled Carbon Nanotubes. *Chem. Mater.*, Vol. 10, No. 9, pp. 2338-2340

Krstic, V.; Roth, S. & Burghard, M. (2000). Phase breaking in three-terminal contacted single-wall carbon nanotube bundles. *Phys. Rev. B*, Vol. 62, No. 24, pp. R16353-R1635

Ksenevich, V. K.; Galibert, J.; Forro, L. & Samuilov, V. A. (2006). Magnetotransport in 2D-arrays of single-wall carbon nanotubes, In: *Carbon Nanotubes: From Basic Research to Nanotechnology*, NATO ASI Series II: Mathematics, Physics and Chemistry, Vol.222, Popov, V. N. & Lambin, P. (Eds.), pp. 183–184, Springer, ISBN 978-1-4020-4572-1, Dordrecth, The Netherlands

Ksenevich, V. K.; Aboltin, R. S.; Dauzhenka, T. A., Galibert, J. & Samuilov, V. A. (2008, a). Charge transport mechanisms in the arrays of multi-walled carbon nanotubes, *Proceedings of the III International conference Materials and structures of modern electronics*, pp. 271-273, ISBN 978-985-518-091-4, Minsk, Belarus, September, 2008, BSU publishers, Minsk

Ksenevich, V.K.; Odzaev, V.B.; Martunas, Z.; Seliuta, D.; Valusis, G.; Galibert, J.; Melnikov, A.A.; Wieck, A.D.; Novitski, D.; Kozlov, M.E. & Samuilov, V.A. (2008, b). Localization and nonlinear transport in single walled carbon nanotube fibers. *J. Appl. Phys.*, Vol 104, No. 7, pp. 073724 1–073724-7

Ksenevich, V.K.; Seliuta, D.; Martunas, Z.; Kasalynas, I.; Valusis, G.; Galibert, J.; Kozlov, M.E. & Samuilov, V.A. (2008, c). Charge Carrier Transport Properties in Single-Walled Carbon Nanotube Fibers. *Acta Phys. Pol. A,*, Vol. 113, No. 3, pp. 1043–1046

Ksenevich, V.; Dauzhenka, T.; Seliuta, D.; Kasalynas, I.; Kivaras, T.; Valusis, G.; Galibert, J.; Helburn, R.; Lu, Q. & Samuilov, V.A. (2009). Electrical transport in carbon nanotubes coatings of silica fibers. *Physica Status Solidi (c),* Vol. 1-3, DOI 10.1002/pssc.200982553 (in press).

Kurobe, A. & Kamimura, H. (1982). Correlations Effects on Variable Range Hopping Conduction and the Magnetoresistance. *J. Phys. Soc. Jap.,* Vol. 51, pp. 1904-1913

Lassagne, B.; Cleuziou, J.-P.; Nanot, S.; Escoffer, W.; Avriller, R.; Rosche, S.; Forro, L.; Raquet, B. & Broto, J.-M. (2007). Aharonov-Bohm Conductance Modulation in Ballistic Carbon Nanotubes. *Phys. Rev. Lett.,* Vol. 98, No. 17, p.p. 176802-1-176802-4

Lee, P. & Ramakrishnan, T.V. (1985). Disordered electronic systems. *Rev.Mod.Phys.,* Vol. 57, No. 2, pp. 287-337.

Liu, H.; Li, J.; Liu, X. & Jiang, S. (2009). A novel multiwalled carbon nanotubes bonded fused-silica fiber for solid phase microextraction–gas chromatographic analysis of phenols in water samples. *Talanta,* Vol. 78, No. 3, pp. 929–935

Nanot, S.; Escoffier, W.; Lassagne, B.; Broto, J.-M. & Raquet, B. (2009). Exploring the electronic band structure of individual carbon nanotubes under 60 T. *C. R. Physique,* Vol. 10, No. 4, pp. 268-282

Nguen, V. L., Spivak, B. Z. & Shklovskii, B. I. (1985). Tunnel hopping in a disordered systems. *Sov. Phys. JEPT,* Vol. 62, No. 5 , pp. 1021-1029

Qian, H.; Bismarck, A.; Greenhalgh, E. S. & Shaffer, M. S. P. (2010). Synthesis and characterisation of carbon nanotubes grown on silica fibres by injection CVD. *Carbon,* Vol. 48, No. 1, pp. 277–286

Pollak, M. & Riess, I. (1976). A percolation treatment of high-field hopping transport. *J. Phys. C,* Vol. 9, No. 12, pp. 2339–2352

Pozela, J. K. (1985). Transport parameters from microwave conductivity and noise measurements, In: *Topics of Applied Physics,* L. Reggiani (Ed.), Vol. 58, pp. 113–147, Springer-Verlag, Berlin

Robertson, J. (2004). Realistic applications of CNTs. *Materials Today,* Vol. 7, No. 10, pp. 46-52

Robertson, J. (2007). Growth of nanotubes for electronics. *Materials Today,* Vol. 10, No. 1-2, pp. 36–43

Schonenberger, C.; Bachtold, A.; Strunk, C.; Salvetat, J.-P. & Forro, L. (1999). Interference and Interaction in multi-wall carbon nanotubes. *Appl. Phys. A,* Vol. 69, No. 3, pp. 283-295

Shaffer, M. & Sandler, J. (2007). Carbon nanotube/nanofibre polymer composites, In: *Processing and properties of nanocomposites,* Advani, S. G., (Ed.), pp. 1–59, World Scientific, ISBN: 978-981-270-390-3, New Jersey, NJ

Sheng, P. (1980). Fluctuation-induced tunneling conduction in disordered materials. *Phys. Rev. B,* Vol. 21, No. 6, pp. 2180–2195

Shklovskii, B. I. & Efros, A. L. (1984). *Electronic properties of doped semiconductors,* Springer, ISBN-10: 0387129952, Berlin, New York

Sivan, U.; Entin-Wohlman, O. & Imry, Y. (1988). Orbital Magnetoconductance in the Variable-Range-Hopping Regime. *Phys. Rev. Lett.,* Vol. 60, No. 15, pp. 1566-1569

Smith, B. W.; Benes, Z.; Luzzi, D. E.; Fischer, J. E. ; Walters, D. A.; Casavant, M. J.; Schmidt, J. & Smalley, R. E. (2000). Structural anisotropy of magnetically aligned single wall carbon nanotube films. *Appl. Phys. Lett.,* Vol. 77, No. 5, pp. 663–665

Stetter, J. R. & Maclay, G. J. (2004). Carbon Nanotubes and Sensors: a Review. In: *Advanced Micro and Nanosystems*, Vol. 1, Baltes, H.; Brand, O.; Fedder, G.K.; Hierold, C.; Korvink, J.G. & Tabata, O. (Eds.), pp. 357–382, Wiley-VCH Verlag GmbH & Co, ISBN-13: 978-3-527-30746-3, Weinheim, Germany

Strunk, C.; Stojetz, B. & Roche, S. (2006). Quantum interference in multiwall carbon nanotubes. *Semicond. Sci. Technol*, Vol. 21, No. 11, pp. S38–S45

Thostenson, E. T.; Li, W. Z.; Wang, D. L.: Ren, Z. F. & Chou, T. W. (2002). Carbon nanotube/carbon fiber hybrid multiscale composites. *J. Appl. Phys.*, Vol. 91, No. 9, pp. 6034–6037

Urbina, A.; Echeverria, I.; Perez-Garrido, A.; Diaz-Sanchez, A. & Abellan, J. (2003). *Phys. Rev. Lett.*, Vol. 90, No. 10, pp. 106603-1-106603-4

Vigolo, B.; Penicaud, A.; Coulon, C.; Sauder, C.; Pailler, R.; Journet, C.; Bernier, P. & Poulin, P. (2000). Macroscopic Fibers and Ribbons of Oriented Carbon Nanotubes. *Science*, Vol. 290, No. 5495, pp. 1331-1334

Vigolo, B.; Poulin, P.; Lucas, M.; Launois, P. & Bernier, P. (2002). Improved structure and properties of single-wall carbon nanotube spun fibers. *Appl. Phys. Lett.*, Vol. 81, No. 7, pp. 1210–1212

Zabrodskii, A. G. (1977). Hopping conduction and density of localized states near the Fermi level. *Soviet Phys. Semicond.*, Vol. 11, pp. 345-348

Zhang, Q.; Qian, W.; Xiang, R.; Yang, Z.; Luo, G.; Wang Y, & Wei, F. (2008). In situ growth of carbon nanotubes on inorganic fibers with different surface properties. *Mater. Chem. Phys.*, Vol. 107, No. 2-3, pp. 317-321

Zhou, W.; Vavro, J.; Guthy, C.; Winey, K. I.; Fischer, J. E.; Ericson, L. M.; Ramesh, S.; Saini, R.; Davis V. A.; Kittrel, C.; Pasqualli, M.; Hauge, R. H. & Smalley, R. E. (2004). Single wall carbon nanotube fibers extruded from super-acid suspensions: Preferred orientation, electrical, and thermal transport. *J. Appl. Phys.*, Vol. 95, No. 2, pp. 649-655

8

Doped Carbon Nanotube Properties

Viktor Griadun
Zaporizhzhia National Technical University
Ukraine

1. Introduction

The last twelve years have seen enormous growth in many aspects of doped carbon nanotubes. The idea of doping carbon nanotubes has come from microelectronics and now develops both in a traditional donor-acceptor direction (Miyamoto et al., 1995; Lee et al., 1997; Rao et al., 1997; Grigorian et al., 1998; Esfarjani et al., 1999) and in a number of new aspects of application. Among recent trends of application of a doping of nanotubes it is necessary to term such as forming for them magnetic properties (Esfarjani et al., 2003; Farajian et al., 2002), change of their optical performances (Bondarev, 2007), embodying of hydrogen storages (Zidan et al., 2003), amplification of catalytic properties (Gong et al., 2009) and improving of emission characteristics (Yao et al., 2002).

In the present chapter we discuss influence of impurities on properties of semiconductor carbon nanotubes with chiral vectors (10,0) and (6,5) which are representative of groups of nanotubes of type "zigzag" and the twirled chirality, accordingly.

At doping nanotubes usually view an intercalation of impurity inside of a nanotube or in space between separate nanotubes. We view doping nanotubes in traditional sense when impurity ranks one of atoms of a nanotube. It allows to hope, that impurity after such localization will not be lightly to migrate in a nanotube or between separate nanotubes. At such approach to a problem of doping of a periodic nanostructure presence in it of vacancies is supposed. Energy of forming of vacancy in a nanotube makes 5...7 eV (Griadun, 2006). It is a little bit more in comparison with energy of forming of vacancy in silicon, but nevertheless leaves probability of substantial existence of vacancies in nanotubes.

Profound knowledges are necessary for an effective utilization carbon nanotubes in nanoelectronic about their local electro-physical performances. In particular extensive data on concentration and parameters of the impurity centers by means of which it is possible to shape fields with the given type and the quantity of an electrical conductivity reproduced by quantities of a lifetime and a diffusion length of non-equilibrium charge carriers are necessary.

Interest to these examinations is called by necessity of development of designs of integrated transistors and logical units for build-up of complete sets of integrated circuits with use semiconductor carbon nanotubes (Martel et al., 1998; Derycke et al., XXXX).

The solution of a problem guesses an opportunity to drive equilibrium concentrations of electrons and holes in semiconductor carbon nanotubes by their doping.

2. Doped carbon nanotube with indices (10,0)

2.1 Method of modeling and base model of the (10,0) carbon nanotube

Models developed within the limits of a method of molecular mechanics model operation MM + (Berkert & Allinger, 1982) with next parameters:

a quadratic and a cubic stretch term in the potential;

atom types;

nonbonded electrostatic interactions are calculated using bond dipole, which values come from the MM+ stretch parameters;

the electrostatic contribution comes from defining a set of bond dipole moments associated with polar bonds;

cutoffs determine the distance limits for calculating nonbonded interactions of the periodic nanostructure;

inner cutoff radius is the maximum interatomic distance for full nonbonded interactions - 10 Å;

outer radius is the minimum distance at which nonbonded interactions are set to zero - 14 Å.

The base model of the nanotube consisting from two hundred of atoms of carbon, is presented on Fig. 1. In models aromatic bonds, providing interacting of valence electrons of carbon atoms are used.

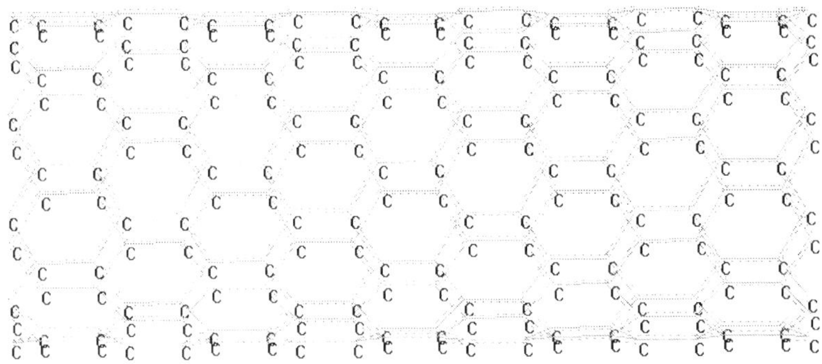

Fig. 1. Basic model of the carbon nanotube with indices (10,0).

The energy-band structure of a nanotube was calculated by the Extended Hückel method (Hoffmann, 1963). The total energy of 200-atomic model of a nanotube has made quantity -323079.99 kcal/mol. On the diagram of its energy structure of a forbidden gap (Fig. 2) is observed a series of levels with energies: -10.60226; -10.37579; -10.37578; -10.36441; -10.36440; -10.30684; -10.27539; -10.26634; -10.26634; -10.24621; -10.24621; -10.19875; -10.19875; -9.6362; -9.6362 eV - matching orbitals with numbers 401 - 415, accordingly.

All from the termed orbitals to a greater or lesser extent match to edges of a nanotube, for example orbital 413 (Fig. 3), and higher orbitals match to a volume part of a nanotube.

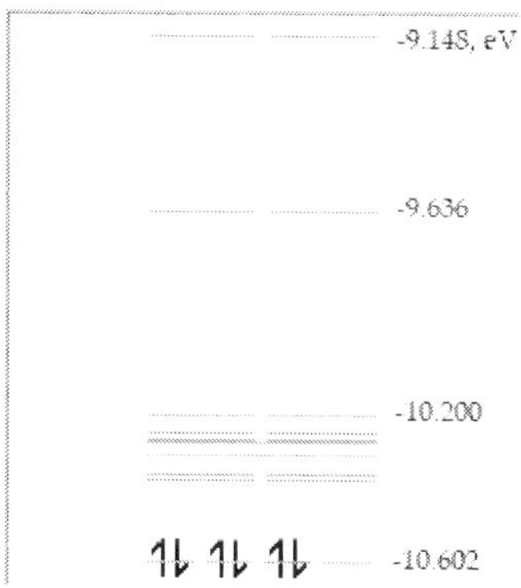

Fig. 2. Structure of a forbidden gap of the nanotube with indices (10,0).

Thus, in a forbidden gap of a nanotube is 30 electronic states matching 15 orbitals. A part from viewed states match to the dangling bonds on edges of a nanotube. For acknowledging this we shall simulate passivation of dangling bonds by atoms of hydrogen.

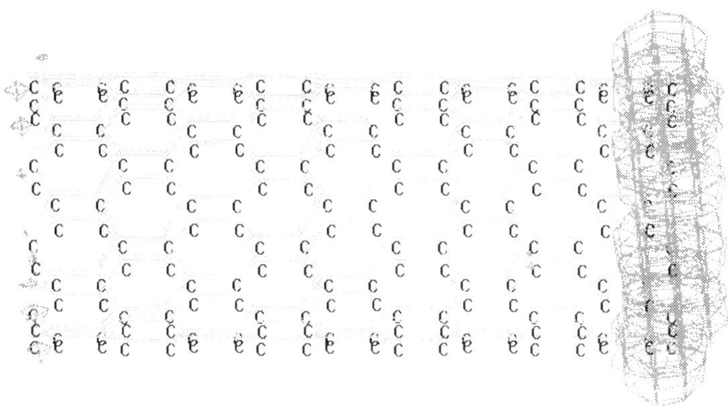

Fig. 3. Orbital 413 of the carbon nanotube with indices (10,0).

On Fig. 4 the model of a 200-atomic nanotube after hydrogen endings and geometrical optimization of nanostructure within the limits of a method of molecular mechanic modeling MM + is figured. Apparently on the model, atoms of hydrogen have filled 20 dangling carbon bonds - on 10 on each edges of a nanotube. Thus, the remained three orbitals feature six states which parent of origination is unknown.

The band of energy diagram in the field of a forbidden gap (Fig. 5) after passivation by hydrogen of the broken off bonds of a nanotube on its edges essentially differs from the diagram of the starting model figured on Fig. 2:

1) The forbidden gap has essentially cleared from energy levels - from 15 remains only 5, that is 10 orbitals that matches to 20 electronic states, were moved in a valence band;
2) Three from five remained energy levels in a forbidden gap should be the acceptor as they are close to a valence band with energies -10.44002 eV (an orbital 411), -10.41608 eV (an orbital 412), and -10.41608 eV (an orbital 413);
3) Passivation by atoms of hydrogen has affected quantity of energy of a ceiling of a valence band (an orbital 410) which has made -10.44002 eV (before passivation was -10.60226 eV);
4) The magnification of energy in item 3 is caused by electronic states of atoms of hydrogen;
5) Passivation by atoms of hydrogen has affected quantity of energy of a bottom of conduction band (an orbital 416) which has made -9.156353 eV (before passivation was -9.147926 eV);
6) The breadth of a forbidden gap of viewed model has made quantity 1.2837 eV (before passivation was 1.4543 eV);
7) The overestimated value of an energy gap in this case is caused by a trace amount of atoms and, accordingly, small length of a nanotube 19.6 Å.

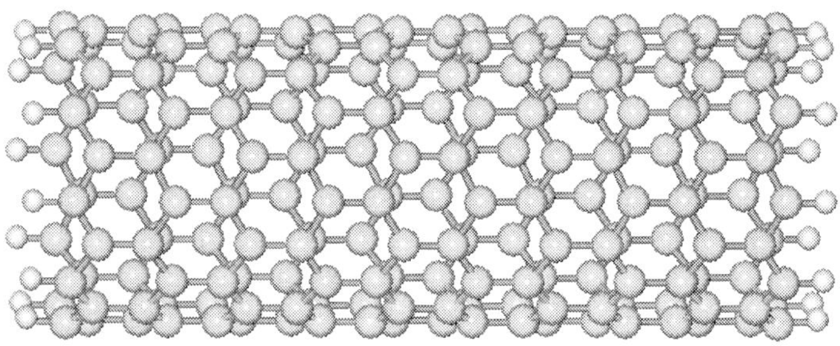

Fig. 4. Hydrogen endings of the carbon nanotube with indices (10,0).

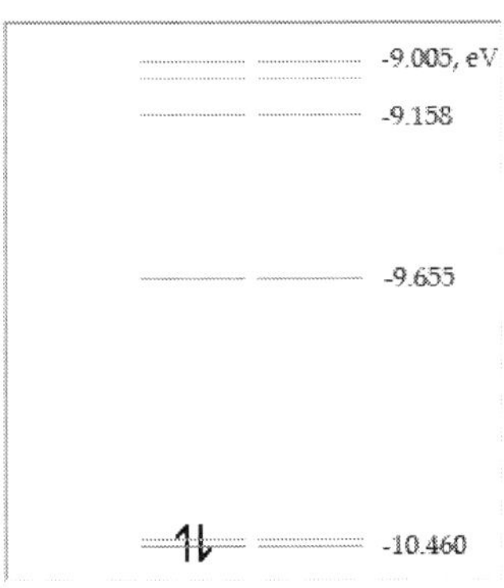

Fig. 5. Structure of a forbidden gap of the (10,0) nanotube after hydrogen endings.

Analogous calculations are lead also for the nanotubes consisting from three hundred and four hundred of atoms of carbon in length ~ 30.15 and 40.63 Å, accordingly. Hydrogen endings of this models also well refines forbidden regions of the energy levels matching broken off bonds. The breadth of a forbidden gap has naturally decreased up to values 1.06 and 0.90 eV, accordingly for 300-atomic and 400-atomic models of nanotubes.

2.2 Doping carbon (10,0) nanotube by atom of boron

Simulation of doping of the nanotube by atom of boron we shall carry out within the limits of a method of MM +, exchanging in the program of basic model the chosen atom of carbon on atom of boron (Griadun, 2007). Thus we can use aromatic bonds of atom of boron with atoms of carbon or exchange them on single bonds, as boron trivalent. We shall consider also at model operation spin multiplicity of a nanostructure.

The geometry of a 200-atomic nanotube doped by atom of boron, is presented on Fig. 6. We see, that at modeling of impurity of boron by aromatic bonds with atoms of carbon the nanostructure distorts feebly. In case of single bonds B-C the nanostructure noticeably distorts in the field of localization of atom of boron. It is related by that single bonds B-C are longer than aromatic on 0.08 Å.

Extended-Hückel energy of the nanostructure with aromatic bonds of atom of boron is less than energy of the nanostructure with single bonds on 2.22 kcal/mol, therefore it is necessary to expect, that in real experiment embodying of the nanostructure with aromatic bonds of atom of boron is more probable.

The energy band diagram of the viewed nanotube doped by atom of boron (Fig. 7a), differs from an energy-band structure of basic model (Fig. 2), however, only quantities of values of energy levels. The expected new acceptor level localized on atom of boron, we do not observe. As well as in basic model three acceptor energy levels, but with other values of energy are observed: -10.60; -10.38 and -10.37 eV. It, visible, is related by that bonds B-C the 1S-electron of atom of boron shares. Except for that removal of degeneration of some orbitals is observed, for example orbitals 414 and 415 with energy -9.636 eV borrow levels with energies -9.641 and -9.611 eV.

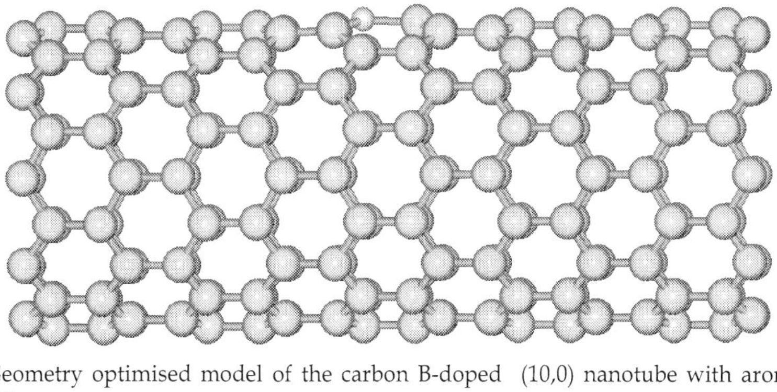

Fig. 6. Geometry optimised model of the carbon B-doped (10,0) nanotube with aromatic bonds B-C.

Hydrogen endings of a nanotube doped by atom of boron practically does not influence geometry of a nanostructure near to atom of boron, however noticeably influences its energy distribution (Fig. 7b). Really, upper occupied level HOMO-0, featured 410-th orbital, has energy -10,46021 eV. Following three orbitals with numbers 411 - 413 are at levels -10.45723, -10.43564, and -10.42292 eV, accordingly. As a whole, such effect is normal as levels of energy of electrons of hydrogen atoms have placed in a valence band and biased its ceiling on 0.15 eV upwards.

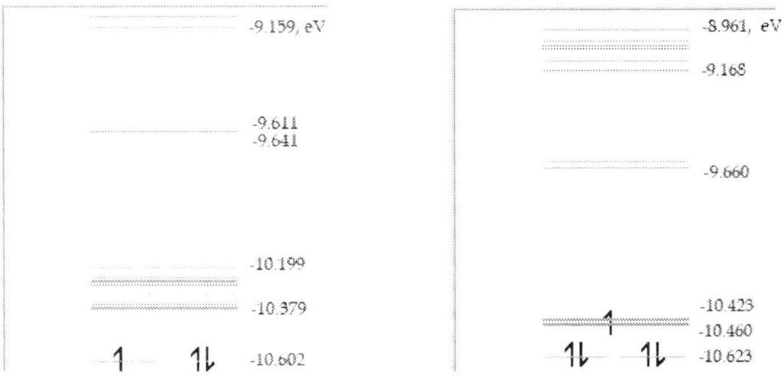

a b

Fig. 7. Energy band structure of the B-doped (10,0) carbon nanotube before (a) and after (b) hydrogen endings.

2.3 Doping of a carbon (10,0) nanotube by atom of nitrogen

The atom of nitrogen also is well inscribed in geometry of a nanotube, not calling essential strain of its nanostructure (Fig. 8). It is caused by that lengths of bonds N-C and C-C differ a little - 1.28 Å and 1.40 Å, accordingly.

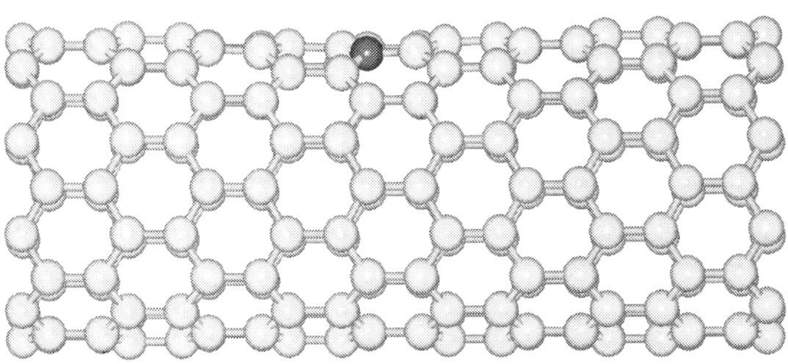

Fig. 8. Model of the nanotube with indices (10,0) doped by atom of nitrogen.

The energy band distribution of a nanotube without hydrogen endings, doped by atom of nitrogen, is presented on Fig. 9a. Under effect of impurity atom in a spectrum of energy levels of a nanostructure there was an orbital filled by an electron the number 401. This

orbital matches to the fifth valence electron of atom of nitrogen and is presented by an energy level equal to quantity -10.6018 eV. The ceiling of a valence band of a model nanostructure is featured by an orbital the number 400 with energy -10.6023 eV. Thus, the extrinsic energy level of atom of nitrogen will not display the donor properties as is near to a valence band, it only will neutralize the inferior acceptor level of a basic nanotube.
In case of nitrogen doping a carbon nanotube with hydrogen endings, the fifth valence electron of atom of nitrogen also occupies the inferior acceptor level near to a valence band (Fig. 9b). Quantity of energy of a level which value makes in this case -10.4579 eV varies only.

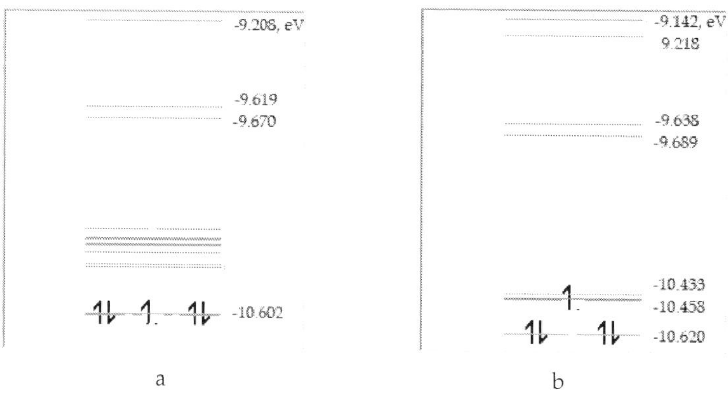

Fig. 9. Effect of impurity atom of nitrogen on energy spectrum of the carbon (10,0) nanotube before (a) and after (b) hydrogen ending.

Thus, doping by atom of nitrogen of a carbon nanotube with indices (10,0) does not result in to formation of a donor centre, and only results in to blocking of one of acceptor levels of a nanostructure.

2.4 Doping of a carbon (10,0) nanotube by atom of aluminium

Atom of aluminium as the device of the third group, well approaches for doping carbon nanotubes. As if to its influence on geometrical and other properties of a carbon nanotube with coefficients (10,0) we shall view on model of a nanostructure in which atom №133 of the basic model of nanotube we shall replace with atom of aluminium. After geometrical optimization of a nanostructure in which aromatic bonds Al-C use, the model gets a view figured on Fig. 10.
We see, that the atom of aluminium is not inscribed in wall of a nanotube, but located in immediate proximity from it apart 1 Å. Besides atoms of carbon with which bound atom of aluminium, are noticeably drifted for limits of a nanotube. In case of use in model of a nanostructure of single bonds Al-C results in to even greater infringements of geometry of a nanotube. The atom of aluminium settles down thus apart 1.2 Å from geometrical wall of a nanotube. The Extended Hückel calculation of energy distribution in area of a forbidden gap of the Al-doped nanotube is presented on Fig. 11.

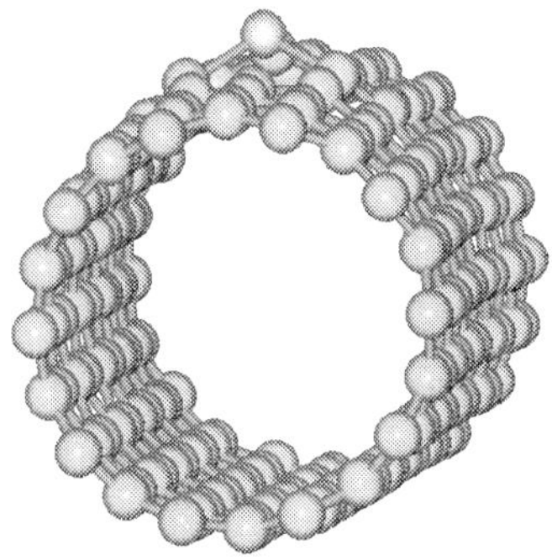

Fig. 10. Optimised geometry of the Al-doped carbon (10,0) nanotube.

It is visible, that the type of bonds of atom of aluminium with atoms of carbon very feebly influences quantity of energy levels. In comparison with a spectrum of basic model of the carbon (10,0) nanotube (Fig. 2) has occurred an acceptor level with energy -10.532 eV. Also there were splits of levels -9.636 and -9.148 eV. The parent of this split is spin multiplicity of the nanostructure doped by aluminium which quantity is equal to two.
It is necessary to note, that the nanostructure with aromatic bonds Al-C should be realized with greater probability as the nanostructure with single bonds Al-C is more on 10 kcal/mol.

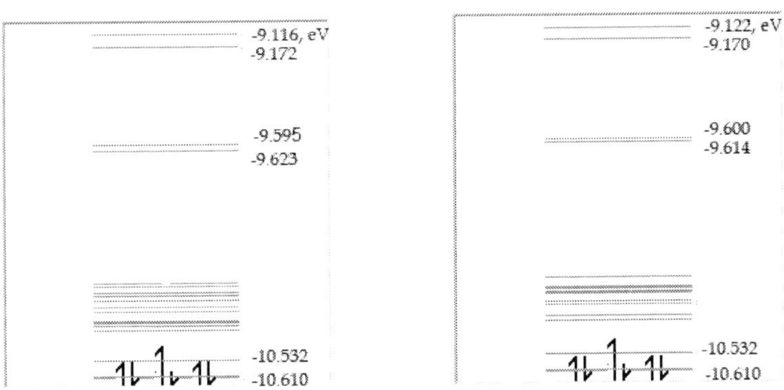

a b

Fig. 11. Energy band structure of a forbidden gap region of the Al-doped carbon (10,0) nanotube: (a) - aromatic bond Al-C; (b) - unary bonds Al-C.

Passivation of the Al-doped carbon (10,0) nanotube by hydrogen endings (Fig. 12a) renders appreciable influence on its energy distribution in the field of a forbidden gap (Fig. 12b). We see, that after passivation of edges of a nanotube by hydrogen in a forbidden region of a nanostructure some level by quantity -10.289 eV which visible is related to the defect created by impurity atom of aluminium was displayed.

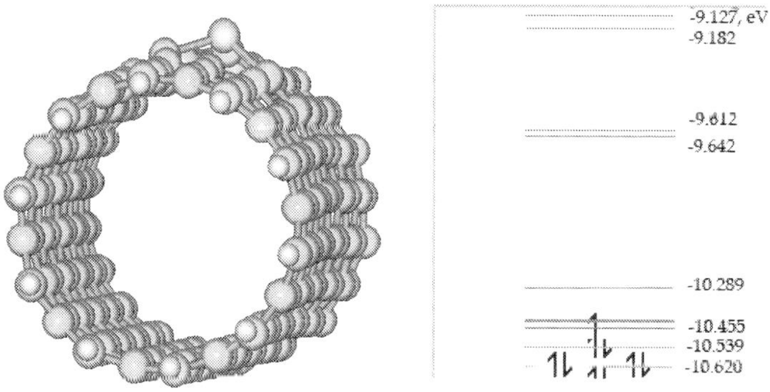

a b

Fig. 12. Al-doped carbon (10,0) nanotube with hydrogen endings: (a) - geometry of nanostructure; (b) - energy band structure of a forbidden gap.

2.5 Doping of a carbon (10,0) nanotube by atom of phosphorus

Doping of the carbon (10,0) nanotube by atom of phosphorus results in appreciable strain of geometry of a nanostructure (Fig. 13). It is related by that bonds C-P, having quantities 1.80, 1.80 and 1.79 Å, longer in comparison to typical length of bonds C-C, equal 1.41 Å. Besides the atom of phosphorus has on 0.3 E greater atomic radius than at atom of carbon that results in to its replacement for limits of a nanotube.

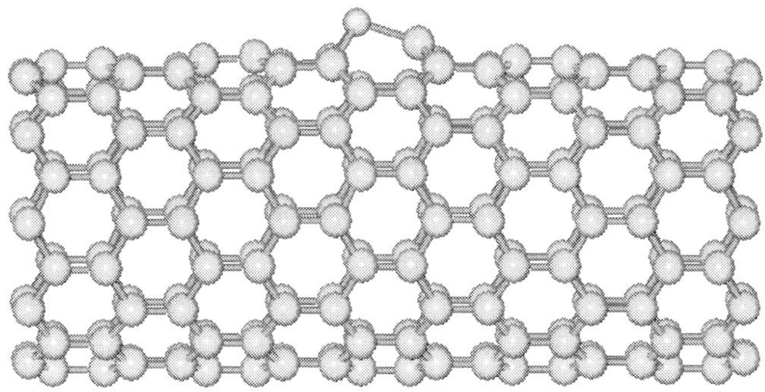

Fig. 13. Molecular mechanics model of the 200-atomic carbon (10.0) nanotube doped by atom of phosphorus.

It is necessary to note, that the atom of phosphorus shapes in a viewed nanostructure an essential lapse rate of mechanical stresses which value equally 0.05 kcal/(Å·mol). In a basic nanotube geometrical optimization is attained at a root-mean-square lapse rate equal 0.000001 kcal/(Å·mol).

The energy diagram of a viewed nanostructure is presented on Fig. 14. We see, that in a forbidden gap of the doped nanotube there was a level (-10.082 eV), which probably matches to the flaw created by atom of phosphorus.

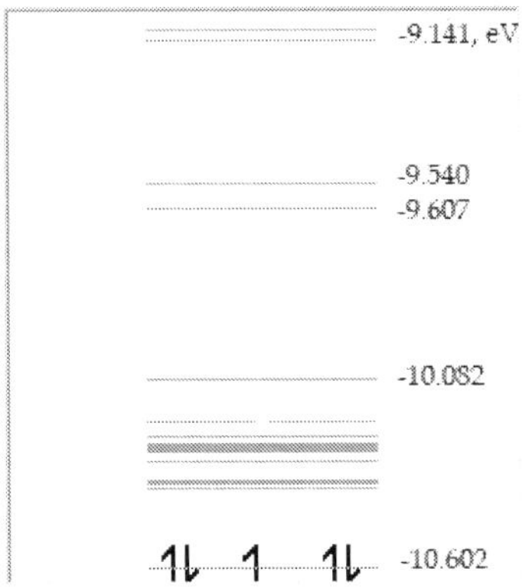

Fig. 14. Energy levels in a forbidden gap of the carbon(10,0) nanotube doped by atom of phosphorus.

The model of a phosphorus-doped (10,0) nanotube with hydrogen passivation of dangling bonds and its energy band structure are presented on Fig. 15. We see, that the geometry of a viewed nanostructure in the field of localization of atom of phosphorus after hydrogen passivation of dangling bonds on its edges has not changed.

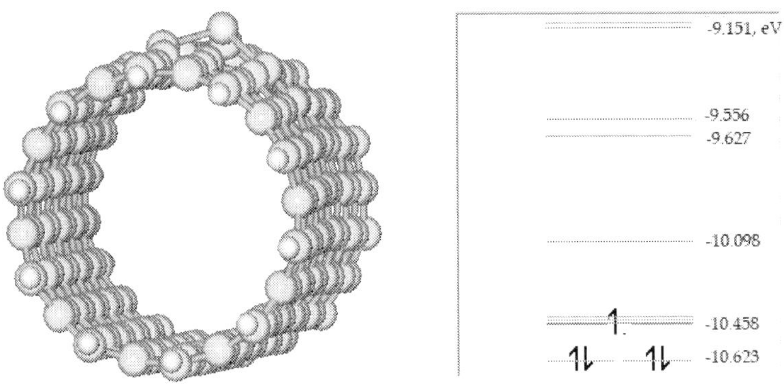

a b

Fig. 15. P-doped carbon (10,0) nanotube with hydrogen endings: (a) - geometry of the molecular mechanics model; (b) - Extended Hückel calculation of energies of orbitals.

In an energy distribution of a forbidden gap there were changes (Fig. 15b):

the upper filled level has risen up to quantity -10.46 эВ;

degeneration of some energy levels is removed;

quantities of energy levels have changed.

It is necessary to note, that the new level with energy -10.098 eV under effect of hydrogen endings has not changed the quantity.
The magnification of number of atoms of carbon in model of a phosphorus-doped nanotube up to 300 pieces results in to diminution of breadth of a forbidden gap and small bias of a new level up to quantity -10.11 eV.

3. Properties of doped carbon nanotube with indices (6,5)

3.1 Basic model of a carbon nanotube with indices (6,5)

There are many methods of model operation of those or other processes and structures. For nanotechnologies the method of the molecular mechanic model operation which basic element is the molecule well approaches. The separate atom too is considered as a molecule. In case of a nanotube the molecule is the nanotube. Atoms of a nanotube are foliated, featured by coordinates and type and also quantity of bonds.
Our model of nanotube consists from two hundred atoms of carbon, which geometrical optimization was led by method MM + with application the block-diagonal of the Newton-Raphson algorithm (Fig 16). The length of a nanotube is equated 22 Å, diameter - 7.3 Å. The distance between atoms is equated 1.41 Å. Chirality of the nanotube which it is well visible on figure, defined by its indices (6,5).
At model operation the applied aromatic bonds which well justify itself when atoms of carbon create rings as in this case. There is one more expedient of the plotting of bonds which total for atom of carbon should equate to four.
To calculation of energy properties of a nanotube we applied Extented Hückel Method (Hoffmann, 1963) without taking into account d-orbitals, with unit spin multiplisity and unweighted Hückel constant equal 1.75. Energy has made quantity equal -323263.1900 kcal/mol.

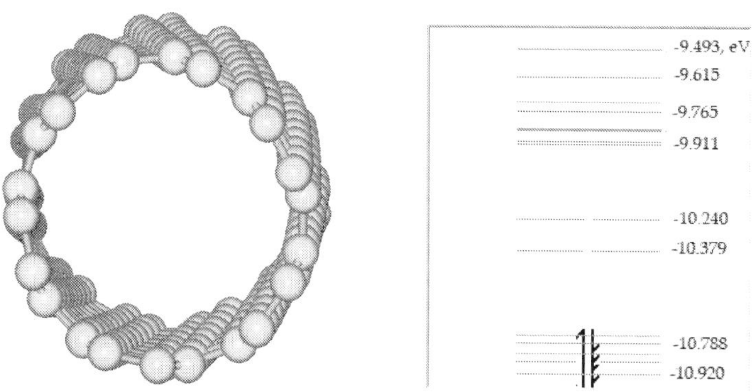

a b

Fig. 16. 200-atoms basic model (a) and energy-band structure (b) of a carbon nanotube with indices (6,5).

The energy-band structure of a basic nanotube (6,5) is presented on Fig. 16b. The upper filled level (HOMO 0), which is featured by orbital number 400, makes quantity -10.788 eV.

Energy of bottom of a conduction band is equated -9.91 eV. In fact the forbidden region has breadth equal 0.88 eV.

In a forbidden gap there is a series of levels which are caused by dangling bonds on edges of the nanotube. So, it is possible to apply atoms of hydrogen or nitrogen to passivation of the broken off bonds. On Fig. 17 passivation by atoms of hydrogen which are optimum in such cases is shown.

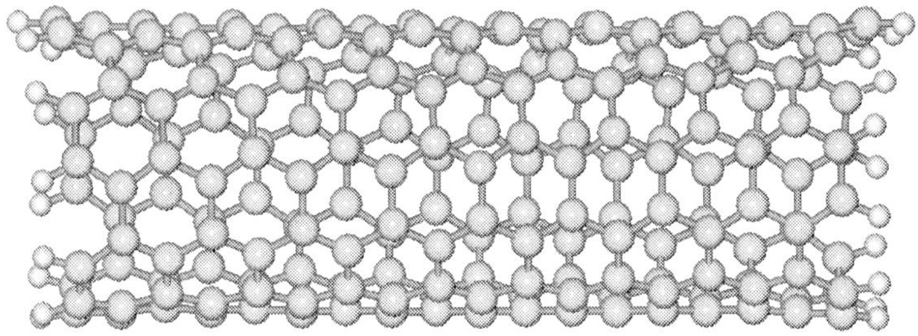

Fig. 17. Atoms of hydrogen fill dangling bonds on edges of the carbon (6,5) nanotube.

The energy of the nanotube with passivated ends has decreased to value -332889.25 kcal/mol that is caused by both H-atoms and a relaxation of a nanotube on edges where bonds have been broken off. The energy-band structure presented on Fig. 18a, considerably has changed in this case. The forbidden gap became clear from energy levels of the dangling bonds. Apparently from this figure, that the forbidden gap quantity is defined by a difference

$$\Delta E = E_{orb412} - E_{orb411} = 1.02 \text{ eV}.$$

The greater width of the gap is caused by a trace amount of carbon atoms.

Fig. 18. The energy-band structure of the base nanotube with indices (6,5) after passivation the dangling bonds by atoms of hydrogen: (a) - 200-atomic model; (b) - 400-atomic model.

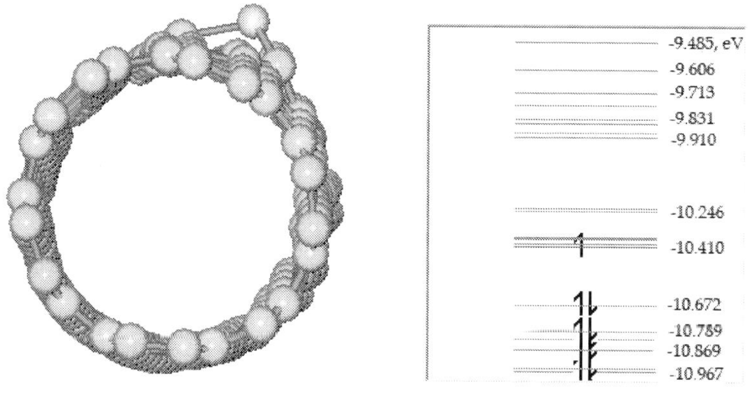

a b

Fig. 25. Al-doped carbon (6,5) nanotube: (a) - a dilatational view; (b) - an energy-band structure.

The nanostructure with hydrogen endings is shown on Fig. 26. Its Extended-Hückel energy is equaled -332121.573 kcal/mol. Electrical dipole moment of a nanotube, is equal 297.4 D - greater enough in comparison with a boron-doped nanotube (6,5).

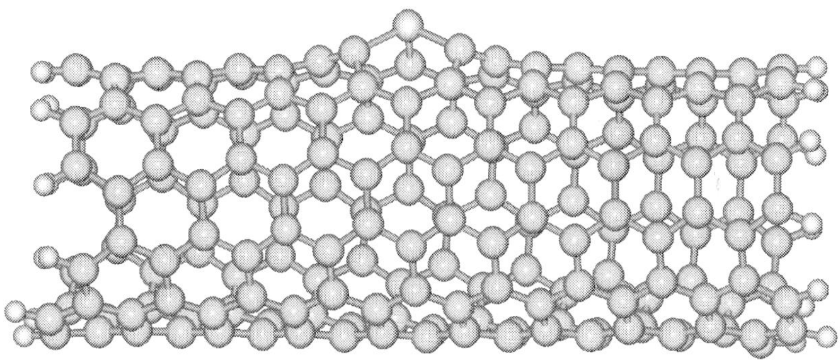

Fig. 26. Doped by atom of aluminium the carbon (6,5) nanotube with hydrogen endings.

Doping by atom of aluminium very strongly influences an energy band spectrum (Fig. 27) nanotubes that is related to its significant strain after doping, and also spin density of a nanostructure. In this case the impurity atom of aluminium creates an acceptor level (-10.373 eV) with an ionization energy equal 0.015 eV.

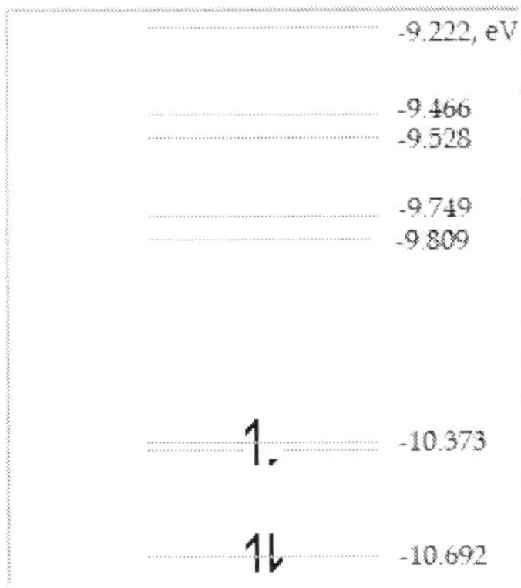

Fig. 27. Energy-band structure of the Al-doped (6,5) nanotube with hydrogen endings.

3.5 Doping by atom of phosphorus of a carbon (6,5) nanotube

The atom of phosphorus has five valence electrons, therefore it is necessary to expect occurrence in a nanotube of a donor centre near to it, as for aromatic bonds with atoms of carbon enough four electrons. We shall exchange in basic model of a 200-atomic nanotube the any atom of carbon with atom of phosphorus, having applied aromatic bonds C-P. For carrying out of geometrical optimization of the constructed P-doped nanostructure we shall use the molecular mechanic method MM + with application of the block-diagonal algorithm of Newton-Raphson. We shall gain model of the P-doped carbon (6,5) nanotube presented on Fig. 28. Apparently on pattern, that the atom of phosphorus has taken a place in exterior area near to a nanotube. It is caused by that bond C-P and atomic radius of phosphorus is more than bonds C-C and atomic radius of carbon on 0.4 and 0.3 Å, accordingly.

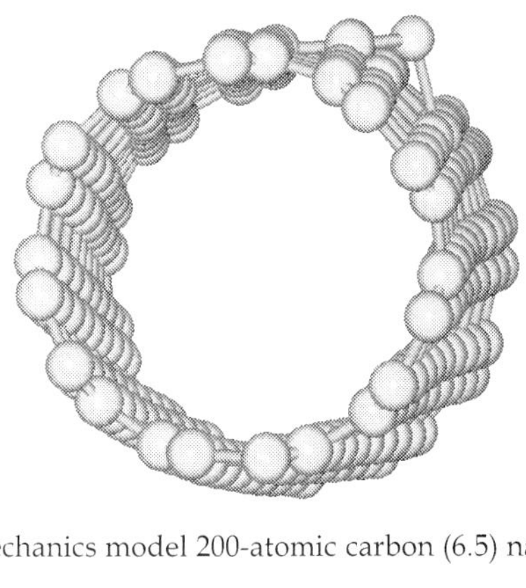

Fig. 28. The molecular mechanics model 200-atomic carbon (6.5) nanotube doped by atom of phosphorus.

In connection with mechanical stresses near to the impurity atom it is necessary to note, that the considered nanostructure is optimized only at RMS gradients of greater 0.015 kcal/(Å·mol), and for a basic nanotube analogous magnitude makes 0.000001 kcal/(Å·mol). The band diagram of a viewed nanostructure is presented on Fig. 29. Comparing the gained diagram with the diagram of a basic nanotube (Fig. 16b), we see, that the fifth electron of atom of phosphorus has occupied the free orbital 401 with an energy -10.758 eV and that in a forbidden region the level with an energy -10.082 eV was generated.

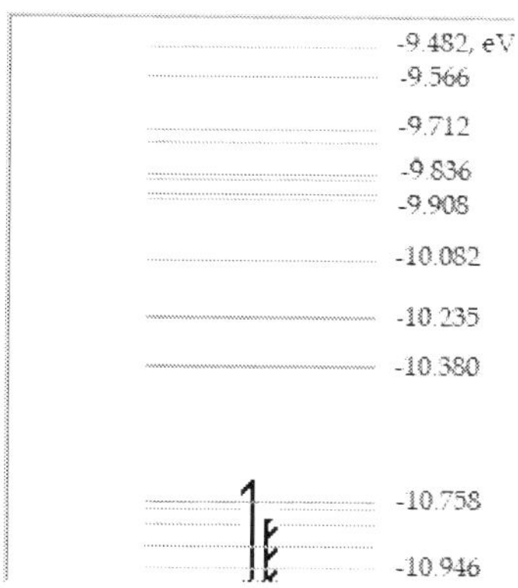

Fig. 29. Energy band structure of the 200-atomic carbon (6.5) nanotube doped by atom of phosphorus.

The models of a phosphorus-doped (6,5) nanotube with hydrogen passivation of dangling bonds is presented on Fig. 30.

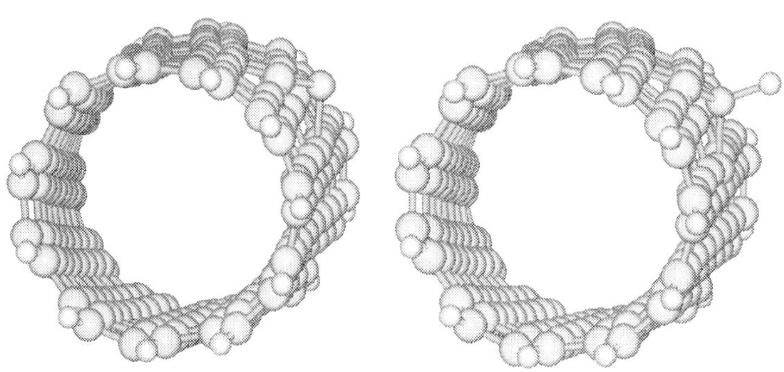

a b

Fig. 30. Molecular mechanics models of the P-doped carbon (6,5) nanotube : (a) - hydrogen endings on edges of a nanostructure; (b) - the complete hydrogen endings.

We see, that the doped nanotube has one broken off bond on atom of phosphorus. If such bond to leave without hydrogen passivation to it any other radical including the same nanotube can be affiliated.
Apparently on Fig. 31, passivation of the dangling bonds has essentially changed an energy distributions of a viewed nanostructure:
the upper filled levels have risen up to magnitudes -10.077 eV (a) and -10.487 eV (b);
degeneration of some energy levels is taken off;
quantities of energy levels have changed.

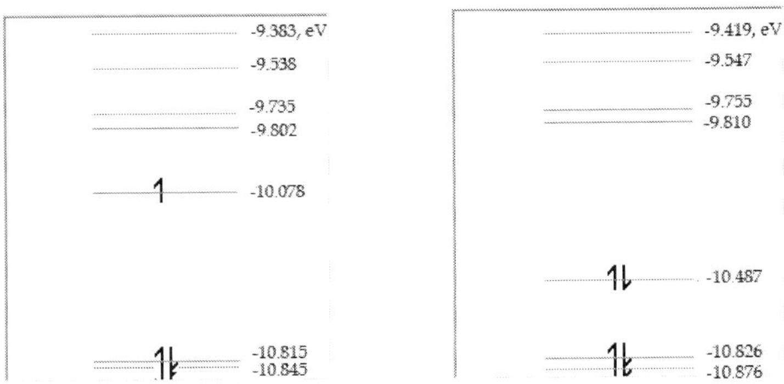

a b

Fig. 31. The phosphorus-doped 200-atomic carbon (6,5) nanotube: (a) - hydrogen endings on edges of a nanostructure; (b) - complete hydrogen endings.

The magnification of number of atoms of carbon in model of a phosphorus-doped nanotube up to 410 pieces results in to diminution of breadth of a forbidden gap and bias of a HOMO 0 levels up to quantities -10.466 eV (Fig. 32a) and -9.852 eV (Fig. 32b).

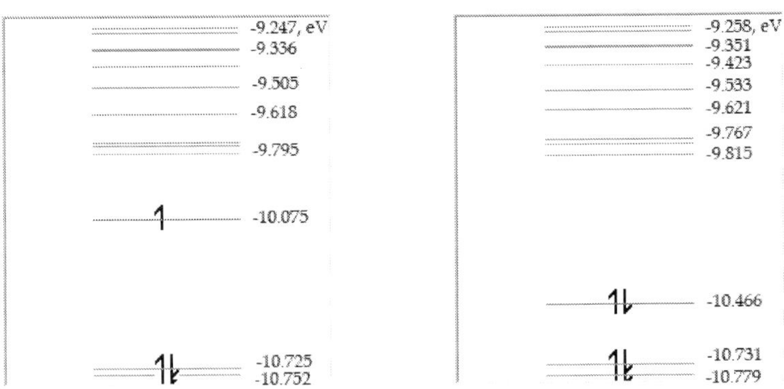

a b

Fig. 32. The phosphorus-doped 410-atomic carbon (6,5) nanotube: (a) - hydrogen endings on edges of a nanostructure; (b) - the complete hydrogen endings.

Thus, the impurity atom of phosphorus without hydrogen ending of its dangling bond can create in a nanotube a deep donor level with an ionization energy equal 0.28 eV. In case of when the free bond of atom of phosphorus is neutralized by atom of hydrogen the recombination centre is formed. Besides in any case, the impurity atom of phosphorus will reinforce emissivity of a nanotube as its valence electron occupies higher energy levels.

4. Conclusion

On models of semiconductor carbon nanotubes with chiral vectors (10,0) and (6,5), consisting of 200 atoms, effect of impurities of atoms Al, B, N and P on performances of their energy-band structure is viewed. The magnification of number of atoms of carbon of the doped nanostructure twice feebly influences its energy distribution, but results in to diminution of a forbidden gap of a nanotube approximately on 0.1 eV.

Process of doping of a carbon nanotube was simulated by replacement of one of its atoms of carbon on an impurity atom. Such doping of a nanotube in experiment can be implemented at existence of vacancies in it. Theories about equilibrium concentration of vacancies in nanotubes are not present, therefore it can become the purpose of the further developments. It is necessary to note, that intercalation atoms in nanotubes can be viewed as interstitial atoms in crystals. Then their presence at a nanotube can determine process of doping also.

It is shown, that the energy distribution of levels in the field of a forbidden gap of the doped nanostructure strongly depends on passivation of its torn off bonds. In the present investigation modeling of process of such passivation was carried out by hydrogen endings. Also it is shown, that doping of a carbon nanotube by atom of phosphorus results in to occurrence of dangling bond on its fifth valence electron, as for support of aromatic bonds with atoms of carbon enough four electrons. Passivation of this bond by atom of hydrogen results in to changes in an energy band structure of the doped nanotube.

Impurity atoms of boron and aluminium are representatives of the third group of a periodic system of elements, therefore they create similar acceptor levels in carbon nanotubes with chiral vectors (10,0) and (6,5):

a) atom of boron creates levels with activation energies 0.17 and 0.08 eV, accordingly;

b) atom of aluminium creates levels with identical activation energies 0.08 eV.

Atoms of nitrogen and phosphorus are representatives of the fifth group of a periodic system of elements, therefore it was expected, that they will create similar donor levels, but it has not proved to be true. The impurity atom of nitrogen creates in a carbon nanotube with a chiral vector (6,5) donor level with an ionization energy 0.07 eV, and in a nanotube (10,0) - very deep donor level near to valence band. The impurity atom of phosphorus creates in a carbon nanotube with a chiral vector (6,5) donor level with an ionization energy 0.28 eV, and in a nanotube (10,0) - very deep donor level near to valence band and a level with energy 10.1 eV.

5. References

Berkert U., Allinger N.L. (1982). *Molecular mechanics.* Amer. Chem. Soc. Mon. 177, Washington, D.C.

Bondarev, I. V. (2007). Quantum Optics Phenomena in Atomically Doped Nanotubes. *Optics and Spectroscopy,* Vol. 103, No 3, (2007) p. 366.

Derycke, V.; Martel, R.; Appenzeller J., Avouris, Ph. (XXXX). Carbon nanotube inter- and intramolecular logic gates. *Nano Letters*, vol. 0, 0, (XXXX) A-D.

Esfarjani, K.; Farajian, A. A.; Hashi, Y.; Kawazoe, Y. (1999). Electronic and transport properties of N-P doped nanotubes. *Appl. Phys. Lett.*, **74**, (1999) p. 79.

Esfarjani, K.; Chen, Z. and Kawazoe, Y. (2003). Electronic properties of magnetically doped nanotubes. *Bull. Mater. Sci.*, Vol. 26, No 1, (January 2003) pp. 105-107.

Farajian, A. A.; Mizuseki, H. and Kawazoe, Y. (2002). Spin-polarized transport through doped nanotube junctions in presence of applied magnetic field, *Proceeding of 10th Int. Symp. "Nanostructures Physics and Technology"*, pp. TN.06p, St Petersburg. Russia, June 17-21 (2002), IOFFE Institute, St Petersburg.

Gong, K.; Du, F.; Xia, Z.; Durstock, M.; Da, L. (2009). Nitrogen-doped Carbon Nanotube Arrays with High Electrocatalitic Activity for Oxygen Reduction. *Science*, Vol. 323, No 5915, (6 February 2009) pp. 760-764.

Griadun V. I. (2006). Vacancies in nanotubes and fullerenes, *Proceedings of 16-th International Crimean Conference "Microwave & Telecommunication Technology" (CriMiCo'2006)*, vol. 2, pp. 669-670, ISBN 966-322-006-6, Sevastopol, September 11-15, 2006, Weber Publ. Co.

Griadun V. I. (2007). Doped carbon nanotube (10,0) characteristics, *Proceedings of 17-th International Crimean Conference "Microwave & Telecommunication Technology" (CriMiCo'2007)*, vol. 2, pp. 573-574, ISBN 978-966-335-013-4, Sevastopol, September 10-14, 2006, Weber Publ. Co.

Grigorian, L.; Sumanasekera, G. U.; Loper, A. L.; Fang, S.; Allen, J. L. and Eklund, P. C. (1998). Transport properties of alkali-metal-doped single-wall carbon nanotubes. *Phys. Rev.*, **B 58**, (1998) R4195-R4198.

Hoffmann, R. (1963). An Extended Hückel Theory. *J. Chem. Phys.*, (1963), vol. 39, No 6, (1963) pp. 1397-1412.

Lee, R. S.; Kim, H. J.; Fisher, J. E.; Thess A.; Smalley, R. E. (1997). Conductivity enhancement in single-walled carbon nanotube bundles doped with K and Br. *Nature*, **388**, (17 July 1997) pp. 255-257.

Martel, R.; T. Schmidt, T.; Shea, H. R.; Hartel, T. and Avouris, Ph. (1998). Single- and multi-wall carbon nanotube field-effect transistors. *Appl. Phys. Lett.*, vol. 73, No 17, (1998) pp. 2447-2449.

Miyamoto, Y.; Rubio, A.; Blase, X.; Cohen, M. L. and Louie, S. G. *(1995)*. Ionic Cohesion and Electron Doping of Thin Carbon Tubules with Alkali Atoms. *Phys. Rev. Lett.*, **74**, (1995) pp. 2993-2996.

Rao, A. M.; Eklund, P. C.; Bandow, S.; Thess, A.; Smalley, R. E. (1997). Evidence for charge transfer in doped carbon nanotube bundles from Raman scattering. *Nature*, **388**, (17 July 1997) pp. 257-259.

Yao, J. B.; Han, J. H.; Choi, S. H.; Lee, T. Y.; Park, C. Y.; Jeong, T. W.; Lee, J. H.; Yu, S.; Park, G.; Yi, W. K.; Baik Y. -J. and Kin, J. M. (2002). Emission Characteristics of boron nitride coated carbon nanotubes. *Physica B: Condensed Matter*, Volume 323, Issues 1-4, (October 2002) pp. 180-181.

Zidan, R.; Rao, A. M.; Au, M. (2003). Doped Carbon Nanotubes for Hydrogen Storage. FY 2003 Progress Report.

9

Fundamental Physical Aspects of Carbon Nanotube Transistors

Zoheir Kordrostami and Mohammad Hossein Sheikhi
Department of Electrical Engineering, Shiraz University
Nanotechnology Research Institute, Shiraz University
Shiraz, Iran

1. Introduction

The increasing demand for ultra-high speed processors, smaller dimensions and lower power consumption of integrated circuits has made the technology scaling of the electronic components a challenging issue for device designers. In past few decades, miniaturization of transistors has always obeyed the moore's law: the number of transistors that can be placed inexpensively on an integrated circuit has doubled approximately every two years. Nanoscale field effect transistors in the sub-10 nm regime, suffer from short channel effects such as direct tunneling from source to drain, increase in gate-leakage current and punch-through effect. These effects have posed severe problems for miniaturized transistors and directed the recent research toward better alternative semiconductors than silicon. Semiconducting carbon nanotubes (CNTs) because of their properties like large mean free path, excellent carrier mobility and improved electrostatics at nanoscales as the result of their non-planar structure, have been known as the best ideal replacement for silicon. In particular, they exhibit ballistic transport over length scales of several hundred nanometers (Heinze, et al., 2006). Absence of the dangling bond states at the surface of CNTs and purely one-dimensional transport properties improve gate control while meeting gate leakage constrains and allows for a wide choice of gate insulators. That's why CNTs suppress the short channel effects in transistor devices. Symmetry of the conduction and valence bands makes CNTs advantageous for complementary applications. CNTs are very attractive for nanoelectronic applications and can be used to achieve high speed ballistic carbon nanotube field effect transistors (CNTFETs). Theoretically, CNTFETs could reach a higher frequency domain (terahertz regime) than conventional semiconductor technologies. CNTFETs, only a few years after the initial discovery of CNTs in 1991 by Sumio Iijima (Iijima, 1991) were first demonstrated in 1998 by Dekker et al. at Delft University (Tans, et al., 1998) and soon after by groups at IBM (Martel, et al., 1998) and Stanford University (Soh, et al., 1999). Intensive research has led to a significant progress in understanding the fundamental properties of CNTs. By using a single-wall CNT as the channel between two electrodes which work as the source and drain contacts of a FET, a coaxial CNTFET can be fabricated. Coaxial devices are of special interest because their geometry allows for better electrostatics because the gate contact wraps all around the channel (CNT) and has a very good control on carrier transport. Type of Metal-CNT contacts plays crucial role in the output characteristics of the transistor. Heavily doped semiconductors because of the ability to form Ohmic contacts can be used as ideal electrodes but they suffer from high parasitic resistance. Existence of potential barrier at the metal-CNT interface, changes the device to a CNTFET resembling to Schottky barrier MOSFETs. However, heavily doped CNT contacts can be used to get to a behavior similar to conventional MOSFETs.

Understanding CNTFETs from electronic point of view requires a deep insight for mesoscopic physics. For modelling a CNTFET, a powerful methodology with the abilities of solving Schrödinger equation under non-equilibrium conditions in the presence of self-consistent electrostatics and treating coupling of the channel to contacts is needed. The non-equilibrium Green's function (NEGF) formalism provides a sound basis for quantum device simulations. In this chapter we aim to introduce different types of CNTFETs, their electrostatics and output characteristics.

The chapter is organized as follows: section 2 introduces different types of CNTFETs and typical design parameters, section 3 explains different physical issues involved with CNTFETs and ultimately in section 4 we briefly describe the NEGF formalism and WKB treatment for simulating CNTFETs. We use these methods to plot the curves for implying physical aspects through the chapter. While explaining the important issues, new challenges for achieving a well-designed CNTFET will be explained. After reading this chapter, the reader should be able to distinguish how a specific type of CNTFETs behaves (both DC analysis and high frequency response) and why.

2. Common CNTFET Designs

2.1 Types of CNTFETs

There are two main types of CNTFETs that are being currently studied, differing by their current injection methods. CNTFETs can be fabricated with Ohmic or Schottky contacts. The type of the contact determines the dominant mechanism of current transport and device output characteristics. CNTFETs are mainly divided into Schottky barrier CNTFETs (SB-CNTFETs) with metallic electrodes which form Schottky contacts and MOSFET-like CNTFETs with doped CNT electrodes which form Ohmic contacts. In SB-CNTFETs, tunneling of electrons and holes from the potential barriers at the source and drain junctions constitutes the current. The barrier width is modulated by the application of gate voltage, and thus, the transconductance of the device is dependent on the gate voltage (Raychowdhury, et al., 2006).

The other type of the CNTFETs takes advantage of the n-doped CNT as the contact. Potassium doped source and drain regions have been demonstrated and the behavior like MOSFETs have been experimentally verified (Javey, et al., 2005). In this type of transistors a potential barrier is formed at the middle of the channel and modulation of the barrier height by the gate voltage controls the current.

2.2 Typical Designs

At this point, we consider two typical designs for two types of the CNTFETs and use them for investigation and comparison of the transistors' behaviors. Let's consider two coaxial CNTFETs with a (13,0) zigzag CNT as the channel which corresponds to a bandgap of about 0.83 eV and a diameter of 1 nm. A 2 nm thick ZrO_2 with a relative permittivity of 25 separates the gate electrode and the CNT. The gate is 10 nm long and wrapped around the channel. The gate thickness is assumed 6 nm. SB-CNTFET employs an intrinsic CNT and 20 nm long metallic contacts as the source and drain. A typical structure and the distribution of the potential energy on the channel are shown in figure 1. The gate voltage makes the barriers near the source/drain thinner and increases the tunneling current.

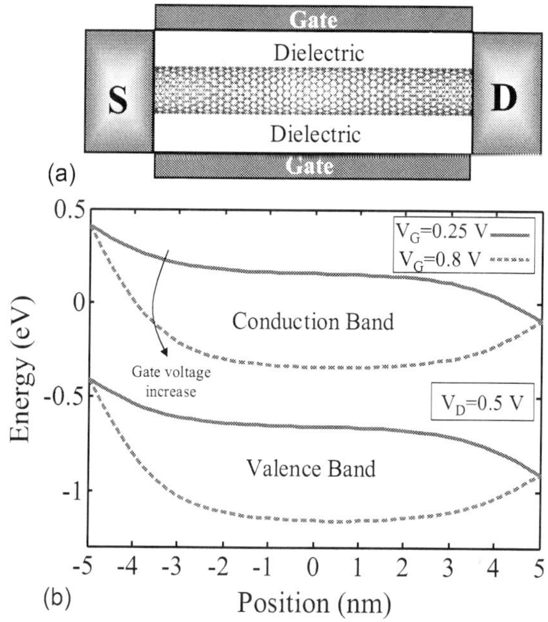

Fig. 1. Schottky Barrier CNTFET, a) 2-D cross section of the coaxial structure with intrinsic CNT as the channel and metal source/drain contacts, b) Energy band diagram obtained from Poisson equation. The metal Fermi level is taken to be at the midgap of the CNT.

On the other hand the MOSFET-like CNTFET benefits from heavily doped ends of the CNT which act as Ohmic source and drain contacts. A typical structure of the MOSFET-like CNTFETs with 20 nm doped sections and its energy band diagram is shown in figure 2. The source/drain doping is 10^{-9} m^{-1} (~0.01 dopant per atom). In the following sections of this chapter these two transistors will be compared to each other and the results can be generalized to other designs. The curves of this chapter have been obtained using the described methods in section 4.

3. Important Aspects of CNTFETs

3.1 Ambipolarity

One of the important aspects of nanotube transistors is the ambipolarity or unipolarity of their current-voltage characteristics. SB-CNTFETs exhibit strong ambipolar behavior. For high enough gate voltages the tunneling probability of electrons through the source

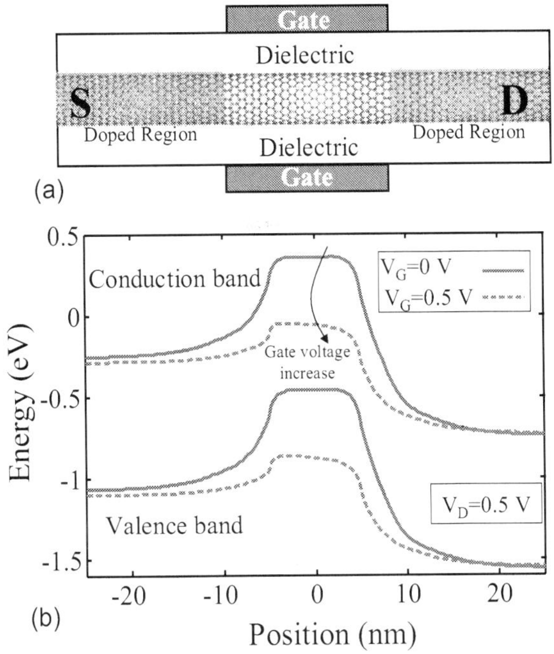

Fig. 2. MOSFET-like CNTFET, a) 2-D cross section of the coaxial structure with intrinsic CNT as the channel and doped CNT sections as source/drain contacts , b) Energy band diagram obtained from Poisson equation.

Schottky barrier in the conduction band is high and for the low and negative gate voltages, a Schottky barrier is formed at the drain in valence band and tunneling of holes increases the current. The energy bands for low and high gate voltages and the Schottky barriers are shown in figure 1(b). However for MOSFET-like CNTFETs only positive gate voltages because of lowering the barrier in the channel increase the current. The energy bands for low and high gate voltages and the potential barrier in the channel are shown in figure 2(b). It is clear from the figures why the characteristics of these two transistors differ. The current-voltage characteristics of the devices have been compared in figure 3.

Ambipolar behavior of the SB-CNTFETs constraints the use of these transistors in conventional CMOS logic families. We have proposed some methods to reduce or eliminate the ambipolarity in CNTFETs. Using asymmetrical contact types we have introduced Schottky-Ohmic CNTFET which has a Schottky barrier at the CNT-metal interface at the source and an Ohmic contact at the drain at the channel-doped CNT interface (Kordrostami, et al., 2008). However, the device still suffers from band to band tunneling. That is, in low or negative voltages the electrons from the valence band can tunnel to the conduction band and contribute to the total current of the transistor and cause increase of the current in negative voltages. We have compared the potential energy distribution of the channel of the transistors in figure 4 (Kordrostami, 2007).

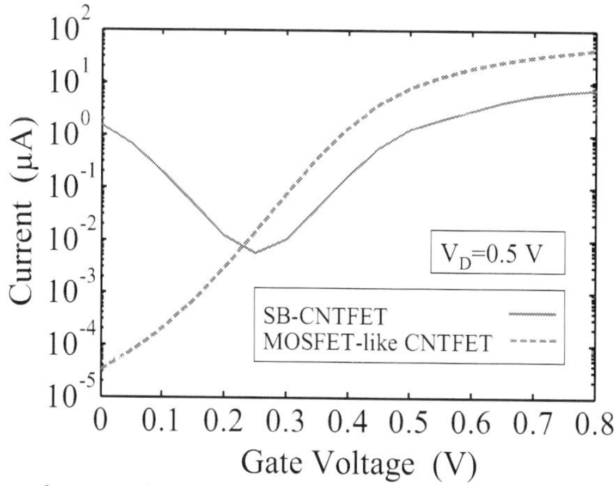

Fig. 3. Comparison between the current-voltage characteristics of the SB and MOSFET-like CNTFETs. Ambipolar behavior of the SB-CNTFET and the unipolar characteristic of the MOSFET-like CNTFET can be seen from the curves.

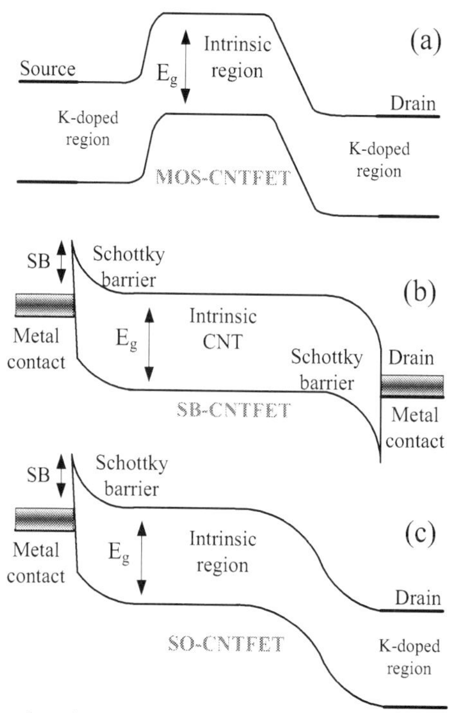

Fig. 4. Potential energy distribution along the carbon nanotube in three different CNTFET structures, a) MOSFET-like, b) Schottky barrier, c) Schottky-Ohmic (Kordrostami, 2007)

Double-gate structures can also reduce the ambipolarity while the CNT is still intrinsic. In the structure shown in figure 5 the first gate controls carrier injection at the source and the second one acts like an electrostatic doping and controls carrier injection at the drain which can be used to suppress the hole tunneling current, for example by applying the same voltage as the drain voltage to the second gate (Pourfath, et al., 2005). In this case, at any drain voltage the band edge profile near the drain contact will be flat like the band diagram in figure 4(c). On the other hand, an advantage of an ambipolar SB-CNTFETs is that they can be used as either an n-type and p-type FET in a CMOS application (Guo and Lundstrom, 2009).

In MOSFET-like CNTFETs with heavily doped source and drain regions, when applying a negative gate voltage, the band to band tunneling may lead to ambipolarity. For suppressing this effect in MOSFET-like CNTFETs we proposed a non-uniform doping profile as shown in figure 6 (Hassaninia, et al., 2008b). This reduces the gradient of the channel potential barrier at each interface between the intrinsic and doped sections of the CNT and suppresses the band to band tunneling and ambipolar conduction.

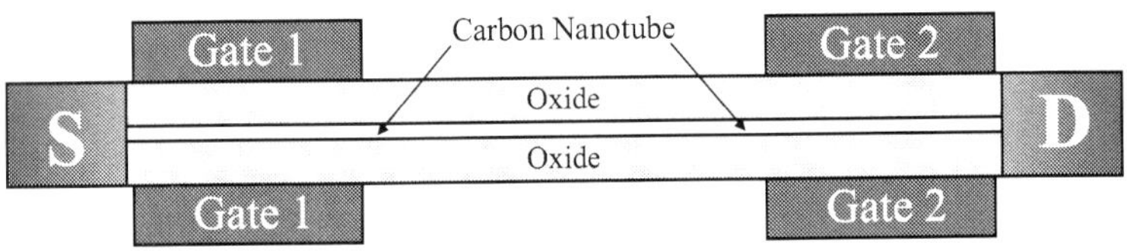

Fig. 5. Double gate CNTFET (Hassaninia, et al., 2008a). The first gate modulates the source tunneling current and the second gate in fact eliminates the Schottky barrier at the drain (path for tunneling of the holes in the valence band).

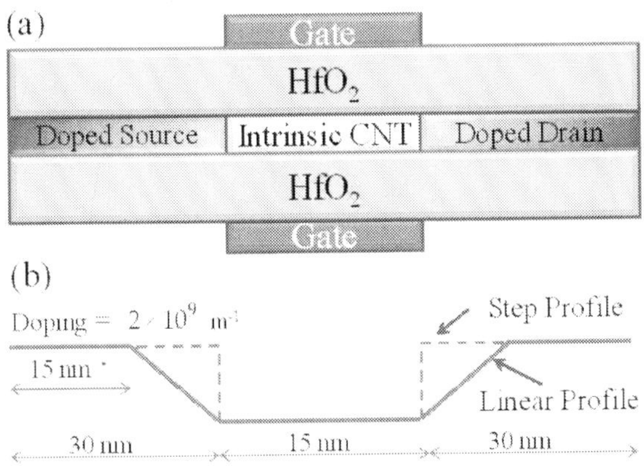

Fig. 6. a) Cross section of the CNTFET. (b) Step and linear doping profiles for the MOSFET-like structure versus position.

3.2 High Permittivity Gate Dielectrics

The relatively low dielectric constant of SiO_2 (at 3.9) limits its use in transistors as gate lengths scale down to tens of nanometers. As the device dimensions approach the 10 nm regime, strong electrostatic coupling of source/drain electrodes arises fundamental challenges especially about the gate control over the channel. Enhancing the gate efficiency requires thinner gate oxides. However, due to excessive direct tunneling leakage currents through the ultra-thin dielectric, the gate dielectric layer is already approaching its limit (~ 1nm) and the only feasible way is to use high-κ gate dielectrics which afford high gate capacitance without relying on ultra-small film thickness (Javey, et al., 2002). The most common gate oxides for nanotransistors are Zirconium[1] (ZrO_2) and Hafnium dioxide (HfO_2) with relative permittivities about 25 and 16 respectively. The current versus the dielectric constant of the oxide in both types of transistors has been shown in figure 7.

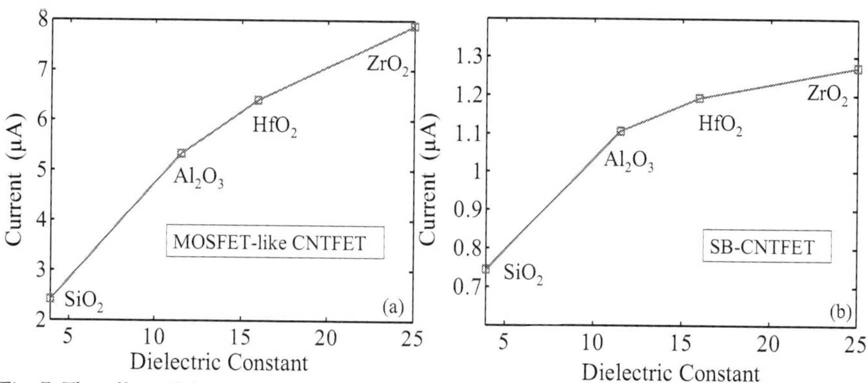

Fig. 7. The effect of the dielectric constant on the on current of the transistors in the on state. (V_G=V_D=0.5 V).

3.3 On/Off Current

Comparison shows that MOSFET-like CNTFETs due to absence of Schottky barriers have a lower off leakage current. On the other hand, in the on state, there is no barrier at the source-channel junction and hence, the device demonstrates significantly higher on current. The minimum current of the SB-CNTFETs occurs when the contribution of the electron and hole tunneling currents to the total current becomes equal, that is when $V_G= V_D/2$ where the energy band diagram is symmetric. For each power supply voltage (V_D), the off-current is defined at the minimal leakage point $V_G(off)=V_D/2$ and the on-current is defined at $V_G(on)=V_G(off)+V_D$. There exists a trade-off: reducing the off-current by lowering the power supply voltage degrades the on-current.

3.4 Transconductance

The transconductance (g_m) of the SB-CNTFETs is severely limited by the Schottky barriers in the on state. The transconductances of the devices as a function of the gate voltage have

---

[1] The name Zirconium originates from the Persian word 'zargun' meaning gold-like.

been compared to each other in figure 8(a). It is seen that transconductance of the MOSFET-like CNTFET is higher than the SB-CNTFET. Figure 8(b) plots CNTFET on state transconductance versus the gate dielectric constant for four types of widely used gate insulators. The high-κ gate insulator improves the CNTFET performance by reducing the self-consistent potential produced by the charge on the tube (Guo, 2004). For the SB-CNTFETs the transconductance tends to saturate when the gate insulator dielectric constant is large. The reason is that the self-consistent potential produced by the charge on the tube is already small and further improving the gate dielectric constant does not help to significantly reduce the Schottky barrier thickness and the transistor performance (Guo, 2004).

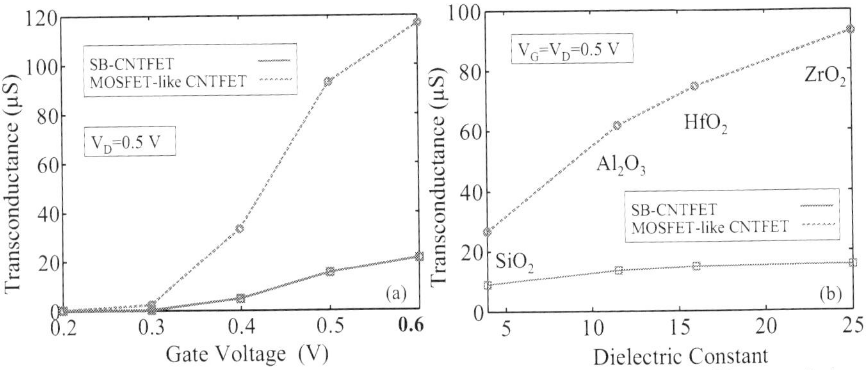

Fig. 8. Transconductance of the CNTFETs, a) g_m versus the gate voltage, b) g_m variations with respect to the dielectric constant.

3.5 Gate Capacitance

For a one dimensional nanotube FET, the total capacitance between the gate and the channel (gate capacitance) depends both on the geometry and the density of states (DOS). If the electrostatic potential on the channel (V_S) is uniform, then the voltage drop over the gate oxide is also uniform and the gate voltage V_G is the summation of the channel potential and the voltage drop over the gate oxide. This can be modeled as the potential division between two capacitances which means the gate capacitance (total capacitance) can be modeled as the series of the electrostatic (geometry dependent) capacitance and the quantum (DOS dependent) capacitance as shown in figure 9. Gate capacitance can be calculated from:

$$C_g = \partial Q / \partial V_G \tag{1}$$

where Q is the total charge on the gate electrode (which has the same magnitude but an opposite sign as the total net charge of the CNT) and V_G is the gate voltage. The gate oxide plays the role of the dielectric between the plates of the oxide (electrostatic) capacitance which can be expressed as:

$$C_{ox} = 2\pi\varepsilon L_g \Big/ Ln\left(\frac{t_{ox} + r_{CNT}}{r_{CNT}}\right) \tag{2}$$

where t_{ox} is the gate dielectric thickness, ε is the dielectric constant of the gate insulator, L_g is the gate length and r_{CNT} is the nanotube diameter. Since the charge on the channel (Q) can be derived from CNT density of states and the Fermi level for a given channel potential, the quantum capacitance can be calculated from:

$$C_q = \partial Q / \partial V_s \tag{3}$$

where V_s is the channel potential. For MOSFET-like CNTFETs V_s and Q can be defined as the potential energy and charge density at the top of the channel potential barrier.

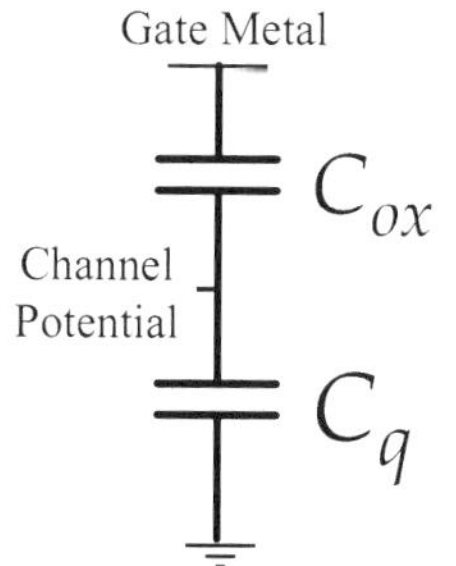

Fig. 9. Series combination of the oxide and quantum capacitances and the voltage drop on the capacitances.

The above derivation is strictly valid only when V_S is position-independent, otherwise the definition of the oxide capacitance breaks down (Guo, et al., 2007). The gate capacitances of the transistors versus the dielectric constant are shown in figure 10.

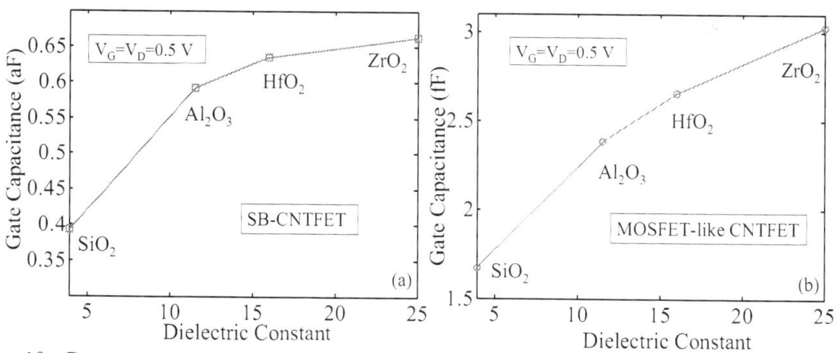

Fig. 10. Gate capacitance versus the dielectric constant a) SB-CNTFET b) MOSFET-like CNTFET.

The energy at the top of the barrier in the channel depends on the gate voltage and on the charge at the top of the barrier. When C_{ox} is small, the gate voltage controls the charge at the top of the barrier (which is independent of the drain voltage) and for large C_{ox}, the gate voltage controls the potential at the top of the barrier, which is independent of the drain

voltage (Guo and Lundstrom, 2006). In bulk devices like conventional MOSFETs, the quantum capacitance is larger than the oxide capacitance, however, CNTFETs usually work in the quantum capacitance limit (C_q<<C_{ox}).
We have compared the effect of the gate thickness on the gate capacitances of a completely gated and a partially gated SB-CNTFET. The results are shown in figure 11 and imply that the thicker gate increases C_g in the partially gated and has no effect in a completely gated SB-CNTFET (Kordrostami and Sheikhi, 2009b).

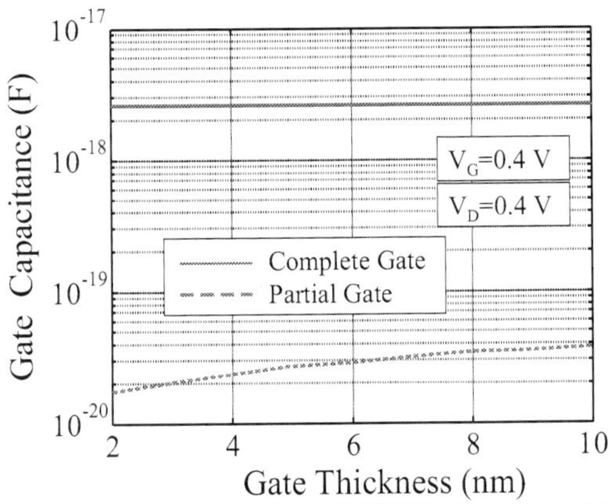

Fig. 11. Gate capacitances for two types of SB-CNTFET. The channel is an intrinsic 50 nm long (19,0) zigzag CNT, the gate length in partially gated SB-CNTFET is 5 nm and the dielectric is SiO_2 with a permittivity of 3.9.

3.6 Fringing Capacitance

The fringing fields between the gate metal and source and drain contacts result in capacitances which are called fringing or parasitic capacitances. Fringing field becomes more important when the channel length of a CNTFET is reduced. This could be comparable to the intrinsic device capacitances, and hence must be considered. In practice, appropriate electrode geometries are required to minimize the fringing field parasitic capacitances so that the parasitic capacitance due to fringing fields become negligible compared to the gate capacitance required to modulate the conductance, for example, by using 1-D metallic electrodes or nanotubes themselves as the electrodes (Yu, et al., 2006). We have investigated the effect of the contact geometry on the electrostatics of both SB-CNTFETs and MOSFET-like CNTFETs which clearly verifies the contribution of the fringing field to the device electrostatics and frequency response (kordrostami and Sheikhi, 2009a).

3.7 Cutoff Frequency

The unity current gain cutoff frequency (the frequency at which the current gain falls to unity) is usually used to describe high-frequency performance of a transistor. The cutoff frequency of the intrinsic CNTFET is called the intrinsic and the cutoff frequency of the CNTFET with inclusion of the parasitic capacitances is called the extrinsic cutoff frequency. When the parasitic capacitances are small the extrinsic cutoff frequency approaches the intrinsic cutoff frequency. We compute the cutoff frequency (f_T) using the quasi-static approximation. The quasi-static treatment works well when the signal varies slowly compared to the time constant determined by the intrinsic gate capacitance and the channel inductance (Guo, et al., 2005). The small-signal circuit model for a CNTFET based on the quasi-static approximation, which includes the equivalent capacitive and resistive elements, but omits the equivalent inductive elements, is shown in figure 12.

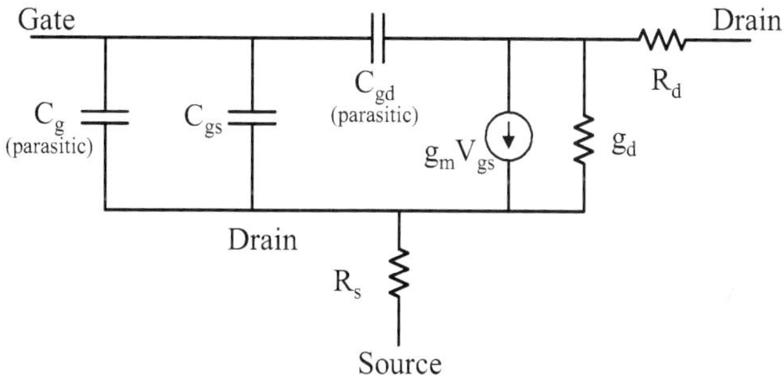

Fig. 12. Small-signal circuit model for a nanotube transistor including intrinsic and extrinsic elements.

By using the circuit model, we can generally write for the cutoff frequency:

$$\frac{1}{2\pi f_T} = (R_S + R_D)C_{gd} + \frac{1}{g_m}(C_g + C_{gs} + C_{gd}) + \frac{g_d}{g_m}(R_S + R_D)(C_g + C_{gs} + C_{gd}) \quad (4)$$

Where g_m is the transconductance, R_S and R_D are the parasitic resistances, C_g is the intrinsic gate capacitance and C_{gs} and C_{gd} are the parasitic capacitances (Burke, 2004). Calculation of the parasitic capacitances between the gate and source (drain) electrode requires separate electrostatic computations. If the parasitic resistances are excluded from the calculations we have:

$$f_T = \frac{1}{2\pi} \frac{g_m}{C_g + C_{gs} + C_{gd}} \quad (5)$$

The cutoff frequency of the intrinsic transistor without parasitic capacitances is:

$$f_T = \frac{1}{2\pi} \frac{g_m}{C_g} \quad (6)$$

In CNTFETs the intrinsic cutoff frequency can be determined in terms of the ratio of the change in the current to the change in the channel charge:

$$f_T = \frac{1}{2\pi} \frac{g_m}{C_g}\bigg|_{V_D=V_D(on)} = \frac{1}{2\pi} \frac{g_m}{C_g} \frac{\partial V_G}{\partial V_G} \approx \frac{1}{2\pi} \frac{\partial I_D}{\partial Q_{Ch}}\bigg|_{V_D=V_D(on)} \quad (7)$$

Where, I_D is the source-drain current and Q_{Ch} is the total charge in the CNT channel and is given as (Yoon, et al., 2006):

$$Q_{Ch} = q\int_{0}^{L_{Ch}} N_e(x)dx \tag{8}$$

where $N_e(x)$ is the electron density as a function of the channel position and L_{Ch} is the channel length. The intrinsic cutoff frequencies of the transistors with respect to the channel length have been compared in figure 13. The shorter the channel the larger the cutoff frequency of the CNTFET.

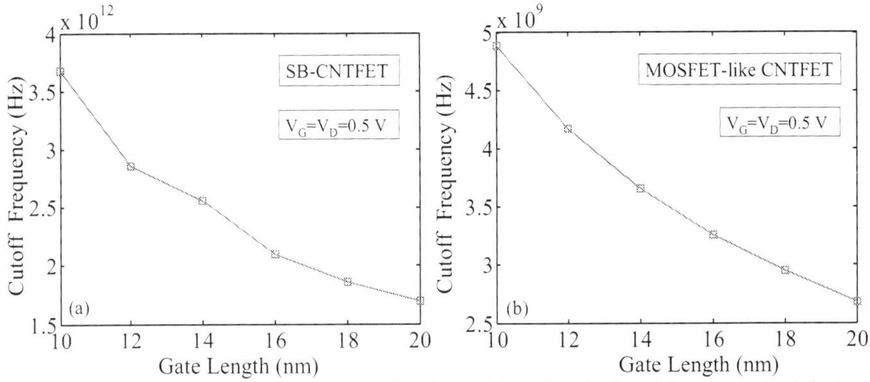

Fig. 13. Intrinsic cutoff frequency versus the channel length calculated for the typical designs in section 2.2, for a) SB-CNTFET, b) MOSFET-like CNTFET.

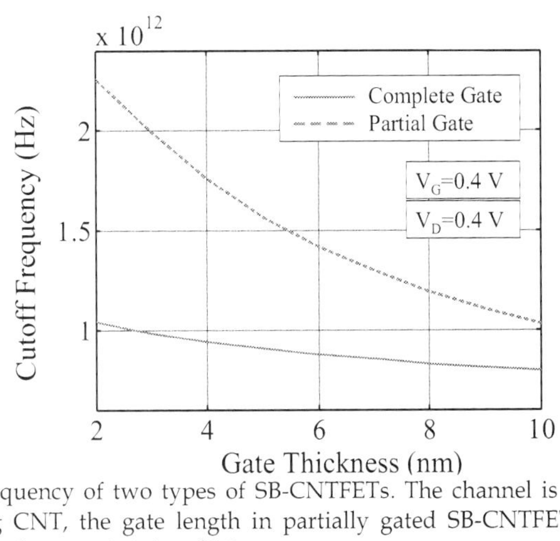

Fig. 14. Cutoff frequency of two types of SB-CNTFETs. The channel is an intrinsic 50 nm long (19,0) zigzag CNT, the gate length in partially gated SB-CNTFET is 5 nm and the dielectric is SiO_2 with a permittivity of 3.9.

From the equation 5 it can be concluded that the cutoff frequency is inversely proportional to the fringing capacitances. For example the partially gated SB-CNTFET has a higher cutoff frequency than the complete gate SB-CNTFETs because the fringing capacitances between metal electrodes are lower in partially gated SB-CNTFETs (Kordrostami and Sheikhi, 2009b). figure 14 shows the reduction of the cutoff frequency when the gate metal is thicker because of the increase of the fringing capacitances.

3.8 Intrinsic Delay

The main limitation to a faster intrinsic delay time is the gate capacitance resulting from fringing fields through the high-κ dielectric directly from the gate to source and gate to drain (Khairul and Roger, 2006). The intrinsic delay is one of the most important performance metrics for digital electronic applications and characterizes how fast a transistor switches. The switching delay can be calculated from:

$$\tau = \frac{Q_{on} - Q_{off}}{I_{On}} \tag{9}$$

Where Q_{on} is the total charge of the channel at on state ($V_G = V_{D(on)}$) and Q_{off} is the total charge of the channel at off state ($V_G=0$, $V_D = V_{D(on)}$) and I_{On} is the on current (Yoon, et al., 2006). The calculated intrinsic delays of the typical transistors are shown in figure 15.

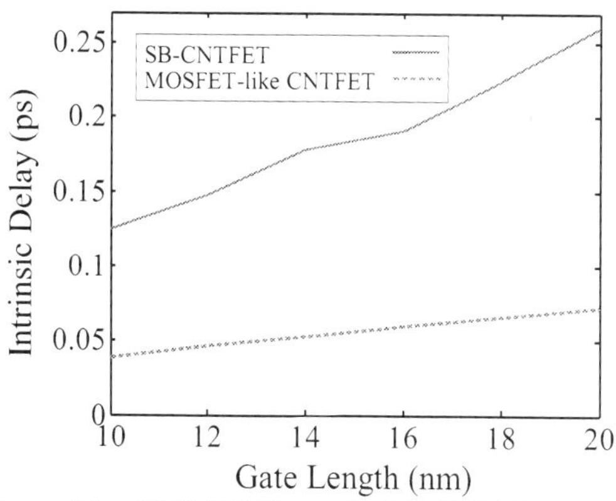

Fig. 15. Intrinsic delays of the SB-CNTFET and the MOSFET-like CNTFET versus the gate length. Longer channel length leads to larger latency. The on state is defined when $V_G=V_D=0.5$ V and the off state is defined as $V_G=0$ V and $V_D=0.5$ V.

4. Physical modelling of the CNTFETs

As electronic devices are being downscaled to nanometer range, the validity of the conventional modeling approaches becomes questionable. The quantitative simulation methods for nanoscale devices should incorporate an understanding of both atomistic structures and quantum mechanical effects.

In the following sections, we describe two common methodologies which can simulate the CNTFETs efficiently: NEGF formalism and WKB method for SB-CNTFETs.

4.1 NEGF Formalism

The non-equilibrium Green's function (NEGF) formalism provides a sound basis for modeling quantum devices, due to the following reasons:

- Atomic-level description of the channel.
- Channel-contact interfaces can be treated by describing open boundary conditions for the Schrödinger equation.
- Dissipative scattering processes and other phenomena like light emission can be modeled.
- Considering quantum mechanical tunneling through Schottky barriers at the metal/nanotube contacts, tunneling and reflection at barriers in the nanotube channel and band to band tunneling.

In this section, we give a brief summary of the NEGF simulation procedure and how to apply the approach to a nanotransistor. The procedure is as follows (Guo and Lundstrom, 2009):

- Identify a suitable basis set and Hamiltonian matrix (H) for an isolated channel.
- Compute the self-energy matrixes (Σ_S and Σ_D).
- Compute the retarded Green's function (G).
- Determine the physical quantities of interest from the Green's function.
- Solve NEGF transport equation iteratively with the Poisson equation until self-consistency is achieved.
- Calculate the drain current.

The procedure starts with an initial guess for the potential of the channel and then the charge density is calculated from the NEGF equations. For a given charge density, the Poisson equation is solved to obtain the electrostatic potential in the nanotube channel. Next, the computed potential profile is used as the input for the NEGF transport equation, and an improved estimate for the charge density is obtained. The iteration between the Poisson equation and the NEGF transport equation continues until self-consistency is achieved. At this time, all the physical quantities are exact and the current can be calculated. The Green's function can be calculated from:

$$G = \left[\left(E + i0^{+}\right)I - H - \Sigma_S - \Sigma_D\right]^{-1} \tag{10}$$

Where H is the Hamiltonian matrix and $\Sigma_{S,D}$ is the self energy matrixes for the source/drain interfaces. The device and the methodology are shown in figure 16.

Table 1 describes the NEGF formalism and how the physical quantities can be calculated. By choosing appropriate self-energy matrixes, the procedure can be implemented to model both types of the CNTFETs. The computationally expensive part of the procedure is calculating the Green's function. Using the real space basis set for calculation of the Hamiltonian matrix for the CNT channel leads to a matrix whose size is the total number of carbon atoms in the nanotube. In CNTFETs the mode space basis which uses the periodic boundary condition around the circumference of the nanotube is the appropriate approach

for calculation of the Hamiltonian of the channel because this approach significantly reduces the size of the Hamiltonian matrix.
After calculation of the transmission coefficient (Table 1), the current of the transistor can be computed from Landauer-Buttiker formula:

$$I = \frac{4e}{h} \int T(E)[f_S(E) - f_D(E)]dE \tag{11}$$

Where $f_{S,D}(E)$ is the Fermi function at the source/drain contact.

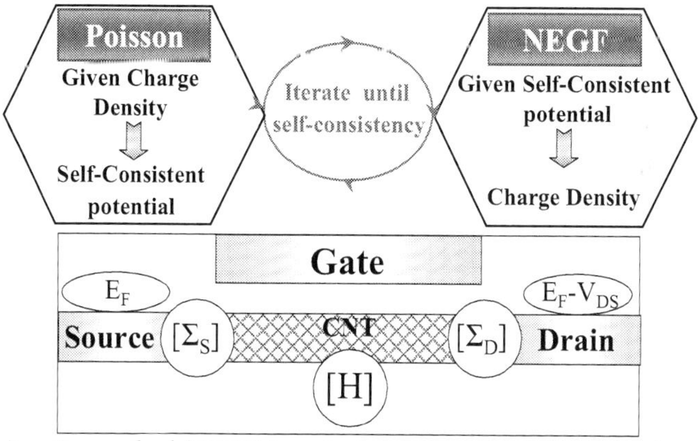

Fig. 16. The iterative method (NEGF) for calculating the potential and the charge density of the channel and the structure of the CNTFET under investigation.

Physical Quantity	Matrix Computation	Description
Broadening function	$\Gamma_S = i[\Sigma_S - \Sigma_S^+]$, $\Gamma_D = i[\Sigma_D - \Sigma_D^+]$	Related to the broadened density of states in the channel.
Spectral function	$[A_S(E)] = G\Gamma_S G^+$ $[A_D(E)] = G\Gamma_D G^+$	Diagonal elements give the LDOS at energy E in that representation: $D(E) = \frac{1}{2\pi} Trace[A(E)]$
Correlation function	$[G^n(E)] = [A_S]f_S + [A_D]f_D$	The matrix version of the electron density per unit energy. From $[G^n(E)]$ density matrix can be calculated.
Density Matrix	$[\rho] = \frac{1}{2\pi} \int [G^n(E)]dE$	Diagonal elements give the electron density: $n(x) = Diag[\rho]$
Transmission Coefficient	$T(E) = Trace[\Gamma_S G \Gamma_D G^+]$	Transmission probability from source to drain.

Table 1. NEGF formalism and the description of physical quantities.

4.2 WKB Treatment of Schottky Barrier CNTFETs

The Wentzel-Kramers-Brillouin (WKB) approximation can be used to solve Schrödinger equation and find the tunneling probability at the source/drain-metal interface. This method is assumed a semi-classical approach. The channel potential distribution can be found by solving Laplace equation. The potential of the channel found from Laplace equation is valid as far as the device works in the quantum capacitance limit. At this limit, the nanotube quantum capacitance is very small and the associated accumulated charge is close to zero (Jiménez, et al., 2006). That's why the Laplace equation is valid under this condition. By using WKB method and solving Schrödinger equation, the transmission coefficient of the channel (probability of tunneling through the Schottky Barriers) is achieved which is a function of the potential energy:

$$T(E) = \exp\left[-2\int_{z_1}^{z_2} k(z)dz\right] \tag{12}$$

Where z_1 and z_2 are classical turning points and $k(z)$ is the wave number which can be calculated from:

$$k(z) = \frac{2}{3aV_0}\sqrt{\left(\frac{E_g}{2}\right)^2 - [E + eV(z)]^2} \tag{13}$$

where a = 0.144 nm, E_g = 0.83 eV and V_0 = 2.5 eV are the Carbon–Carbon bond length, the CNT band gap and the tight-binding parameter, respectively. $V(z)$ is the electrostatic potential along the CNT and is obtained by solving the Laplace equation. The computed potential profile is used as the input for the transmission coefficient and ultimately the current can be calculated from equation 11 (Kordrostami and Sheikhi, 2009c). The Schottky barrier, potential variations and the classical turning points are shown in figure 17.

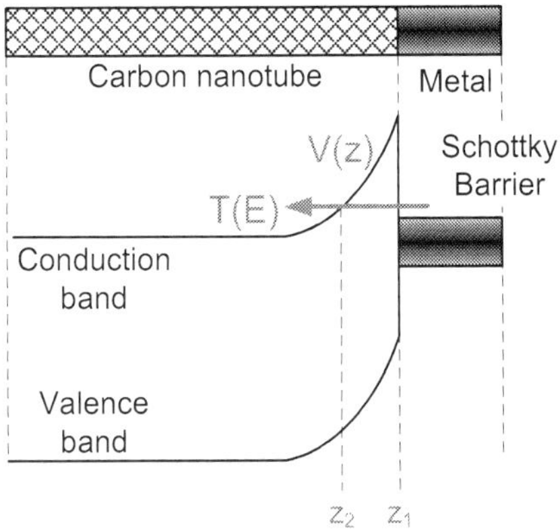

Fig. 17. Schottky barrier, potential variations near the channel-contact interface and the classical turning points which show the tunneling path.

5. Conclusion

The current-voltage characteristic of the SB-CNTFETs is ambipolar and the MOSFET-like CNTFETs exhibit unipolar characteristics. Asymmetric contact types and double gate structures are two ways for suppressing the ambipolarity of the SB-CNTFETs. Linear doping profile in MOSFET-like CNTFETs can reduce the band to band tunneling leakage current in negative voltages. High permittivity dielectrics are needed to ensure the proper control of the gate on the channel at the oxide thickness limit. The cutoff frequency of the CNTFETs severely depends on the gate and fringing capacitances. 1-D contacts can make the fringing capacitances as small as possible. Partially gated SB-CNTFET has smaller fringing capacitances and thus higher cutoff frequency in comparison with completely gated SB-CNTFET. Two simulation methodologies are reliable for modelling CNTFETs: NEGF transport equation self-consistently with Poisson equation for both types of the transistors and semi-classical WKB method for calculating the tunneling current through Schottky barriers in SB-CNTFETs. By using the simulation methods we discussed some trade-offs between different parameters of a particular CNTFET design. In order to achieve a well-designed nanotransistor, a compromise is always needed between different parameters.

6. References

Burke, P. J. (2004). AC performance of nanoelectronics: towards a ballistic THz nanotube transistor. *Solid-State Electronics,* 48, 10-11, 1981-1986, 0038-1101.

Guo, J. (2004). Carbon Nanotube Electronics: Modeling, Physics, and Applications. Purdue University.

Guo, J., Hasan, S., Javey, A., Bosman, G. and Lundstrom, M. (2005). Assessment of high frequency performance potential of carbon nanotube transistors. *IEEE transactions on Nanotechnology,* 4, 6, 715-721, 1536-125X.

Guo, J. and Lundstrom, m. (2006). *Nanoscale Transistors: Device Physics, Modeling and Simulation,* Springer, 978-0-387-28002-8, United States of America.

Guo, J. and Lundstrom, M. (2009). Device Simulation of SWNT-FETs, In: *Carbon Nanotube Electronic,* Javey, A. and Kong J., (Ed.), Springer, 978-0-387-36833-7, New York, USA.

Guo, J., Yoon, Y. and Ouyang, Y. (2007). Gate Electrostatics and Quantum Capacitance of Graphene Nanoribbons. *Nano Letters,* 7, 7, 1935-1940, 1530-6984.

Hassaninia, I., Kordrostami, Z., Sheikhi, M. H. and Ghayour, R. (2008a). Non-Equilibrium Green's Function Calculations for Double Gate Coaxial Carbon Nanotube FETs. *Proceedings of Nanostructures.* March 2008, Kish Island, Iran.

Hassaninia, I., Sheikhi, M. H. and Kordrostami, Z. (2008b). Simulation of carbon nanotube FETs with linear doping profile near the source and drain contacts. *Solid-State Electronics,* 52, 6, 980-985, 0038-1101.

Heinze, S., Tersoff, J. and Avouris, P. (2005). Carbon Nanotube Electronics and Optoelectronics, In: *Introducing Molecular Electronics,* Cuniberti, G. and Richter, K., (Ed.), 381-409, Springer, 978-3-540-27994-5, Netherlands.

Iijima, S. (1991). Helical microtubules of graphitic carbon. *Nature,* 354, 6348, 56-58.

Javey, A., Kim, H., Brink, M., Wang, Q., Ural, A., Guo, J., McIntyre, P., McEuen, P., Lundstrom, M. and Dai, H. (2002). High-[kappa] dielectrics for advanced carbon-nanotube transistors and logic gates. *Nat Mater,* 1, 4, 241-246, 1476-1122.

Javey, A., Tu, R., Farmer, D. B., Guo, J., Gordon, R. G. and Dai, H. (2005). High Performance n-Type Carbon Nanotube Field-Effect Transistors with Chemically Doped Contacts. *Nano Letters,* 5, 2, 345-348, 1530-6984.

Jiménez, D., Cartoixà, X., Miranda, E., Suñé, J., Chaves, A. F. and Roche, S. (2006). A drain current model for Schottky-barrier CNT-FETs *Computational Electronics,* 5, 4, 361-364.

Khairul, A. and Roger, K. L. (2006). Dielectric scaling of a zero-Schottky-barrier, 5 nm gate, carbon nanotube transistor with source/drain underlaps. *Journal of Applied Physics,* 100, 2, 024317.

Kordrostami, Z. (2007). Circuit Modeling of Carbon Nanotube Transistors and Interconnects for Integrated Circuit Applications. *Electrical and Electronic Engineering.* Shiraz University.

Kordrostami, Z., Hassaninia, I. and Sheikhi, M. H. (2008). Unipolar Schottky-Ohmic carbon nanotube field effect transistor. *Proceedings of 3rd IEEE International Conference on Nano/Micro Engineered and Molecular Systems,* 529-531, 978-1-4244-1907-4, Jan. 2008.

kordrostami, Z. and Sheikhi, M. H. (2009a). Contact Geometry dependent Electrostatics of Carbon Nanotube Transistors, *Nanoscience and Technology,* September 2009, Beijing, China.

Kordrostami, Z. and Sheikhi, M. H. (2009b). High Speed Switching Performance of Carbon Nanotube Field Effect Transistors, *Nanotech Europe,* September 2009, Berlin, Germany.

Kordrostami, Z., Sheikhi, M. H. (2009c). Schottky Barrier Field Effect Transistors with a Strained Carbon Nanotube Channel. Journal of Computational and Theoretical Nanoscience, 7, 6, 216–221, 1546-1955.

Martel, R., Schmidt, T., Shea, H. R., Hertel, T. and Avouris, P. (1998). Single- and multi-wall carbon nanotube field-effect transistors. *Applied Physics Letters,* 73, 17, 2447–2449.

Pourfath, M., Ungersboeck, E., Gehring, A., Kosina, H., Selberherr, S., Park, W.-J. and Cheong, B.-H. (2005). Numerical Analysis of Coaxial Double Gate Schottky Barrier Carbon Nanotube Field Effect Transistors. *Journal of Computational Electronics,* 4, 1, 75-78, 1569-8025.

Raychowdhury, A., Keshavarzi, A., Kurtin, J., De, V. and Roy, K. (2006). Carbon Nanotube Field-Effect Transistors for High-Performance Digital Circuits; DC Analysis and Modeling Toward Optimum Transistor Structure. *IEEE Transactions on Electron Devices,* 53, 11, 2711-2717, 0018-9383.

Soh, H. T., Quate, C. F., Morpurgo, A. F., Marcus, C. M., Kong, J. and Dai, H. (1999). Integrated nanotube circuits: Controlled growth and ohmic contacting of single-walled carbon nanotubes. *Applied Physics Letters,* 75, 5, 627–629, 0003-6951.

Tans, S., Verschueren, A. and Dekker, C. (1998). Room-temperature transistor based on a single carbon nanotube. *Nature* 393, 49–52, 0028-0836.

Yoon, Y., Yijian, O. and Jing, G. (2006). Effect of phonon scattering on intrinsic delay and cutoff frequency of carbon nanotube FETs. , *IEEE Transactions on Electron Devices,* 53, 10, 2467-2470, 0018-9383.

Yu, Z., Rutherglen, C. and Burke, P. J. (2006). Microwave nanotube transistor operation at high bias. *Applied Physics Letters,* 88, 23, 233115, 0003-6951.

10

Gate controlled Particle-Wave duality in a single walled carbon nanotube hole-transistor

Takafumi Kamimura
National Institute of Advanced Industrial Science and Technology
Japan Society for the Promotion of Science
CREST-Japan Science and Technology Agency
Japan

Kazuhiko Matsumoto
ISIR, Osaka University
National Institute of Advanced Industrial Science and Technology
CREST-Japan Science and Technology Agency
Japan

1. Introduction

Various electrode materials for single-walled carbon nanotube (SWNT) transistors were investigated. Pd electrodes have been used for Ohmic contacts (Javey et al., 2003). Ti electrodes have been used for Schottky contacts for hole conduction (Heinze et al., 2002; Martel et al., 2001; Kamimura & Matsumoto, 2005), and Mg and Ca electrodes for electron conduction (Nosho et al., 2006). Moreover, in the case of sub-μm order channel length at low temperature, SWNT transistors with Ohmic contacts have shown resonant tunneling transistor (RTT) characteristics (Liang et al., 2001), which are also called as Fabry-Perot characteristics, and SWNT transistors with Schottky contacts have shown single-hole transistor (SHT) characteristics (Suzuki et al., 2001), in which the Schottky barriers act as tunneling barriers. Therefore, electrodes materials should be chosen to obtain the desired characteristics.

In this study, we succeeded in fabricating a multifunctional quantum transistor using the particle nature and wave nature of holes in SWNT. This transistor can operate as an RTT and also as an SHT. An RTT is a device that uses the wave nature of hole and an SHT uses the particle nature of hole in the SWNT. Both devices need tunneling barriers at both sides of the quantum island. The RTT needs strong coupling while the SHT needs weak coupling between the quantum island and the electrodes. Usually, these tunneling barriers are made from thin oxide layers, etc. Therefore, the thickness of the tunneling barriers and the coupling strength cannot normally be controlled in a given device. In the present device, however, the Schottky barriers act as the tunneling barriers between the SWNT quantum island and electrodes. Therefore, the thickness of the tunneling barriers and the coupling strength between the SWNT and electrodes can be controlled by the applied gate voltage V_G.

Moreover, a SWNT is a cylindrical material with a diameter of several nanometers. The small diameter makes it possible to detect an electrical field from even a single-charge. Moreover, by observing the relative energy difference between the conducting carrier and the single-charges to be measured, it is possible to define the potential energy of the single-charges to be measured. However, as described in many reports (Kim et al., 2003; Radosavljevic et al., 2002), SWNT electron devices show hysteresis characteristics in gate voltage-drain current characteristics. The hysteresis characteristics are caused by gate-voltage-dependent charge fluctuation, e.g., adsorption of water molecules around a SWNT (Kim et al., 2003), charging into insulator layer around a SWNT (Radosavljevic et al., 2002), and charging into amorphous carbon around a SWNT (Martel et al., 2001) By eliminating these origins of the hysteresis characteristics, the number of fluctuating charges becomes small and a single-charge can be detected by a SWNT multi-functional quantum transistor.

2. Materials and methods

2.1 Method

We have eliminated the three origins of the hysteresis characteristics of a SWNT field effect transistor mainly pointed out in current reports (Martel et al., 2001; Kim et al., 2003; Radosavljevic et al. 2002; Kamimura & Matsumoto, 2004). To burn out amorphous carbon, we annealed a SWNT at low temperature in oxidizable atmosphere (Kamimura & Matsumoto, et al., 2004). To reduce the number of adsorbed atmosphere molecules, we covered the channel with a silicon dioxide layer. To reduce the number of trap sites in the insulator, we reduced channel length to 73 nm. The SWNT multi-functional quantum transistor fabricated by the process mentioned above shows almost no hysteresis characteristics in the gate voltage rage from -40 to 40 V. Moreover, an abrupt discrete switching of the source-drain current is observed in the electrical measurements of the SWNT multi-functional quantum transistor at 7.3 K. These random telegraph signals (RTS) are attributed to charge fluctuating charge traps near the SWNT multi-functional quantum transistor conduction channel. The current-switching behavior associated with the occupation of individual electron traps is demonstrated and analyzed statistically.

2.2 Sample preparations

A schematic of the sample structure is shown in Fig. 1. SWNT was prepared as follows. An n^+-Si wafer with a thermally grown 300 nm thick oxide was used as a substrate. Layered Fe/Mo/Si (2 nm/20 nm/40 nm) catalysts were evaporated using an electron-beam evaporator under a vacuum of 10^{-6} Pa. These layered catalysts were patterned on the substrate using the conventional photo-lithography process. SWNT was grown by thermal chemical vapor deposition (CVD) using the mixed gases of hydrogen and argon-bubbled ethanol. After the growth of the SWNT, it was purified by burning out the amorphous carbon around the SWNT in an air atmosphere at a temperature of several hundred degrees Celsius (Kamimura and Matsumoto, 2004). Ti (30 nm) electrodes were deposited on the patterned catalysts as the source and drain, and on the back side of the n^+-Si substrate for the gate, using the electron-beam evaporator under a vacuum of 10^{-6} Pa. The distance (L) between the source and drain was 73 nm. Thus, a back gate type multi-functional quantum transistor with an SWNT channel was fabricated that had the functions of an RTT and an

SHT. The single-charge measurement was carried out with the structure that silicon dioxide layer is on the SWNT channel.

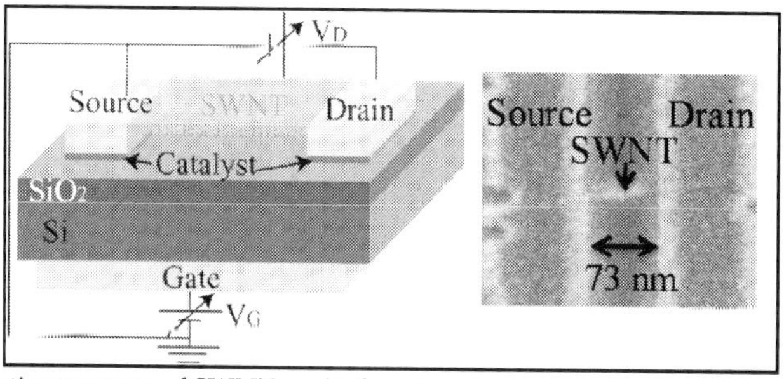

Fig. 1. Schematic structure of SWNT multi-functional quantum transistor covered by silicon dioxide layer. The channel length is 73 nm. The inset shows a SEM image around the channel before silicon dioxide deposition. A few charge storages are fabricated in the SiO_2 layer.

3. Results and discussions

3.1 Particle-Wave duality

Fig. 2(a) shows the differential conductance dI_D/dV_D characteristic as a function of the V_G at 7.3 K, where the drain voltage was set at 8 mV.

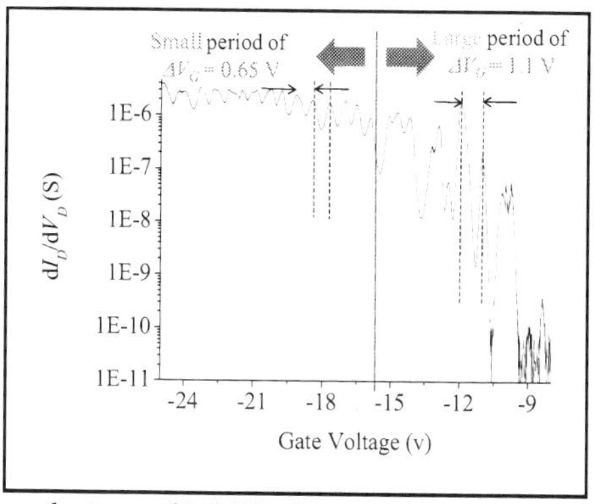

Fig. 2. (a) Differential conductance dI_D/dV_D characteristic as a function of V_G at 7.3 K. Two frequencies of oscillation were observed, which were $\Delta V_G = 1.1$ V and 0.65 V.

An oscillation characteristic with two oscillation periods is also observed in Fig. 2(a). A large oscillation period of $\Delta V_G = 1.1$ V was at $V_G \geq -16$ V and a small oscillation period of $\Delta V_G = 0.65$ V was at $V_G \leq -16$ V. Fig. 2(b) shows the dI_D/ dV_D peak on a linear scale at $V_G \geq -16$ V and $V_G \leq -16$ V. A clear difference in the oscillation period is observed as follows. The peaks of dI_D/ dV_D at $V_G \geq -16$ V are well fitted by the Gaussian, i.e.

$$G \approx \frac{2e^2}{h}\frac{T_L T_R}{\Gamma}\frac{1}{2kT}exp\left(\frac{\varepsilon_F-\varepsilon_0}{2kT}\right), \tag{1}$$

and at $V_G \leq -16$ V by the Lorentzian, i.e.

$$G = \frac{2e^2}{h}\frac{4T_L T_R}{(\varepsilon_F-\varepsilon_0)^2+\Gamma^2} \tag{2}$$

as shown in Fig. 2(c) and (d), respectively, where e is the elementary charge, h is the plank constant, T_L and T_R are the tunneling probabilities at the left and right tunneling barriers, Γ is the full width at half maximum, k is the Boltzmann constant, T is the temperature, ε_F is the Fermi level, and ε_0 is the quantum level.

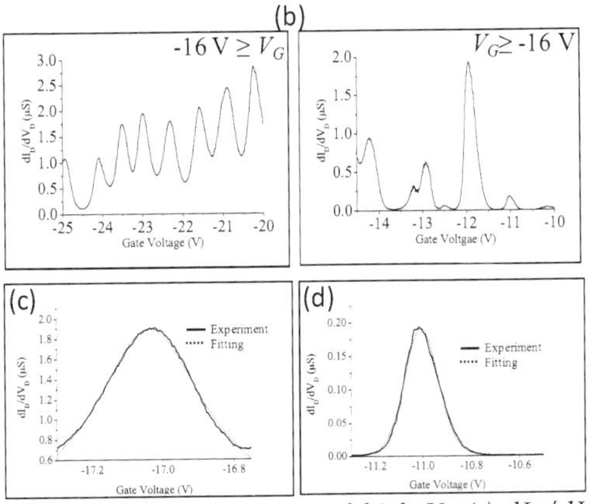

Fig. 2. (b) dI_D/dV_D peak in linear scale at low and high V_G. (c) dI_D/dV_D peak at V_G= -17 V (solid) and Lorentzian fitting line (dotted). The peak is well fitted by the Lorentzian function, which means the shape of the peak can be attributed to energy uncertainty broadening. Therefore, the peak must be a resonant tunneling current peak. (d) dI_D/dV_D peak at V_G = -11 V (solid) and Gaussian fitting line (dotted). The peak is well fitted by the Gaussian function, which means the shape of the peak can be attributed to thermal broadening. Therefore, the peak must be a Coulomb oscillation peak.

The shape of the Coulomb oscillation peaks must be Gaussian, which is attributed to thermal broadening, while the resonant tunneling current peaks must be Lorentzian, which is attributed to energy uncertainty (Radosavljevic et al., 2002). Therefore, the dI_D/dV_D oscillation at $V_G \geq -16$ V in Fig. 2(a) should be Coulomb oscillation peaks, and at $V_G \leq -16$ V in Fig. 2(a), they should be resonant tunneling current peaks. In other words, the device operates in particle nature mode at $V_G \geq -16$ V, and in wave nature mode at $V_G \leq -16$ V.

Fig. 3 shows the dI_D/dV_D characteristic as a function of the drain voltage V_D at 7.3 K and V_G = -12.795 V. The dI_D/dV_D peak spacing of 26 mV is observed in the plot, which corresponds to the quantum energy level separation in the SWNT quantum island. The separation of the quantum energy levels is indicated by

$$\Delta E_Q = \left(\frac{hv_F}{2L}\right)\left[1 + \left(\frac{2L}{3rn}\right)^2\right]^{-\frac{1}{2}}, \tag{3}$$

where r is the radius of the SWNT, L is the length of the SWNT, and v_F is the Fermi velocity (Kamimura & Matsumoto, 2004). When n becomes large, the equation becomes

$$\Delta E_Q = \frac{hv_F}{2L} \tag{4}$$

and shows a constant value independent of n.

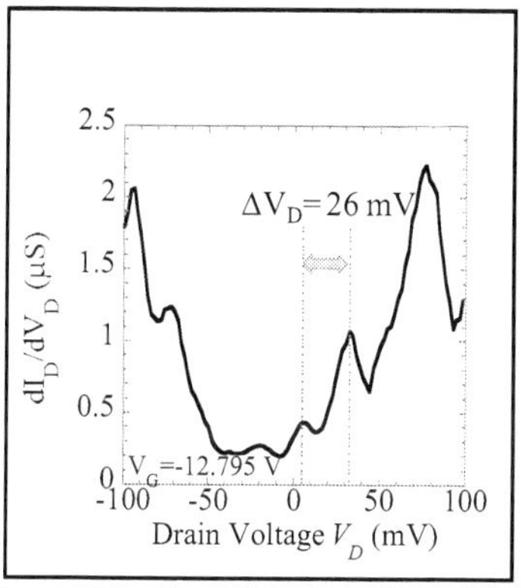

Fig. 3. dI_D/dV_D characteristic as a function of drain voltage V_D at V_G = -12.795 V and 7.3 K. The peaks are attributed to quantum energy levels in the SWNT. The estimated energy separation in 73 nm SWNT was 24 meV, which is in good agreement with the peak spacing of 26 mV in the plot.

The energy separation ΔE_Q of the quantum levels for an SWNT length of 73 nm is calculated to be 24 mV from eq. (2). Because this estimated value of energy separation of 24 mV is in good agreement with the dI_D/dV_D peak spacing of 26 mV in Fig. 3, it can be concluded that the entire SWNT channel acts as a single quantum island.
Fig. 4 shows a contour plot of the dI_D/dV_D characteristic as a function of V_G and V_D at 7.3 K. The characteristic can also be divided into two modes, the particle nature mode and wave nature mode. At $V_G \geq$ -16 V, as shown in Fig. 4(a) and (b), the plot clearly shows the Coulomb diamond structures, which are getting smaller with negatively increasing V_G. Additionally, line shaped quantum levels are observed outside of these Coulomb diamond structures. Therefore, at $V_G \geq$ -16 V, the device operated in the particle nature mode. As shown in Fig. 4(b), at -16 ≥ $V_G \geq$ -20 V, the Coulomb blockade was lifted and the Coulomb diamond structures disappeared at around V_G = -16 V. The quantum levels still remain and the so-called Fabry-Perrot quantum interference pattern is observed at this region. Finally, at $V_G \leq$ -20 V, the quantum levels are getting blurred with negatively increasing V_G, as shown in Fig. 4(c). Therefore, at -16 V ≥ V_G, the device operated in the wave nature mode.

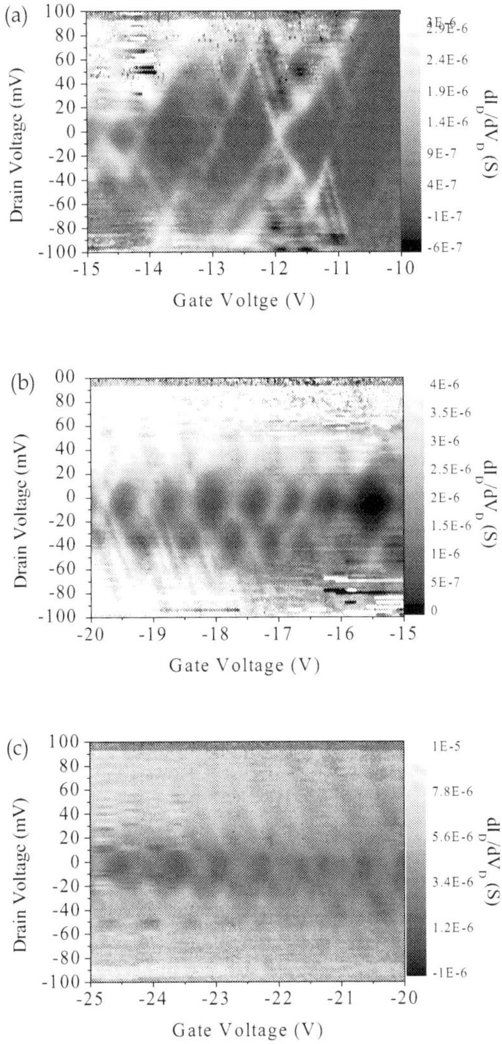

Fig. 4. Contour plot of dI_D/dV_D characteristic as a function of V_G and V_D at 7.3 K. (a) At $V_G \geq -16$ V, Coulomb diamond structures and line shaped quantum levels outside of the Coulomb diamond structures are observed. Moreover, the Coulomb diamonds become smaller with a negatively increasing V_G. (b) At $-16\text{ V} \geq V_G \geq -20$ V, the Coulomb blockade is lifted and the Coulomb diamonds disappear. And the quantum levels become blurred with the negatively increasing V_G. The so-called Fabry-Perrot quantum interference pattern is observed. (c) At $-20\text{ V} \geq V_G$, the blurred quantum levels still remain.

Fig. 5(a) shows the Coulomb charging energy E_C as a function of V_G. E_C is obtained by

$$E_C = \Delta E - E_Q \tag{5}$$

where ΔE is estimated from the size of the Coulomb diamond. E_C drastically decreases with a negatively increasing V_G, almost reaching zero at $V_G \geq$ -16 V. Therefore, the Coulomb diamonds disappear at -16 V $\geq V_G$. Tunneling capacitance C_t and gate capacitance C_G as a function of V_G are shown in Fig. 5(b). C_t and C_G are obtained from

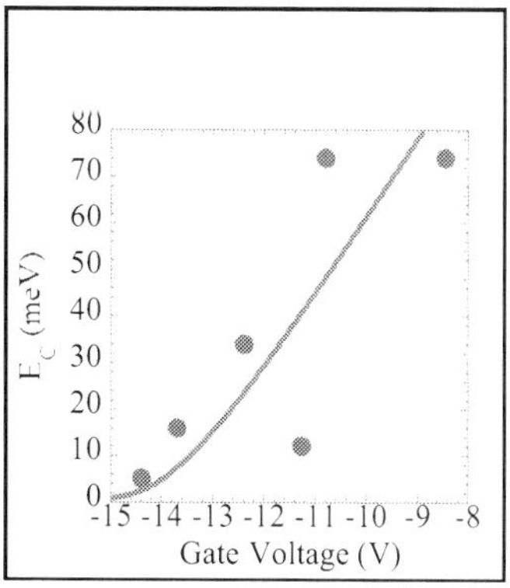

Fig. 5. (a) Coulomb charging energy E_C as a function of V_G. E_C drastically decreases at $V_G \geq$ -16 V and reaches zero at around V_G = -16 V. Therefore, the Coulomb diamonds disappear at the mid-V_G region.

$$C_t = \mathrm{e}/2E_C - C_G \tag{6}$$

and

$$C_C = \mathrm{e}/\Delta V_G \tag{7}$$

respectively. C_t depends on the thickness of the tunneling barrier and C_G depends on the gate structure. C_t drastically increases with a negatively increasing V_G, while C_G is almost constant, independent of the changing V_G. The drastic increase in C_t is attributed to the decrease in the thickness of the Schottky barriers at the contacts between the SWNT quantum island and electrodes (Javey et al., 2003). When the Schottky barriers become thin with negatively increasing V_G, as shown in the inset of Fig. 5(b), C_t drastically increases and the coupling strength of the wave function between the outside and the inside of the Schottky barrier becomes stronger. Therefore, the Coulomb blockade is lifted and E_C becomes zero at -16 V$\geq V_G$.

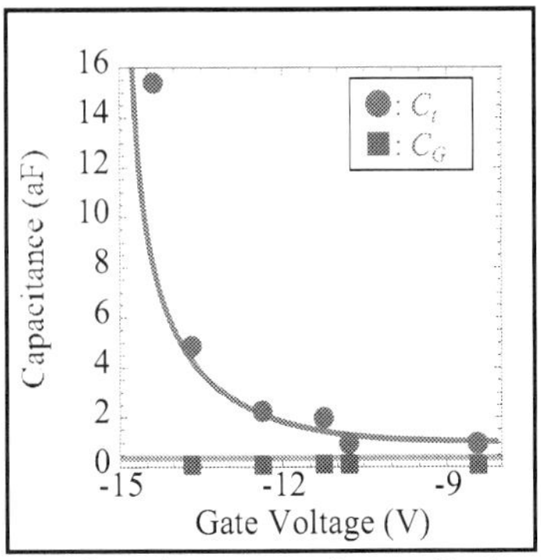

Fig. 5. (b) Tunneling capacitance C_t and gate capacitance C_G characteristics as functions of V_G. C_t drastically increases at $V_G \geq -16$ V. On the other hand, C_G is almost constant, independent of the changing V_G. The drastic increase of C_t is attributed to the change in the thickness of the Schottky barriers at the contact between the SWNT channel and electrodes.

The dependence of the full-width at half maximum (FWHM) of the resonant tunneling current peak characteristic on V_G is shown in Fig. 6(a).

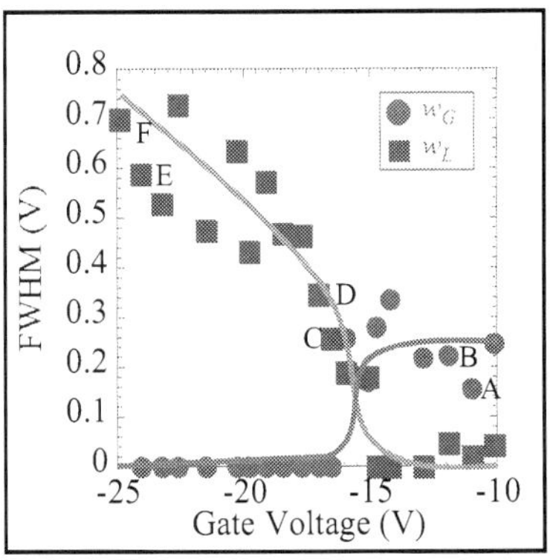

Fig. 6. (a) FWHM of the dI_D/dV_D peak characteristics as a function of V_G. w_G is almost constant, independent of V_G and w_L is negligibly small at $V_G \geq -16$ V. w_G become vanishingly small and w_L increases linearly with a negatively increasing V_G at $-16\text{ V} \geq V_G$.

The FWHM is estimated from an I_D-V_G plot using the Voigt function, which is a convolution of the Gaussian and Lorentzian functions, and can be used to divide the FWHM into the FWHM of the Gaussian w_G and that of the Lorentzian w_L. Fig. 6(b)-(d) shows several curves fitted by the Voigt function, where the solid line is the experimental data, the dotted lines are the fitting curves, the dashed line is the cumulative fitting curves, and each alphabetic marker for peaks A-F corresponds to a marker in the plot of Fig. 6(a).

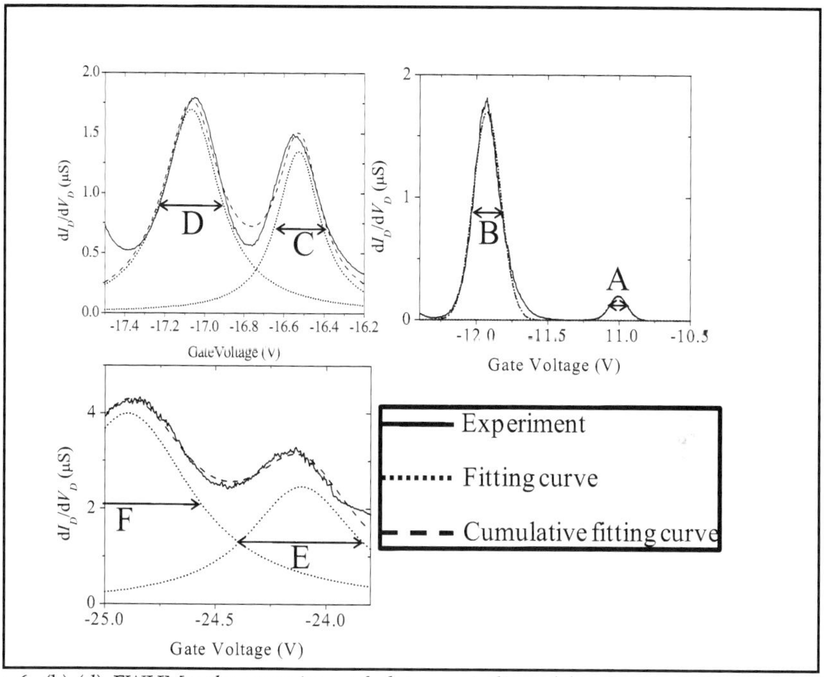

Fig. 6. (b)-(d) FWHMs: the experimental data are indicated by the black line, the fitting curves are indicated by the blue lines, the cumulative fitting curves are indicated by red line, and each alphabetic marker for a peak corresponds to a marker in the plot of Fig. 6(a).

In the particle nature mode of $V_G \geq -16$ V, w_L is negligibly small and the w_G of about 0.2 V is independent of V_G because the FWHM of the Coulomb oscillation mainly depends on the thermal broadening of Fermi dispersion (Foxman et al., 1993). On the other hand, in the wave nature mode of -16 V $\geq V_G$, w_G becomes negligibly small and w_L increases linearly with the negatively increasing V_G. w_L is proportional to the tunneling probability as follows:

$$\alpha w_L = Thv_F/2L, \tag{8}$$

where α is the ratio of the modulated energy to applied V_G, and T is the tunneling probability (Kamimura & Matsumoto, 2006). Therefore, the increase in w_L seen in Fig. 6 represents an increase in the tunneling probability, which is attributed to the decrease in the thickness of the Schottky barriers with a negatively increasing V_G.

The logarithmic dependence of the drain current I_D on the inverse of the temperature is shown in Fig. 7.

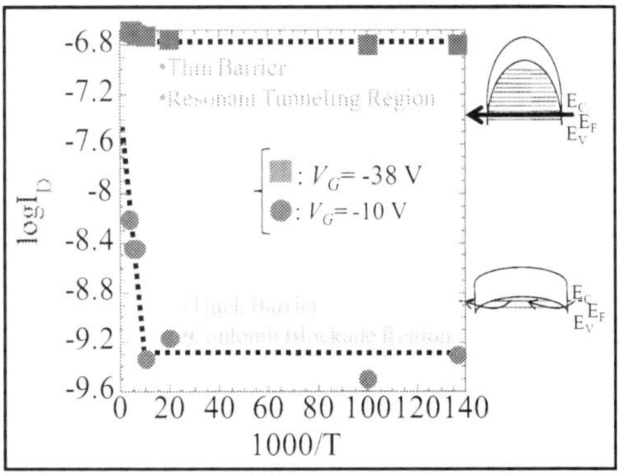

Fig. 7. The logarithmic dependence of the drain current I_D on the inverse of the temperature. At V_G = -38 V of the wave nature mode, I_D is almost constant, independent of the temperature. In the wave nature mode, the Schottky barrier is so thin that the tunneling current becomes dominant. In contrast, at V_G = -10 V of the particle nature mode, I_D drastically increases at the high temperature region, and the Schottky barrier height estimated from the slope is $\Delta\Phi$= 50 mV.

3.2 Hysteresis elimination

Fig. 8 shows the temperature dependence of differential conductance as a function of the gate voltage.

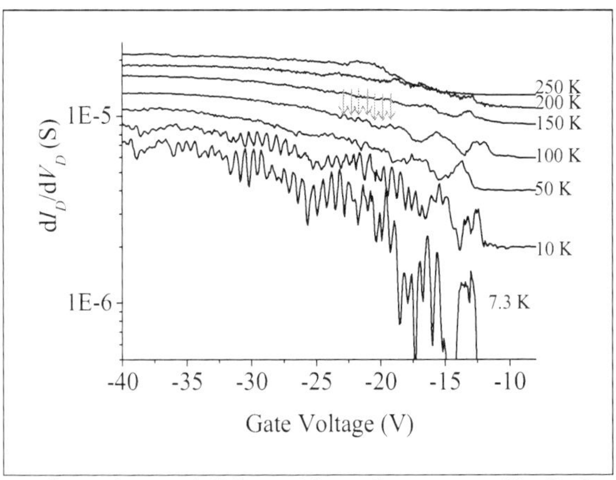

Fig. 8. Temperature dependence of differential conductance on V_G. Each plot is shifted by 3 μS to make it easier to see. The quantum level peaks are becoming blurred with increasing temperature. However, even at T = 100 K, the quantum levels remain, as indicated by the arrows.

The quantum levels peaks became blurred with increasing temperature. However, even at T = 100 K, the quantum levels still remained, as indicated by arrows.

The static characteristic measurement of the devise with silicon dioxide layer on the SWNT channel shown in Fig. 9 was carried out using an Agilent B 1500.

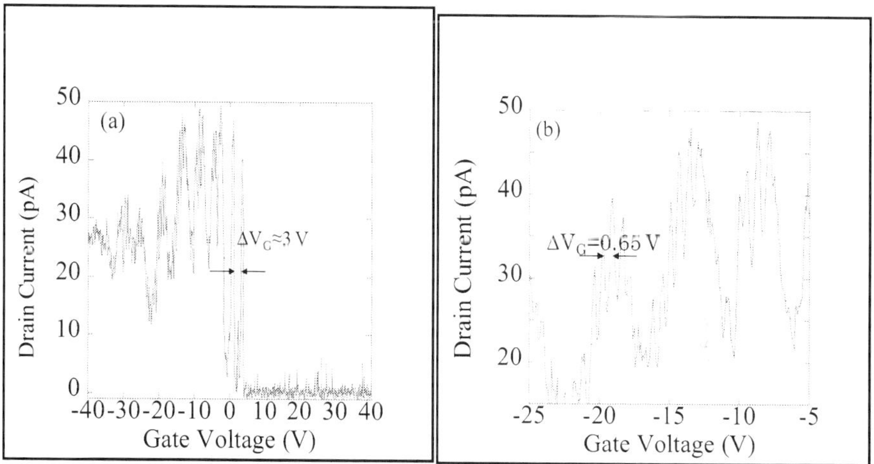

Fig. 9. (a) Coulomb oscillation characteristic at 7.3 K. Drain current was observed only in negative gate voltage regions, which indicates that the measured SWNT is a p-type semiconductor. A large period of ~3 V at an applied gate voltage of about V_G=0 V was observed, which was attributed to Coulomb oscillation characteristic. (b) The small period of ΔV_G=0.65 V of drain current peaks at 7.3 K, which was attributed to the quantum interference property, namely, Fabry-Perot interference of the hole, the coherent length of which was more than 73 nm. Coherent transport in the entire channel was achieved.

In the measurement, the data were integrated for a few seconds to eliminate the effect of noise, which was set by the equipment automatically. After that, the data were recorded. In the drain current I_D-gate voltage V_G characteristics of the SWNT multi-functional quantum transistor under a drain voltage of 11 mV at 7.3 K, the drain current showed a periodic peak-and valley structure with two periods. As shown in Fig. 9(a), I_D was observed only in the negative V_G region, which indicates that the measured SWNT was a p-type semiconductor. Moreover, a large period of ~3 V at around V_G=0 V was observed, which was attributed to the Coulomb oscillation characteristic. A part of Fig. 9(a) is expanded in the horizontal scale and shown in Fig. 9(b), where small period of ΔV_G=0.65 V was observed at a higher V_G, which was attributed to the Fabry-Perot interference property (Liang et al., 2001) of hole, the coherent length of which was more than 73 nm, and coherent transport in the entire channel was achieved. Fig. 9(c) shows a contour plot of the second-order differential conductance as a function of gate and drain voltages, in which a clear Fabry-Perot interference pattern can be seen.

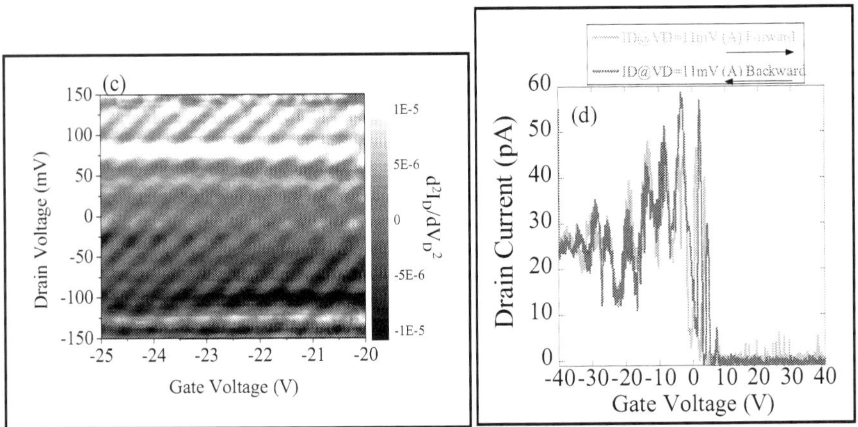

Fig. 9. (c) Contour plot of second-order differential conductance as a function of gate and drain voltages at 7.3 K, which shows a clear Fabry-Perot interference pattern. (d) Drain current-gate voltage characteristics with round trip applied gate voltage sweeping from V_G=40 to -40 V after sweeping from V_G=-40 to 40V at 7.3 K. The two drain current characteristics almost overlapped.

From the small period of drain current oscillation of ΔV_G=0.65 V and the equation (Kamimura & Matsumoto, 2006) of

$$\Delta E_Q = \alpha V_G = \mathrm{h}\nu_F/2Le, \tag{9}$$

where ΔE_Q is the quantum energy separation, a is the gate modulation coefficient, h is Plank's constant, ν_F is the Fermi velocity in graphene, L is the length of the cavity of the hole, and e is the elementary charge, we estimated the width of the quantum well and obtained the length of the cavity of the hole L to be 55 nm. The estimated value is in agreement with the channel length 73 nm of the device which was obtained by scannig electron microscope (SEM) (inset of Fig. 1). The slight difference between the estimated L and the channel length must be attributed to the width of band bending in the SWNT e.g., Schottky barriers.

In Fig. 9(d), the gray (black) line shows drain current vs. applied gate voltage from -40 (40) to 40 (-40)V in a forward (backward) sweep of gate voltage. The SWNT multi-functional quantum transistor showed almost no hysteresis characteristics. The SWNT multi-functional quantum transistor was covered with a silicon dioxide layer, which prevents the SWNT from absorbing and releasing molecules in ambient, because fluctuation of molecules on the SWNT usually induces the hysteresis characteristics in the drain current-gate voltage characteristics. Moreover, the SWNT multi-functional quantum transistor has a small channel length of 73 nm, which reduced the number of trap sites near the SWNT and, at the same time, the number of trapped carriers causing hysteresis characteristics. We believe that the purification was effective to reduce the hysteresis characteristics. Thus, the hysteresis characteristic in the drain current–gate voltage characteristic was eliminated.

3.2 Single charge sensitivity of SWNT multi-functional quantum transistor

Figure 10(a) shows the time dependence of the conductance of the SWNT multi-functional quantum transistor at 7.3 K with a gate bias of V_G=-25.36 V. A short sampling time of 10 ms was set in the dynamic characteristic measurements shown in Fig. 10.

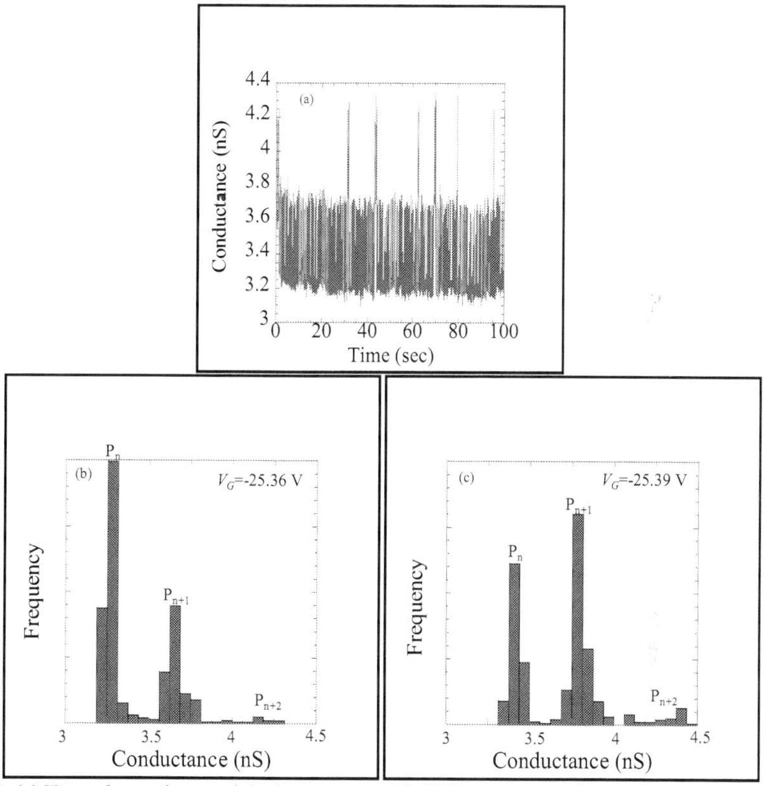

Fig. 10. (a) Time dependence of drain current with RTS at a gate voltage of V_G=-25.36 V. The time dependence of drain current was sampled for 100 s and the sampling time was 10 ms; (b) and (c) show histograms of the conductance levels of RTS at V_G= -25.36 and -25.39 V, respectively. P_n decreases and P_{n+1} and P_{n+2} increase with slightly increasing applied gate voltage from V_G= -25.36 to -25.39 V.

The applied gate voltage was under the Fabry-Perot interference region. The SWNT multi-functional quantum transistor shows RTS, as shown in Fig. 10(a), The RTS showed three levels, n, n+1 and n+2, of the conductance shown in Fig. 10(a). At a lower applied gate voltage, current levels higher than n+2 such as n+3 and n+4 appeared. The multiple levels of RTS are attributed to charge fluctuating charge storages near the conduction channels of the SWNT multi-functional quantum transistor. Moreover, because there was a single-charge storage including multiple energy levels or were some charge storages being at almost the same distances from the conductance channel of the SWNT multi-functional quantum transistor, the RTS appeared. Figures 10(b) and 3(c) show histograms of the conductance levels of RTS at V_G= -25.36 V and V_G= -25.39 V, respectively. The three peaks of conductance of RTS expressed as P_n, P_{n+1}, and P_{n+2} are shown in Figs. 10(b) and 10(c). The conductance levels of the three peaks directly correspond to the conductance levels of RTS. On the other hand, the relative heights of the peaks correspond to occupation probabilities at each conductance level of the RTS. The heights of the peaks depend on applied gate voltage. P_n decreases and P_{n+1} and P_{n+2} increase with slightly increasing applied gate voltage from V_G= -25.36 to -25.39 V, which means that the energy levels in the charge storage are modulated by applied gate voltage.

The gate voltage dependences of the natural log of the ratio between the m_{th} peak and the $(m+1)_{th}$ peak in the conductance histogram P_{m+1}/P_m (m= n, n+1, n+2, ···, n+4) are shown in Fig. 11(a). The natural log of P_{m+1}/P_m linearly depends on applied gate voltage and saturates in each V_G. Each starting point of saturation is marked by an arrow in Fig. 11(a). The charge storage energy levels are floating. Therefore, modulations of energy by V_G may be different at each charge storage. We believe that the reason why the natural log of P_{m+1}/P_m saturates at each V_G may depend on the difference in energy modulation by V_G at each charge storage.

The charge transition is modeled, as shown in Fig. 12(a), in which the energy barrier is between the SWNT and the charge storage. Figures 11(b)-11(e) are enlargement plots of each P_{m+1}/P_m (m= n, n+1, n+2, ···, n+4). The energy differences ΔE_n between the charge storage energy level E_n and the Fermi level E_f are expressed as $\Delta E_n = E_f - E_n$, which is modulated by the applied gate voltage V_G. According to equilibrium statistical mechanics, P_{m+1}/P_m is given by (Peng et al., 2006)

$$P_{m+1}/P_m = (g_f/g_s)\mathrm{e}^{-\beta(E_f-E_n)} = (g_f/g_s)\mathrm{e}^{-\beta\Delta E_n} \qquad (10)$$

where g_f and g_S are the degeneracy of the top of valence band and the charge storage, respectively. g_f/g_S is assumed to be 1. β is 1/kT. E_n includes the contributions of e electrostatic potential induced by V_G, intrinsic energy level in the storage, and Coulomb charging energy. The basis of eq. (1) is the Arrhenius equation. In this model, the height of the barrier is the energy difference between E_f and E_n. Assuming a linear dependence of ΔE_n on V_G, ΔE_n can be written as $\Delta E_n = ae(V_0 - V_G)$, where a is the gate modulation coefficient, and is a constant value of 0.062. a is obtained from the periods of Fabry-Perot interference characteristic on V_D and V_G. V_0 is the offset voltage, and is obtained from the intersecting point of the extrapolating line of the fitting lines and the line of $\ln(P_{m+1}/P_m)$= 0. Therefore, eq. (1) is transformed to

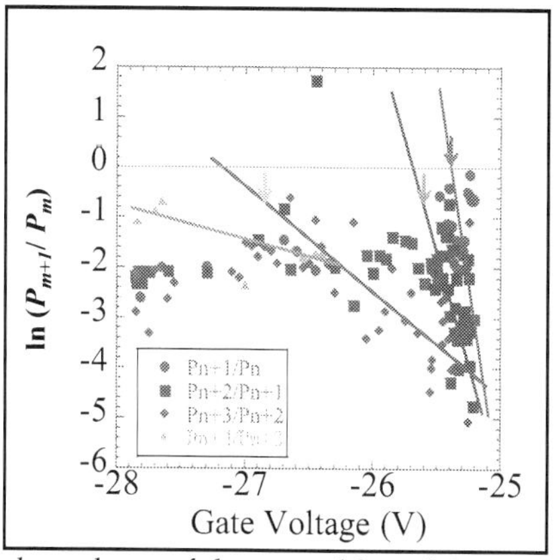

Fig. 11. (a) Gate voltage dependence of the natural log of the ratio between the occupancy probabilities of the mth current levels P_{n+1}/P_n.

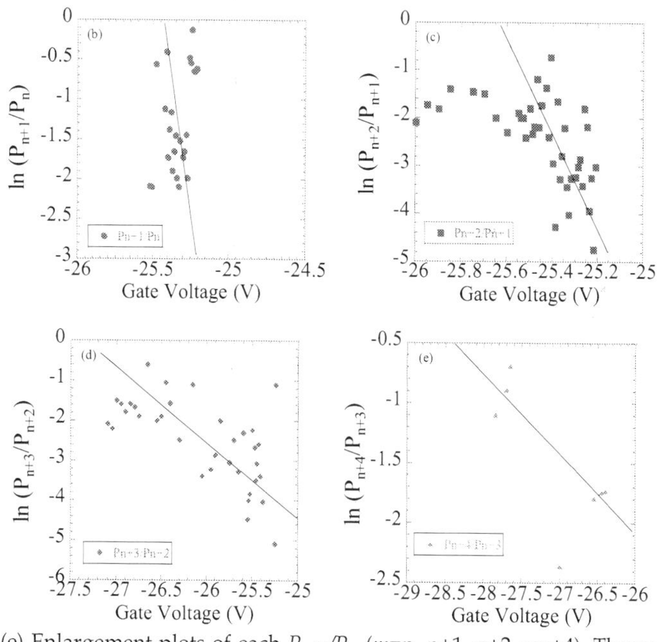

Fig. 11. (b)-(e) Enlargement plots of each P_{m+1}/P_m (m=n, n+1, n+2,··n+4). The natural log of P_{n+1}/P_n was linearly dependent on gate voltage, the slopes of which are -10.4, -4.78, -1.47, and -0.690 V^{-1}, respectively.

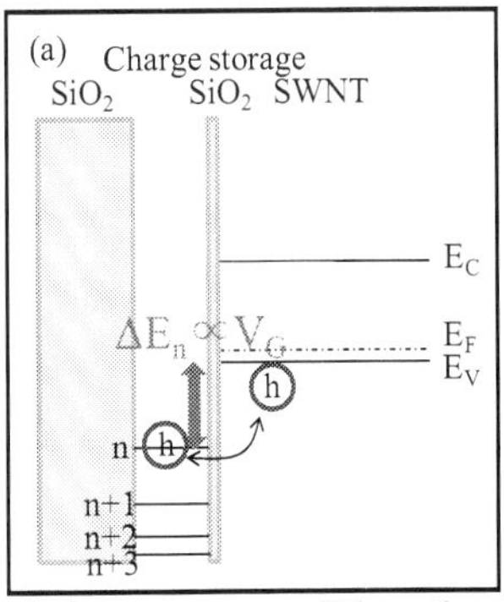

Fig. 12. (a) Schematic model of charge storage. When the charge storage energy level is coincident with the top of the valence band owing to applied gate voltage, the carrier goes and comes between them through the barrier with tunneling. The exiting probabilities of the carrier at the charge storage and the valence band depend on the relative height of their energy levels under equilibrium condition.

$$\ln(P_{m+1}/P_m) = -\beta e\alpha(V_0 - V_G) \tag{11}$$

Equation (2) is the transformed Arrhenius equation, in which V_G is the parameter.
From the dependence of P_{n+1}/P_n on V_G shown in Fig. 11 and eqs. (1) and (2), ΔE_n can be obtained, and is shown in Fig. 12(b). The obtained energy levels are from 1.57 to 1.79 eV.

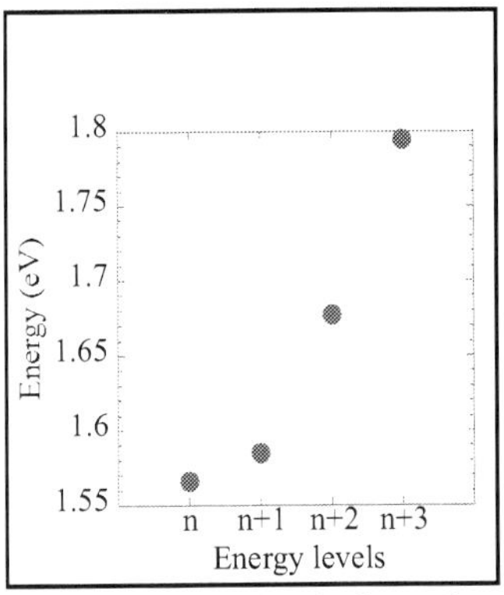

Fig. 12. (b) Estimated charge storage energy levels from the dependence of P_{n+1}/P_n on V_G shown in Figs. 4(b)-4(e) and eq. (1). The energy levels increase from 0.17 to 0.28 eV with increasing number of energy levels in the region of V_G from -25 to -28 V.

4. Conclusion

In summary, we succeeded in fabricating and demonstrating a multi-functional quantum transistor using the particle nature and wave nature of holes in SWNT. This transistor can operate in the wave nature mode as an RTT and in the particle nature mode as an SHT. We were able to reveal that the principle of the characteristic transition from an SHT to an RTT is the modulation of the coupling strength between the SWNT quantum island and the electrodes by the applied V_G.

5. References

Foxman, E. B.; mcEuen, P. L.; Meirav, U.; Wingreen, Ned S.; Meir,Yigal; Belk, Paul A.; & Wind, S. J. (1993). Effects of quantum levels on transport through a Coulomb island. Phys. Rev., Vol. B 47, 10020-10023.

Heinze, S.; Tersoff, J.; Martel, R.; Derycke, V.; Appenzeller, J.; & Avouris, Ph. (2002). Carbon Nanotubes as Schottky Barrier Transistors. Phys. Rev. Lett., Vol. 89, 106801-1-106801-4.

Javey, A.; Guo, J.; Wang, Q.; Lundstrom, M. & Dai, H. (2003). Ballistic carbon nanotube field-effect transistors. Nature, Vol. 424, 654-657.

Kamimura, T. & Matsumoto, K. (2004). Reduction of Hysteresis Characteristics in Carbon Nanotube Field-Effect Transistors by Refining Process. IEICE Trans. Electron., Vol. E-87C, 1795-1798.

Kamimura, T. & Matsumoto, K. (2005). Electrical Heating Process for p-Type to n-Type Conversion of Carbon Nanotube Field Effect Transistors. Jpn. J. Appl. Phys., Vol. 44, 1603-1605.

Kamimura, T. & Matsumoto, K. (2006). Coherent Transport of Hole and Coulomb Blockade Phenomenon in Long p-Type Semiconductor Carbon Nanotube. Jpn. J. Appl. Phys., Vol. 45, 338-340.

Kim, W.; Javey, A.; Vermesh, O.; Wang, Q.; Li, Y.; & Dai, H. (2003). Hysteresis Caused by Water Molecules in Carbon Nanotube Field-Effect Transistors. Nano Lett., Vol. 3, 193-198.

Liang, W.; Bockrath, M.; Bozovic, D.; Hafner, J. H.; Tinkham, M. & Park, H. (2001). Fabry - Perot interference in a nanotube electron waveguide. Nature, Vol. 411, 665-669.

Martel, R.; Derycke, V.; Lavoie, C.; Appenzeller, J.; Chan, K. K.; Tersoff, J. & Avouris, Ph. (2001). Ambipolar Electrical Transport in Semiconducting Single-Wall Carbon Nanotubes. Phys. Rev. Lett., Vol. 87, 256805-1-256805-4.

Nosho,Y.; Ohno, Y.; Kishimoto, S. & Mizutani, T. (2006). Relation between conduction property and work function of contact metal in carbon nanotube field-effect transistors. Nanotechnology, Vol. 17, 3412-3415.

Peng, H. B.; Hughes, M. E. & Golovchenkoa, J. A. (2006). Room-temperature single charge sensitivity in carbon nanotube field-effect transistors. Appl. Phys. Lett., Vol. 89, 243502-243502-3.

Radosavljevic´, M.; Freitag, M.; Thadani,K. V. & Johnson, A. T. (2002). Nonvolatile Molecular Memory Elements Based on Ambipolar Nanotube Field Effect Transistors. Nano Lett., Vol. 2, 761-764.

Suzuki, M.; Ishibashi, K.; Ida, T.; Tsuya, D.; Toratani, K.; & Aoyagi, Y. (2001). Fabrication of single and coupled quantum dots in single-wall carbon nanotubes J. Vac. Sci. Technol., Vol. B 19, 2770-2774.

11

Numerical Modeling of the I-V Characteristics of Carbon Nanotube Field Effect Transistors

Jose Mauricio Marulanda and Ashok Srivastava
Department of Electrical and Computer Engineering
Louisiana State University
Baton Rouge, LA 70803, USA

1. Introduction

Carbon nanotubes (CNTs) were first discovered in 1991 by Sumio Iijima (Iijima, 1991), and they have been a rapid and successful target of development by many researchers (Baughman et al., 2002). Especially with the end of Moore's law in sight (Wind et al., 2002), CNTs are being considered as possible candidates substituting silicon in the fabrication and design of analog and digital integrated circuits (ICs) (Guo et al., 2002; Wong, 2002).

Carbon nanotubes are basically two dimensional graphene sheets rolled into a one dimensional tubular structure (Martel et al., 1998). Their properties are determined by the chiral vector represented by the indices (n,m) (Tanaka et al., 1999; Wallace, 1947; Wildoer et al., 1998). Depending on the number of layers rolled, carbon nanotubes can be either single walled (one layer), or multi walled (two or more layers) (Dresselhaus et al., 2001). Single walled carbon nanotubes (SWNTs) are used in the fabrication of carbon nanotube field effect transistors (CNT-FETs), the first CNT-FETs were implemented in 1998 (Martel et al., 1998; Tans et al., 1998). The structure of a CNT-FET is similar to the structure of a typical MOSFET where the CNT forms the channel between two electrodes that work as the source and the drain of the transistor. The structure is build on top of an insulating layer and a substrate wafer that works as the back gate, (Nihey et al., 2003) (Wind et al., 2002). The basic structure of a CNT-FET based on a SWNT is shown in Fig. 1. Among the CNT-FETs reported (Martel et al., 1998; Wind et al., 2002), a 40 nm gate length transistor (Lin et al., 2005), a multistage complementary logic (Javey et al., 2002), oscillators (Chen et al., 2006), and an 80 GHz operating field effect transistor (Nougaret et al., 2009) have been achieved.

In addition, carbon nanotubes are one-dimensional conductors (1D), which confines the electrons to only back-scattering effects. This property provides a large electron mean free path in metallic carbon nanotubes of usually a few micrometers (White & Todorov, 1998). Carbon nanotubes also exhibit large current capabilities of $\sim 10^{-9}\ A/cm^2$, (Wei et al., 2001; Yao et al., 2000) and it has also been reported that doping can be avoided in the CNT fabrication process, yet still achieving complementary CNT-FETs (Zhang et al., 2007). With carbon nanotubes interconnects being also a major target of research (Koo et al., 2007; Xu et al., 2008) and a completely carbon nanotube based integrated circuit (IC) already reported

(Liang et al., 2009), fabrication of fully integrated carbon nanotubes in custom designed ICs and systems on a chip is no longer a future circuit design promise but an attainable grasp.

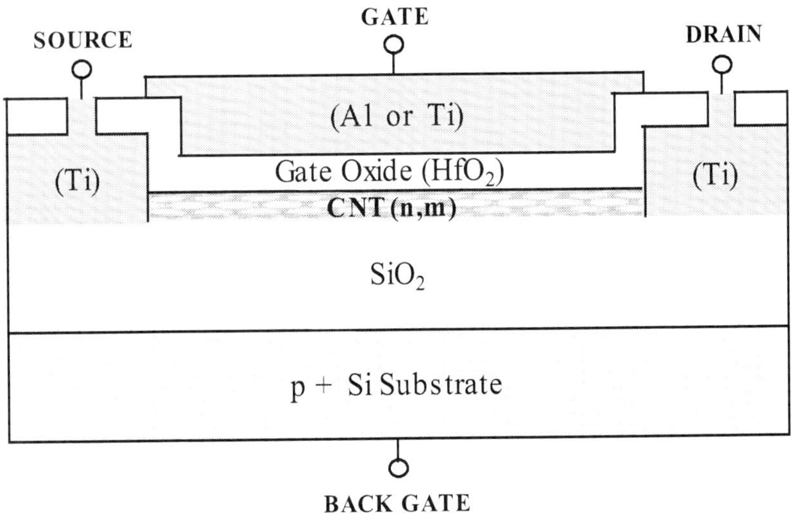

Fig. 1. Cross sectional view of a CNT-FET. Note: HfO_2 (Halfnium dioxide) is a high *k*-dielectric (~17-20) (Javey et al., 2004; Wang et al., 2003).

In this paper, we have used our modeling of the potential and charge balance of CNT-FETs (Marulanda, 2009; Marulanda et al., 2008; Srivastava et al., 2009) to develop a numerical model of the current-voltage relation for a CNT-FET of n-type behaviour. This model uses numerical approximation methods to solve this current and voltage equations and present the I-V characteristics of different CNT-FETs.

2. Current Equation

The current per band in a semiconductor (Datta, 1999; Xia et al., 2004) can be expressed as:

$$I_{ds_i} = \frac{q}{\pi\hbar} \int_{E_{c_i}}^{E_{max}} T_i(E)\left[f_s(E) - f_d(E)\right] dE, \tag{1}$$

where E_{c_i} and $T_i(E)$ are respectively the conduction band energy minimum and transmission coefficient for the i_{th} band, $\hbar$ is Planck's constant divided by 2π, and $f_s(E)$ and $f_d(E)$ are the Fermi distributions at the source and drain, respectively.

Although the total I_{ds} will be the sum of all I_{ds_i} over all bands, the summation in Eq. (1) can be dropped, as only the first band is sufficient. The reason is because the remaining energy bands will have negligible contribution to the current due to the vanishing exponential behaviour of the Fermi distribution for high-energy conduction band values.

By letting the limit of E_{max} go to infinity and the transmission coefficient be a constant equal to unity for the energy range of integration, Eq. (1) becomes (Marulanda et al., 2008):

$$I_{ds} = \frac{qKT}{\pi\hbar}\left[\ln\left(1+e^{\frac{E_{Fs}-E_c}{KT}}\right)-\ln\left(1+e^{\frac{E_{Fd}-E_c}{KT}}\right)\right], \tag{2}$$

where E_{Fs} and E_{Fd} are the Fermi level energies at the source and drain, respectively. This equation for the current can be used in the linear and saturation regions.

3. Gate and Carbon Nanotube Potential Relation

Figure 2(a) shows the basic cross section of a CNT-FET including the charge distributions, Fig. 2(b) shows the corresponding potential distributions between the gate and substrate. In Fig. 2(a), charge distributions are explained as follows: the charge on the gate is Q_g, charges in oxide layers are Q_{01} and Q_{02}, charge inside the CNT is Q_{cnt} and the charge in the substrate is Q_{subs}. In Fig. 2(b), six different potential distributions are shown, which are also described as follows. The voltage between the gate and the substrate (back gate) is V_{gb}, the potential drop across the oxides ψ_{ox1} and ψ_{ox2}, the surface potential in the substrate with respect to the back gate is ψ_{subs}, the potential across the CNT is ψ_{cnt}, and the work function difference between the gate and the substrate materials is ϕ_{ms}. The work function, ϕ_{ms} can be divided in two parts and is expressed as follows,

$$\phi_{ms} = \phi_{mc} + \phi_{cs}, \tag{3}$$

where ϕ_{mc} and ϕ_{cs} are the work function differences between the metal gate and carbon nanotube materials and the carbon nanotube and substrate materials, respectively. We can also combine potentials ψ_{cnt}, ψ_{ox2}, ϕ_{cs}, and ψ_{subs} into a single potential, $\psi_{cnt,s}$, which describes the potential at the interface of the gate oxide and carbon nanotube with respect to the back gate, we can call $\psi_{cnt,s}$ the carbon nanotube surface potential.

Using Kirchoff's voltage law, the potential balance, and the charge neutrality condition, we can write for Fig. 2,

$$V_{gb} = \phi_{ms} + \psi_{ox1} + \psi_{cnt,s}, \tag{4}$$

$$Q'_g + Q'_{01} + Q'_{cnt} + Q'_{02} + Q'_{subs} = 0. \tag{5}$$

The prime in Eq. (4) denotes the charge per unit area. From basic laws of electrostatics (Cheston, 1964; Langmuir, 1961; Shen & Kong, 1995; Thomas, 1972), we can utilize Eqs. (4) and (5) and the charge modeling described in (Marulanda, 2009; Marulanda et al., 2008; Srivastava et al., 2009), to express the gate voltage in a CNT-FET as,

$$V_{gb} = \psi_{cnt,s} - \frac{Q_{cnt}}{C_{ox1}} + V_{fb}, \tag{6}$$

where V_{fb} is the flat band voltage, which is the voltage needed at the gate with respect to the back gate to compensate for the band bending at the gate oxide and carbon nanotube interface and it is given by,

$$V_{fb} = \phi_{mc} - \frac{Q_{01}}{C_{ox1}}. \tag{7}$$

In addition, C_{ox1} and C_{ox2} are the oxide capacitances for the respective oxide regions, T_{ox1} and T_{ox2} and are described for a carbon nanotube of length, L and radius, r as follows (Cheston, 1964; Thomas, 1972),

$$C_{ox} = \frac{2\pi\varepsilon_{ox}L}{\ln\left(\frac{T_{ox} + r + \sqrt{T_{ox}^2 + 2T_{ox}r}}{r}\right)}. \tag{8}$$

In Eqs. (6), (7), and (8), Q_{cnt}, Q_{01} and C_{ox1} are the total charges and capacitance, respectively, which are obtained by multiplying Q'_{01}, Q'_{cnt}, and C'_{ox} with their respective areas.

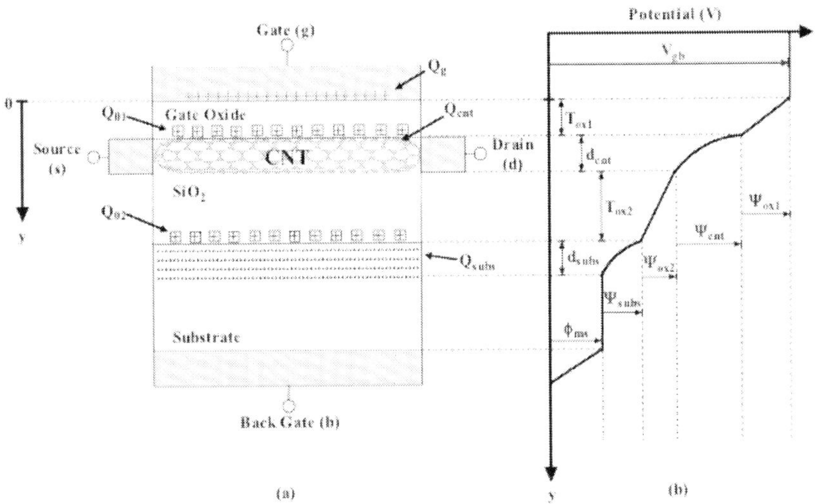

Fig. 2. (a) Plot of the charges from the gate to the substrate and (b) plot of the potential distribution from the gate to the substrate in a CNT-FET.

Using the carrier concentration modeling in (Marulanda & Srivastava, 2008) and charge modeling described in (Marulanda, 2009; Marulanda et al., 2008; Srivastava et al., 2009), we can express Eq. (6) as:

$$V_{gb} = \psi_{cnts} + \delta f(\psi_{cnts}, V_{cb}) + V_{fb}, \tag{9}$$

where

$$f(\psi_{cnt,s},V_{cb})=\begin{cases} Ie^{\frac{\Delta E_F+q(\psi_{cnt,s}-V_{cb}-\phi_0)-E_c}{kT}}; & for\ \psi_{cnt,s}\le V_{cb}+\phi_0-\frac{\Delta E_F}{q}+\frac{E_c}{q}-\frac{kT}{q} \\ \frac{\sqrt{(\Delta E_F+q\psi_{cnt,s}-qV_{cb}-q\phi_0)^2-E_c^2}}{kT}; & for\ \psi_{cnt,s}\ge V_{cb}+\phi_0-\frac{\Delta E_F}{q}+\frac{E_c}{q}+\frac{kT}{q}\end{cases} \tag{10}$$

and

$$\delta=\frac{qLN_c}{C_{ox1}}. \tag{11}$$

In Eq. (9), V_{cb} is the induced potential between the carbon nanotube and the substrate due to the drain and source terminal voltages. The potential, V_{cb} varies from V_{sb} (source to back gate potential) to V_{db} (drain to back gate potential). Furthermore, in Eq. (10), k is Boltzmann constant, T is temperature, ϕ_0 is the carbon nanotube surface potential, $q\psi_{cnt,s}$ when $V_{gb} = V_{fb}$ and $\psi_{cnt} = 0$, ΔE_F is the shift in the Fermi level of the carbon nanotube depending upon doping, , ϕ_0 and ΔE_F are given by,

$$\phi_0=\psi_{ox2}+\psi_{subs}+\phi_{cs}, \tag{12}$$

$$\Delta E_F=\pm kT\ln\left(1+\frac{N}{n_{cnt,i}}\right), \tag{13}$$

where $n_{cnt,i}$ is the intrinsic concentration of a CNT (n,m) and N is the impurity doping concentration. ΔE_F is positive for an n-type carbon nanotube (donors impurity concentration, $N = N_D$) and negative for a p-type carbon nanotubes (acceptors impurity concentration, $N = N_A$).

In Eq. (10), I is an indefinite integral involved in finding the carrier concentration of a carbon nanotube, we have found an analytical solution in (Marulanda, 2009; Marulanda & Srivastava, 2008); however, the general form of the integral, I is given by,

$$I=\frac{1}{\sqrt{kT}}\int_0^{\frac{6E_c}{kT}}\frac{(kTx+E_c)}{x^{1/2}(kTx+2E_c)^{1/2}}e^{-x}dx. \tag{14}$$

The term N_c introduced by Eq. (11) is found from our carrier concentration model for carbon nanotubes (Marulanda & Srivastava, 2007; Marulanda & Srivastava, 2008) and is given by,

$$N_c=\frac{8kT}{\pi V_{pp\pi}a\sqrt{3}}, \tag{11}$$

where $V_{pp\pi}$ is the carbon-carbon (C-C) tight binding overlap energy, a is the lattice constant with a value of $\sqrt{3}\,a_{c-c}$, and a_{c-c} is the nearest neighbor distance between C-C bonds (0.144 nm) (Dresselhaus et al.).

4. I-V Characteristic Simulation

Equation (9) cannot be solved explicitly for the carbon nanotube surface potential, $\psi_{cnt,s}$ in terms of the terminal voltages, hence, a numerical solution technique must be taken in order to find the exact channel potential given a gate voltage input.

In a CNT-FET, when the terminal voltages V_{db}, V_{sb}, and V_{gb} are applied, the two terms $(E_{Fs} - E_c)$ and $(E_{Fd} - E_c)$, referred in Eq. (2) can be replaced by $\Delta E_F + q(\psi_{cnt,s} - V_{sb} - \phi_0) - E_c$ and $\Delta E_F + q(\psi_{cnt,s} - V_{db} - \phi_0) - E_c$, respectively (Marulanda, 2009; Raychowdhury et al., 2004; Srivastava et al., 2009; Tsividis, 1999). V_{sb} and V_{db} are the source to bulk and drain to bulk voltages, respectively; furthermore, $\psi_{cnt,s}$ is calculated using Eq. (9) employing a numerical solution approach. Thus, our Eq. (2) becomes:

$$I_{ds} = \frac{qkT}{\pi \hbar}\left[\ln\left(1+e^{\frac{\Delta E_F + q(\psi_{cnt,s}(0) - V_{sb} - \phi_0) - E_c}{kT}}\right) - \ln\left(1+e^{\frac{\Delta E_F + q(\psi_{cnt,s}(L) - V_{db} - \phi_0) - E_c}{kT}}\right)\right]. \quad (15)$$

The complete process to find the current I_{ds} is as follows:

1. Define the voltages V_{db}, V_{sb}, and V_{gb}.
2. Use a numerical method to find $\psi_{cnt,s}$ from Eq. (9).
3. Compute the current, I_{ds}, using Eq. (15).
4. Repeat steps 1 - 3 for each of the input voltages, V_{db}, V_{sb}, and V_{gb} combination.

Figure 4 shows the I-V characteristics for a CNT-FET with a carbon nanotube of chiral vector (5,3) and diameter of 0.2759 nm, under different overdrive gate voltages. These plots were derived following the four steps just mentioned.

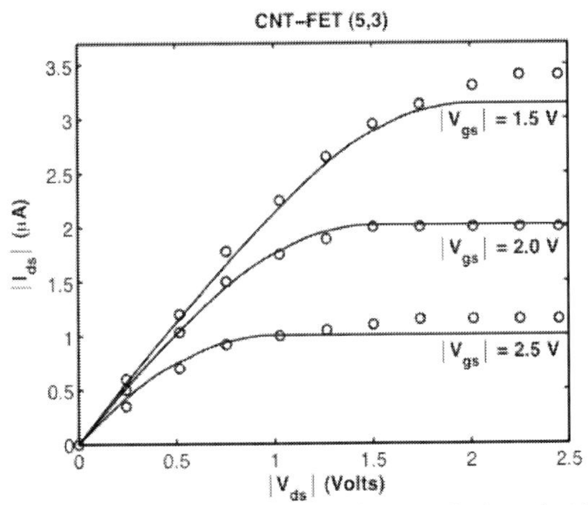

Fig. 3. I-V characteristics for a CNT-FET (5,3) with V_{fb} = 0, ϕ_0 = 0, ΔE_F = 0, and device dimensions of $L = 3\ \mu m$, $T_{ox1} = 500\ nm$, and $T_{ox2} = 1\ \mu m$.

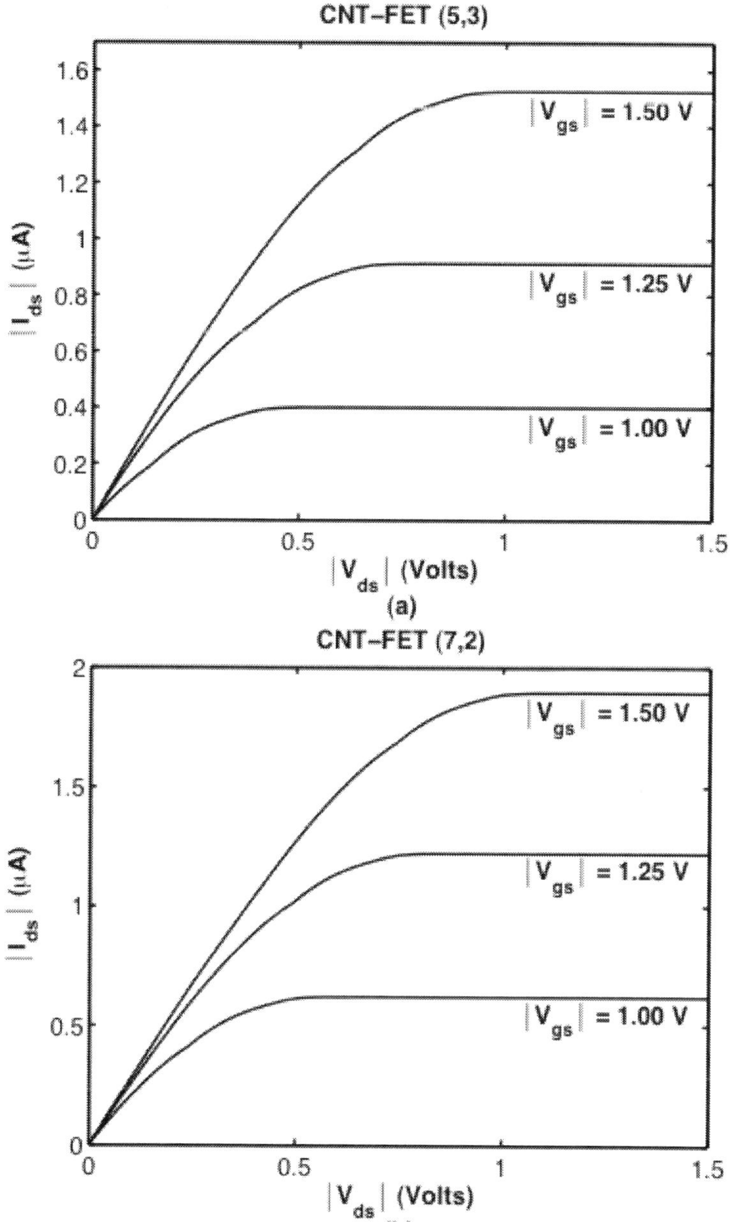

Fig. 4. I–V characteristics for (a) CNT-FET (5,3) and (b) CNT-FET (7,2) with $V_{fb} = 0$, $\phi_0 = 0$, $\Delta E_F = 0$, and device dimensions of $L = 25\ nm$, $T_{ox1} = 40\ nm$, and $T_{ox2} = 400\ nm$.

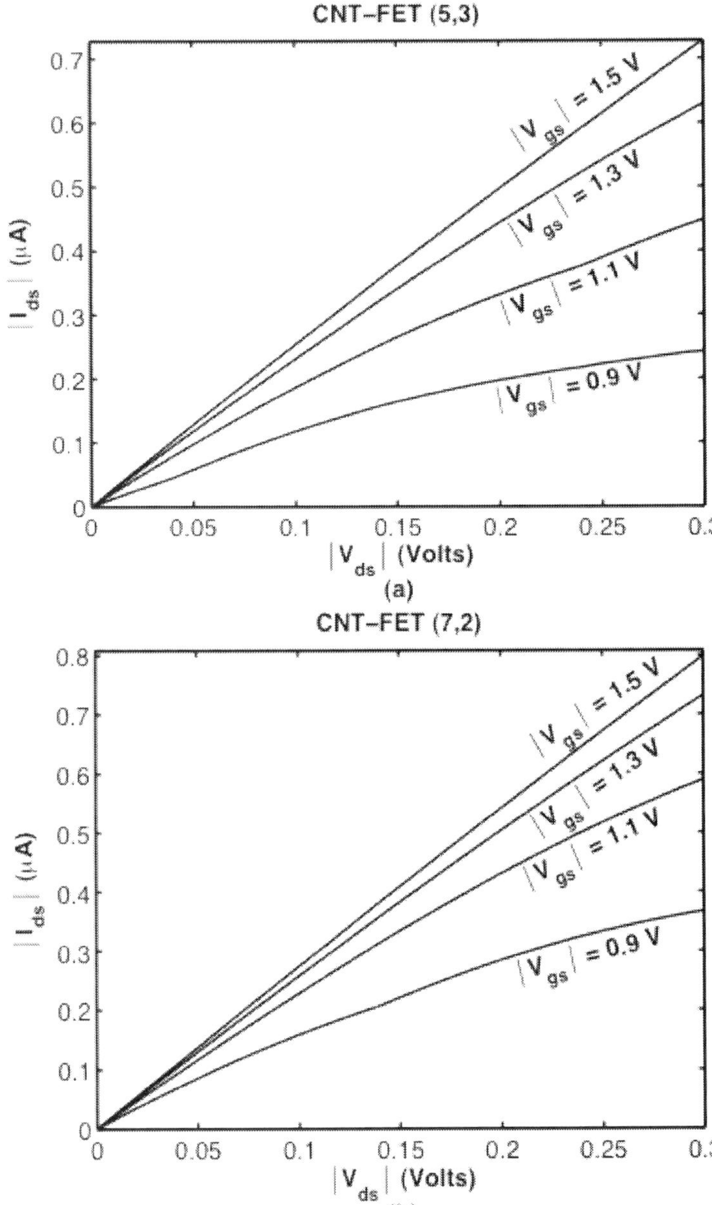

Fig. 5. I–V characteristics under low V_{ds} for (a) CNT-FET (5,3) and (b) CNT-FET (7,2) with $V_{fb} = 0$, $\phi_0 = 0$, $\Delta E_F = 0$, and device dimensions of $L = 25\ nm$, $T_{ox1} = 40\ nm$, and $T_{ox2} = 400\ nm$.

12

Compact Modeling of Carbon Nanotube Transistor and Interconnects

Yu Cao, Saurabh Sinha and Asha Balijepalli
Arizona State University, Tempe
USA

Silicon based devices have dominated mainstream computing for the last four decades. Achieving sustainable scaling of physical dimensions and device performance (Moore, 1965) has been key to their success. However, due to limitations in fundamental physics, materials, and manufacturing limits, this scaling trend has slowed down. Examples of major bottlenecks for continual scaling include short channel effects, high leakage currents(Wann et al., October 1996), excessive process variations(Bowman et al., 2002) and reliability issues(Chen et al., February 1985). These pitfalls are posing dramatic challenges to fabrication of circuits with scaled silicon devices. As we approach these fundamental limits in planar CMOS process, it becomes imperative to search for alternative materials, structures, devices as well as design paradigm to replace silicon transistor as the building block of future nanoelectronics. Novel structures like FinFETs(Hisamoto et al., 2000) and Trigate devices(Doyle et al., 2003), strained channel to enhance carrier mobility(Welser et al., 1994) and high-K/metal gate to reduce gate leakage current(Chau et al., 2004) have been proposed. These innovations have limited potential and will extend the scaling by a generation or two. Amongst more radical search for new devices and materials, carbon nanotube electronics has attracted significant attention owing to their high intrinsic carrier mobility.

For the sake of simplicity, carbon nanotubes can be defined as hollow cylinders made up of one (single-walled) or more (multi-walled) concentric layer of carbon atoms arranged in a hexagonal lattice structure, which is similar to a rolled-up sheet of graphene. With diameters of 1-4 nm and the length extending to several micrometers, carbon nanotube is essentially a one dimensional object possessing unique properties attributed to low dimensional structures, such as 1-D density of state for carriers (McEuen et al., 2002). This enables reduced phase space for scattering and near ballistic transport of carriers when the device dimension is less than the mean-free path for scattering. Depending on the direction in which the graphene sheet is rolled up, single-walled carbon nanotubes is either metallic or semiconducting. Hence CNT transistor and interconnect can be made out of semiconducting and metallic nanotubes, respectively. Functional field effect transistors with semiconducting carbon nanotube channel(Lin et al., 2005; Zhang et al., 2006) and metallic nanotubes as interconnects(Close & Wong, 2007) have been demonstrated. Theoretically, it is possible to get current densities much higher than that of silicon devices with a similar dimension(Raychowdhury et al., 2006) using multiple CNTs in parallel.

To speed up the evolution of this novel alternative technology, parallel efforts in circuit design are essential. For this purpose, the development of compact model is a vitally important

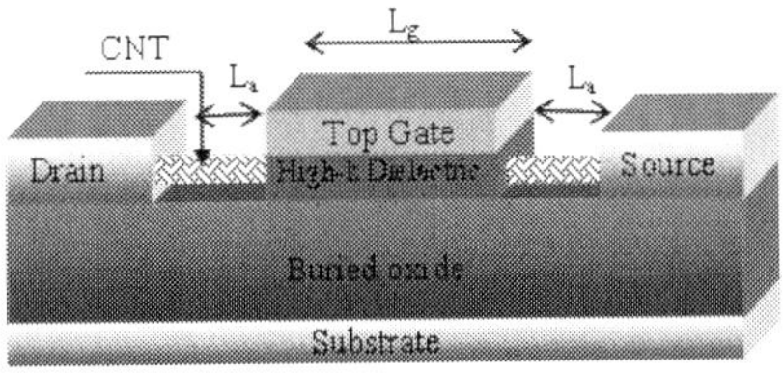

Fig. 1. Cross-section of a generic SW-CNT structure with top gated region as the intrinsic transistor of length L_g and highly doped undated access region of length L_a as the extrinsic part

step that enables circuit simulation and exploration. Early work on carbon nanotube transistor modeling assumed doped source-drain junctions resulting in optimistic performance projections (Guo et al., 2002). The CNT-FET model developed in (Castro et al., 2002) accurately calculated the degradation of current due to presence of Schottky barriers at the contacts. However, it requires self-consistent numerical iterations to calculate the final current and tunneling probability. Currently most of the models developed for carbon nanotube transistors and interconnects employ some kind of numerical approach (Guo et al., 2004; Wong et al., Nov. 2006) to obtain the I-V and C-V characteristics. Though highly physical and accurate, such numerical approaches reduce the computation efficiency and are not suitable for large-scale circuit simulations. Some other modeling approaches include threshold voltage based models(Raychowdhury et al., 2004) and models that resort to SPICE simulator to solve iterative differential equations and compute the surface potential (Deng & Wong, 2007a).

In this chapter, we discuss the development of an integrated compact model for carbon nanotube transistors and interconnects that is non-iterative and SPICE compatible. Initial models concentrated on modeling only the ballistic transport model of the transistor channel. However, the effect of the Schottky barrier at the metal source-drain contacts cannot be decoupled from the channel region. The developed model accounts for the presence of these barriers accurately. The implemented model has been systematically verified with TCAD simulations and measured data. Using this model, we benchmark digital and analog performance metrics and compare them with 22nm CMOS process to explore design potentials with CNTs.

1. Model Development

The cross-sectional view of a typical carbon nanotube transistor is shown in Fig. 1. The basic structure is similar to a conventional FET with the channel replaced by a semiconducting carbon nanotube. The similarity to the structure of CMOS device improves the compatibility with today's process and design infrastructure.

In the ideal case for ballistic transport, the source and drain electrodes would behave as reservoirs that supply and sink mobile carriers without any reflection at the source and drain. This is true only when there are ideal source and drain contacts, i.e., no significant energy barrier between the channel and the contact. There has been extensive work on finding the appropriate contact material for the CNT-FET (Chen et al., 2005) but they all have a finite energy gap when contacting the carbon nanotube forming a Schottky barrier. The device performance is primarily limited by the Schottky contact, depending on the properties of the contact material

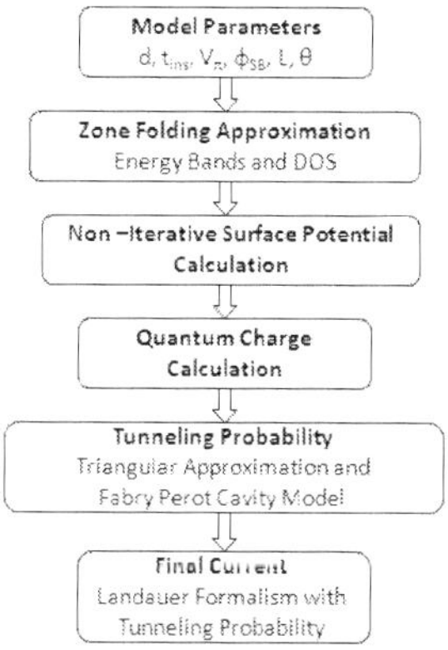

Fig. 2. Flowchart describing the model development

and the nanotube. The energy gap is sensitive to the work function of the contact, the diameter of the nanotube, as well as the chirality.

In this section, we will first discuss the intrinsic channel and relate the physical parameters to the electrical equivalents by the Zone-folding approximation. Using linear approximations, we derive an expression for surface potential. Then the impact of the Schottky barrier on the drain current is studied and integrated into the model. Based on similar modeling principles of the channel and the contact, a compact model of CNT interconnect and CNT-FET are developed. Fig. 2 shows a flowchart describing the CNT-FET compact model. Details of the physical mechanisms and model derivations of the CNT interconnect model are discussed in the next section.

1.1 Zone-folding approximation

We begin with characterizing the structure of single-walled carbon nanotubes (SWCNT) and defining its basic electronic properties such as band-gap, density of states etc. A SWCNT device is essentially a one-dimensional nanowire formed by rolling a two-dimensional graphene sheet. The $2s$, $2p_x$ and $2p_y$ orbitals form σ bonds in graphene. Since the σ bonds are weakly coupled to the $2p_z$ orbitals, they form π bonds, which give rise to the electronics properties of graphene. The E-k values for graphene can be obtained from the tight-binding model given by (1),

$$E_{g2D}(k_x, k_y) = \pm t\left\{1 + 4\cos\left(\frac{\sqrt{3}k_x a}{2}\right)\cos\left(\frac{k_y a}{2}\right) + 4\cos^2\left(\frac{k_y a}{2}\right)\right\}^{1/2} \tag{1}$$

To get the bandstructure of carbon nanotubes, we begin with the bandstructure of graphene given in (1), apply periodic boundary conditions along the circumference of the nanotube. The rolling-up of the honeycomb lattice of the graphene sheet along a specific direction, known as

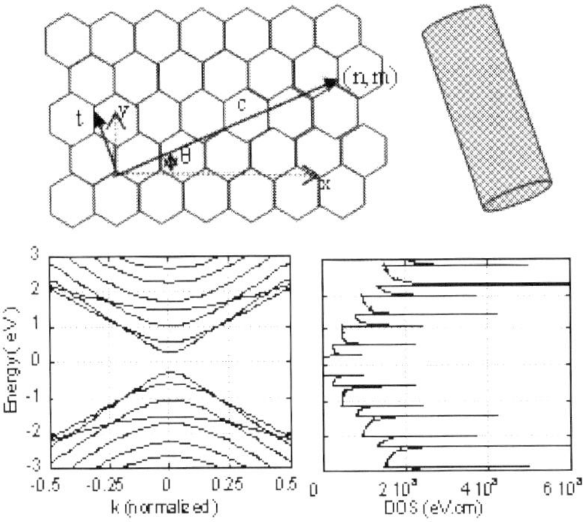

Fig. 3. Honeycomb lattice graphene sheet showing the chiral vectors (n,m) and the corresponding E-k diagram calculated using (1) and DOS using (2)

the chiral vector (shown in Fig. 3), causes the quantization of the wave-vector space along its direction.

A chiral vector can be denoted by the coordinates (n,m). If (n-m) is a multiple of 3, the carbon nanotube is metallic, else it is semi-conducting. To calculate the current, the electron density of states (DOS) near the Fermi level is required. Classical tight-binding models are used to accurately compute the DOS. At low bias, the DOS D(E) at energy E can be approximated as expressed in (2) (Guo & Lundstrom, 2006),

$$D(E) = \frac{D_0|E|}{\sqrt{E^2 - {E_n}^2}} \text{ where } D_0 = \frac{8}{3\pi V_\pi a} \tag{2}$$

All variables used in the above equations are defined in Table 1. For a more detailed discussion on the band-structure of carbon nanotubes, zone-folding approximation and complete derivation the interested reader is referred to (Guo & Lundstrom, 2006).

1.2 Surface-potential based modeling

On applying a gate voltage V_G is applied, the surface potential (ϕ_s) is modulated. The expressions for surface potential and the total charge are as follows :

$$\phi_s = V_{GS} - \frac{|Q_{CNT}|}{C_{ins}} \tag{3}$$

$$\begin{aligned} Q_{CNT} = & N_0 \sum_n \int_{E_n} F(\sqrt{E^2 - E_n^2}, \mu_s) \\ & + F(\sqrt{E^2 - {E_n}^2}, \mu_s - V_{DS}) dE \end{aligned} \tag{4}$$

where

$$F(E,\mu) = \frac{1}{1+e^{(E-\mu)}} \text{(Fermi-Dirac Integral)} \tag{5}$$

and n is the number of sub-bands under conduction.
The conventional method to compute the ϕ_s (using the conduction-band minima and DOS calculated from Table 1 and (2), respectively) involves numerically solving the 1-D poisson equation and the total charge equation self-consistently. In spite of being accurate, this method is computationally intensive and inappropriate for compact modeling and circuit simulations. Additionally, SPICE solvers may encounter convergence errors when loaded with a task of solving complicated numerical functions. Hence, a linearized equation for ϕ_s has been derived. By eliminating numerical iterations, the simulation speed is considerably improved making the model suitable for large scaling circuit simulation.
We derive a closed form linear approximation for surface potential from the fundamental equations of channel charge and bias conditions. From (3) and (4)

$$\begin{aligned}\phi_s =& V_{gs} - \frac{N_0}{C_{ins}} \sum_n \int_{E_n} \frac{1}{1+exp(\frac{E-V_s-E_n-\phi_s}{KT})} \\ &+ \frac{1}{1+exp(\frac{E-V_d-E_n-\phi_s}{KT})} dE \\ =& V_{gs} - \gamma \sum_n \left[ln(1+e^{(\frac{E_f-V_s-E_n-\phi_s}{KT})}) \right. \\ &\left. + ln(1+e^{\frac{(E_f-V_d-E_n-\phi_s)}{KT}}) \right]\end{aligned} \tag{6}$$

For low bias voltages, the first order approximation (6) is linear. Hence, we may write the equation as

$$\begin{aligned}\phi_s =& V_{gs} - \sum_n \gamma \left[\frac{(E_f - V_s - E_n - \phi_s)}{KT} \right. \\ &\left. + \frac{(E_f - V_d - E_n - \phi_s)}{KT} \right]\end{aligned} \tag{7}$$

Moving ϕ_s from RHS on to the LHS and re-arranging the terms, we get the non-iterative closed form expression for surface potential as follows

$$\phi_s = \sum_n \left(\frac{V_t \gamma (\xi_s |\xi_s| + \xi_d |\xi_d|)}{2(1+2\gamma)} \right) - V_{gs} \tag{8}$$

where $\gamma = N_0 / C_{ins}$, and

$$\xi_{s,d} = \frac{(E_f - V_{s,d} - E_{o,p} + V_{gs})}{V_t}; |\xi_{s,d}| = \begin{cases} 1, & \text{if } \xi_{s,d} > 0, \\ 0, & \text{if } \xi_{s,d} < 0. \end{cases}$$

This expression forms the basis of the compact model. All existing models (Castro et al., 2002; Deng & Wong, 2007a; Guo et al., 2004; 2002; Raychowdhury et al., 2004; Wong et al., Nov.

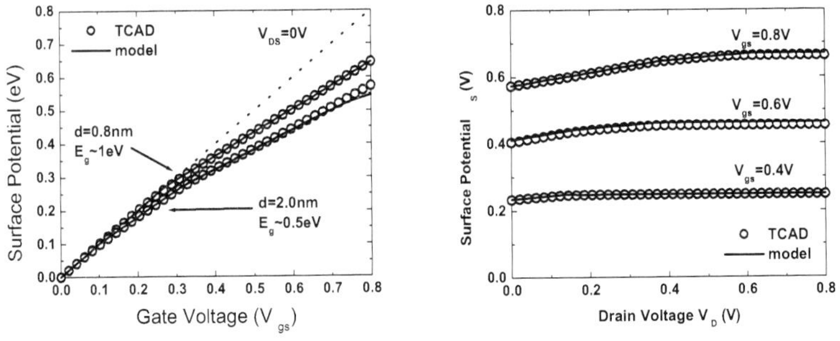

(a) (b)

Fig. 4. ϕ_s as a function of V_{gs} and V_{ds} for d=0.8nm and 2nm The voltage range chosen is the region where there is good gate control and FET-like behavior

2006) use some kind of self-consistent numerical methods to solve for ϕ_s. Fig. 5(b) shows the variation of surface potential as a function of V_{GS} and V_{DS}, for different diameters. At low voltages, the model is in good agreement with the numerical simulations and no regional approximations are required in the expression. The surface potential is a function of the diameter, temperature and gate dielectrics to the first order. At higher voltages, higher sub-bands are filled and therefore the slope of the line in Fig. 5(b) changes and this behavior is modeled by (8).

1.3 Capacitance Model

Surface potential is calculated using (3), a function of Q_{CNT}. However, Q_{CNT} itself is a function of ϕ_s and most other models employ self-consistent iterations to solve for charge and surface potential. As explained in the previous subsection, we approximate the charge in each sub-band to be linear to get a closed form solution for surface potential. With ϕ_s known, the quantum charge Q_{CNT} can be calculated in closed form as well. Fig. 5 shows a plot of quantum charge calculated with respect to varying gate voltage for V_{DS}=0.2 and 0.8V.
Rate of change of quantum charge with respect to gate voltage is given by

$$\frac{\partial Q_{CNT}}{\partial V_G} = \frac{\partial Q_{CNT}}{\partial \phi_s}\frac{\partial \phi_s}{\partial V_G} \tag{9}$$

where the term $\partial Q_{CNT}/\partial \phi_s$ is known as quantum capacitance, C_Q. Since the model shows excellent agreement with TCAD simulations is demonstrated for Q_{CNT} vs. V_G (Fig. 5) and ϕ_s vs. V_G (Fig. 4), the effect of quantum capacitance is implicit and need not be calculated separately. A capacitance branch model similar to (Deng, 2007), shown in Fig. 5(b), is used to calculate the intrinsic voltages that are effective inside the channel.
Extrinsic capacitance such as parasitic fringe capacitance, inter-electrode capacitance and coupling capacitance between adjacent gates in multiple gate/multiple nanotube device geometries can dominate over the intrinsic capacitance and impact the performance of a CNT-FET. A provision for including non-ideal extrinsic capacitance has been incorporated in our compact model in the form of parasitic capacitance, C_p, which is a fitting parameter. The exact value

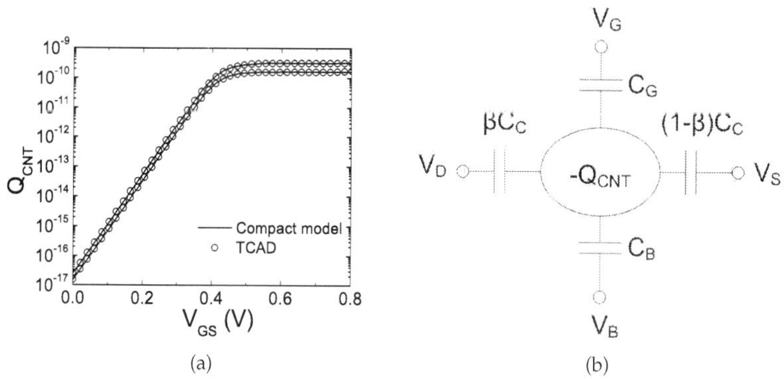

Fig. 5. (a) Q_{CNT} as a function of V_{GS} for V_{DS}=0.2V and 0.8V and CNT diameter 1nm. (b) The capacitance divider network used in the model. C_C and β are fitting parameters.

of C_p is device-geometry dependent and can be estimated from the expressions in (Deng & Wong, 2007b; John & Pulfrey, 2006; Paul et al., 2006).

1.4 Schottky Barrier modeling

The Schottky barrier height ϕ_{SB} depends on the work function difference and the barrier width depends on the insulator thickness. The total current at the junction is the sum of thermoionic emission and the tunneling current through the barrier. Hence it is important to accurately model carrier conduction in a CNT FET through the Schottky barrier. Transmission coefficients for a single barrier are calculated using the WKB approximation (Nakanishi et al., 2002) as follows

$$T(E) = exp\left[-\int_{z_i}^{z_f} k(z)dz\right] \tag{10}$$

The WKB approximation does not take into account reflection between the source and drain junctions. The conductance of the actual device is lower than calculated (Heinze et al., 2002). The Fabry-Perot Cavity model is used to include the effect of reflection given as

$$T(E) = \frac{T_s(E)T_d(E)}{T_s(E) - T_s(E)T_d(E) + T_d(E)} \tag{11}$$

The WKB approximation (D.Jimenez, Jan 2007) has two exponential terms in the E-k space which does not have a closed form solution. Approximating the barrier profile as a triangle allows us to get a closed form solution and results in negligible loss in accuracy. Extending on the derivation in (D.Jimenez, Jan 2007), we have the following

$$E - E(z) = \pm\frac{3a}{2}V_\pi\sqrt{{k_n}^2 + {k_z}^2}$$

where

$E(z)$=energy profile of conduction/valence band

k_n=momentum component of nth sub-band

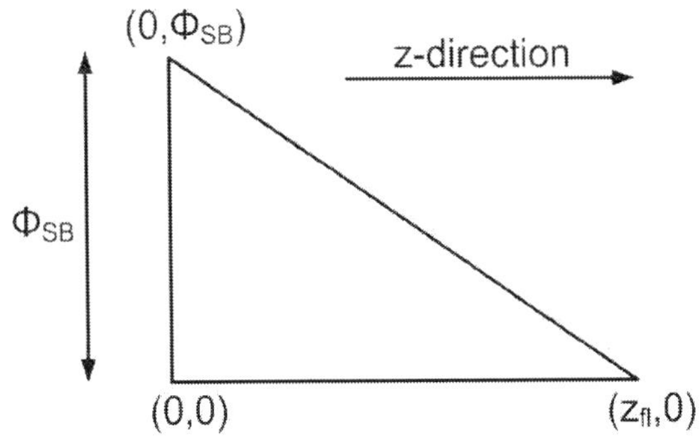

Fig. 6. Triangular Schottky Barrier Height model used in deriving the closed form expression for Tunneling Probability.

k_z=momentum component along direction of electron transport
$a = 0.142nm$, the carbon-carbon bond distance
Hence

$$k_z = \sqrt{\frac{4[E - E(z)]}{(3aV_\pi)^2} - k_n{}^2}$$

Using the value of k_z in (10),

$$T(E) = exp\left[- \int_{z_i}^{z_f} \sqrt{k_n{}^2 - \frac{16[E - E(z)]^2}{(3aV_\pi)^2}} \right] \tag{12}$$

In order to get an expression for $E - E(z)$, a triangular barrier profile as shown in Fig. 6 is used, which gives us

$$E(z) = -\frac{\phi_{sb}}{z_{fl}}(z) + \phi_{sb}$$
$$E - E(z) = E - \phi_{sb} + \frac{\phi_{sb}}{z_{fl}}(z) \tag{13}$$

From (12) and (13), we get a closed form expression for tunneling probability in (14). The limits of the integration are $z_i = 0$ and $z_f = z_{fl}$, where $z_{fl} = [-t_{ins}/2]ln(E/\phi_{sb})$.

$$T(E) = exp\left[\frac{-t_{ins}k_n}{\phi_{sb'}} (E'\sqrt{1 - K'^2} + (E - \phi_{sb'})E_t) \right] \tag{14}$$

where

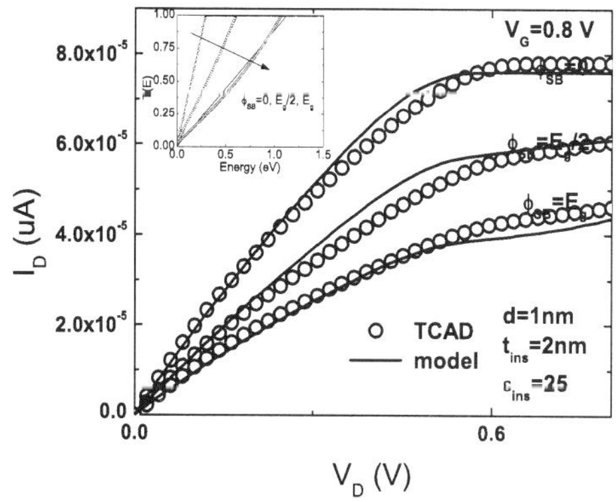

Fig. 7. I_{ds} vs. V_{ds} at V_{gs}=0.8V for three different barrier heights. The inset shows the variation in the tunneling probability (numerical vs. triangular approximation) for the different contacts.

$$
\begin{aligned}
E_t &= \sqrt{(1 - K(E - \phi_{sb'}))^2} - sin^{-1}(-E'\sqrt{1 - KE'^2}) \\
&\quad + sin^{-1}(\phi_{sb'} - E) \\
K &= \left(\frac{q\pi}{4k_n N_0}\right) \\
E' &= (E - \phi_{sb'}) + \phi_{sb'} ln\left(\frac{E}{\phi_{sb'}}\right) \\
\phi_{sb'} &= \mu_{s,d} + \phi_{sb}
\end{aligned} \tag{15}
$$

Fig. 7 demonstrates the excellent agreement between the triangular approximation model and the numerical model for the contact part. The tunneling probability equation given by (14) is solved at the source and drain junctions and (16) is used to compute the final current.

$$
\begin{aligned}
I = \frac{4q}{h} \sum_n \int_{E_n} & sgn(E)T(E)\Big[F(sgn(E)(E, \mu_s)) \\
& + F(sgn(E), (E, \mu_s - V_{ds}))\Big] dE
\end{aligned} \tag{16}
$$

where $sgn(E) = 1$ or -1 for conduction and valence band respectively and $F(\mu, E)$ is as defined in (5).
Using the equations and results discussed above, (summarized in Table. 1), a physics based compact model of CNT FET is completed and implemented in VerilogA. It is computationally efficient and supports transient simulations. The I-V characteristics are presented in Fig.

Physical constants

V_π	C-C Bonding Energy	2.97eV
a	C-C Bonding Length	0.142nm
q	Charge	$1.6e^{-19}$ C
V_t	Thermal Voltage	26mV

Model Parameters

d	Diameter(m)	θ	Chiral Angle (deg)
L	Nanotube length (m)	t_{ins}	Insulator thickness (m)
ϕ_{sb}	Barrier height (eV)	ϵ_{ins}	Insulator Dielectric Constant

Derived Parameters

Energy Gap (eV)	$E_g = 2V_{pi}a/d$
Sub-band Energy Levels (eV)	$E_n = (E_g/8)(6n - 3 - (-1)^n)$
Intrinsic Carrier Conc.	$N_0 = 4q/(3\pi V_\pi a)$
Insulator Capacitance	$C_{ins} = \frac{2\pi\epsilon_r\epsilon_0}{\log((t_{ins}+d/2)/(d/2))}$

Table 1. Constants and Parameters used in the Model

8(a). These results prove that the model is scalable to different diameters and bias conditions. Since we use the surface potential approach, scattering effects that may further affect the I-V characteristics can be easily incorporated in the future.

2. Transistor Model Validation and Extraction

The parameters enlisted in Table 2 comprise the SPICE based circuit model for CNT FET. Running simulations by varying each parameter enables us to gain detailed insight on its impact on the design potential of the CNT FET.

2.1 Extraction procedure

Our compact model can be used to comprehend measurement data in order to gain process-related insight such as parasitics, variations etc. This is achieved by properly tuning the model parameters enlisted in Table 2. A capacitor divider network similar to (Deng, 2007) is assumed in this model, as shown in Fig. 5(b). C_C, the coupling capacitance and β are the two primary fitting paramters from this network.
Scattering effects are not directly incorporated in the model. The fitting parameter *mob* is used to capture the effect of scattering in the nanotube and is multiplied to the final current computed in (16) in the model. *mob* value is dependent on the length of the nanotube and lies between 0 and 1. Appropriate values for this parameter can be obtained from published results (Deng, 2007) .The main fitting steps are

1. Define instance parameters; calculate physical parasitics (C_C is set to a very small value, which is about 1/10 of the insulator capacitance).
2. Csubfit: tuned to fit I_{DS} vs. V_{GS} at low V_{DS} (0.1V) and V_{BS} fixed. This is to match the flat bland voltage.

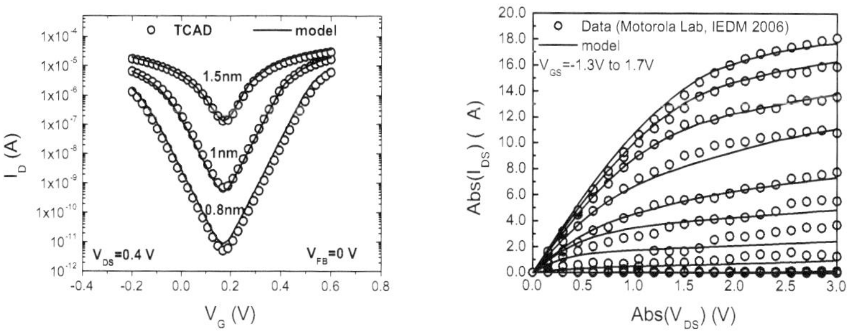

(a) (b)

Fig. 8. (a) I_{ds} as a function of V_{gs} for d=0.8nm, 1nm and 1.5nm. $V_{FB} = 0V$ $t_{ins} = 2nm$ $\epsilon_r = 25$ and $L = 10nm$. (b) Model validation with experimental data. Both scales show absolute values of drain voltages and currents (Amlani et al., 2006).

3. β: tuned to fit I_{DS} vs. V_{DS} at a high V_{GS} to match the saturation region (basically the shape of the I_{DS} vs. V_{DS} curve).
4. C_p: tuned to match I_{DS} vs. V_{GS} in the subthreshold region, at high V_{DS}; ϕ_{SB} also needs to be tuned to match I_{DS} vs. V_{GS} in the saturation region.
5. $R_{D,S}$: tuned to primarily match I_{DS} vs. V_{DS} in the linear region. $R_{D,S}$ also affects the saturation region. Hence, iterations may be required to get the correct fit.
6. mob: used to match the saturated drain current.

2.2 Model Validation

Using the extraction procedure described in the previous section, the model has been validated with published measurement data (Amlani et al., 2006) as shown in Fig. 8(b). An interesting feature of the fitting is the exact replication of the gap in the I-V plot, which is due to the multiple band conduction in carbon nanotubes.
The I-V characteristics in general have the following trends:

- The off current varies exponentially with diameter and barrier height.
- The on current degrades with barrier height and increases linearly with diameter.

These conclusions have been confirmed in previous models. But the new model helps us run SPICE simulations fast enough to benchmark circuits performance. All the results in the next section are generated using the Verilog-A model that supports DC and transient analysis for a single inverter several times faster than numerical simulations in MATLAB.

3. Interconnect modeling

Metallic CNT possess properties of high mechanical and thermal stability, thermal conductivity and high current carrying capabilities (Naeemi et al., Feb. 2005)(Srivastava & Banerjee,

Parameter	Description	Default (units)
Instance Parameters		
d	Diameter	2nm
θ	Chiral angle ($0 \leq \theta < 30^o$)	0
$tins$	Insulator thickness	10nm
$eins$	Dielectric constant of insulator	9
$tback$	Backgate insulator thickness	130nm
$eback$	Dielectric constant of substrate	3.9 (SiO_2)
L	Gate length	100nm
$type$	n-type=1 p-type= -1	1
Model Parameters		
$phisb$	Schottky barrier height	0eV
mob	Mobility parameter	1
R_s	Parasitic Source access resistance	0 ohm
R_d	Parasitic Drain access resistance	0 ohm
β	Coupling Coefficient	1
C_C	Coupling Capacitance	7aF
C_p	Parasitic Capacitance	120aF

Table 2. Spice Model File Parameters

2005) making them candidates for interconnects. Ideally, metallic SWCNT's have a fermi velocity of about $8X10^5$ m/s (M.S. Dresselhaus & Eklund, 1996) but in reality the ballistic motion is degraded by several scattering mechanisms. The scattering mechanisms has been been discussed in several previous publications (Park et al., 2004; Raychowdhury & Roy, 2006). The resistance is normally modeled by a set of piece-wise linear equations. In this work we present a continuous expression for the resistance of the interconnect and the resistance of the contact making them suitable for SPICE simulators. For the sake of simplicity, the effect of temperature on resistance has not been included in this model. For a detailed discussion on the effect of temperature on resistance of metallic CNTs, the reader is referred to (Pop et al., 2007). The circuit model for the interconnect is shown in Fig. 9(b). At high frequencies, the inductance and the capacitance determine the total impedance of the interconnect. The following subsections present the DC and small-signal parameters of the CNT interconnect.

Due to the nature of the band structure, in an ideal ballistic motion regime, the resistance is constant as expressed in (17)

$$R_{ballistic} = \frac{h}{4e^2} = \frac{1}{G_0} \tag{17}$$

However, when the length of the interconnect is much longer that the mean free path (MFP), several scattering mechanisms dominate. At low bias, the predominant mechanism is the acoustic phonon scattering with a MFP of $1\mu m - 1.6\mu m$ Park et al. (2004). As the bias voltage increases, the electrons can scatter from band to band and also within the same band. This

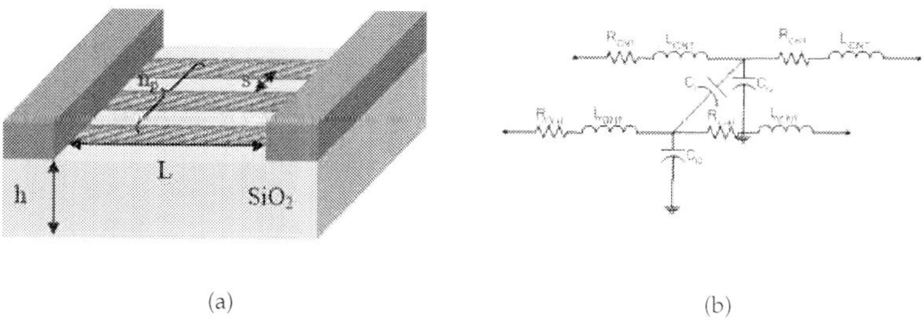

Fig. 9. (a) Cross-section of a generic interconnect using carbon nanotubes. (b) Circuit model for CNT interconnect

leads to optical phonon scattering and zone-boundary scattering. These scattering mechanisms are well-known and have been modeled in the past. In this compact model, we derived a single equation to continuously model all these scattering effects across multiple regions as shown in (18),

$$G(V,L) = G_{op\_zo} + \frac{V_{eff}[G_{acc} - G_{op\_zo}]}{V} \tag{18}$$

where

$$V_{eff} = V_{cr} - \frac{1}{2}\left[(V_{cr} - V - \delta) + \sqrt{(V_{cr} - V - \delta)^2 + 4V_{cr}\delta}\right]$$

3.1 Resistance

Here a piece-wise linear model for resistance (Raychowdhury & Roy, 2006) is modified by including V_{eff} which ensures that mobility is a continuous function of the bias voltage and length facilitating convergence in circuit simulators . The dependence of length and bias voltage on the resistance of a CNT interconnect is shown in Fig. 10(a).

3.2 Capacitance and Inductance

As shown in Fig. 9(a) carbon nanotube interconnects are usually formed by arranging arrays of nanotubes aligned next to each other with the terminals at the ends of the two tubes. The coupling capacitance between two adjacent nanotubes C_C, and the quantum capacitance within the nanotube C_q have considerable effect on its conduction properties. The coupling capacitance has the form

$$C_c = \frac{\pi \epsilon L}{log\left(\frac{d}{S} + \sqrt{\left(\frac{d}{S}\right)^2 + 1}\right)} \tag{19}$$

And the quantum capacitance is given by

$$C_q = \frac{4e^2 L}{\pi h v_f} \tag{20}$$

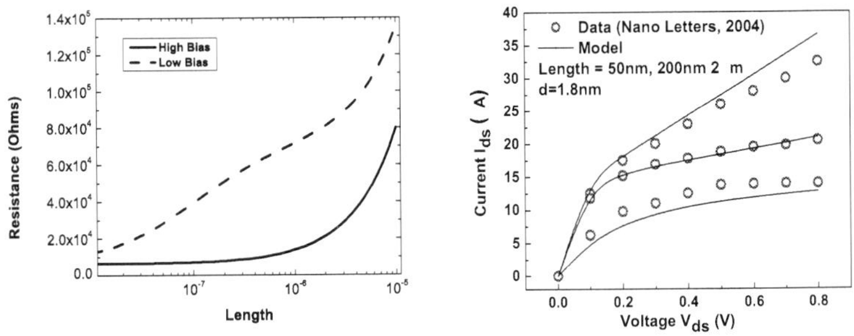

(a) (b)

Fig. 10. (a) Resistance of a carbon nanotube interconnect with varying length for high and low bias across the terminals. (b) Interconnect model validation with measured data for varying length. Diameter of the metallic nanotubes are 1.8nm.

Carbon nanotubes posses two kinds of inductances, the magnetic or mutual inductance and the kinetic or self-inductance. As discussed in detail in (Raychowdhury & Roy, 2006), kinetic inductance dominates mutual inductance in a one-dimensional structure like carbon nanotubes and hence we only consider kinetic inductance in the model. This is given by

$$L_e = \frac{h}{2e^2 v_f} \tag{21}$$

Carbon nanotubes have two modes of propagation, with two electrons in each mode (spin up and spin down). This results in a total of four modes of propagation giving one-fourth of the total inductance calculated in (21) and four times the quantum capacitance given in (20).

3.3 Interconnect model extraction and verification

Three model parameters, V_{crit}, l_{acc} and l_{zb} are used to model the optical phonon scattering, acoustic phonon scattering and zone boundry phonon scattering respectively. The SPICE circuit parameters for the interconnect model are enlisted in Table 3.
The rest of the parameters are geometry dependent.

1. Coupling capacitance is either calculated by external 2D or 3D solvers such as Raphael (*Raphael Interconnect Analysis Program Reference Manual*, n.d.) and entered as an instance parameter, or can be calculated internally by (19).
2. If CNT length ranges between 10nm and 1μm, V_{crit} is tuned in the range of 0.08 to 0.16 to decrease the resistance. If CNT length is greater than 1μm, acoustic phonon scattering dominates and therefore l_{acc} changes the slope of the curve.
3. If the contacts are short and ohmic then R_n and R_p can be ignored.
4. At high current values the ϕ_{SB} value is extracted.

The model is validated against measured data in Fig. 10(b).

Parameter	Description	Default value
Instance		
d	Diameter	1nm
np	Number of CNTs in parallel	1
s	Spacing between CNTs	10nm
e_{ins}	Dielectric constant of insulator	25
Cc	Coupling Capacitance	0
h	Substrate insulator thickness	100nm
L	Gate length	100nm
Model		
$phisb$	Schottky Barrier height	0eV
V_{crit}	Optical-phonon scattering parameter	0.16eV
R_p, R_n	Parasitic Access Resistance	0 ohm
l_{acc}	Mfp for acoustic phonon scattering	1.0 um
l_{zb}	Mfp for zone boundary phonon scattering	20nm

Table 3. Interconnect Model File Parameters

4. Design Insights

The developed compact model for CNT transistors accurately predicts I-V and C-V characteristics. It is scalable to key process and design parameters such as diameter, chirality, gate dielectrics, and bias voltages. Using the model, we explore design possibilities with CNT in order to identify the optimum design space. CNT with L=100nm is compared with 22nm CMOS (PTM) for both analog and digital applications. For consistency we have used $V_{FB} = Vdd/2$(NCNT) and $-Vdd/2$(PCNT). The dielectric material used has $\epsilon_r = 25$. Parasitic capacitances have been lumped into a single parameter based on published values (J. Deng, Sept.2006). Since all the characteristics are mainly dependent on the diameter of the nanotube, our analysis is for varying diameters. Above 1.8nm, the SB-FET has I_{ON}/I_{OFF} less than 50 and is not included in this study.

4.1 Digital Design

Numerous simulations of FO4 inverter comparing CNT FETs with 22nm CMOS have been undertaken to study the effect of Schottky barrier height (Source/Drain contact material), gate dielectric thickness, leakage power, supply voltage scaling and process variations on digital design (Balijepalli et al., 2007). It is found that for smaller diameters of the range of 1-1.5nm and optimum contact materials, up to 10X improvement in speed, power and energy consumption is observed as compared to 22nm CMOS. The speed contours have been plotted for adequate scaling in dielectric thickness to ensure the same performance. By varying the diameter, Fig. 11 shows that up to 10X increase in speed can be achieved when compared to 22nm CMOS.

The reason for diameters of 1-1.5nm being optimal is depicted by the shaded region in the Fig. 11. Larger diameters have higher leakage and are harder to switch off. Smaller diameters have

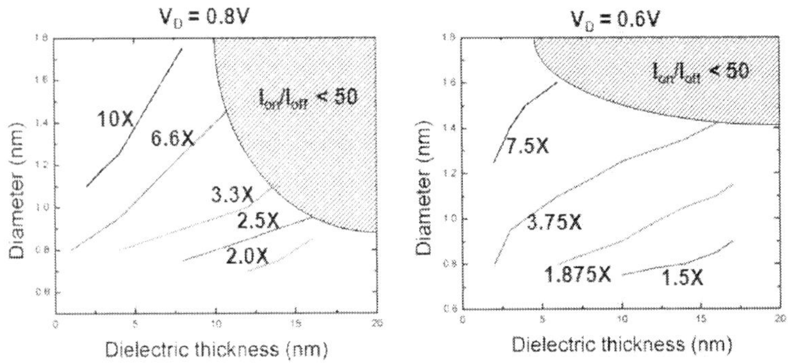

Fig. 11. Speed contours for varying diameters and t_{ins} for V_{DD} = 0.8V and 0.6V

a 5X decrease in speed compared to larger diameter CNTs. Thus, there is a trade-off between speed and power in using CNT FET for digital applications, similar as that of CMOS.

4.2 Analog Design

If parasitic capacitance is reduced, CNTs have another advantage in very low quantum capacitance. The device can have very high cut-off frequency given by (Akinwande et al., 2006).

$$f_T = \frac{g_m}{2\pi C_g} \tag{22}$$

Based on the above equation, cut-off frequency for a 2nm diameter CNT transistor is calculated to be around 120 GHz in our model. An in-depth critique of the high frequency performance of carbon nanotube field effect transistors is given in (Pulfrey & Chen, 2008) which discusses the effect of nanotube chirality (diameter), oxide permittivity and other process parameters on f_T. These effects can be incorporated in our model with minimal effort.

The two major hurdles preventing large scale manufacturing of CNT based devices are lack of measurement technique to characterize analog performance and reducing the parasitic capacitance during fabrication. The AC gain and frequency response are mainly controlled by the transconductance (g_m) and output impedance (R_{out}). Here we plot the variation of output impedance of CNT FET compared to 22nm CMOS (Fig. 12). For a fair comparison, R_{out} is calculated for the same saturation current of both devices. For CMOS, R_{out} vs. V_{DS} is mainly influenced by the triode region, channel length modulation, drain induced barrier lowering (DIBL) and finally substrate current induced body effect (SCBE) with increasing V_{DS} (Huang et al., 1992). In CNTs, R_{out} is affected by the linear, saturation and ambipolar characteristics of the device. As shown in Fig. 12, due to better saturation characteristics in CNTs, a CNT FET can have up to 25X higher R_{out} compared to 22nm CMOS for the same saturation current.

5. Conclusion

In this chapter, a detailed procedure for developing compact models for carbon nanotube transistors and interconnects has been presented. Since the developed model does not use any iteration-based calculations, is scalable with process and design parameters and is highly accurate, it increases the scope of predictive design research. These models have been used for

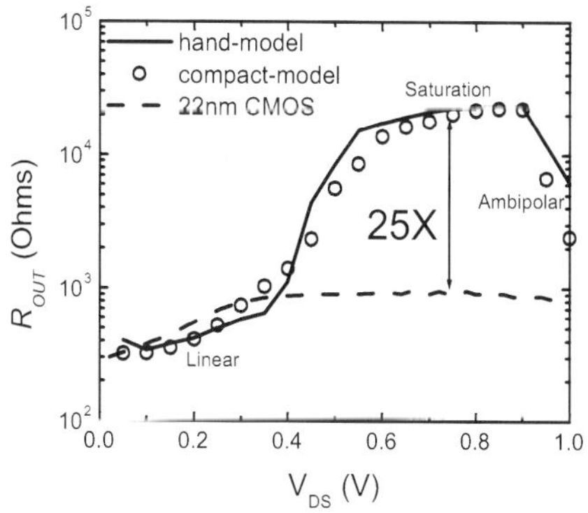

Fig. 12. R_{out} as a function of drain voltage, compared with 22nm CMOS with same saturation current. Regions affecting R_{out} are annotated.

circuit simulation to benchmark the performance of carbon nanotube transistors with 22nm bulk CMOS transistors providing important perspective on the design trade-offs and potential of CNT based devices. Thus, compact modeling serves as one of the most important bridges between CNT process and design giving key insights into the development of carbon nanotube based electronics.

6. References

Akinwande, D., Close, G. F. & Wong, H.-S. P. (2006). Analysis of the Frequency Response of Carbon Nanotube Transistors, *Nanotechnology, IEEE Transactions on* **5**(5): 599 – 605.

Amlani, I., Lewis, J., Lee, K., Zhang, R., Deng, J. & Wong, H.-S. P. (2006). First demonstration of ac gain from a single-walled carbon nanotube common-source amplifier, pp. 1–4.

Balijepalli, A., Sinha, S. & Cao, Y. (2007). Compact modeling of carbon nanotube transistor for early stage process-design exploration, *ISLPED '07: Proceedings of the 2007 international symposium on Low power electronics and design*, ACM, New York, NY, USA, pp. 2 – 7.

Bowman, K. A., Duvall, S. G. & Meindl, J. D. (2002). Impact of die-to-die and within-die parameter fluctuations on the maximum clock frequency distribution for gigascale integration, *Solid-State Circuits, IEEE Journal of* **37**(2): 183 – 190.

Castro, L., John, D. & Pulfrey, D. (2002). Towards a compact model for schottky-barrier nanotube fets, pp. 303–306.

Chau, R., Datta, S., Doczy, M., Doyle, B., Kavalieros, J. & Metz, M. (2004). High-κ/MetalñGate Stack and Its MOSFET Characteristics, *IEEE Electron Device Lett.* **25**(2004): 408 – 410.

Chen, I. C., Holland, S. & Hu, C. (February 1985). Electrical Breakdown in Thin Gate and Tunneling Oxides, *IEEE Trans. Electron Devices* **32**: 413 – 422.

Chen, Z., Appenzeller, J., Knoch, J., Lin, Y.-M. & Avouris, P. (2005). The role of metal-nanotube contact in the performance of carbon nanotube field-effect transistors, *Nano Lett.* **5**: 1497 – 1502.

Close, G. F. & Wong, H.-S. P. (2007). Fabrication and Characterization of Carbon Nanotube Interconnects, *Electron Devices Meeting, 2007. IEDM 2007. IEEE International* pp. 203 – 206.

Deng, J. (2007). A compact spice model for carbon-nanotube field-effect transistors including nonidealities and its application6part i: Model of the intrinsic channel region.

Deng, J. & Wong, H.-S. (2007a). A Compact SPICE Model for Carbon-Nanotube Field-Effect Transistors Including Nonidealities and Its Application6Part I: Model of the Intrinsic Channel Region, *Electron Devices, IEEE Transactions on* **54**(12): 3186 – 3194.

Deng, J. & Wong, H.-S. (2007b). Modeling and analysis of planar-gate electrostatic capacitance of 1-d fet with multiple cylindrical conducting channels, *Electron Devices, IEEE Transactions on* **54**(9): 2377–2385.

D.Jimenez, e. (Jan 2007). A simple drain current model for schottky-barrier carbon nanotube field effect transistors, *Nanotechnology* **18**(2): 025201. ID: 61221.

Doyle, B., Boyanov, B., Datta, S., Doczy, M., Hareland, S., Jin, B., Kavalieros, J., Linton, T., Rios, R. & Chau, R. (2003). Tri-Gate fully-depleted CMOS transistors: Fabrication, design and layout, *VLSI Symp. Tech. Dig.* pp. 133 – 134.

Guo, J., Datta, S, Lundstrom & M (2004). A numerical study of scaling issues for Schottky-barrier carbon nanotube transistors, *Electron Devices, IEEE Transactions on* **51**(2): 172 – 177.

Guo, J. & Lundstrom, M. (2006). *Nanoscale Transistors: Device Physics, Modeling and Simulation*, Springer.

Guo, J., Lundstrom, M. & Datta, S. (2002). Performance projections for ballistic carbon nanotube field-effect transistors, *Applied Physics Letters* **80**(17): 3192–3194.
URL: *http://link.aip.org/link/?APL/80/3192/1*

Heinze, S., Tersoff, J., Martel, R., Derycke, V., Appenzeller, J. & Avouris, P. (2002). Carbon nanotubes as schottky barrier transistors, *Phys. Rev. Lett.* **89**(10): 106801.

Hisamoto, D., Lee, W.-C., Kedzierski, J., Takeuchi, H., Asano, K., Kuo, C., Anderson, E., King, T.-J., Bokor, J. & Hu, C. (2000). FinFET-a self-aligned double-gate MOSFET scalable to 20 nm, *IEEE Trans. Electron Devices* **47**(12): 2320 – 2325.

Huang, J. H., Liu, Z. H., Jeng, M. C., Ko, P. K. & Hu, C. (1992). A physical model for MOSFET output resistance, *Electron Devices Meeting, 1992. Technical Digest., International* pp. 569 – 572.

J. Deng, H.-S. W. (Sept.2006). *A circuit-compatible SPICE model for enhancement mode carbon nanotube field effect transistors*, Synopsys Inc.

John, D. L. & Pulfrey, D. L. (2006). Switching-speed calculations for schottky-barrier carbon nanotube field-effect transistors, **24**(3): 708–712.
URL: *http://link.aip.org/link/?JVA/24/708/1*

Lin, Y.-M., Appenzeller, J., Chen, Z., Chen, Z.-G., Cheng, H.-M. & Avouris, P. (2005). High performance dual-gate carbon nanotube FETs with 40-nm gate length, *IEEE Electron Device Lett.* **26**: 823 – 825.

McEuen, P. L., S, F. M. & Park, H. (2002). Single-walled carbon nanotube electronics, *Nanotechnology, IEEE Transactions on* **1**(1): 78 – 85.

Moore, G. E. (1965). Cramming more components onto integrated circuits, *Electronics* **38**(8).

M.S. Dresselhaus, G. D. & Eklund, P. (1996). *Science of Fullerenes and Carbon Nanotubes*, Elsevier Inc.

Naeemi, A., Sarvari, R. & Meindl, J. D. (Feb. 2005). Performance comparison between carbon nanotube and copper interconnects for gigascale integration (GSI), *IEEE Electron Device Lett.* **26**(2): 84 – 86.

Nakanishi, T., Bachtold, A. & Dekker, C. (2002). Transport through the interface between a semiconducting carbon nanotube and a metal electrode, *Phys. Rev. B* **66**(7): 073307.

Park, J.-Y., Rosenblatt, S., Yaish, Y., Sazonova, V., Ustunel, H., Braig, S., Arias, T. A., Brouwer, P. W. & McEuen, P. L. (2004). Electron-Phonon Scattering in Metallic Single-Walled Carbon Nanotubes, *Nano Lett.* **4**(3): 517 – 520.

Paul, B., Fujita, S., Okajima, M. & Lee, T. (2006). Impact of geometry-dependent parasitic capacitances on the performance of cnfet circuits, *Electron Device Letters, IEEE* **27**(5): 380–382.

Pop, E., Mann, D. A., Goodson, K. E. & Dai, H. (2007). Electrical and thermal transport in metallic single-wall carbon nanotubes on insulating substrates, *Journal of Applied Physics* **101**(9): 093710.
URL: *http://link.aip.org/link/?JAP/101/093710/1*

Pulfrey, D. L. & Chen, L. (2008). Examination of the high-frequency capability of carbon nanotube fets, *Solid-State Electronics* **52**(9): 1324 – 1328. Papers Selected from the 37th European Solid-State Device Research Conference - ESSDERC'07.
URL: *http://www.sciencedirect.com/science/article/B6TY5-4ST45Y8-1/2/35986f4a38a6adca353dfae27fc40cf2*

Raphael Interconnect Analysis Program Reference Manual (n.d.). Synopsys Inc.

Raychowdhury, A., Keshavarzi, A., Kurtin, J., De, V. & Roy, K. (2006). Carbon Nanotube Field-Effect Transistors for High-Performance Digital Circuits-DC Analysis and Modeling Toward Optimum Transistor Structure, *Electron Devices, IEEE Transactions on* **53**(11): 2711 – 2717.

Raychowdhury, A., Mukhopadhyay, S. & Roy, K. (2004). A circuit-compatible model of ballistic carbon nanotube field-effect transistors, *Computer-Aided Design of Integrated Circuits and Systems, IEEE Transactions on* **23**(10): 1411 – 1420.

Raychowdhury, A. & Roy, K. (2006). Modeling of metallic carbon-nanotube interconnects for circuit simulations and a comparison with Cu interconnects for scaled technologies, *Computer-Aided Design of Integrated Circuits and Systems, IEEE Transactions on* **25**(1): 58 – 65.

Srivastava, N. & Banerjee, K. (2005). Performance analysis of carbon nanotube interconnects for vlsi applications, *Computer-Aided Design, 2005. ICCAD-2005. IEEE/ACM International Conference on* pp. 383–390.

Wann, C. H., Noda, K., Tanaka, T., Yoshida, M. & Hu, C. (October 1996). A Comparative Study of Advanced MOSFET Concepts, *IEEE Transactions on Electron Devices* **Vol. 43**(No. 10): 1742 – 1753.

Welser, J., Hoyt, J. L. & Gibbons, J. F. (1994). Electron mobility enhancement in strained-Si n-type metal-oxideñsemiconductor field-effect transistors, *IEEE Electron Device Lett.* **15**(3): 100 – 102.

Wong, H.-S., Deng, J., Hazeghi, A., Krishnamohan, T. & Wan, G. (Nov. 2006). Carbon nanotube transistor circuits - models and tools for design and performance optimization, *IC-CAD* pp. 651–654.

Zhang, G., Wang, X., Li, X., Lu, Y., Javey, A. & Dai, H. (2006). Carbon Nanotubes: From Growth, Placement and Assembly Control to 60mV/decade and Sub-60 mV/decade Tunnel Transistors, *Electron Devices Meeting, 2006. IEDM '06. International* pp. 1 – 4.

Measurement of High-Frequency Characteristics of CNTFETs and Equivalent Circuit Model Analysis

Kaoru Narita
NEC Corporation
Japan

1. Introduction

Carbon nanotube field effect transistors (CNTFETs) are high-mobility devices that operate at very high-speeds. Theoretical analyses suggest that the cut-off frequency (f_T) of an ideal CNT-FET is between 800 GHz and 1.3 THz when its gate length is 0.1 μm (1; 2). Since this frequency is much higher than that of state-of-the-art Si, GaAs, and InP transistors, CNTFETs are promising candidates for future nanoelectronic devices. Singh et al.(3) measured frequency responses of top-gated CNTFETs up to 100 MHz. Li et al.(4) observed 2.6-GHz operation of CNTFETs with an LC impedance-matching circuit. However, as Li et al. pointed out (4), measuring high-frequency performance of high-impedance devices, such as CNTFETs, is quite difficult. This is because their output impedances are much higher ($\sim 10^5$ Ω) than the impedance of the measurement system (50 Ω) using a network analyzer. To perform accurate high-frequency measurements, especially those to determine f_T values of such devices, we must measure S-parameters with a network analyzer even though large impedance mismatches hinder us from obtaining accurate measurement data. Kim et al.(5) measured S-parameters of multi-finger CNTFETs by using a network analyzer and obtained an f_T value of 2.5 GHz. They also concluded a maximum oscillation frequency ($f_{\max}$) of more than 5 GHz was obtained using the maximum stable gain (G_{msg}). Le Louarn et al.(6) obtained intrinsic f_T value of 30 GHz by measuring a CNTFET the channel of which was fabricated using dielectrophoresis to increase the CNT density. They also obtained G_{msg} value of more than 10 dB at 20 GHz.
This chapter will describe a method for accurately measuring and modeling the high-frequency characteristics of CNTFETs, with reference to our experiment and analysis (7). In the experiment, we first decreased the device impedance to be able to measure the S-parameter using network analyzer. This was achieved by developing a high-density multiple-channel CNTFET structure the output impedance of which is much lower than that of the conventional single-channel CNTFETs. Then we used a de-embedding procedure to remove existing errors in measured S-parameters of small-signal devices in order to obtain the current gain and unilateral power gain (U) that can determine accurate f_T and $f_{\max}$ values. For accurate RF modeling of CNTFETs, we developed an equivalent circuit RF model that includes parasitic resistances and capacitances of the CNTFET. Then the expression of the f_T ($f_{\max}$) was derived as a function of them. Not ignoring the higher order parasitic resistances and capacitances neglected in the cases of current RF transistors, an accurate model was obtained that can fully explain the experimental results.

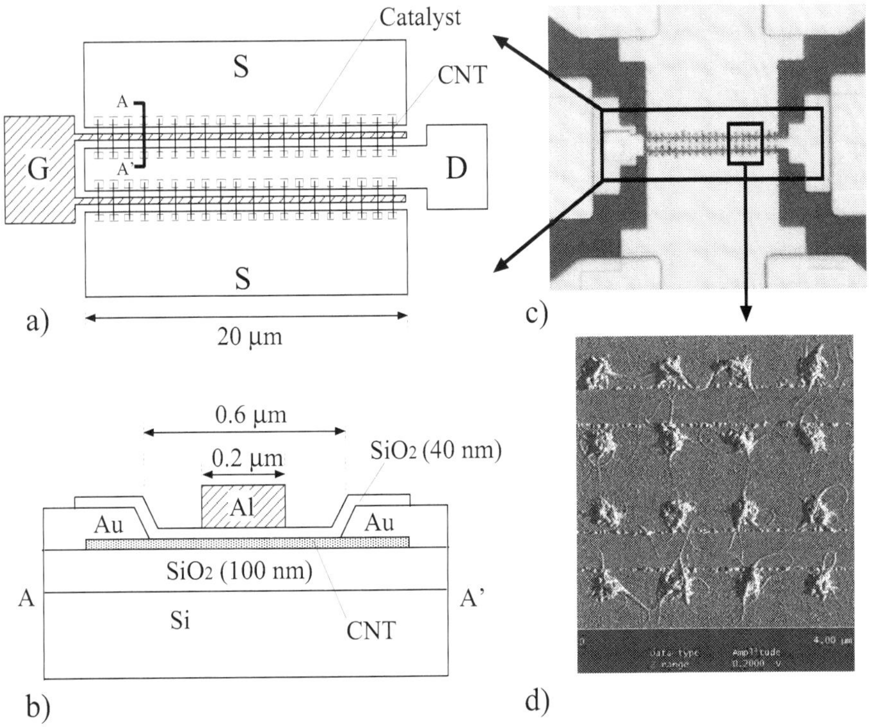

Fig. 1. Multiple-channel CNTFET structure: a) top view, b) cross section, c) optical micrograph, d) atomic force micrograph.

2. Multiple-channel CNTFET Structure

As shown in Figure 1, the evaluated CNTFET was fabricated on a SiO_2 insulator on a highly resistive (10 kΩcm) Si substrate. Iron was deposited for a catalyst and was patterned by electron-beam lithography. Single-walled carbon nanotubes (SWCNTs) were grown from the catalyst islands by chemical vapor deposition. The average density of the SWCNTs was 5 per μm, as observed in the AFM analysis (Figure 1-d). The gate oxide was 40-nm thick SiO_2, which serves as a passivation layer to retain stable characteristics and suppress hysteresis of the CNTFET I-V curve. The top-gated structure was used to reduce parasitic capacitances. The gate consisted of two 20-μm wide fingers. Thus, approximately 200 SWCNT channels were constructed in the total 40-μm gate width. The drain and source electrodes were formed by evaporation of Au, and ohmic contacts were made with CNT channels.

3. De-Embedding Procedure

Using the multiple-channel structure decreases the output impedances of the devices more than those of the single-channel CNTFETs. Therefore, their output signals can be observed

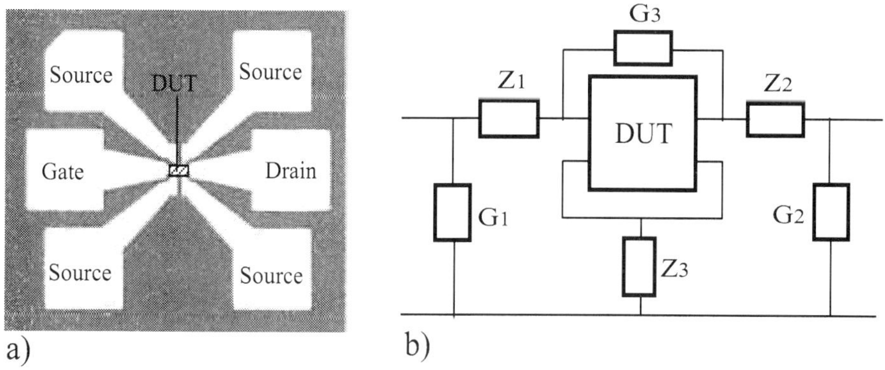

Fig. 2. RF test structure: a) pad layout, b) equivalent circuit.

directly with the network analyzer. However, output impedances of CNTFETs are still higher than those of conventional RF transistors. This means that the output signal of the device is small and easily disturbed or masked by the parasitic elements. The drain, gate, and source electrodes of the CNTFET were connected with pads for RF probe contacts. The dimensions of the pad were 100 $\times$ 100 μm, and its layout is shown in Figure 2-a. The areas of the pad and the connective wiring region are much larger than the transistor area (shown as DUT in Figure 2-a), and this large area forms parasitic elements and causes large errors in CNTFET S-parameters. Therefore, we applied the de-embedding procedure to effectively eliminate the parasitic error matrix, and only the S-parameters of the transistor were extracted using open-short-through standards on the substrate. This method is basically the same as that described in Vandamme et al.(8) and Temeijer et al.(9). The equivalent circuit of the RF test-structure, including pads and CNTFETs, is shown in Figure 2-b. In the figure, parasitic elements $(Z_1, Z_2, Z_3, G_1, G_2, G_3)$ are shown. Z_1, Z_2, Z_3 are parasitic impedances, and G_1, G_2, G_3 are parasitic admittances. To determine the parasitic elements, we made four standard patterns (Open, Short1, Short2, Through) that are the same as the CNTFET measurement patterns but without CNT channels (Figure 3). The equivalent circuits of the four standard patterns are shown in Figure 4. Each standard pattern contains a different combination of the parasitic elements, and so they can be determined by the measured S-parameters of the four standards. Let us transform the measured S-parameters $(s_{ij} : i, j = 1, 2)$ of the four standards to the Y-parameters and express them as $y_{ijop}, y_{ijsh1}, y_{ijsh2}, y_{ijthr} (i, j = 1, 2)$. Here, y_{ijop} is the Y-parameters of the Open standard, y_{ijsh1} is the Y-parameters of the Short1 standard, y_{ijsh2} is the Y-parameters of the Short2 standard, y_{ijthr} is the Y-parameters of the Through standard.

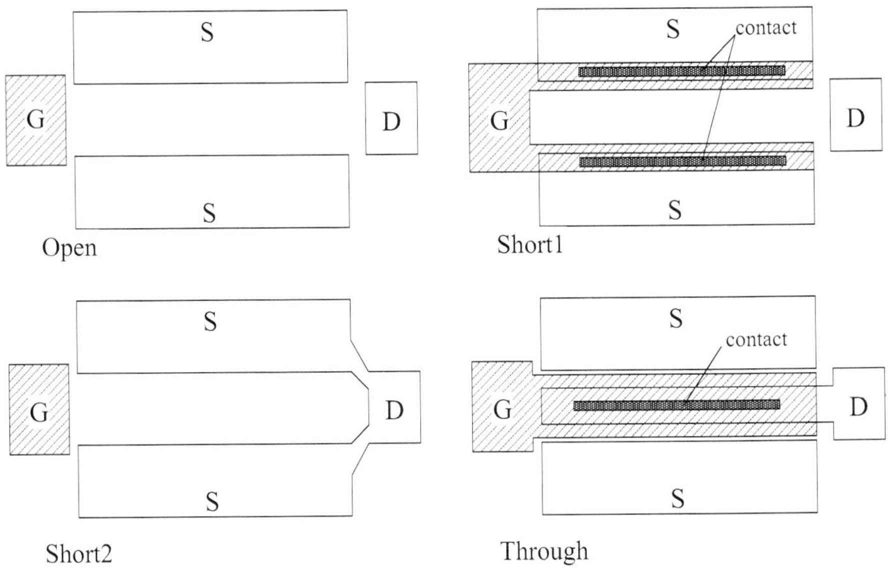

Fig. 3. Standard patterns for de-embedding

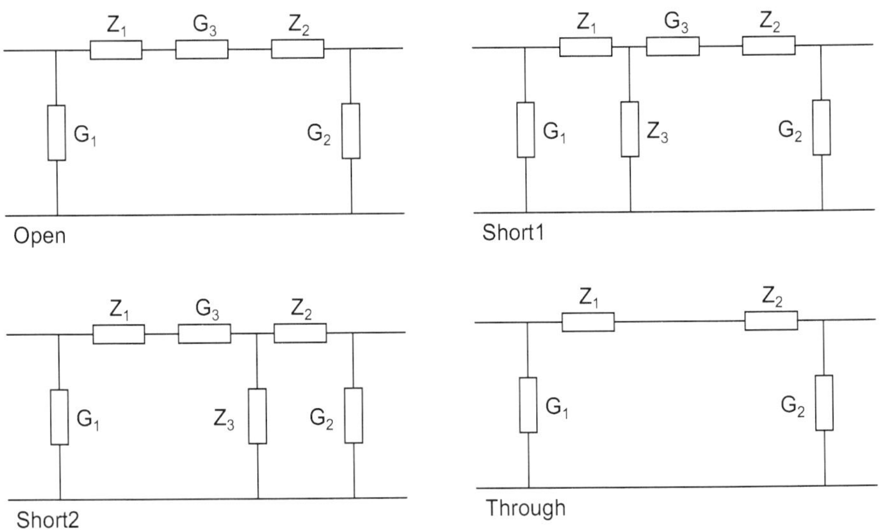

Fig. 4. Equivalent circuit of standard patterns.

Thus the parasitic elements $(Z_1, Z_2, Z_3, G_1, G_2, G_3)$ can be expressed as follows:

$$G_1 = y_{11op} + y_{12op} \tag{1}$$

$$G_2 = y_{22op} + y_{12op} \tag{2}$$

$$G_3 = \left(\frac{-1}{y_{12op}} + \frac{1}{y_{12thr}} \right)^{-1} \tag{3}$$

$$Z_1 = \frac{1}{2}\left(\frac{-1}{y_{12thr}} + \frac{1}{y_{11sh1} - G_1} + \frac{1}{y_{22sh2} - G_2} \right) \tag{4}$$

$$Z_2 = \frac{1}{2}\left(\frac{-1}{y_{12thr}} - \frac{1}{y_{11sh1} - G_1} + \frac{1}{y_{22sh2} - G_2} \right) \tag{5}$$

$$Z_3 = \frac{1}{2}\left(\frac{1}{y_{12thr}} + \frac{1}{y_{11sh1} - G_1} - \frac{1}{y_{22sh2} - G_2} \right) \tag{6}$$

Using the above parasitic elements (parasitic impedance and parasitic admittance), the de-embedded matrix can be obtained by the following procedure. Let us transform the measured S-matrix ($\mathbf{S}_{meas}$) into the Y-matrix and write it as $\mathbf{Y}_{meas}$. First, we subtract G_1, G_2 from $\mathbf{Y}_{meas}$ and obtain $\mathbf{Y}_A$ as follows:

$$\mathbf{Y}_A = \mathbf{Y}_{meas} - \begin{bmatrix} G_1 & 0 \\ 0 & G_2 \end{bmatrix} \tag{7}$$

Transforming the obtained $\mathbf{Y}_A$ to the Z-matrix ($\mathbf{Z}_A$), we next subtract Z_1, Z_2, Z_3 from $\mathbf{Z}_A$ and obtain $\mathbf{Z}_B$ as follows:

$$\mathbf{Z}_B = \mathbf{Z}_A - \begin{bmatrix} Z_1 + Z_3 & Z_3 \\ Z_3 & Z_2 + Z_3 \end{bmatrix} \tag{8}$$

Again, transforming the obtained $\mathbf{Z}_B$ into the Y-matrix ($\mathbf{Y}_B$), we subtract G_3 from $\mathbf{Y}_B$ and obtain $\mathbf{Y}_{DUT}$ as follows:

$$\mathbf{Y}_{DUT} = \mathbf{Y}_B - \begin{bmatrix} G_3 & -G_3 \\ -G_3 & G_3 \end{bmatrix} \tag{9}$$

$\mathbf{Y}_{DUT}$ is the final de-embedded Y-matrix of the DUT part.

4. Measurement Results

4.1 DC Characteristics

The DC characteristics of the multiple-channel CNTFET were measured with a semiconductor parameter analyzer (Agilent 4156C). Figure 5-a shows the drain current (I_d) versus gate voltage (V_g) curve when the drain voltage (V_d) was -2 V. I_d versus V_d curve is shown in Figure 5-b. These characteristics are like p-type FETs but the drain current is not zero even when the gate voltage is small enough. This is due to the metallic carbon nanotubes. Because the metallic carbon nanotubes do not affect the high-frequency characteristics of the device, we did not perform a special removal process such as a burn out procedure. From the DC curve (Figure 5), transconductance ($g_m = \partial I_d / \partial V_g$) of 226 μS and drain conductance ($g_d = \partial I_d / \partial V_d$) of 1 mS (at V_g = 5 V, V_d = -2 V) were obtained. The drain current of our multi-channel CNTFET is more than 200 times larger than that of single-channel CNTFETs. We observed hysteresis in the I-V curves; however, the width of the hysteresis is much smaller ($\Delta V_g < 1$ V, $\Delta V_d < 0.1$ V) than that of non-passivated CNTFETs.

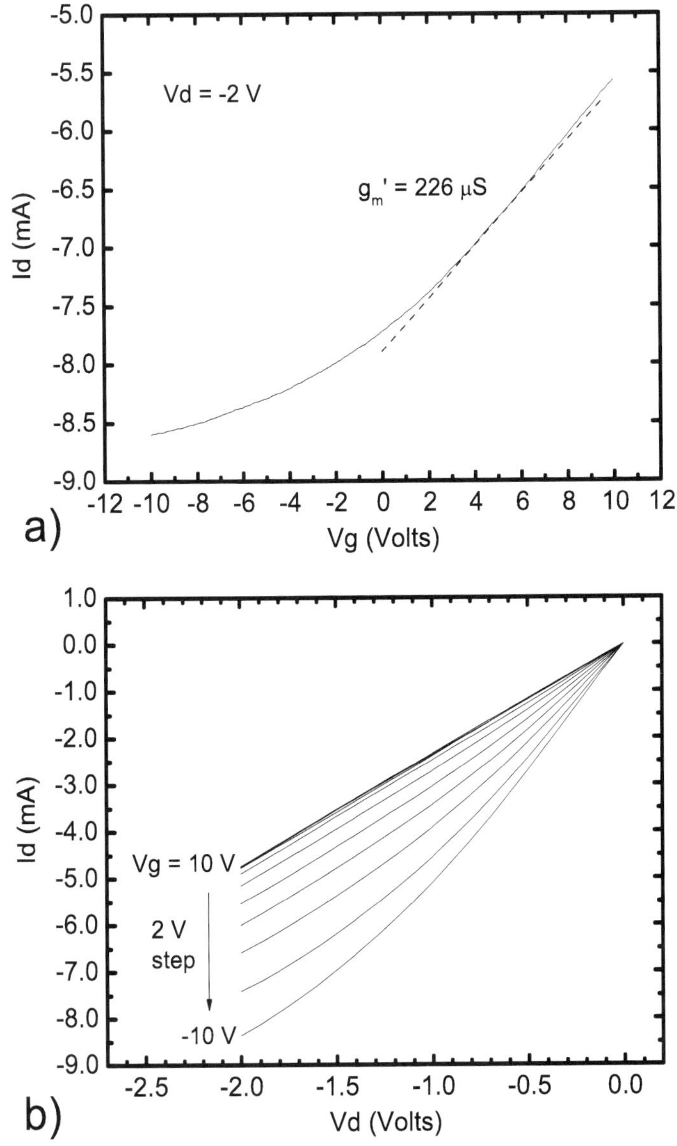

Fig. 5. DC characteristics of multiple-channel CNTFET: a) I_d-V_g curve, b) I_d-V_d curve.

4.2 RF Characteristics

Using RF probes (Cascade Microtech I40-GSG-125), we measured 2-port S-parameters of the device between 100 MHz and 20 GHz with the network analyzer (Agilent PNA N5230A). Standard SOLT-calibration was performed at the probe ends by using the calibration substrate. The measured S-parameters were de-embedded by the previously mentioned error removal procedure, and we obtained current gain ($|h_{21}| = |y_{21}/y_{11}|$) from the de-embedded Y-parameters ($\mathbf{Y}_{DUT}$) (Figure 6-a). In this figure, the measured and de-embedded data are displayed. The de-embedded data was 15 dB larger than the measured data. The f_T value was determined to be 10.3 GHz by obtaining the frequency when the current gain was unity (0 dB). Also, the unilateral power gain (U) was calculated from the de-embedded Y-parameters ($\mathbf{Y}_{DUT}$) by the formula below and plotted as a function of frequency in Figure 6-b.

$$U = \frac{|y_{21} - y_{12}|^2}{4\left[\mathrm{Re}(y_{11})\,\mathrm{Re}(y_{22}) - \mathrm{Re}(y_{12})\,\mathrm{Re}(y_{21})\right]} \tag{10}$$

The $f_{\max}$ value was determined to be 3.5 GHz by obtaining the frequency when the unilateral power gain was unity (0 dB).

5. Equivalent circuit model analysis

Figure 7 shows our proposed equivalent small-signal circuit model for multiple-channel CNTFET. Here, R_s (R_d) is the resistance of the CNT between the source (drain) and gate, and contains the resistance of the CNT extensions and contact resistances. R_s and R_d were extracted from the DC measurements, as described in (10). The extracted values of R_s and R_d were 420 Ω each for our multiple-channel CNTFET. Note that R_s and R_d values of usual RF transistors are negligibly small (a few ohms); however, for CNTFETs, even when using multiple-channel structures, these resistances play a dominant role in the analysis. The g_m and g_d in Figure 7 are intrinsic transconductance and drain conductance and relate to measured transconductance (g'_m) and drain conductance (g'_d), like in Chow and Antoniadis (11), as follows:

$$g_m = \frac{g'_m}{1 - g'_m R_s - g'_d(R_s + R_d)} \quad \text{and} \quad g_d = \frac{g'_d}{1 - g'_m R_s - g'_d(R_s + R_d)}. \tag{11}$$

Using these relations and the DC measurement results, we calculated the intrinsic transconductance and drain conductance as g_m = 3.47 mS and g_d = 15.4 mS. The intrinsic g_m and g_d values are one order of magnitude larger than g'_m and g'_d because of the large R_s and R_d values. The value $C_{g-cnt}(= C_{g-cnts} + C_{g-cntd})$ is the capacitance between the gate electrode and CNTs. We assumed that $C_{g-cnts} = C_{g-cntd}$ because of the symmetry of the device. According to Burke (1), C_{g-cnt} consists of the electrostatic capacitance (C_{ES}) and the quantum capacitance (C_Q) and is given by $C_{g-cnt}^{-1} = C_{ES}^{-1} + C_Q^{-1}$. C_Q is about 100 aF/μm. C_{ES} is calculated from geometry (as shown in Figure 8-a) and is given by $C_{ES} = 2\pi\varepsilon_r\varepsilon_0/\cosh^{-1}(2h/d)$. In our case, h = 40 nm and d = 1 nm, so C_{ES} was estimated to be 42.5 aF/μm. Thus, C_{g-cnt} (one CNT) was calculated to be 30 aF/μm, and taking into account the gate length of 0.2 μm and 200 CNT channels, we calculated C_{g-cnt} = 1.2 fF ($C_{g-cnts} = C_{g-cntd}$ = 0.6 fF). C_{gs} and C_{gd}, as shown in Figure 7, are parasitic capacitances between the gate and source and the gate and drain. $C_{gs} + C_{gd}$ can be approximated by the electrostatic capacitance between a coplanar stripline and ground planes on a dielectric (Figure 8-b), like in Collin (12), as follows:

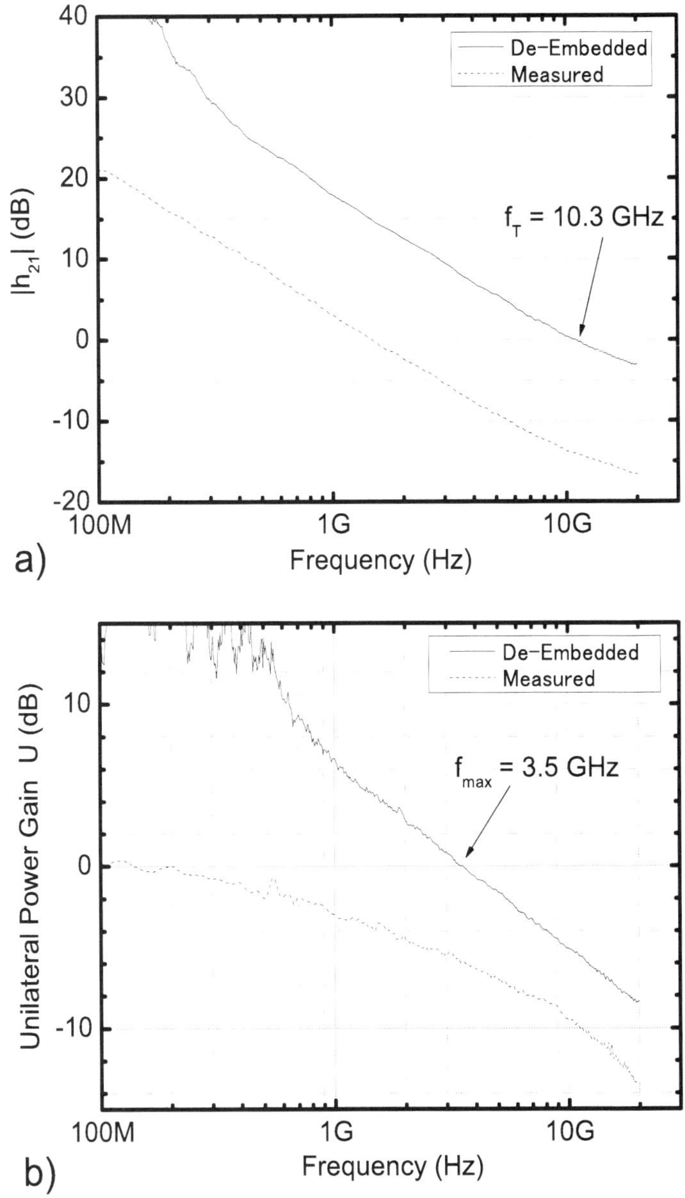

Fig. 6. RF characteristics of multiple-channel CNTFET: a) current gain ($|h_{21}|$), b) power gain (U), bias condition is $V_d = -2$ V, $V_g = 5$ V.

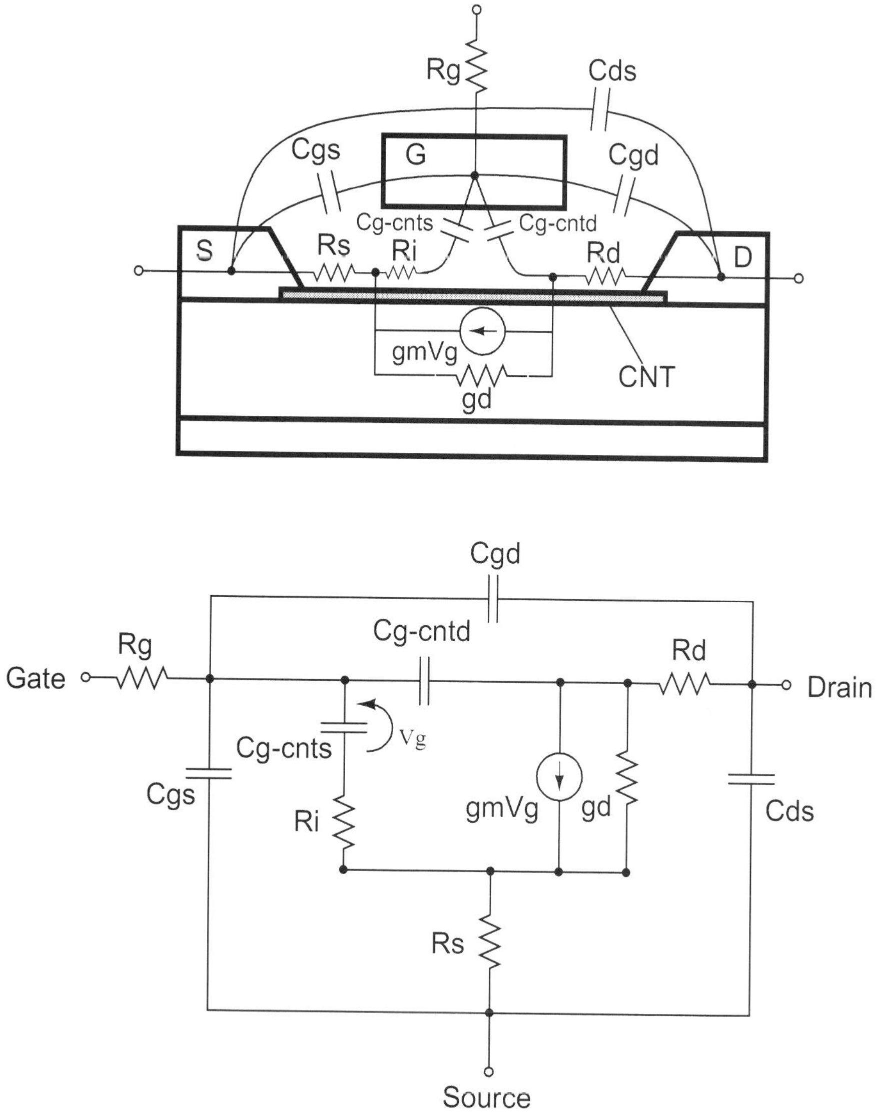

Fig. 7. Equivalent small-signal circuit model for CNTFET.

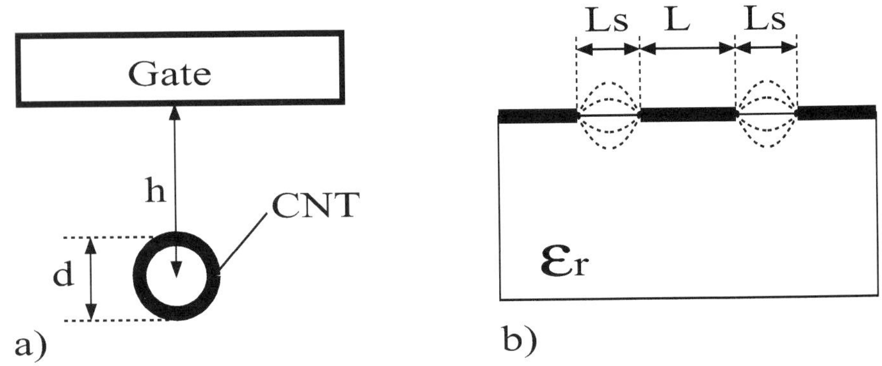

Fig. 8. Capacitance geometry: a) gate-CNT capacitance, b) coplanar stripline on dielectric

$$C_{gs} + C_{gd} = 2(\varepsilon_r + 1)\varepsilon_0 \frac{K(k)}{K(\sqrt{1-k^2})}, \tag{12}$$

where $k = L/(L + 2L_S)$, and K is the complete elliptic integral of the first kind. Considering $L = L_s = 0.2\ \mu$m and $W = 40\ \mu$m, we obtain $C_{gs} + C_{gd} = 2.2$ fF.
Once the equivalent circuit model is constructed like in Figure 7, we can derive f_T by calculating the H-parameters of the circuit (13). f_T is given as

$$\frac{1}{2\pi f_T} = \frac{1}{g_m} C_g + \frac{g_d}{g_m} C_g (R_s + R_d) + C_{g-cntd}(R_s + R_d) + (C_{gs} + C_{gd}) R_s, \tag{13}$$

where $C_g = C_{gs} + C_{gd} + C_{g-cnts} + C_{g-cntd}$. Substituting the parameters into Equation (13), we obtain f_T(model) = 10.6 GHz. This value is consistent with the experimental one: f_T (experiment) = 10.3 GHz.
Similarly, we can derive the expression of $f_{\max}$ by calculating the unilateral power gain U (Equation (10)) from the Y-parameter of the equivalent circuit and solving the equation: $U = 1$. The following formula is the calculation result.

$$f_{\max} = \sqrt{\frac{f_T}{2\pi R_g \left[\left(\frac{g_d}{g_m} C_g + C_{gd} + C_{g-cntd} \right) + \frac{2\pi f_T}{g_m^2 R_g} \Psi \right]}} \tag{14}$$

where C_g and Ψ are defined as follows:

$$C_g = C_{gs} + C_{gd} + C_{g-cnts} + C_{g-cntd} \tag{15}$$

$$\Psi = R_s R_d \left(g_d C_{g-cnts} + g_d C_{g-cntd} + g_m C_{g-cntd} \right)^2 + R_i g_d C_{g-cnts} (C_{g-cnts} + g_d R_s C_{g-cnts} + g_d R_d C_{g-cnts} + g_m R_d C_{g-cntd}) \tag{16}$$

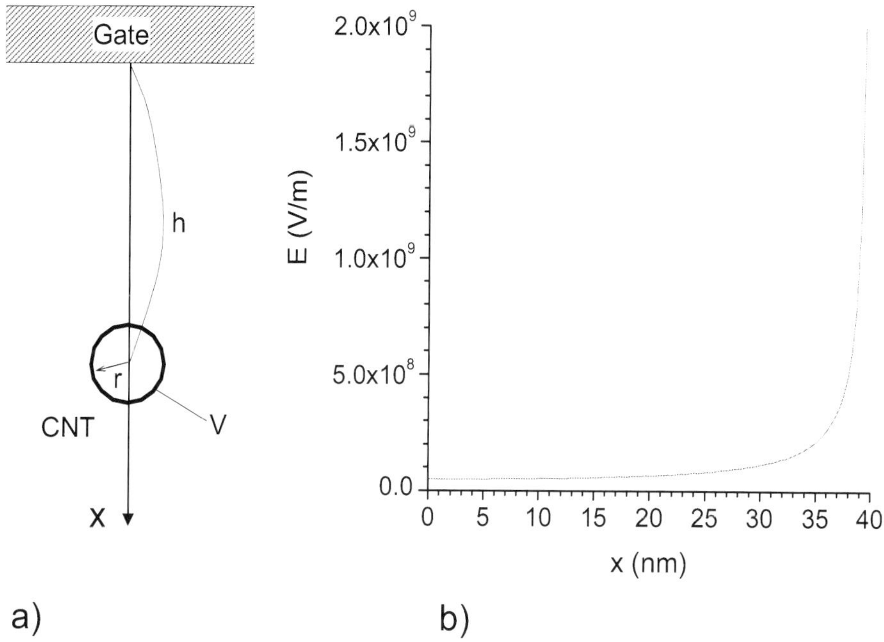

Fig. 9. (a) Geometry of gate electrode and CNT, (b) Electric field at x calculated under the condition of $V = -5$ V, $r = 0.5$ nm, $h = 40$ nm.

The above formula of $f_{\max}$ is more complicated than the expression of f_T (Equation(13)). Note that the expression of $f_{\max}$ contains not only the parameters included in the expression of f_T, but also the parameters R_g and R_i. R_g is the gate resistance and can be estimated from the resistivity of the gate electrode material of aluminum. R_g in our case is about 40 ohms. R_i is equivalent to the channel resistance (R_{ch}) in the case of current transistors and can be expressed as $R_{ch} = \alpha / g_m$. Here, α is a coefficient and has a value less than 0.2 (13). If we assume that $\alpha = 0.2$ and use the g_m of the CNTFET, we obtain $R_i \approx 60$. Using these values of R_g, R_i and the other parameters used for estimation of f_T, we can estimate $f_{\max}$ as $f_{\max}|_{R_i=60}$=19.2GHz. This value is much larger than the experimental value: $f_{\max}$(experiment)=3.5 GHz. To clarify this discrepancy, we should consider the phenomenon peculiar to the CNT channels, and one consistent model was proposed as shown below.
The diameter of the SWCNTs used for the multiple-channel CNTFET is about 1 nm, and S-parameter was measured under the condition that the gate electrode voltage is 5 V. This gate voltage is relatively high enough that the strong electric field exists at the vicinity of the CNTs. To estimate the electric field, we assume that the gate electrode and the CNT are an infinite conductive plane and an infinite length of conductive cylinder that is separate from the plane at the distance h and has a radius of r, as shown in the Figure 9-a. When the potential of the

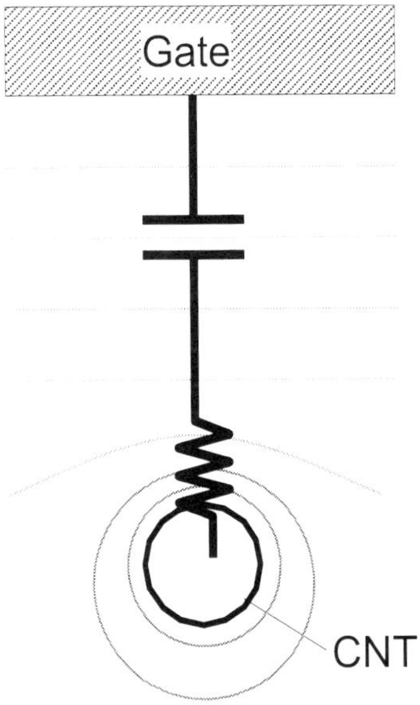

Fig. 10. Equivalent circuit between gate electrode and CNT

CNT is V, the potential at x is written as the expression below (14):

$$\phi = \frac{V}{2\cosh^{-1}(h/r)} \ln\left(\frac{x+\sqrt{h^2-r^2}}{x-\sqrt{h^2-r^2}}\right)^2 \tag{17}$$

Substituting $V = -5$ V, $r = 0.5$ nm, $h = 40$ nm (thickness of the gate oxide) to the above formula, and calculating the electric field $E = -\partial\phi/\partial x$, the result are shown in the Figure 9-b. The result shows that the electric field increases rapidly as the point is getting close to the CNT. The electric field at the CNT surface ($x = 39.5nm$) is 2.0×10^9 V/m. This means that the electric field is extremely strong at the vicinity of the CNT. In this condition, the tunneling current, Fowler-Nordheim(F-N) current, easily flows from the CNT into the gate oxide. This phenomenon is considered also in (15), and they describe that the current density of the F-N current is expressed as below:

$$J_{FN} = aE^2 \exp(-b/E) \tag{18}$$

where a and b are constants and have values $a = 2 \times 10^6 A/MV^2$ and $b = 230MV/cm$. The total F-N current that flows through all CNTs in the transistor can be described as $I = AJ_{FN}$, where A is the total area through which the F-N current flows. Let us consider the F-N current at the CNT surface ($x = 39.5nm$). In this case, the area is upper half surface of the CNT cylinder under the gate electrode. Multiplied by 200 CNT channels, the total area can be obtained as $A = 200 \cdot L \cdot \pi r$, where L is the gate length.
The variation of the potential ($\delta\phi$) corresponding to the variation of the CNT potential (δV) can be written as $\delta\phi = \frac{\partial\phi}{\partial V}\delta V$. Also the variation of the F-N current(δI) can be written as

$\delta I = \frac{\partial I}{\partial V}\delta V$. Thus, the resistance originated from the F-N current can be written as $R_{FN} = \delta\phi/\delta I = \frac{\partial \phi}{\partial V} / \frac{\partial I}{\partial V}$. Calculating R_{FN} using Equations (17) and (18), we obtain $R_{FN} \approx 75k\Omega$. Consequently, we can guess that the equivalent circuit between the gate electrode and CNT consists of the capacitance of the gate oxide and additional serial resistance(R_{FN}), as shown in Figure 10. This adds R_{FN} to R_i in our equivalent circuit model of the CNTFET, so that we can write $R_i = R_{ch} + R_{FN} = 60 + 75k \approx 75k\Omega$. Substituting this into Equation (14), we obtain $f_{\max}$(model)=3.4 GHz. This $f_{\max}$ value is very near to the experimental value: $f_{\max}$(experiment)=3.5 GHz.
From the above discussion, we get a guideline to improve the high-frequency characteristics of the CNTFET. To increase the f_T, the parasitic capacitances $(C_{gs} + C_{gd})$ and the resistances $(R_s + R_d)$ need to be reduced. If $(C_{gs} + C_{gd})$ and $(R_s + R_d)$ are negligible, then Equation (13) becomes $f_T = g_m/2\pi C_{g-cnt}$, and we obtain $f_T \approx 460$ GHz. The parasitic resistance R_s (R_d) is the resistance of CNT extensional part between the source (drain) electrode and the channel. A candidate to decrease this resistance is doping (16). To improve the value of $f_{\max}$, R_g and $R_i(R_{FN})$ should be decreased in addition to reducing the parasitic capacitances and resistances. Because R_{FN} originates from the strong electric field near CNTs, as discussed above, we should develop a CNTFET that operates with low gate voltages.

6. Summary

This chapter described a method for accurately measuring and modeling the high-frequency characteristics of CNTFETs. To directly measure using a network analyzer, we developed a high-density multiple-channel CNTFET structure the output impedance of which is much lower than conventional single-channel CNTFETs. We also focused on accurately measuring S-parameters, e.g., we used a de-embedding procedure that removes existing errors in measured S-parameters of high-impedance devices. Consequently, we obtained a cut-off frequency (f_T) of 10.3 GHz and a maximum oscillation frequency ($f_{\max}$) of 3.5 GHz. We also proposed an equivalent circuit RF model that includes the higher order parasitic resistances and capacitances that are neglected in the case of current RF transistors. The model, therefore, can explain the experimental results very accurately. The analysis also revealed that decreasing the parasitic capacitances of the electrodes and the resistances of the CNT extensions greatly improves the high-frequency performance of CNTFETs.

7. References

[1] P. J. Burke, "AC performance of nanoelectronics: towards a ballistic THz nanotube transistor", Solid-State Electronics, 48, (2004), pp. 1981-1986.
[2] S. Hasan, S. Salahuddin, M. Vaydyanathan, and M. A. Alam, "High-Frequency Performance Projections for Ballistic Carbon-Nanotube Transistors", IEEE Transaction on Nanotechnology, Vol. 5, No. 1, (2006), pp. 14-22.
[3] D.V.Singh, K.A. Jenkins, J. Appenzeller, D. Neumayer, A. Grill, and H.-S.P. Wong, "Frequency Response of Top-Gated Carbon Nanotube Field-Effect Transistors", IEEE Transactions on Nanotechnology, Vol. 3, No. 3, (2004), pp. 383-387.
[4] S. Li, Z. Yu, S. Yen, W.C. Tang, and P.J. Burke, "Carbon Nanotube Transistor Operation at 2.6 GHz", Nano Letters, Vol. 4, No. 4, (2004), pp. 753-756.
[5] S. Kim, T. Choi, L. Rabieirad, J.-H. Jeon, M. Shim, and S. Mohammadi, "A Poly-Si Gate Carbon Nanotube Field Effect Transistor for High Frequency Applications", IEEE MTT-S International Microwave Symposium Digest, 12-17 June, (2005), pp. 303-306.

[6] A. L. Louarn, F. Kapche, J.-M. Bethoux, H. Happy, G. Dambrine, V. Derycke, P. Chenevier, N. Izard, M. F. Goffman, and J.-P. Bourgoin, "Intrinsic current gain cutoff frequency of 30 GHz with carbon nanotube transistors," Appl. Phys. Lett. 90,(2007) 233108.
[7] K. Narita, H. Hongo, M. Ishida, and F. Nihey, "High-frequency performance of multiple-channel carbon nanotube transistors," Physica Status Solidi (a), 204, No.6, (2007), pp.1808-1813.
[8] E. P. Vandamme, D.M.M.-P. Schreurs, and C.V.Dinther, "Improved Three-Step De-Embedding Method to Accurately Account for the Influence of Pad Parasitics in Silicon On-Wafer RF Test-Structures", IEEE Transaction on Electron Devices, Vol. 48, No. 4, (2001), pp. 737-742.
[9] L. F. Temeijer, R. J. Havens, A. B. Jansman, and Y. Bouttement, " Comparison of the "Pad-Open-Short" and "Open-Short-Load" Deembedding Techniques for Accurate On-Wafer RF Characterization of High-Quality Passives", IEEE Transaction on Microwave Theory and Techniques, Vol. 53, No. 2, (2005), pp. 723-729.
[10] P. I. Suciu and R. L. Johnston, "Experimental Deviation of the Source and Drain Resistance of MOS Transistors", IEEE Transaction on Electron Devices, Vol. ED-27, No. 9, (1980), pp.1846-1848.
[11] S. Y. Chou and D. A. Antoniadis, "Relationship Between Measured and Intrinsic Transconductance of FET's", IEEE Transaction on Electron Devices, Vol. ED-34, No. 2, (1987), pp. 448-450.
[12] R. E. Collin, "Foundations for Microwave Engineering (Second Edition)", McGraw-Hill Inc., (1992), p. 175.
[13] W. Liu, "Fundamentals of III-V Devices", Wiley-Interscience, (1999).
[14] S. Ramo, J. R. Whinnery and T. V. Duzer, "Fields and Waves in Communication Electronics (Second Edition)," John Wiley & Sons, (1984), pp.339.
[15] Arnaud Robert-Peillard and Slava V.Rotkin, "Modeling Hysteresis Phenomena in Nanotube Field-Effect Transistors," IEEE Transactions on Nanotechnology, Vol.4, No.2, (2005), pp.284-288.
[16] A. Javey, R. Tu, D. B. Farmer, J. Guo, R. G. Gordon, and H. Dai, "High Performance n-Type Carbon Nanotube Field-Effect Transistors with Chemically Doped Contacts", Nano Letters, Vol. 5, No. 2, (2005), pp.345-348.

14

Carbon Nanotubes Interconnects for Nanoelectronics Circuits

A.G. Chiariello*, A. Maffucci°, G. Miano* and F. Villone°
*DIEL, Università di Napoli "Federico II", Napoli
°DAEIMI, Università di Cassino, Cassino
Italy

1. Introduction

Future nanoelectronics will be enabled only by providing the effective capability of connecting the nanometric devices to the circuit boards, therefore a major challenge is the design and fabrication of the *nano-interconnects*. For nanotechnology applications the limits imposed by physics, materials, assembly and design could not be overcome by simply scaling the conventional metal/dielectric systems: innovation in new materials, new technology and new system integration techniques is required.

Carbon nanotubes (CNTs) are recently discovered carbon structures (Iijima, 1991), which have been soon considered as *emerging research materials* (ITRS, 2007) for nanoelectronics applications (Fig.1), because of their unique properties (Avouris et al., 2003; Saito et al., 2004; Anantram & Léonard, 2006). CNT interconnects are expected to meet many of the requirements for technologies below the 22nm node (ITRS, 2007) in terms of mechanical strength, thermal conductivity and electrical performances.

This Chapter is devoted to CNT interconnects, proposed for wiring and for packaging nanotechnology ICs, with the aim of presenting the state-of-the-art of electrodynamics and circuit modelling of CNTs and to provide performance comparisons between conventional and CNT interconnects. Some case-studies of practical application are carried-out, referring to real-world wiring and packaging problems for future nanoelectronics.

(a) (b) (c)

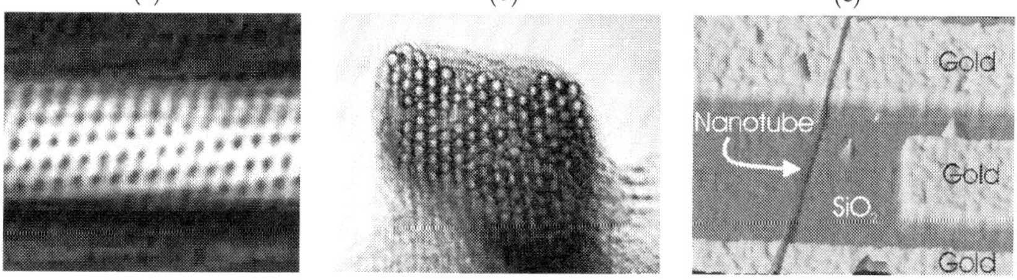

Fig. 1. Some real-world nanotubes: (a) AFM image of chiral tube of 1.3 nm diameter (*Technical University, Delft*); (b) TEM image of a crystalline nanotube bundle (*Rice University*); (c) a single nanotube as interconnect between gold electrodes (*IBM*).

Due to the potential applications of CNTs in nanoelectronics, many efforts have been made in literature to derive models able to describe the electrical propagation. Examples of phenomenological models are found in (Burke, 2002; Burke, 2003), based on the Luttinger liquid theory. Semi-classical approaches describing the CNTs as electron waveguides may be found in (Wesström, 1996; Sarto, 2009), and in (Salahuddin, 2005), where a general model for a quantum wire is derived from the transport theory based on the Boltzmann equation. In this Chapter we adopt the semi-classical model first proposed in (Miano & Villone, 2006), based on a fluid description of the motion of the CNT free electrons. This model has been used to derive transmission line equivalent circuits for metallic CNT interconnects in (Maffucci et al, 2008a; Maffucci et al., 2009), and has been recently extended to CNTs of arbitrary chirality (Forestiere et al., 2009a and 2009b). Note that the same model may be extended to take into account also the effect of CNT curvature (Forestiere et al., 2009b)

Using any of the quoted models, the comparative analysis between CNT and conventional Cu interconnects leads to a quite general result: nano-interconnects made by good quality CNT bundles outperform Cu ones at intermediate and global level, whereas at local level the behaviour is comparable (Li et al., 2009; Maffucci et al., 2008b; Naeemi, 2008). Therefore, the focus is now on the availability of a low-cost fabrication process providing CNT bundles with low contact resistance, good densities, good direction control and compatibility with CMOS technology. Satisfactorily results have been achieved for vertical CNT bundles (ITRS, 2007; Li et al., 2009; Liu, 2009). The same good results, however, have not yet been obtained in the fabrication of CNT interconnects parallel to the substrate, which remains a challenging task. On the other hand, many of the problems related to the integration between the CNT technology and the CMOS one are still open. However, the gap between theoretical predictions and practical applications is reducing faster and faster, and many examples have been recently given of the first real-world applications of CNT interconnects. In the nanopackaging area CNT bumps for flip-chip interconnects have been fabricated, replacing the conventional solder bumps, Fig.2a (Soga, 2008). Recently, a 1.02 GHz oscillator ring circuit using CNT wiring for a conventional chip has been presented, see Fig.2b (Close et al., 2008). These are only few of the many examples of successful integration between CNTs and ICs, testimonials of this technological trend.

(a) (b)

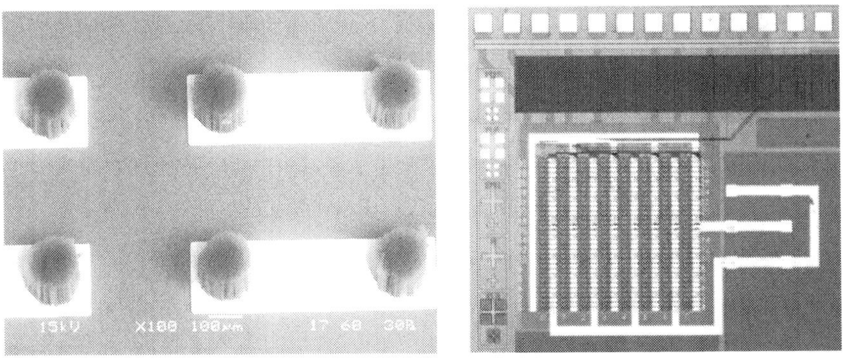

Fig. 2. (a) Vertical CNT bundles as flip-chip bumps (*Fujitsu Labs.*); (b) carbon nanotubes wiring in an oscillator circuit (*Stanford University & Toshiba*).

2. Electromagnetic propagation models for CNTs

A carbon nanotube is realized by rolling-up a sheet of a mono-atomic layer of graphite (*graphene*), see Fig.3a. The rolled graphene may become either metallic or semiconducting, depending on its *chirality*, i.e. the way in which it is rolled up. A single-wall carbon nanotube (SWCNT) is made by a single sheet of graphene (Fig.3b), whereas a multi-wall carbon nanotube (MWCNT) is made by nested sheets (Fig.3c). In the graphene the conduction phenomenon is related to the motion of the so-called π-electron, i.e. the one out of the four valence electrons which is free to move between the positive ion lattice. Both semiclassical as well as quantum-mechanical approaches have been used to describe the CNT electron dynamics.

We study the electrodynamics of a CNT shell, which may either describe an isolated SWCNT or a shell of a MWCNT, in which the direct coupling of electronic states of adjacent shells may be disregarded. The electrical signal propagation along the CNT shell is analyzed assuming a low frequency regime, where only intraband transitions of π-electrons with unchanged transverse quasimomentum occur. This allows using the semi-classical transport theory expounded, for instance, in (Slepyan et al., 1999; Maksimenko et al., 2000; Miano & Villone, 2006: Maffucci et al., 2008a; Forestiere et al., 2009a and 2009b). Using this approach, the longitudinal dynamic conductivity of a CNT shell of arbitrary chirality may be obtained, expressed in terms of the number of effective conducting channels. This leads to the constitutive equation of this innovative material, which could be coupled to Maxwell equations in order to derive electromagnetic and/or circuital models, following the stream of what done in (Miano & Villone, 2006: Maffucci et al., 2008a; Forestiere et al., 2009).

2.1 Constitutive relation of a CNT shell

The chirality of a CNT is described by two integers n and m, which are the coefficients of he basis vectors $\mathbf{a}_1$ and $\mathbf{a}_2$ depicted in Fig.3a. The CNT is metallic if $|n-m|=3q$, where q=0,1,2... A zig-zag CNT is obtained for $n=0$ or $m=0$, whereas an armchair CNT is given by $n=m$, therefore, armchair CNTs are always metallic, whereas zig-zag ones are metallic only if $m=3q$, with q=0,1,2... For $m\to\infty$ or $n\to\infty$ any CNT tends to the graphene sheet. The CNT diameter D is given by:

$$D = 0.0782\sqrt{n^2+nm+m^2}\ \text{[nm]}. \quad (1)$$

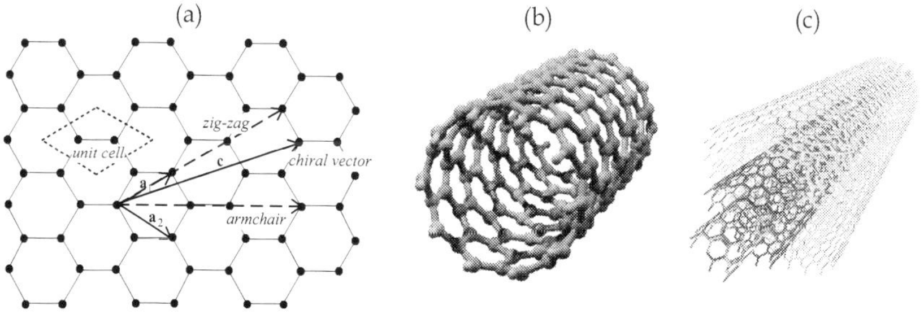

Fig. 3. (a) the graphene lattice; (b) single-wall CNT; (c) multi-wall CNT.

For each segment of the first Brillouin zone of the nanotube we introduce a distribution function for the π electrons in the valence and conduction bands:

$$f_{\mu}^{(\pm)} = f_{\mu}^{(\pm)}(z,k,t), \quad \mu = 0,1,...N, \; -\frac{\pi}{T} < k \leq \frac{\pi}{T}, \tag{2}$$

where k is the longitudinal wavenumber, t is the time, N is the number of graphene exagons in the primitive cell of the CNT, and T is the translational length. By assuming that the electric field is transversally uniform, $f_{\mu}^{(\pm)}$ satisfy the quasi-classic Boltzmann equation

$$\frac{\partial f_{\mu}^{(\pm)}}{\partial t} + v_{\mu}^{(\pm)} \frac{\partial f_{\mu}^{(\pm)}}{\partial z} + \frac{2\pi e}{h} E_z \frac{\partial f_{\mu}^{(\pm)}}{\partial k} = -\nu(f_{\mu}^{(\pm)} - f_{0,\mu}^{(\pm)}), \tag{3}$$

where e is the electron charge, h is the Planck constant, $E_z(z,t)$ is the component of the electric field tangent to the CNT surface, ν is the frequency of collision of the electrons with the ion lattice, $v_{\mu}^{(\pm)}$ are the longitudinal velocities of the electrons in the conduction/valence subbands and $f_{0,\mu}^{(\pm)}$ are the distribution functions at equilibrium. The latter are given by:

$$f_{0,\mu}^{(\pm)}(k) = \frac{1}{\pi^2 D} F\left[E_{\mu}^{(\pm)}(k)\right], \qquad F[E] = \frac{1}{e^{E/k_B T_0} + 1}, \tag{4}$$

being $E_{\mu}^{(\pm)}$ the energy of the generic subband, $F[E]$ the Dirac-Fermi function, k_B the Boltzmann constant and T_0 the CNT absolute temperature. Let us assume time-harmonic electric field and surface current density:

$$E_z(z,t) = \mathrm{Re}\left\{\hat{E}_z e^{i(\omega t - \beta z)}\right\}, \qquad J_z(z,t) = \mathrm{Re}\left\{\hat{J}_z e^{i(\omega t - \beta z)}\right\}. \tag{5}$$

For small perturbations of functions (2) around the equilibrium values (4) we get

$$\hat{E}_z = \hat{\sigma}_{zz}(\beta,\omega)\hat{J}_z, \tag{6}$$

where the spectral domain conductivity $\hat{\sigma}_{zz}(\beta,\omega)$ is given by

$$\hat{\sigma}_{zz}(\beta,\omega) = \frac{2ie^2}{\pi^2 D} \sum_{\mu=0}^{N-1} \int_0^{\pi/T} \left(\frac{1}{\omega - v_{\mu}^{(+)}\beta - i\nu} + \frac{1}{\omega + v_{\mu}^{(+)}\beta - i\nu} \right) v_{\mu}^{(+)2} \frac{dF}{dE_{\mu}^{(+)}} dk \,. \tag{7}$$

For all the subbands that give a meaningful contribution to the conductivity we may assume $v_{\mu}^{(\pm)} \cong v_F$ in (7), being v_F the Fermi velocity. This leads to

$$\hat{\sigma}_{zz}(\beta,\omega) = \frac{4e^2 v_F}{\pi h D} \frac{1}{i\omega + \nu} \frac{1}{1 + v_F^2 \beta^2 / (i\omega + \nu)^2} M(n,m), \tag{8}$$

where M may be regarded as the *equivalent number of conducting channels*:

$$M(n,m) = \frac{h}{\pi v_F} \sum_{\mu=0}^{N-1} \int_0^{\pi/T} v_F^2 \left(\frac{dF}{dF_\mu^+} \right) dk. \tag{9}$$

The *constitutive relation* of the CNT in the frequency domain comes from (6), (8) and (9):

$$(1 + i\omega / \nu) J_z + \frac{v_F^2}{\nu(1 + \nu / i\omega)} \frac{\partial \rho_s}{\partial z} = \sigma_c E_z, \tag{10}$$

where ρ_s is the surface charge density and σ_c is the static limit for the axial conductivity:

$$\sigma_c = \frac{4e^2 v_F}{\nu \pi \hbar D} M(n,m). \tag{11}$$

In the limit $\omega / \nu >> 1$, (11) reduces to the simple fluid model in (Miano & Villone, 2006; Maffucci et al., 2008a) and for $\omega / \nu << 1$ the electrons behave as a cold fluid.

2.2 Approximation of the equivalent number of conducting channels

The number of conducting channels M is a function of the parameters affecting the number of subbands around the CNT Fermi level: CNT chirality, diameter and temperature. Figure 4a shows M for increasing CNT diameter, evaluated at $T = 273$ K. For small diameters it is $M = 2$ for metallic CNTs and $M = 0$ for semiconducting ones, i.e. only metallic CNTs contribute the conduction. For larger diameters the channels of a metallic CNT increase and those associated to semiconducting ones are no longer negligible.

Let us now compute the axial conductivity for the two cases: Fig.4b shows these values, normalized to the axial conductivity for a graphene sheet, given by (Slepyan, 1999):

$$\sigma_\infty = \frac{8\pi \ln 2}{h^2} \frac{e^2 k_B T}{\nu + i\omega}. \tag{12}$$

Figure 4b agrees with the condition that for $D \to \infty$ a CNT tends to the graphene.

To obtain a simple analysis tool for studying MWCNTs or bundles of CNTs, it is useful to fit the behavior of M through a simple approximated function of the diameter D and temperature T. A piecewise linear function able to fit the asymptotic behaviors for $D \to \infty$ and $D \to 0$ may be given by the following expression (the parameters are given in Table 1):

$$M \approx \begin{cases} M_0 & \text{for } D < d_0 / T \\ a_1 DT + a_2 & \text{for } D \geq d_0 / T \end{cases}. \tag{13}$$

Figure 5 shows the fitting results computed at $T = 273$ K and $T = 373$ K.

This result is consistent with similar approximations proposed in literature (Naeemi, 2008), where however M is sligthly overestimated, due to the approximations used in evaluating the number of subbands around the Fermi level.

	Metallic	*Semiconducting*
M_0 [adim.]	2	0
a_1 [1/nmK]	$3.3 \cdot 10^{-4}$	$3.3 \cdot 10^{-4}$
a_2 [adim.]	0.15	–0.20
d_0 [nm·K]	$5.6 \cdot 10^{3}$	$0.6 \cdot 10^{3}$

Table 1. Parameters for the fitting formulas (13).

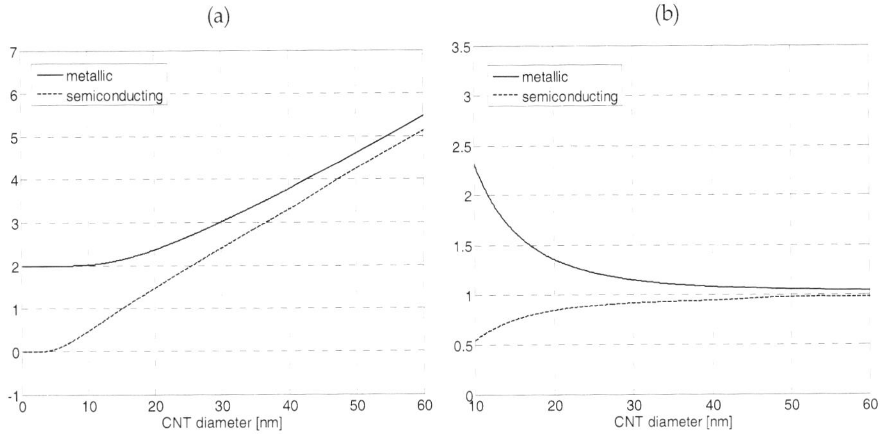

Fig. 4. (a) Number of effective conducting channels versus CNT diameter, at T=273 K; (b) axial conductivity normalized to the graphene value.

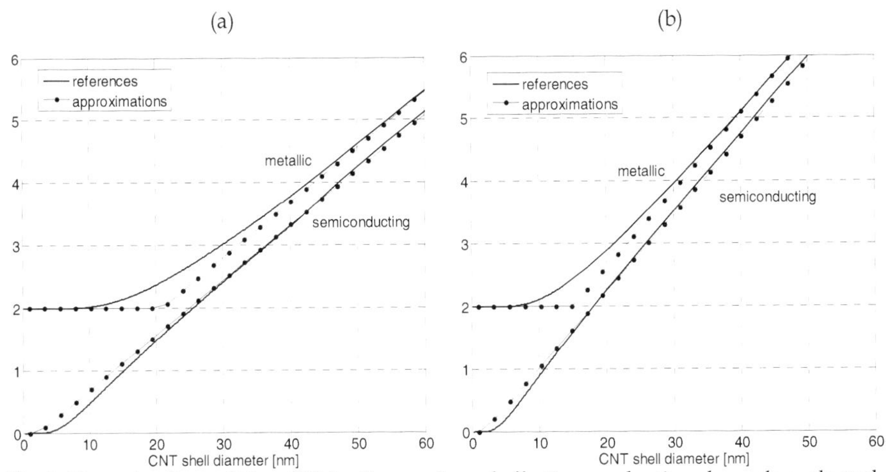

Fig. 5. Piecewise linear fitting (13) for the number of effective conducting channels evaluated at: (a) $T = 273$ K ; and (b) $T = 373$ K .

3. Circuit models for CNT bundle interconnects

3.1 Circuit model for a single CNT above a ground plane

A circuit model for CNT interconnects may be derived in the framework of the transmission line (TL) theory. The problem may be formulated in frequency domain by coupling Maxwell equations to the CNT constitutive relation (10). Then, assuming the propagation to be of quasi-TEM type, a distributed RLC circuit may be derived.

Let us first consider the simple case of a CNT shell of diameter D, located at a distance t above a PEC ground, Fig. 6a. As shown in paragraph 4.3, the distribution of surface currents and charges along the contour of a CNT section may be assumed to be uniform, since the proximity effect is negligible, hence the current and charge at any position along the CNT axis are given by:

$$Q(\omega,z)=\pi D\rho_s(\omega,z), \quad I(\omega,z)=\pi D J_z(\omega,z). \tag{14}$$

Using (14) in (10) we get:

$$(\nu+i\omega)I(\omega,z)+\frac{v_F^2}{(1+\nu/i\omega)}\frac{\partial Q(\omega,z)}{\partial z}=2v_F\frac{M}{R_0}E_z(\omega,z), \tag{15}$$

being $R_0=h/2e^2=12.9\,\mathrm{k\Omega}$ the so-called *quantum resistance*. The longitudinal component of the electric field E_z may be derived from the scalar electric and vector magnetic potentials:

$$E_z(\omega,z)=-\frac{\partial V(\omega,z)}{\partial z}-i\omega A_z, \tag{16}$$

which may be related to charge and current through the classical expressions:

$$V(\omega,z)=\frac{Q(\omega,z)}{C_e}, \quad A_z(\omega,z)=L_m I(\omega,z), \tag{17}$$

being C_e, L_m the per-unit-length (pul) electrostatic capacitance and magnetic inductance.

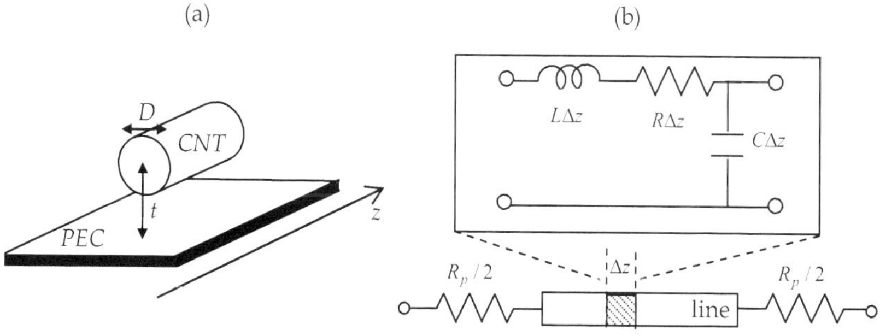

Fig. 6. (a) A CNT shell above a PEC; (b) equivalent circuit: elementary cell (inset) and lumped contact resistances.

Combining (15)-(17) we get the first Telegraphers' equation:

$$-\frac{dV(\omega,z)}{dz} = (R(\omega)+i\omega L(\omega))I(\omega,z), \tag{18}$$

with the pul parameters R and L defined as:

$$L = (L_m + L_k)/a_C, \quad R = \nu L_k / a_C, \quad C = C_e, \tag{19}$$

where the parameter a_C is given by:

$$a_C = 1 + \frac{C_e}{C_q}\frac{1}{1+\nu/i\omega}. \tag{20}$$

In (19) we have introduced the *p.u.l. kinetic inductance* L_k and *p.u.l. quantum capacitance* C_q :

$$L_k = \frac{R_0}{2v_F M} = \frac{h}{4e^2 v_F M}, \quad C_q = \frac{1}{L_k v_F^2} = \frac{2M}{R_0 v_F} = \frac{4e^2 M}{h v_F}. \tag{21}$$

The second Telegraphers' equation is obtained by imposing the charge conservation law:

$$-\frac{dI(\omega,z)}{dz} = i\omega C V(\omega,z), \tag{22}$$

with the pul capacitance C defined as $C = C_e$.

Equations (18) and (22) describe a lossy TL where the quantum effects are combined to the classical electrical and magnetic ones in the definition of the *p.u.l.* parameters R, L, and C. Assuming a frequency operating condition such as $\nu/\omega << 1$, the parameters R and L do not depend on frequency, since (20) becomes:

$$a_C \approx 1 + C_e / C_q. \tag{23}$$

Typical orders of magnitude for the collision frequency ν are $10^{11} \div 10^{12}$ Hz, hence the above condition is satisfied for frequencies up to the order of hundrends of GHz. Assuming this limit, the propagation along a CNT is then described by a simple lossy TL with constant parameters R, L, and C. The elementary cell is depicted in the inset of Fig.6b. The circuit model obtained here is a generalization of other models proposed in literature: for instance in (Burke, 2003; Salahuddin, 2005; Maffucci et al., 2008a) TL models for metallic SWCNTs are obtained. The elementary cell is however slightly different from those used in (Burke, 2003; Salahuddin, 2005), where the voltage variable is derived from the *electrochemical* potential rather than from the electrostatic one. A detailed discussion may be found in (Maffucci et al., 2009). The TL models in the quoted references describe metallic CNTs, where the model presented here is more general, since it deals with both metallic and semiconducting CNTs of any diameter. For instance the values of L_k and C_q used in literature may be obtained using M=2 in (21), i.e. assuming metallic CNT with small radius:

$$L_{k0} = \frac{h}{8e^2 v_F} = \frac{R_0}{4v_F}, \quad C_{q0} = \frac{8e^2}{hv_F} = \frac{4}{R_0 v_F}. \tag{24}$$

Assuming in (24) $c_s \approx v_F \approx 8 \cdot 10^5\ m/s$, we get $L_{k0} \approx 4\,\mathrm{nH}/\mu\mathrm{m}$ and $C_{q0} \approx 400\,\mathrm{aF}/\mu m$. However the actual values deviate from this bulk values as an effect of the increase of the number of channels with the CNT diameter, as shown in Figure 7.

Let us now investigate the influence of the quantistic capacitance and kinetic inductance on the pul parameters (19). Assuming a homogeneous dielectric the classical parameters are:

$$L_m = \frac{\mu}{2\pi}\cosh^{-1}\left(\frac{2t}{D}\right), \quad C_e = 2\pi\varepsilon \Big/ \cosh^{-1}\left(\frac{2t}{D}\right). \tag{25}$$

Let us consider the case of a metallic SWCNT with $D = 4\,\mathrm{nm}$, $\varepsilon_r = 2.2$ and $\mu = \mu_0$. For such a condition $L_k \approx L_{k0}$ and $C_q \approx C_{q0}$. Table 2 shows the ratio between classical and quantistic parameters for different aspect ratios t/D. The kinetic inductance dominates the magnetic one, being 3-4 order of

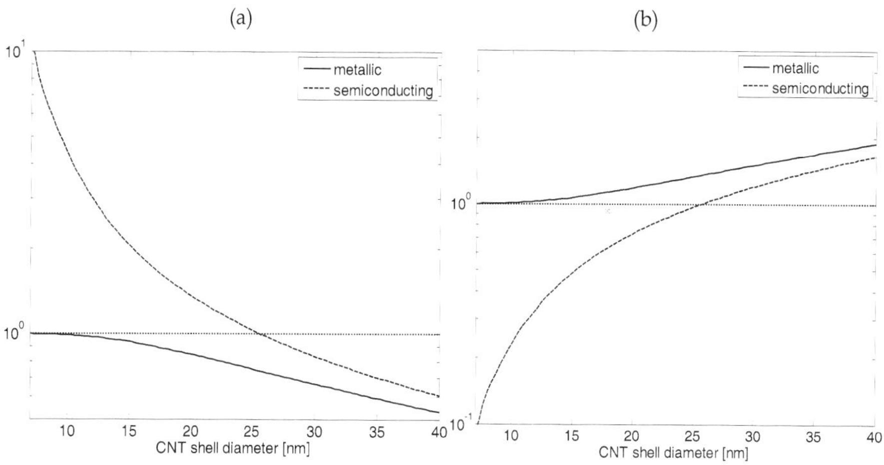

Fig. 7. (a) kinetic inductance, and (b) quantum capacitance for a CNT shell vs diameter, normalized to the bulk values (24).

t/D	L_k/L_m	C_e/C_q	α_C
1.5	$8.1 \cdot 10^3$	0.13	1.13
3	$7.5 \cdot 10^3$	0.14	1.14
5	$5.5 \cdot 10^3$	0.09	1.09
10	$4.6 \cdot 10^3$	0.07	1.07

Table 2. Comparison between classical and quantistic pul TL parameters for a SWCNT.

magnitude higher. As for the capacitance, the quantum capacitance is approximately one order of magnitude smaller than the electrostatic one, hence the parameter α_C is close to 1. A consequence of the huge value of the kinetic inductance is a low propagation velocity and an high characteristic impedance: the values of these two parameters normalized to an ideally scaled PEC interconnect of same dimensions would be

$$\frac{c_{CNT}}{c} \approx \sqrt{\frac{L_m}{L_k}}, \quad \frac{Z_{CNT}}{Z_C} \approx \sqrt{\frac{L_k}{L_m}}. \tag{26}$$

Indeed the lower propagation velocity is not a real limit for the practical application of CNT interconnects. Assuming the same conditions as those used for Table 2, for $t = 3D$ it is $c_{CNT} \approx 2.5 \cdot 10^6$ m/s, instead of $c \approx 2 \cdot 10^8$ m/s, hence at $f = 10\,\text{GHz}$ we have a wavelength $\lambda = 0.25$ mm. This means that at local and semilocal level the line is electrically short and the propagation delay, although higher than that introduced by an ideally scaled conventional interconnect, is negligible compared to the delay introduced by the RC or RLC behavior of the line (e.g., Banerjee, 2008; Maffucci et al., 2008b).
Let us now analyze the p.u.l. resistance, using (19) with $\alpha_C \approx 1$ and assumimg $L_K \approx L_{k0}$:

$$R = \nu L_K = \frac{R_0}{2M}\frac{1}{l_{mfp}}, \tag{27}$$

where the collision frequency ν has been defined as

$$\nu = v_F / l_{mfp}, \tag{28}$$

being l_{mfp} the mean-free-path. The values of l_{mfp} may vary significantly if low or high bias conditions are considered. In the low bias condition l_{mfp} may be assumed constant, whereas for higher bias conditions, l_{mfp} depends on the longitudinal voltage too, hence the parameter R is a non-linear function of the longitudinal voltage, usually described as a piecewise linear function with only two asymptotic values (Raychowdhury & Roy, 2006). A detailed discussion on this point may be found in (Park et al., 2004; Maffucci et al., 2009). In addition, assuming a CNT length greater than the *ballistic* length, a simple linear dependence of l_{mfp} on the diameter D may be considered (Nieuwoudt & Massoud, (2006).

As a final remark we note that experimental evidences and theoretical limits show a bulk value of $R_0/2 \approx 6.45\ \text{k}\Omega$ for the CNT resistance, which is a lower bound even when ballistic transport is considered (*e.g.*, Park et al., 2004). Furthermore, if the contacts between the CNT and the metallic electrodes at its terminations are not ideal, we have to consider an additional parasitic lumped resistance, whose value is strongly dependent on the quality of the contacts. For poor contacts this value may easily rise up to $100\,\text{k}\Omega$, even if recently the fabrication techinques are demonstrated very good quality contacts (Li et al, 2009). We include these phenomena by adding two lumped resistors $R_P/2$ in series to the line (see Fig.6b), taking into account both the bulk resistance and the imperfect contact resistance.

3.2 Circuit model for bundles of CNTs

The huge value of the resistance makes useless an interconnect made by a single CNT shell. For practical application purposes, bundles of SWCNTs or MWCNTs are proposed as material to be used in fabricating interconnect traces (ITRS, 2007): all the CNT shells are fed in parallel, so lowering the total resistance. Following the stream of what done in paragraph 3.1, we can model the propagation along a CNT bundle in the frame of the multiconductor TL theory, starting from the scheme in Fig.8.

Let $\mathbf{V}(\omega,z)=[V_1(\omega,z)....V_N(\omega,z)]^T$ and $\mathbf{I}(\omega,z)=[I_1(\omega,z)....I_N(\omega,z)]^T$ be the vectors of the N voltages and currents at given z, in frequency domain. On the k-th CNT eq. (15) reads:

$$(\nu_k+i\omega)I_k(\omega,z)+\frac{v_F^2}{(1+\nu_k/i\omega)}\frac{\partial Q_k(\omega,z)}{\partial z}=2v_F\frac{M_k}{R_0}E_{zk}(\omega,z). \tag{29}$$

Let us assume that in evaluating the low-energy band structure we can neglect the interactions between the CNT shells, as shown by experimental evidence (Maarouf et al., 2000). This assumption allows using the results obtained for isolated CNT shells in evaluating the parameters M_k and ν_k in (29). Using again (16)-(17) to derive $E_{zk}(\omega,z)$ and imposing the charge conservation law, we get the MTL equations:

$$-\frac{d\mathbf{V}(\omega,z)}{dz}=(R+i\omega L)\mathbf{I}(\omega,z),\quad -\frac{d\mathbf{I}(\omega,z)}{dz}=i\omega C\mathbf{V}(\omega,z), \tag{30}$$

with the pul parameter matrices given by:

$$L=\alpha_C^{-1}(L_M+L_k),\quad C=C_E,\quad R=\alpha_c^{-1}L_k\nu,\quad \alpha_c=I+C_q^{-1}C_e, \tag{31}$$

where I is the identity matrix, and the other matrices are given by:

$$\nu=diag(\nu_k),\quad L_k=diag(L_{kk}),\quad C_q=diag(C_{qk}). \tag{32}$$

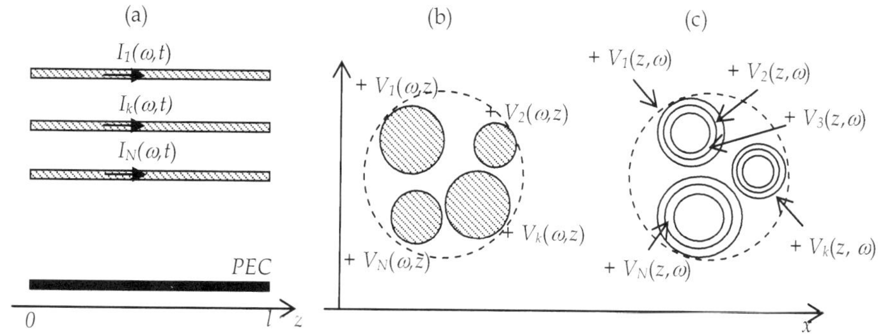

Fig. 8. A bundle of CNTs modeled as a multiconductor interconnect: (a) longitudinal view; transverse section of (b) SWCNTs and (c) MWCNTs bundle.

In practical applications any CNT bundle is used to carry a single signal, hence a CNT bundle above a ground may be described by a two-conductor TL model, which could be rigorously derived from the MTL one (30)-(32) assuming all the CNTs in parallel. Alternatively, we can use approximated expressions for its pul parameters. It is important to stress that statistically a low-cost fabrication process for CNT bundles ends up with 1/3 of metallic CNTs and 2/3 semiconducting ones (Li et al., 2005; Naeemi & Meindl, 2006). In addition we have to recall that typical values of SWCNTs diameters fall in the range $2 \div 10$ nm (Cheung et al., 2002), whereas the external shells of MWCNTs have typically diameters greater than 20 nm (Li, 2005; Naeemi, 2006). Recalling the results of Section 2, we can state that in a bundle of SWCNTs we may assume 1/3 conducting (M=2) and 2/3 non conducting (M=0), whereas all the MWCNTs shells may assumed to be conducting.
Let us first focus on the bundle capacitance. Figure 9 shows two circular bundles made by MWCNTs of external diameter $D_{out} = 20\,\text{nm}$, embbedded in a homogeneous dielectric matrix of SiO_2 with $\varepsilon_r = 4.5$. The bundle diameter is $D_b = 146\,\text{nm}$ and the center-center distance is $s = 300\,\text{nm}$. Figure 9 shows the distribution of the electrostatic potential, assuming a differential mode on the pair, which leads to a computed pul capacitance of 88.7 pF/m. The classical pul capacitance of a pair of solid cylindrical conductors would be:

$$C_e = \pi\varepsilon \Big/ \cosh^{-1}\left(\frac{2s}{D_b}\right). \tag{33}$$

Using (33) would give a value of $92.8\,\text{pF/m}$, with an error of less of 5%. In a SWCNTs only 1/3 of the SWCNTs are metallic, hence the distribution of the potential may sensibly vary assuming random distribution of the fraction of metallic CNTs in the bundle. Figure 10 shows two possible cases, referred to a pair made by two bundles of SWCNTs with CNT diameter $D = 2\,\text{nm}$, bundle diameter $D_b = 18.8\,\text{nm}$, center-center distance $s = 38.2\,\text{nm}$, embedded in a SiO_2 matrix ($\varepsilon_r = 4.5$). Table 3 shows the results obtained for the cases of Fig.10 and assuming all the CNTs metallic (case 3). In this condition approximation (33) would give an error of 20%. This approximation is acceptable or not depending on the values of the load capacitance in practical applications, as will be shown in Section 4.

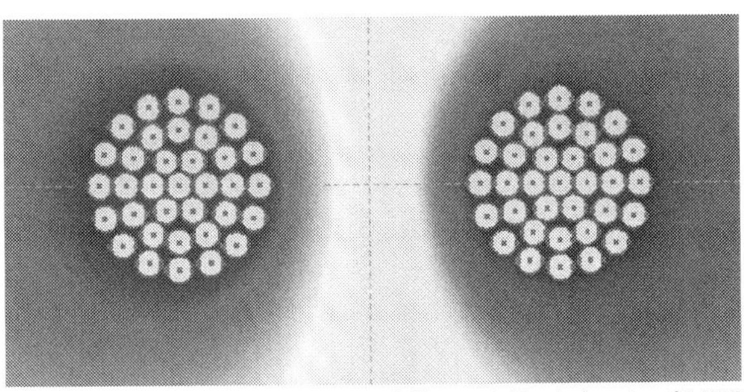

Fig. 9. Distribution of electrostatic potential for a differentially-fed pair of MWCNT bundles.

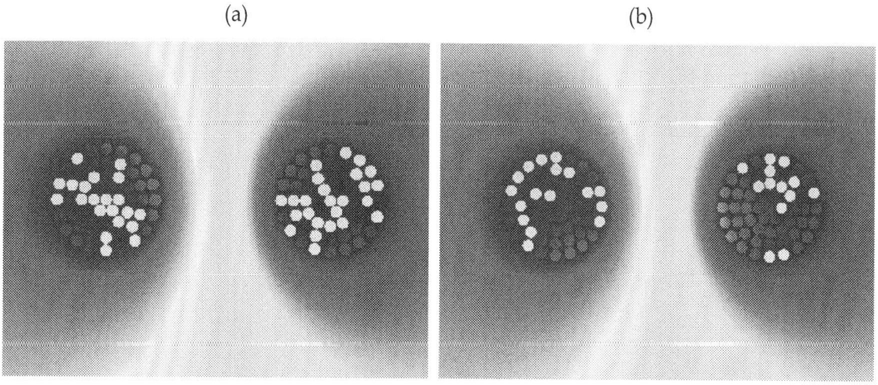

Fig. 10. Distribution of the electrostatic potential for a differentially-fed pair of SWCNT bundles: two random cases obtained assuming that 1/3 of CNTs are metallic (highlighted).

Case	C_b *(exact)* [pF/m]	C_b *(33)* [pF/m]	*% err*
1	76.5	92.8	17.6
2	74.3	92.8	19.9
3	87.9	92.8	5.4

Table 3. Evaluation of the bundle pul capacitance for the SWCNT cases depicted in Fig.10.

Let us now investigate the behavior of the bundle pul inductance and resistance. The magnetic inductance of the bundle may be computed from the vacuum space electrostatic capacitance, since it is $L_m = \mu_0 \varepsilon_0 C_{e0}^{-1}$, where C_{e0} may be computed exactly or may be approximated by (33) if the introduced error is acceptable (which is the case of MWCNT bundles). When considering the kinetic inductance L_{kb} of a bundle of N SWCNTs, we may assume that only 1/3 are metallic, each of them contributing with $L_k = L_{k0}$. In addition, if we consider CNTs with the same diameters, we have a unique value for ν, hence:

$$L_{kb} = \frac{L_{k0}}{N/3}, \qquad R_b = \frac{\nu L_{k0}}{N/3} = \frac{3}{4} \frac{R_0}{l_{mfp} N} . \tag{34}$$

Recalling the results in Table 2, for N up to the order of 10^3 we may still assume that $L_{kb} >> L_m$, and since it is $\alpha_C \approx 1$, we may approximate the total bundle inductance with the first of (34), hence following the simple scaling law obtained in literature (Maffucci et al., 2009). Let us consider the case of a microstrip with effective dielectric constant $\varepsilon_{reff} = 2.2$, where the signal trace of width $w = 27$ nm, thickness $t = 2w$ and distance to ground $2t$ is made by a bundle of N SWCNTs of diameter $D = 2.2$ nm. Figure 11 shows the self admittance of the line computed with the rigorous approach and with the proposed approximation, for N=100 and N=200. Here we have assumed $v_F = 8.7 \cdot 10^5$ m/s and $l_{mfp} = 1\,\mu\text{m}$.

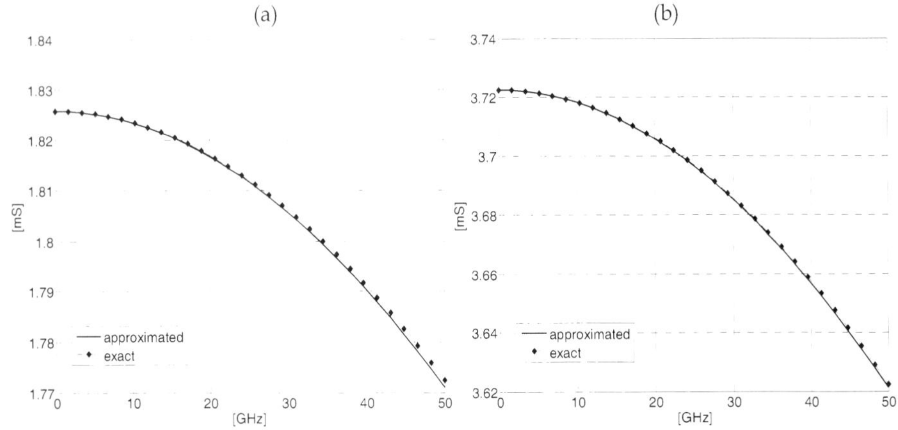

Fig. 11. Self admittance for a bundle of (a) 100 and (b) 200 SWCNTs.

Let us consider MWCNTs bundles. For these structures, the electrostatic capacitance and hence the magnetic inductance may be computed from (33). As for the kinetic inductance, we may introduce a simple approximation based on (13). Let D_{in} and D_{out} be the outer and inner shells diameters, respectively. The inter-shell spacing is the Van-der Waals distance $\delta = 0.34$ nm (Li et al., 2005), hence the number of shells is $N_{shell} = 1 + (D_{out} - D_{in}) / 2\delta$. Once again we assume 1/3 of the shells to be metallic and 2/3 semiconducting. Assuming the above distribution and using approximation (13) we can compute the equivalent number of channels for a single shell of diameter D_s in a MWCNTs through the simple law:

$$M_{shell} \approx \begin{cases} 2/3 & \text{for } D < d_s / T \\ a_{1s} D_s T + a_{2s} & \text{for } D \geq d_s / T \end{cases} \tag{35}$$

where $a_{1s} = 3.3 \cdot 10^{-4}$ 1/nmK, $a_{2s} = -0.08$ and $d_s = 1.9 \cdot 10^3$ nm $\cdot$ K. The kinetic inductance of a bundle of N_{CNT} MWCNTs is then given by:

$$L_{kb} \approx \frac{2L_{k0}}{N_{CNT} N_{shell} M_{shell}}. \tag{36}$$

where L_{k0} is the bulk value defined in (24). Using (36) in (21) we can derive the bundle quantum capacitance. Using $\alpha_C \approx 1$, from (19), (21) and (28) we may derive the bundle resistance:

$$R_b \approx \nu L_{kb} \approx \frac{2\nu L_{k0}}{N_{CNT} N_{shell} M_{shell}} = \frac{1}{2 l_{mfp}} \frac{R_0}{N_{CNT} N_{shell} M_{shell}}. \tag{37}$$

Note that the mean-free-path l_{mfp} may be a function of the diameter D. A simple linear dependence is proposed in (Nieuwoudt & Massoud, 2006).

4. Performance analysis of carbon nanotube interconnects

This Section is devoted to performance analysis of CNT interconnects, compared to conventional Cu ones. The investigated scenarios refer to typical arrangements foreseen for future 22 nm technology node, expected for year 2016 and beyond (ITRS, 2007). In the following the circuit models for CNT interconnects are those presented in Section 3, whereas the conventional ones are described through classical TL models.

4.1 Signal integrity analysis for CNT on-chip interconnects

Let us consider the on-chip interconnect in Fig.12 made by two horizontal traces and a vertical via. The electrical and geometrical parameters given in Table 4 are typical values for the intermediate level at 22 nm technology node (ITRS, 2007). The signal traces on the layers and the via barrel may be either constituted by a solid Cu conductor or by a SWCNT bundle. In particular we consider the case of traces made by bundles of SWCNTs of $D = 0.94\,\text{nm}$, with a fraction of 1/3 metallic. At the ends of the two horizontal tracts, a lumped contact resistance of $50\,\text{k}\Omega$ is considered for each metallic CNT. For the Cu case, the horizontal tracts are described as striplines, whereas the via is modeled through a series impedance *R*-*L* and a pad capacitance to ground (*e.g.*, Chiariello et al., 2009):

$$R_{via} = \frac{\rho l_{via}}{\pi(d_{via}/2)^2},\; C_{via} \approx \frac{1.41\varepsilon_r D_P T}{D_{AP} - D_P}[pF],\; L_{via} \approx 5.08 \cdot l_{via} \cdot \left[\ln\left(\frac{4 l_{via}}{d_{via}}\right) + 1\right]\,[nH], \quad (38)$$

where d_{via} and l_{via} are the diameter and the length of the via barrel, respectively, T is the dielectric thickness, D_P and D_{AP} are the diameters of the pad and the antipad, respectively.

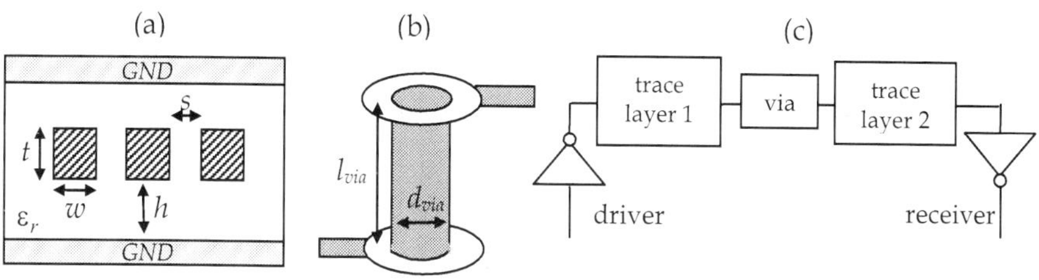

Fig. 12. On-chip interconnect: (a) section of the horizontal tracts; (b) vertical via; (c) circuit.

w [nm]	t [nm]	h [nm]	s [nm]	d_{via} [nm]	l_{via} [μm]	ε_r [SiO_2]	ρ_{Cu} [μΩ·cm]
44	44	39.6	44	60	1.0	4.5	4.235

Table 4. Parameters for the case study in Fig.12.

	R [Ω/μm]	L [pH/μm]	C [fF/μm]	R_{via} [Ω]	L_{via} [pH]
Cu	22.31	0.30	0.17	10.19	0.03
CNT	2.29	2.62	0.17	1.29	1.47

Table 5. Computed pul and lumped parameters.

As for the terminations in Fig.12c, we assume that the driver is modeled as a voltage source of $V_s = 0.8V_{DD}$ and a series resistor of 1.5 $k\Omega$, whereas the receiver is modeled as a capacitor of 1 fF. These are typical values for the 22-nm node, obtained from the minimum size gate values in the case with a size factor of 20x (Li et al., 2009).

Let us investigate the signal integrity performances of these two systems, assuming an high data rate (DR) signal transmission. Figure 13 shows the eye-diagrams obtained for the Cu case (Fig.13a-b) and the CNT case (Fig.13c-d), for values of the data rate *DR* ranging from 3 to 20 Gbit/s. The complete SI analysis is given in Table 6. The jitter is of the same order, whereas the performances of CNTs in terms of eye-opening factor are much better than the Cu ones. For instance at 10 Gbit/s the Cu channel could not be used (Fig.13b), whereas the CNT one still shows good performances (Fig.13c). The behavior obtained here is justified from the consideration that the higher inductance of CNT is compensated by its lower resistance and the SI performance of the overall CNT channel is better than Cu.

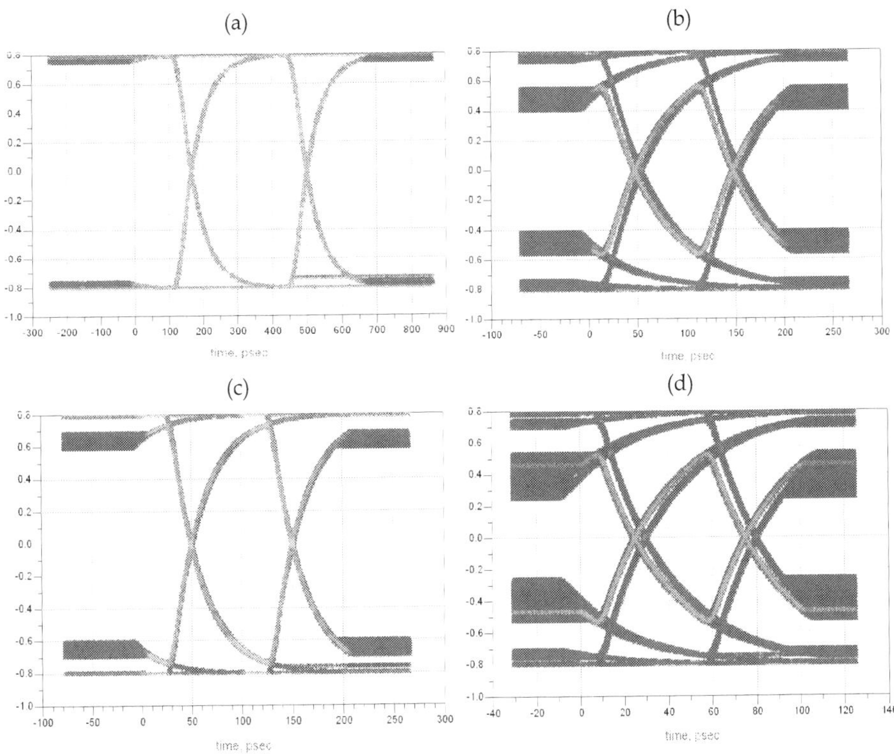

Fig. 13. Eye-diagram analysis for the case in Fig.12: Cu interconnect with (a) DR =3 Gbit/s, t_r=33 ps; and (b) DR =10 Gbit/s, t_r=10 ps; CNT interconnect with (c) DR =10 Gbit/s, t_r=10 ps; and (d) DR =20 Gbit/s, t_r=5 ps.

	DR [Gbit/s]	t_r [ps]	*Eye jitter* [ps - rms]	*Eye opening factor*
CNT	1	100	2.1	1
Cu	1	100	1.6	1
CNT	3	33	166	0.99
Cu	3	33	167	0.97
CNT	5	20	100	0.98
Cu	5	20	100	0.91
CNT	10	10	50	0.89
Cu	10	10	50	0.76
CNT	20	5	24	0.72
Cu	20	5	22	0.47

Table 6. Results of the Eye-Diagram analysis for the case-study in Fig.12.

4.2 EMC behavior of CNT interconnects

As pointed out in Section 3, the parameters of the circuit model for CNT interconnects are strongly affected by the presence of a kinetic inductance and a quantum capacitance. Within the limits of the model used here, these two parameters are independent on frequency, hence the frequency behavior of CNT interconnects may be completely different from that of conventional ones. Here we investigate two typical problems arising when performing an electromagnetic compatibility (EMC) analysis: the high-frequency effects on the line impedance and the crosstalk noise.

The case-study used for EMC analysis is the stripline configuration in Fig.14a, assuming global level parameters: $w = 0.32\,\mu\text{m}$, $t = 0.19\,\mu\text{m}$, $h = 0.15\,\mu\text{m}$ and $\varepsilon_r = 2.2$. Three different realizations of the signal traces are compared: solid copper, a bundle of SWCNTs and a bundle of MWCNTs. At such a scale the copper resistivity is $\rho \approx 2.942\,\mu\Omega \cdot cm$.

Let us first study the high-frequency impedance. The first realization of the signal traces is made a bundle of SWCNTs, of diameter $D = 1.4\ nm$, assuming the area of the trace filled by CNTs with a density of 80%, and considering 1/3 of the total number of CNTs to be metallic. The second realization is made by a bundle of MWCNTs, with the same density, each of them with outer shell diameter of $D_{out} = 40\text{ nm}$, inner shell diameter $D_{in} = 0.5D_{out}$ and inter-shell distance $\delta = 0.34$ nm. The temperature is 300 K.

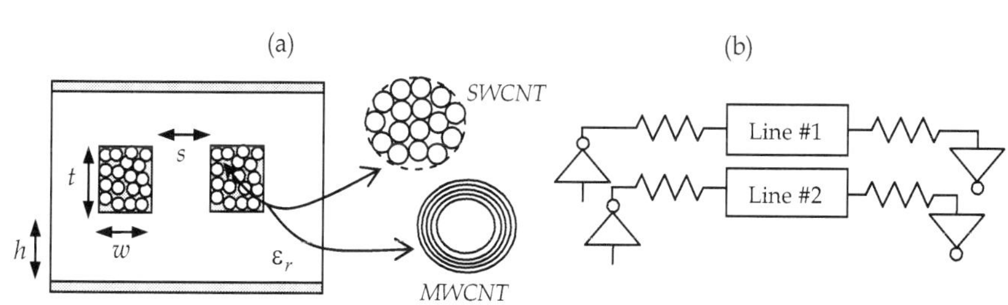

Fig. 14. Interconnect geometry used for EMC analysis: (a) the signal traces are made by either solid Cu or CNT bundles (SWCNT or MWCNTs); (b) circuit for the crosstalk analysis.

The p.u.l. resistances for the three realizations are reported in Fig.15a. For Cu lines this parameter is affected by the skin-effect, whereas for CNTs this effect is not present. For instance, at 200 GHz the penetration depth is $0.19\,\mu\text{m}$, equal to t and lower than w. The CNT interconnects outperform the Cu one and show a resistance parameter insensitive to the skin effect. In particular, the MWCNTs bundle outperforms the other two realizations in the whole frequency range: this is due to the fact that all the shells contribute to the conduction. As for the p.u.l. inductance (Fig.15b), the Cu line shows a slight decrease of this parameters, due to the decrease of the contribution of the internal inductance with increasing frequency, whereas the CNTs bundles are again insensitive to it. The MWCNT bundle interconnect introduces an inductance comparable to the Cu one, whereas the SWCNT realization shows an higher inductance. Since the kinetic inductance of the bundles decreases as (34) or (36), it may be comparable to the magnetic one, as in this case. As a conclusion we may assume that CNT interconnects exhibit parameters independent on frequency, whereas the conventional ones would suffer from this problem.

For the crosstalk analysis we consider a coupled stripline as in Fig.14a, assuming $s=w$, and typical intermediate level parameters: $w = t = 44$ nm, $h = 39.6$ nm and ε_r = 2.2. The considered circuit is shown in Fig.14b. We assume all the resistors of $1450\,\Omega$ assuming that line 1 is fed at near end by a time-harmonic voltage source of amplitude V_{in} = 1 (a.u.). The far-end receivers are modeled as capacitors of 1 fF. For the CNT realizations we consider a density of 80%, a temperature of 300 K and a mean free path $l_{mfp} = 0.5\,\mu\text{m}$. The SWCNTs diameter is $D = 1.4\,nm$, whereas for the MWCNTs the outer shell diameter is $D_{out} = 30$ nm , the inner shell diameter is $D_{in} = 0.5D_{out}$ and the inter-shell distance is $\delta = 0.34$ nm. Because of the large number of CNTs in the bundle, we may disregard the effect of the parasitic contact resistances. Figure 16 shows the computed near-end and far-end crosstalk voltages, normalized to the input voltage V_{in} . The crosstalk noise for the three realizations is of the same order, hence from this point of view we may assume an equivalent EMC performance.

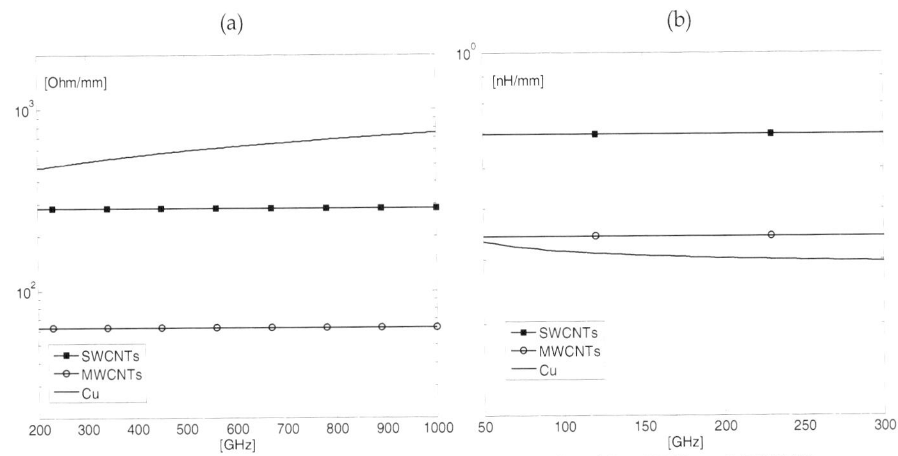

Fig. 15. (a) P.u.l. resistance for a trace made by Cu, a bundle of SWCNTs or MWCNTs; (b) P.u.l. inductance for a trace made by Cu, a bundle of SWCNTs or MWCNTs.

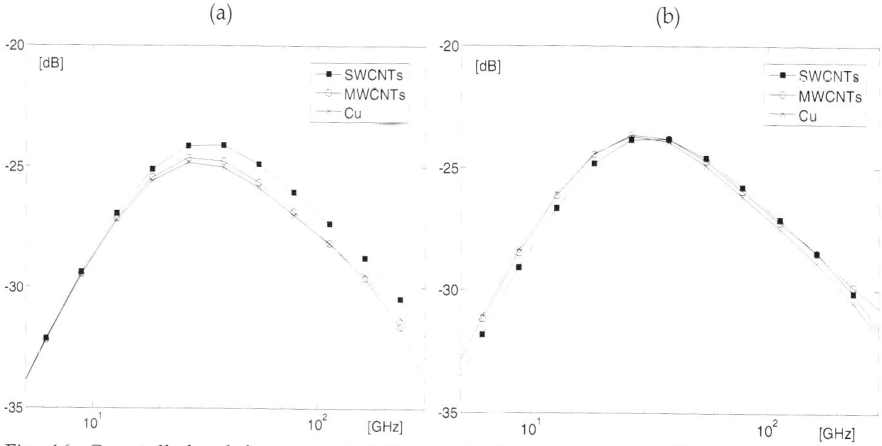

Fig. 16. Crosstalk level for a coupled line made by copper trace, SWCNTs bundle and MWCNTs bundle: (a) near end; (b) far end.

4.3 CNT interconnects as pillars for nanopackaging

Recently the CNTs have been proposed as material for nanopackaging interconnects, since their unique electrical, mechanical and thermal properties allows meeting many of the new requirements for bonding, molding compound, underfill, thermal interface, die attach (Morris, 2008; Maffucci, 2009). One of the main reasons pushing towards the use of CNTs in nanopackaging is the possibility either to achieve good electrical performances and implement new heat removal technologies. The thermal management of future nanoelectronics requires new approaches, since the classical heat removal techniques are inadequate. Using the conventional approach, for instance, the cooler for future Systems in Package would require much greater volume than the semiconductor itself. Recently, CNTs have been proposed as microchannel coolers in thermofluidic cooling approaches and as thermal interface material (TIM). The main limit to the practical use of CNTs is still given by the possibility of integrating a high density of aligned CNTs in a polymer matrix, without degrading thermal conductivity.

In order to investigate the electrical properties of CNT nanopackaging interconnects, let us consider the use of CNTs bundles as chip to package interconnect, following the scheme depicted in Fig.17a: the two vertical pillars are realized using Cu, a bundle of SWCNTs (Fig.17b) or a bundle of MWCNTs (Fig.17c). In terms of electrical properties, the goal of a good packaging interconnect is of course to introduce low parasitics.

As done before, in the following we assume that in the SWCNT bundle only a fraction of 1/3 of the total CNTs is metallic, and similarly in a MWCNT only 1/3 shells are metallic. Using the results in Section 2, since the semiconducting SWCNTs do not give contribution to the conduction, very high density bundles must be fabricated to have low resistance CNT pillars. In MWCNTs, instead, the semiconducting shells give a non-negligible contribution to the conduction and the conduction of the metallic shells themselves is improved, hence the optimal density is a trade-off between the MWCNT outer diameter and its shell number.

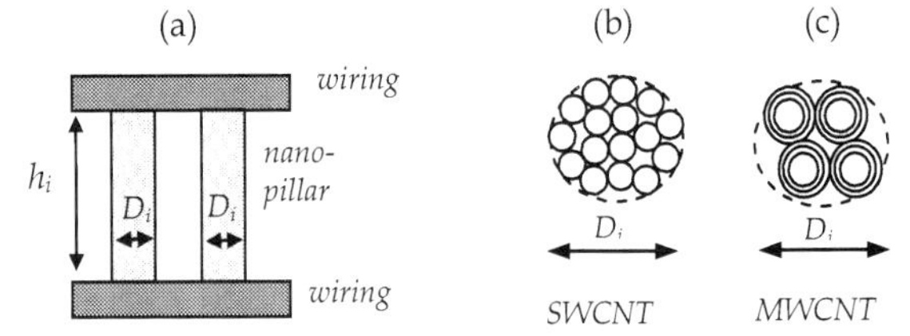

Fig. 17. Chip-to-package interconnect: (a) scheme; (b) SWCNTs and (c) MWCNTs realization

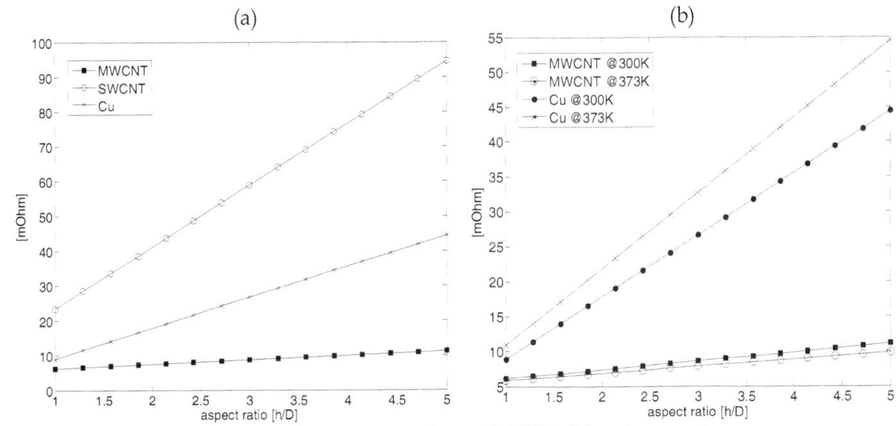

Fig. 18. Parasitic resistance: (a) vs aspect ratio, at T=300K; (b) vs temperature.

Let us consider the case of Fig. 17, assuming the pillars to be cylinders of diameter D_i and height h_i, a wire bond pitch of $20\,\mu m$ and an operating frequency of 10 GHz at a temperature of 300 K. First we consider Cu interconnects with $\sigma_{Cu} = 5.8 \cdot 10^7\ S/m$. Next, we fill the interconnect with a bundle of SWCNTs of mean diameter 2.7 nm, assuming a density of 80% of the total area and assuming 1/3 of metallic CNTs. In addition, a parasitic contact resistance of $20\,k\Omega$ in series to any single CNT is considered. Finally we consider the bundle to be filled by MWCNTs with outer diameter $D_{out} = 30\,\text{nm}$, inner diameter $D_{in} = 0.5 D_{out}$ and inter-shell distance 0.34 nm. We assume again 1/3 of metallic shells and the rest semiconducting, a density of 80% and a contact resistance of $10\,k\Omega$ per shell. The considered mean free path for all CNTs is $l_{mfp} = 0.5\,\mu\text{m}$.

For this case the parasitic inductances are well beyond the maximum allowed of $5-10\,pH$, hence we focus on the resistance. Figure 18a compares the resistance values obtained for aspect ratios ranging from 1 to 5. The lowest values are obtained by MWCNT bundle, whereas SWCNT bundle shows higher resistance. The better performance of MWCNTs is due to the contribution to the conduction given by the semiconducting shells.

Let us now investigate the problem of the compatibility of CNT interconnects with the temperatures of CMOS technology. Let us consider a realistic chip operating temperature of

T=373 K. When the temperature increases the kinetic inductance decreases, as shown in Section 2, and the difference is more pronounced for large diameters, hence for MWCNTs rather than for SWCNTs. Let us neglect the variation of the mean-free path with temperature: in the same conditions described above, the parasitic resistance for the MWCNT bundle and the Cu interconnects would be that shown in Fig.18b. Note that for the copper at 373K we have assumed $\sigma_{Cu} = 4.0 \cdot 10^7 \text{ S/m}$. The MWCNT interconnect has better performances at higher frequencies, which is opposite to the behaviour of Cu ones.

Finally let us focus on the problem of proximity effect, which is a major problem for packaging interconnect, since it leads to non uniform current crowding which cause local thermal hot-spots. Figure 19 shows the distribution of the current density over two CNT bundles at a distance equal to their width. The coupled line is driven by a differential signal (in the figures the underlying PEC ground is omitted). The simulations have been carried out by means of the full-wave 3D numerical simulator SURFCODE (Miano & Villone, 2005). As shown in Figs.19a-b, the CNT bundles do not suffer from the proximity effect, which is a problem for the conventional interconnects, Figs.18c-d. This is again due to the presence in the CNT electrodynamics of a kinetic term which dominates over the magnetic one. Within the limits of validity of our model, this kinetic term is independent from the distance between the bundles, which, on the contrary, is a crucial parameter in determining the proximity effect when conventional interconnects are analyzed.

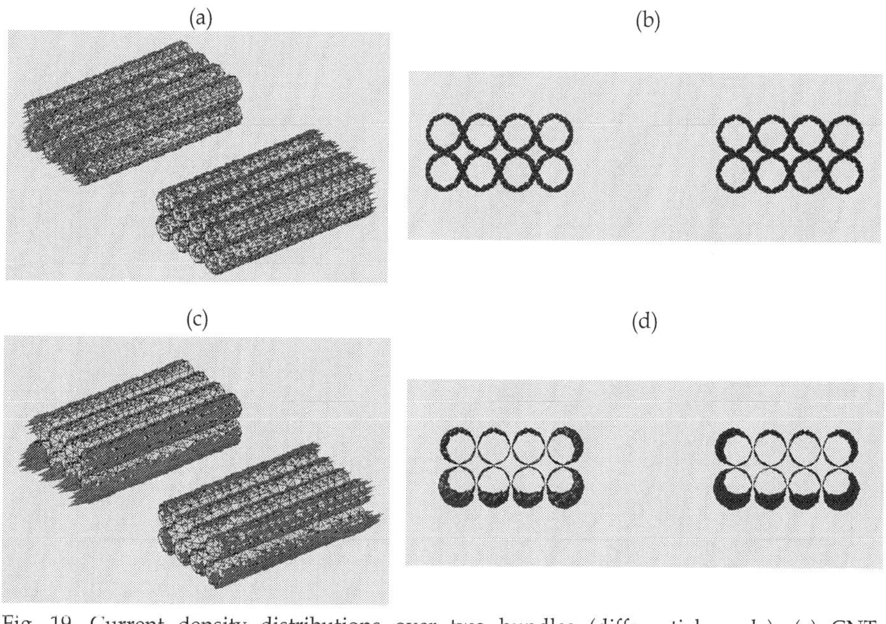

Fig. 19. Current density distributions over two bundles (differential mode): (a) CNT bundles; (b) CNT transverse section; (c) PEC bundles; (d) PEC transverse section.

5. Conclusion

This Chapter has analyzed the use of carbon nanotubes as possible innovative material to fabricate high-speed interconnects at various hierarchical levels of the future nanoelectronics technology.

The electromagnetic model presented here to describe the electrical propagation along CNT interconnects is derived from a semi-classical solution of the transport equation. All the quantistic and kinetic effects affecting the electrodynamics of charge carriers are taken into account through global parameters (*kinetic inductance* and *quantum capacitance*) which have been simply related to the number of effective conducting channel per CNT shell. Both the rigorous approach and a simple approximation have been provided to evaluate this number for CNT shells of arbitrary chirality, hence allowing the modelling of either single-wall and multi-wall CNTs.

The electrodynamical model provides a constitutive equation for the CNT interconnect, which is coupled to the Maxwell equation: the problem is then solved in the frame of the classical transmission line theory, so obtaining simple equivalent RLC circuits for single CNTs and bundles of CNTs. These models are consistent with those used in literature for metallic CNTs, but introduce the possibility to deal with arbitrary chiralities and CNT diameters. The analysis of CNT transmission lines highlights unusual behaviours with respect to the Cu lines, mainly due to the strong influence of the kinetic and quantistic terms over the classical electromagnetic inductance and capacitance parameters.

Some case-studies have been carried out referring to a realistic use of CNTs to build interconnects for wiring and for packaging nanotechnology ICs. In the considered examples typical values of the future 22nm technology nodes have been assumed. A signal integrity analysis is performed with reference to an on-chip interconnect arrangement, where the conventional Cu material is replaced by a bundle of CNTs. Next, the EMC behaviour of two adjacent traces in a stripline is analyzed, checking the high-frequency effects and the crosstalk noise. Finally the use of CNTs as pillars for nanopackaging is studied and the introduced parasitics are compared to conventional material realization.

Carbon nanotube interconnects are shown to have better behaviour with respect to conventional Cu ones in terms of SI and EMC performances, introduced parasitics and response to temperature variations. From a practical point of view, these simulated performances may be achieved provided that good quality bundles are realized, *i.e.* bundles of high density, good direction control and good terminal contacts.

6. References

Anantram, M.P.; & Léonard, F. (2006). Physics of carbon nanotube electronic devices. *Report Progress in Physics*, Vol.69, 507.

Avouris, P.; Appenzeller, J.; Marte, R. & Wind, S.J. (2003). Carbon nanotube electronics, *Proceedings of IEEE*, Vol.91, 1772.

Banerjee, K.; Li, H. & Srivastava, N. (2008). Current status and future perspectives of carbon nanotubes interconnects. *Proc. of IEEE Nanotechonology Conference*, Arlington, TX, USA, August 18-21, 2008.

Burke, P.J. (2002). Luttinger liquid theory as a model of the gigahertz electrical properties of carbon nanotubes, *IEEE Transactions on Nanotechnology*, Vol. 1, 129-144.

Burke, P.J. (2003). An RF circuit model for carbon nanotubes. *IEEE Transactions on Nanotechnology*, Vol.2, 55.

Cheung, C.L. & et al. (2002). Diameter-controlled Synthesis of Carbon nanotubes. *Journ. Phys. Chem. B*, Vol.106, 2429-2433.

Chiariello, A.G.; Maffucci, A. & Miano, G. (2009). Signal Integrity Analysis of Carbon Nanotube on-chip Interconnects. *IEEE Workshop on Signal Propagation on Interconnects SPI 2009*, paper no. S4_1, Strasbourg, France, 12-15 may 2009.

Close, G.F.; Yasuda, S.; Paul, B.; Fujita, S. & Philip Wong, H.-S. (2008). A 1 GHz integrated circuit with carbon nanotube interconnects and silicon transistors. *Nano Letters*, Vol.8, 706–709.

Forestiere, C.; Maffucci, A. & Miano, G. (2009a). A Generalized Model for the Signal Propagation along Single- and Multi-Walled Carbon Nanotubes with Arbitrary Chirality, *Proc. of 9th IEEE Nanotechnology Conference*, paper #64, Genova (Italy), 27-30 Jul 2009.

Forestiere, C.; Maffucci, A.; Miano, G., Maksimenko, S.A. & Slepyan, G.Y. (2009b). Signal propagation in carbon nanotubes of arbitrary chirality. In press on *IEEE Trans. On Nanotechnology*.

ITRS (2007) *International Technology Roadmap for Semiconductors*. http://public.itrs.net. Edition 2007.

Iijima, S. (1991). Helical microtubules of graphitic carbon. *Nature*, Vol.354, 56-58.

Li H.J.; Lu, W. G.; Li, J. J. ; Bai, X. D. & Gu, C. Z. (2005). Multichannel Ballistic Transport in Multiwall Carbon Nanotubes. *Physical Review Letters*, Vol.95, 086601.

Li, H.; Xu, C; Srivastava, N. & Banerjee, K. (2009). Carbon Nanomaterials for Next - Generation Interconnects and Passives: Physics, Status, and Prospects. *IEEE Transactions on Electron Devices*, Vol. 56, 1799-1821.

Liu, Z. (2009). Fabrication and electrical characterization of densified carbon nanotube micropillars for IC interconnection. *IEEE Trans. on Nanotechnology*, Vol.8, 196-203.

Maarouf, A.A.; Kane, C.L. & Mele, E.J. (2000). Electronic structure of carbon nanotube ropes. *Physical Review B*, Vol.61, 11156-11165.

Maksimenko, S.A. & Slepyan, G.Y. (2000). Electrodynamic properties of carbon nanotubes, in *Electromagnetic Fields in Unconventional Materials and Structures*, O. N. Singh and A. Lakhtakia Eds., pp. 217–255, Wiley, New York.

Maksimenko, S.A.; Khrushchinsky, A.A.; Slepyan, G.Y. & Kibisb, O.V. (2007). Electrodynamics of chiral carbon nanotubes in the helical parametrization scheme. *Journal of Nanophotonics*, Vol.1, p.013505.

Maffucci A.; Miano, G. & Villone, F. (2008a). A transmission line model for metallic carbon nanotube interconnects, *Intern. Journal of Circuit Theory and Applic.*, Vol. 36, 31-52.

Maffucci A.; Miano, G. & Villone, F. (2008b). Performance Comparison Between Metallic Carbon Nanotube and Copper Nano-Interconnects, *IEEE Transactions on Advanced Packaging*, Vol. 31, 692 - 699.

Maffucci A.; Miano, G. & Villone, F. (2009). A new Circuit Model for Carbon Nanotube Interconnects with Diameter-dependent Parameters. *IEEE Transactions on Nanotechnology*, Vol.8, 345 - 354.

Maffucci, A. (2009). Carbon Nanotubes in Nanopackaging applications. *IEEE Magazine on Nanotechnology*, Vol.3, no.3, 22-25.

Miano, G. & Villone, F. (2005). A surface integral formulation of Maxwell equations for topologically complex conducting domains. *IEEE Trans. on Antennas and Propagation*, vol. 53, 4001-4014.

Miano, G. & Villone, F. (2006). An Integral Formulation for the Electrodynamics of Metallic Carbon Nanotubes Based on a Fluid Model, *IEEE Transactions on Antennas and Propagation*, Vol.54, 2713-2724, 2006.

Morris, J.E. (2008). *Nanopackaging: Nanotechnologies and electronics packaging*. New-York: Springer.

Naeemi, A. & Meindl, J. D. (2006). Compact Physical Models for Multiwall Carbon-Nanotube Interconnects. *IEEE Electron Devices Letters*, Vol.27, 338-340.

Naeemi, A. & Meindl, J. D. (2008). Performance Modeling for Single- and Multiwall Carbon Nanotubes as Signal and Power Interconnects in Gigascale Systems, *IEEE Trans. on Electron Devices*, Vol. 55, n.10, 2574-2582.

Nieuwoudt, A. & Massoud,Y. (2006). Evaluating the impact of resistance in carbon nanotube bundles for VLSI interconnect using diameter-dependent modeling techniques. *IEEE Trans. Electron Devices*, vol. 53, 2460–2466.

Park, J.Y.; et al., (2004). Electron-phonon Scattering in Metallic Single-Walled Carbon Nanotubes. *Nano Letters*, Vol.4, 517-520.

Raychowdhury, A.; Roy, K. (2006). Modelling of metallic carbon-nanotube interconnects for circuit simulations and a comparison with Cu interconnects for scaled technologies. *IEEE Trans. on Computer-Aided Design for Integrated Circ. and Syst.*, Vol. 25, 58-65.

Rutherglen, C. & Burke, P.J. (2009). Nanoelectromagnetics: Circuit and Electromagnetic Properties of Carbon Nanotubes. *Small*, Vol.5, 884–906.

Saito, R.; Dresselhaus, G. & Dresselhaus, M. S. (2004). *Physical Properties of Carbon Nanotubes*, Imperial College Press, Singapore.

Salahuddin, S.; Lundstrom, M. & Datta, S. (2005). Transport effects on signal propagation in quantum wires. *IEEE Transactions on Electron Devices*, Vol.52, 1734-1742.

Sarto, M.S.; Tamburrano, A. & D'Amore, M. (2009). New Electron-Waveguide-Based Modeling for Carbon Nanotube Interconnects. *IEEE Transactions on Nanotechnology*, Vol. 8, 214-225.

Slepyan, G.Y.; Maksimenko, S. A.; Lakhtakia, A.; Yevtushenko, O. & Gusakov, A.V. (1999). Electrodynamics of carbon nanotubes: dynamics conductivity, impedance boundary conditions, and surface wave propagation. *Physical Review B*, Vol. 60, 17136.

Soga, I.; Kondo, D.; Yamaguchi, Y.; Iwai, T.; Mizukoshi, M.; Awano, Y.; Yube, K. & Fujii T. (2008). Carbon nanotube bumps for LSI interconnect, *Proc. Of Electronic Components and Technology Conf.*, 1390-1394, May 2008.

Wesström, J.J. (1996). Signal propagation in electron waveguides: Transmission-line analogies, *Physical Review B*, Vol. 54, 11484-11491.

Pu, S-N.; Yin, W.-Y.; Mao, J.-F. & Liu, Q.H. (2009). Crosstalk Prediction of Single- and Double-Walled Carbon-Nanotube (SWCNT/DWCNT) Bundle Interconnects. *IEEE Transactions on Electron Devices*, Vol. 56, 560-568.

15

Interconnect Challenges and Carbon Nanotube as Interconnect in Nano VLSI Circuits

Davood Fathi and Behjat Forouzandeh
School of Electrical and Computer Engineering, Faculty of Engineering, University of Tehran, Tehran, Iran;
Email: davfathi@gmail.com

This chapter discusses about the behavior of Carbon Nanotube (CNT) different structures which can be used as interconnect in Very Large Scale (VLSI) circuits in nanoscale regime. Also interconnect challenges in VLSI circuits which lead to use CNT as interconnect instead of Cu, is reviewed. CNTs are classified into three main types including Single-walled Carbon Nanotube (SWCNT), CNT Bundle, and Multi-walled Carbon Nanotube (MWCNT). Because of extremely high quantum resistance of a SWCNT which is about 6.45 kΩ, rope or bundle of CNTs are used which consist of parallel CNTs in order to overcome the high delay time due to the high intrinsic (quantum) resistance. Also MWCNTs which consist of parallel shells, present much less delay time with respect to SWCNTs, for the application as interconnects. In this chapter, first a short discussion about interconnect challenges in VLSI circuits is presented. Then the repeater insertion technique for the delay reduction in the global interconnects will be studied. After that, the parameters and circuit model of a CNT will be discussed. Then a brief review about the different structures of CNT interconnects including CNT bundle and MWCNT will be presented. At the continuation, the time domain behavior of a CNT bundle interconnect in a driver-CNT bundle-load configuration will be discussed and analyzed. In this analysis, CNT bundle is modeled as a transmission line circuit model. At the end, a brief study of stability analysis in CNT interconnects will be presented.

1. Interconnect Challenges in VLSI Circuits

As interconnect feature sizes shrink, copper resistivity increases due to surface and grain boundary scatterings and also surface roughness [1]. Furthermore, wires, especially power and ground lines, are becoming more and more vulnerable to electromigration because of rapid increases in current densities [2]. The resistance of copper interconnects, with cross-sectional dimensions of the order of the mean free path of electrons (~40 nm in Cu at room temperature) in current and imminent technologies [2], is increasing rapidly under the combined effects of enhanced grain boundary scattering, surface scattering and the presence of the highly resistive diffusion barrier layer [3]. The steep rise in parasitic resistance of copper interconnects poses serious challenges for interconnect delay [2] (especially at the global level where wires traverse long distances) and for interconnect reliability [4], hence it

has a significant impact on the performance and reliability of VLSI circuits. In order to alleviate such problems, changes in the material used for on-chip interconnections have been sought even in earlier technology generations, for example the transition from aluminum to copper some years back [3].
Carbon nanotubes (CNTs) exhibit a ballistic flow of electrons with electron mean free paths of several micrometers, and are capable of conducting very large current densities [3]. They are therefore proposed as potential candidates for signal and power interconnections [5], [6]. Because of their extremely desirable properties of high mechanical and thermal stability, high thermal conductivity and large current carrying capacity [7], CNTs have aroused a lot of research interest in their applicability as VLSI interconnects of the future. Depending on their chirality (the direction along which the graphene sheets are rolled up), CNTs demonstrate either metallic or semiconducting properties. Fig. 1 shows different structures depending on the chirality

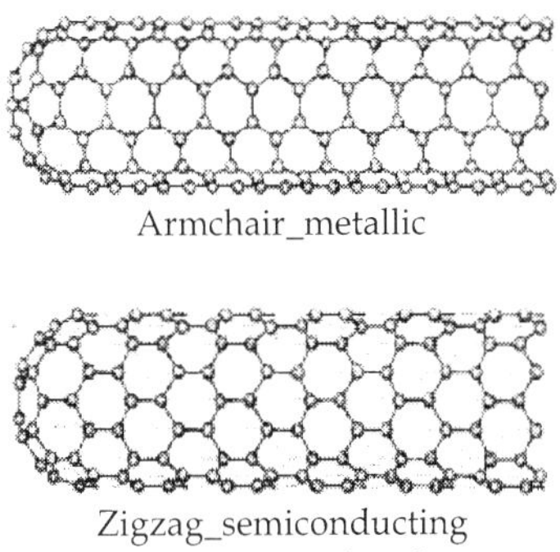

Fig. 1. Two different structures of carbon nanotubes for an armchair-type nanotube and a zigzag-type nanotube [8].

Carbon nanotubes are also classified into single-walled and multi-walled nanotubes.

2. Repeater Insertion as a Technique for the Delay Reduction

With the technology scaling in very deep submicron (VDSM) CMOS circuits, the gate delay decreases rapidly, while the delay of global interconnects tends to increase because of increasing their aspect ratio (thickness to width ratio) with scaling [2], [9], [10]. The repeater (buffer) insertion technique is generally used to reduce the delay of long (semi global) and global interconnects [9], [11]-[16]. An analytical model for obtaining the optimal buffer size and segment length for an equal partitioning network, in which the buffers sizes and segments lengths are constant, has been presented [12], [14], [17].
In [18] we have discussed about the optimization of global interconnects using unequal repeater (buffer) partitioning technique. This method which is discussed and reviewed in this chapter, is based on the segmentation of a long global interconnect into unequal parts, and inserting buffers with unequal sizes between them. The related structure is named as "*Unequal buffer partitioning network*" against "*Equal buffer partitioning network*" that was mentioned above. The optimum delay is a function of various parameters such as the

buffers sizes, the interconnect segments lengths, the load and so on [9], [11], [12], [15]-[17], [19]-[22]. It is shown that for the optimization of a buffer inserted interconnect behavior, the energy-delay product minimization is better than the delay minimization. Thus in this research, the energy-delay product has been chosen as target function for the minimization. There are different algorithms for minimizing a function, which in this chapter, the genetic algorithm (GA) using MATLAB [23] has been used for minimizing the energy-delay product function.

2.1 Equal Buffer Partitioning Network

Fig. 2 shows a global interconnect with the buffer insertion, in which each segment has equal length and all the buffers have the same size

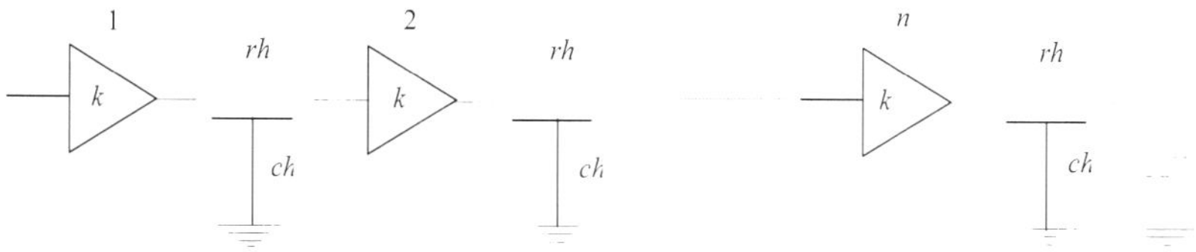

Fig. 2. Equal buffer partitioning network [18]

where r, c, and h are the interconnect resistance per unit length, the interconnect capacitance per unit length, and each segment length of the interconnect, respectively. Also k, n and C_L are the buffers size, the number of buffers, and the load capacitance, respectively. The total time delay of global line interconnect including the buffers and load, using Elmore relation [12], [21] will be

$$Delay_1 = n\left[R_{Bo}(C_{Bo}+ch)+\frac{1}{2}rch^2\right]+(n-1)(R_{Bo}+rh)C_{Bi}+(R_{Bo}+rh)C_L \,. \qquad (1)$$

where $R_{Bo}=r_0/k$, $C_{Bi}=kc_0$, and $C_{Bo}=kc_p$ are the buffers output resistance, the buffers input capacitance, and the buffers output capacitance, respectively. Also r_0, c_0, and c_p are similar parameters of the minimum sized repeater (buffer), respectively. We can express the total energy as

$$Energy_1 = n\left[ch+k(c_0+c_p)+\frac{C_L}{n}\right]V_{dd}^2 \qquad (2)$$

where V_{dd} is the power supply voltage. The energy-delay product will be written as

$$EDP_1 = Energy_1 \times Delay_1 \,, \qquad (3)$$

$$EDP_1 = n^2\left[ch+k(c_0+c_p)+\frac{C_L}{n}\right]\times\left[\frac{r_0}{k}(kc_p+\frac{n-1}{n}kc_0+\frac{C_L}{n}+ch)+rh(\frac{n-1}{n}kc_0+\frac{C_L}{n}+\frac{1}{2}ch)\right]V_{dd}^2 \,. \qquad (4)$$

2.2 Unequal Buffer Partitioning Network

Fig. 3 shows a global interconnect with buffer insertion, in which each segment length is *a* times of the previous segment length, and each buffer size is *f* times of the previous buffer size, respectively

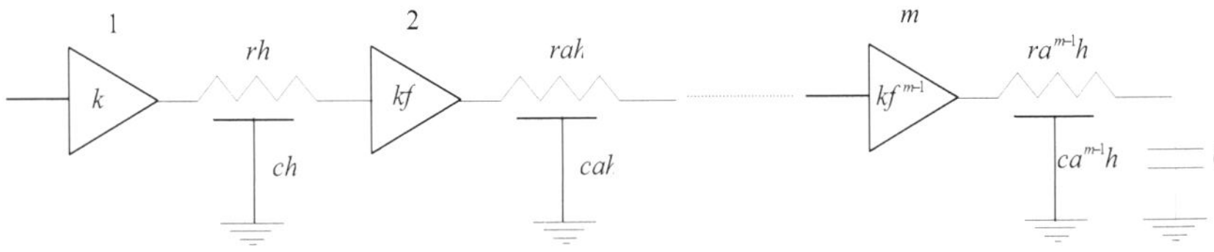

Fig. 3 Unequal buffer partitioning network [18]

where $h=l(1-a)/(1-a^m)$ which l is the total length of line (interconnect), k is the first buffer size, and m is the number of buffers. The other parameters are the same as in Fig.2. The total time delay of global interconnect in this structure, including the buffers and load, using Elmore relation [12], [21] will be

$$Delay_2 = mr_0c_p + (m-1)r_0c_0f + \left(\frac{1-\left(\frac{a}{f}\right)^m}{1-\frac{a}{f}} \right) \frac{r_0}{k} ch + \left(\frac{1-(af)^{m-1}}{1-af} \right) rhkfc_0 + \frac{1}{2}\left(\frac{1-a^{2m}}{1-a} \right) rch^2 + \left(\frac{r_0}{kf^{m-1}} + ra^{m-1}h \right) C_L \quad (5)$$

where r_0, c_0, and c_p are the output resistance, the input capacitance, and the output capacitance of the minimum sized buffer, respectively. We can express the total energy as

$$Energy_2 = \left[cl + \left(\frac{1-f^m}{1-f} \right) k(c_0 + c_p) + C_L \right] V_{dd}^2 . \quad (6)$$

Thus the energy-delay product for "*Unequal partitioning network*" will be written as

$$EDP_2 = Energy_2 \times Delay_2 , \quad (7)$$

$$EDP_2 = \left[cl + \left(\frac{1-f^m}{1-f} \right) k(c_0 + c_p) + C_L \right]$$
$$\times \left[mr_0c_p + (m-1)r_0c_0f + \left(\frac{1-\left(\frac{a}{f}\right)^m}{1-\frac{a}{f}} \right) \frac{r_0}{k} ch + \left(\frac{1-(af)^{m-1}}{1-af} \right) rhkfc_0 + \frac{1}{2}\left(\frac{1-a^{2m}}{1-a} \right) rch^2 + \left(\frac{r_0}{kf^{m-1}} + ra^{m-1}h \right) C_L \right] V_{dd}^2 . \quad (8)$$

2.3 Optimization Procedure

In this section, EDP[1] functions for the two networks "*Equal partitioning network*" and "*Unequal partitioning network*", which are defined and obtained in sections II and III, are minimized using the genetic algorithm (GA) of MATLAB [23]. This procedure is performed on the two networks for the three technology nodes 65, 90, 130 nm, which the specifications of global interconnect and the minimum sized repeater (buffer) in each technology node have been extracted from ITRS[2] [2], [9]. Also the interconnect capacitance per unit length is obtained using the formulations presented in [9]. Moreover, in each step of the optimization (minimization) procedure, the load capacitance has been taken as a parameter which varies from one to hundred times of the minimum sized buffer output capacitance.

In Figs. 4-9, the propagation delay improvement for "*Unequal partitioning network*" respect to "*Equal partitioning network*", versus the global interconnect length, and for three technology nodes 65, 90 , 130 nm, have been plotted whereas the capacitive load varies from one to hundred times of the minimum sized buffer output capacitance (c_p). The genetic algorithm (GA) of MATLAB [23] has been used as a tool for minimizing the energy-delay product (EDP) for the two networks at different technology nodes and various loads. For obtaining the correct results for each step of the minimization procedure, the algorithm has been done 1000 times in each step and the least value has been chosen as the best answer.

It is found from Figs. 4-9 that the improvement of the propagation delay, in unequal partitioning network is more than equal partitioning network. This improvement is obvious for the technology nodes 90, 130 nm and goes high with increasing the load capacitance. Also for technology node 65 nm, the delay improvement will be achieved for the high values of the load capacitance, which is cleared from Figs. 8, 9.

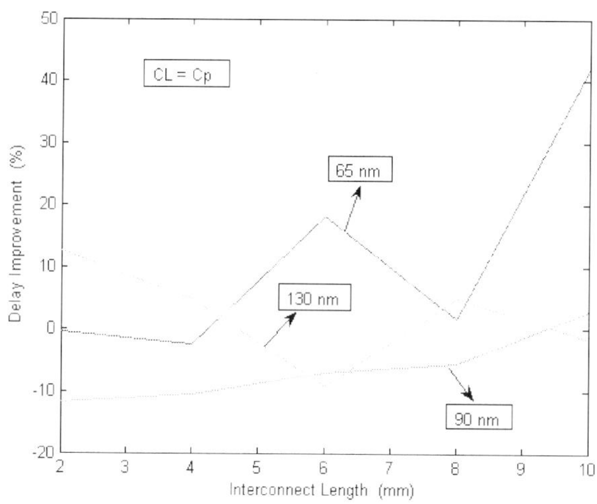

Fig. 4. The delay improvement of unequal partitioning network respect to equal partitioning network, for $C_L=c_p$ [18].

---

[1] Energy-Delay Product

[2] International Technology Roadmap for Semiconductors

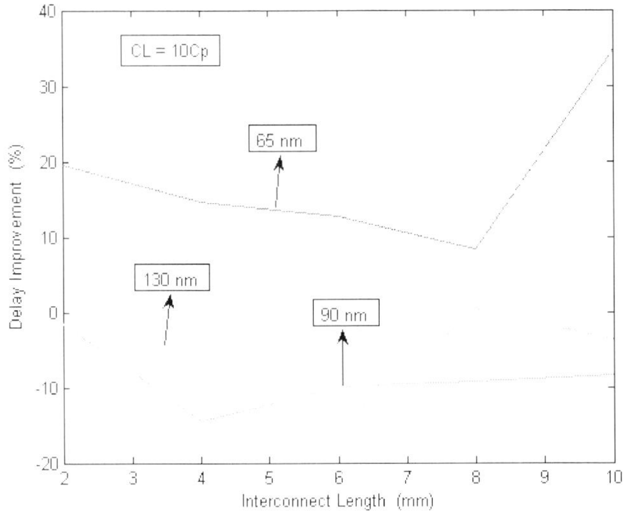

Fig. 5. The delay improvement of unequal partitioning network respect to equal partitioning network, for C_L=10c_p [18].

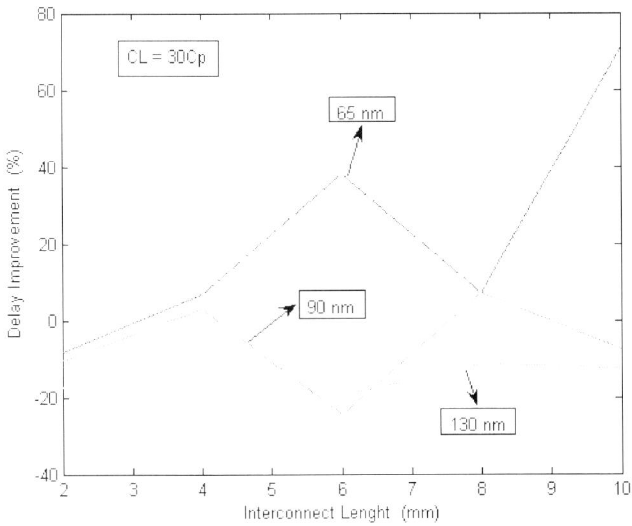

Fig. 6. The delay improvement of unequal partitioning network respect to equal partitioning network, for C_L=30c_p [18].

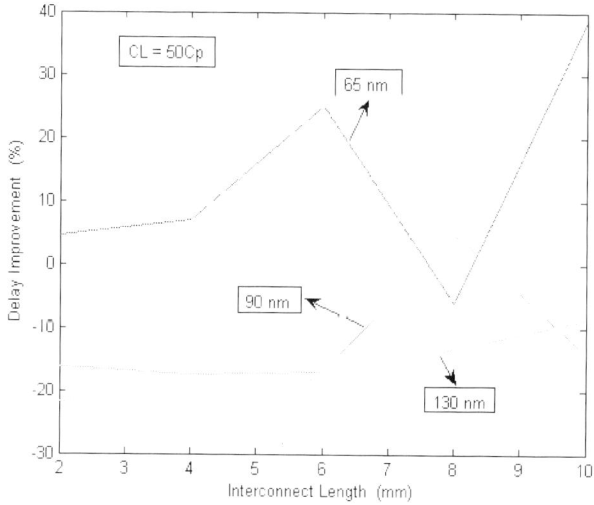

Fig. 7. The delay improvement of unequal partitioning network respect to equal partitioning network, for C_L=$50c_p$ [18].

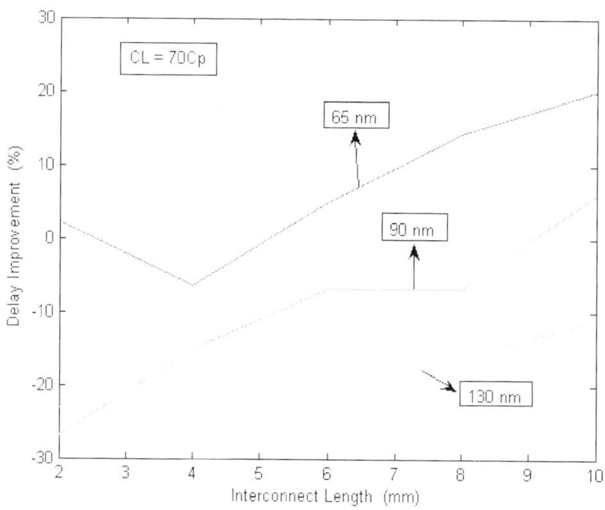

Fig. 8. The delay improvement of unequal partitioning network respect to equal partitioning network, for C_L=$70c_p$ [18].

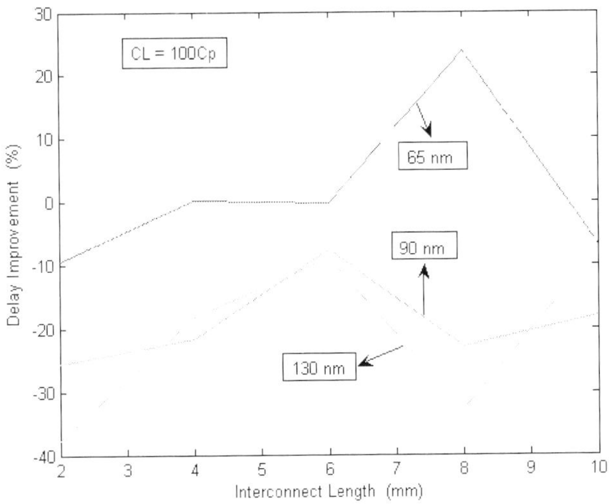

Fig. 9. The delay improvement of unequal partitioning network respect to equal partitioning network, for C_L=$100c_p$ [18].

3. Circuit Model and Parameters for CNT

Fig. 10 depicts the equivalent circuit for an isolated single-walled carbon nanotube (SWCNT) of length less than the mean free path of electrons in a CNT [24], [25].

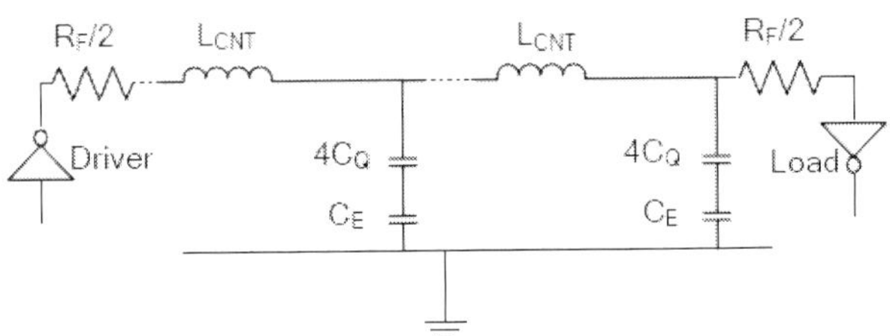

Fig. 10. Equivalent circuit model for an isolated SWCNT of length less than the mean free path of electrons, assuming ideal contacts [25].

In this figure, R_F is the fundamental (quantum) resistance, L_{CNT} is the total inductance, and C_Q and C_E are the quantum and electrostatic capacitances, respectively. In the following subsections, these parameters and their related formulations will be discussed.

3.1. CNT Resistance

Due to spin degeneracy and sub-lattice degeneracy of electrons in graphene, each nanotube has four conducting channels in parallel [3], [26]. Hence, the conductance of an isolated ballistic single-walled CNT (SWCNT) assuming perfect contacts, is $4e^2/h$ = 155 μS, which yields a resistance of 6.45 KΩ [8], [24]. This is the quantum resistance associated with a SWCNT that cannot be avoided [3], [27]. This fundamental resistance, as shown in Fig. 11, is equally divided between the two contacts on either side of the nanotube and can be expressed as [27]

$$R_F = \frac{h}{4e^2} \tag{9}$$

where h is plank's constant and e is electron charge. The mean free path (MFP) of electrons in a CNT is typically 1 μm [3], [8], [28]. For CNT lengths less than λ_{CNT}, electron transport within the nanotube is essentially ballistic [3]. In this case, the resistance of nanotube with ideal coupling to the two metal contacts at its ends is independent of length and is given by $h/(4e^2) \approx 6.45$ KΩ [26], [27]. However, for lengths greater than the mean free path, scattering leads to an additional ohmic resistance which increases with length as [29]

$$R_{CNT} = R_F \frac{\ell}{\lambda_{CNT}} = \left(\frac{h}{4e^2}\right)\frac{\ell}{\lambda_{CNT}} \tag{10}$$

where ℓ and λ_{CNT} are the length and the mean free path of CNT, respectively. This has also been confirmed by experimental observations [3], [28]. It should be noted that this additional scattering resistance would appear as a distributed resistance per unit length [24] in the equivalent circuit, to account for resistive losses along the CNT length

$$R_{CNT}(p.u.l) = \frac{R_F}{\lambda_{CNT}} = \left(\frac{h}{4e^2}\right)\frac{1}{\lambda_{CNT}}. \tag{11}$$

It is necessary to note that there are inconsistent results published in literature, both experimental and theoretical, regarding the dependency of resistance on length [30]. Some of these results indicate an exponential relationship [31], [32]

$$R = R_F \exp\left(\frac{\ell}{2\lambda_{CNT}}\right) \tag{12}$$

and some show a linear dependency [28], [33]

$$R = R_F\left(1 + \frac{\ell}{\lambda_{CNT}}\right). \tag{13}$$

It can be observed from (10)-(13) that the value of mean free path (MFP) plays an important role in determining the resistance of the carbon nanotube. It has been proven that the MFP of a CNT, both for metallic and semiconductor types, is proportional to the diameter [34], [35]. For the MFP of metallic CNTs, we have [34], [36]

$$\lambda_{CNT} = \left(\frac{\sqrt{3}\pi\phi^2}{2\sigma_\varepsilon^2 + 9\sigma_\phi^2} \right).D \tag{14}$$

where D is the diameter of the CNT, φ is the nearest neighbor tight-binding parameter, ε is the on-site energies, and σ_ε and σ_φ are the variances of ε and φ, respectively. For the MFP of semiconducting CNTs, we have [34], [35]

$$\lambda_{CNT} = \left(\frac{v_F}{\alpha T} \right).D \tag{15}$$

where v_F is the Fermi velocity, α is the coefficient of scattering rate, and T is the temperature. For a typical SWCNT with diameter 1 nm, the value of MFP has been reported about 1 µm based on measurements [37], [38]. Thus irrespective of the nature of SWCNTs (shells in an MWCNT), metallic or semiconducting, we can assume $\lambda_{CNT} \approx 1000D$ [34].

Fig. 11 shows the equivalent distributed circuit model of an individual CNT (shell in a multi-walled CNT)

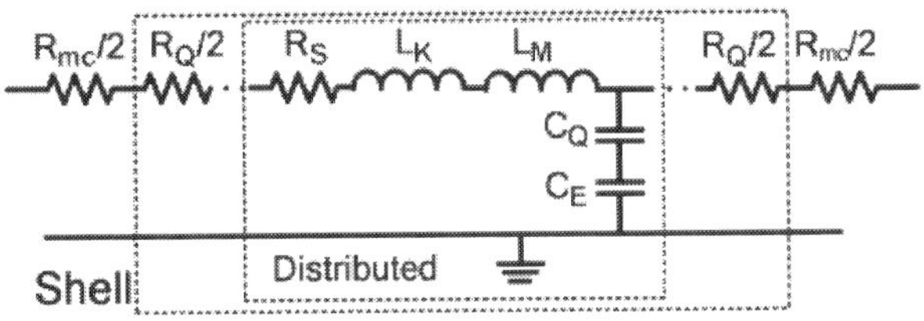

Fig. 11. Equivalent distributed circuit model of an individual CNT [34].

In this figure, R_{mc} is the imperfect contact resistance, R_Q is the quantum resistance (as the fundamental resistance R_F in Fig. 10), R_S is the scattering-induced resistance (as R_{CNT} in (10), (11)), L_K and L_M are the kinetic and magnetic inductances, respectively, and C_Q and C_E are the quantum and electrostatic capacitances, respectively.

The imperfect metal-to-nanotube contacts at each of the two ends of the nanotube, give rise to an additional resistance typically about 100 KΩ in series with the fundamental resistance R_F [24], [39].

3.2. CNT Capacitance

The total capacitance of a CNT arises from two sources: the electrostatic capacitance which is the intrinsic plate capacitance of an isolated CNT, and the quantum capacitance which accounts for the quantum electrostatic energy stored in the nanotube when it carries current [3], [26].

The electrostatic capacitance is calculated by treating the CNT as a thin wire placed away from a ground plane, as shown in Fig. 12, and its value per unit length is given by [3], [26], [30]

$$C_E = \frac{2\pi\varepsilon}{\cosh^{-1}\left(\frac{2y}{d}\right)} \approx \frac{2\pi\varepsilon}{\ln\left(\frac{y}{d}\right)} \tag{16}$$

where ε, d and y are the dielectric permittivity, the CNT diameter, and the distance of CNT from the ground plane, respectively.

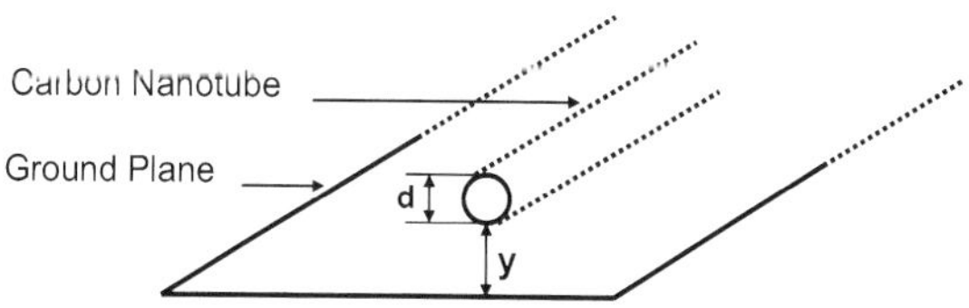

Fig. 12. Structure of an isolated CNT over a ground plane.

The electron cloud in a CNT can be assumed to be a quantum electron gas in one dimension. Hence, this follows Pauli's exclusion principle and it is not possible to add an electron with energy less than the Fermi energy of the system (E_F) [40]. The quantum capacitance is used to model the energy needed to add an electron at an available quantum state above the Fermi level [26]. By equating this energy to that of an effective capacitance, the expression for the quantum capacitance per unit length is obtained as [25]

$$C_Q = \frac{2e^2}{hv_F} \tag{17}$$

where v_F is the Fermi velocity in graphite [40] and is approximately 8×10^5 m/s [25]. Also for a CNT, C_Q is approximately 100 aF/µm [25], [26]. Since a CNT has four conducting channels as described before, the effective quantum capacitance resulting from four parallel capacitances is given by 4 C_Q.

In [34], the following relations for the quantum capacitance per unit length of a shell in a MWCNT have been expressed, according to the result in [25]

$$C_{Q/channel} = 2\times\frac{2e^2}{hv_F} \approx 193 \text{ aF}/\mu\text{m}, \tag{18}$$

$$C_{Q/shell} = C_{Q/channel} \times N_{shell}(D), \tag{19}$$

where

$$N_{shell}(D) \approx a.D + b \qquad D > 3 \text{ nm} \tag{20}$$

is the number of conducting channels (spin degeneracy is already considered) in any shell, D is the diameter of the shell, a = 0.0612 nm^{-1}, and b = 0.425.

On the other hand, in [38], the following relation for the number of conducting channels in any shell has been reported

$$N_{shell}(D) \approx a.D + b \qquad D > 3 \text{ nm}$$
$$\approx \frac{2}{3} \qquad D < 6 \text{ nm}. \tag{21}$$

It should be noted that the error introduced by (20), (21), due to different chiralities, is within 15% for all values of D [34], [38]. Note that the two regions in (21) have an overlap, and for 3 nm < D < 6 nm, both constant and linear functions can be used without any considerable error [38].

3.3. CNT Inductance

The total inductance of a CNT (L_{CNT} in Fig. 10) arises from two sources: the magnetic inductance and the kinetic inductance (L_M and L_K in Fig. 11). In the presence of a ground plane, the magnetic inductance per unit length is given by [25], [40]

$$L_M = \frac{\mu}{2\pi}\cosh^{-1}\left(\frac{2y}{d}\right) \approx \frac{\mu}{2\pi}\ln\left(\frac{y}{d}\right). \tag{22}$$

For a typical situation, the nanotube is placed on top of an insulating substrate (typically silicon dioxide), with a conducting medium below. A typical oxide thickness is between 10 nm and 1 μm with a typical nanotube radius of 1 to 2 nm. It can be noted that the magnetic inductance is a relatively weak function of the factor (y/d) and for typical geometries, it can be estimated to be around 1 pH/μm [40], [41].

In one-dimensional CNT conductors, apart from the magnetic inductance, another inductive component appears due to the kinetic energy of the electrons. The details of its derivation can be obtained in [40]-[42]. The kinetic inductance per unit length can be expressed as [25], [40], [41]

$$L_K = \frac{h}{2e^2 v_F}. \tag{23}$$

It is necessary to note that the four parallel conducting channels in a CNT give rise to an effective kinetic inductance of L_K /4. Also as it has been shown from (23), the kinetic inductance per unit length for a one dimensional CNT conductor is around 16 nH/μm [26], [40], more than 4 orders of magnitude larger than its magnetic counterpart L_M (≈ 1 pH/μm) [30], [40], [41], and will essentially play a vital role in high frequency applications [40].

In [34], the following relations for the kinetic inductance per unit length of a shell in a MWCNT have been expressed, according to the result in [25]

$$L_{K/channel} = \frac{h}{2e^2 v_F} \times \frac{1}{2} \approx 8\ \mathrm{nH}/\mu\mathrm{m}\,, \tag{24}$$

$$L_{K/shell} = \frac{L_{K/channel}}{N_{shell}(D)} \tag{25}$$

where N_{shell} (D) has been defined in (20), (21).

4. Different Structures of CNT as Interconnect

4.1. CNT Bundle as Interconnect

While SWCNTs have desirable material properties, individual nanotubes suffer from an intrinsic ballistic resistance of approximately 6.5 kΩ that is not dependent on the length of the nanotube [43]. As a result, the high resistance associated with an isolated CNT, causes excessive delay for interconnect applications. To alleviate the intrinsic resistance problem, bundles or ropes of CNTs conducting current in parallel, have been proposed and physically demonstrated as a possible interconnect medium for local, intermediate, and global interconnects [3], [43]. Fig. 13 shows a CNT bundle interconnect structure consists of a signal line and two ground return paths

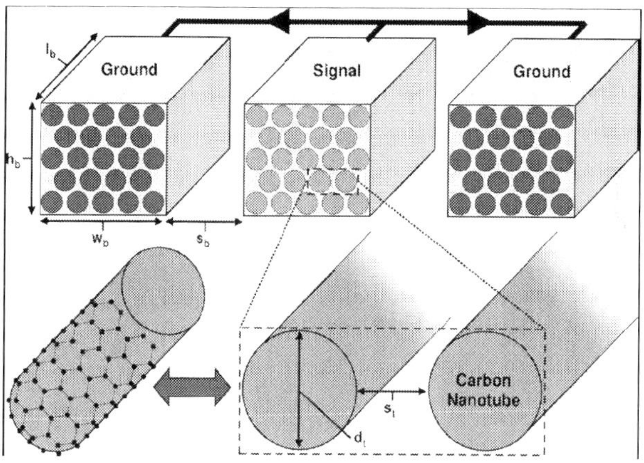

Fig. 13. System of SWCNT interconnect bundles implementing a signal line and two adjacent ground return paths [43].

Due to the lack of control on chirality, any bundle of CNTs consists of metallic as well as semi-conducting nanotubes. The required relations for the parameters of CNT bundles including the magnetic and kinetic inductances, the electrostatic and quantum capacitances, the fundamental and scattering resistances, can be obtained from [5]. In section 5 the time domain behaviour of a CNT bundle as interconnect is discussed based on [44].

4.2. Multi-walled CNT as Interconnect

Fig. 14 shows a geometric structure of a Multi-Walled carbon nanotube (MWCNT) over a ground plane.

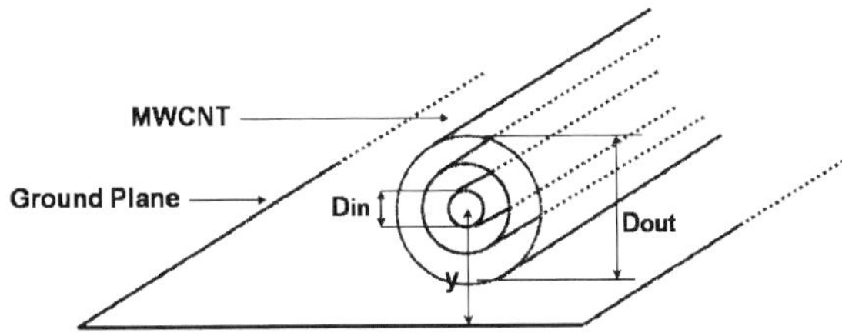

Fig. 14. Structure of a Multi-Walled CNT over a ground plane.

In this figure, D_{in} and D_{out} are the diameter of inner shell and the diameter of outer shell, respectively, and y is the height of inner shell from the ground plane. Recently wide spread studies regarding the benefits of the performance of MWCNTs as interconnect in comparison with CNT bundles and Cu have been performed. In [34] the performance of MWCNT interconnects has been analyzed and their circuit modelling has been discussed. Although MWCNT has an important role in the interconnect applications, the main scope of this chapter which follows in the subsequent section, is dedicated to the analysis of the behaviour of CNT bundle interconnects.

5. Time Domain Response

In [44] we have discussed about the time domain analysis of a CNT bundle interconnect in a driver-interconnect-load configuration and a new relation for the input-output transfer function in the related configuration has been extracted. A review of the discussion presented in [44] is brought in this section. Fig. 15 shows a CNT bundle interconnect with resistance, capacitance and inductance per unit length of R_S, C and L respectively, driven by a repeater of output resistance R_{tr} and output parasitic capacitance C_{out}, and driving an identical repeater with input capacitance equal to the load capacitance C_L.

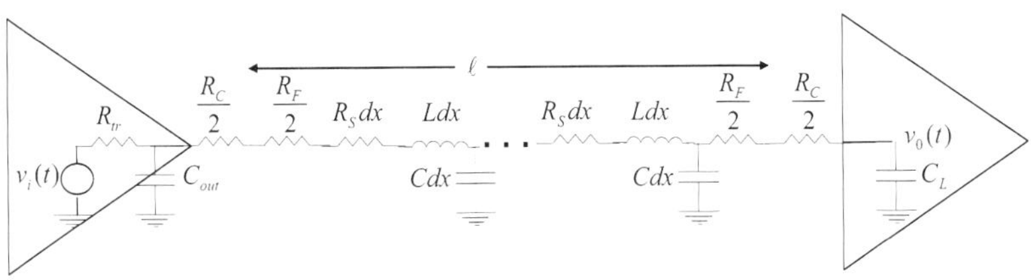

Fig. 15. Equivalent circuit of a driver-CNT bundle-load configuration, based on transmission line modeling [44].

In this figure, the CNT bundle interconnect has been modelled as a transmission line. For calculating the input-output transfer function of the configuration in Fig. 15, we need to derivate the total transmission parameter matrix. Using the ABCD transmission parameter matrix for a uniform RLC transmission line of length ℓ as given in [45]-[47], we can express the total ABCD transmission parameter matrix of the configuration in Fig. 15 as

$$T_{total} = \begin{bmatrix} 1 & R_{tr} \\ 0 & 1 \end{bmatrix} \begin{bmatrix} 1 & 0 \\ sC_{out} & 1 \end{bmatrix} \begin{bmatrix} 1 & R_{ex} \\ 0 & 1 \end{bmatrix} \begin{bmatrix} \cosh(\theta^T \ell) & Z_0^T \sinh(\theta^T \ell) \\ \frac{1}{Z_0^T} \sinh(\theta^T \ell) & \cosh(\theta^T \ell) \end{bmatrix} \begin{bmatrix} 1 & R_{ex} \\ 0 & 1 \end{bmatrix} \tag{26}$$

where $R_{ex} = (R_C + R_F)/2$, $Z_0^T = \sqrt{(R_S + sL)/(sC)}$, $\theta^T = \sqrt{(R_S + sL)sC}$, and $s = j\omega$ is the complex frequency. The elements of matrix T_{total} can be written, using (26), as

$$A_T = (1 + sR_{tr}C_{out})\cosh(\theta^T \ell) + \frac{(R_{tr} + R_{ex} + sR_{tr}R_{ex}C_{out})}{Z_0^T}\sinh(\theta^T \ell), \tag{27}$$

$$B_T = [R_{tr} + 2R_{ex} + 2sR_{tr}R_{ex}C_{out}]\cosh(\theta^T \ell) + [Z_0(1 + sR_{tr}C_{out}) + \frac{R_{ex}(R_{tr} + R_{ex} + sR_{tr}R_{ex}C_{out})}{Z_0^T}]\sinh(\theta^T \ell), \tag{28}$$

$$C_T = sC_{out}\cosh(\theta^T \ell) + \frac{1 + sR_{ex}C_{out}}{Z_0^T}\sinh(\theta^T \ell), \tag{29}$$

$$D_T = (1 + 2sR_{ex}C_{out})\cosh(\theta^T \ell) + [sC_{out}Z_0^T + \frac{R_{ex}(1 + sR_{ex}C_{out})}{Z_0^T}]\sinh(\theta^T \ell). \tag{30}$$

Therefore the input-output transfer function of the configuration in Fig. 15 can be written as

$$H(s) = \frac{V_0(s)}{V_i(s)} = \frac{1}{A_T + sC_L B_T} = \left(\begin{array}{l} [1 + s(R_{tr}C_{out} + R_{tr}C_L + 2R_{ex}C_L) + s^2(2R_{tr}R_{ex}C_{out}C_L)]\cosh(\theta^T \ell) \\ + \left[\begin{array}{l} \frac{R_{tr} + R_{ex}}{Z_0^T} + s\left(\frac{R_{tr}R_{ex}C_{out} + R_{ex}(R_{tr} + R_{ex})C_L}{Z_0^T} + Z_0^T C_L \right) \\ + s^2 \left(\frac{R_{tr}R_{ex}^2 C_{out}}{Z_0^T} + Z_0^T R_{tr} C_{out} \right) C_L \end{array} \right] \sinh(\theta^T \ell) \end{array} \right)^{-1}. \tag{31}$$

For simulation purposes, we need to extract a parametric linear approximation for (31). For this purpose, we need to calculate the equivalent linear terms for $\cosh(\theta^T \ell)$ and $\sinh(\theta^T \ell)/Z_0^T$, and put them in (31). Using the relation for the transfer function of a driver-CNT interconnect-load configuration, as discussed in [47], and rigorous calculations, we can extract the following linear expressions

$$\cosh(\theta^T \ell) = 1 + \left(\frac{R_S C \ell^2}{2!}\right) s + \left(\frac{L C \ell^2}{2!} + \frac{R_S^2 C^2 \ell^4}{4!}\right) s^2$$

$$+ \left(\frac{2 R_S L C^2 \ell^4}{4!} + \frac{R_S^3 C^3 \ell^3}{6!}\right) s^3 + \left(\frac{L^2 C^2 \ell^4}{4!} + \frac{3 R_S^2 L C^3 \ell^6}{6!} + \frac{R_S^4 C^4 \ell^8}{8!}\right) s^4, \tag{32}$$

$$\frac{\sinh(\theta^T \ell)}{Z_0^T} = sC\ell \left[1 + \frac{R_S C \ell^2}{3!} s + \left(\frac{L C \ell^2}{3!} + \frac{R_S^2 C^2 \ell^4}{5!}\right) s^2 + \left(\frac{R_S^3 C^3 \ell^6}{7!} + \frac{2 R_S L C^2 \ell^4}{5!}\right) s^3\right]. \tag{33}$$

With substituting the two terms $\cosh(\theta^T \ell)$ and $\sinh(\theta^T \ell) / Z_0^T$ in (31), by (32) and (33), we can obtain the linear parametric equivalent for the transfer function of (31) as

$$H(s) = \frac{1}{1 + a_1 s + a_2 s^2 + a_3 s^3 + a_4 s^4 + a_5 s^5 + a_6 s^6} \tag{34}$$

where

$$a_1 = R_{tr}(C_{out} + C\ell + C_L) + R_{ex}(C\ell + 2C_L) + R_S \ell \left(\frac{C\ell}{2!} + C_L\right), \tag{35}$$

$$a_2 = R_{ex} R_{tr} (C_{out} C\ell + C_L C\ell + 2 C_{out} C_L) + R_{tr} R_S C \ell^2 \left(\frac{C_{out} C\ell}{2!} + \frac{C^2 \ell^2}{3!} + C_{out} C_L + \frac{C\ell C_L}{2!}\right)$$

$$+ R_{ex} R_S C \ell^2 \left(\frac{C\ell}{3!} + C_L\right) + \frac{R_S^2 C^2 \ell^4}{4!} + \frac{L C \ell^2}{2!} + R_{ex}^2 C\ell C_L + \frac{R_S^2 C \ell^3 C_L}{3!} + L\ell C_L, \tag{36}$$

$$a_3 = \frac{2 R_S L C^2 \ell^4}{4!} + \frac{R_S^3 C^3 \ell^6}{6!} + \left(R_{ex} + R_{tr} + \frac{R_S}{C} C_L\right)\left(\frac{L C^2 \ell^3}{3!} + \frac{R_S^2 C^3 \ell^5}{5!}\right)$$

$$+ [R_{tr}(C_{out} + C_L) + 2 R_{ex} C_L]\left(\frac{L C \ell^2}{2!} + \frac{R_S^2 C^2 \ell^4}{4!}\right)$$

$$+ \frac{R_S C^2 \ell^3}{3!}\left(R_{ex} R_{tr}(C_{out} + C_L) + R_{ex}^2 C_L + \frac{L}{C} C_L + \frac{R_S}{C} R_{tr} C_{outL}\right)$$

$$+ R_{ex} R_{tr} C_{out} C\ell C_L (R_{ex} + R_S \ell) + R_{tr} C_{out} L\ell C_L, \tag{37}$$

$$
\begin{aligned}
a_4 = {} & \frac{L^2C^2\ell^4}{4!} + \frac{3R_S^2LC^3\ell^6}{6!} + \frac{R_S^4C^4\ell^8}{8!} + [R_{tr}(C_{out}+C_L)+2R_{ex}C_L]\left(\frac{2R_SLC^2\ell^4}{4!} + \frac{R_S^3C^3\ell^6}{6!}\right) \\
& + \left(R_{tr} + R_{ex} + \frac{R_S}{C}C_L\right)\left(\frac{2R_SLC^3\ell^5}{5!} + \frac{R_S^3C^4\ell^7}{7!}\right) + 2R_{tr}R_{ex}C_{out}C_L\left(\frac{LC\ell^2}{2!} + \frac{R_S^2C^2\ell^4}{4!}\right) \\
& + \left(R_{tr}R_{ex}(C_{out}+C_L) + R_{ex}^2C_L + \frac{L}{C}C_L + \frac{R_S}{C}R_{tr}C_{out}C_L\right)\left(\frac{LC^2\ell^3}{3!} + \frac{R_S^2C^3\ell^5}{5!}\right) \\
& + \frac{R_SC^2\ell^3}{3!}\left(R_{tr}R_{ex}^2C_{out}C_L + \frac{L}{C}R_{tr}C_{out}C_L\right),
\end{aligned}
\tag{38}
$$

$$
\begin{aligned}
a_5 = {} & [R_{tr}(C_{out}+C_L)+2R_{ex}C_L]\left(\frac{L^2C^2\ell^4}{4!} + \frac{3R_S^2LC^3\ell^6}{6!} + \frac{R_S^4C^4\ell^8}{8!}\right) \\
& + 2R_{tr}R_{ex}C_{out}C_L\left(\frac{2R_SLC^2\ell^4}{4!} + \frac{R_S^3C^3\ell^6}{6!}\right) \\
& + \left(R_{tr}R_{ex}(C_{out}+C_L) + R_{ex}^2C_L + \frac{L}{C}C_L + \frac{R_S}{C}R_{tr}C_{out}C_L\right)\left(\frac{2R_SLC^3\ell^5}{5!} + \frac{R_S^3C^4\ell^7}{7!}\right) \\
& + \left(R_{tr}R_{ex}^2C_{out}C_L + \frac{L}{C}R_{tr}C_{out}C_L\right)\left(\frac{LC^2\ell^3}{3!} + \frac{R_S^2C^3\ell^5}{5!}\right),
\end{aligned}
\tag{39}
$$

$$
\begin{aligned}
a_6 = {} & 2R_{tr}R_{ex}C_{out}C_L\left(\frac{L^2C^2\ell^4}{4!} + \frac{3R_S^2LC^3\ell^6}{6!} + \frac{R_S^4C^4\ell^8}{8!}\right) \\
& + \left(R_{tr}R_{ex}^2C_{out}C_L + \frac{L}{C}R_{tr}C_{out}C_L\right)\left(\frac{2R_SLC^3\ell^5}{5!} + \frac{R_S^3C^4\ell^7}{7!}\right).
\end{aligned}
\tag{40}
$$

Fig. 16 shows the step response of configuration in Fig. 15, for 32 nm technology node, using our extracted linear transfer function of (34), and HSPICE simulation.

The repeater size has been assumed 174 times larger than the minimum sized repeater, which its parameters have been extracted from ITRS 2007 [2]. Also the load capacitance has been considered equal to the input capacitance of repeater. Recall that $\lambda_{CNT} \approx 1000D$ [34] where D is the diameter of each individual CNT, the mean free path of CNT (λ_{CNT}) has been assumed 1 μm. As it is cleared from Fig. 16, there is an excellent match between the result of our extracted parametric transfer function and HSPICE simulation result. Fig. 17 shows the step response of configuration in Fig. 15, using our extracted linear parametric transfer function of (34), for the contact resistance values from 1 kΩ to 50 kΩ. As shown in Fig. 17, the propagation delay increases from 0.138 ns to 5.58 ns, with the increase of contact resistance value from 1 kΩ to 50 kΩ.

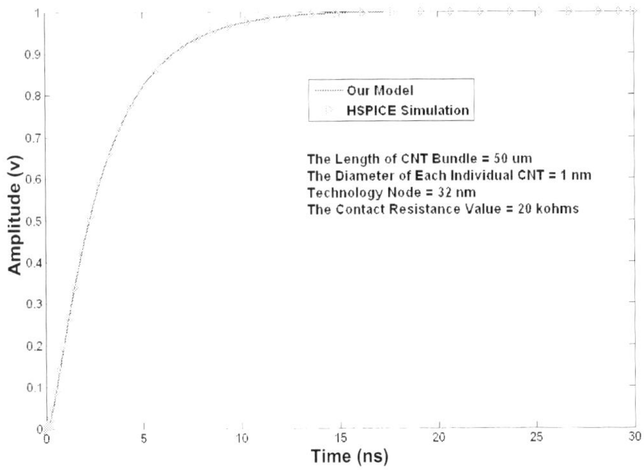

Fig. 16. The step response of configuration in Fig. 15, using our extracted linear parametric transfer function, and HSPICE simulation. In this figure, the number of transmission line model sections for HSPICE simulation, has been considered 400 [44].

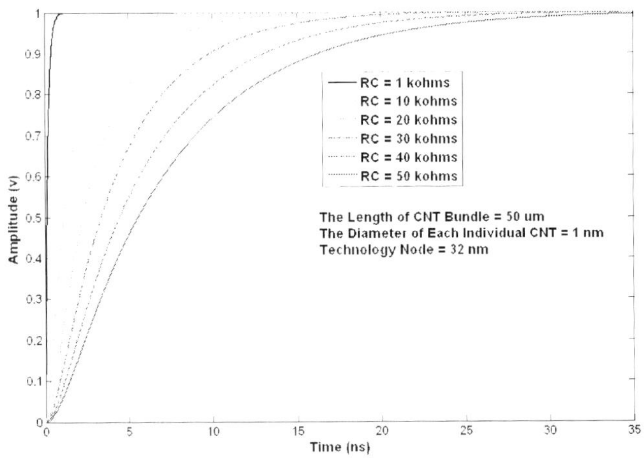

Fig. 17. The step response of configuration in Fig. 15, using our extracted linear parametric transfer function, for various values of the contact resistance [44].

Fig. 18 shows the propagation delay of configuration in Fig. 15, using our extracted linear parametric transfer function of (34), versus the contact resistance value, and for the CNT bundle lengths 50 μm, 200 μm, 500 μm and 1000 μm.

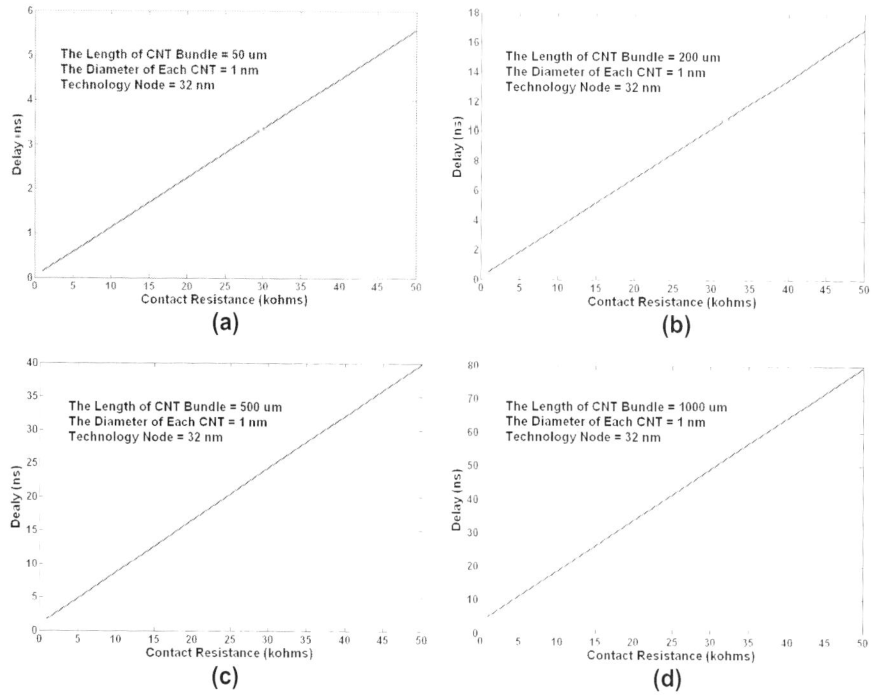

Fig. 18. The propagation delay of configuration in Fig. 15, using our new linear parametric transfer function, versus the contact resistance value, for the CNT bundle lengths: (a) 50 μm, (b) 200 μm, (c) 500 μm, (d) 1000 μm [44].

In this figure, the diameter of each individual CNT has been chosen 1 nm, and therefore as discussed before, the mean free path of CNT will be 1 μm. As shown in Fig. 18, for the length of CNT bundle equal to 50 μm, the propagation delay changes from 0.138 ns to 5.58 ns for the contact resistance values from 1 kΩ to 50 kΩ, i.e. a variation range of 39.43 times the minimum value. The related delay variation ranges for the length values 200 μm, 500 μm, and 1000 μm, are 31.37, 22.61, and 15.42 times the minimum value, respectively. This means that, the impact of the contact resistance on the propagation delay, decreases with the increase of the bundle length. The reason is that, with the increase of the bundle length, the role of scattering resistance which increases with the length [29], would be more important.
In Fig. 19, the nyquist diagrams for a driven CNT bundle interconnect, versus the length of CNT bundle and the diameter of each individual CNT, have been plotted using MATLAB [23].

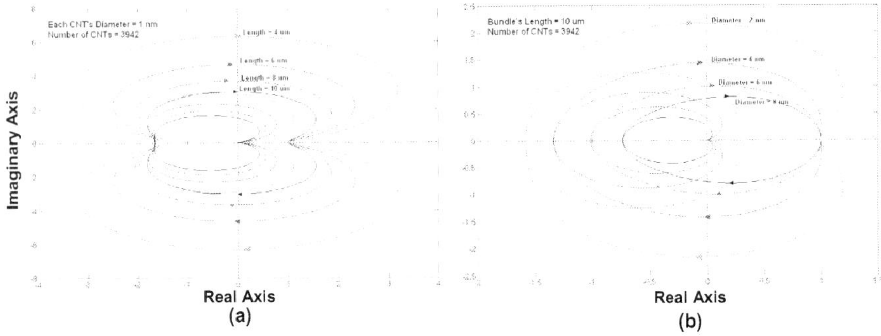

Fig. 19. The nyquist stability diagrams for a driven CNT bundle interconnect; (a) versus length; (b) versus diameter. In this figure, the space between two adjacent CNTs has been chosen as 0.34 nm, and the length, the width and the thickness of CNT bundle, have been considered 10 µm, 50 nm and 125 nm, respectively. Also the load capacitance has been assumed equal to 5 fF [44].

As shown in Fig. 19 (a), by increasing the length of CNT bundle, the complex point (-1,0) goes toward outside the diagram. So, by increasing the length of CNT bundle, the system becomes more stable. As shown in Fig. 19 (b), by increasing the diameter of each individual CNT, the complex point (-1,0) goes toward outside the diagram and then, the diagram goes farther from this point. So, by increasing the diameter of each individual CNT, the system becomes more stable. It should be noted that in simulations of Fig. 19, the driver has been considered ideal with perfect contacts, and all individual CNTs in the bundle have been assumed metallic. A more detailed discussion about the stability analysis in CNT interconnects has been presented in [48].

6. Summary

In this chapter, we have studied interconnect challenges and the behaviour of carbon nanotube (CNT) as interconnect in VLSI circuits. In this review we discussed about the two main structures of CNT, including CNT bundles and MWCNTs, which achieve good performance due to the parallel SWCNTs in a bundle or the parallel shells in a MWCNT. These optimized configurations give the better characteristics including decreased delay time in comparison with SWCNTs, which is a vital parameter for the application as interconnect. The repeater (buffer) insertion technique that is used for the reduction of delay time in the global interconnects, has been discussed. Also in this chapter, we analyzed the time domain response of CNT bundle interconnect in a driver-interconnect-load configuration, based on the formulations and discussions we have presented in the reference [44]. At the continuation, we discussed briefly about the stability concept in CNT bundle interconnects, versus the length and diameter of each CNT in a CNT bundle.

7. References

[1] A. Naeemi, R. Sarvari, J. D. Meindl, " Performance Comparison Between Carbon Nanotube and Copper Interconnects for Gigascale Integration (GSI)", *IEEE Electron Device Letters*, vol. 26, No. 2, pp. 84-86, Feb. 2005.

[2] International Technology Roadmap for Semiconductors (ITRS), 2007.

[3] N. Srivastava, K. Banerjee, "Performance Analysis of Carbon Nanotube Interconnects for VLSI Applications", *IEEE Int. Conf. on Computer-Aided Design (ICCAD)*, pp. 383-390, CA, Nov. 2005.

[4] N. Srivastava, K. Banerjee, "A Comparative Scaling Analysis of Metallic and Carbon Nanotube Interconnections for Nanometer Scale VLSI Technologies", *Proc. 21st Intl. VLSI Multilevel Interconnect Conf.*, pp. 393-398, 2004.

[5] K. Banerjee, N. Srivastava, "Are Carbon Nanotubes the Future of VLSI Interconnections?", *ACM Design Automation Conf. (DAC)*, pp. 809-814, CA, Jul. 2006.

[6] J. Li. et al., "Bottom-Up Approach for Carbon Nanotubes Interconnects", *Appl. Phys. Lett.*, vol. 82, pp. 2491-2493, Apr. 2003.

[7] A. Naeemi, J. D. Meindl, "Design and Performance Modeling for Single-Walled Carbon Nanotubes as Local, Semiglobal, and Global Interconnects in Gigascale Integrated Systems", *IEEE Trans. Electron Devices*, vol. 54, No.1, pp. 26-37, Jan. 2007.

[8] P. L. McEuen, et al., "Single-Walled Carbon Nanotube Electronics", *IEEE Trans. Nanotechnology*, vol. 1, No. 1, pp. 78-85, 2002.

[9] Min. Tang, Jun-Fa Mao, "Optimization of Global Interconnects in High Performance VLSI Circuits", *Proc. 19th Int. Conf. VLSI Design, 2006.*

[10] D. Fathi, B. Forouzandeh, "Accurate Analysis of Global Interconnects in Nano FPGAs", *Nano*, pp. 171-176, Vol. 4, No. 3, 2009.

[11] V. Prasad, M.P. Desai, "Interconnect delay minimization using a novel pre-mid-post buffer strategy", *Proc. 16th Int. Conf. VLSI Design*, pp. 417-422, Jan. 2003.

[12] Jan M. Rabaey, Anantha Chandrakasan, "Digital Integrated Circuit", 2nd Edition, Chapters 4 & 9, 2006.

[13] A.Naeemi, R.Venkatesan J.d.mendlS, "SYSTEM-ON-CHIP GLOBAL INTERCONNECT OPTIMIZATION", *Proc. ASIC/SoC Conf., Sep. 2002, pp. 399-403.*

[14] M. Mui, K. Banerjee, "A Global Interconnect Optimization Scheme for Nanometer Scale VLSI With Implications for Latency, Bandwidth, and Power Dissipation", *IEEE Trans. ELECTRON DEVICES, vol. 51, NO. 2, pp. 195–203, Feb.* 2004.

[15] K. Banerjee, Amit Mehrotra, "A Power-Optimal Repeater Insertion Methodology for Global Interconnects in Nanometer Designs", *IEEE Trans. Electron Devices, vol. 49, No. 11, pp. 2001-2007, Nov. 2002.*

[16] R. Venkatesan, J. A. Davis, J. d. Mendil, "Compact Distributed RLC Interconnect Models-Part IV: Unified Models for Time Delay, Crosstalk, and Repeater Insertion", *IEEE Trans. ELECTRON DEVICES, vol. 50, NO. 4, pp. 1094–1102, Apr. 2003.*

[17] B. Bakoglu, "Circuits, Interconnections and Packaging for VLSI", *Addison-Wesley, 1990.*

[18] D. Fathi, B. Forouzandeh, A New Repeater Insertion Technique for Optimization of Global Interconnects in Nano VLSI", *NANO*, To be published.

[19] Shyh-Chyi Wong, Gwo-Yann Lee, "Modeling of interconnect capacitance, delay, and crosstalk in VLSI", *IEEE Trans. Semiconductor Manufacturing*, vol. 13, Issue 1, pp. 108-111, Feb. 2000.

[20] A. B. Kahng, K. Masuko, S. Muddu, "Analytical delay models for VLSI interconnects under ramp input", *Proc. ACM/IEEE Int. Conf. Computer-Aided Design, pp. 30-36, Nov. 1996.*

[21] W. C. Elmore, "The Transient Analysis of Damped Linear Networks with Particular Regard to Wideband Amplifiers", *J. Appl. Phys.*, vol. 19, No. 1, pp. 55-63, Jan. 1948.

[22] X. Li, J. Mao, H. Huang, Y.Liu, "Global Interconnect Width and Spacing Optimization for Latency, Bandwidth and Power Dissipation," *IEEE Trans. Electron Devices, vol. 52, No. 10, pp. 2272-2279, Oct. 2005.*

[23] Mathematics Laboratory (MATLAB), 2007.

[24] N. Srivastava, R. V. Joshi, K. Banerjee, "Carbon Nanotube Interconnects: Implications for Performance, Power Dissipation and Thermal Management", *IEEE Int. Electron Devices Meeting (IEDM)*, Washington DC, pp. 257-260, Dec. 2005.

[25] P. J. Burke, "Luttinger Liquid Theory as a Model of the Gigahertz Electrical Properties of Carbon Nanotubes", *IEEE Trans. Nanotechnology*, vol. 1, No. 3, pp. 129-144, Sep. 2002.

[26] K. Banerjee, N. Srivastava, "Are Carbon Nanotubes the Future of VLSI Interconnections?", *ACM Design Automation Conf. (DAC)*, San Francisco, CA, pp. 809-814, Jul. 2006.

[27] S. Datta, "Electrical Resistance: An Atomistic View", *Nanotechnology*, vol. 15, S433-S451, 2004.

[28] J. Y. Park, et al., "Electron-Phonon Scattering in Metallic Single-Walled Carbon Nanotubes", *Nano Letters*, vol. 4, No. 3, pp. 517-520, 2004.

[29] S. Datta, "Electronic Transport in Mesoscopic Systems", *Cambridge Univ. Press*, 1995.

[30] A. Naeemi, R. Sarvari, J. D. Meindl, "Performance Comparison Between Carbon Nanotube and Copper Interconnects for Gigascale Integration (GSI)", *IEEE Electron Device Letters*, vol. 26, No.2, pp. 84-86, Feb. 2005.

[31] P. J. de Pablo, et al., "Nonlinear Resistance Versus Length in Single-Walled Carbon Nanotubes", *Phys. Rev. Lett.*, vol. 88, pp. 036804/1-036804/4, Jan. 2002.

[32] A. Andriotis, et al., "Non-Linear Resistance Dependence on Length in Single-Wall Carbon Nanotubes", *Nano Lett.*, vol. 3, pp. 131-134, 2003.

[33] S. Li, et al., "Electrical Properties of 0.4 cm Long Single-Walled Carbon Nanotubes", *Nano Lett.*, vol. 4, pp. 2003-2007, 2004.

[34] H. Li, et al., "Circuit Modeling and Performance Analysis of Multi-Walled Carbon Nanotube Interconnects", *IEEE Trans. Electron Devices*, vol. 55, No. 6, pp. 1328-1337, Jun. 2008.

[35] X. Zhou, et al., "Band Structure, Phonon Scattering, and the Performance Limit of Single-Walled Carbon Nanotube Transistors", *Phys. Rev. Lett.*, vol. 95, No. 14, p. 146805, Sep. 2005.

[36] J. Jiang, et al., "Universal Expression for Localization Length in Metallic Carbon Nanotubes", *Phys. Rev. B, Condens. Matter*, vol. 64, No. 4, p. 045409, Jul. 2001.

[37] A. Naeemi, J. D. Meindl, "Design and Performance Modeling for Single-Walled Carbon Nanotubes as Local, Semiglobal, and Global Interconnects in Gigascale Integrated Systems", *IEEE Trans. Electron Devices*, vol. 54, No.1, pp. 26-37, Jan. 2007.

[38] A. Naeemi, J. D. Meindl, "Compact Physical Models for Multiwall Carbon-Nanotube Interconnects", *IEEE Electron Device Lett.*, vol. 27, No. 5, pp. 338-340, May 2006.

[39] A. Naeemi, J. D. Meindl, "Mono-Layer Metallic Nanotube Interconnects: Promising Candidates for Short Local Interconnects", *IEEE Electron Device Letters*, vol. 26, No. 8, pp. 544-546, Aug. 2005.

[40] A. Raychowdhury, K. Roy, "Modeling of metallic Carbon-Nanotube Interconnects for Circuit Simulations and a Comparison With Cu Interconnects for Scaled Technologies", *IEEE Trans. Computer-Aided Design of Integrated Circuits and Systems*, vol. 25, No. 1, pp. 58-65, Jan. 2006.

[41] P. J. Burke, "An RF Circuit Model for Carbon Nanotubes", *IEEE Trans. Nanotechnol.*, vol. 2, No. 1, pp. 55-58, Mar. 2003.

[42] M. W. Bockrath, "Carbon Nanotubes: Electrons in One Dimension", *Ph.D. Dissertation*, Dept. Phys., Univ. California, Berkeley, 1999.

[43] A. Nieuwoudt, Y. Massoud, "Evaluating the Impact of Resistance in Carbon Nanotube Bundles for VLSI Interconnect Using Diameter-Dependent Modeling Techniques", *IEEE Trans. Electron Devices*, vol. 53, No. 10, pp. 2460-2466, Oct. 2006.

[44] D. Fathi, B. Forouzandeh, S. Mohajerzadeh, R. Sarvari, "Accurate Analysis of Carbon Nanotube Interconnects Using Transmission Line Model", *Micro & Nano Letters*, vol. 4, No. 2, pp. 116-121, 2009.

[45] K. Banerjee, A. Mehrotra, "Analysis of On-Chip Inductance Effects for Distributed RLC Interconnects", *IEEE Trans. Computer-Aided Design of Integ. Circuits and Systems*, vol. 21, No. 8, pp. 904-915, Aug. 2002.

[46] A. K. Palit, et al., "Reduced Order Long Interconnect Modeling", *15. ITG-GI-GMM Workshop Testmethoden und Zuverlässigkeit von Schaltungen und Systemen*, Timmendorfer Strand, pp. 42-47, Mar. 2003.

[47] D. Fathi, B. Forouzandeh, "Time Domain Analysis of Carbon Nanotube Interconnects Based on Distributed RLC Model", *Nano*, vol. 4, No. 1, pp. 13-21, Feb. 2009.

[48] D. Fathi, B. Forouzandeh, "A Novel Approach for Stability Analysis in Carbon Nanotube Interconnects", *IEEE Electron Device Letters*, vol. 30, No. 5, pp. 475-477, May 2009.

16

Affinity of CNT for metal – Its importance to application: Molecular dynamics approach

Shuhei Inoue
Hiroshima University
Japan

1. Introduction

As we know, carbon nanotube (especially single-walled carbon nanotube: SWCNT) possesses outstanding properties such as high thermal conductivity, ultra high physical strength, durability, and so forth. Therefore, we expect SWCNT to be a candidate for future material that can realize all of our hopes. However, thinking about its application, the affinity for metal and/or the bonding to metal species are inevitably important problems and those still remain to be unsolved. If there are some discontinuities between SWCNT and metal species, these discontinuities must result in thermal resistance and electrical resistance; that mean we loose the reason and advantage of using SWCNT. If SWCNT can be coated with metal, some new properties may arise, but there is no guarantee that SWCNT can maintain its outstanding properties. These kinds of new approach for application have been partially reported experimentally; however, their mechanisms are not solved. For example, some metal species were deposited on single-walled carbon nanotube forests (Ishikawa et al., 2007) and an isolated single-walled carbon nanotube was coated by metals using e-beam (Zhang et al., 2000). But to investigate the properties of functionalized-SWCNT (metal coated SWCNT) experimentally is extremely difficult. In this chapter molecular dynamics simulation (MD) easily tells us the answers for above-mentioned questions.

The classical MD is merely solving the Newtonian equation according to the forces affected among atoms, but provided the potential function is appropriate, it gives us quite proper perspectives. In this chapter I mention how to build up the appropriate potential function and the coincidence with experimental results of metal deposition on SWCNT. Furthermore, I refer to the perspectives of physical properties of the functionalized-SWCNT.

2. Computational methods

2.1 Driving the potential function

The Brenner potential was applied for carbon-carbon interaction and a classical many-body potential was applied for metal-metal and carbon-metal interactions as a function of the bond order potential. The parameter sets for nickel-nickel and carbon-nickel were derived as a function of coordination number (Shibuta & Maruyama, 2007) while those for gold-gold

atom	S	β (Å^{-1})	D_{e1} (eV)	D_{e2} (eV)	C_D	R_{e1} (Å)	R_{e2} (Å)	C_R	R_1 (Å)	R_2 (Å)
Fe–Fe*	1.3	1.2173	0.4155	0.8392	0.8730	2.627	0	–	2.7	3.2
Ni–Ni*	1.3	1.570	0.4217	1.0144	0.8268	2.4934	0.1096	0.3734	2.7	3.2
Ti–Ti	1.3	2.331	0.6500	0.2001	0.3700	3.819	2.411	0.2357	2.7	3.2
Au–Au	1.3	1.750	0.5290	1.3510	1.5251	3.3043	0.7573	0.2939	2.9	3.5

atom	De (eV)	S	β (Å^{-1})	R_e (Å)	R_1 (Å)	R_2 (Å)	b	δ
Fe–C*	3.3249	1.3	1.5284	1.7304	2.7	3.0	0.0656	-0.4279
Ni–C*	2.4673	1.3	1.8706	1.7628	2.7	3.0	0.0688	-0.5351
Ti–C	3.6240	1.3	1.466	1.8000	2.7	3.0	0.0600	-0.5000
Au–C	2.1840	1.3	2.0745	1.9030	2.9	3.3	0.0970	-0.5350

Table 1. Potential parameters for MD simulations

and carbon-gold were derived by density functional theory (DFT) calculations. Gaussian 03 was employed in this study, and Becke's three-parameter exchange functional with the Lee-Yang-Parr correlation functional (B3LYP) (Beche, 1993 and Lee et al. 1998) was applied. The Los Alamos effective core potential with DZ (LANL2DZ) (Hay et al. 1985) was used as the basis set.

Because the method for deriving parameters is similar to abovementioned paper, it is briefly explained here. First, we assume the symmetrical structure of Aux and AuCx clusters that means we do not take into account the Jahn-Teller distortion (Castro et al., 1997) and calculate their total energy. The binding energy is calculated by subtracting isolated energies from the total energy and dividing its result by the number of bonds, as shown in Eq. 1:

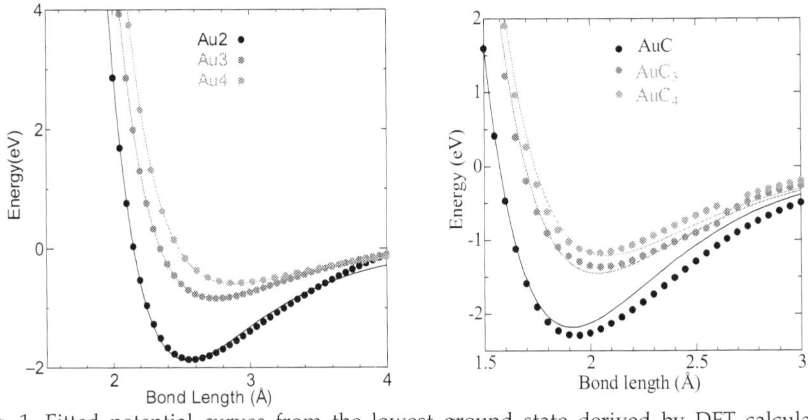

Fig. 1. Fitted potential curves from the lowest ground state derived by DFT calculation (Gaussian03, B3LYP/LANL2DZ). Dots show the binding energies, and the fitted curves are the bond order potential curves. In each case, various spin states are examined and their minimum values are employed.

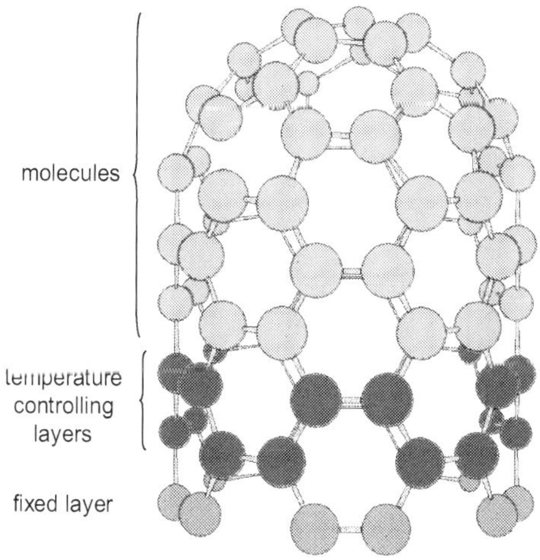

Fig. 2. Surface consisting of a capped carbon nanotube. The calculated system consists of 6 × 6 units (2880 atoms) and has dimensions of 60 Å × 52 Å × 60 Å.

$$De = \left(E^{total}_{min.} - E^{isolated}_{m,min.} - N_c \cdot E^{isolate}_{c,min.}\right) / N_b \tag{1}$$

where $E^{total}_{min.}$ denotes the total energy at a certain spin state, $E^{isolated}_{m,min.}$ is the isolated energy for the metal, N_c is the coordination number of carbon, $E^{isolate}_{c,min.}$ is the isolated energy for carbon, and N_b is the bond number. In each case, several spin states have been calculated, and their minimum values corresponding to some distance are employed for the fitting. The clusters might take various spin states, but we employ the spin state with the minimum energy. Figure 1 shows the binding energies of gold-gold and carbon-gold derived from DFT calculation and fitted by the potential functions of the bond order potential shown in Eqs. (2–9).

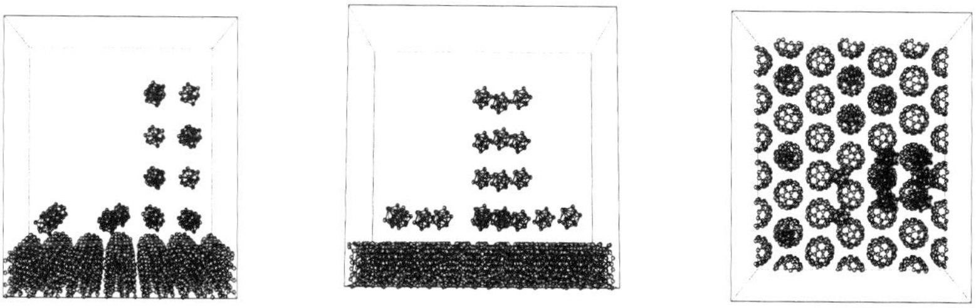

Snapshot from x direction   Snapshot from y direction   Snapshot from z direction

Fig. 3. Cluster configuration onto the flat CNT surface. The center of gravity of each cluster is above the center of the carbon nanotube or its hollow site. To express continuous deposition, some clusters are placed at distances of 10 Å vertically.

$$E = V_R - V_A \tag{2}$$

$$V_R = \frac{D_e}{S-1}\exp\left\{-\beta\sqrt{2S}\left(r-R_e\right)\right\} \tag{3}$$

$$V_A = B^* \frac{D_e S}{S-1}\exp\left\{1+b\left(N^C-1\right)\right\}^{\delta} \tag{4}$$

$$D_e = D_{e1} + D_{e2}\exp\left\{-C_D\left(N^M-1\right)\right\} \tag{5}$$

$$R_e = R_{e1} - R_{e2}\exp\left\{-C_R\left(N^M-1\right)\right\} \tag{6}$$

$$f(r)=\begin{cases} 1 & (r<R_1) \\ 0.5\cdot\left(1+\cos\dfrac{r-R_1}{R_2-R_1}\pi\right) & (R_1<r<R_2) \\ 0 & (R_2<r) \end{cases} \tag{7}$$

$$N_i^C = 1 + \sum_{carbon} f\left(r_{ik}\right) \tag{8}$$

$$N_i^M = 1 + \sum_{metal} f\left(r_{ik}\right) \tag{9}$$

where r denotes the distance between two atoms and V_R and V_A are the Morse-type potential (Morse, 1929) repulsion and attraction terms, respectively. R_e represents the equilibrium bond length, and D_e is the potential depth at $r = R_e$. S means the ratio of the effective repulsive and attractive terms. N_i^C and N_i^M are the coordination numbers, which are derived by using the cut-off function of $f(r)$. All these parameters are shown in Table 1.

2.2 Physical vapor deposition onto vertically aligned single-walled carbon nanotube

The vertically aligned single-walled carbon nanotube (VA-SWCNT) surface is expressed by a capped short carbon nanotube with a (5, 5) chirality, as shown in Fig. 2. In this paper, the calculated system consists of 6 × 6 units with dimensions of 60 Å × 52 Å × 60 Å, and their configuration resembles the FCC lattice (111). Atoms interact with each other within their cut-off distances. The first layer is a fixed layer, while the second and third layers are temperature-controlling layers. These layers maintain their temperature by velocity scaling. The system has periodic boundary conditions in each direction and the velocity Verlet algorithm is employed. Assuming a realistic metal evaporation in a vacuum chamber, physical vapor deposition (PVD) onto a VA-SWCNT surface is expressed by the collision of each M_{13} cluster with small evaporation energy (kinetic energy, K.E.) of 10.0 meV. Each metal cluster is fully annealed above its melting point (Ni: 1400 K, Au: 1000 K) and set as shown in Fig. 3. The VA-SWCNT is also annealed at 300 K. The melting point of the bulk expressed by these bond order potentials is higher than these temperatures, but the M_{13} cluster is very small and therefore its melting temperature is less than that of the bulk (Ni: 1728 K, Au: 1338 K) (Kittel, 2004).

2.3 Coating nanotube with metal

The dimensions of the calculated system are 60 Å × 100 Å × 100 Å, and a periodic boundary condition is applied in each direction. As shown in Fig. 4, an isolated SWCNT is 60 Å in

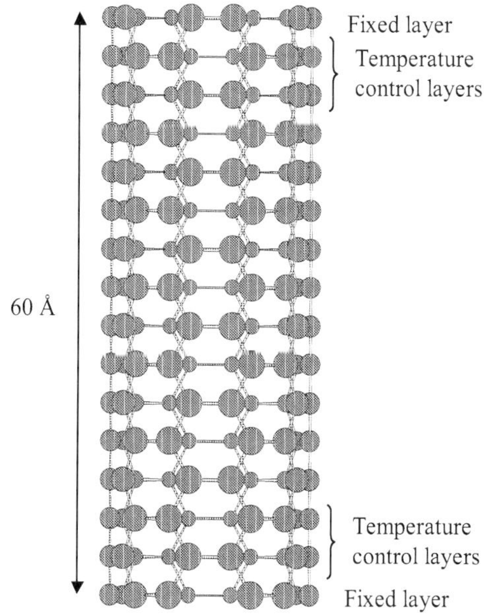

Fig. 4. Isolated single-walled carbon nanotube. The both ends are fixed layer and following two layers work to control the temperature.

length with (5, 5) chirality; both its ends are fixed, and the next two layers works to control its temperature. Before coating this carbon nanotube with a metal, it is annealed at 300 K by velocity scaling technique. The molecules of the coating metal are expressed as M_{13} clusters; this expression is suitable for actual experiment because in the conventional metal evaporation process, the average metal particle size is approximately a few angstroms in diameter. This metal cluster is also fully annealed at 1200 – 1400 K; the annealing temperature is clearly higher than their melting temperature but less than their boiling temperature. The melting temperatures in the bulk form expressed by these potential parameters are approximately 1800 – 2000 K; however, in this paper, the metal clusters under study are too small to exhibit bulk properties. In order to achieve continuous metal evaporation coating, the fully annealed metal clusters are distributed at appropriate distances (15 – 20 Å) and made to collide with the carbon nanotube at low evaporation energy of 10 meV or 30 meV.

3. Results and discussion

3.1 PVD on VA-SWCNT surface

The initial configuration of the metal clusters is above mentioned and shown in Fig. 3. Each cluster is placed just above the carbon nanotube, or its hollow site for expressing various collision models. In order to express continuous deposition, the clusters are placed at distances of 10 Å. Figure 5 shows the result of PVD onto a VA-SWNTs surface. In both cases, some of the metal clusters collided with each other before arriving at the surface and formed a larger cluster, while others arrived at the surface without clustering. In the case of nickel deposition, irrespective of clustering, the deposited nickel atoms spread over the VA-SWCNT surface and formed a smooth surface; however, in the case of gold clusters, small clusters of gold could spread over the VA-SWCNT surface, but large clusters could not, and they formed a grain-like structure. On comparing the potential curves of both metals, the equilibrium bond length is slightly different, but the binding energy appears to be almost similar. The point at which they differ the most is the potential depth at which their coordination number is one. The potential depth of gold suddenly decreases as the coordination number increases. This could be a reason for the formation of the grain-like structure, because the contact angle is closely related to the wettability, which is well explained by solid-liquid affinity (ε_{sl} / ε_{l}) (Maruyama et al., 1998).

According to their results the contact angle tends to become larger as the solid-liquid affinity becomes smaller. In the condensed phase the coordination number should be larger, so that the affinity becomes smaller comparing to that of nickel. The electron configuration of an isolated gold atom is [$4f^{14}$, $5d^{10}$, $6s^1$] and that of a nickel atom is [$3d^8$, $4s^2$]. Therefore, it is speculated that the number of electrons that greatly contribute to binding should be one for gold and more than two for nickel. The order of energy levels of 6s, 4f, and 5d are this order but they are not so far with each other; thus sometimes this order can be reversal for taking a closed shell. In case of gold the 4f-orbit and the 5d-orbit take a closed shell as they

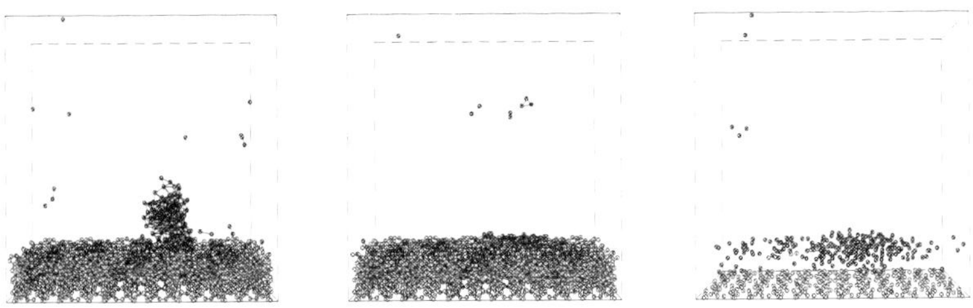

(a) Deposition of nickel clusters.

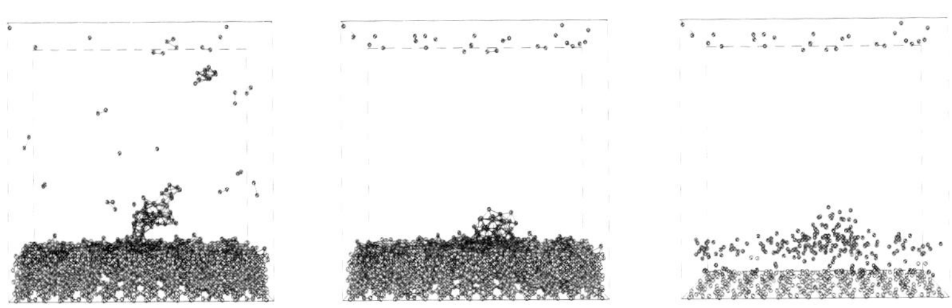

(b) Deposition of gold clusters.

Fig. 5. Metal PVD onto VA-SCNT surface. In both cases, small metal clusters collide with each other before arriving at the VA-SWCNT surface and form large clusters. The carbon atoms except for those in the fixed layer are hidden in the third figure to identify the metal atoms.

are; therefore, lone electron is one. Thus, the gold atoms have a strong binding energy in the form of a dimmer. On the other hand, the nickel atom has some free electrons, so that their binding energy is smaller than that of the gold atom. Compared to nickel deposition, gold deposition may require stringent conditions to form a flat surface; however, this can be achieved by carrying out deposition under a high vacuum condition and at a low deposition rate. Small clusters of gold can spread over the VA-SWCNT surface, while some of the gold clusters collide with each other before arriving at the surface and form a grain-like structure; it is thus essential to avoid clustering.

3.2 Coating

Figure 6 shows the simulation result of continuous metal evaporation coating ((a) Ti, (b) Fe, (c) Ni, (d) Au). In this simulation, 24 M_{13} clusters were gently collided with one SWCNT. The impact direction was perpendicular to the axis of the carbon nanotube, and at each impact point, about four clusters collided simultaneously. In case of titanium and iron clusters (Fig. 6(a), (b)), the metal atoms seemed to firmly combine with carbon atoms; thus, the metal atoms entered the carbon nanotube and distorted its structure. The strong interaction between the metal atoms and carbon caused the structural deformation of the nanotube; moreover, the continuous impact in a short time for reducing the calculation load (10 ps/layer; usually 1-250 ms/layer in evaporation method for pure gold in experiment) resulted in unrealistic phenomena. In the case of nickel clusters (Fig. 6(c)), the nickel atoms

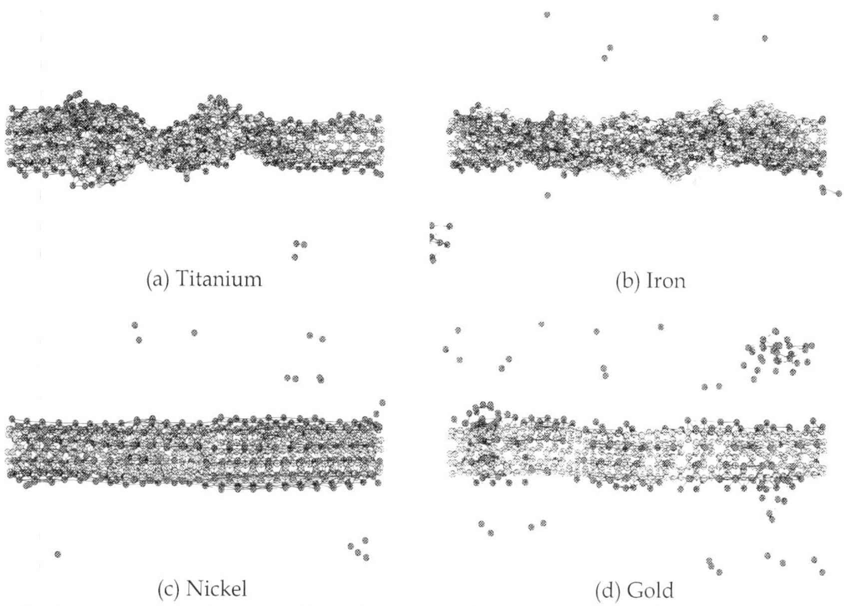

(a) Titanium (b) Iron (c) Nickel (d) Gold

Fig. 6. Continuous metal evaporation. Blue is metal atom, green is carbon atom with three bonds (sp^2), orange is carbon atom with two bonds (one dangling bond), and red is carbon atom with one bond (two dangling bonds).

smoothly and seamlessly covered the carbon nanotube. The equilibrium position of the nickel atoms coincided with the center of the hexagonal carbon network. When the first nickel cluster collided with the carbon nanotube, a grain structure was temporarily formed; however, the nickel atoms immediately found a stable position and moved into it. Once they reached the equilibrium position, they remained stationary at this temperature. Next, the second cluster reached the surface, acquired a grain structure instantly, and then moved to the most stable site. When all the clusters reached the surface, the equilibrium position on the carbon network was saturated; therefore, the nickel clusters formed a second metal layer on the carbon nanotube. Even at this stage, the nickel atoms did not form a grain structure but formed a smooth and seamless layer. Figure 6(d) shows the result of gold cluster coating. According to Zhang et al., gold atoms form a grain structure and exhibit a highly discontinuous coating; however, according to our simulation, the gold atoms can be coated smoothly but discontinuously. The gold atoms initially formed a grain structure, as reported by Zhang et al.; this was because the gold clusters were accidentally concentrated in the same area on the carbon nanotube. This indicates that if the evaporation rate decreases or the experiment is conducted under a higher vacuum condition, the formation of a grain structure can be avoided. Zhang et al. explained this result in terms of the weak interaction between the gold atoms and the SWCNT. However, according to our DFT calculations, the binding energy of the Au-C bond is not low; on the contrary, it is slightly larger than the binding energy of the Au-Au bond. The value of the binding energy is not considered to be significant; however, the ratio of the binding energies, i.e., $\varepsilon_{M-C} / \varepsilon_{M-M}$, should be substantial. Because these condition each atom has many coordination number, the values of this ratio for titanium, iron, nickel, and gold are 2, 3, 1.5 (approximately), and 1 (approximately), respectively. Thus, in the case of titanium and iron coating, the carbon atoms experience an outward pull and the nanotube structure gets deformed. In the case of nickel, a very smooth coating is achieved. On the other hand, in the case of gold, though the Au-C interaction is slightly stronger than the Au-Au interaction, a grain structure is formed. In some areas, the gold atoms are deposited smoothly but not seamlessly. The reason for this phenomenon is attributed to the equilibrium bond length. As mentioned above, the most stable site for the first layer metal atoms is the center of the hexagonal carbon network, of which diameter is 2.5 Å. Coincidently, the equilibrium bond lengths of nickel and iron are comparable, but that of gold is slightly greater (2.6-3.3 Å, depending on the coordination number). The gold atoms must have a shorter bond length so that they can get accommodated in the stable site of the carbon nanotube, thereby creating a discontinuous coating that reduces the distortion energy. Thus, even gold atoms can be coated smoothly, provided the distances between them is sufficient to reduce the distortion for each fragment of coating (in our simulation, the fragment of coatings is approximately 10-15 nm in length).
Owing to the strong C-Ti and C-Fe interactions, the structure of the carbon nanotube is distorted under a high coating rate in our simulation. In order to confirm this, a lower coating rate is applied and the results are examined. In this case, each M_{13} cluster collides with the center of the carbon nanotube in a sequence. For the fully annealing, this sequential impact took an enough interval. Figure 7 shows this sequential impact of the metal clusters on the carbon nanotube. A full annealing duration for each impact step results in a smooth and continuous coating and no defects are observed in the carbon nanotube. A schematic diagram of the iron cluster impact is shown in Fig. 3(a). Similar to the nickel atom, the iron atom finds a stable position and moves into it. However, in this simulation, because each

cluster reaches almost the same position, all the iron atoms do not spread over the surface

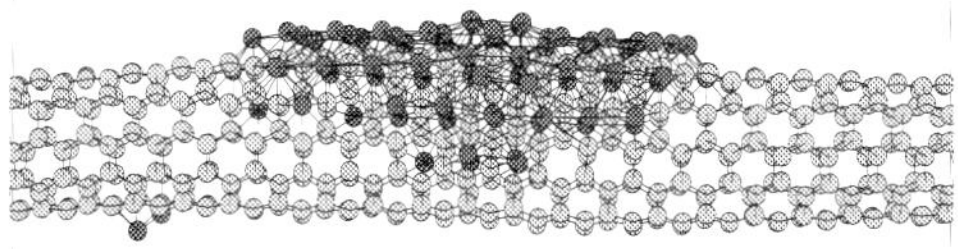

(a) Sequential Fe coating. Totally six Fe_{13} clusters collide.

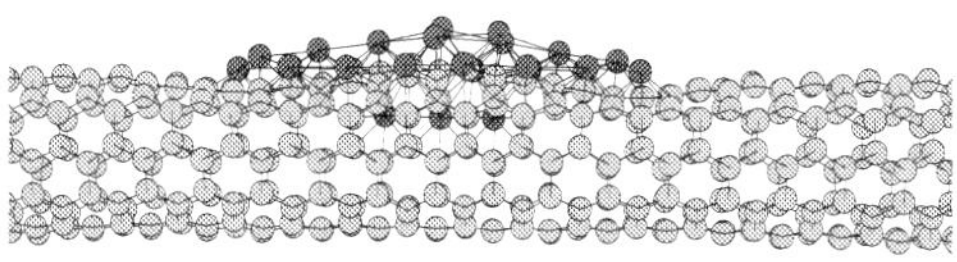

(b) Sequential Ti coating. Totally three Ti_{13} clusters collide.

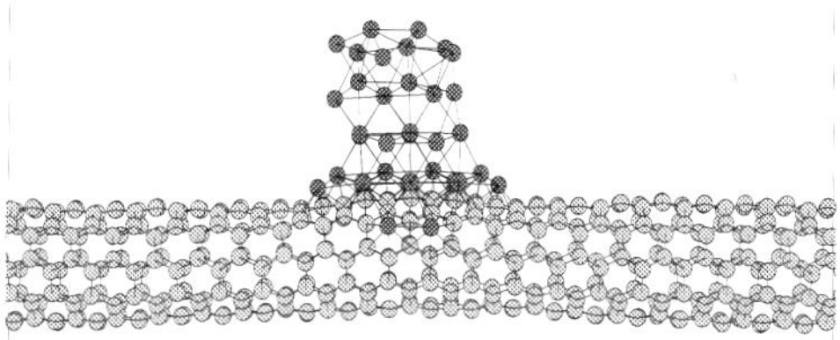

(c) Sequential Au coating. Totally three Au_{13} clusters collide.

Fig. 7. Sequential impacts of Fe, Ti, and Au cluster on the carbon nanotube surroundings with enough annealing duration after every impact.

but form a second layer. This can occur in the nickel system as well. When too many clusters arrive at the same point, each atom seeks the stable site and tends to move into it, but the temperature reduces before the diffusion is completed. As a result, the iron atoms form a second layer. A similar behavior is observed in titanium coating, as shown in Fig. 7 (b). However, in the case of titanium clusters, when the impact energy is small, a different phenomenon is observed. In our simulation, first three clusters were able to reach the surface even at a low evaporation energy, but the next cluster could notreach. This can be attributed to the difference in the equilibrium bond lengths. According to Eq. 6, if the number of coordinates is sufficient, the equilibrium bond length must be equal to R_{e1} for the metal-metal potential and the metal-carbon equilibrium bond length is simply determined as a function of R_e (Table 1). A comparison reveals that the difference in this value for the titanium coating system are larger than those for the iron coating system. This is because if there is a titanium layer on the carbon nanotube, the cluster having a small impact energy cannot overcome the repulsive energy. Thus, the cluster is reflected before it is attracted to the carbon atoms.

This proves that at a high evaporation energy (e.g., 1 eV), the titanium cluster can arrive at the surface and produce a smooth coating. This might be a reason why the titanium coating is smoother than the nickel coating in the experimental results. [15] In the case of gold coating, as shown in Fig. 6(c), the first cluster was able to spread over the surface but the second and the third could not; therefore, they formed a grain structure. This indicates that if the gold cluster reaches the same place, as in the case of the experiment with a high deposition rate, the grain formation is not avoidable, as inferred from Fig. 7(d). From the point of wettability the contact angle is closely related to the wettability, which is mentioned above section explained by solid-liquid affinity (ε_{sl} / ε_l). According to their results the contact angle tends to become larger as the solid-liquid affinity becomes smaller that means the gold cluster tends to form a grain structure.

3.3 Physical properties

Figure 8 shows the stress-strain (force-strain) curves of coated and uncoated SWCNT. In this simulation one end of SWCNT was fixed and the other end was pulled slowly enough to be annealed that avoid unintentionally breaking of SWCNT. Owing to the disturbance of perfect sp^2 bonding of SWCNT, it might be easily expected that the physical strength becomes much weaker, but the decrease is not so large; or rather, the force constant becomes larger. The rupture is induced by metal atom that tends to take the bonds from the carbon. Because the bond length of carbon-metal is longer than that of carbon-carbon, the interaction between carbons becomes suddenly weaker than that of carbon-metal when the SWCNT pulled.

Figure 9 indicates the thermal diffusivity of coated and uncoated SWCNT. The thermal diffusivity is derived by fitting the one-dimensional non-Fourier heat conduction equation with two relaxation timescales (τ_1, τ_2) shown below equation.

$$\tau_1 \frac{\partial^2 T}{\partial t^2} + \frac{\partial T}{\partial t} = \alpha \left(\nabla^2 T + \tau_2 \frac{\partial}{\partial t} \nabla^2 T \right) \tag{10}$$

The dominant heat carrier in SWCNT is phonon and this equation model is fully described in other's paper (Shiomi & Maruyama, 2006).

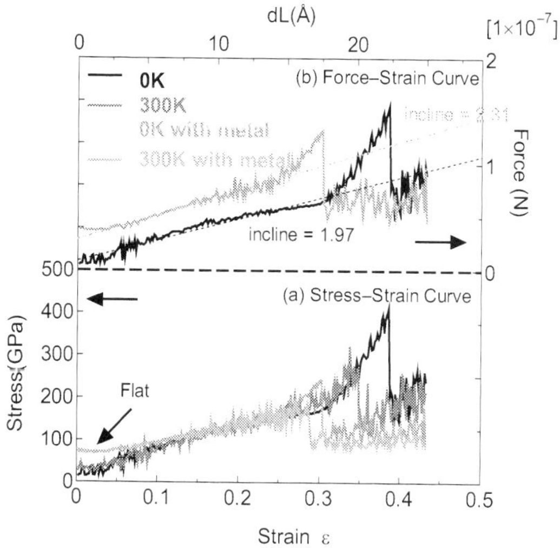

Fig. 4

Fig. 8. The stress–strain curve (a) and force–strain curve (b). The stress is defined by assuming their cross section with 6.93 Å for the uncoated SWCNT and 8.32 Å for the metal-coated SWCNT in diameter. The metal-coated SWCNT meets earlier rupture point but has a larger force constant. The metal-coated SWCNT has a residual stress that makes the inclination flat in a small strain (displacement) range.

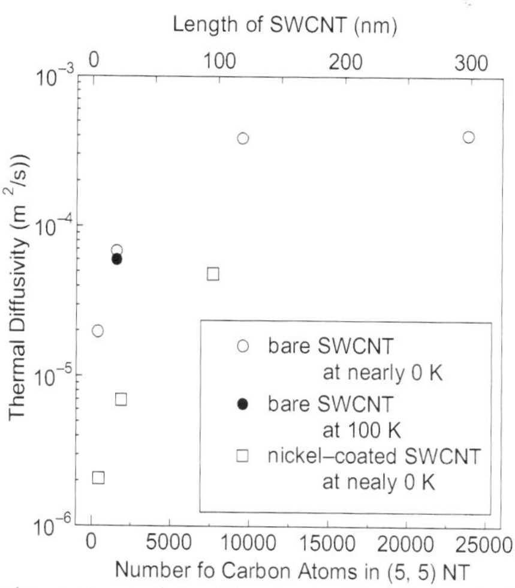

Fig. 9. Thermal diffusivity estimated in this study. The thermal diffusivity of the nickel-coated SWCNT decreases by 90%. The estimated thermal conductivity (λ) is expressed as follows: $\lambda = 1 \times 10^6\ \alpha - 3 \times 10^6\ \alpha$. It depends on the definition of the density and specific heat capacity (C_v) of the SWCNTs is assumed to be $3R$, where R is a gas constant.

4. Conclusion

In this chapter MD approach for CNT analysis was explained. The classical MD merely solves the Newtnian equation, but if the potential function is fully appropriate, the results must be convincive. In case of PVD and coating this simulation shows good agreement with the experimental results; on the other hand, for the physical properties, the adequate results are given and also the mechanism should become clear.

5. References

Beche AD. (1993) Density-Functional Thermochemistry .3. The Role of Exact Exchange. *J. Chem. Phys.* 98, 1993, 5648-5652.

Castro M, et al. (1997) Structure, bonding, and magnetism of small Fe-n, Co-n, and Ni-n, clusters, n<=5. *Chem. Phys. Lett.* 271 1997, 133-142.

Hay PJ & Wadt WR. (1985) Abinitio Effective Core Potentials for Molecular Calculations - Potentials for the Transition-Metal Atoms Sc to Hg. *J. Chem. Phys.* 82, 1985, 270-283.

Ishikawa K. et al. (2007) Extended abstracts, ASME-JSME Thermal Eng., (2007) HT2007-32783.

Kittel C. (2004)Introduction to Solid State Physics. John Wiley & Sons Inc. 2004.

Lee CT, et al. (1998) G. Development of the Colle-Salvetti Correlation-Energy Formula Into a Functional of the Electron-Density. *Phys. Rev. B* 37, 1998, 785-789.

Maruyama S. et al. Liquid droplet in contact with a solid surface. *Micro Thermophys. Eng.* 2, 1998, 49-62.

Morse PM. (1929) Diatomic Molecules According to the Wave Mechanics. II. Vibrational Levels. *Phys. Rev.* 34, 1929, 57-64.

Shibuta Y, Maruyama S. (2007) Bond-order potential for transition metal carbide cluster for the growth simulation of a single-walled carbon nanotube. *Comput. Mat. Sci.* 39, 2007, 842-848.

Shiomi J & Maruyama S. (2006) Non-Fourier heat conduction in a single-walled carbon nanotube: Classical molecular dynamics simulations. , 273 2006, 205420.

Zhang Y. et al. *Phys. Lett.*, 331 (2000) 35.

17

Carbon nanotube field emitters

Alexander Zhbanov[1,2], Evgeny Pogorelov[1] and Yia-Chung Chang[1]
[1]*Research Center for Applied Sciences, Academia Sinica, Taiwan*
[2]*Department of Mechatronics, Gwangju Institute of Science and Technology, Korea*

1. Introduction

Application of various one-dimensional nanostructure materials as field emission sources has attracted extensive scientific efforts. Elongated structures are suitable for achieving high field-emission-current density at a low electric field because of their high aspect ratio. Area of its application includes a wide range of field-emission-based devices such as flat-panel displays, electron microscopes, vacuum microwave amplifiers, X-ray tube sources, cathode-ray lamps, nanolithography systems, gas detectors, mass spectrometers etc.

Since the discovery of carbon nanotubes (CNTs) (Iijima, 1991; Iijima & Ichihashi, 1993; Bethune et al., 1993) and experimental observations of their remarkable field emission characteristics (Rinzler et al., 1995; de Heer et al., 1995; Chernazatonskii et al., 1995), significant efforts have been devoted to the application of using CNTs for electron sources.

One of the main problems for design such field emission emitter is the difficulties in estimation of the electric field on the apex of nanotubes. Only a few works considered forces acting on nanoemitters under electric field. Thus far, there is no analytical formula which provides a good approximation to the total current generated by the nanoscale field emitter. In this chapter, we theoretically consider the electric field strength, field enhancement factor, ponderomotive forces, and total current of a metallic elliptical needle in the form of hemi-ellipsoid in the presence of a flat anode. Also we shortly review the history CNT cold emitters and technology of their fabrication. Furthermore we consider the application areas of CNT electron sources.

2. Historical preview

Field emission is an emission of electrons from a solid surface under action of external high electric field E. Field emission was experimentally discovered in 1987 by R.W. Wood (Wood, 1897). In 1929 R.A. Millikan and C.C. Lauritsen established linear dependence of the logarithm of current density on $1/E$ (Millikan & Lauritsen, 1929). Field emission was explained by quantum tunneling of electrons through the surface potential barrier. This theory was developed by R.H. Fowler and L.W. Nordheim in 1928 (Fowler & Nordheim, 1928).

According to the Fowler–Nordheim theory, the current density of the field emission j is determined by the following expression

$$j \approx \frac{C_1 E^2}{\phi} \exp\left(-\frac{C_2 \phi^{3/2}}{E} \right), \tag{1}$$

where j denotes the emission current density in Acm^{-2}, E is local electric field at the emitting surface in Vcm^{-1}, ϕ is work function in eV, and the first and second Fowler–Nordheim constants are $C_1 = 1.56 \times 10^{-6}$ $AeVV^{-2}$, $C_2 = 6.83 \times 10^7$ $VeV^{-3/2}cm^{-1}$, respectively. The electric field E at the CNT tip increases compared with the average field E_0. Substituting in Eq. (1) the expression $E = \beta E_0$, where β is a field enhancement factor, we shall write the Fowler-Nordheim dependence in the form:

$$j \approx \frac{C_1 (\beta E_0)^2}{\phi} \exp\left(-\frac{C_2 \phi^{3/2}}{\beta E_0} \right). \tag{2}$$

Thus current-voltage characteristics of the field electron emission in the Fowler-Nordheim coordinates $\left(\log j / E_0^2, 1/E_0\right)$ are presented by straight lines. It was assumed that $\varphi = 4.8$ eV for nanotubes. The field enhancement factor β varied from 300 to 3000 depending on the tube size.

Theory of field emission considered in details in recent books (Fursey, 2005; Ducastelle et al., 2006).

Strong electric fields ($E \sim 10^7$ Vcm^{-1}) near a surface are necessary to obtain the appreciable field emission current from pure metals. Therefore, the emitters in early investigations were produced in the form of thin spike with radiuses of curvature on the ends about 1 micron.

Development of lithographic techniques allowed fabricating so called "Spindt tips" in which the field emitters are small sharp molybdenum microcones. One of the first papers describing such technology has appeared in 1968 (Spindt, 1968). Essential efforts have been spent by several companies for development of the Spind-type field emission display, but no large-screen production has been forthcoming.

The new potential in designing field emitters and devices on their basis has appeared after discovery of carbon nanotubes.

Field emission of carbon nanotubes was for the first time reported by Fishbine (Phillips Lab.) (Fishbine et al., 1994), Gulyaev (Institute of Radio-engineering and Electronics, Russia) (Gulyaev et al., 1994), and Rinzler (Rice University) (Rinzler et al., 1994) in 1994.

Four first journal papers (Gulyaev et al., 1995; Chernozatonskii et al., 1995; Rinzler et al., 1995; de Heer et al., 1995) dedicated to this problem were published in 1995. As is known, several papers have appeared in the two subsequent years: two works (Chernozatonskii et al., 1996; Collins & Zettl, 1996) were published in 1996 and seven works (Collins & Zettl, 1997; Gulyaev et al., 1997; Sinitsyn et al., 1997; de Heer et al., 1997; Saito et al., 1997a; Saito et al., 1997b; Lee et al., 1997) were published in 1997. Starting from 1998, interest in field-emission properties of CNT was increasing explosively all over the world. Today we can speak of thousands of published papers.

Recently, field emission from metals (Lee et al., 2002), metal oxides (Li et al., 2006; Banerjee et al., 2004; Jo et al., 2003; Seelaboyina et al., 2006), metal carbides (Charbonnier et al., 2001), and other elongated nanostructures have also been explored. It is now possible to control the diameter, height, radius of curvature of the tip, and basic form of emitters during growth. Elongated structures of different shapes such as nanotubes, nanocones, nanofibers,

nanowires, nanoneedles, and nanorods have been successfully grown (Li et al., 2006; Li et al., 2007; Jang et al., 2005; Hu & Huang, 2003).
Promising new materials for field-emission sources are B- and N-doped CNTs. Terrones et al. (Terrones et al., 2004; Terrones et al., 2008) have reviewed the field emission properties of B- and N doped CNTs and nanofibres. B-doped multi-wall CNTs could exhibit enhanced field emission (turn on voltages of ~1.4 V/μm) when compared to pristine multi-wall CNTs (turn on voltages of ~3 V/μm). N-doped CNTs are able to emit electrons at relatively low turn-on voltages (2 V/μm). This phenomenon arises from the presence of B atoms (holes) or N atoms (donors) at the nanotube tips.

3. Carbon nanotubes field emitters

3.1 Physical properties of carbon nanotubes suitable for cold emission

From the practical application point of view CNTs are preferable field emitters due to their low threshold voltage, good emission stability and long emitter lifetime.
CNTs possess these advantages due to the large aspect ratio, high electric and thermal conductivity, highest flexibility, elasticity, and Young's modulus. Their strong covalent bonding makes them chemically inert to poisoning and physically inert to sputtering during field emission. They can also carry a very high current density of order 10^9 A cm^{-2} before electromigration. Nanotubes have a high melting point and preserve their high aspect ratio over time. CNTs emits electrons under conditions of technical vacuum. They are chemically inert to poisoning due to strong covalent bonding. Measuring of field emission properties (Kung et al., 2002) and theoretical ab-initio calculations (Park et al., 2001) shows that emission currents are significantly enhanced when oxygen is adsorbed at the tip of carbon nanotubes.

3.2 Manufacturing techniques for CNT-based field-emission cathodes

Many technologies for fabrication of CNT-based field-emission cathodes were offered. We shall consider only some of them.
Individual CNT field emitters have a large potential for application in electron guns for scanning electron microscopes. To investigate the emission properties of individual CNTs de Jonge et al (de Jonge & Bonard, 2004) improved the mounting method using a piezo-driven nanomanipulator. For the mounting of an individual CNT on a tungsten tip, a tungsten wire was fixed by laser-welding on a titanium (or tungsten) filament.
Field emission CNT-based cathodes are manufactured either as a bulk solid containing nanotubes or as a film with thickness from hundreds of nanometers to tens of microns.
Bulk cathodes are known to be manufactured by two methods. The Alex Zettl team from California University, Berkley, USA used a technology in accordance with which the ready material of unsorted randomly aligned nanotubes is mixed into a compound, baked, and surface ground. Flexible and elastic nanotubes are not broken during the grinding. In accordance with the technology used by the Yahachi Saito team of Mie University (Japan), the graphite-electrode material processed by an electric arc is cut to pellets and glued to a stainless-steel plate by silver paste.
Film technologies are used in all other cases. Film cathodes are basically manufactured by two methods: either preliminary synthesized tubes are attached to a substrate or the tubes are grown directly on the substrate.

In the two methods, different technologies yield films of both well oriented and strongly entangled tubes.

N.I. Sinitsyn group from Institute of Radio-engineering and Electronics (Saratov, Russia) used CVD methods for synthesizing films of both regularly grown nanotubes (Fig. 1) and "felt" of entangled fibers. Strips were obtained using a catalyst deposited through a template (Zhbanov et al., 2004).

Fig. 1. Photograph of strips of oriented CNTs synthesized on a substrate. The strip width is 20 μm and the gap between the strips is 5 μm (Zhbanov et al., 2004).

The Jean-Marc Bonard team of Lousanne Polytechnical School (Switzerland) developed the technology of microcontact printing of catalytic precursor for growing oriented tubes arranged in accordance with a specified pattern on a substrate (Bonard et al., 2001b). The catalyst, the so-called "ink", was applied to the stamp surface. The ink was a solution containing from 1 to 50 mM of $Fe(NO_3)_3 \cdot 9H_2O$. The duration of contact during the printing was 3 s. Nanotube deposition was by the CVD method in a standard flow reactor at a temperature of 720 °C.

In the case of low concentration of catalyst (1 mM, Fig. 2*a*), several single nanotubes are randomly distributed over the printing region. The catalyst-concentration growth is accompanied by formation of films of entangled tubes, as is shown in Figs. 2*b* and 2*c*. For concentrations about 50 mM, clusters of nanotubes oriented normally to the surface are formed. Figure 2*d* shows that the sides of the walls are flat, and not a single tube is hanging outward. For concentrations above 60 mM, growth of nanotubes is retarded, and the printed template is covered by amorphous carbon particles.

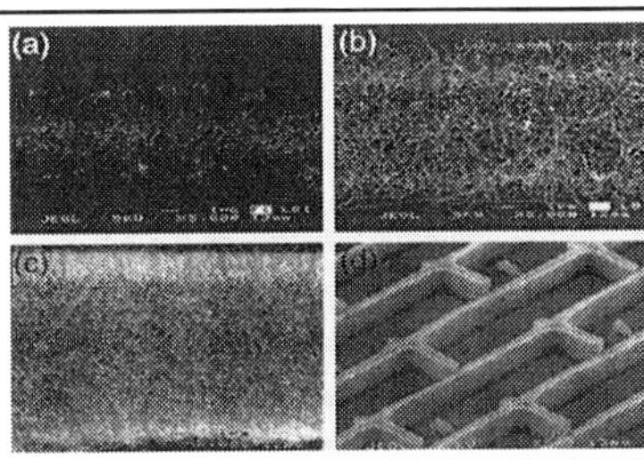

Fig. 2. Nanotube growth for various concentrations of catalytic ink used for the precursor application. Catalyst concentration in the solution was 1 mM (a), 5 mM (b), 40 mM (c), and 50 mM (d). The figure is taken from (Bonard et al., 2001b).

Hongjie Dai team of Stanford University, USA, used the following technology for obtaining arrays of well-oriented carbon nanotubes. First, porous silicon was formed on the surface of a silicon substrate by anode etching and then the ferrum film was deposited on the latter through the shadow mask by electron-beam evaporation (Fan et al., 1999). Then nanotubes were grown as a result of acetylene decomposition in argon flow at 700 °C.
E. F.Kukovitsky team of the Kazan Physics-Technical Institute (Russia) developed the technology of synthesis of oriented nanotubes with conical layers (Fig. 3) (Musatov et al., 2001; Kukovitsky et al., 2003). The first stage of the process involves polyethylene pyrolysis in the first oven at a temperature of 600 °C. Then, by the helium flow, the gaseous products of pyrolysis are transferred to the second oven where nanotubes grow on the nickel foil catalyst at a temperature of 800 to 900 °C. For the obtained specimens, the current density was 10 mA/cm^2 for the electric field from 4 to 4.5 V/μm.

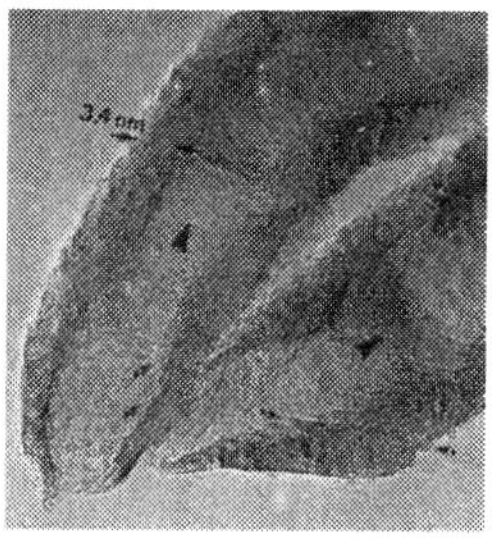

Fig. 3. High-resolution electron microscope image of nanotubes with conical layers. Graphene layers are marked by arrows with points, and the CNT growth direction is marked by large arrows (Musatov et al., 2001).

As is obvious from the literature analysis, almost all CNT-based cathodes show high emission irrespective of the fact whether the tubes are multi-wall or single-wall, well-oriented or entangled. Bamboo-shaped aligned carbon nanotubes (Srivastava et al., 2006; Ghosh et al., 2008) as well as carbon nanocones (Yudasaka et al., 2008) demonstrate high field electron emission.
Let's note, that not only elongated carbon nanotubes, but also pyramids from fullerenes are used as cold cathodes. Formation and characteristics of fullerene coatings on the surface of tungsten tip field emitters and emitters with ribbed crystals formed on their surface are studied by group of Sominskii from St. Petersburg State Technical University (Russia) (Tumareva et al., 2002; Tumareva et al., 2008). Methods of creating microprotusions on the surface of the coatings that considerably enhance the electric field have been developed and tested. Emitters with a single microprotrusion demonstrated emission current densities up to 10^6–10^7 A/cm^2. It was shown that single micron-sized emitters can stably operate at currents up to 100 μA.

3.3 Electric field and field enhancement factor in diode configuration

The field enhancement factor is very important parameter for characterization of CNT emitters.

The model of a hemisphere on a post for CNT emitters is widely used in analytical approximations and numerical simulations (Fig. 4). To calculate the electric-field intensity and the field enhancement factor on the nanotube tips, the following assumptions are usually done:

1) Nanotubes are regularly located on a flat substrate in a "honeycomb-like" order. A nanotube is a cylinder with height h and diameter 2ρ capped by a hemisphere of ρ radius. Total height of closed nanotube is H, the distance from cathode to anode is L, the gap between anode and nanotube tip is l, and the distance between the nearest neighbors is D.

2) A nanotube obeys the laws of continuous medium, is perfectly conducting, and the cathode potential is maintained on its entire surface.

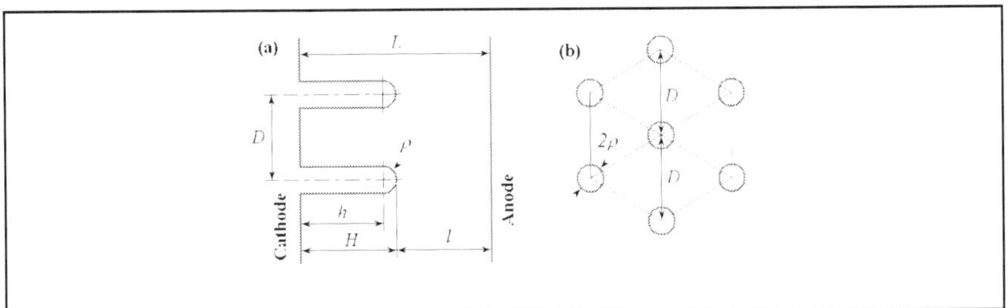

Fig. 4. Scheme of aligned nanotube film, the model of a hemisphere on a post: (a) side view; (b) top view.

Let us introduce dimensionless parameters for the geometrical characterization of model. The dimensionless height of emitter, the dimensionless gap between anode and emitter tip, and the dimensionless distance between individual emitter are the following:

$$\eta = \frac{\rho}{h},\ \lambda = \frac{\rho}{l},\ \delta = \frac{\rho}{D}. \tag{3}$$

Until now the analytical solution for the model of a hemisphere on a post is unknown. There is no even a solution for the individual cylindrical nanotube closed by hemispherical cap in a uniform electric field.

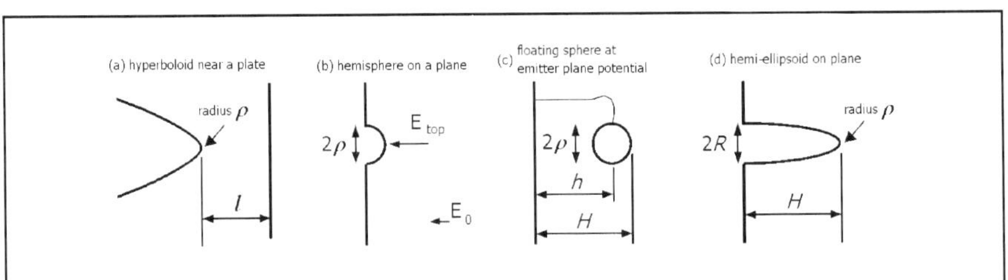

Fig. 5. Schemes of simplest models for field enhancement factor estimation: (a) hyperboloid near a plate; (b) hemisphere on a plane; (c) floating sphere at emitter-plane potential, and (d) hemi-ellipsoid on a plane.

Numerical simulations were reported in many papers (Edgcombe & Valdrè, 2001; Edgcombe & Valdrè, 2002; Read & Bowring, 2004; Musatov et al., 2001). Calculation difficulties in these numerical methods arise due to the large nanotube aspect ratio and very long distance between cathode and anode in comparison with emitter height. Usually, these numerical results were generalized and simple fitting formulas of field enhancement factor for individual nanotube (Edgcombe & Valdrè, 2001; Edgcombe & Valdrè, 2002; Read & Bowring, 2004; Shang et al., 2007), for nanotube in space between parallel cathode and anode planes (Bonard et al., 2002a; Filip et al., 2001; Nilsson et al, 2002; Smith et al., 2005), and for a nanotube surrounded by neighboring nanotubes with a screening effect (Jo et al., 2003; Glukhova et al., 2003; Nilsson et al., 2000; Read & Bowring, 2004; Wang et al., 2005) were suggested. The main problem for such algebraic fitting formulas is the lack of a definitive proof of their accuracy.

Four of the simplest models are the "hyperboloid near a plate" model, the "hemisphere on a plane" model, the "floating sphere at emitter-plane potential" model, and the "hemi-ellipsoid on plane" model. We follow to the classification suggested by Forbes *et al.* (Forbes et al., 2003).

These models allows analytical solutions, they are illustrated in Fig.5. We will use dimensionless parameters Eq. (3) to define geometry of these models.

3.3.1 Hyperboloid near a plate model

We introduce the prolate spheroidal coordinates σ and τ to consider the model of a hyperboloid near a plate (Fig. 6).

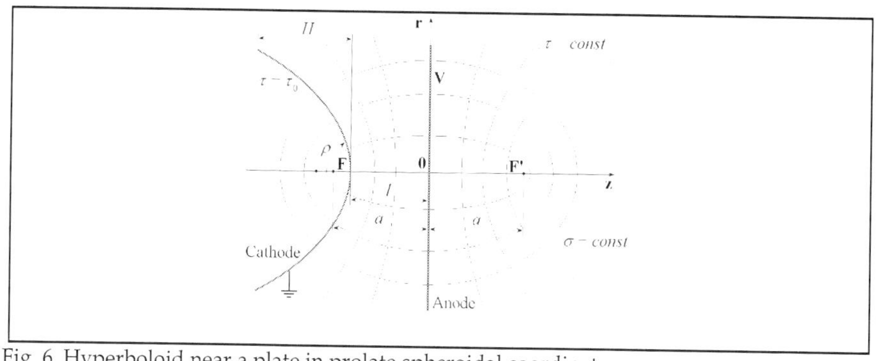

Fig. 6. Hyperboloid near a plate in prolate spheroidal coordinates.

The equation of prolate spheroid is:

$$\frac{r^2}{a^2(\sigma^2-1)}+\frac{z^2}{a^2\sigma^2}=1\,;\;\sigma\geq 1\,. \tag{4}$$

The equation of hyperboloid of two sheets is:

$$\frac{r^2}{a^2(\tau^2-1)}+\frac{z^2}{a^2\tau^2}=1\,;\;1\geq\tau\geq-1\,. \tag{5}$$

Points F (0;-a) and F′ (0; a) are the foci of the hyperboloids and spheroids. The cathode represents a hyperboloid of revolution τ_0 = *const* and the anode is a plane $\tau = 0$. They are show in Fig. 6 by solid red lines. The radius of hyperboloid curvature of the tip is ρ.
The electric field is calculated according to the formula:

$$E=\frac{V}{a\sqrt{(\sigma^2-\tau^2)(1-\tau^2)}\operatorname{arctanh}(l/a)}, \tag{6}$$

where function arctanh is inverse hyperbolic tangent, V is the voltage applied across a gap between anode and cathode.
The model of a hyperboloid near a plate is suitable to describe the interaction of individual CNT field emitter with surface in scanning electron microscopes. Usually in cases important for practice we have $\rho << l$ and $l << H$.
If $\rho << l$ the maximal value of the module of intensity is approximated by the formula (Drechsler & Müller, 1953):

$$E_{top}\approx\frac{V}{\rho\ln(4l/\rho)}. \tag{7}$$

If we define the macroscopic field by $E_0 = V/l$ then we can write the field enhancement factor

$$\beta=\frac{\sqrt{1+\lambda}}{\lambda\cdot\operatorname{arctanh}\sqrt{1/(1+\lambda)}}. \tag{8}$$

Let us estimate the electric force acting on the surface of ellipsoid. The electrostatic force acting on the elementary area, s of the external surface is given by

$$\vec{F}=\int_S\frac{\varepsilon_0}{2}E^2\vec{n}ds, \tag{9}$$

where ε_0 is the electric constant, $\vec{n}$ is a vector normal to the surface.
Taking into account that the infinitesimal surface element is $ds=2\pi a^2\sqrt{(1-\tau_0^2)(\sigma^2-\tau_0^2)}d\sigma$, we can analytically integrate the force acting on the top of a hyperboloid surface of height H (see Fig. 6). It is clear that r-component of force equals to zero, $F_r = 0$. For the z-component we have

$$F_z(H)=\frac{\pi\varepsilon_0V^2}{2\operatorname{arctanh}^2\sqrt{1/(1+\lambda)}}\ln\left(1+\frac{2H}{\rho}+\frac{2H}{l}+\frac{H^2}{l\rho}+\frac{H^2}{l^2}\right). \tag{10}$$

The total current is calculated by integration of current density from Eq. (1) over a hyperboloid surface

$$J_{total} \approx \int jds. \tag{11}$$

We note that the exchange and correlation effect is ignored in the basic equation (1). Thus the Fowler-Nordheim theory is suitable only for approximate calculations. Nevertheless this theory is widely used for analysis of field emission current from elongated nanostructures. After substitution of field distribution over the sphere surface and an infinitesimal surface element we have

$$\int jds = \frac{2\pi C_1 V^2}{\phi\sqrt{1-\tau_0^2}\,\mathrm{arctanh}^2\tau_0}\int^{\infty} \frac{\exp\left(-C_2 V^{-1}\phi^{3/2} a\sqrt{(1-\tau_0^2)(\sigma^2-\tau_0^2)}\,\mathrm{arctanh}\,\tau_0\right)}{\sqrt{\sigma^2-\tau_0^2}} d\sigma. \tag{12}$$

If $\rho << l$ then $\tau_0 \approx 1$ and $\sqrt{\sigma^2-\tau_0^2} \approx \sqrt{\sigma^2-1} \approx \sqrt{2(\sigma-1)}$. This approximation allows us to reduce our integral to another one $\int^{\infty} \frac{\exp\left(-a\sqrt{x-1}\right)}{\sqrt{x-1}} dx = \frac{2}{a}$.

Thus the total field emission current is

$$J_{total} \approx \frac{2\pi C_1 V^3 \sqrt{1+\lambda}}{C_2 \phi^{5/2} \rho\, \mathrm{arctanh}^3 \sqrt{1/(1+\lambda)}}, \tag{13}$$

where the total current, J_{total} is measured in A; the radius of curvature, ρ is measured in cm.

3.3.2 Hemisphere on a plane

The metallic sphere in a uniform electric field E_0 (Fig. 5(b)) was considered in many papers (for example Refs. (Forbes et al., 2003; Wang et al., 2004; Pogorelov et al., 2009)). We can replace the sphere by point electric dipole. If the electric dipole moment is p_0 then the dipole potential is

$$\varphi_{dip} = -\frac{p_0}{4\pi\varepsilon_0}\frac{z}{\left(z^2+r^2\right)^{3/2}}. \tag{14}$$

Equation of circle is $\varphi_{dip} + zE_0 = 0$. From this equation we can find the relation between the electric dipole moment and the sphere radius: $p_0 = 4\pi\varepsilon_0 E_0 \rho^3$. The electric field on the top of hemisphere reaches $E_{top} = p_0 / 2\pi\varepsilon_0\rho^3 + E_0 = 3E_0$. The field enhancement factor is $\beta = E_{top} / E_0 = 3$. The field distribution over the sphere surface have the form $E = 3E_0 \cos\theta$, where θ is polar angle.

Pogorelov et al. (Pogorelov et al., 2009) have shown that the total current emitted from the hemisphere surface is

$$J_{total} = \frac{2\pi\rho^2 C_1 \phi^{7/2} C_2^3}{3E_0}\left[\frac{\mathrm{E}_1(C_3)}{6} + \left(\frac{1}{3C_3^3} - \frac{1}{6C_3^2} + \frac{1}{6C_3}\right)\exp(-C_3)\right], \quad (15)$$

where $C_3 = C_2\phi^{3/2}/3E_0$ and $\mathrm{E}_1(x) \equiv \int \exp(-xt)/t\,dt$ is the exponential integral.

Due to small β for the hemisphere we need to use very strong electrical field to produce slightly visible current in experiment.

3.3.3 Floating sphere at emitter-plane potential

The "floating sphere at emitter-plane potential" model has no "body" of the field emitter and possesses only its "head". This model gives too high estimation of electric field on the apex of nanotube but plausibly reproduce tendencies of change of the field enhancement factor. Approximate analytical solution for the "floating sphere at emitter-plane potential" model is well known (for example Refs. (Forbes et al., 2003; Wang et al., 2004)). To solve this problem the method of images (Jackson, 1999) is usually used.

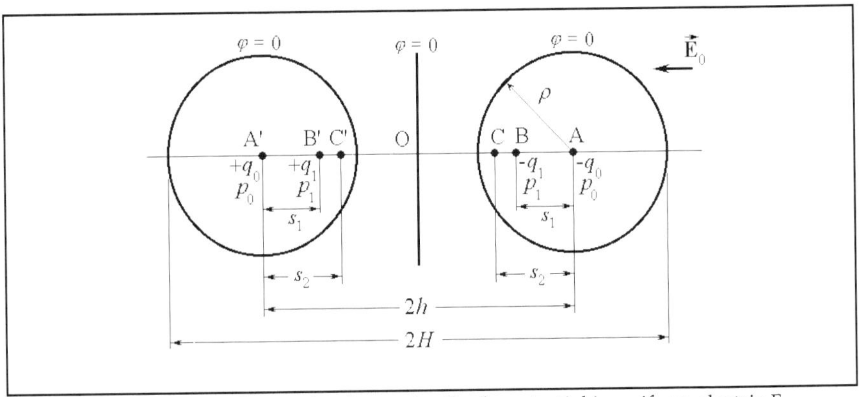

Fig. 7. Two conducting spheres of radius ρ at cathode potential in uniform electric E_0.

The charge $-q_0 = -4\pi\varepsilon_0 hE_0\rho$ and the electric dipole $p_0 = 4\pi\varepsilon_0 E_0\rho^3$ placed at point A (Fig. 7) create a sphere of radius ρ and potential $\varphi = 0$ in uniform external electric field. The charge q_0 and dipole p_0 cause a potential variation across the emitter plane. To correct this we have to place an image-charge q_0 and image-dipole p_0 at point A′ behind the emitter plane. The image-charge and image-dipole will distort the surface of sphere. To restore the shape we should place additional charge $-q_1 = -q_0\rho/2h$ and dipole $p_1 = p_0\rho^3/8h^3$ at point B on the distance $s_1 = \rho^2/2h$ from the center of sphere (see Fig. 7).

Next we have to put q_1 and p_1 at point B′, after to put $-q_2$ and p_2 at C and so on. Neglecting terms of higher smallness in this series of approximation we find the electric field on the top of floating sphere

$$E_{top} = \frac{1}{4\pi\varepsilon_0}\frac{q_0}{\rho^2} + \frac{2p_0}{4\pi\varepsilon_0}\frac{1}{\rho^3} + \frac{1}{4\pi\varepsilon_0}\frac{q_1}{(s_1+\rho)^2} + E_0 \approx E_0\left(\frac{h}{\rho} + 3.5\right). \quad (16)$$

Thus the field enhancement factor is

$$\beta = \frac{1}{\eta} + 3.5 = \frac{H}{\rho} + 2.5\,. \tag{17}$$

We can provide more accurate calculations. Recurring formulas for the distance s_{i+1}, the charge q_{i+1}, and the dipole moment p_{i+1} through s_i, q_i, and p_i are the following $s_{i-1} = \frac{\rho^2}{2h - s_i}$, $p_{i-1} = p_i \frac{\rho^3}{(2h - s_i)^3}$, and $q_{i-1} = q_i \frac{\rho}{2h - s_i} - p_i \frac{\rho}{(2h - s_i)^2}$, where the initial distance is zero: $s_0 = 0$.

Series expansion of the field enhancement factor is

$$\beta = \eta^{-1} + \frac{7}{2} - \frac{1}{2}\eta + \frac{1}{8}\eta^2 + \frac{7}{16}\eta^3 - \frac{25}{32}\eta^4 + \frac{25}{32}\eta^5 + O(\eta^6). \tag{18}$$

As the next step of approaching to CNT film, consider an assembly of floating spheres and a screening of the individual emitter by neighbors. The view from above of the sphere surrounded by another one is shown in Fig. 8. Large red circles in this picture are the floating spheres. Small black circles mark places where charges are located. Numbers "0" show initial charges in the center of balls. Numbers "1" specify image charges induced only by nearest neighbors. Numbers "2" concern to secondary image charges.

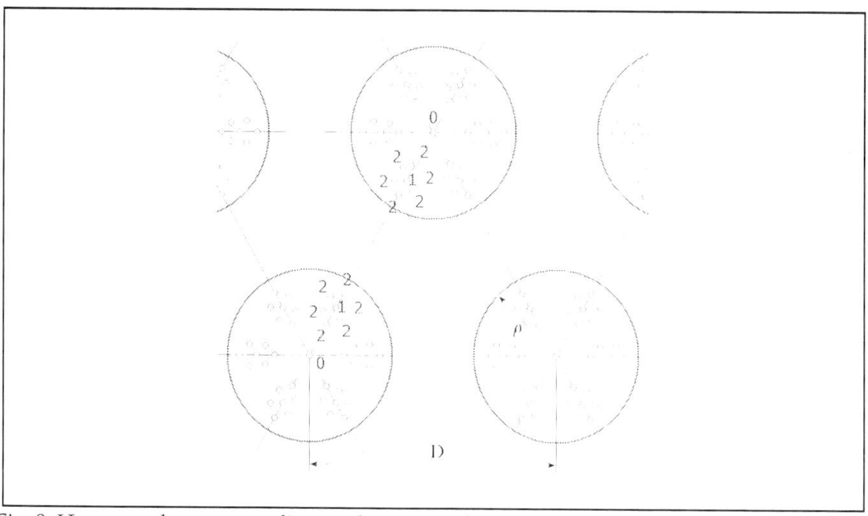

Fig. 8. Honeycomb structure, distance between spheres is D, sphere radius is ρ.

If the distance between spheres, D is large enough ($D \gg \rho$) then all image charges collect on small area around the center of sphere. In that case we can combine all charges inside the ball into its center. Also we will neglect influence of image dipoles.

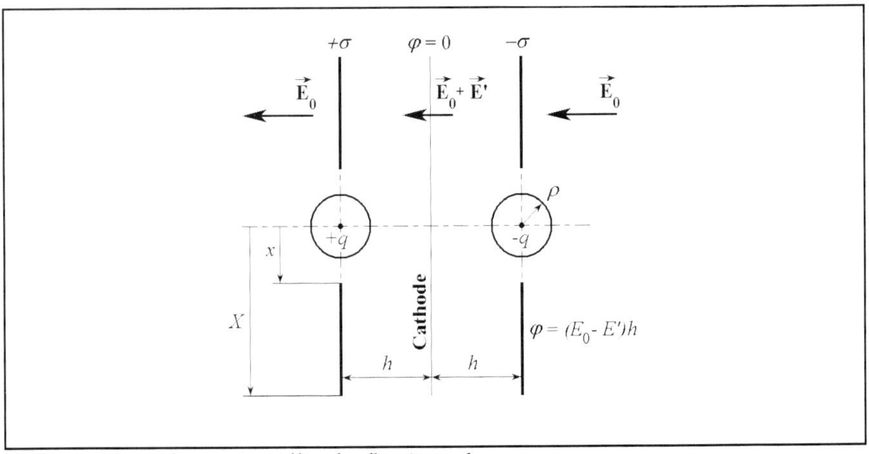

Fig. 9. Modeling of screening effect for floating spheres.

The set of floating spheres produces an idealized surface charge density $\sigma = 2q/\sqrt{3}D^2$. Positively and negatively charged surfaces form the parallel plate capacitor (Fig. 9). The electric field between two large parallel plates is given by $E' = \sigma/\varepsilon_0$.

From the equation $h\left(E_0 - \frac{2}{\sqrt{3}\varepsilon_0}\frac{q}{D^2}\right) = \frac{1}{4\pi\varepsilon_0}\frac{q}{\rho}$ we can find the total charge in the center of each sphere $q = \varepsilon_0 h E_0 \left(\frac{1}{4\pi\rho} + \frac{2h}{\sqrt{3}D^2}\right)^{-1}$.

Thus we can find the maximal field on the surface of floating sphere

$$E_{top} = \frac{hE_0}{\rho}\frac{\sqrt{3}D^2}{\left(\sqrt{3}D^2 + 8\pi\rho h\right)} + E_0 \tag{19}$$

and the field enhancement factor

$$\beta = \frac{\sqrt{3}}{\sqrt{3}\eta + 8\pi\delta^2} + 1 . \tag{20}$$

More accurate approximation

$$\varphi_{\sigma+} = 2\pi\int_x^X \frac{\sigma}{4\pi\varepsilon_0}\frac{r}{\sqrt{(2h+\rho)^2 + r^2}}dr\ ,\ \varphi_{\sigma-} = -2\pi\int_x^X \frac{\sigma}{4\pi\varepsilon_0}\frac{r}{\sqrt{\rho^2 + r^2}}dr\ ,\ x = \sqrt{\frac{\sqrt{3}}{2\pi}}D . \tag{21}$$

$$\varphi_{P''} = \lim_{X\to\infty}\left(\varphi_{\sigma+} + \varphi_{\sigma-}\right) + \frac{1}{4\pi\varepsilon_0}\left(\frac{q'}{2h+\rho} - \frac{q'}{\rho}\right). \tag{22}$$

Solving equation $\varphi_{P''} = E_0 h$ we find

$$\beta = \frac{6(1+\eta)^2}{4\sqrt{6\pi}\,\delta(1+\eta)\left(\sqrt{2\pi\delta^2(2+\eta)^2+\sqrt{3}\eta^2}-\eta\sqrt{2\pi\delta^2+\sqrt{3}}\right)+3\eta(2+\eta)}. \tag{23}$$

Eq. (23) is transformed to Eq. (20) after neglect in values of higher order of smallness.
On the one hand the field enhancement factor and the current density on nanotube apex reach its maximum if the distance between emitters is very large. On the over hand in this case the current density on the anode will be very small. Clearly we can find optimum distance between emitters. As an approximation, assume that the emitting surface of each sphere equals $\pi\rho^2$ and that the electric field is a constant on this surface. The anode current density takes the form

$$\mathbf{j}_{\mathbf{anode}} \approx \frac{2\pi}{\sqrt{3}}\delta^2\frac{\mathbf{C}_1(\beta\mathbf{E}_0)^2}{\varphi}\mathbf{exp}\left(-\frac{\mathbf{C}_2\varphi^{3/2}}{\beta\mathbf{E}_0}\right). \tag{24}$$

If $h >> \rho$ and $D >> \rho$ then $\beta \approx \sqrt{3}/\left(\sqrt{3}\eta + 8\pi\delta^2\right)$. Let's use this relation for the simplicity.
Solving the equation

$$\frac{\partial \mathbf{j}_{\mathbf{anode}}}{\partial\delta} = 0 = 24\pi\mathbf{E}_0\delta^2 - 3\sqrt{3}\mathbf{E}_0\eta + 24\pi\varphi^{3/2}\delta^2\mathbf{C}_2\eta + 64\sqrt{3}\pi^2\mathbf{C}_2\varphi^{3/2}\delta^4 \tag{25}$$

we find the optimal dimensionless distance between emitters in honeycomb structure

$$\delta_{opt} = \left(\frac{3\mathbf{E}_0\eta}{4\pi\left(\sqrt{3}\mathbf{E}_0 + \sqrt{3}\mathbf{C}_2\varphi^{3/2}\eta + \sqrt{3\mathbf{E}_0^{\,2} + 18\mathbf{C}_2\mathbf{E}_0\varphi^{3/2}\eta + 3\mathbf{C}_2^2\varphi^3\eta^2}\right)}\right)^{1/2}. \tag{26}$$

After neglect terms of higher smallness we can write the simplification

$$\delta_{opt} \approx \sqrt{\frac{\sqrt{3}\mathbf{E}_0\eta}{4\pi\left(\mathbf{E}_0 + 2\mathbf{C}_2\varphi^{3/2}\eta\right)}}. \tag{27}$$

Fig. 10 illustrates the dependence of anode current density from geometrical parameters of emitter. We have assumed that the work function is ϕ = 4.8 eV, the external field is E_0 = 60000 Vcm^{-1}, the dimensionless height is η = 0.001 (for Fig. 10a), and the dimensionless distance between emitters is δ = 0.002 (for Fig. 10b).

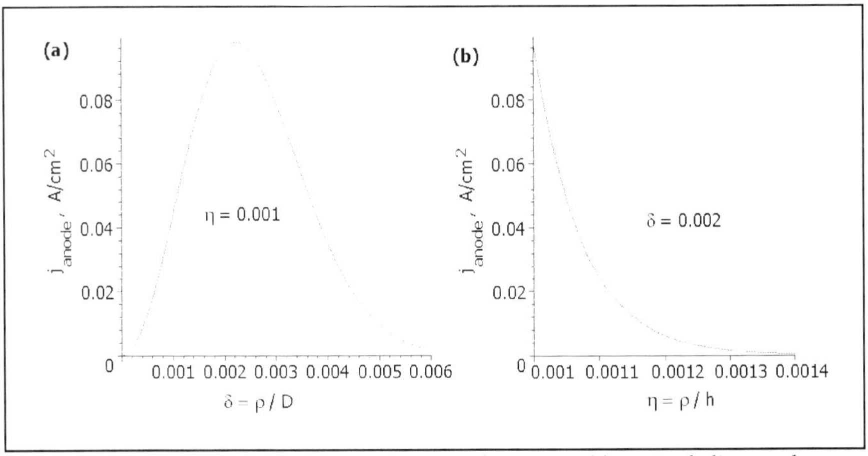

Fig. 10. Anode current density versus dimensionless sizes: (a) optimal distance between emitters if the height is fixed; (b) influence of emitter height on anode current if the density of emitting centers is constant.

Let's consider influence of the limited anode-cathode distance (Fig. 11.) on the field enhancement factor.

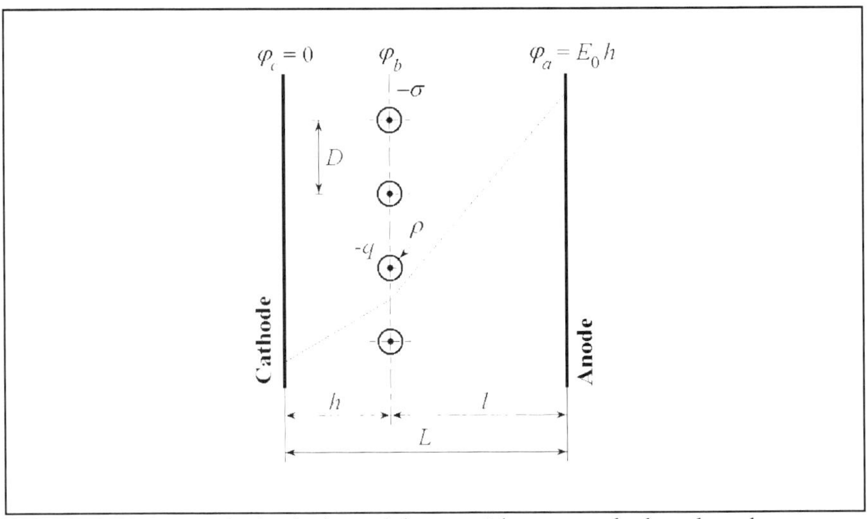

Fig. 11. Geometrical model for the limited distance, L between cathode and anode.

The cathode has zero potential $\varphi_c = 0$; $\varphi_a = E_0 L$ is the anode potential.

As before, assume that average charge per each conductive ball, q is concentrated at its center. Equation for average potential φ_b on plane with grounded conductive balls is

$$-\frac{\varepsilon_0}{h}\varphi_c + \left(\frac{\varepsilon_0}{h} + \frac{\varepsilon_0}{l}\right)\varphi_b - \frac{\varepsilon_0}{l}\varphi_a = -\sigma \tag{28}$$

Solving the equation $-\frac{2hl}{\sqrt{3}L\varepsilon_0 D^2}q + E_0 h = \frac{1}{4\pi\varepsilon_0\rho}q$, we find the charge in the center of ball

$$q = \frac{4\sqrt{3}\pi\varepsilon_0 E_0 h\rho D^2 L}{\sqrt{3}D^2 L + 8\pi h\rho L - 8\pi h^2\rho} \tag{29}$$

Thus the maximal field on the surface of floating sphere is

$$E_{top} = \frac{E_0 h}{\rho}\left[1 + \frac{8\pi}{\sqrt{3}}\frac{h\rho}{D^2}\left(1 - \frac{h}{L}\right)\right]^{-1}. \tag{30}$$

The field enhancement factor is

$$\beta = \left[\frac{\rho}{h} + \frac{8\pi}{\sqrt{3}}\left(\frac{\rho}{D}\right)^2\left(1 - \frac{h}{L}\right)\right]^{-1} = \left[\eta + \frac{8\pi}{\sqrt{3}}\delta^2\left(1 - \frac{\lambda}{\eta}\right)\right]^{-1}. \tag{31}$$

Let us note here that the model of floating sphere and the method of images allow considering field emission not only on flat anode but also on spherical anode.

3.3.4 Hemi-ellipsoid on a plane

Consider a prolate metallic spheroid in a uniform electric field. We can replace the spheroid by a linearly charged thread as we show in our recent paper (Pogorelov et al., 2009). The thread is a green line in Fig. 12 and the linear charge distribution is represented by a red line. The length of a thread is $2h$. The electrostatic potential produced by the charged thread is

$$\varphi(z,r) = -\int_{-h}^{h}\frac{1}{4\pi\varepsilon_0}\frac{\tau z' dz'}{\sqrt{(z'-z)^2 + r^2}}, \tag{32}$$

where (r; z) denotes the in-plane radial and z coordinates, τz is the linear charge density at point (0; z), h is half of the thread length. The solution is independent of the azimuthal angle. We assume the coefficient of linear charge density, τ to be positive.

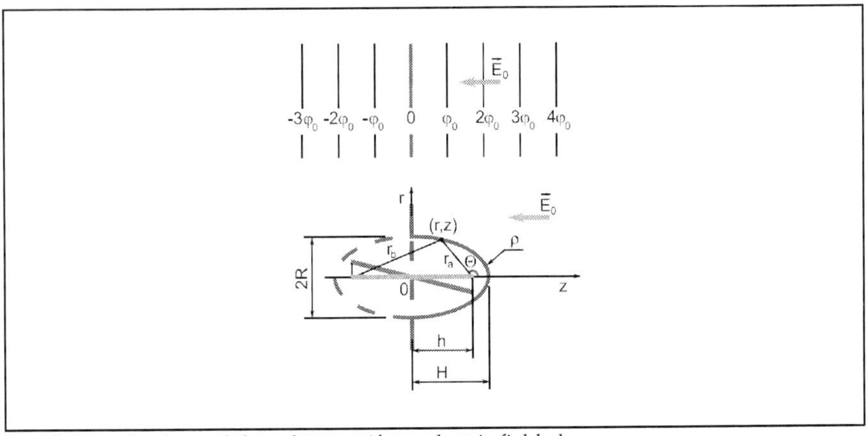

Fig. 12. Linearly charged thread in a uniform electric field along z.

The shape of metallic spheroid is given by the solution to the equation.

$$\varphi(z,r) + E_0 z = 0 \,. \tag{33}$$

Using coordinates on the spheroid surface $r_a = \sqrt{(z-h)^2 + r^2}$ and $r_b = \sqrt{(z+h)^2 + r^2}$ and a dimensionless parameter, the eccentricity $\xi = \dfrac{2h}{r_a + r_b}, (0 < \xi < 1)$, we can rewrite Eq. (33) in the form

$$\frac{4h}{r_a + r_b} - \ln\left(\frac{r_a + r_b + 2h}{r_a + r_b - 2h}\right) = -C \text{, where } C = 4\pi\varepsilon_0 \frac{E_0}{\tau} = \ln\left(\frac{1+\xi}{1-\xi}\right) - 2\xi \,. \tag{34}$$

The zero equipotential which represents the metallic hemi-ellipsoidal cathode on a plate is shown in Fig. 12 by solid blue line. Points (0, -*h*) and (0, h) are the foci of the ellipse, r_a and r_b are distances between (r; z) and the two foci.

If ξ is close to 1, the ellipse becomes elongated. If $\xi \to 0$ the ellipse turns into a circle. Therefore, by changing the coefficient of linear charge density, τ we may modify the shape of the ellipse.

We can also adjust other geometrical parameters of the ellipse: the length of semi-major axis or height *H*; the length of semi-minor axis or base radius, *R* at $z = 0$; and radius of curvature, ρ at point (0, *H*) (see Fig. 13).

$$H = \frac{h}{\xi} = \frac{r_a + r_b}{2}, \quad R = \frac{h\sqrt{1-\xi^2}}{\xi}, \quad \rho = \frac{R^2}{H} \,. \tag{35}$$

We can calculate components of the electric field on the surface of the metallic spheroid:

$$E_z = -\frac{E_0}{C}\frac{h(r_b - r_a)^2}{r_a r_b (r_b + r_a)},\; E_r = -\frac{E_0}{C}\frac{h(r_b - r_a)\sqrt{4h^2 - (r_b - r_a)^2}}{r_a r_b \sqrt{(r_b + r_a)^2 - 4h^2}}. \quad (36)$$

Thus the modulus of the electric field is

$$E = \sqrt{E_z^2 + E_r^2} = \frac{E_0}{C}\frac{4h^2(r_b - r_a)}{(r_b + r_a)\sqrt{r_a r_b [(r_b + r_a)^2 - 4h^2]}}. \quad (37)$$

Eqs. (36), (37) allow determining the electric field strength on the surface of the half ellipsoid at an arbitrary point. The field enhancement factor at the apex of the ellipsoid is as follows:

$$\beta = \frac{2\xi^3}{(1-\xi^2)C} = \frac{2\xi^3}{(1-\xi^2)\left(\ln\frac{1+\xi}{1-\xi} - 2\xi\right)}. \quad (38)$$

Analytical expressions for field strength on the z-axis and for field enhancement factor on the tip of the half ellipsoid obtained previously (Forbes et al., 2003; Kosmahl, 1991; Latham, 1981; Latham, 1995) are in agreement with our result. Here, by taking gradient of Eq. (33) we can obtain the field strength at any point we desire. In the limit $\xi \to 0$ we have a metallic half sphere and the field enhancement factor $\beta = 3$. If $\xi \to 1$ then for the elongated metallic needle, we have

$$\beta \approx 2\frac{H}{\rho}\frac{1}{\ln(4H/\rho) - 2}. \quad (39)$$

Ponderomotive forces. Let us estimate the electric force acting on the surface of ellipsoid. We can calculate the force acting on the spheroid surface between circles $r_a = A$ and $r_a = B$ (see Fig. 13).

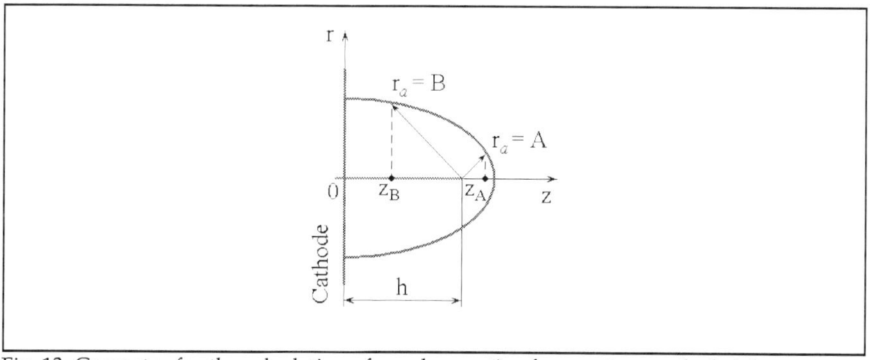

Fig. 13. Geometry for the calculation of ponderomotive force acting on the belt between $r_a = A$ and $r_a = B$.

It is obvious that r-component of force on that surface is equal to zero. For the z-component, after routine operations we obtain

$$F_z(z_A, z_B) = \frac{\pi\rho^2\varepsilon_0(\beta E_0)^2}{2\left(1-\frac{\rho}{H}\right)^2}\left[1-\left(1-\frac{\rho}{H}\right)\frac{z^2}{H^2}-\ln\left\{1-\left(1-\frac{\rho}{H}\right)\frac{z^2}{H^2}\right\}\right]\Bigg|_{z_A}^{z_B}, \quad (40)$$

where z varies from z_A to z_B (see Fig. 13). It gives us the value of net detaching force acting on the surface of the spheroid integrated over the surface between planes $z = z_A$ and $z = z_B$. The total detaching force acting on the ellipsoidal needle is

$$F_{total} = \frac{\pi\rho^2\varepsilon_0(\beta E_0)^2}{2\left(1-\frac{\rho}{H}\right)^2}\left(\frac{\rho}{H}-1-\ln\frac{\rho}{H}\right). \quad (41)$$

In Fig. 14a we show the relative detaching force $F_z(H,z)/F_{total}$ as a function of coordinate z, where $F_z(H,z')$ is the force acting on the surface between plane $z = z'$ and the tip of the spheroid $(z = H)$. Distribution depends only on the parameter H/ρ. The major part of the detaching force is concentrated near the tip when H/ρ is large. Fig. 14b shows the total force isolines on the $(\rho, H/\rho)$ plane. When we chose logarithmic scale for ρ and H/ρ with logarithmic steps we obtain a set of nearly straight isolines with equal distances in the plot.

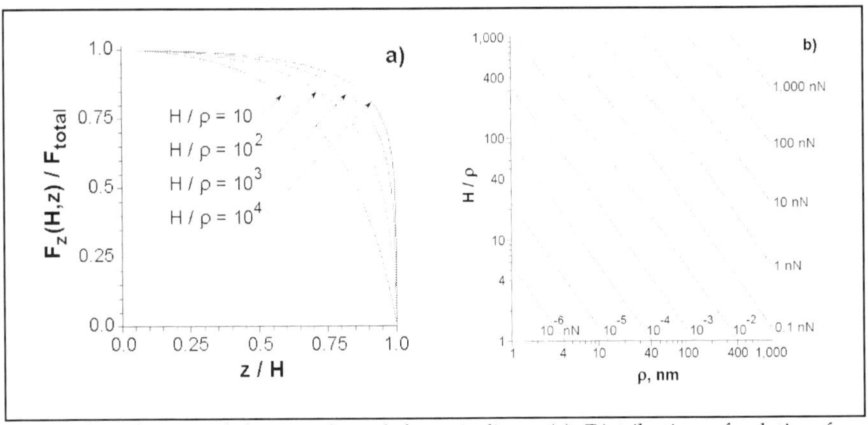

Fig. 14. Distribution of force and total force isolines. (a) Distribution of relative force $F_z(H,z)/F_{total}$ over the axis of needle. (b) Isolines of total force on the $(\rho, H/\rho)$ plane.

We think that under the action of ponderomotive forces in the external electric field, carbon nanotubes that are even chaotically located on the substrate straighten and become oriented (Musatov et al., 2001; Glukhova et al., 2003).

Field emission from individual needle. The current emitted from the surface bounded by $\theta \in (0, \Theta)$ can be expressed in terms of the E_1-function

$$J(\Theta) = \frac{2\pi \rho^2 C_1 (\beta E_0)^2}{\phi} E_1 \left[\frac{C_2 \sqrt{2} \phi^{3/2}}{\beta E_0 \sqrt{1+\lambda}} \right] \Bigg|_{\lambda=\cos\Theta}^{1} . \tag{42}$$

where $\lambda = \cos\theta$, θ is angle between axis z and r_a as shown in Fig. 12.
By comparing with accurate numerical integration of (1) we find that formula (42) is accurate up to the first four digits for $H/\rho > 100$. So formula (42) is a good approximation for investigating the current emitted from an area of surface depending on angle Θ. The total current with similar accuracy can be written approximately as

$$J_{total} = \frac{2\pi \rho^2 C_1 (\beta E_0)^2}{\phi} E_1 \left[\frac{C_2 \sqrt{2} \phi^{3/2}}{\beta E_0 \sqrt{1+\lambda}} \right] \Bigg|_{-\xi}^{1} \approx \frac{2\pi \rho^2 C_1 (\beta E_0)^2}{\phi} E_1 \left[\frac{C_2 \phi^{3/2}}{\beta E_0} \right] \tag{43}$$

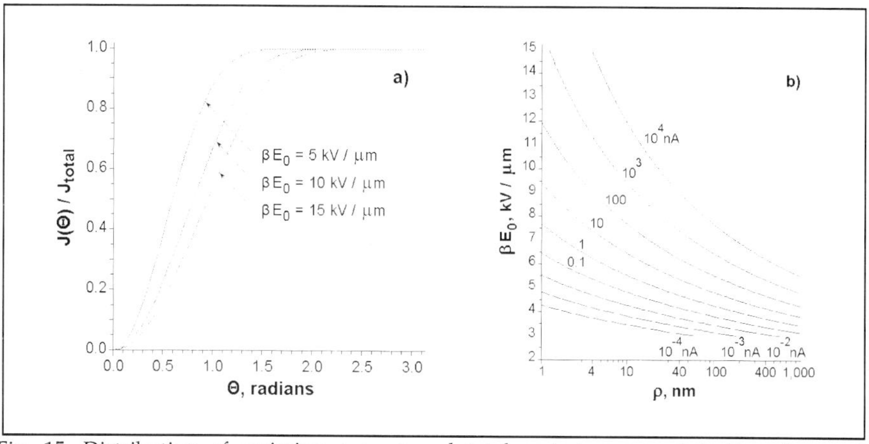

Fig. 15. Distribution of emission current and total emission current isolines for work function φ=4.8 eV. (a) Distribution of relative emission current over Θ. (b) Total emission current isolines on the (βE_0, ρ) plane.

The emission current has much sharper distribution on the tip than the detaching force due to exponential dependence of Fowler-Nordheim formula (1). In Fig. 15a we plot the relative emission current, $J(\Theta)/J_{total}$ as a function of Θ (angle between r_a and axis z), where $J(\Theta)$ is emission from the tip on the surface of the ellipsoidal needle. From (43) we see that total emission current depends only on the area of the hemisphere, $2\pi\rho^2$, work function, φ, and the electric field on the tip, βE_0. So in Fig. 15b we show the total emission current isolines within a reasonable range (from 10^{-4} to 10^4 nA) with logarithmic steps on the (βE_0, ρ) plane, where the work function φ=4.8 eV is assumed. If the electric field is $E_0 \approx 1..5$ V/μm, then we have to use an enhancement factor $\beta \approx 10^3..10^4$ to get an appreciable current.

Simulation of planar anode by "image" charges. The presence of a flat anode placed at a distances comparable with the length of the nanoneedle has strong influence on the emission current and the detaching force. The basic idea of calculation is to replace the cathode by a linearly charged thread in a uniform electric field and to use a set of "image" charges to reproduce the anode as shown in Fig. 16. We put infinite set of "images" of the linearly charged thread with the same spacing, $2L$ $(L > H)$. It is clear that planes $z = 0$ and $z = L$ are planes of symmetry for distributed charges and will have potentials 0 on cathode and $V = E_0 L$ on anode.

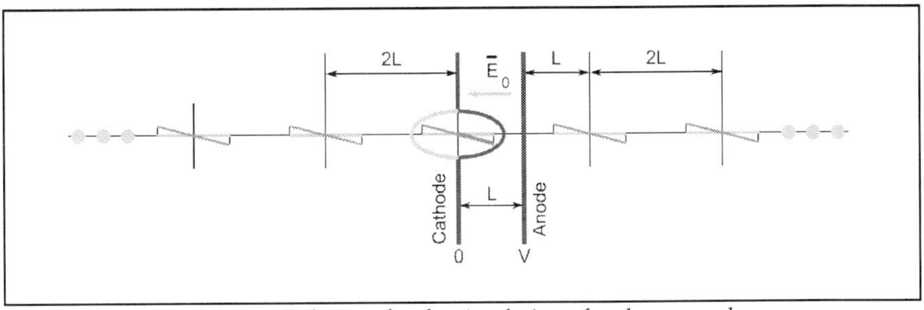

Fig. 16. Infinite set of "image" charges for the simulation of a planar anode.

With the same τ we will get different geometry of cathode due to additional potential of "images". We assume that the new form of the elongated needle will be approximately described by ellipsoid especially near the tip. On the surface of the thin ellipsoid, we have $r << h$ and $z < H < L$. So, we can describe the image potential φ_i as a function of z.

$$\varphi_i(z) = -\frac{E_0}{C^*}\sum_{n=1}^{\infty}\left[(z-2nL)\ln\frac{2nL-z+h}{2nL-z-h} + (z+2nL)\ln\frac{2nL+z+h}{2nL+z-h}\right], \tag{44}$$

where $C^* \equiv 4\pi\varepsilon_0 \frac{E_0}{\tau^*}$, $\quad L = \mu H, \quad \mu > 1$. The thread potential is

$$\varphi = \frac{zE_0}{C}\left[\frac{4h}{r_a + r_b} - \ln\left(\frac{r_a + r_b + 2h}{r_a + r_b - 2h}\right)\right]. \tag{45}$$

Near the tip $z \approx H$ we have the following equation for describing the shape of the ellipsoid:

$$\varphi(r,z) + E_0 z + \varphi_i(z) \approx \varphi(r,z) + E_0 z + \frac{z}{H}\varphi_i(H) = 0\,. \tag{46}$$

So we have to change constant C on constant C^* as following

$$C^* = \ln\left(\frac{1+\xi}{1-\xi}\right) - 2\xi - P\,, \tag{47}$$

where

$$P = \sum_{n=1}^{\infty}\left[(2n\mu-1)\ln\frac{2n\mu-1+\xi}{2n\mu-1-\xi}-(2n\mu+1)\ln\frac{2n\mu+1+\xi}{2n\mu+1-\xi}\right]. \quad (48)$$

Therefore we have the field emission factor in the form

$$\beta(L/H,H/\rho)=\frac{2\xi^3}{(1-\xi^2)C^*}. \quad (49)$$

Finally for the force and emission current we may use the above derived formulas. Instead of (38) we should use (47), (48), and (49). So, the parameter $\mu = L/H$ in the force and emission current formulas (40), (41), (42), and (43) should be included only through the field enhancement factor $\beta=\beta(L/H,H/\rho)$.

What distance between the anode and cathode is large enough to assert that the elongated metallic spheroid is placed in a uniform electric field? In experiment we may measure distance L between the anode and cathode and the distance, L - H between the anode and the needle apex. We can determine the applied electric field both by $E_0=V/L$ and by $E_0'=V/(L-H)$. In our calculations we move the anode and increase the voltage V so as to keep E_0 or E_0' immutable. It is clear that for a large distance (in the limit $L\to\infty$) the difference between definitions of E_0 and E_0' disappears. Fig. 17 shows the total current emitted from the needle versus the anode-cathode distance.

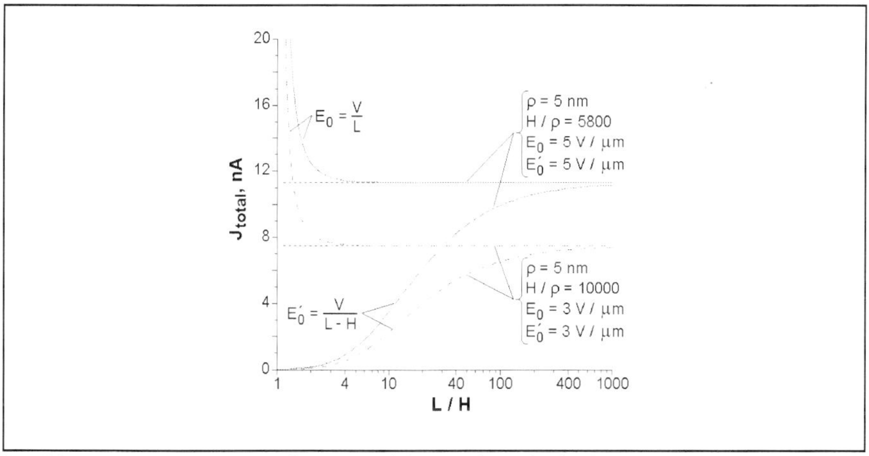

Fig. 17. Emission current as a function of the anode-cathode distance parameter $\mu = L/H$ for constant electric field $E_0 \equiv V/L$ (red line) or $E_0' \equiv V/(L-H)$ (blue line). Radius of curvature is fixed at $\rho = 5$ nm. For solid curves, $H/\rho = 5800$ and for dash curves, $H/\rho = 10000$. For red and blue solid curves $E_0 = 5\,\mathrm{V/\mu m}$, for red and blue dash curves $E_0 = 3\,\mathrm{V/\mu m}$.

We set for blue lines $E_0' = \text{const}$ and for red lines $E_0 = \text{const}$. With $E_0 = E_0' = \text{const}$ the total currents have the same limit for $\mu = L / H \to \infty$. But currents with $E_0' = \text{const}$ tend to approach the limit much slower than the corresponding currents with $E_0 = \text{constant}$. If we define the applied electric field as E_0 and the anode-cathode distance is ten times more than the needle height, then we can neglect the influence of the anode and assume that the metallic needle is placed in a uniform field. In contrast, if the applied electric field is defined as E_0', then the anode-cathode distance must exceed one thousand times of the needle height for the above statement to be valid.

3.3.5 The model of a hemisphere on a post. Numerical approximations

The model of a hemisphere on a post allows only the numerical solution. There are many numerical results obtained by various researchers which have been generalized by simple algebraic formulas of field enhancement factor for an individual nanotube and assembly of nanotubes.

From our point of view the most accurate formula belongs to Edgcombe et. al. (Edgcombe & Valdrè, 2001; Edgcombe & Valdrè, 2002; Forbes et al., 2003)

$$\beta_0 = 1.2\left(2.5 + \frac{H}{\rho}\right)^{0.9} . \tag{50}$$

This formula accurately describes the field enhancement factor of individual nanotube in a uniform electric field. Comparison of field enhancement factors for the "floating sphere at emitter-plane potential" model (green line), the "hemi-ellipsoid on plane" model (red line), and fitting formula for the "hemisphere on a post" model (blue line) is shown in Fig. 18.

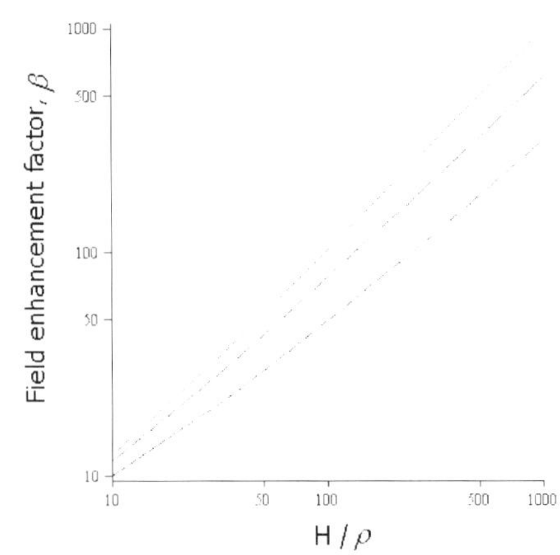

Fig. 18. Comparison the field enhancement factors for three models: floating sphere (green line), hemi-ellipsoid (red line), and hemisphere on a post (blue line).

For a nanotube in space between parallel cathode and anode planes we prefer approximation (Bonard et al., 2002)

$$\beta = \beta_0 \left[1 + 0.013 \left(\frac{L-H}{L} \right)^{-1} - 0.033 \left(\frac{L-H}{L} \right) \right]. \tag{51}$$

For a nanotube surrounded by neighboring nanotubes with a screening effect we recommend (Jo et al., 2003)

$$\beta = \beta_0 \left[1 - \exp\left(-2.3172 \frac{D}{H} \right) \right]. \tag{52}$$

3.4 Further investigations

Let us shortly mention the further work that should now be carried out on the field-emission properties of CNTs.
1) Developing of field emission theory for CNT emitters
2) Temperature effects in CNTs: Joule heating; Peltier and Nottingham effects; carbon nanotube heat radiation; thermo-field emission from nanotubes
3) Action of electrostatic forces on CNT field emitters: elongation and straightening of nanotube; pulling out of nanotubes in strong electric field.
4) Reliability of CNT field emitters: degradation mechanism of field emission; lifetime of nanotube emitters.

4. Applications

Field emission is the most promising application areas of carbon nanotubes. We shall consider only a few samples of working devices.

4.1 Field emission displays

Flat-panel displays with CNT-based cathodes are proposed as alternative to other displays with film emitters. The first diode-type display consisting of 32 × 32 matrix-addressable pixels was manufactured by Wang et al. in 1998 (Wang et al., 1998). At present, flat-panel displays based on CNT field-emission cathodes are developed in hundreds of laboratories, and engineering samples have been already manufactured.
A 4.5-inch full-color CNT display developed by Choi et al. (Choi et al., 1999) from Samsung Company was demonstrated at several exhibitions.
A full-color addressable display developed jointly by the limited liability company "Volga-Svet" (Saratov, Russia) and "CopyTele Inc.} (New York, USA) (Abanshin et al., 2002) was demonstrated at the conference IveSC'02. Emission in the proposed design was from thin edges of carbon nanocluster films, which are slightly hanging from the supporting pedestals (Fig. 19). The anode covered by a phosphor layer is shown on the top in Fig. 19. Gating/blocking of emission current is performed by a metal control electrode located on the substrate between the pedestals. The figure shows the results of the trajectory analysis for an option of the structure (Zhbanov et al., 2004).

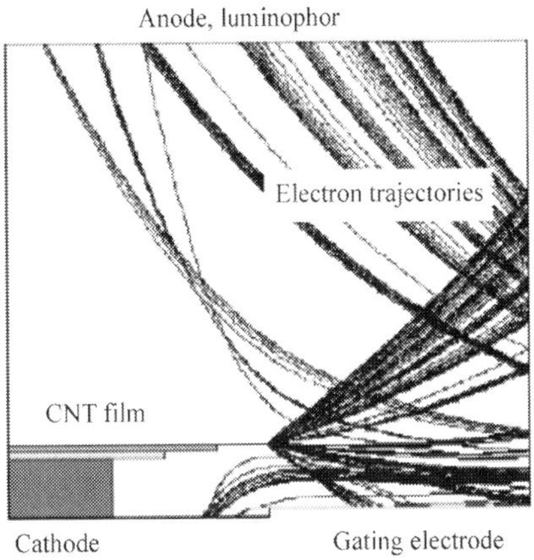

Fig. 19. Trajectory analysis for the flat-panel display produced by the limited liability company "Volga-Svet" and "CopyTele Inc." (Zhbanov et al., 2004).

4.2 X-ray tubes

A compact X-ray tube with a field emitter based on carbon nanotubes was developed by Musatov's group from Institute of Radio-engineering and Electronics (Russia) (Musatov et al., 2007). Over a long time interval, the X-ray tube maintains an anode current of 300 μA, an anode voltage of 10 kV, and the stable characteristics of the field emitter (Fig. 20).

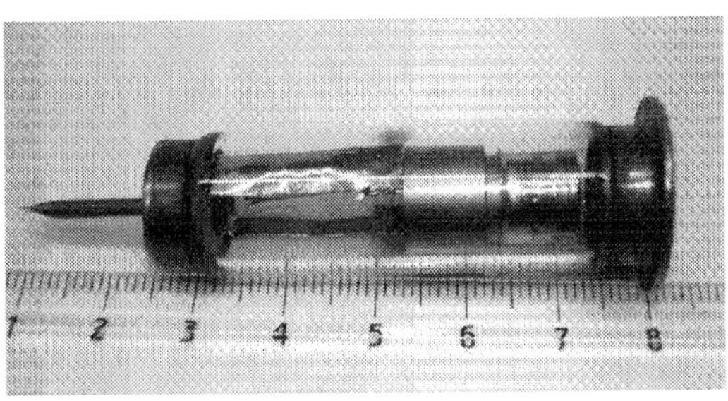

Fig. 20. Compact X-ray tube. The figure is taken from (Musatov et al., 2007).

4.3 Light elements

The Bonard team developed the cathodoluminescent light-emitting element of cylindrical geometry (Bonard et al., 2001a). The cathode in the form of a metal rod with deposited CNTs is located on the axis of a glass tube covered by phosphor from the inside. The operating voltage is 7.5 kV, the current density on the cathode is 0.25 mA/cm^2, the current density on the anode is 0.03 mA/cm^2, and the lamp luminance is 10^4 cd/m^2.

CNT light-emitting elements of various colors in the form of a filamentary electric lamp are proposed by Saito team in (Bonard et al., 2002b). The operating grid voltage is from 0.2 to 1.2 kV, the current density at the cathode with an area of 2 mm^2 is 10 mA/cm^2, the average electric field intensity is 1.5 $V/\mu m$, and the luminance of elements of various colors ranges from $1.5 \cdot 10^4$ to $6.3 \cdot 10^4$ cd/m^2.

Samples of the light-emitting elements were demonstrated by A.N. Obraztsov (Obraztsov et al., 2002) and E. P. Sheshin (Leshukov et al., 2002) at the Saratov 4th International Conference on Vacuum Electrons Sources (IveSC'02).

4.4 Microwave devices

The electron gun consisting of edge or cylindrical emitters with flat faces; a grid control electrode placed above the emitters; and a three-anode focusing system which is common for all the emitters (Fig. 21) was designed by Zakharchenko's group (Zakharchenko et al., 1996). The flat face surfaces of the edge emitter are covered with a film made of carbon nanotubes of diameter 30-100 Å. A substantial density of emitting centers being as great as 10^8-10^{10} tips/cm^2 provides a high current density at low accelerating voltages.

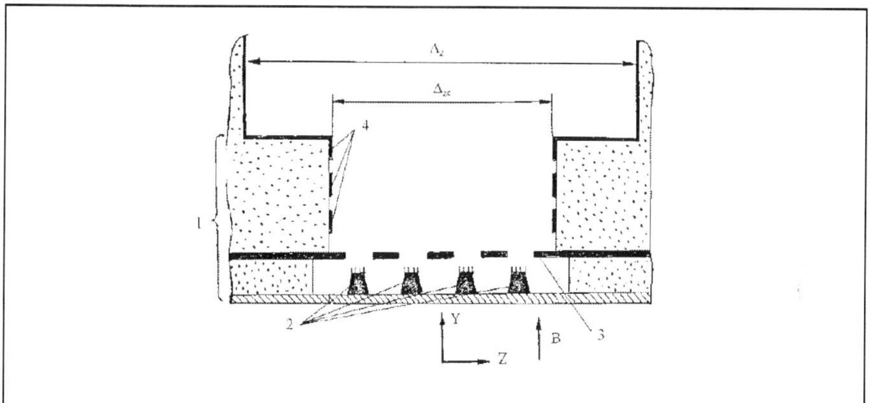

Fig. 21 Schematic of the electron gun for an amplifier. Magnetic field B=0.08 T, voltage V=400 V, and current I=1 A/cm^2. The electron gun (1) consists of edge or cylindrical field emitters (2) covered with CNT film, a control electrode (3), and three anodes (4) (Zakharchenko et al., 1996).

5. Conclusions and Acknowledgements

In this chapter we theoretically investigate the field emission from carbon nanotube field emitters in diode configuration between a flat anode and cathode. Exact analytical formulas of the electrical field, field enhancement factor, ponderomotive force, and field emission current are found. Applied voltage, height of the needle, radius of curvature on its top, and the work function are the parameters at our disposal. The field enhancement factor, total force and emission current, as well as their distributions on the top of the needle for a wide range of parameters have been calculated and analyzed. Also we review the technology of fabrication and the application areas of CNT electron sources.

We have right to conclude that carbon nanotubes are excellent field emitters. Now CNTs conquer appreciable positions as electron sources in compact X-ray tubes, lighting elements, scanning electron microscopy electron guns, and electron guns for microwave devices. We think CNTs have good potential in the field emission display market.

We gratefully acknowledge support through the National Science Council of Taiwan, Republic of China, through the project NSC 95-2112-M-001-068-MY3.

6. References

Abanshin, N.; Muchina, E. & Nikishin, N. et al. (2002). *Proc. IVeSC'2002 Supplement,* Saratov, Russia, July 15-19, p.13.

Banerjee, D.; Jo, S. & Ren, Z. (2004). Enhanced Field Emission of ZnO Nanowires. *Advanced Materials,* 16, 2028.

Bethune, D.; Kiang, C. & de Vries, M. et al. (1993). Cobalt-Catalysed Growth of Carbon Nanotubes with Single-Atomic-Layer Walls. *Nature,* 363, 605.

Bonard, J-M.; Stöckli, Th. & Noury, O. et al. (2001a). Field emission from cylindrical carbon nanotube cathodes: Possibilities for luminescent tubes. *Applied Physics Letters,* Vol. 78, No. 18, P. 2775-2777.

Bonard, J.-M.; Kind, H. & Stöckli, Th. et al. (2001b). Field emission from carbon nanotube: the first five years. *Solid State Electronics,* Vol. 45, P. 893-914.

Bonard, J. -M.; Dean, K. & Coll, B. et al. (2002a). Field Emission of Individual Carbon Nanotubes in the Scanning Electron Microscope. *Physical Review Letters,* 89, 197602.

Bonard , J.-M.; Gaal, R. & Garaj, S. et al. (2002b). Field emission properties of carbon nanohorn films. *Journal of Applied Physics,* 91, 10107–10109.

Charbonnier, F.; Mackie, W. & Hartman, R. et al. (2001). *The Journal of Vacuum Science Technology B,* 19(3), 1064.

Chernozatonskii, L.; Gulyaev, Yu. & Kosakovskaja, Z. et al. (1995). Electron field emission from nanofilament carbon films. *Chemical Physics Letters,* Vol. 233, P. 63-68.

Chernozatonskii, L.; Kosakovskaya, Z. & Gulyaev, Yu. et al. (1996). Influence of external factors on electron field emission from thin-film nanofilament carbon structures. *The Journal of Vacuum Science Technology B,* Vol. 14(3), P. 2080-2082.

Choi, W.; Chung, D. & Kang, J. et al. (1999). Fully sealed, high-brightness carbon-nanotube field-emission display. *Applied Physics Letters,* Vol. 75, No. 20, P. 3129-3131.

Collins, P. & Zettl, A. (1996). A simple and robust electron beam source from carbon nanotubes. *Applied Physics Letters,* Vol. 69, No. 13, P. 1969-1971.

Collins, P. & Zettl, A. (1997). Unique characteristics of cold cathode carbon-nanotube-matrix field emitters. *Physical Review B,* Vol. 55, No. 15, P. 9391-9399.

Drechsler, M. & Müller, E. (1953). Zur Feldelektronenemission und Austrittsarbeit Einzelner Kristallflachen. *Zeitschrift Fur Physik,* Vol. 134, Issue 2, P. 208-221.

Ducastelle, F.; Blase, X. & Bonard, J.-M. et al. (2006). *Electronic Structure,* Lecture Notes of Physics 677, 199–276.

Edgcombe, C. & Valdrè, U. (2001). The enhancement factor and the characterization of amorphous carbon field emitters. *Solid-State Electronics,* Vol. 45, No. 6, pp. 857-863.

Edgcombe, C. & Valdrè, U. (2002). Experimental and computational study of field emission characteristics from amorphous carbon single nanotips grown by carbon contamination I. Experiments and computation. *Philosophical Magazine B*, Vol. 82, No. 9, P. 987-1007.

Fan, Sh.; Chapline, M. & Franklin, N. et al. (1999). Self-Oriented Regular Arrays of Carbon Nanotubes and Their Field Emission Properties. *Science*, Vol. 283, P. 512-514.

Filip, V.; Nicolaescu, D. & Tanemura, M. et al. (2001). Modeling the electron field emission from carbon nanotube films. *Ultramicroscopy*, Vol. 89, P. 39–49.

Fishbine, B.; Miglionico, C. & Hackett, K. et al. (1994). Graphene nanotubule cold field emission electron sources. *Materials Research Society Symposium – Proceedings*, Vol. 349, pp. 319-324.

Forbes, R.; Edgcombe, C. & Valdrè, U. (2003). Some comments on models for field enhancement. *Ultramicroscopy*, Vol. 95, pp. 57–65.

Fowler, R. & Nordheim, L. (1928). Electron Emission in Intense Electric Fields. *Proceedings of the Royal Society (London) A*, 119, pp. 173-181.

Fursey, G. (2005). *Field emission in vacuum microelectronics*. Kluwer Academic / Plenum Publishers, New York, ISBN: 0-306-47450-6.

Ghosh, P.; Tanemura, M. & Soga, T. et al. (2008). Field emission property of N-doped aligned carbon nanotubes grown by pyrolysis of monoethanolamine. *Solid-State Communications*, Vol. 147, Issue 1-2, pp. 15-19.

Glukhova, O.; Zhbanov, A. & Torgashov, I. et al. (2003). Ponderomotive forces effect on the field emission of carbon nanotube films. *Applied Surface Science*, Vol. 215, pp. 149-159.

Gulyaev, Yu.; Nefyodov, I. & Sinitsyn, N. et al. (1994). Simulation of the emission characteristics of field emitter arrays based on nanofilament carbon structures. *7-th International Vacuum Microelectronics Conference*. ATRIA, World Trade Center, Grenoble, France, July 4-15, pp. 319-321.

Gulyaev, Yu.; Chernozatonskii, L. & Kosakovskaja, Z. et al. (1995). Field emitter arrays on nanotube carbon structure films. *The Journal of Vacuum Science Technology B*, Vol. 13(2), pp. 435-436.

Gulyaev, Yu.; Sinitsyn, N. & Torgashov, G. et al. (1997). Work function estimate for electrons emitted from nanotube carbon cluster films. *The Journal of Vacuum Science Technology B*, Vol. 15(2), pp. 422-424.

de Heer, W. ; Châtelain, A. & Ugarte, D. (1995). A Carbon Nanotube Field-Emission Electron Source. *Science*, Vol. 270, No. 5239, pp. 1179-1180.

de Heer, W.; Bonard, J.-M. & Fauth K. et al. (1997). Electron Field Emitters Based on Carbon Nanotube Films. *Advanced Materials*, Vol. 9, No 1, pp. 87-89.

Hu, Y. & Huang, C.-H. (2003). Computer simulation of the field emission properties of multiwalled carbon nanotubes for flat panel displays. *The Journal of Vacuum Science Technology B*, Vol. 21(4), pp. 1648-1654.

Iijima, S. (1991). Helical microtubules of graphitic carbon. *Nature*, Vol. 354, pp. 56-58.

Iijima, S. & Ichihashi, T. (1993). Single shell carbon nanotubes of one nanometer diameter. *Nature*, Vol. 363, pp. 603-605.

Jackson, J. (1999). *Classical Electrodynamics*. John Wiley & Sons, Inc. - 808 p. ISBN 0-471-30932-X.

Jang, H.; Kim, D.-H. & Lee, H.-R. et al. (2005). Field emission from cone-like single crystalline indium tin oxide nanorods. *Materials Letters*, Vol. 59, pp. 1526–1529.

Jo, S.; Tu, Y. & Huang, Z. et al. (2003a). Effect of length and spacing of vertically aligned carbon nanotubes on field emission properties. *Applied Physics Letters*, Vol. 82, No. 20, pp. 3520-3522.

Jo, S.; Lao, J. & Ren, Z. et al. (2003b). Field-emission studies on thin films of zinc oxide nanowires. *Applied Physics Letters*, Vol. 83, No. 23, pp. 4821-4823.

de Jonge, N. & Bonard, J.-M. (2004). Carbon nanotube electron sources and applications. *Philosophical Transactions of the Royal Society (London) A*, Vol. 362, pp. 2239-2266.

Kosmahl, H. (1991). Analytic Evaluation of Field Emission Enhancement Factors for Ellipsoidal Cones and Elliptic Cross-Section Wedges. *IEEE Trans. Electron Devices*, Vol. 38, No. 6, pp. 1534-1537.

Kukovitsky, E.; L'vov, S. & Sainov, N. et al. (2003). *Applied Surface Science*, Vol. 215, pp. 201–208.

Kung, S.; Hwang, K. & Lin, I. (2002). Oxygen and ozone oxidation-enhanced field emission of carbon nanotubes. *Applied Physics Letters*, Vol. 80, pp. 4819-4821.

Latham, R. (1981). *High Voltage Vacuum Insulation: The Physical Basis*, Academic Press, London.

Latham, R. (1995). *High Voltage Vacuum Insulation: Basic Concepts and Technological Practice*, Academic Press, London.

Lee, Y.; Kim, S. & Tománek, D. (1997). Field-induced unraveling of carbon nanotubes. *Chemical Physics Letters*, Vol. 265, pp. 667-672.

Lee, Y.-H.; Choi, Ch.-H. & Jang, Y.-T. et al. (2002). Tungsten nanowires and their field electron emission properties. *Applied Physics Letters*, Vol. 81, No. 4, pp. 745-747.

Leshukov, M.; Baturin, A. & Chadaev, N. et al. (2002). *Proc. IVeSC'2002 Supplement*, Saratov, Russia, July 15-19, 2002, p. 246.

Li, S.; Lee, Ch. & Lin, P. et al. (2006). Gate-controlled ZnO nanowires for field-emission device application. *The Journal of Vacuum Science Technology B*, Vol. 24(1), pp. 147-151.

Li, J.; Wang, Q. & Gu, C. (2007). Growth and field emission properties of tubular carbon cones. *Ultramicroscopy*, Vol. 107, pp. 861–864.

Millikan, R. & Lauritsen, C. (1929). Dependence of Electron Emission from Metals upon Field Strengths and Temperatures. *Physical Review*, Vol. 33, pp. 598 - 604.

Musatov, A.; Kiselev, N. & Zakharov, D. et al. (2001). Field electron emission from nanotube carbon layers grown by CVD process. *Applied Surface Science*, Vol. 183(1-2), pp. 111-119.

Musatov, A.; Gulyaev, Yu. & Izrael'yants, K. et al. (2007). A Compact X-ray Tube with a Field Emitter Based on Carbon Nanotubes. *Journal of Communications Technology and Electronics*, Vol. 52, No. 6, pp. 714–716.

Nilsson, L.; Groening, O. & Emmenegger, C. et al. (2000). Scanning field emission from patterned carbon nanotube films. *Applied Physics Letters*, Vol. 76, pp. 2071-2073.

Nilsson, L.; Groening, O. & Kuettel, O. et al. (2002). Microscopic characterization of electron field emission. *The Journal of Vacuum Science Technology B*, Vol. 20(1), pp. 326-337.

Obraztsov, A.; Petrushenko, Yu. & Satanovskaya, O. et al. (2002). *Proc. IVeSC'2002 Supplement*, Saratov, Russia, July 15–19, p. 254.

Park, N.; Han, S. & Ihm, J. (2001). Effects of oxygen adsorption on carbon nanotube field emitters. *Physical Review B*, Vol. 64, Issue 12, No. 125401.

Pogorelov, E.; Zhbanov, A. & Chang, Y. (2009). Field enhancement factor and field emission from hemi-ellipsoidal metallic needle. *Ultramicroscopy*, Vol. 109, Issue 4, pp. 373-378.

Read, F. & Bowring, N. (2004). Field enhancement factors of random arrays of carbon nanotubes. *Nucl. Instr. Meth. Phys. Res. A*, Vol. 519, pp. 305-314.

Read, F. & Bowring, N. (2004). Field enhancement factors of one-dimensional and two-dimensional arrays of nanotubes. *Microelectronic Engineering*, Vol. 73-74, pp. 679-685.

Rinzler, A.; Hafler, J. & Nikolaev, P. et al. (1994). Field Emission and Growth of Fullerene Nanotubes. *MRS proceeding*, Vol. 359, pp. 61-68 (December 1994).

Rinzler, A.; Hafner, J. & Nikolaev, P. et al. (1995). Unraveling Nanotubes: Field Emission from an Atomic Wire. *Science*, Vol. 269, pp. 1550-1553.

Saito, Y.; Hamaguchi, K. & Nishino, T. et al. (1997a). Field Emission Patterns from Single-Walled Carbon Nanotubes. *Jpn. J. Appl. Phys.*, Vol. 36, Part 2, No. 10A, P. L1340-L1342.

Saito, Y.; Hamaguchi, K. & Hata, K. et al. (1997b). Conical beams from open nanotubes. *Nature*, Vol. 389, No. 6651, pp. 554-555.

Seelaboyina, R.; Huang, J. & Park, J. et al. (2006). Multistage field enhancement of tungsten oxide nanowires and its field emission in various vacuum conditions. *Nanotechnology*, Vol. 17, pp. 4840-4844.

Shang, X. ; Wang, M. & Qu, S. et al. (2007). A model calculation of the tip field distribution for a single carbon nanotube. *The Journal of Appied. Physics*, Vol. 102, 054301.

Sinitsyn, N.; Gulyaev, Yu. & Torgashov, G. et al. (1997). Thin films consisting of carbon nanotubes as a new material for emission electronics. *Applied Surface Science*, Vol. 111, pp. 145-150.

Smith, R.; Carey, J. & Forrest, R. et al. (2005). Effect of aspect ratio and anode location on the field emission properties of a single tip based emitter. *J. Vac. Sci. Technol. B*, Vol. 23(2), pp. 632-635.

Spindt, C. (1968). A Thin-Film Field-Emission Cathode. *The Journal of Applied Physics*, Vol. 39, Issue 7, 3504.

Srivastava, S.; Vankar, V. & Rao, D. et al. (2006). Enhanced field emission characteristics of nitrogen-doped carbon nanotube films grown by microwave plasma enhanced chemical vapor deposition process. *Thin Solid Films*, Vol. 515, pp. 1851-1856.

Terrones, M.; Jorio, A. & Endo, M. et al. (2004). New direction in nanotube science. *Materials Today*, Vol. 7, No. 10, pp. 30-45.

Terrones, M; Souza, A. & Rao, A. (2008). Doped carbon nanotubes: Synthesis, characterization and applications. *Carbon Nanotubes Book, Series: Topics in Appl. Phys.*, vol. 111, pp. 531-566.

Tumareva, T.; Sominskiĭ, G. & Efremov, A. et al. (2002). Tip Field Emitters Coated with Fullerenes. *Technical Physics*, Vol. 47, No. 2, pp. 244–249.

Tumareva, T.; Sominski, G. & Svetlov, I. et al. (2008). Fullerene-Coated Field Emitters Activated by a Potassium Ion Flux in High Electric Fields. *Technical Physics*, Vol. 53, No. 11, pp. 1504–1507.

Wang, Q.; Setlur, A. & Lauerhaas, J. et al. (1998). A nanotube-based field-emission flat panel display. *Applied Physics Letters,* Vol. 72, No. 22, pp. 2912-2913.

Wang, X.; Wang, M. & He, P. et al. (2004). Model calculation for the field enhancement factor of carbon nanotube. *J. Appl. Phys.,* Vol. 96, No. 11, pp. 6752-6755.

Wang, X.; Wang, M. & Ge, H. et al. (2005). Modeling and simulation for the field emission of carbon nanotubes array. *Physica E,* Vol. 30, pp. 101-106.

Wood, R. (1897). A New Form of Cathode Discharge and the Production of *X*-Rays, together with Some Notes on Diffraction. Preliminary Communication" *Phys. Rev.* (Series I) , Vol. 5, pp. 1 - 10.

Yudasaka, M.; Iijima, S. & Crespi, V. (2008). Single-wall carbon nanohorns and nanocones. *Carbon Nanotubes Book, Series: Topics in Appl. Phys.,* vol. 111, pp. 605-629.

Zakharchenko, Y.; Torgashov, G. & Gulyaev Y. et al. (1996). Two-stage distributed amplifier on field emitter arrays. *J. Vac. Sci. Technol. B,* Vol. 14(3), pp. 1982-1985.

Zhbanov, A.; Sinitsyn, N. & Torgashov, G. (2004). Nanoelectronic Devices Based on Carbon Nanotubes. *Radiophysics and Quantum Electronics,* Vol. 47, pp. 435-452.

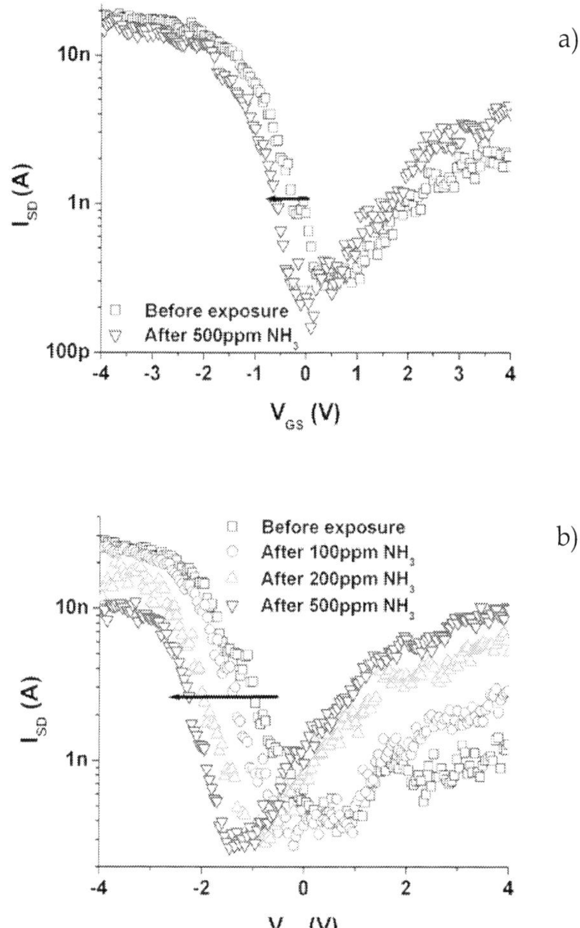

Fig. 7. The transfer characteristics for Device 1A with the contacts passivated before and after exposure to NH_3 at (a) T=150°C and (b) T=200°C, respectively.

The transfer curve started to shift towards negative gate voltage after NH_3 exposure at 150°C and above, see Fig. 7 (a). Since the contacts were fully isolated from NH_3, this parallel shift in the transfer curve suggests that NH_3 could adsorb on the CNT wall and donate electrons to the CNT. Consequently, the Fermi level of the CNT moves towards the conduction band edge so that the threshold voltage V_{TH} becomes more negative. When the testing temperature reached 200°C, this phenomenon became more prominent. Progressive shift of the transfer curve in accordance with NH_3 concentrations is shown in Fig. 7 (b).

Under a first order estimation, the total charge transferred $\Delta Q = C_G \Delta V_{TH}$, where the gate capacitance $C_G = \dfrac{2\pi\varepsilon\varepsilon_0 L}{\cosh^{-1}(h/r)}$, the SiO_2 dielectric constant ε=3.9 and thickness h=200nm. For a SWNT bundle with a length L~5μm and a radius r~5nm, C_G≈0.25 fF. Thus, ΔQ is approximately 0.625fC or about 4000 electrons, at 200°C with $\Delta V_{TH} = 2.5V$ for 500ppm NH_3. If the cross-sectional area of NH_3 A~0.13nm², the length of the exposed CNT channel L'~3μm, the coverage θ~0.07 (interpolated from the Langmuir plot for 500ppm NH_3 on CNT(Bradley, et al., 2003)), the charge transfer rate $f \approx \Delta QA / q\theta\pi rL'$ is about 0.02 electron. This value is reasonably consistent with typical theoretical predictions.

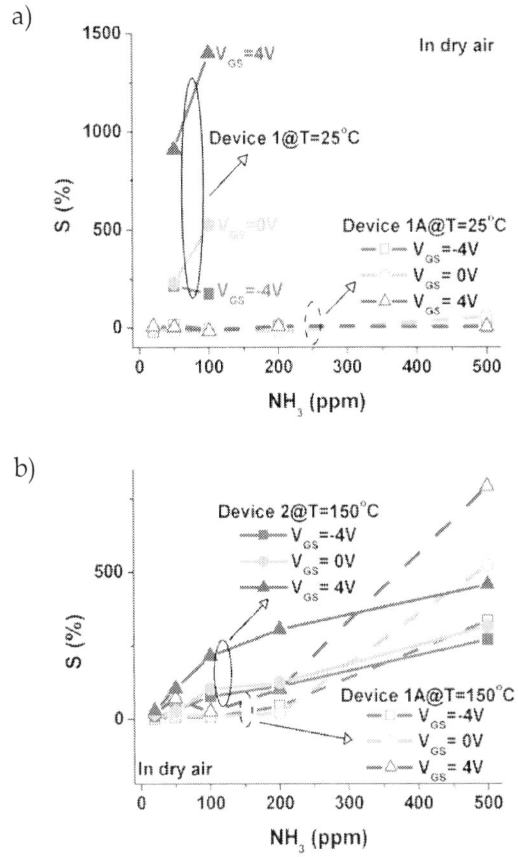

Fig. 8. Extracted sensitivities for (a) Device 1 and Device 1A at T=25°C; (b) Device 2 and Device 1A at T=150°C, respectively.

The extracted sensitivity $S = \Delta R / R_0$ of Device 1 and Device 1A under three gate voltages at T=25°C are shown in Fig. 8 (a). For Device 1, a very high sensitivity and significant gate modulation were observed. When the gate voltage is varied from negative to positive, the dominant carrier injection process changes from tunneling to thermionic emission, and the source-drain current becomes very sensitive to the SB height. A small change in the contact SB height due to NH_3 adsorption will be prominently reflected in the source-drain current. Our device structure with a CNT on top of metal electrodes could also enhance the SB modulation effect. Once the contacts passivation is carried out in Device 1A, it essentially does not respond to NH_3, implying that the CNT channel is not active to NH_3 at room temperature. Fig. 8 (b) compares the sensitivities for Device 1A and Device 2 at T=150°C. At small NH_3 concentrations, a low coverage of NH_3 on the CNT channel and poor charge transfer efficiency result in a small sensitivity in Device 1A. When the NH_3 concentration is increased, the sensitivity for Device 2 becomes saturated, probably due to limited interaction area in the CNT/Au contacts.

4.3 Effect of oxygen on NH_3 sensing

Theoretical studies suggest that, NH_3 interacts weakly with pristine CNTs with little charge transfer (Kong, et al., 2000, Bauschlicher & Ricca, 2004, Peng & Cho, 1999). Existence of a large activation barrier prevents adsorption of NH_3 on perfect CNTs even at high temperatures. However, the adsorption of the gas molecules on defective CNTs could be much easier (Feng, et al., 2005, Robinson, et al., 2007). In addition, the adsorption barrier of NH_3 on a defective CNT can be further lowered by pre-dissociated oxygen atoms, leading to an enchanted charge transfer rate, as pointed out by Andzelm et al (Andzelm, et al., 2006). In order to study the influences of oxygen on NH_3 adsorption onto the CNT wall, we changed the background gas from dry air to N_2. Device 1A was first annealed in N_2 environment at 350°C for 2 hrs to degas the adsorbed oxygen. Note that during the high-temperature annealing, remaining oxygen molecules at the contacts can be further desorbed and the device became more n-type. The transfer curves before and after exposure to 500ppm NH_3 for 1000s from T=25°C to T=150°C were shown in Fig. 9, and real-time sensing results at T=200°C are shown in Fig. 10. No detectable changes due to NH_3 exposure were observed. Comparing with the sensing response observed in dry air environment, we can confirm that the adsorption of NH_3 is facilitated by environmental oxygen. It was also found that the sensitivity was restored after the background was changed to dry air again. Our results are consistent with Andzelm *et al*'s predictions. NH_3 could preferentially attach to the defect sites on CNT with pre-dissociated oxygen, as illustrated in Fig. 11.

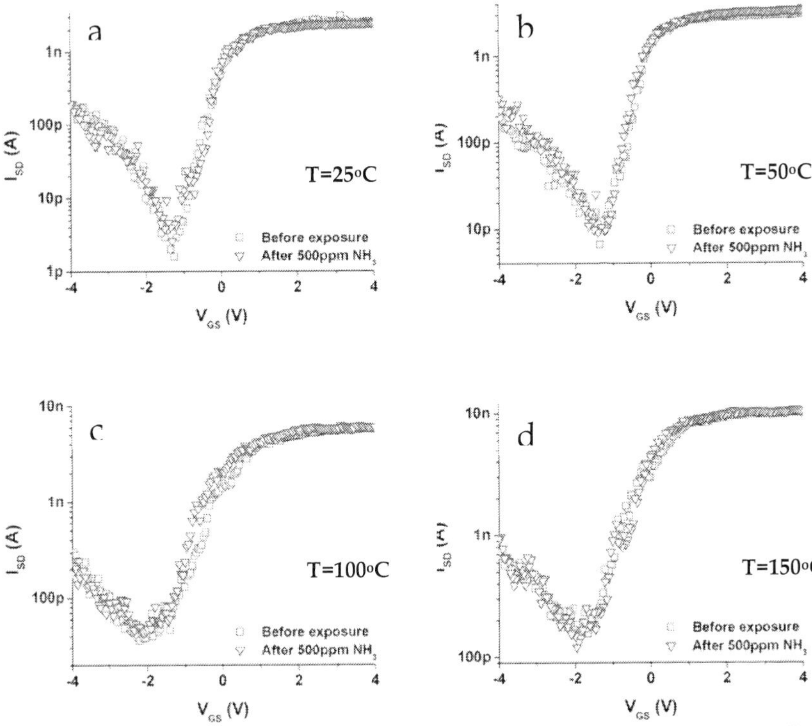

Fig. 9. Transfer characteristics of Device 1A in N_2 before and after exposure to NH_3 at (a) T=25°C, (b) T=50°C, (c) T=100°C and (d) T=150°C respectively.

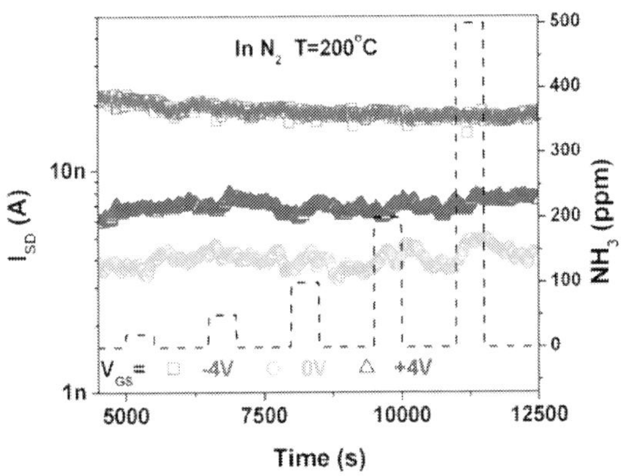

Fig. 10. Response of I_{SD} to various concentrations of NH_3 in N_2 at T=200°C for Device 1A.

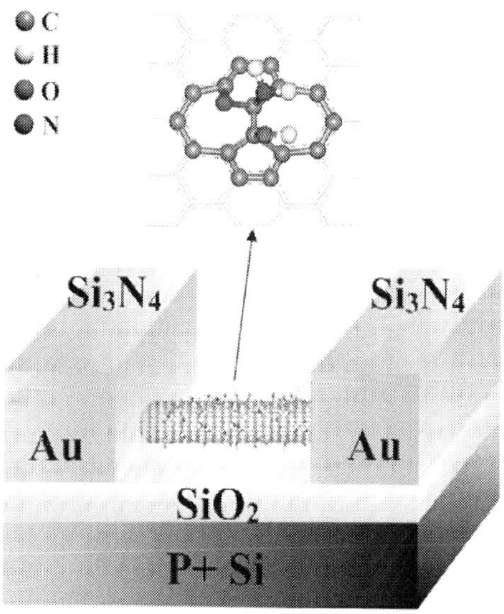

Fig. 11. Schematic of NH_3 adsorption on Device 1A. Here, we illustrate a NH_3 molecule adsorbs on a CNT with Stone-Wales defect with pre-dissociated oxygen atoms, as suggested in by Andzelm et al.

4.4 Comparisons of the sensing mechanisms in CNT based gas sensors

From our results, we are able to rule out several possibilities of indirect interactions between NH_3 and CNT. Firstly, as the testing environment was totally dry, NH_3 adsorption through water layer is not applicable here. Secondly, If NH_3 could interact through the SiO_2 substrate or adsorb inside the CNT bundles, a reduced sensitivity should have been observed after contacts passivation. However, our observation that Device 1A is totally insensitive to NH_3 at room temperature does not support this hypothesis. In fact, charge transfer and the SB modulation are the two mechanisms in our CNT sensors.

At room temperature, the weak adsorption of NH_3 on the CNT wall does not induce any measurable effect on the source-drain current. The sensing signal mainly arises from the contacts. When NH_3 molecules are adsorbed on the CNT/Au interface, the work function of the Au electrode is reduced (Bilic, et al., 2002) and/or the electrostatic charge balance between the CNT and Au is disturbed by the dipoles of NH_3 molecules,(Yamada, 2006) leading to an increased SB for hole injection. The sensitivity reflected in the source-drain current are, however, gate voltage dependant. As illustrated in Figure 12 (b), when a negative gate voltage bends the energy band of the CNT upwards, the SB width becomes very narrow and holes could tunnel through the barrier, even when the SB height is increased by NH_3. In contrast, at a positive gate voltage, the SB width is too thick for tunneling process. Thus, hole injection is only through thermionic emission over the SB height. The source-drain current is then expressed as:(Appenzeller, et al., 2004) $I_{DS} \sim T^2 \exp(-q\phi_B / K_B T)$, where ϕ_B is the Schottky barrier height, K_B is the

Boltzmann's constant and T is the temperature in Kelvin. In this case, the sensitivity for the SB modulation is $S = \Delta R / R_0 \approx \exp(q\Delta\phi_B / K_B T) - 1$. Owing to the exponential dependence on the SB height change $\Delta\phi_B$, extremely high sensitivity can be achieved. Our results are consistent with Yamada's theoretical prediction that the SB modulation is most significant when CNT is operating in the depletion mode.(Yamada, 2006)

One disadvantage of the SB dominated sensors is a typical long recovery process at room temperature. Once the operating temperature is increased, the sensitivity degrades shapely due to its exponential dependence with *1/T*. The nanoscale CNT/metal contacts could also be prone to subtle environmental noises, causing fluctuations during detection. In contrast, using the CNT channel as a sensing element can avoid the problems. In addition, the large surface to volume ratio of CNT offers another merit. At T>150°C, when the adsorption barrier is further lowered by oxygen, NH_3 adsorption on the defect sites of the CNT becomes favored. Upon the consequent charge transfer, the Fermi level of the exposed central CNT channel moves upwards and the energy band shifts downwards with respect to those passivated parts, a potential barrier is therefore created and impedes the current flow, see Fig. 12 (d). Here, we would like to emphasize that, the charge transfer effect can be studied only when the contacts are fully protected. Otherwise, as temperature increases, the sensitivity enhancement from the charge transfer effect and the degradation due to SB modulation counteract each other.

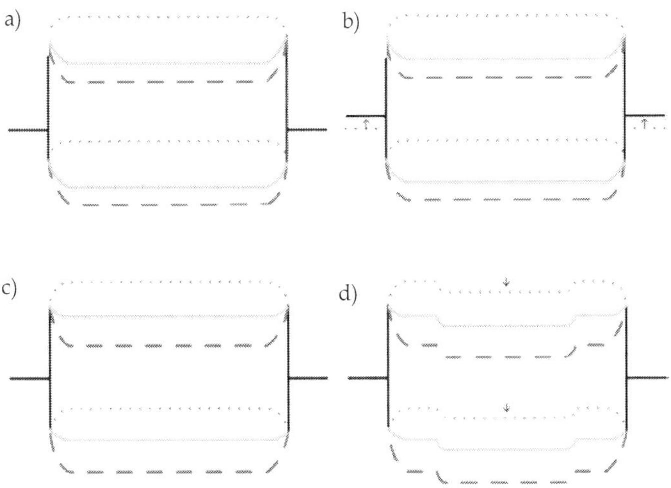

Fig. 12. Schematic energy band diagram for: Device 1 (a) before and (b) after the NH_3 exposure (An intrinsic CNT is considered. The work function of the source/drain electrodes is initially near the valence band edge of the CNT and is reduced after NH_3 exposure); Device 1A (c) before and (d) after the NH_3 exposure (After passivation, the work function of electrodes aligns near the midgap of CNT. The Fermi level of the exposed central CNT channel shifts upwards due to electron-doping from NH_3). Legend: Red dotted for $V_{GS}<0$; Green solid for $V_{GS}=0$; Blue dashed for $V_{GS}>0$.

Table 1 compares the SB modulation and charge transfer mechanisms. The sensors with the SB modulation usually demonstrate a high sensitivity at room temperature. Their sensing performance can be adjusted through the gate voltages. In contrast, the sensors under the charge transfer mechanism require high working temperatures, showing a good reversibility.

Mechanism	SB modulation	Charge transfer
Sensitivity	Extremely high in depletion mode	Low
Gate dependence	Strong	Weak
Operating Temperature	Room temperature	T>150°C
Reversibility	Low, can be improved at positive gate voltage	Good

Table 1. Comparisons between sensors operated by SB modulation and charge transfer.

5. Conclusions and Future Work

In this chapter, we discussed in details the sensing mechanisms of CNT based NH_3 detection. By selective Si_3N_4 passivation, We clearly show that the SB modulation at the CNT/metal contacts dominates the sensing performance at room temperature, and the sensor exhibits high sensitivity and good tunability under appropriate gate voltages. At higher temperatures, say 150°C or above, NH_3 molecules start to adsorb on the CNT wall and the charge transfer process from the adsorbed NH_3 molecules to the CNTs contributes to the sensing signal. As the mechanisms are identified, the next step is how to improve the sensing performance, and more chanllegingly, how to differentiate between different gas species. One promising way could be functionalizaton method (Feng, et al., 2005, Qi, et al., 2003). For example, using metal catalysts in the form of nanoparticles to decorate CNT, which promotes the interaction with speicific gas speicies. With a combination of several metal nanoparticle, the sensing performance to various gas molecules could be compared and contrasted.

6. Reference

Andzelm, J.; Govind, N. & Maiti, A. (2006). Nanotube-based gas sensors - Role of structural defects. *Chemical Physics Letters,* Vol. 421, No. 1-3: 58-62, 0009-2614

Appenzeller, J.; Radosavljevic, M.; Knoch, J. & Avouris, P. (2004). Tunneling versus thermionic emission in one-dimensional semiconductors. *Physical Review Letters,* Vol. 92, No. 4, 0031-9007

Bauschlicher, C. W. & Ricca, A. (2004). Binding of NH_3 to graphite and to a (9,0) carbon nanotube. *Physical Review B,* Vol. 70, No. 11, 1098-0121

Bilic, A.; Reimers, J. R.; Hush, N. S. & Hafner, J. (2002). Adsorption of ammonia on the gold(111) surface. *Journal of Chemical Physics,* Vol. 116, No. 20: 8981-8987, 0021-9606

Bradley, K.; Gabriel, J. C. P.; Briman, M.; Star, A. & Gruner, G. (2003). Charge transfer from ammonia physisorbed on nanotubes. *Physical Review Letters,* Vol. 91, No. 21, 0031-9007

Bradley, K.; Gabriel, J. C. P.; Star, A. & Gruner, G. (2003). Short-channel effects in contact-passivated nanotube chemical sensors. *Applied Physics Letters,* Vol. 83, No. 18: 3821-3823, 0003-6951

Chang, H.; Lee, J. D.; Lee, S. M. & Lee, Y. H. (2001). Adsorption of NH_3 and NO_2 molecules on carbon nanotubes. *Applied Physics Letters,* Vol. 79, No. 23: 3863-3865, 0003-6951

Derycke, V.; Martel, R.; Appenzeller, J. & Avouris, P. (2002). Controlling doping and carrier injection in carbon nanotube transistors. *Applied Physics Letters,* Vol. 80, No. 15: 2773-2775, 0003-6951

Feng, X.; Irle, S.; Witek, H.; Morokuma, K.; Vidic, R. & Borguet, E. (2005). Sensitivity of ammonia interaction with single-walled carbon nanotube bundles to the presence of defect sites and functionalities. *Journal of the American Chemical Society,* Vol. 127, No. 30: 10533-10538, 0002-7863

Heinze, S.; Tersoff, J.; Martel, R.; Derycke, V.; Appenzeller, J. & Avouris, P. (2002). Carbon nanotubes as Schottky barrier transistors. *Physical Review Letters,* Vol. 89, No. 10, 0031-9007

Heller, I.; Janssens, A. M.; Mannik, J.; Minot, E. D.; Lemay, S. G. & Dekker, C. (2008). Identifying the mechanism of biosensing with carbon nanotube transistors. *Nano Letters,* Vol. 8, No. 2: 591-595, 1530-6984

Kaminishi, D.; Ozaki, H.; Ohno, Y.; Maehashi, K.; Inoue, K.; Matsumoto, K.; Seri, Y.; Masuda, A. & Matsumura, H. (2005). Air-stable n-type carbon nanotube field-effect transistors with Si_3N_4 passivation films fabricated by catalytic chemical vapor deposition. *Applied Physics Letters,* Vol. 86, No. 11, 0003-6951

Kong, J.; Franklin, N. R.; Zhou, C. W.; Chapline, M. G.; Peng, S.; Cho, K. J. & Dai, H. J. (2000). Nanotube molecular wires as chemical sensors. *Science,* Vol. 287, No. 5453: 622-625, 0036-8075

Li, J. Q.; Zhang, Q.; Peng, N. & Zhu, Q. (2005). Manipulation of carbon nanotubes using AC dielectrophoresis. *Applied Physics Letters,* Vol. 86, No. 15, 0003-6951

Li, J.; Lu, Y. J.; Ye, Q.; Cinke, M.; Han, J. & Meyyappan, M. (2003). Carbon nanotube sensors for gas and organic vapor detection. *Nano Letters,* Vol. 3, No. 7: 929-933, 1530-6984

Liu, X. L.; Luo, Z. C.; Han, S.; Tang, T.; Zhang, D. H. & Zhou, C. W. (2005). Band engineering of carbon nanotube field-effect transistors via selected area chemical gating. *Applied Physics Letters,* Vol. 86, No. 24, 0003-6951

Peng, N.; Zhang, Q.; Li, J. Q. & Liu, N. Y. (2006). Influences of ac electric field on the spatial distribution of carbon nanotubes formed between electrodes. *Journal of Applied Physics,* Vol. 100, No. 2, 0021-8979

Peng, S. & Cho, K. J. (1999). Chemical control of nanotube electronics. *7th Annual Foresight Conference on Molecular Nanotechnology*, 57-60, Santa Clara, California.

Qi, P.; Vermesh, O.; Grecu, M.; Javey, A.; Wang, O.; Dai, H. J.; Peng, S. & Cho, K. J. (2003). Toward large arrays of multiplex functionalized carbon nanotube sensors for highly sensitive and selective molecular detection. *Nano Letters,* Vol. 3, No. 3: 347-351, 1530-6984

Robinson, J. A.; Snow, E. S. & Perkins, F. K. (2007). Improved chemical detection using single-walled carbon nanotube network capacitors. *Sensors and Actuators a-Physical,* Vol. 135, No. 2: 309-314, 0924-4247

Wei, C.; Dai, L. M.; Roy, A. & Tolle, T. B. (2006). Multifunctional chemical vapor sensors of aligned carbon nanotube and polymer composites. *Journal of the American Chemical Society,* Vol. 128, No. 5: 1412-1413, 0002-7863

Yamada, T. (2006). Equivalent circuit model for carbon nanotube Schottky barrier: Influence of neutral polarized gas molecules. *Applied Physics Letters,* Vol. 88, No. 8, 0003-6951

Yamada, T. (2006). Equivalent circuit model for carbon nanotube Schottky barrier: Influence of neutral polarized gas molecules. *Applied Physics Letters,* Vol. 88, No. 8: 083106, 0003-6951

Zhang, J.; Boyd, A.; Tselev, A.; Paranjape, M. & Barbara, P. (2006). Mechanism of NO_2 detection in carbon nanotube field effect transistor chemical sensors. *Applied Physics Letters,* Vol. 88, No. 12, 0003-6951

Zhao, J. J.; Buldum, A.; Han, J. & Lu, J. P. (2002). Gas molecule adsorption in carbon nanotubes and nanotube bundles. *Nanotechnology,* Vol. 13, No. 2: 195-200, 0957-4484

CNT/substrate system. In other words, cooling by gas occurs not only at the CNT surface but at much higher surface of a SiO_2 substrate which is in thermal contact with the CNT. This explains strongly enhanced sensitivity to gases for supported CNTs. On the other hand, the observed strong effect of substrate could be used to control the sensitivity of pressure measurements just by changing the area of thermal contact between the CNT and substrate, or by choosing substrates with different thermal properties.

4. Decorated MWCNTs and selectivity to gases

4.1 Supported MWCNTs

As commented above, the technologies of nanotubes decoration by nanoparticles that were developed last years can be used to promote selectivity to gases (Gelamo et al., 2009a) Here, we used decoration by Ti nanoparticles to detect selectively O_2 and the result using this device in a presence of N_2, Ar and O_2 at different pressures is shown below, see Fig. 10.

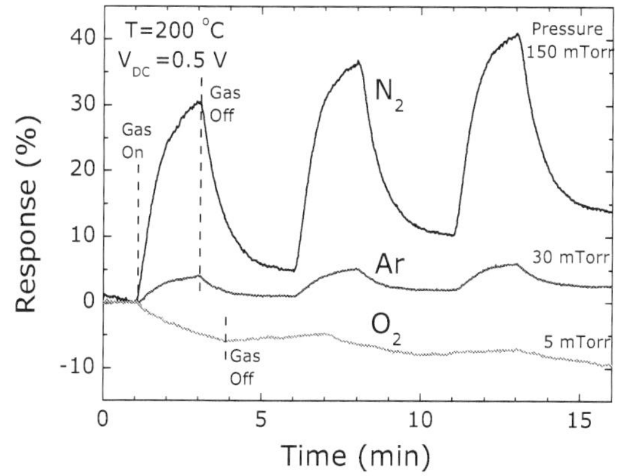

Fig. 10. Sensor response to pulses of different gases.

Tests varying O_2 pressure showed that two mechanisms (chemical and electrothermal or ET) can be observed at the same time in a CNTs/NPs system, see Fig. 11. For O_2 pressures ≤10 mTorr, the chemical effect at NPs (current decrease) is notable, while for pressures >30 mTorr, the ET effect starts to prevail. For intermediate O_2 pressures, a superposition of positive and negative contributions of two effects can give results that might appear confusing (e.g., these contributions in some cases can be mutually canceled for a specific pressure), though characteristic times for these effects are much different. In particular, interesting behavior can be seen in Fig. 11, when the current drops abruptly after O_2 pumping out (at t =15 min) for pressures higher than 70 mTorr. This is a result of fast rise of the CNTs/NPs system temperature after the gas removal (no more gas cooling), and corresponding decrease of CNTs resistance. Note that apparently much smaller rise of current is observed when gas is injected (t =1 min), since the corresponding current rise is mostly compensated by its almost simultaneous decrease due to the chemical effect. As a result, in contrast to N_2 when a good linearity with pressure is observed, the signals for oxygen (measured at t =5 min) tend to saturate for small pressures (~10 mTorr), see Fig. 12.

When Ti decorated MWCNTs are exposed to N_2, only ET effect is observed, with the response similar to that for pristine MWCNTs. Figure 12 shows the device responses to injection of O_2 at various pressures. For O_2, tendency for saturation at very low pressures can be seen, while for N_2 the response is linear for the pressure range used.

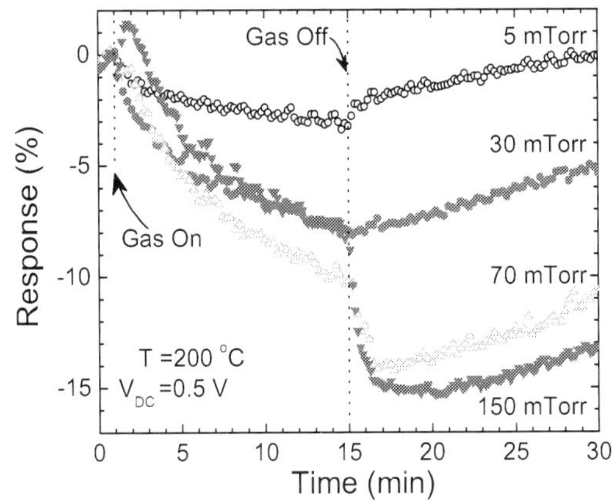

Fig. 11. Response to O_2, supported MWCNTs: effect of pressure variation, pulsed gas injection. Gas is on at t = 1 min and off at 15 min.

4.2 Suspended MWCNTS

As previously discussed above, there is a significant heat transfer from the CNTs self-heated by current to the substrate (oxidized Si). Thus it is clear that by eliminating the heat losses to substrate, significant increase of nanotube (and attached nanoparticles) temperature can be achieved. To prepare suspended nanotubes, first a FIB system was used to mill 5 µm deep and 1 µm wide cuts between electrodes and after this MWCNTs were DEP deposited. Thin Ti film (1-2 nm) was then deposited by sputtering on the surface of nanotubes (Fig. 5).

In this case, due to much stronger self-heating, selective gas measurements turn out to be possible even at a *room* substrate temperature (Fig. 13), with greatly enhanced sensitivity (by an order of magnitude, at much lower substrate temperature), as compared with supported CNTs. Faster sensor response can be also clearly seen. Some reports give values of Joule heating for suspended MWCNTs exceeding 1000 °C (Cai, Akita and Nakayama 2004), but this seems to be unlikely for the present conditions, as heating above 600 °C at the presence of oxygen would eventually burn out the nanotubes that was not observed in our experiments. However, local CNTs/NPs heating under present conditions evidently can reach very high values (up to 300-400 °C or even more, Gelamo et al., 2009b) under low pressure conditions when cooling by gases is negligible. This makes possible very intense interaction of oxygen with partially oxidized Ti NPs.

Besides the enhanced device response under exposition to O_2, other interesting parameter that changes greatly is the response time, see Fig. 14 where results for both supported and suspended configurations are presented. This parameter can give important information

about characteristic times of processes occurring during chemical interaction of nanoparticles and gas, and can useful to analyze the underlying mechanisms.
For supported MWCNTs, longer response times (near 5 min.) are indication of slow processes of heat transfer to the substrate. For suspended nanotubes, the response time is an order of magnitude smaller (0.5-0.8 min.), and its value is still higher than that residence time for gas molecules in the chamber (~0.2 min.). Probably, this time is determined by chemical interaction between O_2 and Ti particles. Careful study of the interaction mechanisms is subject of future work.

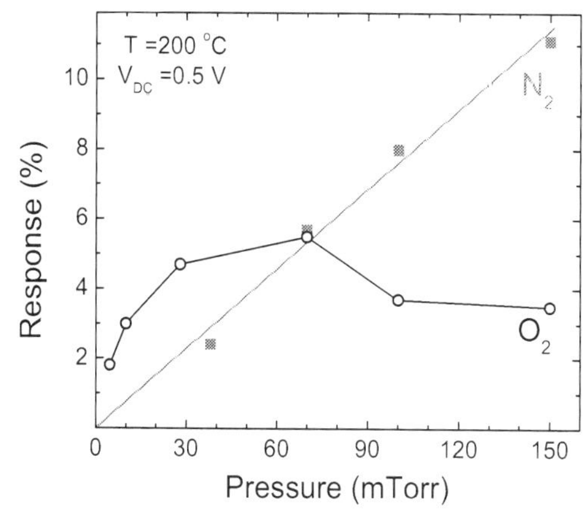

Fig. 12. Dependence of response on pressure for O_2 and N_2 using Ti decorated supported MWCNTs.

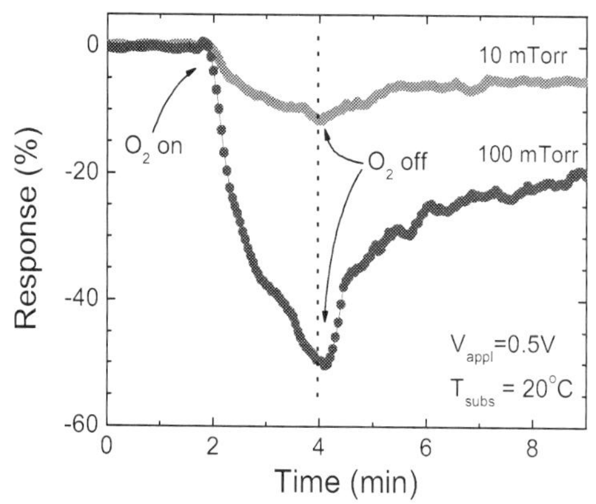

Fig. 13. Response to O_2, suspended MWNCTs, O_2 peak pressures of 10 mTorr and 100 mTorr, respectively.

5. Conclusions

The study of gas sensors based on short MWCNTs in a chemical resistor configuration (where nanotubes can be heated considerably by current) has revealed two main effects occurring under exposure to simple low-pressure gases like Ar, N_2 and O_2, and manifestation of the effects is significantly different for supported (over SiO_2) and suspended nanotubes (Gelamo et al., 2009a):

i) thermoelectric effect, due to cooling of nanotubes upon injection of gas. This cooling results in a current increase, indicating positive TCR values for MWCNTs used, and happens both with pristine and decorated nanotubes. For the latter, nanoparticles should have the same temperature as nanotubes. No significant difference was observed for the cooling efficiency between three gases studied (Ar, N_2 and O_2);

ii) chemical effect due to interaction of Ti nanoparticles with O_2, resulting in a current decrease.

iii) for supported nanotubes, heating appears to be much weaker due to strong heat losses to the substrate, in this process the substrate surface layer also attains high temperature. Upon injection of gas, the nanotube and substrate surface layer that are in a thermal contact are cooled down together, and this increases the sensitivity to gas pressure, due to effectively larger area of interaction with gas;

iv) for suspended nanotubes, self-heating is much stronger (no heat losses to the substrate, and losses to the metal electrodes and by radiation are smaller). On the other hand, the area of interaction with gas (the nanotube surface area) is much smaller as compared with supported nanotubes, so that nanotubes cooling by gas, under the same conditions, is much smaller for suspended ones (at least, two orders of magnitude). That means that suspended nanotubes can be used for pressure sensing at much higher gas pressures (>1 Torr), as cooling rate is proportional to the gas pressure;

v) for suspended nanotubes decorated with Ti nanoparticles, much higher sensitivity to O_2 (chemical effect) is achieved as compared with supported ones, due to much stronger self-heating of nanotubes (and attached nanoparticles), so that measurements at room substrate temperature are possible.

Finally, a CNT/NP hybrid material has been applied for low-pressure gas sensing applications in chemical resistor configuration. In this configuration, multi-wall carbon nanotubes serve as a conductive channel (for electrical signal acquisition), a heating element (for local heating of attached nanoparticles), and a substrate for NPs deposition (for selective gas sensitivity), whereas nanoparticles are employed to provide selective sensitivity to specific gases.

Self-heating by Joule effect, especially strong for suspended nanotubes, provides enhanced sensitivity to gases using CNT/NP material. In contrast to sensors based on self-heated nanowires, where functions of heating and gas sensing are coupled in the same element, in the CNT/NP system these functions are associated with two different elements, that can be beneficial for the sensor performance. It is a flexible configuration allowing pressure and gas sensing in a wide dynamic range using arrays of sensors (suspended or supported, decorated or pristine CNTs) at room substrate temperature and very low power ($\sim 10^{-5}$ W) consumption.

Measurements have been realized for gases O_2, Ar and N_2, with gas pressures varying from a few to 150 mTorr due to current experimental limitations, however, the range of measurable pressures (and gases) can be further expanded.

Finally, the opportunity to perform experiments at locally elevated temperatures using CNT/NP structures can be interesting for studies of temperature activated processes such as catalysis.

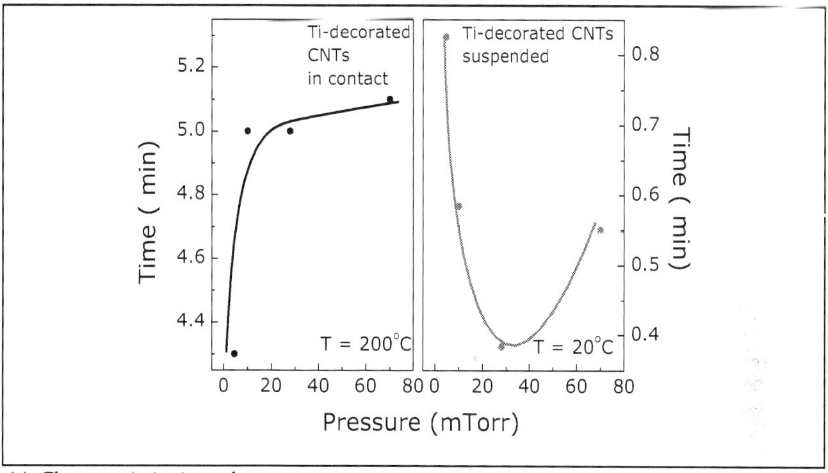

Fig. 14. Characteristic times for sensor response upon gas injection.

Acknowledgments

Authors greatly acknowledge the help of Dr. Carla Veríssimo, Dr. Alfredo R. Vaz, and Dr. Mário A. Bica de Moraes for the images and fruitful discussions; FAPESP, CNPq and NAMITEC for financial support.

6. References

Anchell, J. L. & A. C. Hess (1996) H_2O dissociation at low-coordinated sites on (MgO)(n) clusters, n=4,8. *Journal of Physical Chemistry,* 100, 18317-18321.

Ayyub, P., M. Multani, M. Barma, V. R. Palkar & R. Vijayaraghavan (1988) Size-induced structural phase-transitions and hyperfine properties of microcrystalline Fe_2O_3. *Journal of Physics C-Solid State Physics,* 21, 2229-2245.

Ayyub, P., V. R. Palkar, S. Chattopadhyay & M. Multani (1995) Effect of crystal size-reduction on lattice symmetry and cooperative properties. *Physical Review B,* 51, 6135-6138.

Bandow, S., A. M. Rao, K. A. Williams, A. Thess, R. E. Smalley & P. C. Eklund (1997) Purification of single-wall carbon nanotubes by microfiltration. *Journal of Physical Chemistry B,* 101, 8839-8842.

Berber, S., Y. K. Kwon & D. Tomanek (2000) Unusually high thermal conductivity of carbon nanotubes. *Physical Review Letters,* 84, 4613-4616.

Bockrath, M., J. Hone, A. Zettl, P. L. McEuen, A. G. Rinzler & R. E. Smalley (2000) Chemical doping of individual semiconducting carbon-nanotube ropes. *Physical Review B,* 61, 10606-10608.

Cai, X. Y., S. Akita & Y. Nakayama (2004) Current induced light emission from a multiwall carbon nanotube. *Thin Solid Films,* 464, 364-367.

Chopra, S., A. Pham, J. Gaillard, A. Parker & A. M. Rao (2002) Carbon-nanotube-based resonant-circuit sensor for ammonia. *Applied Physics Letters,* 80, 4632-4634.

Currie, L. A. (1995) Nomenclature In Evaluation Of Analytical Methods Including Detection And Quantification Capabilities (Iupac Recommendations 1995). *Pure and Applied Chemistry,* 67, 1699-1723.

Dai, H. J. (2002) Carbon nanotubes: Synthesis, integration, and properties. *Accounts of Chemical Research,* 35, 1035-1044.

Duesberg, G. S., M. Burghard, J. Muster, G. Philipp & S. Roth (1998a) Separation of carbon nanotubes by size exclusion chromatography. *Chemical Communications*, 435-436.

Duesberg, G. S., J. Muster, V. Krstic, M. Burghard & S. Roth (1998b) Chromatographic size separation of single-wall carbon nanotubes. *Applied Physics a-Materials Science & Processing,* 67, 117-119.

Endo, M., K. Takeuchi, S. Igarashi, K. Kobori, M. Shiraishi & H. W. Kroto (1993) The production and structure of pyrolytic carbon nanotubes (PCNTS). *Journal of Physics and Chemistry of Solids,* 54, 1841-1848.

FernandezGarcia, M., J. C. Conesa & F. Illas (1996) Effect of the Madelung potential value and symmetry on the adsorption properties of adsorbate/oxide systems. *Surface Science,* 349, 207-215.

Ferrari, A. M. & G. Pacchioni (1995) Electronic-structure of F and V centers on the MgO surface. *Journal of Physical Chemistry,* 99, 17010-17018.

Garvie, R. C. & M. F. Goss (1986) Intrinsic size dependence of the phase-transformation temperature in zirconia microcrystals. *Journal of Materials Science,* 21, 1253-1257.

Gelamo, R. V., F. P. Rouxinol, C. Verissimo, A. R. Vaz, M. A. B. de Moraes & S. A. Moshkalev (2009) Low-temperature gas and pressure sensor based on multi-wall carbon nanotubes decorated with Ti nanoparticles. *Chemical Physics Letters,* 482, 302-306.

Haddon, R. C. & S. Y. Chow. 1999. Hybridization as a metric for the reaction coordinate of the chemical reaction. Concert in chemical reactions. *Pure & Appl. Chem.*, Vol. 71, 289-294.

Holzinger, M., A. Hirsch, P. Bernier, G. S. Duesberg & M. Burghard (2000) A new purification method for single-wall carbon nanotubes (SWNTs). *Applied Physics A-Materials Science & Processing,* 70, 599-602.

Iijima, S. (1991) Helical microtubules of graphitic carbon. *Nature,* 354, 56-58.

Iijima, S. & T. Ichihashi (1993) Single-shell carbon nanotubes of 1-nm diameter. *Nature,* 363, 603-605.

Kong, J., M. G. Chapline & H. J. Dai (2001) Functionalized carbon nanotubes for molecular hydrogen sensors. *Advanced Materials,* 13, 1384-1386.

Lee, R. S., H. J. Kim, J. E. Fischer, J. Lefebvre, M. Radosavljevic, J. Hone & A. T. Johnson (2000) Transport properties of a potassium-doped single-wall carbon nanotube rope. *Physical Review B,* 61, 4526-4529.

Li, J., Y. J. Lu, Q. Ye, M. Cinke, J. Han & M. Meyyappan (2003) Carbon nanotube sensors for gas and organic vapor detection. *Nano Letters,* 3, 929-933.

Liu, J., S. S. Fan & H. J. Dai (2004) Recent advances in methods of forming carbon nanotubes. *MRS Bulletin,* 29, 244-250.

Lu, Y. J., J. Li, J. Han, H. T. Ng, C. Binder, C. Partridge & M. Meyyappan (2004) Room temperature methane detection using palladium loaded single-walled carbon nanotube sensors. *Chemical Physics Letters,* 391, 344-348.

Lucas, E., S. Decker, A. Khaleel, A. Seitz, S. Fultz, A. Ponce, W. F. Li, C. Carnes & K. J. Klabunde (2001) Nanocrystalline metal oxides as unique chemical reagents/sorbents. *Chemistry-A European Journal,* 7, 2505-2510.

Moriarty, P. (2001) Nanostructured materials. *Reports on Progress in Physics,* 64, 297-381.

Moshkalev, S. A. & C. Verissimo (2007) Nucleation and growth of carbon nanotubes in catalytic chemical vapor deposition. *Journal of Applied Physics,* 102, 044303.

O'Connell, M. J., S. M. Bachilo, C. B. Huffman, V. C. Moore, M. S. Strano, E. H. Haroz, K. L. Rialon, P. J. Boul, W. H. Noon, C. Kittrell, J. P. Ma, R. H. Hauge, R. B. Weisman & R. E. Smalley (2002) Band gap fluorescence from individual single-walled carbon nanotubes. *Science,* 297, 593-596.

Pacchioni, G., A. M. Ferrari & P. S. Bagus (1996) Cluster and band structure ab initio calculations on the adsorption of CO on acid sites of the TiO2(110) surface. *Surface Science,* 350, 159-175.

Park, J. W. & P. L. McEuen (2001) Formation of a p-type quantum dot at the end of an n-type carbon nanotube. *Applied Physics Letters,* 79, 1363-1365.

Pengfei, Q. F., O. Vermesh, M. Grecu, A. Javey, O. Wang, H. J. Dai, S. Peng & K. J. Cho (2003) Toward large arrays of multiplex functionalized carbon nanotube sensors for highly sensitive and selective molecular detection. *Nano Letters,* 3, 347-351.

Penza, M., G. Cassano, R. Rossi, M. Alvisi, A. Rizzo, M. A. Signore, T. Dikonimos, E. Serra & R. Giorgi (2007) Enhancement of sensitivity in gas chemiresistors based on carbon nanotube surface functionalized with noble metal (Au, Pt) nanoclusters. *Applied Physics Letters,* 90, 3.

Pohl, H. A. (1978). *Dielectrophoresis: The Behavior of Neutral Matter in Nonuniform Electric Fields.* Cambridge: Cambridge University Press.

Pop, E., D. Mann, J. Cao, Q. Wang, K. Goodson & H. J. Dai (2005) Negative differential conductance and hot phonons in suspended nanotube molecular wires. *Physical Review Letters,* 95, 4.

Pop, E., D. A. Mann, K. E. Goodson & H. J. Dai (2007) Electrical and thermal transport in metallic single-wall carbon nanotubes on insulating substrates. *Journal of Applied Physics,* 101, 093710.

Rodriguez, J. A. (2002) Orbital-band interactions and the reactivity of molecules on oxide surfaces: from explanations to predictions. *Theoretical Chemistry Accounts,* 107, 117-129.

Samsonov, V. M., N. Y. Sdobnyakov & A. N. Bazulev. 2003. On thermodynamic stability conditions for nanosized particles. Surface Science, 526, 532-535.

Sayago, I., E. Terrado, M. Aleixandre, M. C. Horrillo, M. J. Fernandez, J. Lozano, E. Lafuente, W. K. Maser, A. M. Benito, M. T. Martinez, J. Gutierrez & E. Munoz (2007) Novel selective sensors based on carbon nanotube films for hydrogen detection. *Sensors and Actuators B-Chemical,* 122, 75-80.

Shelimov, K. B., R. O. Esenaliev, A. G. Rinzler, C. B. Huffman & R. E. Smalley (1998) Purification of single-wall carbon nanotubes by ultrasonically assisted filtration. *Chemical Physics Letters,* 282, 429-434.

Someya, T., J. Small, P. Kim, C. Nuckolls & J. T. Yardley (2003) Alcohol vapor sensors based on single-walled carbon nanotube field effect transistors. *Nano Letters,* 3, 877-881.

Star, A., V. Joshi, S. Skarupo, D. Thomas & J. C. P. Gabriel (2006) Gas sensor array based on metal-decorated carbon nanotubes. *Journal of Physical Chemistry B,* 110, 21014-21020.

Suehiro, J., G. B. Zhou & M. Hara (2003) Fabrication of a carbon nanotube-based gas sensor using dielectrophoresis and its application for ammonia detection by impedance spectroscopy. *Journal of Physics D-Applied Physics,* 36, L109-L114.

Terrones, M. (2004) Carbon nanotubes: synthesis and properties, electronic devices and other emerging applications. *International Materials Reviews,* 49, 325-377.

Thess, A., R. Lee, P. Nikolaev, H. J. Dai, P. Petit, J. Robert, C. H. Xu, Y. H. Lee, S. G. Kim, A. G. Rinzler, D. T. Colbert, G. E. Scuseria, D. Tomanek, J. E. Fischer & R. E. Smalley (1996) Crystalline ropes of metallic carbon nanotubes. *Science,* 273, 483-487.

Valentini, L., L. Lozzi, C. Cantalini, I. Armentano, J. M. Kenny, L. Ottaviano & S. Santucci. (2003). Effects of oxygen annealing on gas sensing properties of carbon nanotube thin films. Thin Solid Films, 436, 95-100.

Varghese, O. K., P. D. Kichambre, D. Gong, K. G. Ong, E. C. Dickey & C. A. Grimes (2001) Gas sensing characteristics of multi-wall carbon nanotubes. *Sensors and Actuators B-Chemical,* 81, 32-41.

Wei, J. Q., H. W. Zhu, D. H. Wu & B. Q. Wei (2004) Carbon nanotube filaments in household light bulbs. *Applied Physics Letters,* 84, 4869-4871.

Zhu, J., M. Yudasaka, M. F. Zhang & S. Iijima (2004) Dispersing carbon nanotubes in water: A noncovalent and nonorganic way. *Journal of Physical Chemistry B,* 108, 11317-11320.

Zhu, X., R. Birringer, U. Herr & H. Gleiter (1987) X-ray-diffraction studies of the structure of nanometer-sized crystalline materials. *Physical Review B,* 35, 9085-9090.

20

Suspended Carbon Nanotubes: Applications in Physical Sensors and Actuators

Anupama B. Kaul and Larry Epp
Jet Propulsion Laboratory, California Institute of Technology, Pasadena, CA 91109 U.S.A.

1. Introduction

Carbon is truly an extraordinary material with physical structures spanning three-dimensional (3D) graphite, two-dimensional (2D) grapheme and zero-dimensional (0D) buckyballs or buckminster fullerine spheres. It is not surprising that the structural characteristics of carbon thus yield band diagrams displaying a diverse array of physical properties. When a 2D graphene sheet is rolled into a cylinder, a one-dimensional (1D) or quasi-1D form of carbon results, namely carbon nanotubes (CNTs), which have been one of the most extensively studied materials since their discovery (Ijima, 1991). A single rolled-up sheet of graphene results in a single-walled nanotube (SWNT) with a typical diameter of 1 - 2 nm. Multi-walled nanotubes (MWNTs) consist of concentric cylinders with an interlayer spacing of 0.3 - 0.4 nm, and diameters that are at least an order of magnitude larger than SWNTs, between 10 - 30 nm. The exceptional thermal, mechanical, electronic and optical properties of nanotubes (Dresselhaus, Dresselhaus, Avouris, 2001) has created a surge of applications, ranging from the use of CNTs as interconnects (Li *et al.*, 2003), heat transport materials (Yu *et al.*, 2006), novel transistors (Bachtold, *et al.* 2001), as well as optical materials (Homma *et al.*, 2009). The focus of this chapter is on the nanoelectronic applications of suspended carbon nanotubes, in particular their use as physical sensors and actuators.

Unlike 3D materials, when nanotubes are dispersed on a substrate, their properties are intimately influenced by the tube-to-substrate interactions, particularly those of SWNTs. For example, when diameter and helicity of SWNTs are controlled such that semiconducting tubes result (Odom *et al.*, 1998), no luminescence is detected for SWNTs lying on a substrate eventhough semiconducting SWNTs have a direct band gap (Lefebvre *et al.*, 2003). In addition, van der Waals interactions between CNTs and the substrate cause radial and axial deformations (Hertel *et al.*, 1998) which affect the electron transport properties of the tubes. The presence of the substrate beneath the tube can also influence heat dissipation mechanisms, which is an underlying motivation for using suspended CNTs as thermal conductivity based pressure sensors, and will be described in Section 2.2. Such sensors are important for vacuum-based microcavity applications (vacuum microelectronics, micro-eletromechanical-systems (MEMS) such as gyroscopes and RF MEMS switches). The high modulus and low mass of the tubes also leads to resonators that operate at high frequency and exhibit a high quality factor (Q), which provides the basis for another physical sensing mechanism, namely mass sensing that will be described in Section 2.3. Other types of physical sensors, derived primarily from the integration of SWNTs in a MEMS platform, have also been fabricated and reviewed (Hierold *et al.*, 2007).

Suspended nanotubes can also be utilized as actuators, which serve as the active elements of many nano-electro-mechanical-systems (NEMS), and at length scales smaller than typical MEMS structures. The extremely high directional stiffness of CNTs (Salvetat *et al.*, 1999), their unique ability to accommodate very large mechanical strains (Yu *et al.*, 2000), coupled with their low mass and chemical inertness, collectively suggest that CNTs are ideal materials for NEMS. Nanotube based NEMS have already been demonstrated in applications ranging from nanotweezers (Kim & Lieber, 1999), memory devices (Rueckes *et al.*, 2000), and nanorelays (Dujardin *et al.*, 2005). In Section 3 and 4, the use of CNTS as NEMS actuators based on *laterally*, as well as *vertically* oriented tubes will be illustrated, respectively. These actuators can act as switches, potentially for low power, low leakage memory and logic applications, and due to their mechanical construct, such switches are also potentially more tolerant to radiation compared to solid-state transistors based on Si.

Throughout this chapter, we will present results from our own research conducted at the Jet Propulsion Laboratory (JPL), California Institute of Technology, where suspended or free-standing quasi-1D carbon nanostructures have been applied to both physical sensing (b. Kaul *et al.* 2009; d. Kaul 2009), as well as NEMS actuators (a. Kaul *et al.*, 2009; c. Kaul *et al.*, 2009; Kaul *et al.*, 2006).

2. Physical Sensor Applications

2.1 Methods of Suspension

The two primary approaches for forming suspended CNT structures are: a) post-growth release (Walters *et al.*, 1999), or b) where the tubes are grown across pre-fabricated trenches on a wafer using chemical vapor deposition (CVD) (Tombler *et al.*, 2000; Minot *et al.*, 2003). In both approaches, the AFM tip was used to mechanically strain the suspended tube(s) in order to understand how structural deformation affects the elastic strain in SWNT ropes (Walters *et al.*, 1999) and the electronic transport properties of individual SWNTs (Tombler *et al.*, 2000; Minot *et al.*, 2003). In our work at JPL, we have implemented both approaches to form suspended nanotube structures, but such structures have been exclusively used for device applications, specifically for physical sensing and NEMS actuators.

Post-growth release was utilized with critical point drying to form suspended SWNTs that were used for pressure sensing (b. Kaul *et al.*, 2009; d. Kaul, 2009). The operational mechanism of such sensors will be illustrated in Section 2.2. For the post-growth release approach, the SiO_2 beneath the SWNTs was etched in 10:1 Buffered-HF (BHF), which was followed by critical point drying. Simple drying techniques have been employed after HF etching to release < 1 μm long tubes by using the reduced surface tension of IPA (21.7 mN/m) (Nygard *et al.*, 2001), compared to that of DI water (72.8 mN/m), whereas the even lower surface tension of tetramethylsilane (10.2 mN/m) has also been utilized in other reports (Walters, 1999). For the long SWNTs used in our sensors (5 - 10 μm), which are more susceptible to surface tension stress upon wet-release, added precautions were taken that employed critical point drying. The electrical yield after release -- defined by the number of electrically conducting devices -- was > 90%, but the resistance was always higher after release. The rise in resistance could be associated with the contacts where the tubes are likely to exhibit kinks that suppress conductance. It could also be due to HF damaging the SWNTs; clearly more work is necessary to fully characterize the extent to which HF damage is also involved.

We have also implemented the second technique to suspend tubes, where SWNTs have been grown across prefabricated trenches. Such structures were utilized for lateral NEMS actuators that will be discussed in Section 3.1. In earlier reports tubes were grown across prefabricated trenches, where the suspended tubes were over SiO_2/Si (Minot *et al.*, 2003) substrates. In contrast, we formed our suspended SWNTs over refractory Nb electrodes, which served as the pull-electrode to transiently charge the SWNTs in order to induce electromechanical switching. In general, the choice of the approach used to suspend tubes depends in part on the aspect ratio of the trenches, and on the quality of the tubes desired. For example, tubes that do not undergo post-growth processing are likely to be less defective, but such a technique is usually not feasible for forming suspended structures for low aspect ratio trenches below ~ 1:10; if the aspect ratio of the trench is too small, the tubes are likely to be undesirably draped on the substrate within the trench. This arises from the large amplitude vibrations that are thermally induced in the tubes from the high temperature CVD growth dynamics, increasing the likelihood of tubes attaching to the underlying substrate if the aspect ratio is too small.

2.2 Application in Pressure Sensors

Among the diverse variety of applications of CNTs, recently, CNTs have also been applied to pressure sensing based on the mechanism of piezoresistance (Grow *et al.*, 2005; Stampfer *et al.*, 2006). In these recently reported CNT pressure sensors (Stampfer *et al.*, 2006), metallic SWNTs were placed on membranes of atomic-layer-deposited (ALD) alumina, and a pressure differential across the membrane caused it to bulge, inducing strain in the overlying SWNT. Strain-induced conductivity changes were detected in non-vacuum environments (760-900 Torr).

Unlike the mechanism of piezoresistance, here we present a CNT pressure sensor that operates on the principle of thermal exchange between a voltage-biased CNT and the surrounding gas (b. Kaul *et al.*, 2009; d. Kaul, 2009). A recent article by Kawano (Kawano *et al.*, 2007) presents a sensing mechanism similar to that reported here, which utilized MWNTs grown using a localized synthesis process (Kawano *et al.*, 2006). Here we report on the device application of sensors based on SWNTs that were grown using conventional thermal CVD. Typical thermal conductivity gauges, such as Pirani or thermocouple gauges are physically large (volume ~ 10 cm^3) prohibiting their integration with many vacuum-encased Si-based micro-cavity applications (vacuum microelectronics, MEMS such as gyroscopes and RF MEMS switches). As a result of their large thermal mass, they are inherently slow, and also operate at high powers and temperatures.

With the advent of MEMS technology, miniaturized, low power pressure sensors were developed, representing the first application of bulk Si-micromachining (Peterson, 1982). Many of these Si-based transducers operate on the piezoresistive or capacitive principle (Esashi *et al.*, 1998) and obtaining wide-dynamic range has been challenging, which is especially important in micro-cavity applications. Due to the small volumes in microcavities, outgassing can cause large pressure changes over short times, affecting device performance.

We present a miniature SWNT-based thermal conductivity gauge (volume ~ 10^{-4} cm^3) that operates at low power (nW-μW) and low temperature, and exhibits a wide dynamic range (760 Torr-10^{-6} Torr). Due to the reduced dimensionality for thermal conduction and phonon relaxation in 1D materials, transport is strongly influenced by environmental factors. The high surface area of the 1-D SWNT, along with the high temperature coefficient of resistivity (TCR) of diffusively contacted 5-10 μm long tubes, greater pressure sensitivity is enabled compared to thin-film or bulk conductors. Moreover, when the CNT is suspended, heat dissipation via the substrate is minimized, increasing sensitivity at even lower pressures. We have carried out an investigation of such devices for their feasibility as vacuum gauges.

A schematic of the device is shown in Fig. 1a where the starting substrate was a thermally oxidized Si <100> wafer. Patterned CVD growth of SWNTs using Fe-catalyst was performed, which was followed by e-beam deposition of Au/Cr electrodes (250 nm/5 nm) patterned by liftoff. In order to release the devices, the SiO_2 beneath the SWNTs was etched in BOE which was followed by critical point drying described in Section 2.1. A low magnification SEM image of a completed device is shown in Fig. 1b, which also shows the wire-bonds for electrically contacting the devices. The high magnification SEM image in Fig. 1c indicates a tube traversing the electrode, where typical tube lengths were 5 - 10 μm.

(a)

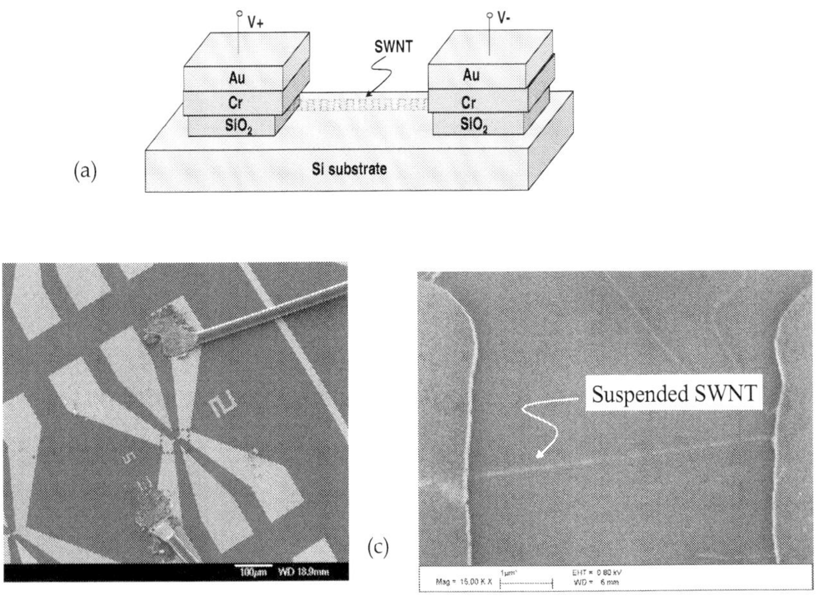

(b) (c)

Fig. 1. a) A schematic of the CNT pressure sensor where the SWNT is suspended. b) Low magnification SEM image of wire-bonded CNT vacuum gauge. Area highlighted by dashed lines indicates the region observed in the high magnification image in (c), where the typical SWNT lengths were 5-10 μm. Trench depth was ~ 310 nm (b. Kaul *et al.*, 2009).

Though the number of tubes that nucleate from the catalyst site can be controlled, to some extent, by the size of the catalyst island in thermal CVD, multiple tubes are shown in the SEM image of Fig. 1c. However, the growth of some of the tubes is seen to terminate prior to the mid-gap region (bottom of right electrode), while other tubes are oriented in a growth direction that does not allow them to bridge the gap to the opposite electrode (top of right electrode). The image in Fig. 1c thus, shows a single tube traversing the gap between the left and right electrodes. However, it is likely the presence of multiple tubes bridging the gap may enhance sensor robustness, and hence performance for practical applications. In conjunction with measuring the pressure response of unreleased and released CNT devices, a thin film meander resistor was also fabricated and pressure response measured for comparative purposes.

The current that passes through a voltage biased wire lying on a substrate induces joule heating. This heat can dissipate via several mechanisms, including the substrate (E_s), radiation (E_r) or the gas (E_g) (Alvesteffer *et al.*, 1999) as shown by Eqn. (1) below; the input power supplied to the bias element is represented by an energy E_t where,

$$E_t = E_s + E_r + E_g \tag{1}$$

The heat dissipated through the substrate is

$$E_s = (K\Delta TA)/L \tag{2}$$

where A is the area through which heat transfer occurs, K is the thermal conductivity of the SiO_2, ΔT is the temperature difference between the current carrying element and the substrate and L is the distance from the element to the substrate. Radiative losses are given by (3) where σ is the Stefan-Boltzmann radiation constant, ε is the thermal emissivity of the element, and T_h and T_a are the temperatures of the element and ambient, respectively.

$$E_r = \sigma\varepsilon(T_h^4 - T_a^4)A \tag{3}$$

Finally, heat dissipation by the gas is given by E_g,

$$E_g = \alpha_r \Lambda_t AP(T_h - T_a)\sqrt{\frac{273}{T_h}} \tag{4}$$

where α_r is the accommodation coefficient, Λ_t is the free molecule thermal conductivity, A is the surface area of the element, and P is the pressure. While the radiative and substrate losses become more significant at low pressures, the thermal conductivity of the gas is a predominant loss mechanism in the viscous flow regime ($0.01 < K_n < 10$, where K_n is Knudsen's number). The heat dissipation in this regime depends linearly on pressure as illustrated by Knudsen's formula (Eqn. 4).

Shown in Fig. 2a, is the normalized current as a function of pressure for the thin-film meander resistor and a CNT device, both of which are biased at a few watts of power. The thin film resistor initially shows a conductance increase (up to about 200 Torr), a plateau region of constant current, and then current decreasing below about 20 Torr; the percent change in current is < 1%. In contrast, the CNT device, shows a rapid and continuous decrease in current from ambient to ~ 40 mTorr, with more than 35 % change in conductance.

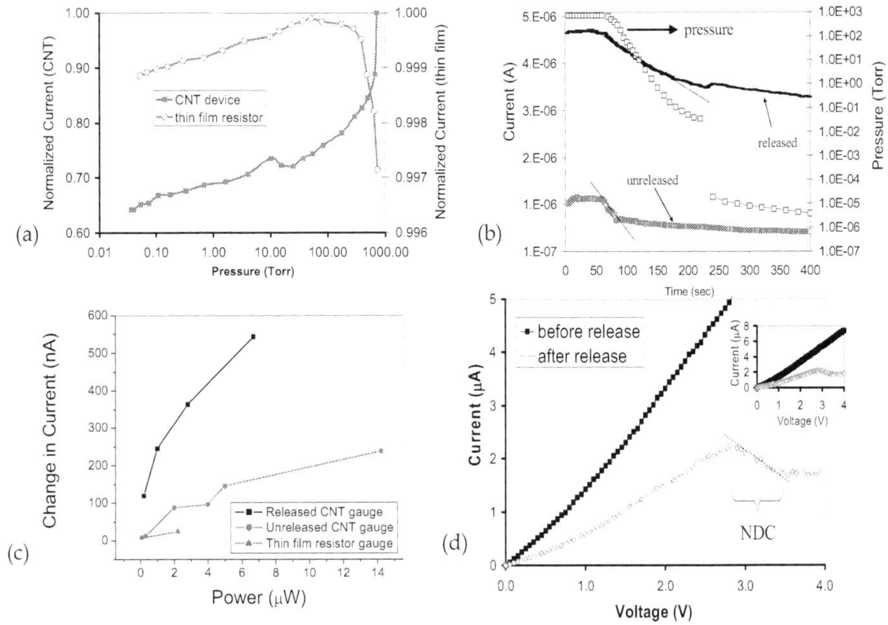

Fig. 2. a) Normalized current-pressure response at 2 μW for thin film resistor and unreleased CNT device. The conductance change is smallest for thin film resistor (< 1 %) while the conductance continues to decrease for the CNT device by about 35% at 35 mTorr. b) Absolute current variation for released and unreleased device vs. time/pressure at ~2 μW. The released device shows a higher rate of conductance change down to ~10^{-6} Torr possibly due to minimizing heat loss through the substrate. c) Net current change from ~ 5 x 10^{-6} Torr to ~ 8 x 10^{-7} Torr for released and unreleased CNT sensors as a function of bias power. The thin film resistor gauge was the least sensitive. The largest change in conductance is for released device at high power levels. d) The conductance of an unreleased SWNT device increases monotonically as the bias is increased, but after substrate removal, current saturation and a negative differential conductance (NDC) regime is observed in this device at high biases, which is attributed to the high optical phonon density in suspended tubes. Inset shows a higher current scale (b. Kaul *et al.*, 2009; d. Kaul, 2009).

At low pressures (< 100 mTorr), conduction through the gas becomes less significant and other pressure independent mechanisms such as radiative losses and solid-state conduction through the substrate tend to dominate, as seen by Eqns. 2 and 3. Radiative losses are significant only when temperatures exceed > 200 °C, and although radiative losses may be minimal at the low operational temperatures (or power) of our devices, the losses through the substrate have been calculated to be significant in the low vacuum regime. We implemented post-growth release to remove the oxide beneath the tubes as described, in order to suspend the tubes and measured their pressure response.

Shown in Fig. 2b is the pressure response of a device before and after the CNT was released from the SiO_2 substrate. As it can be seen, the released CNT device has a continued decrease in conductance well into the 10^{-5} Torr range while the unreleased device shows little variation after about ~ 100 sec (~ 1 Torr), in this case. This data appears to confirm that by removing the substrate underneath the current carrying element, sensitivity is increased to the lower pressure ranges.

The increase in sensitivity after substrate removal is perhaps better exemplified by the data in Fig. 2c which shows the net current change (ΔI) for released and unreleased device in the pressure range of ~ 5 x 10^{-6} Torr to ~ 8 x 10^{-7} Torr. In both cases, ΔI increases with power but the released device has a three times larger change than that of the unreleased device (at a power of ~ 6 μW, $\Delta I_{released}$ ~ 550 nA, compared to $\Delta I_{unreleased}$ ~ 150 nA). The relationship of power effecting sensitivity appears to be in agreement with Fig. 2c, and the combined results suggest that the greatest sensitivity is expected for the released devices operated at high power.

While the increased current sensitivity with substrate removal can be explained on the basis of heat minimization through the substrate, the reduced dimensionality for phonon scattering in 1-D systems, in particular suspended SWNTs, can cause unique effects to arise at large bias voltages and power. At high fields, a large nonequilibrium optical phonon population exists in suspended SWNTs and their long relaxation times result in non-isothermal conditions along the length of the tube. The I-V characteristic of both suspended and unsuspended SWNTs has been iteratively calculated (Pop *et al.*, 2005), by modeling the occupation probability of both optical and acoustic phonons at high biases, according to $I = \frac{V}{R(V,T)}$ where,

$$R(V,T) = R_c + \frac{h}{4e^2}\frac{\lfloor L + \lambda_{eff}(V,T) \rfloor}{\lambda_{eff}(V,T)} \tag{5}$$

Here, R_c is the contact resistance, L is the length of the nanotube and λ_{eff} is the sum of the acoustic, optical (emission) and optical (absorption) mean free paths derived from the Bose-Einstein phonon distribution. This analysis showed that in unsuspended tubes the I-V characteristic increases monotonically at high voltages suggestive of isothermal conditions, since the substrate facilitates in the relaxation of optical phonons emitted through electron scattering. In contrast, the current in the suspended tube saturates and a negative differential conductance (NDC) regime is encountered, which cannot be explained by velocity saturation (at ~ 5 kV/cm).

We have observed NDC in our suspended SWNT devices as shown by the I-V characteristic in Fig. 2d, where such measurements were taken in air and at room temperature. The unreleased device shows a monotonic rise in conductance, while the current after release for the same device saturates at high bias, ~ 2.8 V in this case, and enters a NDC regime. A peak conductance I_{peak}, where

$$I_{peak}(\mu A) = \frac{10}{L(\mu m)} \tag{6}$$

has been calculated and depends inversely on the length of the tube (Pop *et al.*, 2005). In the device shown in Fig. 2d, the experimentally observed I_{peak} was ~ 2.1 μA which appears to be in rough agreement with the expected value since the electrode spacing (and hence tube length) for this device was ~ 6 μm. Measurements show pressure sensitivity increases as the bias voltage or power are increased, but the released devices showed a more pronounced effect with power, as shown in Fig. 2c. The large optical phonon density in suspended SWNTs at high fields, with their long lifetimes may play an important role in determining the rate of temperature rise in the tubes, which can be exploited for their utility as thermal conductivity-based pressure or gas sensors.

2.3 Application in Mass Detection

While the pressure sensing mechanism described in 2.2 relied on suspended tubes that are stationary, the tubes in such bridge-type structures can also be driven into vibration or resonance by an external pertubation, forming the basis for yet another type of sensing mechanism, namely mass sensing. Much like a guitar string, the vibration frequency of the tube or suspended beam is a sensitive function of its total mass. The smaller the total mass, the more sensitive it will be to the external peturbation, and the higher is its resonance frequency. The addition of any adsorbed mass on the tube or beam will cause a shift in its resonance frequency, which can be measured and implemented for mass sensing.

Since sensitivity increases as beam dimensions shrink, there has been a constant push to drive the device dimensions to smaller and smaller length scales. For example, Si MEMS beams that resonate at RF frequencies, were used to detect species as small as bacteria and viruses as they adsorbed on a micromachined beam and caused a shift in its resonance frequency (Ilic *et al.*, 2004). More recently, lithographically fabricated NEMS beams with their even smaller masses, high frequency operation, and high Q's, were successfully used for the detection of protein molecules and nanoparticles (Naik *et al.*, 2009). Theoretical calculations suggest that mass sensitivity below a single dalton could be achieved by making NEMS beams even smaller; this approach may be used for weighing individual atoms and for performing mass spectrometry, for example in proteomics, using a large-scale array of such sensors.

Resonance phenomena has also been observed in SWNTs (Sazonova *et al.*, 2004; Peng *et al.*, 2006), which offer even more advantages than top-down fabricated MEMS and NEMS structures due to the remarkably high Young's modulus and light structure of SWNTs. Such suspended SWNTs have been shown to act as oscillators, where the resonance frequency was measured using an indirect mixing technique with a lock-in amplifier. More recently, resonance was also observed in metallic MWNTs by using a direct on-chip transmission measurement technique where an RF displacement current was detected as the tube was driven into resonance (Eriksson *et al.*, 2008). Suspended CNTs are attractive as mechanical resonators and have significant promise for mass sensing applications.

3. Actuators Based on Laterally Suspended Tubes

3.1 Actuators Based on Single Tubes

Besides physical sensing applications, suspended CNTs can also serve as actuators for NEMS. Such NEMS-based actuators can serve as switches exhibiting reduced leakage currents and power dissipation, when compared to Si transistors that are now facing performance limitations as a result of continued miniaturization. Due to the capacitive nature of switching in solid-state transistors, such devices are also susceptible to radiation and extremes in temperature. Due to the "mechanical" construct of NEMS switches, they should be inherently more robust when operated in harsh environments. In this section, NEMS switches that are based on tubes oriented laterally or parallel to the substrate are discussed. For laterally-oriented tubes, switches based on single tubes are described, and then Section 3.2 follows with an overview of a technology where monolayer thick fabrics of tubes are used.

Electromechanical switching in CNTs was first observed by (Rueckes T. *et al.*, 2000) where single-walled-nanotubes (SWNTs) were mechanically manipulated to form crossed structures with an air-gap. Switching has been demonstrated in deposited MWNT cantilever structures (Lee *et al.*, 2004) that were fabricated using an AC electrophoresis technique. In addition, deposited MWNTs have also been used to demonstrate electrostatic switching, where individual tubes were located by SEM for subsequent e-beam and thin film processing (Cha *et al.*, 2005). To date, switching in both SWNTs and MWNTs has been reported for the case of deposited tubes. Here the work on SWNT NEMS switches (Kaul *et al.*, 2006) is presented, where the tubes are grown on-chip with patterned catalysts using materials that are compatible with the high temperature CVD synthesis of SWNTs.

As described in Section 2.1, besides post-growth release (Walters *et al.*, 1999), nanotubes have also been grown across pre-fabricated trenches (Tombler *et al.*, 2000; Minot *et al.*, 2003) on a Si wafer. Unlike the latter structures where the tubes were suspended directly over SiO_2/Si substrates, in the application of laterally suspended SWNTs for NEMS applications presented here, suspended SWNTs were formed over a refractory metal electrode made from niobium (Nb). The Nb serves as the pull electrode to transiently charge the tube in order to induce electrostatic switching.

The process of formation is shown in Fig. 3a. In the first masking layer, active device regions in the PECVD SiO_2 were patterned to thin down the oxide from 200 nm to ~ 20 nm, which was comparable to the air gap of the switch. Nanotrenches that were as small as 130 nm were then defined using e-beam lithography. The nanotrenches were patterned using CF_4/O_2 RIE to etch the PECVD SiO_2 down to the Nb layer. For nanotube growth, catalyst was patterned by photolithography and liftoff of 0.5 nm Fe which was deposited by e-beam evaporation. The sample, with patterned Fe, was then placed in a CVD furnace for nanotube growth at 850 °C for 10 min using CH_4 and H_2 at flow rates of 1500 sccm and 50 sccm, respectively, where SWNT growth predominates. After CNT growth, the metal electrodes (Au/Ti) were patterned to contact the CNTs, using a bilayer AZ 5214/PMMA lift-off process, which results in easy lift-off of metal films due to an undercut in the PMMA layer. The top electrode metals Au/Ti were then deposited in an e-beam evaporator and lifted off in acetone. The high magnification SEM image of Fig. 3b depicts a nanotube crossing the trench.

The conductance between the left and right electrodes was measured and is shown in Fig. 4a, which yields a resistance of ~ 200 kΩ, much of which can be attributed to the high contact resistance between the CNT and Au/Ti electrodes. In general, these resistances typically ranged in value from tens of kΩ to tens of MΩ's, where the presence of multiple tubes can also contribute to the differences in resistance.

The actuation voltages were measured by applying a DC voltage between either the left or right electrode and the Nb pull electrode beneath the tube. As transient charge develops on the tube with increasing bias voltage, the resulting electrostatic force is sufficient to overcome the elastostatic force and deflects the suspended tube down toward the pull electrode.

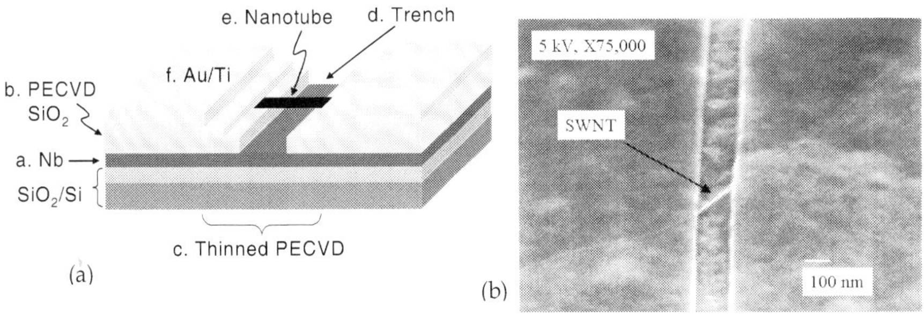

Fig. 3. a) Schematic shows a nanotube switch composed of (a) a 200 nm Nb film deposited using DC magnetron sputtering onto a thermally oxidized Si wafer followed by (b) a 200 nm PECVD SiO_2 layer. (c) Active device regions are etched in the PECVD SiO_2 layer to a thickness of ~ 20 nm, the height of the air gap for the switch. (d) Nanotrenches are defined using e-beam and etched to the Nb layer. (e) SWNT are grown from patterned Fe in a CVD furnace using CH_4/H_2 (1500 sccm/50 sccm) at 850 °C. (f) Electrodes are deposited using e-beam evaporated Au/Ti (220 nm/12 nm) which are patterned using a bilayer resist/PMMA process for easy lift-off. b) High magnification SEM micrograph showing a single nanotube bridging the 130 nm wide trench (Kaul *et al.*, 2006).

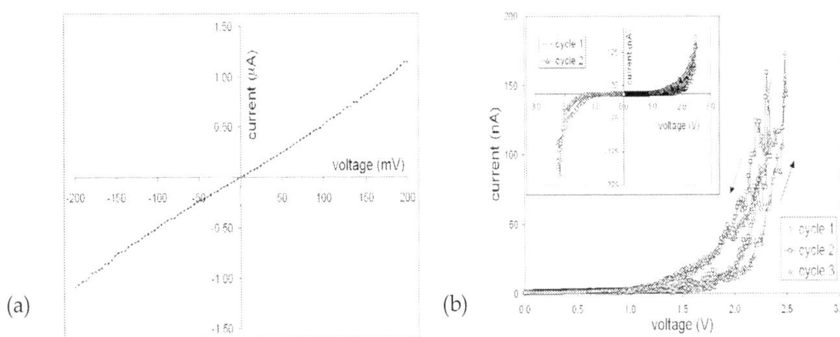

Fig. 4. a) Conduction measurements from the left Au/Ti electrode to the right electrode for a typical device gives R ~ 200 kΩ (trench width ~ 250 nm, 2 μm x 5 μm catalyst area). b) I-V characteristic of a device actuated over multiple cycles (250 nm trench width). The inset shows the ON state voltage to be similar in the forward-bias (pull electrode grounded) and reverse-bias (pull electrode positive) regimes, indicating that field emission is an unlikely possibility at these voltages (Kaul *et al.*, 2006).

Shown in Fig. 4b is an I-V characteristic of a device that was actuated over several cycles. Turn-on occurs at ~ 2.4 V in this case, with a slight variation with cycling that is also reported in other CNT (Lee, 2004) and MEMS switches. The rapidly rising current regime arises in both the forward-biased (pull electrode grounded) and reverse-biased (pull electrode positive) cases, as indicated by the inset of Fig. 4b, although the exact switching voltages are slightly different in the two cases, ~ 2.4 V (forward-biased) and ~ 2.2 V (reverse-biased). The differences in turn-on voltage can perhaps arise from the random distribution of metallic and semiconducting tubes observed in current SWNT growth processes, but still suggest that the differences in resistance between the ON and OFF states far outweigh any differences that may arise from contact resistances. As shown by the inset of Fig. 4b, this switching behavior is polarity independent, as would be expected for electrostatic actuation, and rules out field emission as a likely mechanism at these voltages. In general, the magnitude of the switching voltages in these air-bridge devices was a few volts and the switching times were determined to be a few nanoseconds.

3.2 Actuators Based on Fabrics of Tubes

While the previous section was focused on the electrostatic switching in NEMS devices comprising of individual SWNTs or MWNTs, monolayer thick fabrics of nanotubes are also being investigated and are a cornerstone of Nantero's mechanical memories (http://www.nantero.com/) that are currently under development. In this application, monolayer thick fabrics of CNTs are formed by dispersing a prepared solution of SWNTs in a solvent onto a substrate and evaporating the solvent. The resulting 1-2 nm thick film is then patterned using conventional lithography and monolithically integrated with other CMOS processes to form laterally suspended monolayer thick fabrics that can be switched electrostatically. Companies such as Nantero are developing this monolayer thick fabric fabric technology for making non-volatile memory elements for replacing Static-Random-Access-Memory (SRAM), Dynamic-Random-Access-Memory (DRAM) and Flash memory, for the defense, space, and commercial markets. Applications targeted are the formation of military and space-rated radiation-hard CMOS-CNT hybrid devices and circuits, and non-volatile memory products for the commercial marketplace.

4. Actuators Based on Vertical Tubes

4.1 Growth of Vertically Aligned Carbon Nanofibers

The discussion in Section 2 and 3 has been on the sensor and actuator application of tubes oriented parallel to the substrate. We now describe results from our research in forming tubes oriented perpendicular to the substrate, which are being developed for actuator applications for 3D electronics (a. Kaul *et al.*, 2009; c. Kaul *et al.*, 2009). The formation of vertically oriented tubes was enabled by implementing plasma-enhanced (PE) CVD, which enhances the alignment of the tubes, compared to CNTs synthesized using thermal CVD (c. Kaul *et al.*, 2009). Thermal CVD, which is a heavily utilized technique for the synthesis of CNTs, specifically MWNTs, results in 'bundles' of tubes largely perpendicular to the substrate. Inspection of MWNTs in such bundles reveals individual tubes growing in random directions and forming a 'matted' array. Often, this poor vertical tube alignment does not maximally utilize the exceptional 1D thermal, mechanical, electrical or optical properties of the tubes, and reduces performance for certain applications, such as field emitters for flat panel displays or thermal interface materials for integrated circuits.

A technique which has emerged in recent years to ensure excellent vertical tube alignment is PECVD (Ren *et al.*, 1998), where the inherent E-field in the plasma allows tube growth in a direction parallel to the field. Shown in Fig. 5a is a DC glow discharge produced in our PECVD growth chamber at the JPL Microdevices Laboratory, where a uniform and stable discharge is seen over an area as large as 75 mm. Analysis of crystallinity of individual tubes synthesized using PECVD reveals graphitic structures where the graphene layers are inclined to the central axis; these structures are commonly referred to as carbon nanofibers (CNFs) and control over their physical orientation with the plasma is excellent. Various plasma sources have been employed for tube or CNF growth, such as microwave, inductively coupled plasma (ICP), DC and DC with hot filament (Melechko *et al.*, 2005).

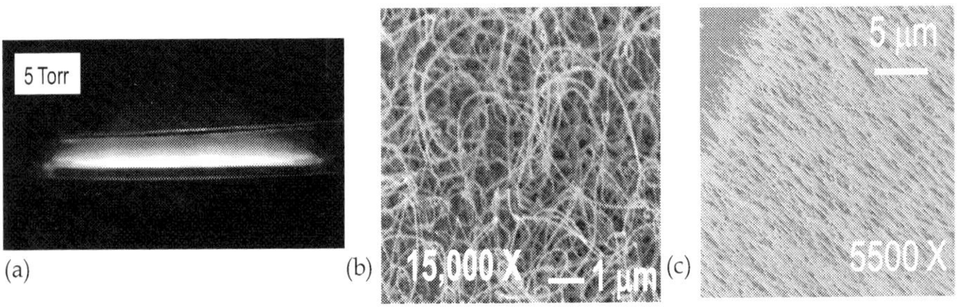

Fig. 5. a) The inset at the top showing a stable and uniform plasma over a diameter spanning ~ 75 mm. b) Unaligned tubes result when C_2H_2 and NH_3 are used in the absence of an E-field, where the substrate was degenerately doped Si with ~ 4 nm thick Ni catalyst. c) A high areal density of vertically aligned tubes was achieved using a 15 nm thick Ni film on degenerately doped Si, with 53 sccm of C_2H_2 and 90 sccm of NH_3 at 5 Torr and 300 mA (c. Kaul *et al.*, 2009). The image is taken where the SEM beam was at ~ 45° relative to the holder on which the sample was mounted.

When only C_2H_2 and NH_3 were present at a temperature of ~ 700 °C in the absence of an electric field, unaligned tubes resulted, as shown in Fig. 5b. In the presence of a stable plasma, a 15 nm thick Ni catalyst resulted in a large areal density of vertically aligned tubes, as shown in the SEM in Fig. 5c (sample holder is tilted at 45° with respect to the SEM beam). Although the growth temperature here was fixed to 700°C, prior work suggests there is a possibility for lowering this to temperatures more compatible with CMOS or low-temperature substrates (Hoffman *et al.*, 2003). In general, elevated temperatures cause the thin Ni catalyst film to dissociate into fine nanoclusters as a result of surface energy driving forces, where the clusters then serve as nucleation sites for tube growth. The plasma may also aid in catalyst particle fragmentation directly via ion bombardment, which may also contribute to heating the substrate surface indirectly.

The physical characteristics of the tubes were controlled to some extent with this bottom-up technique, by adjusting th growth parameters, such as growth pressure, catalyst thickness, and power during dc PECVD synthesis of the tubes. As an example, the SEM in Fig. 6 shows tubes of diameters ranging from 200 nm down to only 40 nm, which resulted from varying growth parameters. Also, as is evident from this figure, a tip growth mechanism was operative, where the nominally pear-shaped nickel cap, remains on the tip as the tube grows.

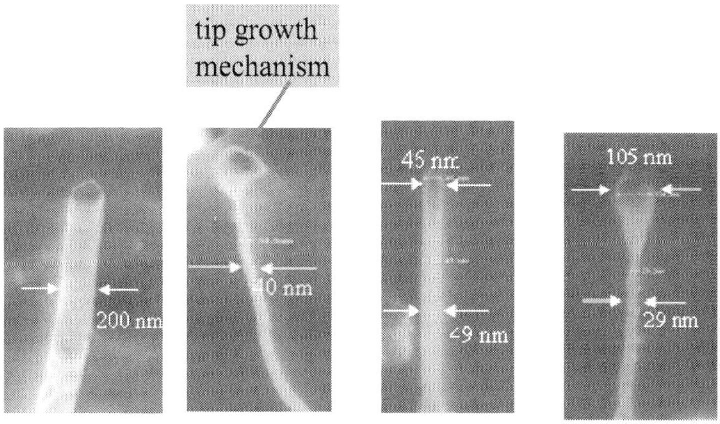

Fig. 6. Tube characteristics were engineered to some extent, via CNT growth parameters (pressure, catalyst thickness and power) and depict very wide ~ 200 nm tubes down to tubes only ~ 29 nm wide. A tip growth mechanism was also seen to be operative for Ni/Si (c. Kaul *et al.*, 2009).

4.2 Actuation in Vertical Carbon Nanofibers

By lithographically patterning the Ni catalyst sites, it is possible to produce a single, vertically aligned tube at a pre-defined location using DC PECVD, as described in Section 4.1. We now describe the electrostatic switching characteristics of such vertical NEMS structures which has the potential to increase integration density for 3D-electronics (a. Kaul *et al.*, 2009). Switching between vertically oriented tubes arranged in a 3-terminal configuration was recently reported (Jang *et al.*, 2005). In this work, a nanoprobe inside an SEM was mechanically manipulated in order to actuate a single, vertically oriented CNF. Nanomanipulation was also used to interrogate the electrical conduction in individual, as-grown CNFs grown on Si, as well as refractory nitride underlayers.

The tubes were synthesized with C_2H_2:NH_3 = 1:4 at 5 Torr and 700°C with Ni catalyst. Tubes were grown directly on Si <100> substrates with resistivity $\rho \sim$ 1–5 m Ω-cm, and NbTiN. The ~ 200 nm thick refractory NbTiN was sputtered, with $\rho \sim$ 113 $\mu\Omega$-cm, and was also chemically compatible with CNF synthesis. The sample was mounted on a 45° beveled holder inside the SEM. A nanomanipulator probe stage (Kammrath and Weiss) was placed inside the scanning-electron-microscope (SEM) (FEI Quanta 200F), where tungsten probes were used to make *in situ* electrical measurements with an HP 4155C parameter analyzer.

The nanoprobe physically contacted an individual CNF grown on NbTiN, as shown by the SEM in the inset of Fig. 7 (ground probe was on substrate). Since the probe-to-tube contact length was < 100 nm for this 2-terminal measurement, conduction was dominated by the large contact resistance. The work function φ for tungsten (W) $\varphi_W \sim$ 4.5 eV < $\varphi_{CNF} \sim$ 5.0 eV (Ahmad *et al.*, 2007), and suggests a Schottky barrier may arise at this interface, and also possibly at the tube-to-substrate interface; $\varphi_{NbN} \sim 3.92$ eV and like most transition metal nitrides with low φ (Saito *et al.*, 1999), it is likely $\varphi_{NbTiN} < \varphi_{CNF}$. A sub-gap region with suppressed conductance at low biases was seen in both polarities, and may have arisen from a native oxide on the W probes; if a small semiconducting junction (e.g. Schottky) also exists, an asymmetry in the I-V characteristic would arise, as observed.

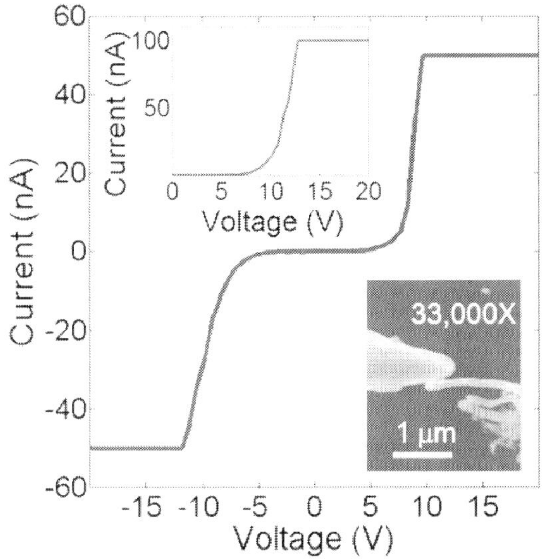

Fig. 7. Electrical continuity measurements for a single CNF grown on NbTiN. This tube is oriented vertically, so the image is rotated by 90°counter-clockwise. A nanoprobe is in contact with CNF, as indicated by the SEM image. The inset shows compliance increased to 100 nA.

Actuation measurements were performed for CNFs on NbTiN, where a nanoprobe was manipulated to within a few hundred nm of a single CNF. The electrostatic force per unit length F_{Elec} increases as $F_{Elec} \prec V^2$, where V is the voltage, and the elastostatic force per unit length F_{Elasto} increases as $F_{Elasto} \prec EI$, where E and I are the elastic modulus and moment of inertia of the nanotube, respectively (Dequesnes *et al.*, 2002). With increasing V ($F_{Elec} > F_{Elasto}$) the tube deflects closer to the probe, and a tunneling current is detected which increases exponentially, and results in a sudden or sharp change in slope at turn-on. The switching I-V in Fig. 8 shows currents rising sharply at $V_{pi} \sim 18$ V. The turn-off occurred at ~ 16 V and was dominated by the large tube-to-probe contact resistance since the tube remained stuck to the probe; thus, the turn-off response was similar to Fig. 7. The inset in Fig. 8 captures another switching event for a different tube, where turn-on and turn-off occurred at ~ 14 V and 10 V, respectively, and also illustrates the abruptness of the turn-on transition.

21

Study of Carbon NanoTube Field Effect Transistors for NEMS

Hasina F. Huq[1], Bashirul Polash[1], Oscar Machado[2] and Nora Espinoza[1]
[1]*The University of Texas-Pan American,* [2]*The University of Texas El Paso*
USA

1. Introduction

The recent developments of Carbon NanoTube Field Effect Transistor (CNTFET) technology indicate the perspective of the Nanoelectromechanical systems (NEMS). Carbon nanotubes (CNT) are ideal candidates for NEMS due to their chemical and physical structures, low masses and exceptional stiffness. Study of NEMS devices in the light of quantum mechanics requires understanding the interplay between the physical, geometrical and electrical parameters of the system (Dang et al., 2006). An analytical representation of (CNT) based field effect transistor is developed for high frequency NEMS applications to examine the characteristics observed from the fabricated devices. The analytical models enable us to gain deep insights of the device performance and behavior. The developed analytical model of CNTFETs represents its viability into transistor applications for NEMS switches, RF circuits, memory cells, field emission displays, biomedical instruments etc (Polash & Huq, 2008).
The metal-nanotube contacts in the CNTFETs are treated as Schottky barriers and analyzed by means of a ballistic model (Natori et al., 2005). The famous Landauer formula is used to calculate the conductance of the tube by relating the energy dependant transmission probability within the tight binding approximation of the CNTFET (Datta, 2000). Transmission function of the CNT is expressed in terms of the Green's functions of the conductors and the coupling of the conductor leads. The Green's function is incorporated with the transfer Hamiltonian approach to calculate the tunneling currents. The non-equilibrium Green's function transport equation is solved iteratively along with a 2D Poisson equation to improve the numerical convergence. The charge density is calculated by integrating the 1D universal density-of-states along with the source-drain Fermi-Dirac distribution function over energy within the energy gap of the CNT. The calculations show that the proposed device can perform stable operation at high current levels (670 μA/μm). Upper limits of device characteristics are considered for the model. Degradation in measured data is observed due to the limitations in device fabrication technology and imperfect contact placement on the CNT. Commercialization of CNTs NEMS/sensors is a great challenge due to the price and size of such measuring equipments (Fujita et al., 2007) . When small bias current is used, resistance measurement of CNTs sensor becomes difficult and high accuracy current source and analogue to digital converter (ADC) are required to maintain a reasonable signal to noise ratio (SNR). Controlled growth of CNT is yet to be

mastered (Javeyet al., 2005). In situ coupling of source/drain to CNT mechanism needs to be idealized for better approximating the NEMS performance. Replacing the metal-CNT SB contact with highly doped CNT source/drain (ohmic like) contact could improve the NEMS performance.

2. CNTFETs and NEMS

Carbon nanotubes are nano size (10^{-9}) Carbon atom made tubes with nanostructure and are wrapped into the form of a cylinders with graphene sheet. There are two basic tube structures: SWNT-Single-Walled NanoTube and MWNT- Multi-Walled NanoTube (Reich et al., 2004). Carbon nanotubes are being used in day to day applications. Advances in medicine are being contributed to Carbon nanotubes in some way (Kam & Wong, 2005). One of the advances that researchers are working on, is CNT based devices such as nanoshells, NEMS, and quantum dots and their properties. Using these electronics devices such as NEMS (nano-electro-mechanical-systems); it will be possible to carry DNA to cancerous cells and destroy them with minimal damage to surrounding healthy cells through methods. CNT based FET devices propose that by working with smart NEMS, it is possible to destroy only cancerous cells leaving healthy untouched. This devices will be able to know and identify threats of cancer cells.

CNT based micro/nano objects or devices can be comprised of a range of miniature structures, including cantilevers and diaphragms, static structures, chemically sensitive surfaces and electrical devices (resistors and FETs). Devices used for NEMs are valves, mixers and pumps and CNTFET can act as an active valve or controlled switch.

Reasons for minaturization of NEMS using CNTFETs can be summerised as follows (i) reducing the sensor element to the scale of the target species and hence providing a higher sensitivity (ii) reduced reagent volumes and associated costs, (iii) reduced time to result due to small volumes resulting in higher effective concentrations,(iv) amenability of portability and miniaturization of the entire system (v) point-of-care diagnostic, (vi) Multi-agent detection capability (Gruner., 2006).

Materials requirements are different from NEMS to NEMS and it is even more important for CNTFET based-NEMS. Desired properties of NEMS materials shows biocompatible, chemically modifiable, easy to fabricate, economic and soft compliable. Because of their dimensions and the good electrical, mechanical, chemical and physical properties CNTFETs make great emitter of electrical fields. These devices exhibit a wide range of favorable electrical properties (Hoenlein et al., 2003); some of them are presented below:

- High transconductance. This property determines the performance of any FET. given that a higher transconductance results in greater gain or amplification.
- Superior threshold voltage.
- Superior subthreshold slope. This property is very important for low power applications.
- High mobility.
- Ballistic transport. This property results in high speed devices .
- High current density.
- High on / off current ratios.

Electrical applications are arising each day, and not only in integrated circuits but nanotubes are also being used in auto industry, medical science, mechanical system, electronics, and recreational purposes.
Researchers in nano technology have encounter barriers that slow down it's progress especially in the medical field. We have seen that this study could lead to a potential cure for cancer however studies have shown that the toxicity of carbon can be lethal to humans. According to an article publish in the U.S National Library of Medicine and the National Institute of Health, they examine the impact of CNTs under different conditions, nanotubes could cross membrane barrier and reach organs and could induce harmful effects as inflammatory and fibrotic reactions in vital organs (Kam & Nadine 2005).

3. CNTFETs device structure

Carbon nanotube field effect transistors (CNTFETs) exhibited promising characteristics through experiments, yet the underlying mechanism is not fully understood (Lin et al., 2005). Theoretical analysis of CNTFETs with Schottky barriers (SB) is represented by means of a ballistic transport within quantum transport regime (Hasan et al., 2006). Due to outstanding electrical and mechanical characteristics, high-speed operation of two terminals or three terminal CNTFET-NEMS switches have been demonstrated. They are expected to use in memory circuits or RF switches.
Carbon nanotubes are single atomic layer thick sheet of graphite rolled into seamless cylinders with nanometer dimensions. Since the invention in 1991 by S. Ijima, CNTs have drawn great attention of researchers because of their extraordinary chemical, physical, and electrical properties. CNT can be metallic or semiconducting based on the direction of graphene sheet roll (Iijima et al., 1992). Understanding in detail the nature of the direction of graphene sheet roll is therefore an important ingredient in the development of reliable CNT NEMS. Minimal series resistance and quantum effects in CNT based NEMS switches govern the electron flow which causes ballistic transport.
The channel of the proposed device is formed with a (13, 0) CNT which has a diameter of 1nm and length is 15nm. Source and drain electrodes are placed at the ends of nanotube length while the gate electrode is co-axially placed along the tube axis. High-κ (~25) dielectric zirconium oxide (ZrO2) is used for gate insulation. The drain is biased with external voltage source and the source is connected to the ground. The electronic properties of a nanotube derived from the dispersion relation of a graphite sheet with wave vectors (*kx, ky*) is given by (Khan et al., 2007);

$$E(k_x, k_y) = \pm\gamma\sqrt{1 + 4\cos(\frac{\sqrt{3}k_x a}{2})\cos(\frac{k_y a}{2}) + 4\cos^2(\frac{k_y a}{2})} \quad (1)$$

γ is the nearest neighbor-hopping parameter and a is lattice constant.
γ = 2.5~3.2 from different measurements and a = 0.246 nm.
Imposing boundary condition along the tube circumference quantizes the two dimensional wave vector

$$(k_x, k_y)\cdot \mathbf{C} = 2\pi q \quad (2)$$

$$\mathbf{C} = n\mathbf{a}_1 + m\mathbf{a}_2 \quad (3)$$

where **C** is the chiral vector of the nanotube having and *n,m,q* are integers. This leads to the condition at which metallic conductance occurs which is

$$(n-m)=3q \tag{4}$$

The above condition suggests that 1/3rd of the tubes are metallic and 2/3rd are semi conducting. It has also been found that when m=0, the energy density of states is the highest which reveals the best semi conducting properties [Someya et al., 2003]. Figure 1 shows the schematic representation of a coaxially gated CNTFET with Non-equilibrium Green's function (NEGF) quantities (Rahman et al., 2003).

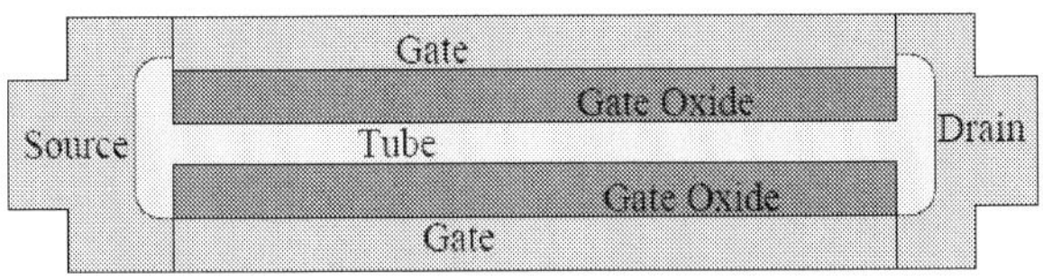

Fig. 1. Schematic representation of a coaxially gated CNTFET with NEGF quantities

4. Modelling of CNTFETs

Energy gap for a semiconducting nanotube is dependent on tube diameter and is given by

$$E_g = \frac{2d_{cc}\gamma}{D} \tag{5}$$

For a (13, 0) nanotube, the tube diameter is 1nm, dcc = 0.1421 nm and γ = 3.00 gives the energy gap of 0.8526 eV. It is reported that there exist characteristic end states at the end of single-wall CNTs. The parameters used in the proposed model and calculations are listed in Table 1.

Energy Gap	0.75 eV	Temperature	300K
Gate Oxide Thickness	2nm	Source Fermi level	-0.32eV
Gate Dielectric constant	25	Gate Control Parameter	0.88eV
CNT Diameter	1nm	Drain Control Parameter	0.035eV

Table 1. parameters of the proposed cntfet

Conductance between two leads source and drain is defined in terms of current and voltage: I=GV.

Using Landauer formula, conductance is expressed by the following equations:

$$G= 2\, q^2T/h \qquad (6)$$

Where q is the charge of electron and h is the Planck's constant.
T is known as the transmission function in terms of energy that represents the probability of an electron injected at one end of a conductor will emit at the other end. T can be expressed as

$$T= \text{trace}\,(\Gamma_S G_0^r \Gamma_D G_0^a) \qquad (7)$$

G_0^r G_0^a represents the retarded and advanced Green's function of the nanotube and $\Gamma_{D,S}$ are the coupling of the CNT to the source and the drain. The retarded Green's function is calculated by NEGF formulation:

$$G_0^r = (EI - H - \Sigma_S - \Sigma_D)^{-1} \qquad (8)$$

Where E is Fermi energy, I is the identity matrix, H is the Hamiltonian of the nanotube. $\Sigma_{D,S}$ is the self energy terms at the source and drain coupling of the contacts are the calculated using the broadening function of the self energy terms at the source and drain:

$$\Gamma_{S,D} = i\,(\Sigma_{S,D} - \Sigma^+_{S,D}) \qquad (9)$$

The nanotube behaves as the conducting channel in the CNTFET from the source to drain; it depends on the current density of the tube. The current density is a measure of the density of flow of an electric current per unit area across a section. The current density is an area density described by *J*. The current through an area *A* is simply the flux of the current density through that area as show below:

$$\bar{I} = \int \bar{J} \bullet d\bar{A}$$

If the flow of the current is through a uniform area $I=JA$

Using the charge density within the device, the NEGF transport equation is solved teratively with the posison equation until self-consistant potential distribution is found. Finally the current is calculated using the Landauer Böuttiker expression:

$$I_d = \frac{4d \int T(E)[f_S(E) - fD(E)]dE}{h} \qquad (10)$$

T is the transmission probability across the source/drain; f_S and f_D are the source/drain Fermi-Dirac distribution functions consistent potential; The equation is solved simultaneously to evaluate and characterize the performance of these devices.

5. Simulation results

The nanoHUB is a rich, web-based resource for research, education and collaboration in nanotechnology. NanoHUB.org was created by the National Science Foundation and the funded the Network for Computational Nanotechnology (NCN). The NCN is a network of universities with a vision to pioneer the development of nanotechnology, from science to manufacturing through innovative theory, exploratory simulation, and novel cyber infrastructure. The reserch work go from Nano-bio, nano-electromechanical systems, nano-electronics to special projects, providing information for nanoHUB to grow. This simulation tool is available at nanoHUB.org and creates a 3D image of the nanotube and gives the "Energy vs. Axial Wave Vector", the "Lowest Sub-band" and the "Density of States (DOS) vs. Energy".

5.1 I-V Characteristics

The FETToy is a tool that simulates the ballistic I-V characteristics of MOSFET. It simulates from single gate MOSFET to double gate MOSFET to Silicon Nanowire MOSFET and Carbon Nanotube MOSFET. Using the simulation tool the characteristics of CNTFET is investigated with different device stucture and applied voltages.

The FETToy simulates the I-V characteristics under a certain biased condition where the Gate Control Parameter and the Drain Control Parameter should be less than or equal to 1. The simulation models the drain voltage V_D at 1 volt. The thickness of the oxide layer and the carbon nanotube diameter determine the performance of the CNTFETs. The thicker the oxide the less voltage will go through the insulator as well as the bigger the diameter the more current density it could develop.

The gate insulator is varied from 1.5nm to 4.5nm, the gate controlled parameter and drain control parameters are varied from 0 to 1 and the diameter of CNT is varied from 0.1nm to 10nm. If the gate control parameter and the drain control parameters are increased, the drain current increases significantly. This increases the drain current by 24.3uA. This current also depends on the thickness of the oxide layer. The current reduces by a factor of 6.7uA at the saturation region where the drain voltage is 1 volt. At a diameter of 0.1nm, the results show a very low drain current figure 2. At a diameter of 1nm, the current coming out of the drain is drastically changing from 3uA to approximately 28uA. The drain current reaches approximately 78uA at the diameter of 10 nm figure 3. The bigger diameter allows higher drain current. The fact is that the triode region and the saturation region are depending on the diameter, unlike the MOSFET depending of the channel pinch off.

5.2 Transconductance Behavior

In order to achieve a relatively large transconductance the CNT must have large diameter. The larger the transconductance, the greater the gain it will deliver. However the increase of g_m at larger V_{GS} has the disadvantage of reducing the allowable voltage signals swing at the drain. As the diameter gets smaller this reduces the carrier mobility, changing the transconductance. The transconductance behavior is obtained at 1nm diameter, with different gate and drain voltage. The transconductance varies by a factor of 10/V depending on the amount of voltage applied to the gate (figure 4,5). However, the increase of V_G will reduce the allowed voltage signal through the drain.

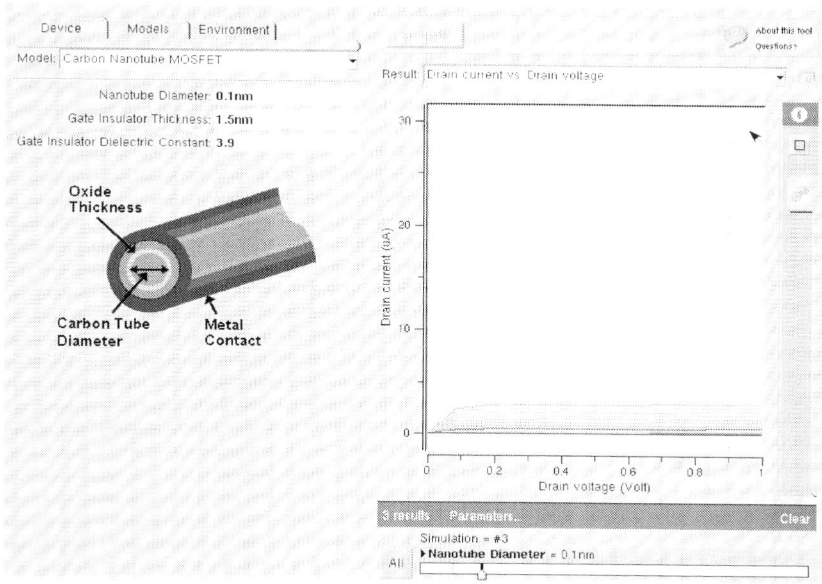

Fig. 2. I-V charecteristics of CNTFET at a diameter of 0.1nm

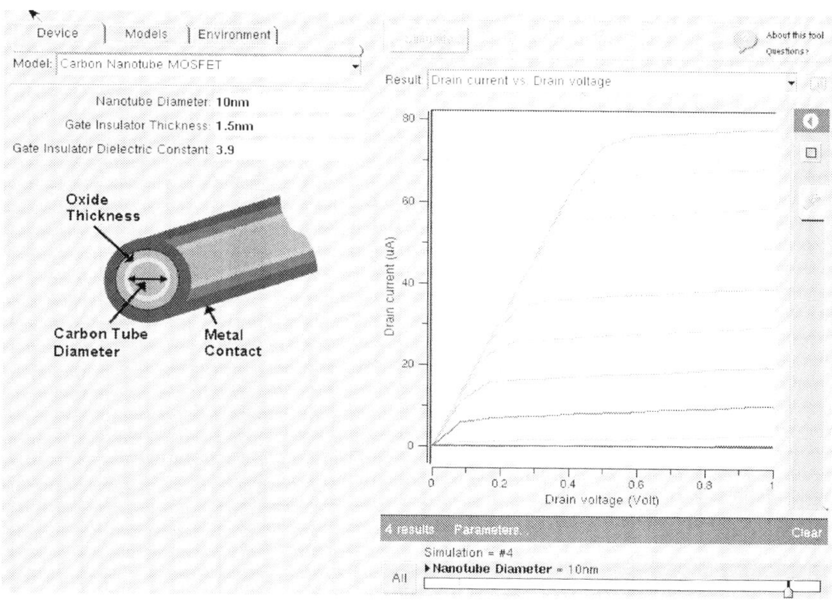

Fig. 3. I-V charecteristics of CNTFET at a diameter of 10 nm

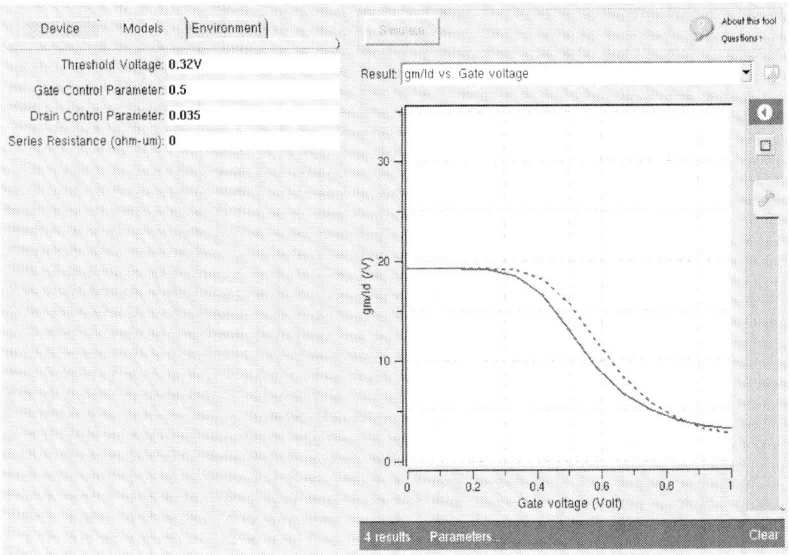

Fig. 4. Transconductance behavior of CNTFET (V_D= 0.5V)

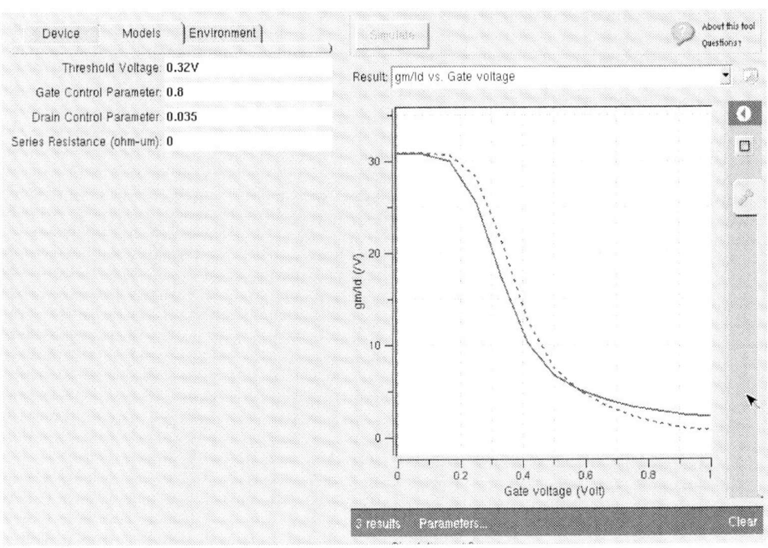

Fig. 5. Transconductance behavior of CNTFET (V_D= 0.8V)

6. Results analysis

Figure 6 shows the density of state (DOS) of different types of nanotubes with respect to energy gap. The simulation is done by using 'CNTbands 2.0' available from www.nanohub.org. The simulated end states are within the energy gap of semiconducting CNTs, implying that the end states are a 1-D analogy with conventional surface states. At metal-semiconductor (M-S) CNT junctions, however, no interface state was experimentally manifested, whereas some electronic states of semiconducting CNT diffuse into the metallic CNT across an M-S CNT. The calculated band gaps are in good agreement with those calculated within local density approximation. The band gaps of CNTs are small (from 0.2 to 2.0 eV), so CNTs are either metallic or semiconductive. The energy band structures of carbon atom C provides an occupied energy level in the band gap depending upon the DOS and types of CNT. The (13, 0) nanotube acts as semi conducting material since it has energy gap between conduction and valence band. The (10, 10) nanotube and the (10, 5) nanotube act as conducting material as the valance and conduction bands are overlapping. The electron can be excited more easily from valance band level to conducting band than that from filled band. This excited electron leads this kind of nanotube to be conductive as that of N-type semiconductor. The energy gap of this kind of CNT is small, so the valance level can easily provide electrons to the conducting band. Therefore, it is the energy gap, DOS and types of CNT that determines the conductivity of the material. The FETToy tool available in the website www.nanohub.org is useful to observe the characteristics of carbon nanotube FET, which is considered having cylindrical geometry. Using the simulation tool the different characteristics of the CNTFET are obtained.

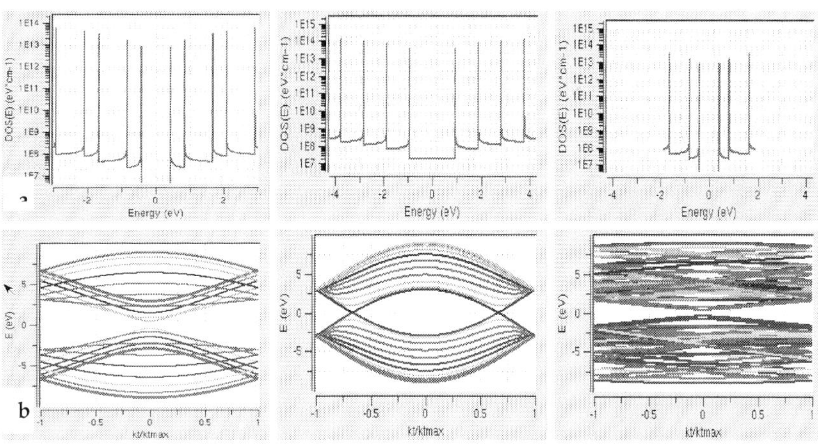

(13, 0) nanotube | (10, 10) nanotube | (10, 5) nanotube

Fig. 6. (a): Energy Density of state (DOS); (b): Energy gap of different types of nanotubes

The input parameters used in the simulation are presented in Table 2. Figure 7(a) shows the output characteristics of the CNTFET. The output characteristics indicate MOSFET like behavior. The channel allows the current flow when the gate voltage is greater than 0.26V.

As a result, the on-current is 9.660e-05 A at Vg=1.20V and Vd=1.20V, and the off-current is 8.665e-11 A at Vg=0.00V and Vd=1.20V.

The FETToy tool also produced six more different plots which are useful to understand the behavior of this transistor. Figure 7(b) presents the average velocity versus the gate voltage at saturation region. Figure 8(a) show the mobile charge behavior as a function of gate voltage. In Figure 8 (b) the mobile charge versus the drain voltage is shown at different gate voltages. It is noticed that increasing the drain voltage beyond a specific value has no longer an effect on the shape of the curves since the mobile charge remains constant. It is also observed that low drain voltage produces higher mobile charge and high drain voltage produces lower mobile charge. Figure 9 (a) shows the quantum capacitance versus the gate voltage at different drain voltages. It is noted that a higher quantum capacitance can be reached at a gate voltage greater than 0.3V. Lower drain voltage shows significant capacitance effect. Figure 9 (b) shows the transconductance behavior as a function of the gate voltage with different drain voltages.

Gate insulator thickness:	2e-09 (m)	Temperature:	300 (K)
Insulator dielectric constant:	25	Initial gate voltage:	0 (eV)
Threshold voltage:	0.32	Final gate voltage:	1.2 (eV)
Gate control parameter:	0.88	Number of bias points (gate):	15
Drain control parameter:	0.035	Initial drain voltage:	0 (eV)
Series Resistance:	0 (ohms)	Final drain voltage:	1.2 (eV)
NT diameter:	1e-09 (m)	Number of bias points (drain):	30

Table 2. input parameters for simulation

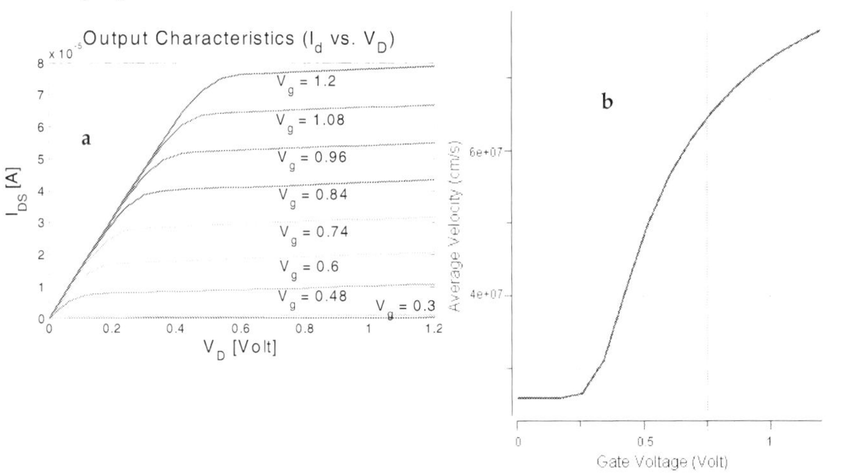

Fig. 7. (a): Output characteristics indicate MOSFET like behavior; (b): Average velocity vs. Gate voltage

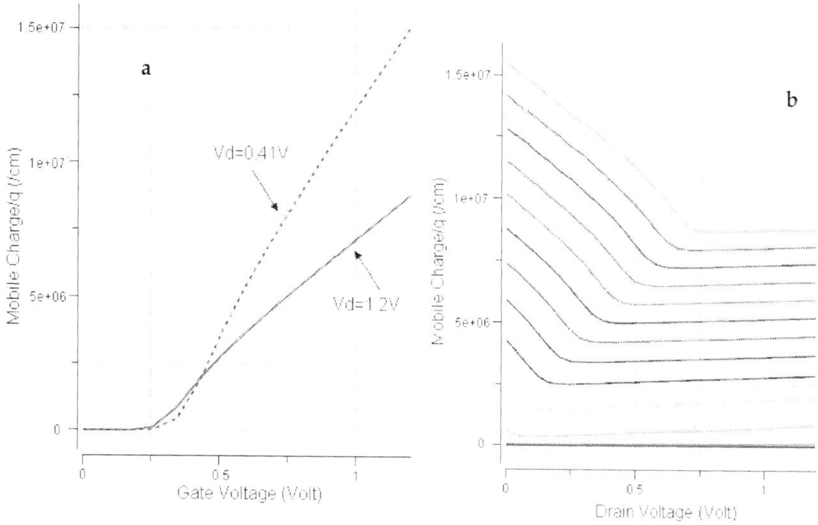

Fig. 8. (a): Mobile charge vs. Gate voltage; (b): Mobile charge vs. Drain voltage

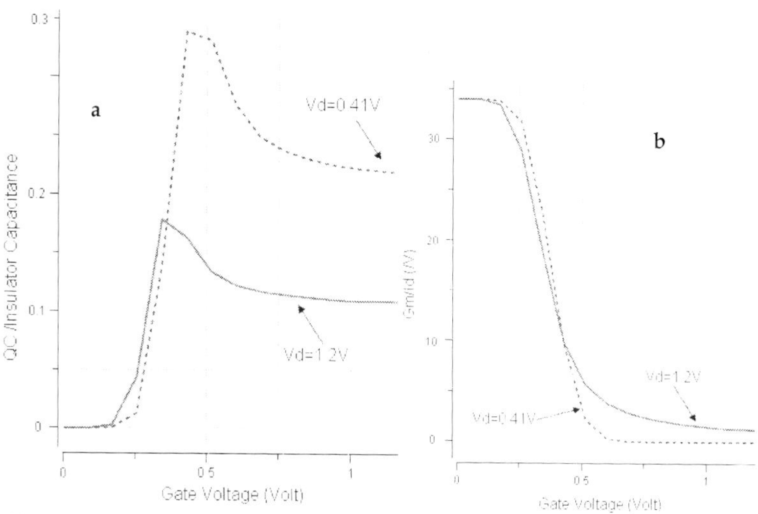

Fig. 9. (a): QC/Insulator Capacitance vs. Gate voltage; (b): Tranceconductance behavior as a function of Gate voltage

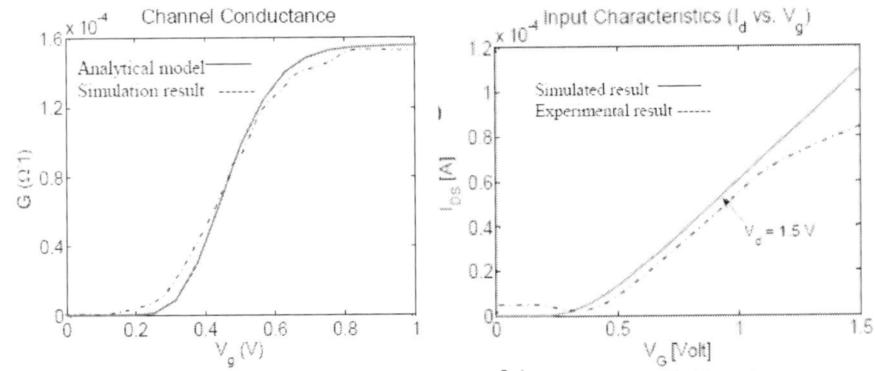

Fig. 10. (a): Channel conductance is saturated after certain gate voltage; (b): Transfer characteristics of the CNTFET shows on current for small bias voltage

Preliminary results from the analytical model and the measured data at room temperature presented in (Someya et al., 2003) demonstrate the feasibility of the developed model. The channel transconductance behaviors as a function of gate voltage are presented in Figure 10(a). The simulated results are compared with the analytical model. The transfer current voltage characteristics are compared with the experimental data in Figure 10(b) at room temperature. The channel conductance approaches the quantum conductance *4q2/h* (1.5x10-4 S) indicates the ballistic transport in the channel. The conductance is largely controlled by the electric field at the contact rather than the electrostatic potential. The SB height is determined by the metal work function which governs the current flow in the channel. Current saturates slower for a larger SB height. The device saturates at a small bias voltage. CNTFET model input characteristics indicates that the transistor is *n*-type. The device becomes *p*-type in presence of oxygen as the Fermi level at the metal-nanotube interface is shifted toward the valence band, allowing for hole conduction. If the same device annealed in vacuum, the Fermi level shifts toward the conduction band and favors electron injection, thus producing an *n*-type FET. Analytical formulation of CNTFET properties using ballistic transport properties with SB is presented. The calculations show that the proposed device can perform stable operation at high current levels (~70 μA). Upper limits of device characteristics are considered for the model. Degradation in measured data is observed due to the limitations in device fabrication technology and imperfect contact placement on the CNT.

7. Conclusion

The development and the dimensions of carbon nanotubes make the transistor a challenge to develop and to control the aspects of it, such as threshold voltage, maximum drain current, maximum transconductance. The investigated characteristics of CNTFETs represent its viability into transistor applications for NEMS switches, memory cells, field emission displays, biomedical instruments etc (Singh, 2005). Even with pessimistic assumptions,

CNTFET nanoelectronics can achieve significantly greater performance at a fraction of resource than of silicon technology. The understanding of CNTFET device physics should prove useful in optimizing device designs (Javey et al., 2002). The electronic conductivity and the thermal conductivity of CNTs are as good as or even better than those of noble metals (Mizutaniet al,. 2008). The mechanical strength of CNTs is extremely high. Significant system benefits are anticipated from CNT based electronics devices; however, there are many technological and material challenges for them. Localized oxidation on the CNT is critical for developing such devices (Tsang, 2003). Controlled growth of CNT is yet to be mastered. In situ coupling of source/drain to CNT mechanism needs to be idealized for better approximating the device performance. Replacing the metal-CNT SB contact with highly doped CNT source/drain (ohmic like) contact could improve the device performance (Javey et al., 2002).

It is possible to explore CNTFETs, especially in the area of optoelectronic devices, where electron and holes carriers are being recombined in a variety of different mechanisms. CNTs are direct-gap material and, as such, they directly absorb and emit light, thus possibly enabling a future optoelectronics technology based on SWCNTs. (Tenne, 2008); so it can be used as light emitters or a light detector depending on the biasing. The American Institute of Physics states that the ambipolar behavior of CNTFETs, even though tends to create a current leakage, if modified can lead to be recombined and give off the recombination energy in the form of light.

Silicon contains dangling bonds. This bonds are defects that affect the flow of electrons, however, carbon does not have surface dangling bonds as silicon does and so there is no need to use silicon dioxide as the gate insulator, the electrons will flow from the source to the gate without dangling bonds; this shows better performance in CNTFETs. CNTFET involves no doping and also allows control of the emission intensity and the position of the emitting spot.

Although major research is needed to achieve the goal of modifying the behavior of CNTFET, the progress made up to this date envisions a future for CNT based electronics (Singh et al., 2005). The different characteristics that carbon nanotubes have by altering their coefficient are amazing. The armchair characterizes lead to a metallic conductive tube, however the flexibility of the Zig-zag and chiral could be defined as both with respect to the different n and m. The nanoHUB community creates friendly user simulation environment that are not only makes it easy to use but makes the student learn by altering the different outcomes to deal with during the simulation.

8. References

Dang T., Anghel L. and Leveugle, R. (2006) "CNTFET basics and simulation," *Design and Test of Integrated Systems in Nanoscale Technology*, pp. 28-33,.

Datta S., (2000) "Nanoscale device modeling: the Green's function method" *Seperlattices and Microstructures*, 28-4, 253-278

Fujita, S. Nomura, K. Abe K., Lee T. H. (2007)'3-D Nanoarchitectures With Carbon Nanotube Mechanical Switches for Future On-Chip Network Beyond CMOS Architecture' *IEEE Transactions on Circuit and System:* 54,. 11,

Gruner G., (2006) "Carbon nanotube transistors for biosensing applications," *Analytical and Bioanalytical Chemistry*, vol. 384-2, 322-335,.

Hasan S., Salahuddin S., Vaidyanathan M. and Alam M. A (2006) "High-Frequency Performance Projections for Ballistic Carbon-Nanotube Transistors," *IEEE Transactions on Nanotechnology*, Vol. 5, No. 1, pp. 14-22,

Hoenlein W., Kreupl F., Duesberg G. S., Graham A. P., Liebau M., Seidel R. and Unger E., (2003) "Carbon nanotubes for microelectronics: status and future prospects," *Current Trends in Nanoscience*, 23, 663-669,

Iijima S., (1991) "Helical Microtubules of Graphitic Carbon,", *Nature* 354, 56 - 58

Iijima Sumio , Ajayan P. M. , and Ichihashi T. (1992) 'Growth model for carbon nanotubes'*Physics Revie . Letter.* 69, 3100–3103

Javey, H. Kim, M. Brink, Q. Wang, A. Ural, J. Guo, P. McIntyre, P. McEuen, M. Lundstrom, and H. Dai, (2002)"High Dielectrics For Advanced Carbon Nanotube Transistors and Logic," *Nature Materials,* 1, 241,

Javey, R. Tu, D. Farmer, J. Guo, R. Gordon, and H. Dai, (2005)"High Performance n-Type Carbon Nanotube Field-Effect Transistors with Chemically Doped Contacts," *Nano Letters,* 5, 345-348,

Kam Shi, Wong Nadine (2005)"Carbon nanotubes as multifunctional biological transporters and near-infrared agents for selective cancer cell destruction." *Proceedings of the National Acedemy of Sciences.* 16

Khan, Aurangzeb et al 2007. "Modeling and Simulation of n-Type Carbon Nanotube Field
Effect Transistor using Ca as contact Electrodes" Mater. Res. Soc. Symp. Proc. Vol. 1018.. Material Research Society

Lin Y.-M., Appenzeller J., Knoch J., and Avouris, P. (2005). "High-performance carbon nanotube field-effect transistor with tunable polarities," *Nanotechnology, IEEE Transactions*, 4, 481-489,

Mizutani T. , Nosho Y. , and Ohno Y. , (2008) "Electrical properties of carbon nanotube FETs," *Journal of Physics: Conference Series,* 109,.

Natori K., Kimura Y. and Shimizu T., (2005) "Characteristics of a carbon nanotube field-effect transistor analyzed as a ballistic nanowire field-effect transistor," *Journal of Applied Physics* 97,.

Polash B. and Huq H. F. (2008) 'Analytical Model of Carbon Nanotube Field Effect Transistors for NEMS Applications", *2008 IEEE Midwest Symposium on Circuits and Systems, MWSCAS,* Knoxville, TN;Aug.10-13,.

Rahman, J. Guo, S. Datta, and M. Lundstrom, (2003)"Theory of Ballistic Nanotransistors", *IEEE Transactions on Electron Devices*, 50, 1853-1864,.

Reich S., Thomsen C., Maultzsch J., (2004) *'Carbon Nanotubes BasicConcepts and Physical Propertie's* Wiley-VCH, Weinheim,

Singh D.V., Jenkins K.A. and Appenzeller, J. (2005) "Direct measurements of frequency response of carbon nanotube field effect transistors," *Electronics Letters,*. 41,. 280-282,

Someya T., Kim P., Nuckolls C. (2003) 'Conductance measurement of single-walled carbon nanotubes' in aqueous environment *Applied. Physics.* Lett.,. 82, 233-8

Tenne R.(2008) "Topics in Applied Physics; v.111, Carbon Nanotubes." Inorganic Nanotubes and Fullerene-like Structures (IF). *New York: Springer-Verlag* 631- 635.

Tsang, S.C., Harris, P.J. F. & Green, M. L. (1993) 'Thinning and opening of carbon nanotubes by oxidation using carbon dioxide' *Nature* 362, 520–522

Solid phase (micro)extraction tools based on carbon nanotubes and related nanostructures

Juan Manuel Jiménez-Soto, Rafael Lucena,
Soledad Cárdenas and Miguel Valcárcel*
*Department of Analytical Chemistry, University of Córdoba
Spain*

1. Introduction

The evolution of Analytical Sciences has been marked by several turning points which have changed the analytical laboratories. They have defined a variety of stages dominated by one or two key trends in Analytical Chemistry with clear influence in the "products" yielded in those years. This evolution started with the massive use of instruments in 1950s and continues with the growing importance of miniaturization and simplification in recent years. Nowadays, the analytical trends are also influenced by the development in other scientific and technological areas. Concretely, nanotechnology can be considered as a new, increasingly important trend in science in general and in analytical chemistry in particular. The outstanding properties of the nanoparticles and nanostructured materials have opening up a wide variety of challenging possibilities in this area. It should be expected that the evolution of this analytical trend will be markedly affected by the nanotechnological developments.

Nowadays there are a large variety of nanoparticles commercially available or easily synthesizable in the laboratory such as gold nanoparticles, quantum dots and carbon nanoparticles, among others. Carbon nanotubes (CNTs) have received special research attention since their discovery on account of their unique and outstanding properties which results in a wide variety of applications in different scientific and technological areas.

Carbon nanotubes can be considered as a graphene sheet that has been rolled up to form as extended tube-like structures (Harris, 1999). They range between a few tens of nanometers and several micrometers in length with an outer diameter from ca. 2.5 to 30 nm. They can consist of a single graphene sheet (singlewalled carbon nanotubes, SWCNTs) or a series of concentric layers (multiwalled carbon nanotubes, MWCNTs).

Since their discovery in 1991 by Iijima (Iijima, 1991), carbon nanotubes have been the target of theoretical and practical research with the aim of dilucidating their outstanding properties at the nanoscale level. The main difficult in this context is the marked tendency to aggregation of this nanomaterial, which makes difficult to distinguish between supramolecular assemblies and isolated carbon nanotubes properties. Apart from their thermal stability, excellent mechanical and electric properties, as well as their highest tensile strength among known materials, three characteristics which will play a crucial role in their sorption capacity can be highlighted (Valcárcel et al., 2007):

- They present nonpolar bonds and high length to diameter ratio. This result in a low water and commonly used organic media solubility and high aggregation tendency in these media as result. This negatively affects the effective surface area available for interactions with analytes.
- They can be covalently functionalized. Although they are not especially reactive, they can incorporate hydroxylic or carboxylic groups on their side walls under strong chemical conditions. It permits the immobilization of carbon nanotubes on solid supports such as controlled pore glass, silica or steel.
- They can establish non-covalent interactions with analytes which include hydrogen bond, π-π stacking, dispersion forces, dipole-dipole interactions and hydrophobic effect.

2. Sorbent properties of carbon nanotubes

As it has been commented on, the hydrophobic surface of carbon nanotubes permits the establishment of strong interactions with organic compounds. Indeed, several examples can be found in the literature dealing with the use of carbon nanotubes as effective sorbents in water treatment as regards octadecyl silica (RP-C18) or activated carbon (Liu et al., 2004) (Pyrzynska et al., 2007). Recently an excellent critical review dealing with the adsorption mechanisms of organic chemicals on carbon nanotubes have been published (Pan & Xing, 2008a). The authors stated in their article that the chemical adsorption on carbon nanotubes cannot be easily described using a single adsorption model. It can be ascribed to the fact that the presence of surface defects, functional groups or even interstitial regions can generate preferential adsorption sites. Readers interested in a detailed explanation on these mechanisms are referred to the article by Pan and Xing. Initially, the hydrophobic interactions between the carbon nanotubes surface and the organic analytes were considered as the sole mechanism for adsorption. If this is so, the interactions can be predicted by means of the octanol-water distribution constant, K_{ow}. However, further investigations demonstrated that it was not possible to stablish an explicit relationship between the adsorption coefficient of the organic compounds and the corresponding K_{ow} value (Chen et al., 2007) (Pan et al., 2008b). Three additional mechanisms have been proposed to explain the adsorption of the analytes on the carbon nanotubes such as the π-π interactions that can be established between the aromatic moiety of the molecule and the π-system of the nanotubes; the hydrogen bonds with the functional groups of the nanostructures; and electrostatic interactions due to the residual charge on the nanotubes surface (Lin & Xin, 2008) (Chen et al., 2008). As several factors affect these interactions, the relative contribution of each mechanism cannot be easily attributed.

Probably, one of the variables that will determine the availability of the adsorption equilibrium is directly related to the carbon nanotubes properties, viz: aggregates and functional groups. Moreover, the chemical environmental conditions (e.g. pH, ionic strength and surfactants) can also affects the analyte-CNTs adsorption.

The insolubility of carbon nanotubes in common aqueous and organic media results in the formation of aggregates of different sizes and stability. As a result, the superficial sites available for analytes interaction decreased as, the interstitial and inner pores adsorption is sterically hindered. The procedure selected for the synthesis of the carbon nanotubes can also influence the adsorption capability of the material as residues of amorphous carbon and metal catalyst could also block the inner pores (Yang et al., 2006).

Different procedures have been described to reduce or even eliminate these negative effects on analytes adsorption. The sorbent capacity of surfactant coated carbon nanotubes has been deeply studied in the literature (Carrillo-Carrión et al., 2007a). This evaluation was conducted using toluene and n-undecane as model compounds for the calculation of the corresponding solid-liquid and liquid-liquid isotherms. In order to obtain relevant conclusions, solid-liquid single component isotherms were calculated under three different experimental conditions: solid multiwalled carbon nanotubes; surfactant coated carbon nanotubes and surfactant media (sodium dodecyl sulphate, SDS). In all cases, the concentration of the nanoparticles and surfactant were maintained constant. For the liquid-liquid isotherms, the distribution of the analytes between an organic solvent (heptane) and an aqueous phase containing solid nanotubes, surfactant coated nanotubes and surfactant was considered.

The behaviour observed for the solid-liquid and liquid-liquid isotherms was analogous. The authors concluded that the maximum extractant capacity for toluene corresponds to dispersed nanotubes, with negligible contribution of the SDS solution. This fact permits one to assess that the whole adsorption can be attributed to the presence of the carbon nanoparticles. If benzene is considered instead of toluene as analyte, the enrichment factor obtained is lower which can be explained by the fact that the presence of methyl group in the toluene molecule would facilitate the adsorption on the MWCNTs surface due to an additional interaction with the hydrocarbon chain of the surfactant. On the contrary, the n-undecane extraction was maximum with the surfactant, with the lowest yields obtained with non-dispersed carbon nanotubes. The hydrocarbon molecules are located in the core of the micelles, and when surfactant coated carbon nanoparticles are used, there are fewer sites available, which reduces the extraction efficiency. However, the n-undecane can substitute the carbon nanoparticle from the micelles as the stability of the suspension is lower, providing an intermediate extraction capability.

From the above, it can be concluded the adsorption properties of carbon nanotubes are affected by the nanoparticles state, being possible to increase this property by the use of surfactant, which clearly reduces the natural aggregation tendency of these nanoparticles, increasing the active surface of CNTs.

Adsorption of organic compounds on carbon nanotubes can also be affected by the synthesis procedure and further purification step. The raw material is mainly hydrophobic and in general, the adsorption can be increased after heating the carbon nanotubes due to the removal of amorphous carbon and oxidation of residual catalyst moieties. The increasing amount of oxygenated groups will lead to a higher interaction with polar compounds due to an increased hydrophilicity of the carbon nanotubes. Other functional groups such as -OH, -COOH and -CO can be added by acid treatment (typically a mixture of HNO_3 and H_2SO_4). The presence of these moieties can be used for carbon nanotubes immobilization on a variety of surfaces. The presence of isolated nanostructures can increase the adsorption

through a reduced aggregation. Finally, the hydrophobicity of the carbon nanotubes can be restored by using extreme heating conditions (viz 2200 °C).
Depending on the conditions used, the most appropriated carbon nanotubes state can be selected for the given application. The next sections are devoted to the reported applications of carbon nanotubes and related materials in solid phase (micro) extraction.

3. Carbon nanoparticles in solid-phase extraction

Solid-phase extraction (SPE) is a well-known, worldwide used sample treatment technique. Its favourable features as regards analytes preconcentration factors, interferents removal, solvent consumption, sample processing time, automation and potential sorbent reusability, justifies its use against its counterpart liquid-liquid extraction. Moreover, the variety of sorbents currently available permits the preconcentration of analytes of different polarity and even charged species.
Among the new sorbents proposed for SPE, carbon nanostructures and more specifically fullerenes and carbon nanotubes have been successfully used in this context (Valcárcel et al., 2008). The applicability of these nanoparticles in this separation technique relies on the sorbent properties described in the previous section of this chapter. Here, the main applications of carbon nanostructures in SPE will be presented. The selected examples have been classified according to the extraction unit in which the nanoparticles can be presented, namely, commercial cartridges and minicolumns. Furthermore, their used immobilized on a solid support and dispersed with surfactants will be commented on.
Carbon nanotubes have been used packed in conventional SPE cartridges for the determination of a variety of organic compounds in different matrices. The isolation and preconcentration of organic pollutant from environmental samples is the most reported application of this configuration. By way of sample, Table 1 lists analytical information related to these references.

Sample	Analyte	Amount of sorbent	Sample volume	Eluent	Detection technique	Reference
Environmental waters	4-n-nonylphenol 4-tert-octylphenol	500 mg MWCNTs	500 mL	MeOH, 2.5 mL	HPLC-FL	Cai et al., 2003
Environmental waters	Chlorophenols	300 mg MWCNTs	200 mL	MeOH, pH 10, 6 mL	HPLC-UV	Cai et al., 2005
Environmental waters	Atrazine Simazine	100 mg MWCNTs	500 mL	ACN, 4 mL	HPLC-DAD	Zhou et al., 2006a
Environmental waters	Cyanazine Chlortoluron Chlorbenzuron	100 mg MWCNTs	250 mL	CH_2Cl_2, 5 mL	HPLC-UV	Zhou et al., 2007a
Waters	Sulphonylurea Herbicides	100 mg MWCNTs	500 mL	ACN 1% acetic acid, 12 mL	HPLC-DAD	Zhou et al., 2007b
Waters	Dichlorophenyltrichloroetane and metabolites	100 mg MWCNTs	500 mL	CH_2Cl_2, 5 mL	HPLC-UV	Zhou et al., 2006b
Pork	Barbital, amobarbital and phenobarbital	250 mg MWCNTs	10 mL	Acetone:Ethylacetate 3:7 (v:v), 5 mL	GC/MS/MS	Zhao et al., 2007
Urine	Antidepressants	30 mg MWCNTs	10 mL, diluted in water 1:1	ACN:acetic acid 97:3, 2 mL	HPLC-UV	Cruz-Vera et al., 2008
Virgin olive	Pesticides	30 mg c-SWCNTs	3 mL, diluted in water 1:1	Ethylacetate, 0.5 mL	GC-MS	López-Feria et al., 2009

Table 1. Main analytical features of the procedure developed using carbon nanotubes packed cartridges.

The first sign of the adsorption capacity of carbon nanotubes was revealed by Long et al. (Long & Yang, 2001) when they proposed their use for dioxin removal. Two years later, the first analytical application of packed carbon nanotubes was reported (Cai et al., 2003). The authors employed a commercial polypropylene SPE cartridge from which the sorbent was replaced by 500 mg of multiwalled carbon nanotubes. The upper and low frits were maintained at both ends to hold the nanotubes in place during the extraction. The cartridge was evaluated for the enrichment of bisphenol A, 4-n-nonylphenol and 4-tert-octylphenol from waters. The high affinity of the carbon nanotubes for the endocrine disruptors is demonstrated by the large volume of sample that can be preconcentrated (up to 1 L) without analytes losses. Moreover, the low volume of organic eluent required (2.5 mL of methanol) permits limits of detection as low as 0.018 ng/mL (for 4-n-nonylphenol). The analytical figures of merit provided by the nanostructured material resulted better than those of RP C_{18} and XAD-2 for these compounds. The applicability of carbon nanotubes to solve environmental-related analytical problems was extended in the following years by determining chlorophenols (Cai et al., 2005), herbicides (Zhou et al., 2006a) (Zhou et al, 2007a) (Zhou et al., 2007b) and dichlorodiphenyltrichloroetane and metabolites (Zhou et al., 2006b) in waters. In all cases, multiwalled carbon nanotubes were preferred to singlewalled carbon nanotubes as the presence of concentric sheets resulted in an enhanced interaction with the analytes. The amount of sorbent used can be reduced to 100 mg and in some of the applications, the cartridge can be reused up to 200 times without efficiency loss (Zhou et al., 2007a). The eluent used was different depending on the analyte and, in some cases the use of mixtures is needed. Moreover, the eluent volume required for quantitative elution can be as high as 6 mL which is an inconvenience for sensitivity. An evaporation-redissolution step can be used to overcome this limitation.

In addition to waters, the use of commercial cartridges packed with MWCNTs has been extended to the determination of barbiturates in porks (Zhao et al., 2007), antidepressants in urine (Cruz-Vera et al., 2008) and pesticides in virgin olive oil (López-Feria et al., 2009). In the last two articles, the amount of carbon nanotubes was reduced to 30 mg and, despite the matrix complexity, the cartridge reusability was demonstrated by the authors by the mere inclusion of a washing step between samples.

Recently, a novel carbon nanostructure, carbon nanocones, has been proposed for its use in SPE cartridge (Jiménez-Soto et al., 2009). The commercial product contains 20%wt carbon nanocones, 70%wt carbon disks and 10%wt amorphous carbon. The presence of the amorphous solid results in irreversible interactions with the analytes and lower recovery values, as result. This negative effect can be reduced by maintaining the commercial solid at 450 °C for 20 min. This treatment dramatically reduced the amorphous carbon content with negligible influence on the nanostructured material. In the model application developed for the determination of chlorophenols in waters, 20 mg of the thermally treated sorbent were packed in a 3-mL commercial SPE cartridge. Limits of detection as low as 0.3 ng/mL were achieved by preconcentrating 2 mL of water sample and eluting the retained compounds with 200 μL of hexane. As compared with carbon nanotubes, the conical nanostructures showed to be more efficient in the preconcentration process as similar analytical features were obtained with lower sorbent amount. This fact can be explained taking into account that the carbon nanocones show lower aggregation tendency than the carbon nanotubes and, therefore, the effective surface available for analytes interaction is higher with the former.

Solid-phase extraction has also been implemented in a continuous fashion by using flow system configurations. This approach permits the miniaturization and automation of the whole process with the inherent advantages. Fullerenes have been extensively used in this context. The general configuration used is depicted in Figure 1. As can be seen, an amount of ca. 80 mg of fullerene C60 are packed in a PTFE minicolumn fitted with cotton wool beds at both ends to prevent sorbent losses. The column is placed in the loop of an injection valve to allow the preconcentration of large volumes of sample. Once the sample has been completely aspirated into the system, a small plug of eluent (in the microliter range) was passed through the C60 column for analytes elution. As the continuous unit is on-line coupled to the atomic absorption spectrometer, a transient analytical signal is obtained when the eluent, containing the analytes, reach the flame. By using this configuration, a large variety of metals and their alkyl derivatives have been determined in waters. Table 2 summarized the main applications developed in this context. As can be seen, the adsorption capacity of fullerene C60 for organic compounds is rather limited while they show great potential retaining organometallic compounds. The study carried out by Ballesteros et al. (Ballesteros et al., 2000) also pointed out that the efficiency of the extraction process decreased when the polarity of the analytes increased. Despite this behaviour, some references dealing with the determination of organic molecules using fullerenes as sorbent have been reported. Serrano et al. used fullerene C60 for the extraction of benzene, toluene, ethylbenzene and xylene isomers from waters (Serrano & Gallego, 2006). In this case, the preconcentration step was followed by the gas chromatographic-mass spectrometric determination of the volatile organic compounds. The sensitivity and precision of the method were better than those obtained with conventional sorbents such as RP-C18 and Tenax TA. Moreover, the selectivity was also enhanced as fullerenes were not affected by the low pH selected for retention. As far as metallic species is concerned, they required previous derivatization to form neutral chelates that could be effectively retained on fullerenes. In order to obtain discriminated information on the sample composition, a chromatographic separation can be implemented after the analysis of the sample by an atomic technique to estimate the total concentration of the target metals in the sample. Such is the case of the determination of lead and alkyllead compounds in rainwaters (Baena et al., 2000) or organometallic compounds in environmental samples (Muñoz et al., 2005).

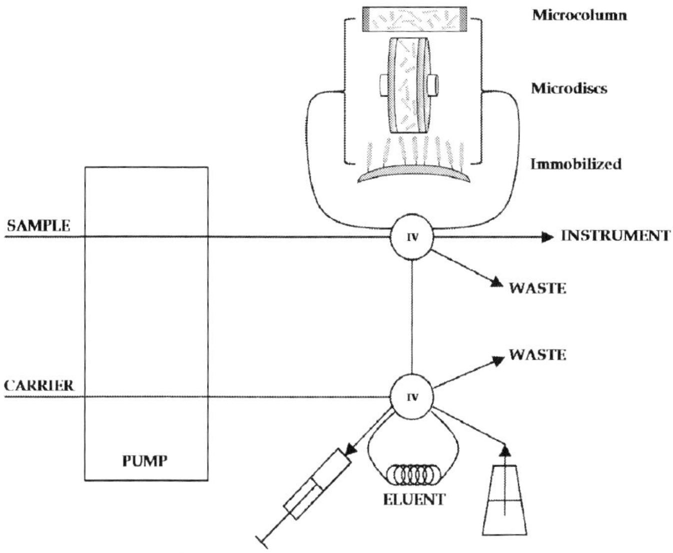

Fig 1. Flow configuration for the use of carbon nanotubes as sorbent in solid phase extraction procedures.

Sample	Analyte	Sorbent	Eluent	Instrumental technique	Reference
Environmental samples	Cd, Mn, Ni	MWCNTs	HNO_3 0.1 M	ICP-AES	Liang et al., 2004
Waters	Cu	MWCNTs	HNO_3 0.5 M	FAAS	Liang et al., 2005
Egg and pork	Sulfonamides	MWCNTs	Mobile phase MeOH:H_2O 22:78	HPLC-UV	Fang et al., 2006
Waters	Pb (II)	c-MWCNTs	HNO_3 1 M	FAAS	Barbosa et al., 2007
Waters	BTEXs	Fullerene C_{60}	Ethyl acetate	GC-MS	Serrano & Gallego, 2006
Waters	Pb	Fullerene C_{60}	IBMK	FAAS	Gallego et al., 1994
Wheat flour	Co	Fullerene C_{60}	IBMK	ETAAS	González et al., 1999
Rainwaters	Pb and alkyllead derivates	Fullerene C_{60}	IBMK	FAAS	Baena et al., 2000

Table 2. Representative examples of the applicability of carbon nanotubes and fullerenes in continuous flow separation techniques.

Reducing the dimensions of the extraction units for carbon nanotubes was also aimed at minimizing the aggregation of the nanoparticles which negatively affects to the retention efficiency. Therefore, Suárez et al. proposed the package of 6 mg of MWCNTs in the minicolumn, which was further located in the loop of the injection valve. It permits the retention and elution processes to be carried out in opposite direction thus reducing the compacting of the sorbent. This approach has been used for the determination of tetracycline residues in waters (Suárez et al., 2007a). In this case, the minicolumn presents a larger diameter, (9 mm i.d.) than that reported for fullerenes, which also helps to reduce the overall pressure in the system. A similar configuration has been on-line coupled to a liquid chromatography for the simultaneous determination of 10 sulfonamides in egg and pork (Fan et al., 2006).

Suárez et al. used immobilized carboxylated singlewalled carbon nanotubes (c-SWCNTs) on inert controlled pore glass and used the combined solid as sorbent in SPE (Suárez et al., 2007b). An accurately weighed amount of 15 mg of the combined solid was packed in a PTFE tube (1.2 mm i.d.) which was placed in a continuous flow device. The automated system was used for the determination of non steroidal anti-inflammatory drugs (NSAIDs) in urine samples. The authors compared the adsorption capacity of the same amount of immobilized c- SWCNTs, SWCNTs and MWCNTs for tetracyclines and NSAIDs [Figure 2]. The best results were obtained with the immobilized sorbent, with the additional consideration that the amount of nanoparticles is lower than in the other cases as the weighed material referrers to both, the c-SWCNTS and the inert glass. It can be ascribed to the fact that the immobilization process results in a vertical alignment of the c-SWCNTs on the particle surface, which clearly enhanced the molecular interactions of the analytes with the nanotube surface.

The determination of inorganic species using carbon nanotubes requires the previous derivatization of the nanoparticles to introduce functional groups (carbonyl, carboxylic and hydroxyl) that allows the interaction with the metallic species. Such is the case of the extraction of cadmium, manganese, and nickel (Liang et al., 2004), Cu (II) (Liang et al., 2005) or Pb (II) (Barbosa et al., 2007) from waters. In these applications, the pH of the sample resulted to be a critical variable as it will define the surface charge of the sorbent, which should be negative to interact with the cations. The microcolumns of variable dimensions and containing and average sorbent amount of 50 mg were inserted in a continuous flow configuration that can be on-line coupled to the flame atomic absorption or inductively coupled plasma atomic emission spectrometers.

As it was indicated in the previous section, surfactants have been reported to disperse carbon nanoparticles in aqueous media, facilitating their use for analytical purposes. The aqueous suspension obtained is stable under optimal conditions (the amount of nanoparticles and surfactant concentration being critical) as the nanoparticles are located inside the micelles. The surfactant coated carbon nanoparticles have been used as pseudophase for liquid-liquid extraction (Carrillo-Carrión et al., 2007a). The interactions between the organic compounds and the dispersed carbon nanotubes differ from that observed for the raw material. As no aggregates exist in the dispersed phase, more adsorption sites are available for retention, which has a positive effect on extraction. This methodology has been successfully applied for the extraction of benzene, toluene, ethylbenzene and xylene isomers form olive oil samples with a clearly sensitivity and selectivity enhancement in comparison with the direct headspace analysis (Carrillo-Carrión et al., 2007b). The procedure is schematically depicted in Figure 3.

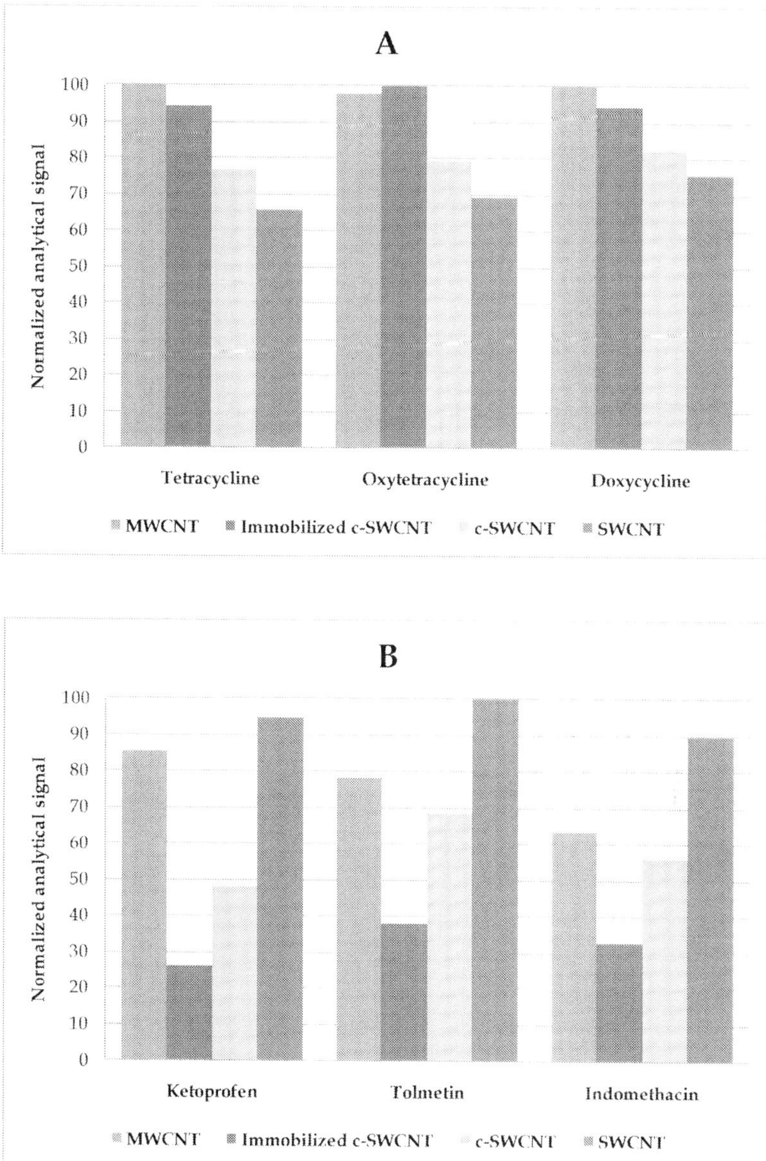

Fig. 2. Comparison of sorption capacity of different types of carbon nanotubes

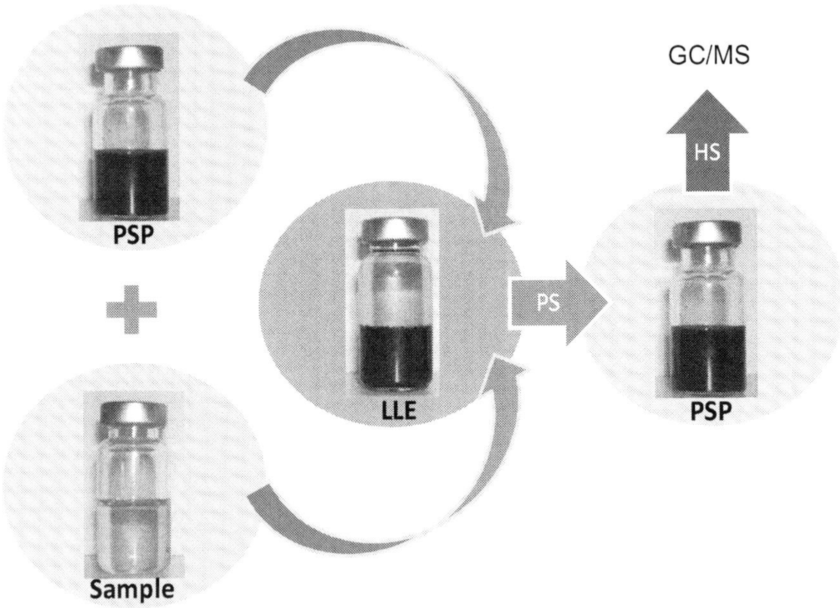

Fig. 3. Analytical procedure for the liquid-liquid extraction (LLE) of BTEXs from olive oil samples using surfactant coated nanotubes as pseudophase (PSP)

Finally, and considering the thermal stability of the carbon nanotubes, they can be used as preconcentration units on-line coupled to a gas chromatograph for retention of low molecular weight compounds (methyl-ethyl ketone, toluene and dimethyl methyl phosphonate) from a vapour stream by maintaining the unit at low temperature. For desorption, the temperature was rapidly increased and the analytes directly transferred to the injection port of the instrument (Zheng et al., 2006).

4. Solid phase microextraction based on carbon nanoparticles

Solid-phase microextraction (SPME) appeared as an evolution of the solid phase extraction. It was introduced by Arthur and Pawliszyn in 1990 (Arthur & Pawliszyn, 1990) and it was conceived as an alternative to traditional methods for sample preparation. SPME overcomes the main drawbacks of classical preconcentration techniques, regarding the number of steps involved, extraction time or the consumption of toxic organic solvents. Solid-phase microextraction has gained importance in recent years because it is an efficient and fast methodology which combines in a single step, the sampling, preconcentration, matrix removal and injection in the chromatographic system. Owing to its favourable features, SPME has widely been applied in various fields including environmental, food, natural products, pharmaceutical, biology, toxicology and forensic (Pawliszyn, 1999).

SPME is based on the distribution of the analytes between the sample (or its headspace) and the stationary phase immobilized on the solid support. The analytes transference starts

$$I = I_0 + I_D + I_M \tag{3}$$

$$\text{where } I_D = \frac{I_0}{2}\left(1+\frac{B}{E_0}+\frac{B^2}{2E_0^2}\right)\left[\left(\frac{E_1}{F_0}\right)^2+\left(\frac{E_2}{F_0}\right)^2\right] \tag{4}$$

$$\text{and } I_M = I_0\left(1+\frac{B}{E_0}+\frac{B^2}{2E_0^2}\right)\left(\frac{E_1}{E_0}\right)\left(\frac{E_2}{E_0}\right)\cos\left[\left(\omega_1-\omega_2\right)t\right] \tag{5}$$

Equations (3)-(5) show that the two applied sinusoidal fields increase the dc current (optical rectification) and also generate a mixer current at the difference frequency.

With optical radiation, where the effects of the photon energy may not be neglected, the major correction to these equations is the effect of a resonance between tunneling electrons and a radiation field (Hagmann, 1995A). This resonance, which is caused by reinforcement of the wave function through virtual photon processes, was later confirmed by others (Mayer & Vigneron, 2000; Mayer et al., 2002, 2003). Figures 1 and 2 show the gain in the power flux density, which increases E_1 and E_2 in Eqs. (4) and (5), for the resonance with tungsten emitters (Mayer & Vigneron, 2000) and (5, 5) armchair (metallic) CNT emitters (Mayer et al., 2002).

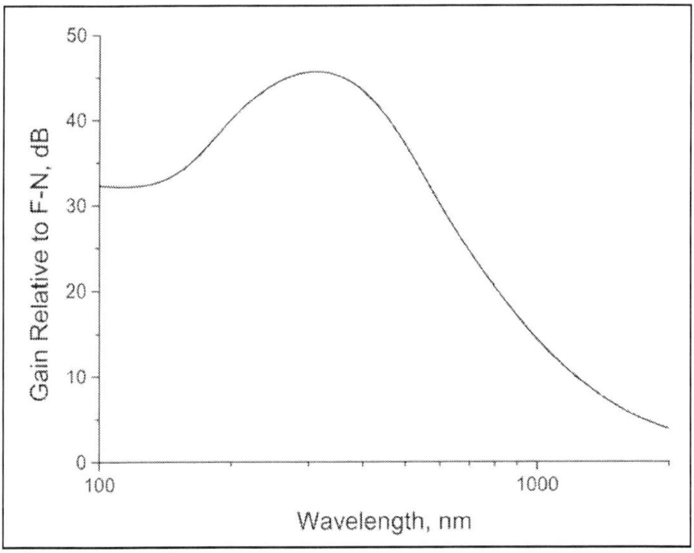

Fig. 1. Gain for tungsten emitters.

Modifying Eqs. (4) and (5) by allowing for the gain G (in dB), as shown in Figs. 1 and 2, the peak value of the mixer current in laser-assisted field emission is given by

$$\frac{I_D}{I_0} = \eta \cdot 10^{G/10} \left(1 + \frac{B}{E_0} + \frac{B^2}{2E_0^2} \right) \frac{(P_1 + P_2)}{E_0^2} \quad (6)$$

$$\frac{I_M}{I_0} = 2\eta \cdot 10^{G/10} \left(1 + \frac{B}{E_0} + \frac{B^2}{2E_0^2} \right) \frac{\sqrt{P_1 P_2}}{E_0^2} \cos\left[(\omega_1 - \omega)t \right] \quad (7)$$

where P_1 and P_2 are the power flux densities of the two lasers at the field emitter, and η is the impedance of free-space (≈ 377 Ω).

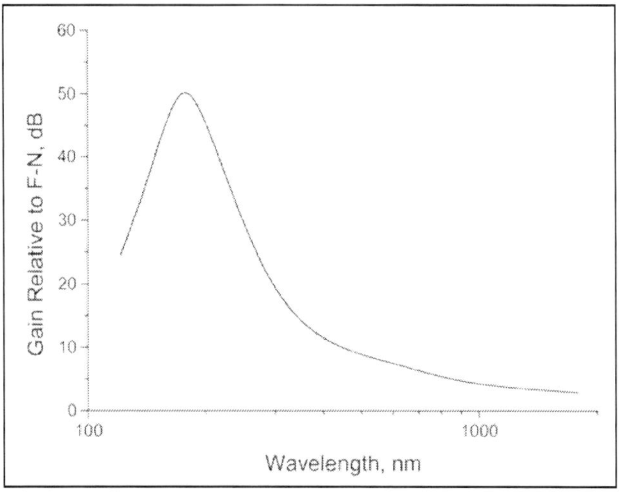

Fig. 2. Gain for (5,5) metallic CNT emitters.

Our prediction of the resonance (Hagmann, 1995A) and first suggestions for microwave and terahertz devices based on LAFE (Hagmann, 1995B) were made at a time when the general consensus was that the increase in field emission current caused by a laser is a small effect due to heating of the emitter that has a time constant of 10 - 1000 μs (Lee & Robbins, 1989). Our measurements of optical rectification in LAFE with a sealed field emitter tube (Leybold AG model 55460) confirm that there is a thermal effect but they also show a non-thermal effect that is in agreement with our analysis and shows the effect of the resonance (Hagmann & Brugat, 1999). We measured a time constant of 300 μs for the decay of the thermal processes, and also found that the cathode-to-anode circuit for the sealed field emitter tube acts as a low-pass filter (Hagmann et al., 2004).

Microwave prototypes have been built and tested using emitters of tungsten and molybdenum (Hagmann, 2004). Two lasers are focused on the emitter to generate current oscillations at their difference frequency by LAFE, as described by Eq. (7). We do not specifically use the emitted electrons as such in our method, but instead we cause the current oscillations to generate a transverse-magnetic (TM) surface wave that propagates on extensions of the emitter (Alonso & Hagmann, 2001). This surface wave is coupled directly to the load or emitted as radiation from antennas that are formed on the emitter (Hagmann, 2004) in order to avoid the low-pass filter in the cathode-to-anode circuit (Hagmann et al., 2004). Our group was the first to suggest the use of surface plasmons to propagate THz radiation on metallic wires (Hagmann, 1998B; Alonso & Hagmann, 1999), which was later implemented by others (Wang & Mittleman, 2004).

It was already shown that photomixing in LAFE can provide an unusually large tunable bandwidth. However, the maximum power that we have obtained with microwave protypes was only 50 pW. This limitation can be understood because the mixer current, which is a few percent of the dc field emission current (30 μA) is coupled to a 50 Ω load or an antenna having a similar value for its radiation resistance. Increasing the size of the emitter is not an effective means to obtain the higher current that is required for greater output power because surface imperfections (Fujieda et al., 2005) cause the dc emitted current to be proportional to $A^{1/4}$, where A is the area of the emitter (Okawa et al., 1988). In the rest of this chapter it will be seen that much greater output power may be obtained without sacrificing the the large tunable bandwidth by using carbon nanotubes (CNT) in LAFE because of a synnergy in which the CNT increase both the dc current and the impedance which is seen by the mixer current.

3. Carbon Nanotubes as High Impedance Transmission Lines

Equation (7) shows that photomixing in LAFE requires incident electric fields at two different frequencies in order to generate a mixer current at a third frequency which is the difference of the first two frequencies. The output power that is coupled to the TM surface waves that propagate on extensions of the emitter is given by ½ $Z_0 I_{MP}^2$, where Z_0 is the (real) characteristic impedance for the transmission line which is a wire extension of the emitter and I_{MP} is the peak value of the mixer current which is given by Eq. (7). Thus, the output power is given by the following equation, and it is proportional to Z_0 which shows the motivation for increasing this impedance in order to obtain greater output power.

$$P_{OUT} = 2\eta^2 Z_0 \left(1 + \frac{B}{E_0} + \frac{B^2}{2E_0^2}\right)^2 \frac{I_0^2}{E_0^4} 10^{G/5} P_1 P_2 \tag{8}$$

The interaction of the mixer current with the impedance Z_0 creates an electric field at the mixer frequency, which has a peak value that is given by $E_{MP} = Z_0 I_{MP}$. Thus, for consistency, it would be necessary to include this electric field in Eq. (2), which reduces the current from optical rectification as well as the mixer current. This negative feedback sets an upper limit on the output power which is consistent with the power that is supplied by the two lasers. However, the effect of this negative feedback is generally negligible unless Z_0 exceeds 1 MΩ which does not occur with carbon nanotube transmission lines (Hagmann, 2008).

Others have simulated single-wall (Burke, 2002, 2003) and multiwall (Tarkiainen et al., 2001; Sonin, 2001; Bachtold, 2001) carbon nanotubes (SWCNT and MWCNT) on a conducting substrate as transmission lines by using a hybrid approach in which quantum theory is combined with classical distributed element transmission line models (Collin, 1990). This method allows for the effects of the size and shape of the carbon nanotube (CNT) and the insulation separating it from the conducting substrate, as well as the electronic properties of the CNT from quantum theory. These studies show that the characteristic impedance is on the order of 5 kΩ, which may be understood because the distributed kinetic inductance of the CNT is approximately10^4 times the inductance for metals. The high value for the kinetic inductance of MWCNT was verified by measurements, and it has been suggested that CNT on conducting substrates could be used as high-impedance transmission lines in nanoelectronics (Ahlskog et al., 2001).

We have extended the previous simulations which were made by others in order to study CNT that are isolated, removed from the substrate, as transmission lines (Hagmann, 2005). The previous work by others was based on the classical distributed element model for a two-conductor transmission line that is formed by the CNT and its image in the substrate, for which the propagation is in a transverse electromagnetic (TEM) wave. However, as a wire is moved away from a conducting surface, there is a transition from TEM propagation (Barlow & Cullen, 1953) to a TM surface wave which is the dominant mode for an isolated wire (Stratton, 1941). Thus, we developed a classical distributed-element model for propagation of the TM surface wave and then added the quantum inductance and the quantum capacitance to this model as it has been done by others (Hagmann, 2005). Highlights of the results of our simulations of isolated CNT as transmission lines follow:

(1) Only the axially-symmetric TM surface wave mode will propagate on an isolated SWCNT. This is consistent with the observation that higher order modes cannot propagate on a metal cylinder having a circumference which is less than one free-space wavelength (Savard, 1967). Single-mode propagation is necessary to prevent dispersion of the propagating electromagnetic fields.

(2) At THz frequencies all of the metallic tubes in an isolated MWCNT can take part in the propagation so that for N shells there are N - 1 propagating coaxial modes in addition to the axially-symmetric TM surface wave mode. Thus, in order to prevent dispersion, only SWCNT should be considered for this application.

(3) The axially-symmetric TM surface wave mode in an isolated SWCNT has a characteristic impedance of approximately 44 kΩ. This value is much greater than that for a CNT on a conducting substrate because the capacitance to the substrate has been eliminated.

(4) The axially-symmetric TM surface wave mode in an isolated SWCNT has a phase velocity of approximately 2.8×10^6 m/s, which is 0.9 percent of the velocity of light in vacuum. The reduced value of the phase velocity is consistent with the increased value of the characteristic impedance.

(5) The axially-symmetric TM surface wave mode in an isolated SWCNT has negligible attenuation for propagation over the lengths that will be used in this application. Low ohmic loss is also found with surface waves on a single conducting cylinder, but the energy is loosely bound to the cylinder so it is necessary to keep other objects at a distance to limit radiative loss (Goubau, 1950). Our applications fall under what has been called the "low-loss regime" for propagation on CNT (Burke at al., 2006).

4. Measurements with Individual CNT

4.1 Description of the prototypes

In 2004, we began our study of LAFE with single SWCNT and MWCNT, funded by the National Science Foundation (NSF-DMI-0338928). Four custom sealed field emission tubes with CNT were made for us by Xintek (Chapel Hill, NC, U.S.A.), having the same outer structure which is shown in Fig. 3. The copper anode is at the right, and the CNT emitter is mounted on a tungsten wire that is attached to the metal cylinder at the left. Figure 4 shows an SEM image of one of the emitters, taken with a JEOL JEM 6300 before it was placed in one of these tubes.

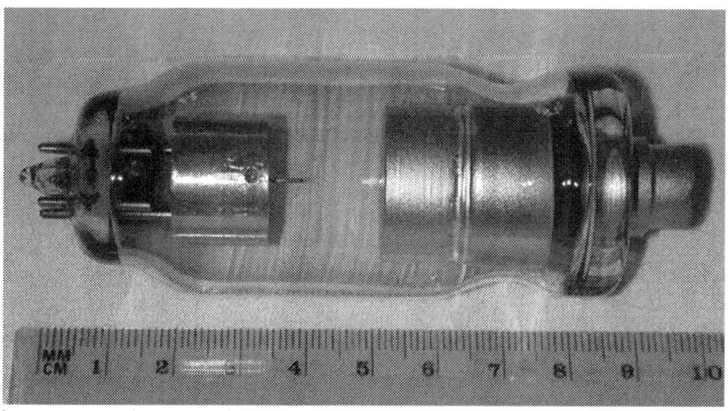

Fig. 3. Field emission tube made for us by Xintek.

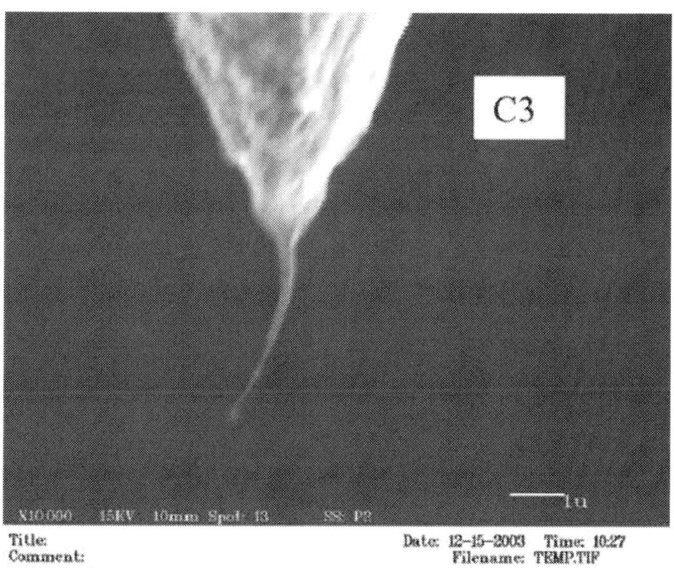

Fig. 4. SEM image of the CNT emitter in tube C-3.

Most of the measurements of field emission from CNT pertain to the dc characteristics of films (Zhu et al., 1999), arrays (Lim et al., 2001), and composites (Wang et al., 1998), which only provide a statistical evaluation of the properties with large numbers of CNT. Furthermore, the electric field, and thus the total emitted current, is reduced when the CNT are close together (Hii et al., 2006). Careful measurements with individual CNT have shown that the field enhancement factor and work function varies with the structure of the CNT and its surface conditions (Xu et al., 2005). Other measurements with single CNT have shown that field emission is affected by irregularities in the structure of the graphitic sheets (Wang et al., 2005), and time-dependent deformations due to reconstruction at the apex increase the electric field at specific sites which are where the field emission occurs (Kuzumaki et al., 2004). Therefore, we chose to begin our studies LAFE with tubes having individual CNT as the emitters.

The two tubes labeled M-1 and M-4 have a single MWCNT as the emitter, and tubes C-3 and C-6 each have a single SWCNT as the emitter. The CNT are in bundles with diameters of 10 to 30 nm that are shaped so that field emission is only from the single CNT at the end of each bundle where the electric field is most intense. A sealed field emitter tube, with an etched tungsten single crystal tip as the emitter (Leybold AG model 55460) was used for comparison in the measurements with the four tubes from Xintek. The characteristics of these five tubes are described in the following sections:

4.2 Fowler-Nordheim characterization of the prototypes

First, the dc current-voltage characteristics were measured for each of the five field emission tubes, and then these data were analyzed using the following procedure that is based on the simplified Fowler-Nordheim equation which was shown as Eq. (1).

The measured field emission current I is proportional to the current density J, and the measured voltage V is proportional to the applied electric field E, so Eq. (1) is equivalent to the following expression that may be used with the measured data:

$$I = CV^2 e^{-D/V} \tag{9}$$

Equation (9) may be written in the following form:

$$\ln\left(\frac{I}{V^2}\right) = \ln(C) - \frac{D}{V} \tag{10}$$

Thus, the values of the parameters C and D for each field emission tube may be determined from the slope and intercept of a graph in which the ordinate and abscissa are $\ln(1/V^2)$ and $1/V$, respectively. The values of the parameters A and B in Eq. (1) are calculated from the work functions φ = 4.5 eV for tungsten, and 4.9 eV for graphene in the CNT. Then the following two parameters may be calculated: $S = CD^2/AB^2$; $R = D/B$. Here S is called the effective area of the emitter, and R is called the effective radius of the emitter.

During the measurements a ballast resistor of 100 MΩ was placed in series with each field emission tube to protect it by limiting the maximum current. The tungsten emitter is mounted on a filament so it was cleaned by electrical heating before each measurement. It is not possible to clean the CNT, which is probaby the cause for the "switch-on" effect in which the supply voltage must be momentarily increased beyond the operating point to initiate field emission with the CNT (Liu & Fan, 2005).

Figure 5 is a Fowler-Nordheim Plot of the data for tube C-6, showing that a straight line is an excellent fit of the data which is consistent with Eq. (10). Linear regressions were made to obtain the values of C, D, S, and R for each field emission tube, which are shown in Table I. Typically the correlation was approximately - 0.998, the standard variance was 0.08, and the probability for the null-hypothesis that no linear relationship exists, was less than 0.0001.

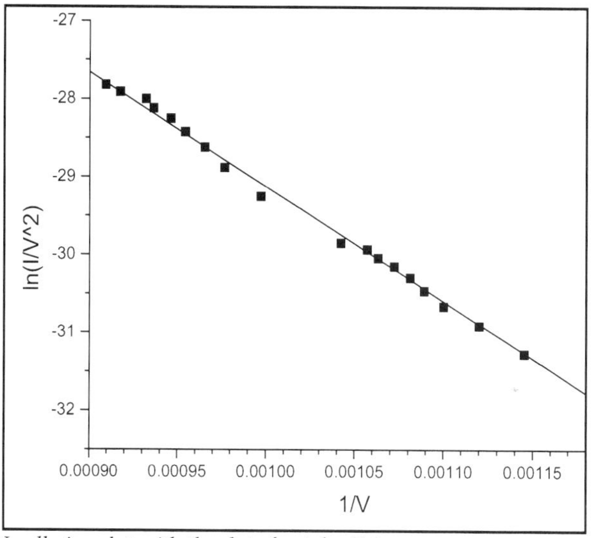

Fig. 5. Fowler-Nordheim plot with the data for tube C-6.

	Leybold	M-1	M-4	C-3	C-6
Emitter	Tungsten	MWCNT	MWCNT	SWCNT	SWCNT
Φ, eV	4.5	4.9	4.9	4.9	4.9
A,A/V^2	3.42×10^{-7}	3.14×10^{-7}	3.14×10^{-7}	3.14×10^{-7}	3.14×10^{-7}
B, V/m	6.52×10^{10}	7.41×10^{10}	7.41×10^{10}	7.41×10^{10}	7.41×10^{10}
C, A/V^2	2.15×10^{-7}	2.20×10^{-9}	3.47×10^{-9}	7.72×10^{-11}	1.27×10^{-8}
D, V	5.89×10^{4}	8.02×10^{3}	7.92×10^{3}	4.84×10^{3}	5.60×10^{3}
S, m^2	5.14×10^{-13}	8.19×10^{-17}	1.26×10^{-16}	1.05×10^{-18}	2.31×10^{-16}
R, m	9.03×10^{-7}	1.08×10^{-7}	1.07×10^{-7}	6.53×10^{-8}	7.55×10^{-8}
R^*, m	2.86×10^{-7}	3.61×10^{-9}	4.48×10^{-9}	4.08×10^{-10}	6.06×10^{-9}
R/R^*	3.16	29.9	23.9	167	12.5

Table 1. DC measurements and Fowler-Nordheim analysis of the data.

In Table 1, R* is an alternative effective radius of the emitter that is calculated from the effective area of the emitter S, assuming that the shape is a hemisphere. The parameters R and S correspond to the ideal case in which the current density is constant over the area S and zero elsewhere, and the radius of curvature in the area S is equal to R. Leybold states that their emitters of etched single-crystal tungsten have a radius of 100 to 200 nm, which is reasonable agreement with thevalues in the table. The values of R/R^* suggest that the field emission from the CNT may have come from an extended length. The unusually large value for tube C-3 suggests that multiple SWCNT may contribute to the current, which is consistent with the SEM images that show the bundle of SWCNT is wider than it is for C-6.

4.3 Measurements of mixing at audio frequencies

We have made rigorous quantum simulations of LAFE (Hagmann, 1999B) which show that the radiation from two lasers increases the dc current (optical rectification) and also causes harmonics and mixing terms with frequencies that are given by $n_1f_1 + n_2f_2$, where f_1 and f_2 are the frequencies of the two lasers and the integers n_1 and n_2 may be positive, zero, or negative. Closed-form expressions have been derived for these terms by using an adiabatic approximation, as was done to obtain Eqs. (1)-(5) which only address optical rectification and mxing at the difference frequency. The terms at frequencies higher than these two are not of immediate interest for photomixing because they would be highly attenuated. However, we made measurements with the five field emission tubes at audio frequencies, where the circuit effects of the sealed tubes do not severely impede the response (Hagmann et al., 2004), to confirm our derivations and demonstrate the action of these devices as mixers.

Two sinusoidal signals, at the frequencies f_1 = 1.67 kHz and f_2 = 1.10 kHz, were superimposed on the high voltage that is fed to a field emission tube, and components of the field emission current at the frequencies f_1, f_2, $2f_1$, $2f_2$, $f_1 + f_2$, and $f_1 - f_2$, as well as the rectified current. The six frequencies correspond to 1.67, 1.10, 3.34, 2.20, 2.77 and 0.57 kHz, respectively. The high voltage path of the measurement circuit consited of the high-voltage power supply, a 100 MΩ ballast resistor, the secondary windings of two transformers, the field emission tube, a dc microammeter, and a 1 MΩ resistor to ground which was a shunt for measuring the current with a digital oscilloscope. The two transformers were used to couple two floating battery-operated Wein bridge oscillators in order to superimpose signals at each of the two frequencies with a potential of 120 V on the dc high voltage. Capacitive shunts across the high-voltage power supply, the ballast resistor, and the dc microammeter were used to simplify the ac equivalent circuit.

With the Leybold tube the currents at the fundamental frequencies f_1 and f_2 were each within 5% of the predicted values, and the rectified current and the currents at each of the other 4 frequencies were each within 10% of the predicted values. However, it was necessary to use fourth order terms (proportional to the fourth derivative) in the Taylor's series to obtain this accuracy. The rectified current and the currents at the 6 frequencies, were each within a factor of 2 of the predicted values for tubes M-4 and C-6, and within a factor of 3 for tube M-1. However, tube C-3 was too unstable to permit measuring the currents at any of the six frequencies. As noted earlier, it is not possible to clean the CNT, and this causes the parameters in Table 1 to be less reproducable than it is for the Leybold tube. We attribute the larger errors in the measurements with the CNT to this effect. Nevertheless, the rectified current and the mixer current were measured with three of the four prototypes using individual CNT, and the values are in reasonable agreement with our analysis.

4.4 Measurements of optical rectification

It was not possible to measure photomixing with the five tubes because they do not contain the special structures that are required to output microwave or THz energy in order to make such measurements (Alonso & Hagmann, 2001). However, we did measure the rectified current that is caused by a single laser diode (20 mW, 658 nm), and Eqs. (6) and (7) show that if two lasers with the same power were used the peak value of the mixer current would be two times the rectification current that we measured. Figure 6 is a block diagram showing the experimental configuration that was used to measure optical rectification.

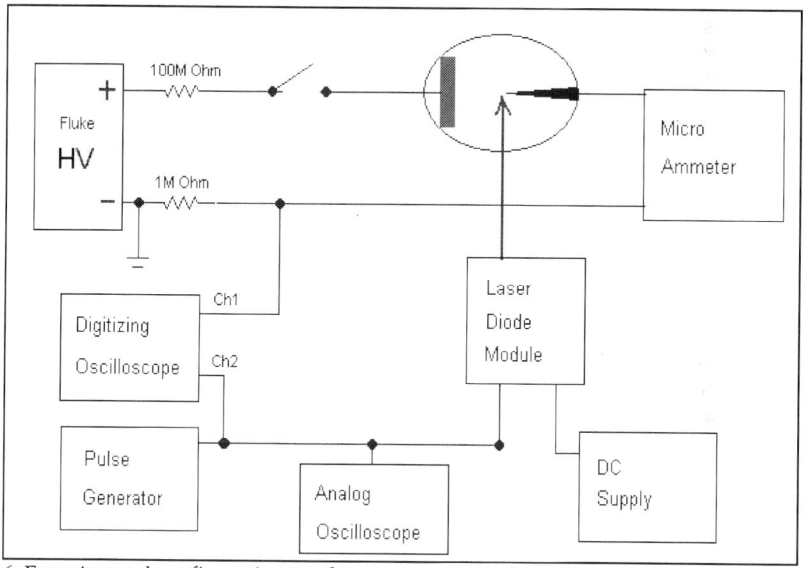

Fig. 6. Experimental configuration used to measure optical rectification.

The laser diode was amplitude-modulated with a square-wave envelope (TTL) by the pulse generator, and a digitizing oscilloscope was used to measure the time dependence of the response of the current to the laser, as shown in the block diagram. The laser diode was maximally-focused to provide a measured Gaussian profile with a power flux density of approximately 10^7 W/m^2 at the emitter. The four prototype tubes were designed to have a long cylindrical glass window with the emitter on the axis to permit the laser radiation to be normal to the glass for minimum distortion, and to reduce the optical path to limit divergence of the beam. However, tube C-6 could not be used in this test because ripples in the glass envelope prevented proper focusing of the laser radiation.

Figure 7 shows the response of the current in tube M-4 as a function of the frequency at which the laser was modulated. The decay in the response is exponential, consistent with the equivalent circuits for the tubes (Hagmann et al., 2004) and the measurement circuit, and it does not show a limitation to the speed of the process of laser-assisted field emission. Table 2 shows values for the time constant τ, and the true value of the rectification current I_D, that were determined from the data by least-squares regression.

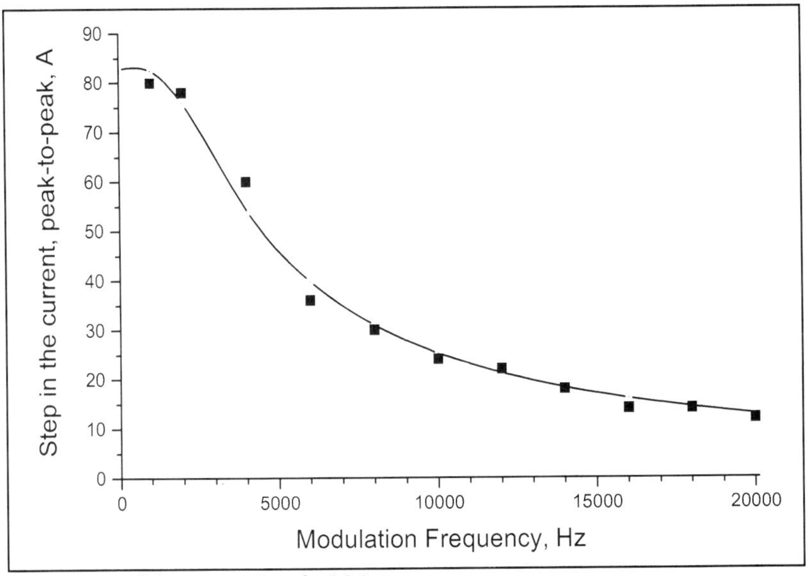

Fig. 7. Response of the current in tube M-4.

	Leybold	M-1	M-4	C-3
I_0, μA	8.0	1.0	1.0	1.0
V_0, V	4600	980	840	920
τ, μs	510	110	80	86
I_D, pA	16	56	83	48
I_D / I_0, %	0.20	5.6	8.3	4.8

Table 2. Measurements of optical rectification.

In Table 2, I_0 is the dc field emission current and V_0 is the dc voltage that is applied across each tube. This table shows that the mean increase in the dc current is 6.2 % for the 3 tubes with CNT emitters, as compared with 0.20 % for the Leybold tube. This shows that if two lasers with the same power were used for photomixing, the peak value of the mixer current would be 12 % of the dc current for the tubes with CNT, as compared with 0.40 % for the Leybold tube.

Two papers must be considered because they describe much larger changes in the field emission current being caused by laser radiation. An increase of the field emission current by a factor of 19 was measured when a CW laser (10 mW, 633 nm) was focused on a CNT film (Cheng et al., 2004). However, the time constant for the increase in the current was 6 minutes, and the values are not consistent with our analysis of optical rectification, so it appears that this effect was caused by heat-related processes. An argon ion laser increased the field emission current from a single CNT by a factor of 300, or decreased by a factor of 50, depending on etching with oxygen, but this was also explained as a thermal process (Colbert & Smalley, 1995). These phenomena that are described in the two papers do not appear to appropriate for generating microwave and terahertz radiation.

5. Photomixing with Clusters of Carbon Nanotubes

Consider a cluster of N transmission lines that are connected across a single load with an impedance Z_L. The Ith transmission line has characteristic impedance Z_I, propagation constant β_I, length Λ_I, and is fed by a constant current source with a peak value of I_I. The phase shift that is caused by propagation over the full length of the Ith line is θ_I. Figure 8 shows this system for only the two transmission lines for N = 2. Photomixing, caused by the radiation radiation from two lasers that is focused on a field emitter at the free end of the Ith transmission line, generates a mixer current I_I which flows through this transmission line.

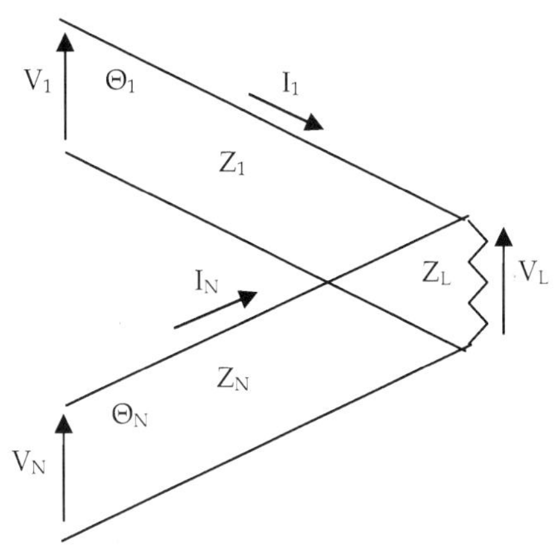

Fig. 8. Branching transmission lines with load.

For the special case where each transmission line has the same characteristic impedance Z_0 and the load has a real impedance R, a broadband impedance match to the load occurs when $N = Z_0/R$. If the mixer current in each transmission line has a peak value that is equal to I_M, then the total current that is delivered to the load is equal to N I_M. Thus, the total power that is delivered to the load, $P_{OUT} = (N\ I_M)^2R/2 = I_M^2Z_0^2/2R$.

For example, let Z_0 = 44 kΩ from Section 3 of this chapter, with R = 50 Ω, so that the total number of CNT in the cluster is given by N = 880. Assume that the dc field emission current from each CNT is I_0 = 10 μA. Then the the total dc field emission current is 8.8 mA and from section 4.4 the mixer current in each CNT is I_M = 0.12 I_0, or 1.2 μA peak. The total mixer current in the load would be 1.06 mA peak, for an output power of 28 μW. This is 14 dB greater than that from photomixing in LTG GaAs (Verghese et al., 1997), but the most significant advantage of photomixing with clusters of CNT is the possibility of obtaining a much greater tunable bandwidth.

More generally, when there is not a perfect impedance match, reflected waves propagate on each transmission line and the voltage and current on the Ith transmission line are given by

$$V_I(z) = A_I e^{-j\beta_I z} + B_I e^{j\beta_I z} \tag{11}$$

$$I_I(z) = \frac{A_I}{Z_I} e^{-j\beta_I z} - \frac{B_I}{Z_I} e^{j\beta_I z} \tag{12}$$

where the free end of the Ith line is at z = 0. Thus, the total current in the load is given by

$$I_L = \sum_{I=1}^{N} \frac{A_I}{Z_I} e^{-j\theta_I} - \frac{B_I}{Z_I} e^{j\theta_I} \tag{13}$$

The voltage across the load must be given by the following expression for all values of I:

$$V_L = A_I e^{-j\theta_I} + B_I e^{j\theta_I} \tag{14}$$

But this voltage is related to the total current through the load by $V_L = Z_L I_L$., which leads to a matrix equation that may be solved to determine the A and B coefficients in Eqs. (11)-(14). For the special case where the characteristic impedance and phase delay of each transmission line are equal to Z_0 and θ, respectively, and the mixer current is the same in each line, the power that is delivered to the load is γ multiplied by what it would be for a single field emitter, where the multiplying factor γ is given by

$$\gamma = \frac{N^2}{\left(1 + \left(N^2 \frac{Z_L^2}{Z_0^2} - 1\right)\sin^2(\theta)\right)} \tag{15}$$

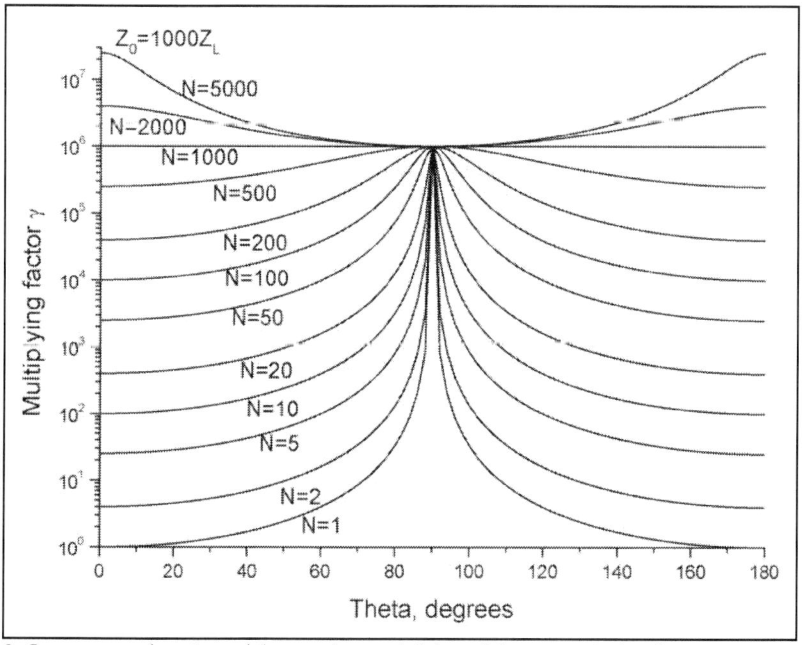

Fig. 9. Gamma as a function of the number and delay of the transmission lines.

Figure 6 shows the multiplying factor γ as a function of the number and the phase delay of the transmission lines, where $Z_0 = 1000\ Z_L$. The horizontal line for N = 1000 corresponds to the broadband match for $N = Z_0/R$ which has already been described. The following behavior is also seen from Eq. (15) and illustrated in Fig. 6:

(1) For small values of θ, $\gamma = N^2$, regardless of the values of Z_0 and Z_L, because the N currents feed directly to the load. This is, it is best to have short transmission lines.

(2) For $\theta = 90^0$, corresponding to quarter-wavelength transmission lines, $\gamma = (Z_0/Z_L)^2$, regardless of the value of N, because the load projects an impedance Z_0^2/NZ_L to each field emitter.

(3) When $Z_0 > Z_L$, if $N > Z_0/Z_L$ the gain decreases from N^2 to $(Z_0/Z_L)^2$ as θ varies from 0° to 90°.

(4) When $Z_0 > Z_L$,, if $N < Z_0/Z_L$ the gain increases from N^2 to $(Z_0/Z_L)^2$ as θ varies from 0° to 90°.

Consideration of some of the design requirements:

(1) The lengths of the SWCNT must be less than one-quarter of the wavelength at the mixer frequency, 7.5 μm at 10 THz, so the mixer currents are generated in phase at the free ends of the SWCNT. The spread in the lengths must not exceed 100 nm at 10 THz so the mixer currents add in phase after propagating to the load on the SWCNT.

(2) The common junction for the SWCNT may be the rounded end of a wire having a radius as small as 30 nm, depending on the size of the SWCNT (Verema et al., 2000).

(3) The contact resistance between the SWCNT and the wire (Tersoff, 1999) must be mitigated by adding gold (Dockendorf et al., 2007) or nickel (Ribaya et al., 2008), or by other means (Kim et al., 2008) at the common junction.

(4) The wire from the common junction may be used as a transmission line or a traveling wave antenna (Alonso & Hagmann, 2001) as we have previously done in microwave prototypes (Hagmann, 2004). The tunable bandwidth is limited only by the transmission line or the antenna so that the ratio of the upper frequency limit to the lower frequency limit could be 10:1 or even somewhat greater. This will be treated in more detail in the following section.
(5) The SWCNT will fan out radially from their common junction during field emission because of electrostatic forces (Kim et al., 2006).
(6) Discrepencies in measurements of luminescence of CNT (Bonard et al., 1998; Rinzler et al., 1995) suggest that the mechanism for electron emission may shift from field emission to thermal processes at high currents so it may be necessary to limit the current to avoid the slower thermal effects.

6. Transmission Lines and Antennas for the Output

6.1 Carbon nanotube transmission lines

The wire from the common junction could be used as a CNT transmission line to propagate energy to the load. In this case the characteristic impedance of the transmission line must be equal to the load impedance or be tapered to provide a broadband impedance match to the load. In section 3 of this chapter we noted that others have shown that CNT on a conducting substrate are TEM transmission lines with a characteristic impedance of approximately 5 kΩ, and isolated SWCNT have a characteristic impedance of approximately 44 kΩ. However, a much lower impedance, such as 50 Ω, is required because the output power varies inversely with the impedance.

6.2 Other types of transmission lines

The wire from the common junction could also be used as another type of transmission line to propagate energy to the load. Again, the characteristic impedance of the transmission line must be equal to the load impedance or be tapered to provide a broadband impedance match to the load. Our group has studied the use of single metallic wires to propagate THz radiation as TM surface waves with enhanced confinement because of surface plasmons (Hagmann, 1998B; Alonso & Hagmann, 1999), and used metallic wires for this purpose in a microwave prototype (Hagmann, 2004). However, the characteristic impedance of this type of transmission line is approximately equal to the impedance of free-space ($\approx$ 377 Ω). A much lower impedance is requried in order to provide greater output power. One possibility is the parallel-plate plasmonic transmission line which can provide an impedance of 50 Ω for TEM mode propagation at THz frequencies with a cut-off frequency of zero Hz (Ghamsari et al., 2008). The common junction of the CNT could be attached to one plate of this transmission line with the other plate grounded.

6.3 Carbon nanotube antennas

The wire from the common junction could be used as a CNT antenna to generate THz radiation. In this case the radiation resistance and reactance of the antenna would constitute the load impedance. The electrical conductivity of CNT is several times larger than copper, but the diameter is small so the resistive lossses are high (Burke et al., 2006). Thus, it may be practical to use short CNT in dipoles, but structures that must be much larger than a wavelength, such as traveling wave antennas and other types of antennas having a large bandwidth, would have too much loss.

6.4 Other types of antennas

The wire from the common junction could be used as another type of antenna to generate THz radiation. Again, the radiation resistance and reactance of the antenna would constitute the load impedance. A variety of different types of antennas have been used at THz frequencies including dipole and bow tie (Yano et al., 2005), spiral (Verghese et al, 1997), and log-periodic structures (Mendis et al., 2005). We have also studied the zigzag antenna for broadband applications of LAFE at THz frequencies (Alonso et al., 2001).

Parallel-plate transmission lines have been used with lenses to obtain highly focused THz radiation comparable to that expected for a 3-D optical element in free-space (Dai et al., 2004). This method shows promise for coupling energy from the common junction because it would be possible to obtain an impedance of 50 Ω for TEM mode propagation at THz frequencies with a cut-off frequency of zero Hz (Ghamsari et al., 2008). As was already noted in section 6.2, the common junction of the CNT could be attached to one plate of this transmission line with the other plate grounded.

7. Conclusions

Photomixing in laser-assisted field emission shows considerable promise as a means to generate microwave or THz radiation with an extremely large tunable bandwidth. However, our simulations and measurements with microwave prototypes show that the output power is quite limited because of the small value for the dc field emission current. This chapter presents further simulations, and measurements with prototypes in which the field emission is from individual CNT, which suggest that it may be able to obtain an output power of 10 μW over a tunable bandwidth ratio of at least 10:1. We are continuing to develop such advanced sources with the objective of providing new devices that are needed for the many different applications of THz radiation (Davies et al., 2002). We also hope that this effort will continue to lead to a more fundamental understanding of the process of quantum tunneling (Hagmann, 1992; Hagmann et al, 1993; Hagmann, 1995A).

8. References

Ahlskog, M., Hakonen, P., Paalanen, M., Roschier, L., and Tarkiainen, R. (2001). Multiwalled carbon nanotubes as building blocks in nanoelectronics. *J. Low Temp. Phys.* 124, 335-352.

Alonso, K., and Hagmann, M.J. (1999). Use of Goubau line to couple microwave signals generated by resonant laser-assisted field emission. *Ultramicroscopy.* 79, 175-179.

Alonso, K., and Hagmann, M.J. (2001). Comparison of three different methods for coupling of microwave and terahertz signals generated by resonant laser-assisted field emission. *J. Vac. Sci. Technol.* B 19, 68-71.

Bachtold, A., de Jonge, M., Grove-Rasmussen, K., McEuen, P.L., Buitelaar, M., and Schonenberger, C. (2001). Suppression of tunneling into multiwall carbon nanotubes. *Phys. Rev. Lett.* 87, 166801 (4 pp).

Barlow, H.M., and Cullen, A.L. (1953). Surface waves. *Proc. IEE.* 100 (III), 329-347.

Bonard, J.-M., Stockli, T., Maier, F., de Heer, W.A., Chatelain, A., Salvetat, J.-P., and Forro, L. (1998). Field-emission-induced luminescence from carbon nanotubes. *Phys. Rev. Lett.* 81, 1441-1443.

Burke, P.J. (2002). Luttinger liquid theory as a model of the Gigahertz electrical properties of carbon nanotubes. *IEEE Trans. Nanotechnology.* 1, 129-144.

Burke, P.J. (2003). An RF circuit model for carbon nanotubes. *IEEE Trans. Nanotechnology.* 2, 55-58.

Burke, P.J., Li, S., and Yu, Z. (2006). Quantitative theory of nanowire and nanotube antenna performance. *IEEE Trans. Nanotechnology.* 5, 314-334.

Carr, G.L., Martin, M.C., McKinney, W.R., Jordan, K., Neil, G.R. and Williams, G.P. (2002). Very high power THz radiation at Jefferson Lab. *Phys. Med. Biol.* 47, 3761-3764.

Cheng, H.-F., Hsieh, Y.-S., Chen, Y.-C., and Lin, I.-N. (2004). Laser irradiation effect on electron field emission properties of carbon nanotubes. *Diam. Relat. Mater.* 13, 1004-1007.

Colbert, D.T., and Smalley, R.E. (1995). Electric effects in nanotube growth. *Carbon.* 33, 921-924.

Collin, R.E. (1990). *Field Theory of Guided Waves.* IEEE Press, New York, 2nd ed.

Dai, J., Coleman, S., and Grischkowsky, D. (2004). Planar THz quasioptics. *Appl. Phys. Lett.* 85, 884-886.

Davies, A.G., Linfield, E.H., and Johnston, M.B. (2002). The development of terahertz sources and their applications. *Phys. Med. Biol.*, 47, 3679-3689.

Dockendorf, C.P.R., Steinlin, M., Poulikakos, D., and Choi, T.-Y. (2007). Individual carbon nanotube soldering with gold nanoink deposition. *Appl. Phys. Lett.* 90, 193116 (3 pp).

Fujieda, T., Hidaka, K., Hayashibara, M., Kamino, T., Ose, Y., Abe, H., Shimizu, T., and Tokumoto, H. (2005). Direct observation of field emission sites in a single multiwalled carbon nanotube by Lorentz Microscopy. *Jpn. J. Appl. Phys.* 44, 1661-1664.

Ghamsari, B.G., and Majedi, A.H. (2008). Terahertz transmission lines based on surface waves in plasmonic waveguides. *J. Appl. Phys.* 104, 083108 (9 pp).

Gomer, R. (1993). *Field Emission and Field Ionization.* American Institute of Physics, New York.

Goubau, G. (1950). Surface waves and their application to transmission lines. *J. Appl. Phys.* 21, 1119-1128.

Hagmann, M.J. (1992). Quantum Tunneling Times: A New Solution compared to Twelve other Methods. *Int. J. Quant. Chem.* 44, 299-309.

Hagmann, M.J. (1995A). Mechanism for resonance in the interaction of tunneling particles with modulation quanta. *J. Appl. Phys.* 78, 25-29.

Hagmann, M.J. (1995B). Simulations of the interaction of tunneling electrons with optical fields in laser-illuminated field emission. *J. Vac. Sci. Technol.* B 13, 1348-1352.

Hagmann, M.J. (1997). Simulations of laser-assisted field emission within the local density approximation of Kohn-Sham density-functional theory. *Int. J. Quant. Chem.* 65, 857-865.

Hagmann, M.J. (1998A). Stable and efficient numerical method for solving the Schrödinger Equation to determine the response of tunneling electrons to a laser pulse. *Int. J. Quant. Chem.* 70, 703-710.

Hagmann, M.J. (1998B). Simulations of the generation of broadband signals from DC to 100 THz by photomixing in laser-assisted field emission. *Ultramicroscopy*. 73, 89-97.

Hagmann, M.J. (199A). Simulations of photon-assisted field emission: Their significance in basic science and device applications. *Ultramicroscopy*. 79, 115-124.

Hagmann, M.J. (1999B). Single-photon and multi-photon processes causing resonance in the transmission of electrons by a single potential barrier in a radiation field. *Int. J. Quant. Chem.* 75, 417-427.

Hagmann, M.J. (2004). Photomixing in resonant laser-assisted field emission—A new technique for wideband-tunable terahertz sources. *IEEE Trans. Microwave Theory Tech.* 52, 2361-2365.

Hagmann, M.J. (2005). Isolated carbon nanotubes as high-impedance transmission lines for microwave through terahertz frequencies. *IEEE Trans. Nanotechnology*. 4, 289-296.

Hagmann, M.J. (2008). Possibility of generating terahertz radiation by photomixing with clusters of carbon nanotubes. *J. Vac. Sci. Technol.* B 26, 794-799.

Hagmann, M.J., and Brugat, M. (1999). Measurements of modulation of the total emitted current in laser-assisted field emission. *Ultramicroscopy* 79, 181-188.

Hagmann, M.J. and Zhao, L. (1993). Experiments pursuant to determining the barrier traversal time for quantum tunneling," *Int. J. Quant. Chem.* 48, 807-814.

Hagmann, M.J., Mousa, M.S., Brugat, M., Sheshin, E.P., and Baturin, A.S. (2004). Large-signal and small-signal electronic equivalent circuits for a field electron emitter. *Surf. Interface Anal.* 36, 402-406.

Hii, K.-F., Vallance, R.R., Chikkamaranahalli, S.B., Menguc, M.P, and Rao, A.M. (2006). Characterizing field emission from individual carbon nanotubes at small distances. *J. Vac. Sci. Technol.* B 24,

Hommelhoff, P., Sortais, Y., Aghajani-Talesh, A., and Kasevich, M.A. (2006). Field emission tip as a nanometer source of free electron femtosecond pulses. *Phys. Rev. Lett.* 96, 077401 (4 pp).

Kim, C.-D., Jang, H.-S., Lee, S.-Y., Lee, H.-R., Roh, Y.-S., Rhee, I-S., Lee, E.-W., Yang, H.-S., and Kim, D.-H. (2006). In situ characterization of the field-emission behaviour of individual carbon nanotubes. *Nanotechnology*. 17, 5180-5184.

Kim, S., Kim, J., Berg, M., and de Lozanne, A. (2008). Robust ohmic contact junctions between metallic tips and multiwalled carbon nanotubes for scanned probe microscopy. *Rev. Sci. Instrum.* 79, 103702 (4 pp).

Kuzumaki, T., Horiike, Y., Kizuka, T., Kona, T., Oshima, C., and Mitsuda, Y. (2004). The dynamic observationof the field emission site of electrons on a carbon nanotube tip. *Diamond Relat. Mater.* 13, 1907-1913.

Lee, M.J.G. and Robins, E.S. (1989). Thermal relaxation of a laser illuminated field emitter. *J. Appl. Phys.* 65, 1699-1706.

Lim, S.C., Jeong, H.J., Park, Y.S., Bae, D.S., Choi, Y.C., Shin, Y.M., Kim, W.S., An, K.H., and Lee, Y.H. (2001). Field-emission properties of vertically aligned carbon-nanotube array dependent on gas exposures and growth conditions. *J. Vac. Sci. Technol.* A 19, 1786-1789.

Lin, M.-C., and Lu, P.-S. (2007). Interaction mechanism of a terahertz wave generator using a field emission cathode. *J. Vac. Sci. Technol.* B 25, 631-635.

Liu, Y., and Fan, S. (2005). Field emission properties of carbon nanotubes grown on silicon nanowire arrays. *Solid State Commun.* 133, 131-134.

Maiwald, F., Lewen, F., Ahrens, V., Beaky, M., Gendriesch, R., Koroliev, A.N., Negirev, A.A., Paveljev, D.G., Vowinkel B., and Winnewisser, G. (2000). Pure rotational spectrum of HCN in the terahertz region: use of a new planar Schottky diode multiplier. *J. Mol. Spectroscopy.* 202, 166-168.

Makishima, H., Miyano, S., Imura, H., Matsuoka, J., Takemura, H., and Okamoto, A. (1999). Design and pereformance of traveling-wave tubes using field emitter array cathodes. *Appl. Surf. Science.* 146, 230-233.

Mayer, A., and Vigneron, J.-P. (2000). Quantum-mechanical simulations of photon-stimulated field emission by transfer matrices and Green's functions. *Phys. Rev.* B 62, 16138-16145

Mayer, A., Miskovsky, N.M., and Cutler, P.H. (2002). Photon-stimulated field emission from semiconducting (10,0) and metallic (5,5) carbon nanotubes. *Phys. Rev.* B 65, 195416 (6 pp).

Mayer, A., Miskovsky, N.M., and Cutler, P.H. (2003). Three dimensional simulations of field emission through an oscillating barrier from a (10,0) carbon nanotube. *J. Vac. Sci. Technol.* B 21, 395-399.

Mendis, R., Sydlo, C., Sigmund, J., Feiginov, M., Meissner, P., and Hartnagel, H.L. (2005). Spectral characterization of broadband THz antennas by photoconductive mixing: Toward optimal antenna design. IEEE Antennas Wireless Propag. Lett. 4, 85-88.

Mross, M., Lowell, T.H., Durant, R., and Mimmitt, M.F. (2003). Performance characteristics of a Smith-Purcell Tunable terahertz source. *J. Biol. Phys.* 29, 295-302.

Okawa, M., Shiori, T., Okubo, H., and Yanabu, S. (1988). Area effect on electric breakdown of copper and stainless steel electrodes in vacuum. *IEEE Trans. Electr. Insul.* 23, 77-81.

Petukhov, A.V., Brudny, V.L., Mochan, W.L., Maytorena, J.A., Mendoza, B.S., and Rasing, T. (1998). Energy conservation and the Manley-Rowe relations in surface nonlinear-optical spectroscopy. *Phys. Rev. Lett.* 81, 566-569.

Ribaya, B.P., Leung, J., Brown, P., Rahman, M., and Nguyen, C.V. (2008). A study on the mechanical and electrical reliability of individual carbon nanotube field emission cathodes. *Nanotechnology.* 19, 185201 (8 pp).

Rinzler, A.G., Hafner, J.H., Nikolaev, P., Lou, L., Kim, S.G., Tomanek, D., Nordlander, P., Colbert, D.T., and Smalley, R.E. (1995). Unraveling nanotubes: field emission from an atomic wire. *Science* 269, 1550-1553.

Ryskin, N.M., Han, S.T., Jang, K.H., and Park, G.S. (2007). Theory of the microelectronic traveling wave klystron amplifier with field-emission cathode array. *Phys. Plasmas.* 14, 093106 (7 pp).

Savard, J.Y. (1967). Higher-order cylindrical surface-wave modes. *IEEE Trans. Microwave Theory Tech.* 15, 151-155.

Schwoebel, P.R., Spindt, C.A., and Holland, C.E. (2005). High current, high current density field emitter array cathodes. *J. Vac. Sci. Technol.* B 23, 691-693.

Sonin, E.B. (2001). Tunneling into 1D and quasi-1D conductors and Luttinger-liquid behavior. *J. Low Temp. Phys.* 124, 321-334.

Stratton, J.A. (1941). *Electromagnetic Theory.* McGraw-Hill, New York.

Tarkiainen, R., Ahlskog, M., Penttila, J., Roschier, L., Hakonen, P., Paalanen, M., and Sonin, E. (2001). Multiwalled carbon nanotube: Luttinger versus Fermi liquid. *Phys. Rev.* B 64, 195412 (4 pp).

Teresoff, J. (1999). Contact resistance of carbon nanotubes. *Appl. Phys. Lett.* 74, 2122-2124.

Verema, L.C., Meunier, V., Lambin, P., and Dekker, C. (2000). Atomic structure of carbon nanotubes from scanning tunneling microscopy. *Phys. Rev.* B 61, 2991-2996.

Verghese, S., McIntosh, K.A., and Brown, E.R. (1997). Optical and terahertz power limits in the low-temperature-grown GaAs photomixers. *Appl. Phys. Lett.* 71, 2743-2745.

Wang, K. And Mittleman, D.M. (2004). Metal wires for terahertz wave guiding. *Nature.* 432, 376-379.

Wang, M.S., Peng, L.-M., Wang, J.Y., and Chen, Q. (2005). Electron field emission characteristics and field evaporation of a single carbon nanotube. *J. Phys. Chem.* B 109, 110-113.

Wang, Q.H., Setlur, A.A., Lauerhaas, J.M., Dai, J.Y., Seeling, E.W. and Chang, R.P.H. (1998). A nanotube-based field-emission flat panel display. *Appl. Phys. Lett.* 72, 2912-2913.

Xu, Z., Bai, X.D., Wang, E.G., and Wang, Z.L. (2005). Field emission of individual carbon nanotube with in situ tip image and real work function. *Appl. Phys. Let.* 87, 163106 (3 pp).

Yano, R.., Gotoh, H., Hirayama, Y., Miyashita, S., Kadoya, Y., and Hattori, T. (2005). Terahertz wave detection performance of photoconductive antennas: Role of antenna structure and gate pulse intensity. *J. Appl. Phys.* 97, 103103 (6 pp).

Yokoo, K., and Ishihara, T. (1997). Field emission monotron for THz emission. *Int. J. Infrared Millimeter Waves.* 18, 1151-1159.

Zhang, X.-C. (2002). Terahertz wave imaging: horizons and hurdles. *Phys. Med. Biol.* 47, 3667-3677.

Zhu, W., Bower, C., Zhou, O, Kochanski, G., and Jin, S. (1999). Large current density from carbon nanotube field emitters. *Appl. Phys. Lett.* 75, 873-875.

Liquid crystal dispersions of carbon nanotubes: dielectric, electro-optical and structural peculiarities

L. Dolgov[1*], O. Koval'chuk[2], N. Lebovka[3], S. Tomylko[2,4], and O. Yaroshchuk[2]
[1]*Institute of Physics, University of Tartu, 51014 Tartu, Estonia*
[2]*Institute of physics, NASU, prospekt Nauki 46, 03028 Kyiv, Ukraine*
[3]*Institute of Biocolloidal Chemistry, NASU, Vernadskii Prosp. 42, 03142 Kyiv, Ukraine*
[4]*Taras Shevchenko National University, Physical Faculty, prospekt Glushkova 2, 03022 Kyiv, Ukraine*

1. Introduction

Liquid crystals (LCs) turn out to be excellent hosts for carbon nanotubes (CNTs). Having molecular structure similar to CNTs, LCs perfectly incorporate CNTs into own structure. Particularly, the liquid crystalline orientational order can be imposed on CNTs so that aligned ensembles of these particles can be attained (Dierking et al., 2004). This alignment can be patterned by pattering alignment of LC host. Furthermore, the alignment axis of CNTs can be easily driven by the LC reorientation in the external field (Dierking et al., 2008); CNTs follow reorientation of LC director demonstrating guest-host effect known for molecular solutions and dispersions of anisotropic nanoparticles in LC hosts (Blinov & Chigrinov, 1996). Finally, LC can be removed and thus pure aligned CNTs can be obtained (Lynch & Patrick, 2002). This altogether means that LC gives unique opportunity for controllable alignment of CNTs.

On the other hand, CNTs bring a number of improvements to LC layers used in electro-optic devices (Qi & Hegmann, 2008). The LC doping by CNTs reduces response time (Huang et al., 2005; Chen et al., 2007; Lee et al., 2008) and driving voltage (Lee et al., 2004), suppresses parasitic back flow and image sticking typical for LC cells (Lee et al., 2004; Baik et al., 2005; Chen & Lee, 2006).

The LC-CNTs systems are not limited to nematic matrices. A series of unique LC-CNTs composites based on thermotropic and lyotropic materials with different LC mesophases is developed and characterized (Weiss et al., 2006; Lagerwall et al., 2006, Lagerwall et al., 2007; Cervini et al., 2008; Podgornov et al., 2009).

A symbiosis of LC and CNTs rouses rapidly increasing interest. The number of publications on this subject grows in geometrical progression. The major results of these studies are summarized in several recent reviews (Lagerwall & Scalia, 2008; Rahman & Lee, 2009).

---

* on leave of absence from *Institute of physics, NASU, prospect Nauki 46, 03028 Kyiv, Ukraine*

A present chapter is focused on remarkable dielectric, electro-optical and micro-structural peculiarities of LC-CNTs dispersions, their correlation and mutual influence. It is mainly based on authors' original results obtained within recent years. The structure of this chapter is the following. The introductory part (section 1) gives short introduction to LC-CNTs composites and elucidates benefit of combination of LC and CNTs. It also outlines a field of questions further considered. A section 2 gives details of our samples and experimental methods. The next three sections correspondingly consider dielectric, electro-optical and structural peculiarities of LC-CNTs composites. Each of these topical sections starts with a short review and lasts with the authors' original results. The final, conclusion part (Section 6), summarizes most interesting properties of LC-CNTs suspensions, their application perspectives and mention some exciting problems for further investigations.

2. Materials and methods

2.1 Liquid crystalline media and chiral dopant

Nematic LCs EBBA (Reakhim, Russia), 5CB, MLC6608, and MLC6609 (Merck, Germany) were used as LC hosts. EBBA was purified by fractional crystallization from the n-hexane solution, 5CB, MLC6608, MLC6609 were used as obtained. Some characteristics of these LCs are presented in Table 1.

LC	Nematic mesophase	dielectric anisotropy $\Delta\varepsilon$, optical anisotropy Δn	Reference
EBBA	308.9 - 350.6 K	$\Delta\varepsilon$=-0.13, Δn=0.25 at 313 K	Goncharuk et al., 2009
5CB	295.5- 308.3 K	$\Delta\varepsilon$=13, Δn=0.177 at 298 K	Blinov & Chigrinov, 1996
MLC6608	clearing point at 363 K	$\Delta\varepsilon$=-4.2, Δn=0.0830 at 293 K	Licristal®, 2002
MLC6609	clearing point at 364.5 K	$\Delta\varepsilon$=-3.7, Δn=0.0777 at 293 K	Licristal®, 2002

Table 1. Characteristics of LC media.

S811 (Merck, Germany) was used as chiral dopant.

2.2 Multiwalled carbon nanotubes

The multiwalled carbon nanotubes (SpetsMash Ltd., Ukraine) were prepared from ethylene by the chemical vapor deposition method (Melezhyk et. al., 2005). Typically, these CNTs have an outer diameter of about 12–20 nm and the length of about 5-10 μm. The specific electric conductivity σ of the powder of compressed CNTs was about 10^3 S/m along the compression axis.

2.3 Preparation of LC-CNTs composites

The all LC-CNTs composites were prepared by 20 min stirring of LC and CNT mixtures using the ultrasonic mixer equipped with a cup horn, at the frequency of 22 kHz and the output power of 150 W. The concentration of CNTs, *c*, was varied in the range 0-2 wt %. Doping of CNTs has not influenced essentially the phase transition temperatures of LC-CNTs composites.

2.4 Cells

The cells for electro-optical and dielectric measurements were made from glass substrates containing patterned ITO electrodes and aligning layers of polyimide. The polyimides AL2021 (JSR, Japan) and SE5300 (Nissan Chemicals, Japan) were used for homeotropic alignment of LC EBBA, MLC6608 and MLC6609, while the polyimide SE150 (Nissan Chemicals, Japan) was utilized for planar alignment of LC 5CB. The polyimide layers were rubbed by a fleecy cloth in order to provide a uniform planar alignment of LC in either field-on state (EBBA, MLC6608 and MLC6609) or a zero field (5CB). The cells were assembled so that the rubbing directions of the opposite aligning layers were antiparallel. A cell gap was maintained by the glass spacers of appropriate size (16 μm, if not otherwise stated). Finally, these cells were filled capillary with neat or CNTs doped liquid crystals heated to isotropic state. In some dielectric measurements the cells without alignment layers were utilized. The structure of LC-CNTs composites was monitored by observation of the filled cells placed between crossed polarizers, both by naked eye and in an optical polarizing microscope.

2.5 Electro-optical measurements

The electro-optical measurements were carried out using the experimental setup described in (Koval'chuk et al., 2001a). The cell was set between two crossed polarizers so that the angle between the polarizer axes and the rubbing direction was 45°.
The sinusoidal voltage 0-60 V (at frequency f=2 kHz) was applied to the cell. The voltage was stepwise increased from 0 to 60 V and then decreased back to 0; the total measuring time, i.e., time of voltage application, was about 1 min. The transmittance of the samples was calculated as $\eta=(I_{out}/I_{in})*100\%$, where I_{in} and I_{out} are intensities of the incident and transmitted light, respectively.

2.6 Dielectric measurements

The dielectric spectra in the frequency region between $5\cdot10^{-2}$ and 10^{6} Hz were recorded by measurement of the frequency f dependences of resistance R and capacitance C using the oscilloscopic method (Twarowski & Albrecht, 1979; Koval'chuk, 1998). The voltage signal applied to the tested cell had a triangle form with amplitude of 0.25 V. The parallel connected R-C circuit was used as an equivalent scheme of the cell and the values of the real (capacitive) ε' and imaginary (ac conductance) ε'' components of the permittivity $\varepsilon=\varepsilon'+i\varepsilon''$ of a composite were then calculated. Based on ε'' data the sample conductivity σ was determined using a formula:

$$\sigma = 2\pi\varepsilon_0\varepsilon''f, \qquad (1)$$

where ε_0 is a dielectric constant.
The ac electrical conductivity was also measured by a LCR meter 819 (Instek, 12 Hz–100 kHz) in the non-relaxation frequency range preliminarily determined from the dielectric spectra. All the measurements were repeated at least 5 times in order to calculate the average values and errors.

3. Dielectric studies of LC composites doped with carbon nanotubes

The interest to the electrophysical behavior of LC composites doped with carbon nanotubes is continuously growing (Lebovka et al., 2008; Lu & Chien, 2008; Koval'chuk et al., 2008). Dispersed nanotubes essentially influence the concentration and spatial distribution of charges in LC cells and thus determine the actual electric field applied to the composites and their electro-optic response (Lee et al., 2004, Huang et al., 2005). These composites demonstrate huge changes in electrical conductivity σ with a small change of CNT concentration c (c<0.5 wt %) (Lisetski et al., 2007; Dierking et al., 2008; Lisetski et al., 2009; Zhao et al., 2009). Besides, the σ vs. c curve shows clear percolation behavior (Lebovka et al., 2008; Koval'chuk et al., 2008; Goncharuk et al., 2009). The huge conductivity increase may result in dielectric breakdown and local heating effects (Jayalakshmi & Prasad, 2009). Monitoring of electrical conductivity identified two time scales in reorientation dynamics of liquid crystal-nanotubes dispersions. These scales are associated with the reorientation of the liquid crystal texture (the short time scale) and with the reorientation of the carbon nanotubes (the long time scale) (Dierking et al., 2008). The electrical conductivity and the dielectric constant of LCs doped with carbon nanotubes demonstrate exctraordinary large changes in electric and magnetic field driven reorientation experiments (Dierking et al., 2004; Jayalakshmi & Prasad, 2009). In spite of these extensive studies, many aspects of LC-CNTs composites remained unclear. In particular, it was not well clarified the nature of electrical conductivity and charge transfer in different phase states of LC, at different concentrations of CNTs and in relations with the magnitude of applied voltage and frequency. The present paragraph is focused on these problems.

3.1 Dielectric spectra

Figure 1 presents typical dielectric spectra of 5CB-CNTs (a) and EBBA-CNTs (b) composites. The measurements were done at the temperatures falling within nematic phase of LC medium, T=297 K (5CB) and T=313 K (EBBA). The LC alignment is planar in case of 5CB and homeotropic in case of EBBA. According to Fig. 1, three frequency ranges, namely, $f<10^2$ Hz (A), 10^2 Hz$<f<10^5$ Hz (B) and $f>10^5$ Hz (C) can be distinguished. As is commonly accepted (Craig, 1995), the frequency range (A) reflects the near electrode barrier layers processes, the frequency range (B) corresponds to the bulk polarization and charge transfer and the frequency range (C) reflects the relaxation process caused by transition from the electronic and dipole polarization to only electronic polarization of LC phase.

In (A) and (C) ranges, the noticeable dispersions of the real ε' and imaginary ε'' components were observed. The data analysis has shown that Cole-Cole approximation (Haase & Wrobel, 2003; Chelidze et al., 1977) can be applied for estimation of corresponding low- and high-frequency times of dielectric relaxation, τ_A and τ_C:

$$\varepsilon^* = \varepsilon_\infty + (\varepsilon_o - \varepsilon_\infty)/(1 + i2\pi f\tau). \quad (2)$$

Here, ε^* is a complex dielectric permittivity, ε_0 and ε_∞ are the limiting values for the frequencies $f = 0$ and $f = \infty$.

Fig. 1 demonstrates that CNT dopant substantially modifies dielectric spectra of LC. These changes in different frequency ranges are further analyzed.

a

b

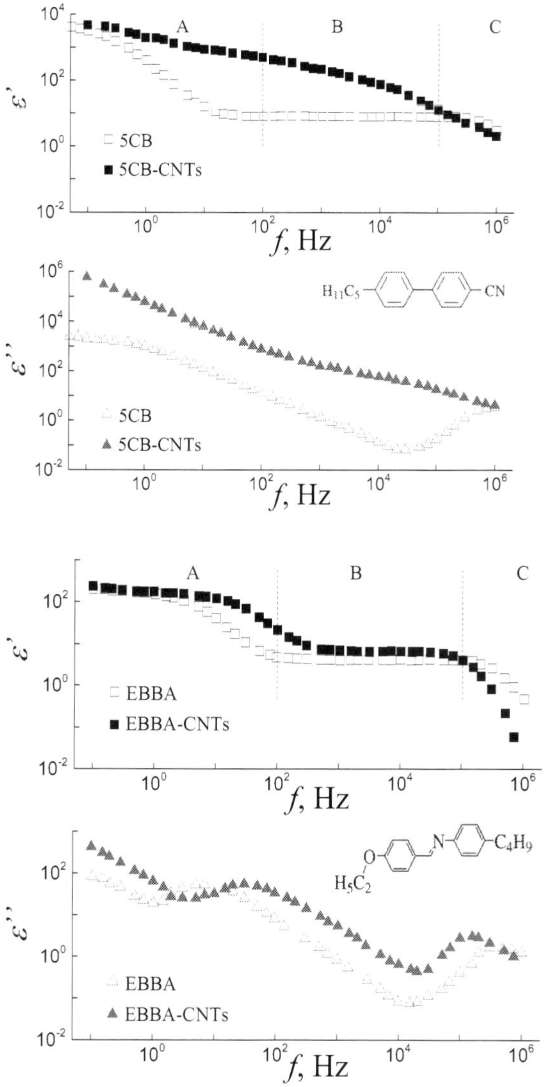

Fig. 1. Complex permittivity components ε' and ε'' versus frequency f for pure LC and LC-CNTs composites of 5CB (a) and EBBA (b). The concentration of CNTs c is 0.25 wt % (a) and 0.5 wt % (b). Chemical formulas of 5CB and EBBA are given on corresponding $\varepsilon''(f)$ graphs.

3.2 Electrical conductivity: Intermediate frequency range

We begin with the intermediate frequency range (B). In this range, large active resistances of the double electric layers (DEL) formed near the cell electrodes are shunted by their capacitance diminishing with a field frequency f (Koval'chuk, 1998; Koval'chuk, 2000; Barbero & Olivero, 2002; Koval'chuk, 2001b). Because of this, the range (B) characterizes volume properties of samples.

Based on formula (1), alternating current conductivity of the composite bulk can be estimated. In a general case, the alternating current conductivity σ of LC-CNTs samples can be represented as a sum of the frequency-independent ionic conductivity σ_i associated with LC, and the frequency-dependent electronic hopping conductivity σ_e, associated with CNTs, $\sigma=\sigma_i+\sigma_e$. These two contributions to the electric conductivity can be easily separated because of considerable frequency dependence of σ_e. The σ_e component becomes essential at CNT concentrations comparable and higher than a threshold concentration c_c corresponding to conductivity percolation discussed in the next subsection. The frequency-independent component σ_i is typical for ionic conductivity of liquids (Frenkel, 1955).

3.2.1 Percolation behavior

Different composites filled with CNTs typically demonstrate percolating behavior of the electrical conductivity σ, when sharp transition from the prevailing ionic to the prevailing charge hopping conductivity occurs at some threshold concentration c_c (Grossiord et al., 2006; Mamunya et al., 2008). At this concentration, the CNTs form a percolation network or cluster that spans the whole system (Stauffer & Aharony, 1992; Torquato, 2002). A theory predicts an inverse proportionality between c_c and the aspect ratio r of the conductive filler particles, $c_c \sim 1/r$ (Balberg et al., 1984; Foygel et al., 2005).

For CNTs with very high aspect ratio, typically, $r \approx 500$-1000, this theoretical estimation results in extremely low values of the percolation threshold, $c_c < 0.1$ wt %, which is in full correspondence with available experimental data (Lisunova et al., 2007; Mamunya et al., 2008; Lebovka et al., 2009).

The percolation behavior of electrical conductivity in different LC-CNTs composites was recently reported (Lisetski et al., 2007; Lebovka et al., 2008; Koval'chuk et al., 2008; Lisetski et al., 2009; Goncharuk et al., 2009). Fig. 2 demonstrates electrical conductivity σ versus filler concentration c curve for 5CB-CNTs samples, which is typical for the LC-CNTs composites. Here, the 5CB is planar oriented in a 16 μm cell. The measurements were carried out at 100 Hz, when frequency dependence of σ was practically absent and thus the ionic contribution to σ was dominating. The threshold increase of σ was observed for the values of CNT concentration c between 0.02 and 0.2 wt %. The experimental data were analyzed using the least-square fitting to scaling equation (Stauffer & Aharony, 1992)

$$\sigma \propto (c - c_c)^t . \qquad (3)$$

As a result, the percolation threshold concentration c_c and conductivity transport exponent t were determined. The insert in Fig. 2 shows c_c and t versus temperature dependencies. There are evident the considerable temperature dependencies for both c_c and t.

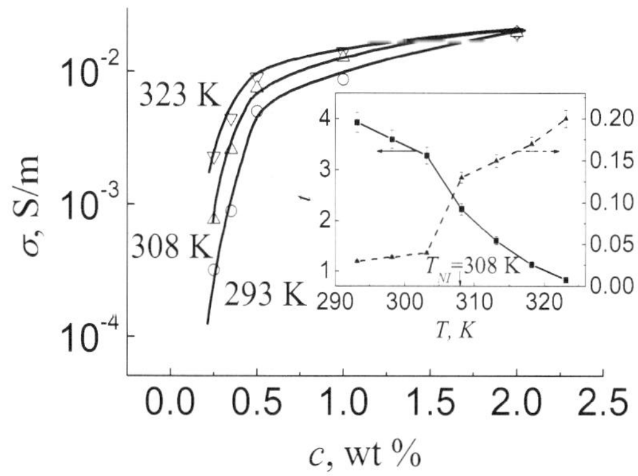

Fig. 2. Electrical conductivity σ versus concentration of nanotubes c in 5CB-CNTs composites for different temperatures T. Insert shows the conductivity exponent t and percolation threshold concentration c_c versus temperature T. The samples were planar oriented in a 16 μm cell, f=100 Hz.

Increase of the percolation threshold c_c with a temperature T is inessential within nematic mesophase ($T<T_{NI}\approx308$ K), however, it becomes noticeable in isotropic phase. At T=325 K the threshold concentration reaches $c_c\approx0.2$ wt %. Such character of $c_c(T)$ curve may be caused by number of effects such as change in the Brownian motion intensity, degree of the orientational order of nanotubes, etc. These factors influence degree of connectedness between the nanotubes and, as result, the percolation threshold (Kyrylyuk & van der Schoot, 2008). The effects of nanotubes' alignment on the percolation behavior were discussed recently, however, they still are not completely clear (Du et al., 2005; Behnam et al., 2007; Kyrylyuk & van der Schoot, 2008). More perfect alignment of CNTs inside the nematic phase is expected. However, it can be assumed that highly aligned nanotubes practically don't intersect each other and fail to create percolation paths. The highest electrical conductivity was experimentally observed for slightly aligned, rather than isotropic, composites (Du et al., 2005). Recent Monte Carlo simulations have also demonstrated that conductivity may be strongly dependent on the measurement direction and the degree of nanotube alignment (Behnam et al., 2007). Along with inducing alignment, nematic LC host stabilizes structure of CNTs damping their Brownian motions. Together with imperfect orientational order of CNTs it may explain the lower concentration threshold of CNTs in the nematic phase comparing with that in the isotropic state (see inset in Fig. 2).

Temperature decrease in the conductivity exponent t (see formula (3)), possibly, reflects changes in the conductivity mechanism and structure of the percolating clusters. The values of conductivity transport exponents $t\approx4/3$ and $t\approx2$ are characteristic for the ordinary 2d (two-dimensional) and 3d (three-dimensional) random percolations, respectively. Note that for the studied systems the 2d-3d crossover percolation behavior (Muller et al., 2003;

Lebovka et al., 2002) with 4/3<t<2 was expected, because the restricted width of the cell (d≈16 μm) was comparable with the length of CNTs (l≈5-10 μm). Moreover, it was found that for the 3d composite filled with anisotropic particles t decreases substantially with increase in the aspect ratio (Foygel et al., 2005). The noticeable deviations from predictions of the standard percolation theory were also observed for systems with diverging local conductances, when distances between adjacent particles are broadly distributed (Johner et al., 2007). The reported transport exponent t often exceeds its classical values, reaching even the values as large as t=5–10 (Johner et al., 2008). Generally, the value of t extracted from the percolation data for CNT-filled composites may be dependent on the distribution function of distances between adjacent conducting particles (Mdarhri et al., 2008).

Note that $c_c(T)$ and $t(T)$ behave discontinuously in the vicinity of nematic-isotropic transition, T_{NI}≈308 K. It reflects strong dependence of the percolating characteristics on the phase state of LC host.

3.2.2 Temperature dependence and energy of activation

In the investigated range of temperature the electrical conductivity σ of the LC-CNTs composites increases with a temperature that is characteristic for the nonmetallic behavior. Moreover, temperature dependence of σ can be satisfactorily described by Arrhenius relationship (Lebovka, 2008):

$$\sigma \propto \exp(-W / kT), \tag{4}$$

where W is the activation energy, k is Boltzman's constant.

Figure 3 shows typical Arrhenius plots for 5CB-CNTs (a) and EBBA-CNTs (b) composites corresponding to different concentrations of nanotubes, c.

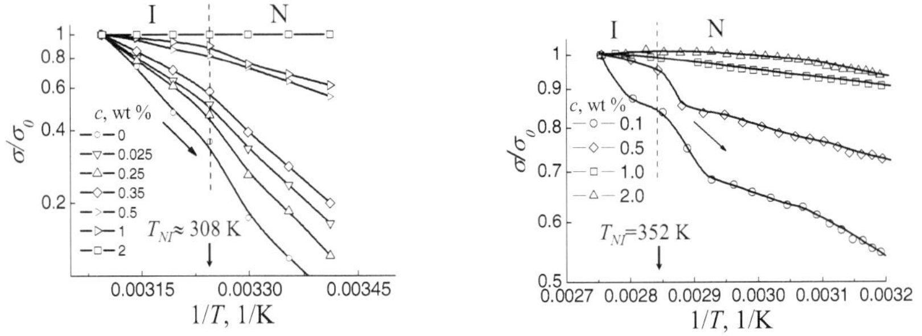

a b

Fig. 3. Relative conductivity σ/σ_0 versus inverse temperature $1/T$ for planar oriented 5CB-CNTs (a) and unoriented EBBA-CNTs (b) composites at different concentrations of nanotubes c. Here, σ_0 corresponds to the electrical conductivity at some reference temperature, which was 323 K in case (a) and 363 K in case (b). The cell gap d was 16 μm (a) and 500 μm (b), the frequency f was 100 Hz (a) and 1000 Hz (b). Arrows mark direction of temperature change.

In these experiments, the conductivity data were recorded in the cooling regimes accounting for the possible influence of the thermal pre-history. The obtained Arrhenius plots of $\sigma(T)$ curves were rather linear within the temperature ranges corresponding to nematic or isotropic phase; however, some deviations from linear behavior were observed near nematic-to-isotropic transition points. Moreover, the slopes were always larger in nematic phase than in isotropic phase; it corresponds to higher activation energy of the electrical conductivity in nematic phase. This can be explained by the restriction of the charge mobility in the nematic phase that is related to distortion of the LC director field in the vicinity of charge carriers (Belotskii et al., 1980). This distortion sphere surrounding the charged particle arises as a result of LC molecule orientation in the electric field generated by this particle.

Figure 4 presents examples of concentration dependencies of the activation energies determined from the Arrhenius plots in nematic phases of 5CB-CNTs and EBBA-CNTs composites.

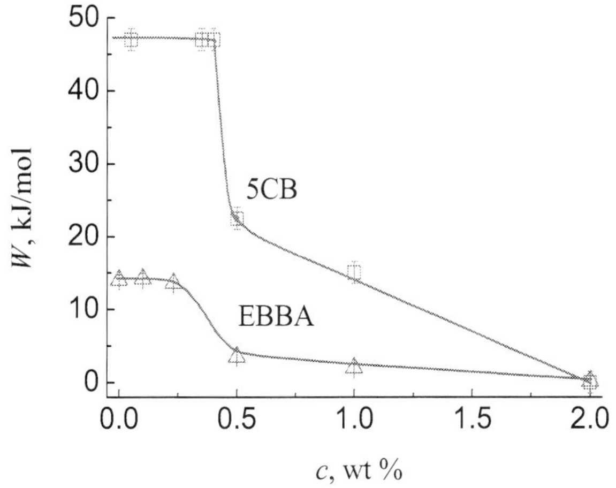

Fig. 4. Activation energy W of electrical conductivity versus concentration of CNTs c for planar oriented 5CB-CNTs and unoriented EBBA-CNTs composites. The values of W were determined from data corresponding to temperature ranges of nematic phases of 5CB and EBBA. The cell gap d was 16 μm (5CB) and 500 μm (EBBA), the frequency f was 100 Hz (5CB) and 1000 Hz (EBBA), the measuring voltage U was 0.25 V (5CB) and 1 V (EBBA).

Though values of W were determined for different LC media, for planar oriented (5CB) and unoriented (EBBA) composites, using different cell gaps and different protocols for σ measurements, similar tendencies were observed in $W(c)$ dependencies: W was constant at small concentrations below the percolation threshold and started to decrease above the percolation threshold approaching $W \approx 0$ kJ/mol at $c \geq 2$ wt %.

This behavior reflects the dominating role of ionic transport mechanism at concentrations below the percolation threshold with activation energy $W_i \approx$ 47 kJ/mol (5CB) and $W_i \approx$ 15 kJ/mol (EBBA). So, below the percolation threshold, the conducting LC-CNTs composites behave as semiconductors. This behavior is also typical for polymer-CNTs composites, where electrical conductivity is an increasing function of temperature, which was explained by thermally assisted hopping or charge tunneling between the conducting particles (Barrau et al., 2003). The decrease of W above the percolation threshold might be explained by the increasing role of charge transport through the CNT structure. Comparing with ionic conductivity of LC, electronic conductivity of CNTs is characterized by low temperature coefficient (Eletskii, 2009). This might decrease the apparent activation energy W of LC-CNTs composites and this tendency should essentially enhance with concentration of CNTs.

3.2.3 Hysteretic behavior and effect of positive temperature coefficient

It was previously demonstrated that temperature affects the spatial arrangement of CNTs in the LC matrix and changes percolation characteristics. The LC-CNTs composite typically displays also the electrical conductivity heating-cooling hysteresis and the pronounced effect of positive temperature coefficient of resistivity (PTC effect) (Lebovka et al., 2008; Goncharuk et al., 2009). Figure 5 presents examples of the electrical conductivity σ heating-cooling hysteresis for EBBA-CNTs unoriented composites in the thick cell ($d \approx 500$ μm). In these experiments, the composites that were initially solid were heated from room temperature up to 363 K and then cooled. The total time of the heating-cooling cycle was about 1 h. The hysteretic loops were most pronounced for CNT concentrations in the vicinity of the percolation transition ($c \approx 0.1$ wt %) and became inessential at higher values of c (figure 5).

The heating-cooling hysteretic behavior of electrical conductivity reflects strong agglomeration and rearrangement of nanotubes during the thermal curing. The abrupt decrease of the $\sigma(T)$ curve in the vicinity of melting point T_{SN} may be caused by several factors. Firstly, CNTs align within the nematic domains. It lowers degree of connectivity of the nanotubes. Besides, the initial network is also partially damaged by thermal expansion of the host EBBA matrix (Lebovka et al., 2008; Lebovka et al., 2009). At high concentration of CNTs the network formed in LC is rather strong because of high density of connectedness between the nanotubes. This strong network cannot be easily destroyed by the Brownian motion and LC reorientation intensified by heating and phase transitions. This explains weakening of hysteretic and PTC effects at $c > c_c$. Note that these effects are especially strong in thick cells, when the length of individual nanotubes, $l \approx 5$-10 μm, is small comparing with a cell gap d. In this case, the Brownian motion of individual nanotubes is unrestricted by nanoparticle interaction with the cell walls and so the Brownian motion and the LC orientational rearrangement are dominating factors in the observed reconstruction of CNTs network during thermal processing.

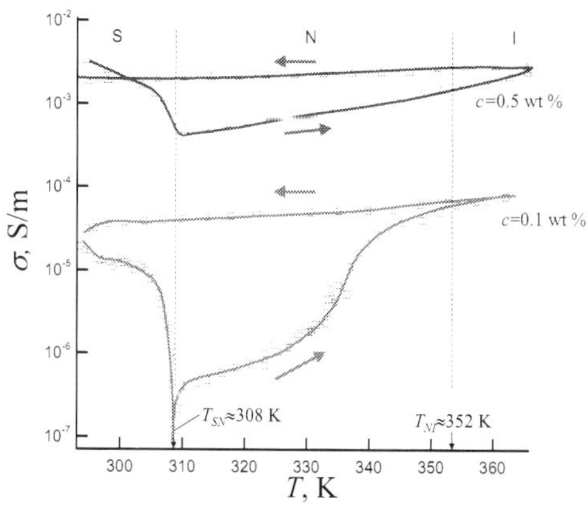

Fig. 5. Heating–cooling hysteresis of electrical conductivity σ for unoriented EBBA-CNTs composites. The measurements were made in a thick 500 μm cell, f=1 kHz, U=1V. The arrows show directions of temperature changes.

3.3 Frequency and voltage dependence

The frequency dependent contribution of electrical conductivity is common for disordered and/or highly alloyed solids (Gantmaher, 2005; Shklovskii & Efros, 1984). It was explained by correlated barrier hopping, tunneling or percolation model (Shklovskii & Efros, 1984; Mott & Davis, 1971, Pike, 1972). Typically, the power law dependences

$$\sigma \propto f^m \text{ or } \varepsilon'' \propto f^{m-1} \tag{5}$$

were observed. Here, m is a frequency exponent.

In the hopping transport model, the current carrier is assumed to hop over a potential barrier between neighboring localized sites and this model predicts that m is slightly lower than 1 and it decreases with rise of temperature. It is different from the tunneling model where the charge carrier is assumed to tunnel across the potential barrier. The tunneling model predicts smaller values of the frequency exponent ($0.4<m<0.8$) and temperature independent value of m. In the percolation theory, the critical exponent m is considered to be universal and determined only by statistical properties of the percolation clusters. The values of m obtained from numerical studies of random composites were in the range of 0.6–0.8 (Straley, 1977). The experimentally found m values near the percolation threshold were reported to be about 0.8–0.9 for carbon black composites (Jager et al., 2001), and 0.66 (Kim et al., 2003) or 0.85-0.9 (Liu et al., 2007) for carbon nanotube dispersions in polymers.

The studied dispersions of CNTs on the base of LCs demonstrated $\sigma(f)$ curves strongly depending on CNT concentration and temperature (Koval'chuk et al., 2008). The value of σ was practically independent on f at small concentrations of CNTs ($c<0.1$ wt % for 5CB) and became a power function of the frequency f at larger c. Examples of $\sigma(f)$ dependences for 5CB-CNTs composites are presented in the insert in Fig. 6a. It is evident that increase of CNT concentration c above c_c results in increase of both ionic σ_i and electronic σ_e contributions, while below $c_c \approx 0.1$ wt % the ionic contribution is dominating. The increase of the ionic contribution σ_i with c can be partially caused by the increased concentration of ions introduced with impurities presented in CNTs. However, this factor is hardly a dominating one because of thorough purification of CNTs used in our studies. The more likely reason is facilitation of ionic transport through the charge exchange mechanism assisted by nanotubes. Such mechanism may be efficient due to the large dielectric constant of the CNTs, which results in strong attraction between impurity ions and nanotube surface (Jagota et al., 2005). This conclusion is supported by presence of double electric layer (DEL) shunting effect assisted by CNTs (see section 3.5). At CNTs concentrations between 0.1 and 0.35 wt %, σ_i =const and σ_e=f^m, which means that σ_i and σ_e can be easily separated. Above c=0.35 wt %, the component σ_e becomes frequency independent that complicates its separation. The frequency independence of σ_e may be the evidence of the transition from the hopping charge transfer to the quasimetallic electronic transfer typical for CNTs. This implies very tight network with direct connections between the nanotubes so that the conductivity barrier between them is rather low.

Figure 6a also presents frequency exponent m vs. the concentration c at 297K. The $m(c)$ dependence is a non-monotonic function with a maximum $m \approx 0.32$ near the concentration $c \approx 0.25$ wt %. Increase of m with c at $c < 0.25$ wt % reflects initial process of CNT cluster formation. In this interval of concentrations, the mechanism of electrical transport is mixed: electrical transport is governed by electron hopping/tunneling mechanism inside CNT clusters and by ionic mechanism between clusters. Above the percolation threshold, at high concentrations of CNTs (c>0.25 wt %), agglomeration of different clusters occurs and contacts between them are multiple. This results in formation of homogeneous ohmic conducting structure, when electron hopping/tunneling mechanism is inessential and electrical conductivity becomes frequency independent.

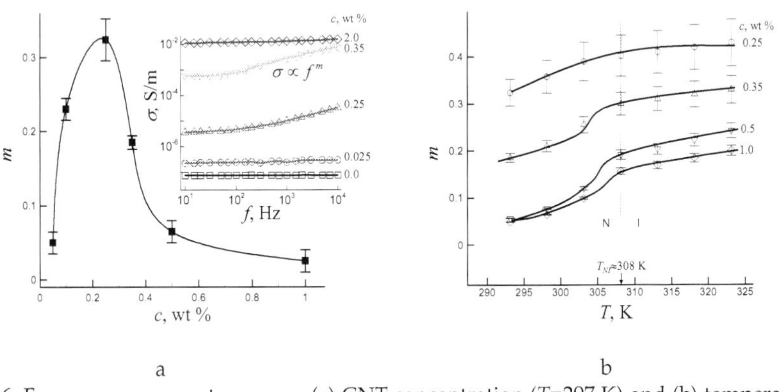

Fig. 6. Frequency exponent m versus (a) CNT concentration (T=297 K) and (b) temperature for planar oriented 5CB-CNTs composites. Insert shows the conductivity σ versus frequency f plots for different concentrations of CNTs c.

The temperature dependencies of frequency exponents m for 5CB-CNTs composites are presented in Fig. 6b. The value of m is an increasing function of temperature. The evident changes in slope of $m(T)$ curves can be observed in the vicinity of nematic-isotropic transition, T_{NI}≈308 K. The obtained data evidence a noticeable influence of the phase state of LC matrix on the charge transport mechanism. This effect can be easily explained accounting for the expected alignment of CNTs inside the nematic matrix (Dierking et al., 2004).

Note that observed behavior of the frequency exponent (rather small values of m (m<0.3) and positive thermal coefficient (dm/dT>0)) contradict with the classical hopping and tunnelling transport theories, predicting dm/dT<0 (Mott & Devis, 1971; Pike, 1972). The percolation theory also predicts rather high values of m, in the range of 0.6–0.8 (Straley, 1977). Note that critical exponent m may vary significantly depending on peculiarities of microstructure and morphology of the composite (Liu et al., 2007). However, at the moment, there is no theory available for explanation of the anomalously small values of m observed in 5CB-CNTs composites.

Large positive thermal coefficient dm/dT>0 obtained for nematic mesophase indicates enhancing of the frequency dependent contribution to electrical conductivity σ. This behavior, possibly, reflects the temperature changes in the percolating CNT networks inside the LC-CNTs composites. As was earlier discussed, temperature increase results in damage of the connectivity in percolation structures and, hence, in decrease of the "effective" filling concentration. Naturally, it resulted in increase of m with temperature T for concentrations above 0.25 wt % (Fig. 6b). This conclusion supports also the observed increase of the percolation threshold concentration c_c with temperature T (insert in Fig. 2).

A noticeably nonlinear behavior was also observed for current–voltage characteristics of LC-CNTs composites (Lebovka et al., 2008). Figure 7 presents voltage dependencies of electrical conductivity σ/σ_0 for EBBA-CNTs composites at different values of temperature T and concentration of nanotubes c.

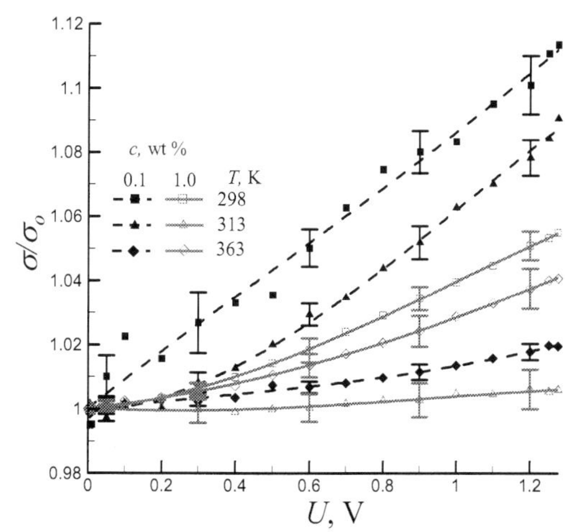

Fig. 7. Relative electrical conductivity σ/σ_0 versus measuring voltage U for composite EBBA-CNTs at different values of temperature T and concentration of nanotubes c. Here, σ_0 is electrical conductivity in the limit of U=0 V. The samples were randomly aligned in the 500 μm cells, f=1 kHz.

Such nonlinear behavior can be explained on the basis of hopping-tunneling model of transport disruption across the insulating LC regions between the CNTs, which predicts the following field dependence for electrical conductivity (Mott & Davis, 1971)

$$\sigma \propto \sinh(e\lambda_h E / kT), \tag{6}$$

where e is the elementary charge, λ_h is the average hopping distance, E is the applied electric field, and k is the Boltzmann constant.

The theory predicts weakening of the non-linear behavior with temperature T in full correspondence with experimental observations (Fig. 7). The most pronounced non-linear behavior was observed in the vicinity of the percolation threshold ($c_c \approx 0.05$-0.1 wt %). It was becoming less explicit above the percolation threshold, where multiple direct contacts between different CNTs were formed. The observed weakening of the current-voltage nonlinear behavior above the percolation threshold (Fig. 7) correlates with decrease of electrical conductivity activation energy (Fig. 4).

3.4 Dielectric relaxation and near-electrode processes: Low frequency range (A)

The dielectric spectra in the most low frequency range (A) ($f<10^2$ Hz) are mainly determined by the near-electrode processes and electron exchange between electrodes and ions (Koval'chuk, 1998; Koval'chuk, 2000; Barbero & Olivero, 2002; Koval'chuk, 2001c). In this range, the most important changes, provoked by presence of CNTs, were observed for the imaginary (ac conductance) component ε'', while changes for real (capacitive) component ε' were smaller. At low frequencies ($f<0.5$ Hz), increase of the component ε'' was observed by doping LC (both 5CB and EBBA) with CNTs (Fig. 1).

This behavior can be explained by enhancement of the electron component in electrical transport through the near-electrode layers, governed by the presence of CNTs in a composite. In fact, the CNTs serve as shunts of the double electrical layers providing paths for the electron exchange between electrodes and impurity ions inside LC.

The shunting effect observed at low frequencies ($f<0.5$ Hz) and related with the presence of CNTs was characterized in terms of the dielectric loss tangent, $\tan \delta_s$, which is given by

$$\tan \delta_s = \varepsilon'' / \varepsilon' . \tag{7}$$

Note that $\tan \delta_s$ characterizes dielectric loses in the near electrod layers and thus is a surface parameter.

Figure 8 presents $\tan \delta_s$ as a function of CNT concentration c for 5CB-CNTs composites.

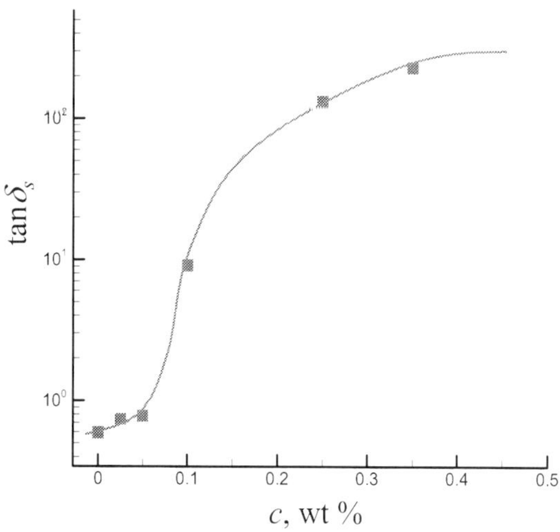

Fig. 8. Dielectric loss tangent tan δ_s versus CNTs concentration c for planar oriented 5CB-CNTs composites in a 16 μm cell at T=293 K, f=0.1 Hz.

At small concentrations of CNTs (c<0.05 wt %), tan δ_s is rather small and dielectric parameters of the near-electrode layers are determined by the properties of 5CB itself. However, the abrupt rise of tan δ_s in the interval of c values between 0.05 and 0.1 wt % evidences about formation of effective shunting paths as elements of a percolation network of CNTs characterized by the same concentration threshold.

The width of the near-electrode layers λ_s can be estimated (Koval'chuk, 2001a; Koval'chuk, 2001c):

$$\lambda_s \approx d\varepsilon_o^C / 2\varepsilon_o^A, \tag{8}$$

where ε_o^A and ε_o^C are the limiting values of ε for the frequency ranges (A) and (C). For example, for pure EBBA, the estimated width of near-electrode layer λ_s is about 0.3 μm and it is practically independent from the cell thickness d (Fig. 9). The doping of EBBA by CNTs (0.5 wt %) results in a noticeable increase of λ_s. This might be caused by increased concentration of ionic impurities or structural modification of DELs in presence of CNTs.

The low-frequency relaxation process characterized by the time τ_A we assign to dipolar polarization relaxation of LC molecules. These molecules can be anchored to the surfaces of the cell and to a skeleton of CNTs acting as a spatially distributed surface in LC host. For pure EBBA, τ_A is an increasing linear function of a cell gap d. The linear increase of the dipole relaxation time with d in the near-electrode zone τ_A was also observed for the planar oriented LC media (Yaroshchuk et al., 2005). It was established that value of τ_A decreases with the increase of electric conductivity. The similar tendency observed for EBBA doped with CNTs (Fig. 9) may be explained by high shunting efficiency of CNTs in the near-electrode zones and the ionic impurities introduced in LC together with CNTs.

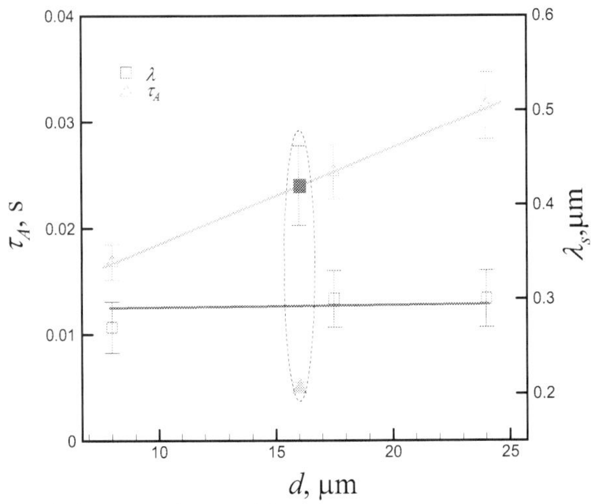

Fig. 9. Near-electrode layer width λ_s and low-frequency relaxation time τ_A vs. a cell gap d for pure EBBA (open symbols) and EBBA-CNTs composite (c=0.5 wt %) (filled symbols). The EBBA media was homeotropically oriented.

3.5 High frequency dielectric relaxation: High frequency range C

The high frequency dielectric relaxation time τ_C was practically independent of d for pure EBBA ($\tau_C \approx 0.67$ μs), and doping by CNTs (0.5 wt %) resulted in increase of τ_C ($\tau_C \approx 2.5$ μs). The similar tendency of τ_C to increase was observed also for 5CB-based composites, where $\tau_C \approx 0.56$ μs for pure 5CB and $\tau_C \approx 0.94$ μs for 5CB-CNTs (0.5 wt %) composite.

4. Electro-optic studies of LC-CNTs composites

Electrically controlled birefringence is a major property of LCs utilized in liquid crystal displays (LCDs), optical shutters, LC lenses and other LC devices. The present section considers influence of CNTs on the electro-optic properties of LC layers. It consists of three subsections. Subsection 4.1 refers to recent improvements of reversible LC response achieved owing to CNT addition. Subsection 4.2 is based on our original results concerning the irreversible electro-optical response of the LC-CNTs systems named as electro-optic memory (Dolgov et al., 2008; Dolgov et al., 2008a). At last, subsection 4.3 describes enhancement of the memory effect in the systems with induced chirality (Yaroshchuk et al., 2010).

4.1 Reversible electro-optic response

Generally, nematic LC devices utilize reversible response of LC layers on the applied voltage (Blinov & Chigrinov, 1996). The characteristics of this response, such as controlling voltage, switching on and off times, contrast ratio, etc., are quite important operational parameters of the LC devices. The major trend in improvement of these parameters is associated with synthesis of new mesogenic compounds and development of new eutectic

LC mixtures on their base. The other direction recently arose due to intensive development of nanotechnologies. Some types of nanoparticles turned out to be very useful fillers fundamentally expanding the range of mechanical, dielectric, magnetic, and optical characteristics of LCs (Qi & Hegmann, 2008).

Among nanoparticles as fillers for LCs, carbon nanotubes take special place. Due to strong shape anisotropy they possess strong anisotropy of polarizability. Therefore embedding even small amount of CNTs (c<0.05 wt %) into liquid crystal host may essentially increase (case of LC with $\Delta\varepsilon>0$) (Lee et al., 2004) or decrease (case of LC with $\Delta\varepsilon<0$) (Huang et al., 2005) dielectric anisotropy. Driving voltage is inversely proportional to the module of $(\Delta\varepsilon)^{1/2}$ and so it may be essentially influenced by carbon nanotubes.

As it was mentioned in section 3.2.1, carbon nanotubes (at $c\geq0.01$ wt %) generally enhance conductivity of LCs. However, at minute amount of nanotubes (~0.001 wt %), this effect is inessential. At the same time, the nanotubes may affect significantly the number of free-moving ions and their distribution within the cell by means of ion adsorption and shunting of double electric layers (see section 3). As a result, a minute doping by carbon nanotubes allows to suppress a number of parasitic ionic effects peculiar to LC electro-optic devices, such as a field screening (Lee et al., 2004), image sticking (Baik et al., 2005), transient current (Chen & Lee, 2006), back flow (Huang et al., 2005; Chen et al., 2007) and hysteresises of capacitance and transmittance (Lee et al., 2004; Baik et al., 2005). This results in lowering of driving voltage (Lee et al., 2004) and shortening of response times (Chen et al., 2007). Along with nematic LCs, CNT dopant may improve properties of LCDs based on ferroelectric LCs. Specifically, it fastens response of the deformed helix ferroelectric liquid crystal displays (Prakash et al., 2009). The changes in structure and resistivity of double electric layers and the composite's rotational viscosity due to carbon nanotubes are discussed as the possible reasons of such speeding-up of the electro-optic response.

4.2 Irreversible electro-optic response: effect of electro-optical memory

The irreverible response is not typical for nematic LCs commonly used in LC devices. However, the embedded nanoparticles grant this property to LCs (Kreuzer et al., 1992; Glushchenko et al., 1997). Regarding LCs filled with CNTs, interesting memory effect has been recently observed in isotropic phase of LC material (Basu & Iannacchione, 2008). It consists in irreversible change of dielectric constant of LC-CNTs composite under the applied voltage and is explained by formation of pseudonematic domains near the CNTs reorienting in the electric field. Since there is no restoring force in the isotropic phase, after the field is off, these domains keep the orientation induced in the electric field.

In what follows, a different memory effect is considered. It is peculiar only to the nematic mesophase of LCs with $\Delta\varepsilon<0$ homeotropically aligned before the action of the electric field. Besides, to realize this effect, the concentration of CNTs in the LCs should be considerably higher (c>0.01 wt %) then in the case of composites with the reversible electro-optic response (c~0.005 wt %).

To elucidate this memory effect, let us consider transmittance η of the samples placed between pair of crossed polarizers as a function of the applied voltage U. The $\eta(U)$ curves for the EBBA-CNTs samples are presented in Fig. 10.

In these experiments the voltage was smoothly increased to 60 V and then decreased to 0 V. One can notice that the $\eta(U)$ curves have oscillating character. The $\eta(U)$ oscillations mean that optical phase incursion during the switching of LC is much more higher than $\pi/2$. The saturation of this curve implies that, together with the bulk, the surface fraction of LC is reoriented in the field. It is evident that the neat LC demonstrates reversible electro-optic response (Fig. 10a). A small hysteresis (1–2 V) can be explained by screening of the applied field with the DELs formed near the cell electrodes (Baik et al., 2005). A very similar character of $\eta(U)$ curves is observed for the LC-CNTs composites with a small content of CNTs ($c<0.01$ wt %).

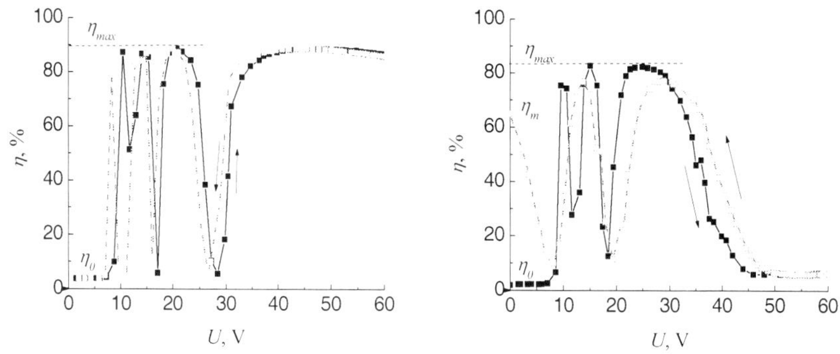

a b

Fig. 10. Dependences of light transmittance η on applied voltage U for EBBA doped with different concentrations of CNTs: a) 0 wt %; b) 0.05 wt %. Arrows indicate the increase and drop of voltage.

The situation, however, changes with a further increase of c. In this case, the oscillations are less pronounced for the reverse part of $\eta(U)$ curve. In addition, transmittance of the sample after the voltage decrease to zero, η_m, is sufficiently higher than the initial transmittance η_0 (Fig. 10b). This means that, to some extent, the cell memorizes transmittance realized in the electric field. This effect named as the effect of electro-optical memory can be characterized by the memory efficiency parameter M (Dolgov et al., 2008; Dolgov et al., 2009):

$$M=(\eta_m-\eta_0)/(\eta_{max}-\eta_0)*100\ \%, \tag{9}$$

where η_{max} is the maximal value of light transmittance (Fig. 10).

For the EBBA-CNTs composites under consideration, dependence of the memory parameter M on concentration of CNTs c is nonmonotonic. The $M(c)$ curve rapidly grows, reaches maximum at nanotube concentrations 0.02-0.05 wt % and then gradually decreases (Fig. 11, curve 1).

The initial rapid growth of this curve has percolation origin, which will be discussed below. The decay is apparently connected with a decrease in the effective voltage applied to the composite layer due to a marked increase in conductivity of the sample at high concentrations of carbon nanotubes.

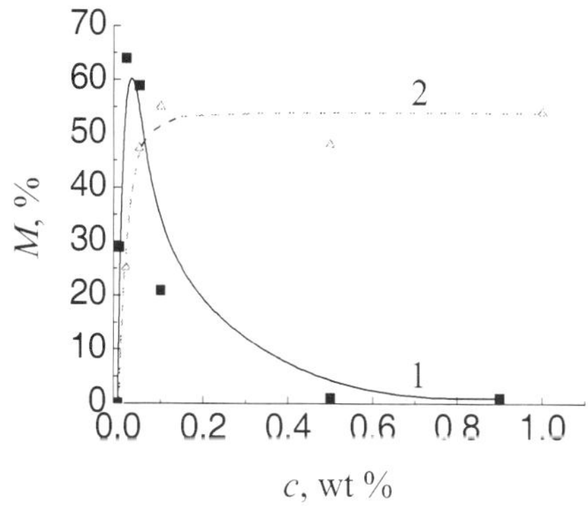

Fig. 11. Memory efficiency M as a function of the weight concentration of nanotubes c for the mixtures based on different liquid crystals: (1) EBBA-based series; (2) MLC6608-based series. The curves (1) and (2) are measured at temperatures 313 K and 297 K, respectively.

Generally, the memory efficiency M grows and reaches saturation with voltage application time τ. The memory reaches the state of saturation slower at lower driving voltages. For instance, at U=50 V the saturation time is 60 s, while at U=20 V it is 270 s. No electro-optical memory is observed at voltages U<10 V.

The effect of electro-optical memory is so pronounced that can be easily seen with a naked eye by sample observation through crossed polarizers (Fig. 12).

1

2

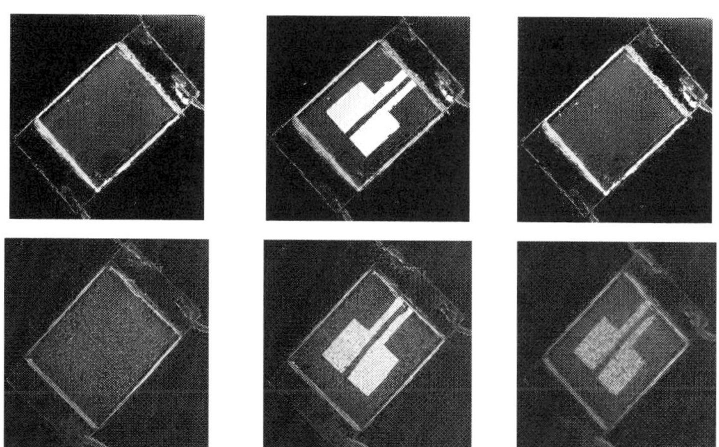

a b c

Fig. 12. Photographs of cells placed between the crossed polarizers and filled with neat EBBA (row 1) and EBBA doped with 0.02 wt % (row 2) of carbon nanotubes: a) initial off-state; b) U=60 V, f=2 kHz; c) off-state after field application. Voltage was applied to the central rectangular area of the sample.

One can see that neat LC cell switches homogeneously from the initial homeotropic to the planar state under the applied voltage. After the voltage switch-off LC returns to the initial state with homeotropic alignment. Similar behaviour is observed for the LC with minute (<0.01 wt %) concentrations of CNTs. LC suspensions with higher amount (c>0.01 wt %) of CNTs switch in the field to the random planar state. This state remains after switching off the voltage and causes high residual transmittance of the sample, i. e. electro-optical memory. Note that the memory can be completely or partially erased by applying to the sample mechanical stress, low frequency voltage (f = 10–50 Hz, U > 30 V) or by the transition of a system to crystalline or isotropic states with subsequent return to the mesophase.

It was further found that, besides EBBA, the described memory is peculiar to MLC6608 and MLC6609, LC mixtures with $\Delta\varepsilon$<0 and nematic mesophase at room temperatures. So there is no need in keeping samples at elevated temperatures during electro-optic tests. It simplifies measuring process in comparison with the EBBA-based mixtures. Below we present results for the MLC6608-based suspensions, but the results for the MLC6609-based samples are quite similar.

The $\eta(U)$ curves for the neat MLC6608 and MLC6608-based suspension are presented in Fig. 13. It is evident that these curves have smaller number of pulsations than the corresponding curves for EBBA-CNTs samples. This is caused by a smaller value of birefringence of MLC6608 (Δn=0.083 vs. Δn=0.25 for EBBA, Table 1).

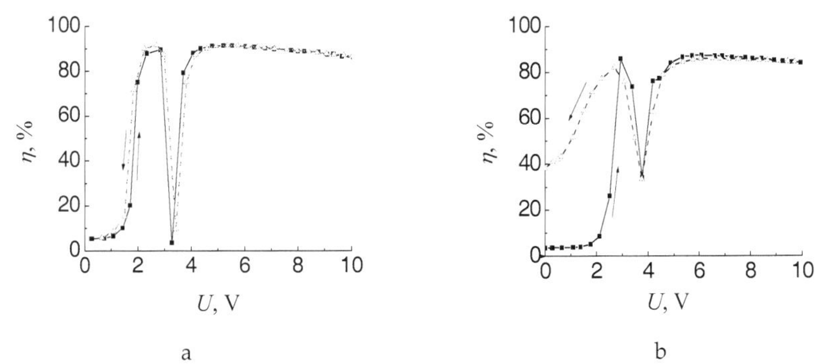

Fig. 13. Transmittance η vs. the voltage U curves for the MLC6608-CNTs composites with different concentrations of nanotubes: (a) 0 wt %; (b) 0.05 wt %. Arrows mark the plots obtained for the increasing and decreasing voltages.

The absolute value of dielectric anisotropy is much higher for MLC6608 than for EBBA (-4.2 vs. -0.13, Table 1). This results in much smaller values of saturation voltages for MLC6608-based mixtures (4–5 V) comparing with those for EBBA-based samples (35–40 V).

The maximal memory efficiency for MLC6608-based suspensions is similar to that for the EBBA-based samples (M=55-60 %). At the same time, the concentration dependences of M are essentially different. While the $M(c)$ curve for EBBA-CNTs series goes through the maximum (Fig. 11, curve 1), the corresponding curve for MLC6608-CNTs series rapidly grows and saturates at c~0.1 wt % (Fig. 11, curve 2). As discussed later, this is caused by different concentrations of ionic impurities and different structuring of CNTs in these LCs.

According to Fig. 14b, the appearance of the cells filled with MLC6608-CNTs composites is similar to that of analogous EBBA-based samples (Fig. 12); they contain areas with uniform homeotropic alignment not subjected to electric field and areas with the random planar alignment realized after the field cycle application. The memorized alignment state typically consists of islets of LC in a random planar state surrounded by the aggregates of CNTs.
It is important to note that for the suspensions based on LCs with $\Delta\varepsilon>0$, particularly for the 5CB, the memory effect is absent. The dependence of memory on a sign of dielectric anisotropy of LC will be elucidated in subsection 5.3.

4.3 Enhancement of memory effect by chiral dopant

The electro-optic memory effect described above is of considerable interest for applications. It suggests new principle for information displaying and storage in the LC based systems and thus can be designed for application in erasable memory cells, bistable displays, etc. These applications require essential improvements of the operational characteristics of LC-CNTs composites, first of all memory efficiency, erasure and recording times.

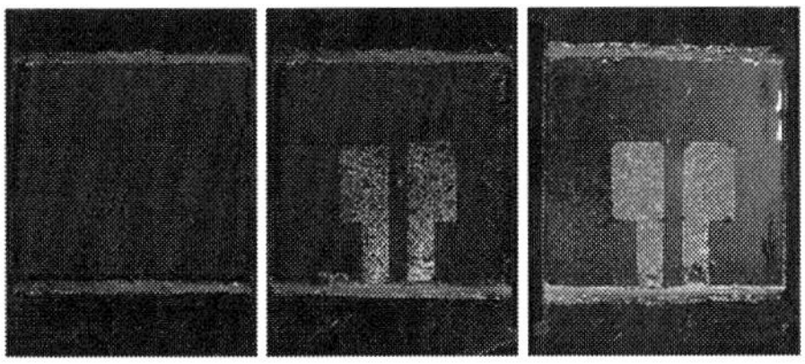

a b c

Fig. 14. The cells filled with LC MLC6608 (a), MLC6608-CNTs (c=0.02 wt %) (b) and MLC6608-ChD-CNTs (c=0.02 wt %, c_{ch}=0.1 wt %) (c) composites viewed between a pare of crossed polarizers. A cycle of electric field (30 V, 1 min) was applied to rectangular areas in the middle of the cells.

This paragraph describes the method of essential improvement of the memory efficiency M. The improvement is achieved by inducing chirality in LC host. This chirality causes additional force stabilizing the state of planar alignment realized in the electric field. Using this principle, efficiency of electro-optic memory can be practically doubled.
In our experiments the chirality of LC MLC6608 was induced by doping it with a small amount of chiral dopant (ChD) S811 from Merck. These studies were carried out in two stages. First stage was aimed at an optimization of chiral dopant concentration, c_{ch}. The increase of c_{ch} strengthens twisting tensions in the MLC6608. It was revealed that at $c_{ch}\leq0.1$ wt % anchoring forces satisfactorily balance these twisting tensions thus stabilizing homeotropic alignment. At $c_{ch}\geq0.15$ wt %, the anchoring forces can not restrain the twisting forces anymore that leads to a formation of various helical structures; at c_{ch}=0.15 wt % the filamentary texture is formed, which transforms in the fingerprint texture at higher concentrations. The concentration c_{ch}=0.1 wt % was selected for further preparation of LC-ChD-CNTs samples. It was the maximal value at which a uniform homeotropic alignment was preserved.
The $\eta(U)$ characteristics for the LC-ChD-CNTs sample, as well as for the reference LC-ChD sample are given in Fig. 15.

As is evident, the LC-ChD sample demonstrates reversible response. In turn, transmittance of LC-ChD-CNTs sample changes irreversibly showing high residual value in a zero field. The memory parameter estimated according to (9) is M=82 %. The corresponding value estimated for the LC-CNTs counterpart with the $\eta(U)$ curve presented in Fig. 13b is 0.44. This means that chiral dopant increases memory efficiency almost by factor 2. The strengthening of the memory effect in the samples containing chiral dopant can be seen even by naked eye (Fig. 14). The observation in polarizing microscope demonstrates that the planar state has an islet structure in the LC-CNTs samples, and a continuous structure in the LC-ChD-CNTs samples. This explains the increased memory efficiency of LC-ChD-CNTs samples.

The enhanced affinity of LC-ChD-CNTs samples to planar alignment might be explained by the enhancement of forces resulting in planar alignment. In the LC-ChD-CNTs samples, the force associated with a CNT network is magnified by a twisting force, which eventually destroys homeotropic alignment.

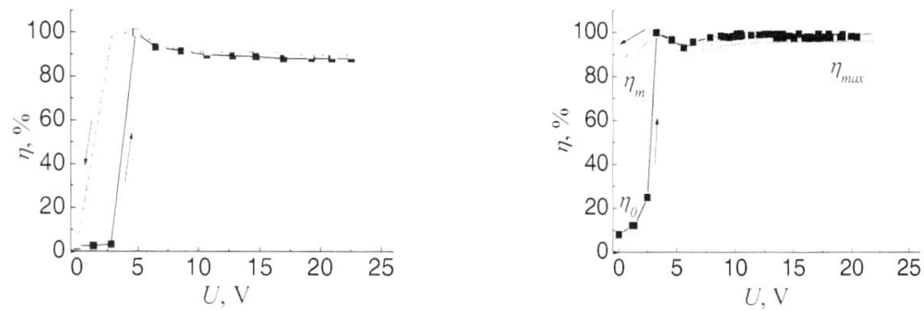

a b

Fig. 15. The transmittance η vs. applied voltage U curves for the cells filled with MLC6608-ChD (c_{ch}=0.1 wt %) (a) and MLC6608-ChD-CNTs (c=0.02 wt %, c_{ch}=0.1 wt %) (b) compositions. The arrows point directions of voltage ramping up and down.

It worth mentioning that, in spite of the memory enhancement of LC-CNTs samples, the twisting force by itself does not cause a memory effect (the case of LC-ChD samples, Fig. 15a). This suggests that the described memory effect is an intrinsic feature of samples containing CNTs.

5. Structural peculiarities of LC-CNT composites and structural transformation under the applied field

This paragraph consists of three subsections. Subsection 5.1 is introductory. It briefly refers to structuring of CNTs in LC hosts and methods of stabilization of CNTs in the LCs. In addition, several known effects connected with the influence of electric field on the CNT aggregates are reviewed. Subsection 5.2 is based on original results and considers microstructuring of the memory type LC-CNTs composites. Particularly it is shown how the electrohydrodynamic flows developed in LC host change the structure of CNTs and promote formation of the memory state. Finally, in subsection 5.3, physical model of electro-optic memory is suggested based on the microstructural observations and results of electro-optical and dielectric measurements.

5.1 Structural transformations in LC-CNTs composites without electrohydrodynamic flows

CNTs strongly attract with each other by van der Waals forces. Therefore they form more or less developed system of aggregates in a liquid host. Weak aggregation of CNTs is observed in the polar aprotic solvents like dimethylformamide (Liu et al., 1999), *N*-methyl-2-pyrrolidone (Giordani et al., 2006), the chlorinated hydrocarbon dichloroethane (Baik et al., 2005), and the polar protic compound ethyl alcohol (Huang & Pan, 2006).

In LC matrices the nanotubes aggregate quite intensively. This is a challenge for the majority of applications demanding LC-CNTs composites with high and uniform dispersion of CNTs. To achieve this, several methods are proposed.

In one of these methods strong aggregation of CNTs was prevented by their initial dispersion in organic solvents with next addition of LC (Baik et al., 2005). After evaporation of solvent one can obtain CNTs dispersed in LC medium. However, one of the drawbacks of this dispersion technique is the undesired residual solvent effect, which may influence quality of LC phase.

Another method lies in modification of CNT surface for enhanced compatibility with LCs. In this case CNTs bear on their surface small molecules, polymers or inorganic species (Trushkevych et al., 2008). But such modification can affect intrinsic mechanical, electrical and optical properties of CNTs (Bahr et al., 2001).

The more simple technique is just ultrasonication of CNTs in LCs without special dispersing agents. Centrifugation and decantation separate the suspension from big aggregates and may be used as additional methods after ultrasonication of CNTs (Chen et al., 2007).

Influence of electric field on CNT aggregates dispersed in LCs turned out to be quite manifold. The electric field may influence structure of CNTs directly or indirectly. One of examples of direct action is a drift of individual CNTs in LC under strong electric fields due to electrophoretic forces (Baik et al., 2006; Chen et al., 2008a). Movement of CNTs in the uniformly aligned LC may be traced due to specific light scattering appearing from the moving nanotube. It was also revealed that the internal forces appearing due to polarization of nanotubes in the electric field may cause structural changes of CNT aggregates. For example, the in-plane electric field of 5.67 V/μm at 60 Hz causes reversible four times elongation of CNT aggregates in the superfluorinated nematic LC (Jeong et al., 2007).

A striking example of the indirect action is the control of CNTs alignment by LC reorientation in electric field (Dierking et al., 2004; Dierking et al., 2008). As a rule, this mechanism does not involve essential LC flows. At the same time, at certain experimental conditions, the electrohydrodynamic (EHD) flows are highly intensive in LC cells (Blinov & Chigrinov, 1996). They can also be realized in the LCs doped by CNTs (Chen et al., 2008b). The next section elucidates structural reconstruction of CNT aggregates under the EHD flows resulting in effect of electro-optic memory. It is based on our recent results published in (Dolgov et al., 2008; Dolgov et al., 2009).

5.2 Structural transformations in the LC-CNT composites with electrohydrodynamic flows

Similarly to other LC-CNTs composites (Dierking et al., 2004; Jeong et al., 2007; Lee et al., 2004; Baik et al., 2005), the memory type EBBA-CNTs composites contain CNTs in aggregated state. Big aggregates are clearly visible in a polarizing microscope. The size of aggregates increases up to tens and even hundreds of microns with increase in the CNT concentration. The single aggregates observed at $c<0.02$ wt % are assembled into a continuous network, when $c>0.1$ wt %.

The evolution of the samples structure under the action of the high frequency electric field (f=2 kHz) depends essentially on the concentration of nanotubes. In the neat EBBA and EBBA doped with a very small amount of CNTs ($c<0.002$ wt %), the initial homeotropic alignment state (Fig. 16a) switches to the uniform planar state (Fig. 16b) at voltages ~10 V. At higher voltages, the classical electrohydrodynamic instabilities develop in these samples. Laminar flows, revealing itself in the form of the Kapustin-Williams domains, arise at U=80 V (Fig. 16c) (Blinov & Chigrinov, 1996). As the voltage increases, the flow patterns become complicated and transform to turbulence patterns at voltages 110–120V (Fig. 16d). The EBBA samples return to the initial homeotropic state after switching off the field (Fig. 16a). In the samples with higher concentrations of nanotubes (c~0.02–0.05 wt %), the development of EHD instabilities is different. At U~10 V, LC switches from the homeotropic to the planar state (Figs. 17a, b). As the voltage increases, EHD flows appear near the aggregates of CNTs. The further increase in the voltage leads to the broadening of flow areas (Fig. 17c), which overlap (at U~40 V) and finally occupy the whole volume of the sample. In this process, the Kapustin-Williams domains appear in some areas of the sample.

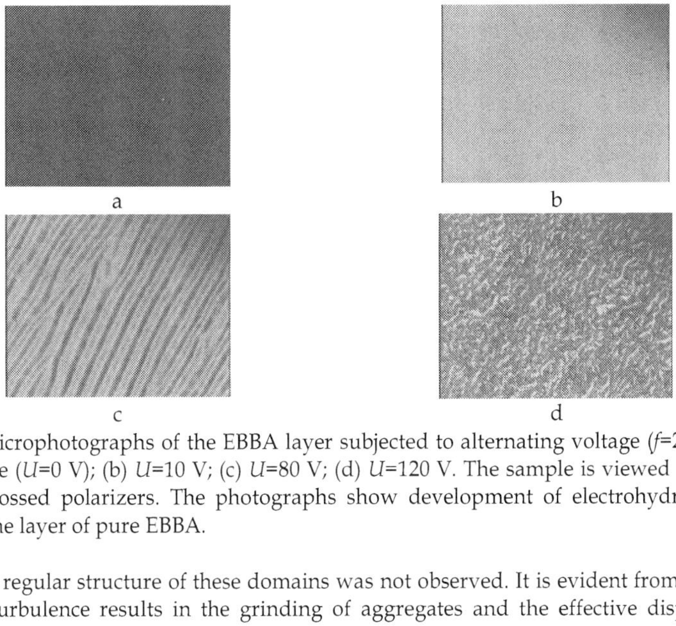

Fig. 16. Microphotographs of the EBBA layer subjected to alternating voltage (f=2 kHz): (a) initial state (U=0 V); (b) U=10 V; (c) U=80 V; (d) U=120 V. The sample is viewed between a pair of crossed polarizers. The photographs show development of electrohydrodynamic flows in the layer of pure EBBA.

However, regular structure of these domains was not observed. It is evident from Figure 17 that the turbulence results in the grinding of aggregates and the effective dispersion of CNTs.

Note that in samples with higher concentration of nanotubes ($c>0.1$ wt %) hydrodynamic motions arose too, but they did not influence essentially the morphology of aggregates. This is caused by lowering of the actual voltage and sample heating due to the essentially increased conductivity (Shah et al, 2008).

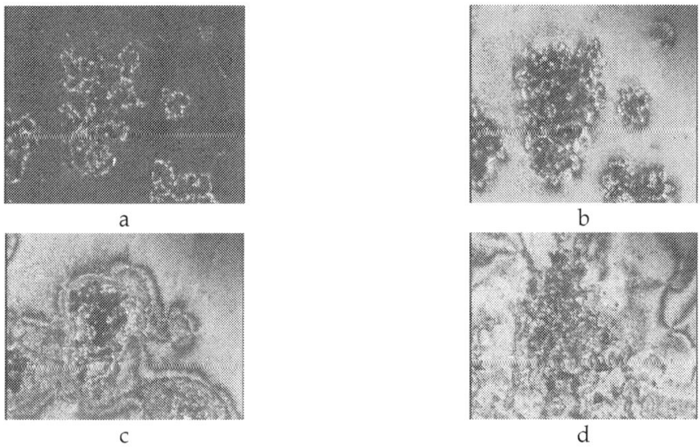

Fig. 17. Microphotographs of the layer of EBBA-CNTs composite (c=0.05 wt %) subjected to alternating voltage (f=2 kHz): (a) initial state (U=0 V); (b) U=10 V; (c) U=20 V; (d) after removing the field (U=0 V). The sample is viewed between a pair of crossed polarizers. The photographs show development of electrohydrodynamic flows in the layer of EBBA-CNTs composite.

The grinding of CNT aggregates and their motion in the EHD flows differs from the earlier described effects of the CNT structure reorganization related to the electrostatic and electrophoretic forces (Jeong et al., 2007; Baik et al., 2006). The dispersion process described above was observed only in the LCs with pronounced EHD instabilities.

In contrast to samples with c<0.002 wt %, the samples with 0.02<c<0.5 wt % remain in the random planar state and have the schlieren microscopic texture after the field switch-off (Fig. 17d). This texture caused the residual transmittance T_m and thus the memory effect.

Comparing with EBBA-based samples development of EHD instabilities in MLC6608 based samples has some differences. In contrast to EBBA, MLC6608 does not reveal any hydrodynamic flows up to 100 V. This is caused by the fact that the ionic conductivity of MLC6608 is one order of magnitude less than that of EBBA. It is well known (Blinov & Chigrinov, 1996), that the sufficient amount of ionic impurities in LC is needed for the development of hydrodynamic instabilities.

The EHD flows in MLC6608 appear only after addition of carbon nanotubes. As in EBBA-based composites, two steps of the LC response to the applied field can be selected: at first, at U=4–9 V, the Friedericksz transition occurs, and then, at higher voltages, hydrodynamic flows develop. In contrast to EBBA-based composites, hydrodynamics in MLC6608-based samples occurs in the form of turbulent flows without the stage of Kapustin-Williams domains. These flows begin to develop just after the Friedericksz transition in the vicinity of CNT aggregates. The areas of turbulence grow with the applied voltage and occupy the whole volume of the sample at 20–40 V. The developed flows result in the grinding of the aggregates of nanotubes. This effect can be considered as a practical method for in situ dispergation of CNTs in a LC medium.

5.3 Percolation structures of CNTs and mechanism of electro-optic memory

According to results of dielectric studies (section 3), the conductivity of LC-CNTs composites demonstrates pronounced percolation behavior. A steep increase of conductivity with concentration of CNTs is caused by self organization of these conductive particles resulting in formation of continuous spatial network percolating through the LC layer.

The results of structural studies of LC-CNTs composites with EHD instabilities (subsection 5.2) lead us to conclusion that the networks of CNTs formed before and after the application of an electric field are different. Before switching on the field, CNTs agglomerate in big clusters forming a network structure at high concentrations (c>0.5 wt %). A much finer structure occurs after the application of an electric field. The massive aggregates of nanotubes crushed in the EHD flows are assumed to form a fine network stabilizing the planar state of LC. This is confirmed by the two-order increase of sample conductivity after development of EHD instabilities.

The decisive role of hydrodynamic flows in the formation of the fine CNT network and thus the memory is confirmed by other experiments. First, the memory effect was not obtained in the composites based on LC 5CB ($\Delta\varepsilon$>0), in which EHD effects are not realized. Second, the memory effect was not observed neither in EBBA- nor in MLC6608-and MLC6609- based composites reoriented from the homeotropic to the planar state by the magnetic field. It is well known that a magnetic field causes no hydrodynamic flows in LC.

The formation of fine CNT network is closely related to memory effect. As was clarified above, the electro-optic memory is caused by the metastable planar LC alignment formed after the field is off. This alignment is stabilized by the network of CNTs acting as a spatially distributed alignment surface for LC. The alignment force of this network overcomes corresponding force of the aligning substrates restricting suspension layer. The analogous mechanism was earlier considered for LC-aerosil composites (Kreuzer et al., 1992; Glushchenko et al., 1997). It is known that such a system is characterized by the pronounced electro-optical memory effect caused by orientational LC transition from the spatially random to oriented state. The oriented state was metastable in a zero field because it was maintained by the network of aerosil particles.

To resist elastic tensions and so maintain planar alignment of LC phase in the LC-CNTs composites, the CNT network should be sufficiently strong. In view of this, the mechanical rigidity percolation should be considered additionally to conductivity percolation. Generally, the rigidity percolation, corresponding to the sol-gel transition, is characterized by a threshold concentration c_m higher than c_c. According to rough estimations, $c_m \approx 1.6c_c$ (Sahimi, 1998). Assuming that this correlation is valid for the studied system and that $c_c \approx 0.05$ wt %, one can obtain that $c_m \approx 0.08$ wt %. Note that this value is by one order of magnitude lower than the value $c_m \approx 1$ wt % estimated for LC-aerosil suspensions (Glushchenko et al., 1997). This might be caused by extremely high aspect ratio of CNT particles capable to form connected structures at so low concentrations.

In summary, the following mechanism is responsible for the effect of electro-optical memory. Initially, LC is homeotropically aligned and CNTs are well aggregated. Electric field application leads to homeotropic-to-planar reorientation and development of electro-hydrodynamic flows in the LC phase. These flows crush bulky CNT aggregates, thus opening way for formation of fine CNT network supporting the planar LC alignment after the turning off the electric field. This mechanism is effective in the limited range of CNT concentrations. On the one hand, c should be higher than rigidity percolation threshold c_m.

On the other hand, it should not be rather high to prevent high electric conductivity of a system which causes critical lowering of the actual voltage applied to suspension and its essential heating.

6. Conclusions

Combination of liquid crystals and carbon nanotubes gives a class of unique composites with fascinating electrical, optical, electro-optical, nonlinear optical and structural properties. The present chapter describes a number of new-found interesting features of these composite materials. In general, a strong correlation between structural, electrical and electro-optic characteristics of LC-CNTs composites is observed.

It is shown that the CNTs shunt double electric layers in the LC cells and, in this way, change essentially a spatial distribution of the electric field applied to the cells. This explains reduction of controlling voltage and response time of LC layers doped by CNTs.

Same as in other liquids and polymers, CNTs demonstrate percolation behavior in LC hosts. At concentration of CNTs $c<0.01$ wt % in the LC-CNTs composites, the nanotubes exist in the form of individual aggregates. The further increase of c results in connection of isolated aggregates and, finally, in formation of continuous network of CNTs permeating LC matrix. The formed CNT network radically increases conductivity of LC samples and changes conductivity mechanism. At $c<0.01$ wt % ionic conductivity, typical for neat LCs, prevails. At $0.01<c<0.4$ wt %, the ionic charge transport is essentially enhanced by a charge hopping transport associated with a CNTs' skeleton. At higher CNT concentration, the conductivity mechanism typical for highly connected CNTs fully dominates. This change in the charge transport mechanisms with the CNTs concentration results in decay of conductivity activation energy from ~10 kJ/mol to ~0 kJ/mol.

The LC-CNTs composites based on the LCs with $\Delta\varepsilon<0$ demonstrate effect of electro-optic memory. It is connected with a metastable alignment state of LC phase stabilized by the network of CNTs acting as a spatially distributed alignment surface. This effect suggests new operation mode for LC devices and can be employed in the systems of information displaying and storage.

It is shown that the electrohydrodynamic flows developing in the LC phases may essentially influence structure of CNTs. Specifically, they crush aggregates of CNTs promoting formation of their finer structure. This suggests a unique method for in situ dispergation of CNTs in LCs. It is especially important for the CNTs non-grafted with special hydrophobic fragments facilitating dispergation of CNTs inside the LC matrix.

We strongly believe that the next studies will open new marvelous effects in these composites and their intriguing applications. Among the problems worthy of future investigation is influence of CNTs on the specific properties of LC materials, such as electrical, optical, thermal and mechanical anisotropy, molecular ordering and variety of phase transitions. It would be interesting to study fundamentally the peculiarities of structuring and percolation of CNTs in mesophases of different symmetry and origin, in the areas of phase transitions. Finally, it is worthwhile to investigate composites based on nanotubes with unusual properties and structure intensively generated by modern nanoscience and nanotechnology.

Acknowledgements

These studies were supported by European Social Fund grant GLOFY0102J, NAS of Ukraine (grant 10-07-H) and "Dnipro" program of Ukrainian-French scientific cooperation (grant M/16-2009).

7. References

Bahr, J. L.; Yang, J.; Kosynkin, D. V.; Bronikowski, M. J.; Smalley, R. E. & Tour, J. M. (2001). Functionalization of carbon nanotubes by electrochemical reduction of aryl diazonium salts: a bucky paper electrode. *J. Am. Chem. Soc.*, Vol. 123, No. 27, June 2001, pp. 6536-6542, DOI: 10.1021/ja010462s.

Baik, In-Su; Jeon, Sang Youn; Lee, Seung Hee; Park, K. Ah; Jeong S. H.; An, K. H. & Lee, Y. H. (2005). Electrical-field effect on carbon nanotubes in a twisted nematic liquid crystal cell. *Appl. Phys. Lett.*, Vol. 87, No. 26, December 2005, pp. (263110-1)-(263110-3), DOI: 10.1063/1.2158509.

Baik, In-Su; Jeon, S. Y.; Jeong, S. J.; Lee, S. H.; An, K. H.; Jeong, S. H. & Lee, Y. H. (2006). Local deformation of liquid crystal director induced by translational motion of carbon nanotubes under in-plane field. *J. Appl. Phys.*, Vol. 100, No. 7, October 2006, pp. (074306-1)-(074306-5), DOI: 10.1063/1.2355535.

Balberg, I.; Anderson, C. H.; Alexander, S.; & Wagner, N. (1984). Excluded volume and its relation to the onset of percolation. *Phys. Rev. B*, Vol. 30, No. 7, October 1984, pp. 3933–3943, DOI: 10.1103/PhysRevB.30.3933.

Barbero, G. & Olivero, D. (2002). Ions and nematic surface energy: belong the exponential approximation for the electric field of ionic origin. *Phys. Rev. E*, Vol. 65, February 2002, pp. (031701-1)-(031701-5), DOI: 10.1103/PhysRevE.65.031701.

Barrau, S.; Demont, P.; Peigney, A.; Laurent, C. & Lacabanne C. (2003). DC and AC conductivity of carbon nanotubes-polyepoxy composites. *Macromolecules*, Vol. 36, Iss. 14, June 2003, pp. 5187-5194, DOI: 10.1021/ma021263b.

Basu, R. & Iannacchione, G. S. (2008). Carbon nanotube dispersed liquid crystal: A nano electromechanical system. *Appl. Phys. Lett.*, Vol. 93, No. 18, November 2008, pp. (183105-1)-(183105-3), DOI: 10.1063/1.3005590.

Behnam, A.; Guo, J. & Urala, A. (2007). Effects of nanotube alignment and measurement direction on percolation resistivity in single-walled carbon nanotube films. *J. Appl. Phys.*, Vol. 102, Iss. 4, August 2007, pp. (044313-1)-(044313-7), DOI: 10.1063/1.2769953.

Belotskii, E. D.; Lev, B. I. & Tomchuk, P. M. (1980). Effective ion mass in a liquid crystal. *JETP Lett.*, Vol. 31, No. 10, May 1980, pp. 539-541.

Blinov, L. M. & Chigrinov, V. G. (1996). *Electrooptic Effects in Liquid Crystal Material*, Springer, ISBN: 0387947086, 9780387947082, New York.

Cervini, R.; Simon, G.; Ginic-Markovic, M.; Matisons, J.; Huynh, C. & Hawkins, S. (2008). Aligned silane-treated MWCNT/liquid crystal polymer films. *Nanotechnology*, Vol. 19, March 2008, pp. (175602-1)-(175602-10), DOI: 10.1088/0957-4484/19/17/175602.

Chelidze, T.L.; Derevyanko, A. I. & Kurilenko, O. D. (1977). *Electric Spectroscopy of Heterogeneous Systems*, Naukova Dumka, Kiev (in russian).

25

Functionalization of carbon nanotubes with luminescent silicon nanocrystals upon nanosecond laser processing in liquid media

Vladimir Švrček

Novel Si Material Team, Research Center for Photovoltaics, National Institute of Advanced Industrial Science and Technology (AIST), Central 2, Umezono 1-1-1, Tsukuba, 305-8568, JAPAN

1. Introduction

Due to unique properties, potential in basic scientific research and technological applications (Xia, 2003, Hu, J. 1999) a discovery of carbon nanotubes (CNTs) (Iijima, 1991) tubular nanostructure has fascinated and activated research in different fields. During recent years nanotubes of many compounds, such as semiconductors, polymers, metals, oxides, sulfides, lipids have been synthesized (Wei et al., 2003; Cao et al., 2003; Mo et al., 2002; Zhang et al., 2002; Nath et al., 2001; Shimizu, 2002; Kobayashi et al., 2002) and extensively investigated (Schnur, 1993). Particularly utilization of nanotube hollow cylinder that can be easily varied within several nanometer-sized has drawn a lot of attention (Schnur, 1993; Tsang, et al., 1994; Shimizu, 2002; Kobayashi, 2002). For instance functionalization of human body more compatible lipid nanotubes (LNTs) cavity could be promising for biological applications since the glycolipid nanotubes sugar headgroup have high biocompatibility and specificity to a certain protein (Shimizu, 2002; Kobayashi, 2002).

During last decade especially CNT has proofed to be a nanomaterial with exceptional physical and chemical properties to many chemical reactions (Pham-Huum, 2002). Concave character of inner-mechanically stable CNTs with confined space offer the possibility to generate metal particles with enhanced structural and chemical stability (Ajayan and Iijima, 1993; Ugarte et al., 1996). Additionally to that the CNTs are electrically and thermally conductive, allowing application in electro- and heterogenous catalysis such hybrid nanomaterials (Tessonnier et al., 2005). The prospects application as templates for nano-wires fabrication and one-dimensional nanocomposite materials through the filling has been envisaged (Ajayan and Iijima, 1993; Ugarte et al., 1996; Thamavaranukup et al., 2004; Tsang, 1994; Lago et al., 1995; Yang, 2004).

The key aspect to realize application based on encapsulated nanoparticles in nanotube cavity relays on an insertion of particles selectively (Tessonnier, 2005). Many techniques has been investigated and applied to efficiently encapsulate the metallic or oxyde particles within especially the CNTs cavity (Ajayan and Iijima, 1993; Ugarte et al., 1996; Tessonnier et al., 2005; Tsang et al., 1994). One approach is that the CNTs are refluxed in concentrated

containing precursor salts, by which CNTs can be opened and filled by the precursors (Lago et al., 1995) or by impregnation (Pham-Huu et al., 2002). However, due to the high surface tension of semiconductor materials and introduction of semiconductor particles is quite difficult (Svrcek, 2006a). Particularly silicon in the form of quantum dots, i.e., silicon nanocrystals (Si-ncs) in sizes less than 10 nm where the quantum confinement can take place, offer attractive optoelectronic features (Canham, 1990; Kanemitsu, 1995; Wolkin et al., 1998). In fact, due to the maturity of silicon technology and simple Si-nc integration with available technologies, Si-ncs has been widely investigated (Hirsmman et al., 1996; Pavesi et al., 2000; Canham, 1990; Kanemitsu, 1995; Wolkin et al., 1998; Švrček et al., 2004; Švrček et al., 2009a). In addition, since Si-nc have low toxicity and fully compatible with a human body, Si-nc may also be preferable as a fluorescent agent in biological systems when compared to other currently used materials (Švrček et al, 2008b). Fictionalization of the Si-ncs with nanotubes (Švrček, et al, 2006a Švrček, et al, 2006b, Švrček et al., 2008a, Švrček et al, 2008b) might open innovative routes for development new design for advanced hybrid materials.
Silicon nanocrystals dispersed in colloidal sulutions with low surface tension could be a way how to introduce them into the nanotube cavity (Švrček, et al., 2006a). One alternative method that has produce colloidal Si-ncs and at the same time allow nanotube filling is pulsed laser ablation and fragmentation in liquid media (Švrček et al., 2006c; Švrček, et al., 2006d, Švrček, et al., 2008a). The main advantage of the technique is an easy control of the experimental conditions, the ambience and the possibilities of using variety of the solutions (water, ethanol, polymer). Synthesis allows of tuning the surface tension of surfactant-free Si-ncs, permits entering of freshly formed Si-ncs into carbon nanotubes (CNT) cavity (Švrček, et al., 2008a, Švrček, et al., 2008b) and allows unique materials such as inorganic-organic nanocomposite structures fabrication (Švrček 2008b). Due to an interaction of ablated material with surfactant (Mafuné et al., 2003, Mafuné et al., **2000,** Usui et al., 2005) it might results into novel nanotube/nanocrystals nanocomposites with peculiar properties while the Si-ncs properties could be greatly enhanced.

In this chapter we discuss and compare the filling processes of carbon and lipid nanotubes with surfactant-free Si-ncs prepared in colloidal solutions. Influence of colloidal suspension on mechanical and chemical stability of the nanotubes is addressed. Since the small value of zeta potential in water results agglomeration that inhibits the filling process the laser processing in ethanol /polymer prevents Si-ncs agglomeration and favorites encapsulation in CNTs cavity. However, drawback of the LNTs is low mechanical and chemical stability. Weak resistance against the ethanol based solutions do not allow simply cavity filling by capillary forces with Si-ncs dispersed in organic (ethanol, methanol. polymer) based solutions. Furthermore, the CNTs showed also stronger physical properties (e.g. Young`s modulus), mechanical stability that permits the laser processing in silicon technology compatible polymer, which results surface passivation and encapsulation of luminescent Si-ncs at room temperature. One step functionalization CNTs cavity by Si-ncs during the laser processing in liquid protects freshly formed nanocrystals from pollutions, inhibits agglomeration, and induces the cavity entrance. We overview a fabrication method of nanocomposites based on Si-ncs and CNTs/LNTs upon nanosecond laser processing in liquid media. The shock waves generated during Si-ncs formation in liquid media by nanosecond laser processing assure the encapsulation of Si-ncs in CNTs cavities and incorporation within LNTs wall. Involved processes as shock waves generation, consequent filling and wall incorporation are discussed in detail.

2. Experimental details

Since the CNTs with inner diameter about 50 nm were commercially purchased at Pyrograf Products Inc., the LNTs were fabricated by following way. The most common way to form LNTs is to make use of the unsymmetrical monolayer lipid membrane made of heteroditopic I, ω-amphiliphiles having two headgroups differing in size (Shimizu, 2002). The LNTs we prepared from N-(11-cis-octadecenoyl)-β-D-glucopyranosylamine as constructing glycolipid molecule. Particularly 10 mg of lipid powder was dissolved in 1 ml of the ethanol and sonicated for 10 minutes then added 100 ml of de-ionised water. The solution was then heated at 90 °C and let it heated for 30 minutes and then naturally cooling down to 25 °C. As reported elsewhere the glycolipid molecules self-assembled in water have ability to form mono-dispersed and well-defined LNTs at high yields (more than 98 %) determined by chiral molecular packing (Kamiya et al., 2005). In benefit for the filling of the LNT cavity the most of the ends are naturally opened. However, this is not the case of CNTs and an additional thermal treatment is necessary (Švrček, et al, 2006b) to open their ends. For this work the CNTs were opened by thermal annealing at temperature 580 °C in oxygen atmosphere.

The Si-ncs used for capillary force induced filling were prepared by electrochemical etching of silicon wafer (p-type <100>, resistivity 0.1 ohm·cm, thickness 0.525 mm). Porous silicon wafers were mechanically scratched from the surface and then dispersed in liquid media (Švrček et al., 2006b). The Si-ncs were ultrasonically mixed with methanol and spin on glass (SOG, Filmtronics, Inc. USA). In order to decrease the viscosity of the SOG polymer we added the methanol at ratio methanol:SOG of 2:1 It has to be noted that the homogeneous Si-ncs distribution within the water has been achieved by wetting of the Si-ncs surface by ~20 drops ethanol. Then opened CNTs were directly introduced into solutions (Si-ncs/methanol, Si-ncs/SOG) and kept in an ultrasonic bath for 30 min. In the case of the LNTs prior to introduction in solution, the water from inner cavity has been removed by freeze draying process. Namely, the suspension of nanotubes was freezed with liquid nitrogen becoming completely glassy. Indeed, the water evaporated from solid state was pumped in order to avoid condensation and decomposition of the nanotube tubular structure. It has to be noted, an introduction of LNTs in Si-ncs/ethanol (methanol/SOG polymer) leads to a complete disaggregation of the LNTs structure therefore only capillary induced filling of LNTs of Si-ncs dispersed in water was possible to perform only.

In order to stabilize freshly prepared Si-ncs by laser ablation processing, the experiment was conducted as follows. In 30 ml of de-ionized water, 0.01 mg of CNTs/LNTs was dissolved. At the same time the Si-ncs production and CNT cavity filling were assured in situ in water. Si wafer immersed and fixed on the bottom of a glass vessel in CNTs/water (CNTs/water) suspensions was used as source for Si-ncs. A third harmonic Nd:YAG laser (Spectra Physics LAB-150-30, 355 nm, 30 Hz, 8 ns) was irradiated onto the target immersed in 5 ml of suspension at room temperature for 2 h. The laser beam was focused on the target plate with a beam size of about ~1-1.5 mm in diameter.

For filling CNTs by laser fragmentation processing, the silicon micrograins were prepared by electrochemical etching of silicon wafers. The micrograins were collected by precipitation when Si-ncs were harvested from supernatant parts of solution. Before laser induced fragmentation the porous Si micrograins precipitated on the bottom of vessel had sizes that exceed several micrometers. Precipitated micrograins contain Si-ncs as well and PL at room temperature could be still observed. Those in 30 ml water together with CNTs (0.03 mg)

were homogenously dispersed. We prepared 0.01 wt.% colloidal solutions of Si micrograins in CNT/water solution for the laser fragmentation experiment. Similarly to the capillary induced filling, in order to achieve homogeneous grain distribution within the CNT/water solution, we have wetted the micrograins surface with a few drops of ethanol (20 drops) prior to adding the CNT/water solution (Švrcek, 2008a).
For the laser fragmentation processing in the transparent polymer (Si-59000, Tokyo Ohka Kogyo Co., Ltd.) in 30 ml of polymer solution, 0.01 mg of the CNTs was homogenously dispersed, and 0.01 wt % of Si micrograins was added. It has to be noted that the polymer is a mixture of ethylpolysillicate ($C_2H_5O[SiO(C_2H_5O)_2]_n C_2H_5$), ethanol, and ethylacetate, and homogeneous distribution of micrograins is naturally achieved. To proceed the laser processing filling 5 ml of the solution (micrograins/CNTs/water,micrograins/CNTs/polymer) was placed into a glass container and irradiated by a pulsed Nd:YAG laserat a laser fluence ranging from 0-23.3 mJ/pulse for 2 h at room temperature (Švrcek, 2008a).
The photoluminescence (PL) measurements of colloidal solutions were carried out at room temperature using fluorophotometer (Shimadzu corporation, RF-5300PC) with excitation by Xe lamp at 300 nm. A small droplet of obtained colloidal solution was deposited onto a copper grid with carbon film for high resolution transmission electron (HR-TEM) and scanning electron microscope (SEM) observations. The SEM images were done with a Hitachi SI 4800 microscope with 30 kV acceleration voltage. The HR-TEM studies were performed on a microscope with 200 kV acceleration voltage (JEOL, JEM 2010). An electron diffraction was employed to perform more localized elemental analyses of Si nanoparticles. In order to perform a localized analysis of the providing Si nanoparticles, a convergent incident electron beam was employed in the diffraction measurements convergent beam electron diffraction.

3. Capillary forces filling of nanotube cavities by room temperature luminescent Si-ncs

Theories based on molecular chilarity are commonly applied to describe the LNTs formation (Spector et al., 1996). A chiral interaction is most important factor for formation of nanotubes and cause constituent molecules to pack at a nonzero angle with respect to their nearest neighbor. The chirality of the molecule induces particular orientation to be energetically favorable and results the formation of a cylindrical hollow (Spector et al., 1996). As the LNTs consist of the unsymmetrical monolayer lipid membrane, the geometry of the LNT can be easily tuned and control by determinative way (Yang et al., 2004). The nanotube length can be varied over large range varying from nanometers to hundreds of micrometers due to the changing of the solvent and preparation conditions (Ratna et al., 1992). As also the diameter of the cavity in LNT can be easily controlled than it can be in principle achieve a selection of particle size. Figure 1 shows typical optical image of LNT formed in glycolipid suspension. As dispersed on glass substrate we observe that the nanotubes are well separated and agglomerated structures are rarely observed. The inner and outer diameter vary and giving approximately 40 and 100 nm in average. The lengths of LNTs vary from 2 μm to 40 μm with average size of about 25 μm.

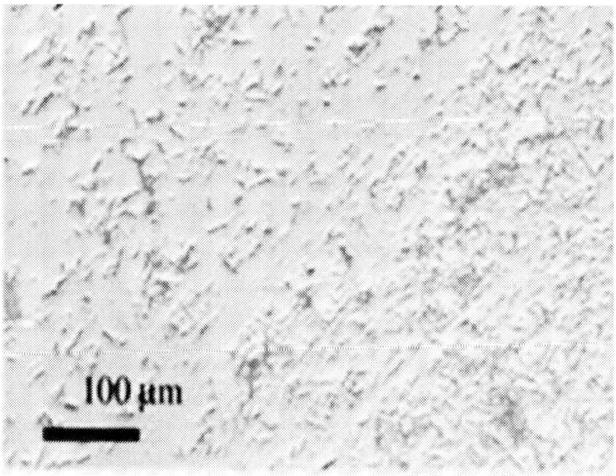

Fig. 1. Optical images of lipid nanotubes (LNTs) formed in glycolipid suspension and dispersed on glass substrate with average length of about 25 µm.

It has been demonstrated that opened LNTs are able to encapsulate different kind of materials in an efficient manner (Yang et al., 2004). It has been pointed out that in order to insert nanoparticles into LNTs at enough high concentrations the cavity has to be empted. (Yang et al., 2004). Similar approach was also applied here. After freeze dried process the SEM and TEM images showed the empty-tubular structure of LNTs. Homogenous hollow cylinders with completely opened ends were observed in majority of the LNTs (~90 %). We recall that the inner hollow of the lipid nanotube is covered with carbony headgroups while outer surface with sugar head group. Presence of carbony headgroups in principle facilitates introduction of the particles dispersed in aqueous solution into hollow cavity (Yui et al., 2005). Therefore it is expected that hydrophilic surface of the inner cavity and capillary forces are favorable to introduce Si nanoparticles dispersed in aqueous solution
Figure 2 (a) demonstrates typical SEM image freeze-dried lipid nanotubes filled with Si-nc in aqueous solution. Transmision electron microscopy images showed the presence of the nanoparticles in tubular structure. Corresponding EDS analysis (Fig 2 (b)) performed within of the nanotube confirmed an increase of the Si concentration within the LNTs cavity. It has to be noted that only small percentage (~1%) of LNTs was filled. Most of the Si-ncs were agglomerated out of the nanotubes in form of spherical particles Agglomerated Si-ncs spheres exceeded several tenths of nanometers in diameter. Fig 2 (c) shows representative SEM image of such particles. Corresponding EDS analysis taken at the particle confirmed presence of Si (Fig 2 (d)).

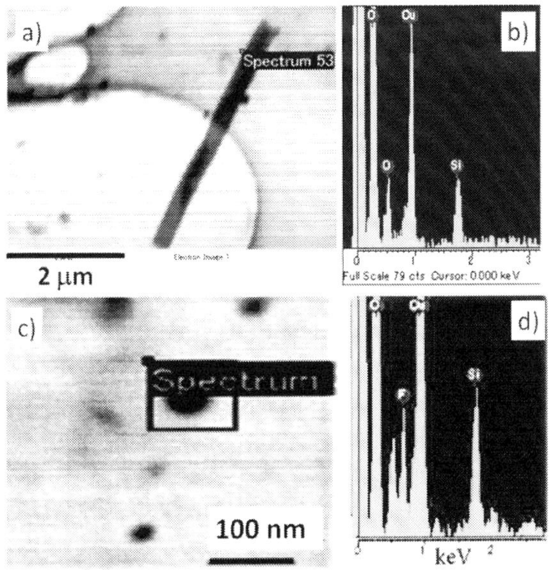

Fig. 2. (a) Freeze-dried lipid nanotubes (LNTs) filled with Si-nc dispersed in aqueous solution. The transmission electron microscopy (TEM) image shows the tubular structure a homogenous hollow cylinder with completely opened end and filled with Si nanoparticles (b) Corresponding EDS taken in the LNTs tubular structure. (c) TEM image of spherical Si particles located outside of LNTs cavity. (d) Corresponding EDS analysis of spherical particle.

The Si-ncs show typical emission at room temperature in spectral range from 500 to 800 nm with maximum around 680 nm. What about the PL properties of the Si-ncs incorporated within LNT tubular structure? Corresponding PL spectra from the solution is shown in Figure 3 (red line). Visible-red PL could be observed in the region where the Si-ncs emission is located only. However, after filling process the colloidal solution showed a decrease in PL intensity. Thus indirectly suggest that some of the Si-ncs could be encapsulated in cavity. It is expected that encapsulation in LNTs cavity, hinders the Si-ncs emission. In order to verify this hypothesis shorter LNTs were prepared with length 3 μm in average. Colloidal solution contained of Si-ncs introduced in shorter LNTs showed stronger PL emission (Fig. 3, blue line). It is probable that as the cavity is shorter also the emission from the Si-ncsLNTs solution could be stronger. Indeed, the longer nanotubes also induce stronger capillary force and Si-ncs introduced deeper within cavity show also weaker emission from Si-ncs. It has to stress that LNTs only did not show up any PL in this region (Fig. 3, open circles).

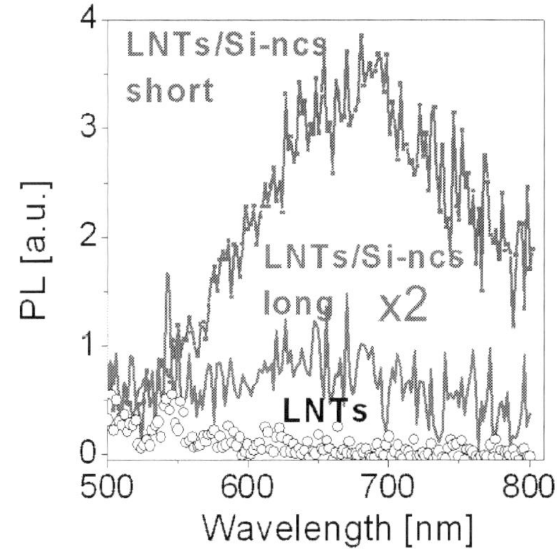

Fig. 3. Photoluminescence (PL) spectra as a function of wavelength for freeze-dried long (red line) and short (blue line) lipid nanotubes (LNTs) filled by silicon nanocrystals (Si-ncs) dispersed in water. The PL spectrum of LNTs only is shown for comparison (open circles).

However, drawback of the Si-ncs dispersed in water is small value of zeta potential, which leads to an agglomeration process (Švrcek, 2009b). As shown in figure 2 (c) agglomerated particles with sizes exceeding of about 50 nm could be find. As the inner diameter of the LNTs is this range than Si-ncs agglomerates cannot in principally enter into LNTs cavity. One possible approach how to increase the zeta potential and allow better Si-ncs dispersion is dispersion in ethanol or methanol. However, in this case the LNTs the tubular structure is destroyed. Figure 4 (a) represents SEM image of freeze-dried lipid nanotubes introduced into Si-nc/ethanol solution. As we can see complete dissociation of lipid tubular structure is experienced.

Contrary to that the CNTs are chemically stable and easily resist many common organic solutions. As results a filling by capillary forces with Si-ncs dispersed in methanol/ethanol can be achieved (Švrcek et al., 2006a). The typical transmission electron microscopy images showed the tubular structure and hollow cylinder of CNTs. Most of the CNTs (~90%) used in this work had two completely opened ends with inner diameter about 50 nm in average. Figure 4 shows a plan-view HRTEM image of an open CNT filled with a Si-nc/dispersed in methanol. Contrasted darker areas are observed in the cavity of the CNT. The location of those Si-ncs inside the CNT channel was examined by tilting the sample along the tube axis (Švrcek et al., 2006a). When the electron beam was focussed on the walls of the starting nanotubes the interplanar distance characteristic of graphite (0.337 nm) and contrasted Si-ncs were evaluated (Švrcek et al., 2006a). Parallel stripes corresponding to atomic columns with an interplanar distance 0.198 nm (inset Fig. 4(b)). This is an agreement with the expected interplanar distance of 0.192 nm in the <220> orientation of silicon. It is worth noting that the HR-TEM image of the Si-ncs was not straightforward due to the very low contrast between the Si-ncs and the grapheme shells. Those were further confirmed by selected area electron diffraction technique (Švrcek et al., 2006a, Švrcek et al., 2006d). The selected area electron diffraction pattern showed Si crystal structure close to a <111> and the JEMS simulation clearly witnesses the presence of Si-ncs with planes with <220> , <422> , <440> orientations (Švrcek et al., 2006d).

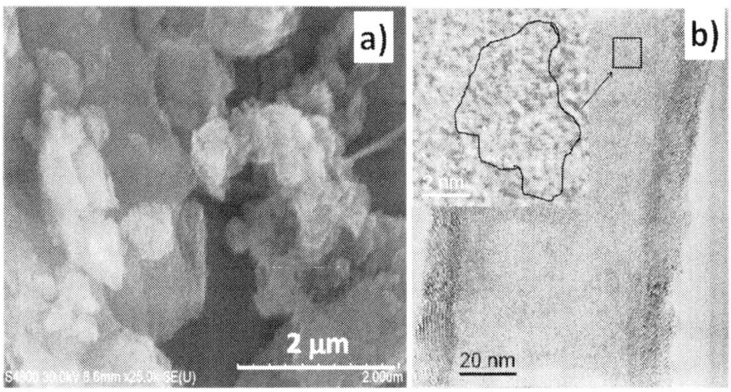

Fig. 4. (a) SEM images of freeze-dried lipid nanotubes (LNTs) introduced into Si-ncs/ethanol solution. Due to the presence of ethanol complete dissociation of LNTs is observed. (b) A plan-view HRTEM image of an open CNT filled with a Si-ncs dispersed in methanol. In set detailed HR-TEM image of the hollow part of the tube with contrasted Si-ncs inside.

One approach that we applied is first the formation of the Si-ncs directly in water by laser ablation (Svrcek et al., 2006c) and then introduction the freeze-drayed LNTs. Induced capillary then can accomplish filling LNT tubular structure. Figure 5 shows SEM images of LNTs in Si-ncs aqueous solution prepared by laser ablation. The SEM image shows the tubular structure a homogenous hollow cylinder of LNT with opened end and inner diameter about 40 nm. However due to the spherical agglomeration of the Si-ncs particles (Svrcek et al., 2009b) a simple dissolution did not allow efficient filling of LNT cavity. Most of the agglomerated Si-ncs nanoparticles with diameter that exceeds the inner diameter of the nanotube (~ 40 nm) remain located out of the cavity (Fig.5, indicated by arrows). It has to be noticed that some parts of the LNT contained darker areas containing smaller sized Si-ncs.

CNT surface can be wetted and filled if the substance having low surface tensions. Similar to the case of LNTs it is expected (M. R. Pederson and J. Q. Broughton, 1992) that liquid solution having a surface tension below < 200 mN/m allows to enter dispersed Si-ncs into the hollow cavity of the CNTs through the capillarity (Durjardin, et al., 1994; Švrcek et al., 2006a). This happened after the surface tension of Si-ncs dispersed in methanol/ethanol droped and then filling of Si-ncs dispersed in liquids into the channel of the nanotubes occurred.

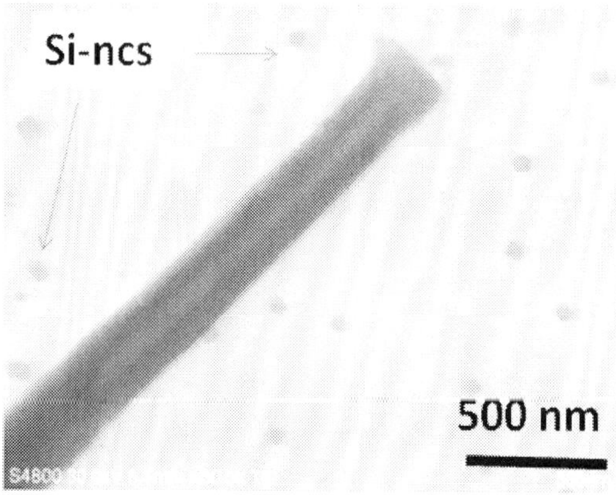

Fig. 5. SEM transmission electron microscopy image of freezed-dried lipid nanotube (LNT) dissolved with Si-nc aqueous solution prepared by laser ablation in water. The image showed the tubular structure a homogenous hollow cylinder with opened end. The presence of Si-ncs spherical clusters outside of LNTs are indicated by arrows.

4. Filling nanotubes through the laser processing in liquid media

4.1. Filling carbon nanotubes during nanosecond laser induced fragmentation of silicon micrograins

It is observed that the rates of filling of CNTs and LNTs by Si-ncs through the capillary forces are quite low. Thus is due to the many factors that act at the same time (i.e. agglomeration of Si-ncs). Recently, some alternative possibilities by nanosecond laser processing in liquid media were shown (Švrceka, 2008a). Namely when Si-ncs micrograins are dispersed in liquid media and the nanosecond laser fragmentation occur generated shock waves - which propagate through the liquid – allow filling nanotube cavity. Shock wave can have enough energy to introduce newly generated Si-ncs particles within carbon nanotubes cavities (Švrcek, 2008a). However, in our experience a weak mechanical strengthens of LNTs leads to dissociation and braking of the LNT tubular structure when nanosecond laser irradiation is applied.

This is not case of mechanically stronger CNTs. In present experiment - in order to exclude the filling CNTs by capillary forces - prior irradiation by ns laser we dispersed CNTs in water and then mixed with wetted Si-ncs micrograins. Figures (2a-b) show photos of the vessels with colloidal suspensions before laser irradiation and after irradiation at fluence 23.3 mJ/pulse. SEM and TEM analysis showed that the micrograins with sizes of several micrometers do not enter into cavity without fragmentation process. Our experiments also showed that at higher laser fluences (>20 mJ/pulse) along with the presence of CNTs, the yellow color of the suspension disappeared. As the irradiation is continuing conducted the solution becomes more transparent (Figure 6 (b)).

Contrary to CNTs absorption only the Si micrograins have a quite significant absorption cross-section at 355 nm wavelength (Švrcek, 2008a) at which the irradiation is conducted. Therefore the majority of the laser energy goes into Si-ncs grain and mostly the grain fragmentation occurs. Corresponding PL spectra of aqueous colloidal CNTs/micrograins solutions fragmented by different laser fluences are depicted in Fig. 6 (c). Similarly to the capillary induced filling, as the laser irradiation power increase, a decrease in PL emission is recorded. At increased irradiation intensity the rate of introduction of Si-ncs in the CNTs cavity is enhanced. Generated reflux increased the Si-ncs concentration within the cavity, which results the PL emission decreases. It has to be noted that the CNTs only dispersed in water did not show any PL in this spectral region at room temperature.

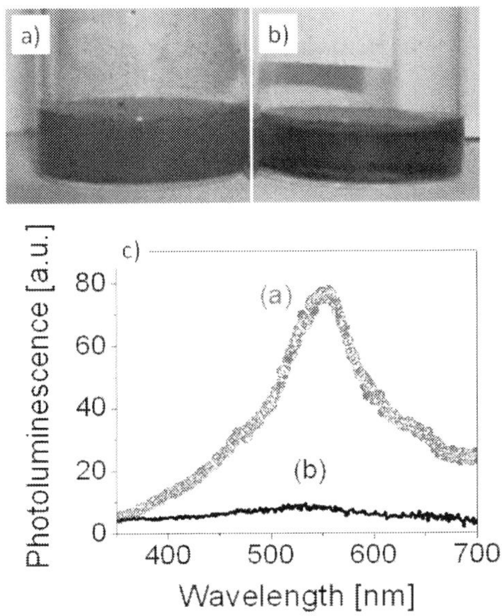

Fig. 6. (a) Photos of 0.01% silicon micrgrains in 5 ml CNT/water solution in glass vessel before and (b) after nanosecond laser induced fragmentation at laser fluence of 23.3 mJ/pulse. (c) Corresponding photoluminescence spectra of aqueous colloidal CNTs/micrograins solutions before (red symbols) and after fragmentation (black line) at laser fluence of 23.3 mJ/pulse are shown.

The question arises how the laser beam interact with the CNTs present in solution during the irradiation and induced filling process. If no ablation is performed the surface of nanotubes is flat and the cavity is mostly empty (Fig 7 a, b). However, when the irradiation is applied even at low laser fluence on the CNT surface are attached same particles or CNT surface is occasionally damaged. Figure 7 c-d show the SEM and corresponding TEM image of nanotube when the solution was irradiated at 5 mJ/pulse. As the power irradiation is increased surface of CNTs is more damaged and also larger amount of nanotubes is found to be broken. At the same time in the cavity higher concentration of silicon particles can be seen. As the fluence is increased smaller particles in diameter could be find in CNTs cavity. Figure 7 (e, f) shows CNT image when the laser fragmentation process was conducted at 23.3 mJ/pulse. Detailed HR-TEM analysis showed that the smaller sized Si-ncs with diameter less then CNTs diameter were localized within the cavity. A detailed HR-TEM analysis has confirmed that the crystalline nanoparticles with quantum confinement size effect are introduced into CNT cavities (Švrcek, 2008a). Results further showed presence of Si-ncs with an interplanar distance of 0.315 nm corresponding to a diamond lattice with different plane orientation (Švrcek, 2008a).

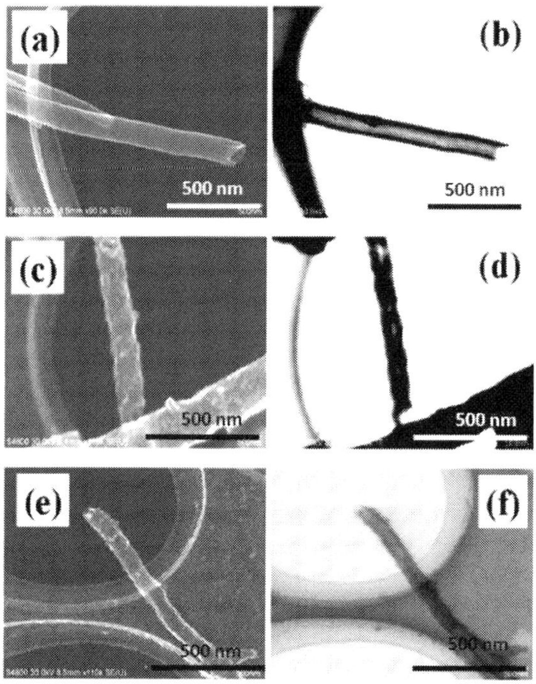

Fig. 7. SEM images of fragmented silicon grains with presence of CNTs a) just dispersed and c) and e) after irradiation at laser fluence at 5 mJ/pulse and 23.3 mJ/pulse respectively. Corresponding TEM images (b, d, f) are shown for comparison to visualize the nanotube cavity, indeed.

An introduction of the Si-nc into silicon technology compatible SOG-transparent polymer allows the stabilization and enhanced of the Si-ncs luminescence properties (Švrcek et al., 2004, Švrcek et al 2008c). Since the CNTs resist ethanol based polymer and mechanical strength also stronger nanocomposites based on Si-ncs/SOG could be envisaged. Furthermore, incorporated Si-ncs within the cavity with efficiently stabilized surface by polymer could be 1D luminescent composite. Dissolution of Si-ncs in liquid SOG modifies the surface tension of the Si-ncs which promotes the filling of CNTs. In addition, compared to water, the SOG polymer is an efficient liquid medium for micrograin fragmentation (Švrcek et al., 2006d). Because of the more efficient fragmentation in SOG solution one can apply lower irradiation intensities and avoid CNT damage. However at higher fluencies the damage of the CNTs was observed (Fig. 8). After performing of the laser processing for 2 h, it is observed that nanotubes cavities are almost fully filled with Si-ncs/polymer composite. Similar to defragmentation and filling in water, it is observed by naked eye, that the solution loses yellowish color when the fragmentation occurred (>4 mJ/pulse). SEM together with EDS analysis showed an excess and increase of silicon content. In addition to elemental analysis, the corresponding electron diffraction pattern in HR-TEM analysis the tube showed the presence of Si-ncs with crystal structure of Si cubic phase (Svrcek, 2008b). Silicon diffraction rings could be assigned to the lattice planes <111>, <220>, and <311>. Importantly corresponding PL spectrum of solidified of a polymer/micrograin/CNT composite showed shift of the PL spectra as a function applied laser fluence. This is similar to the as fragmented Si micrograins in polymer and is due to formation of smaller sized Si-ncs (Švrcek et al., 2006d). It has to be stressed that filling LNTs with Si-ncs/SOG polymer nanocomposites is impossible due to the methanol based polymer solution which dissolve the LNTs tubular structure.

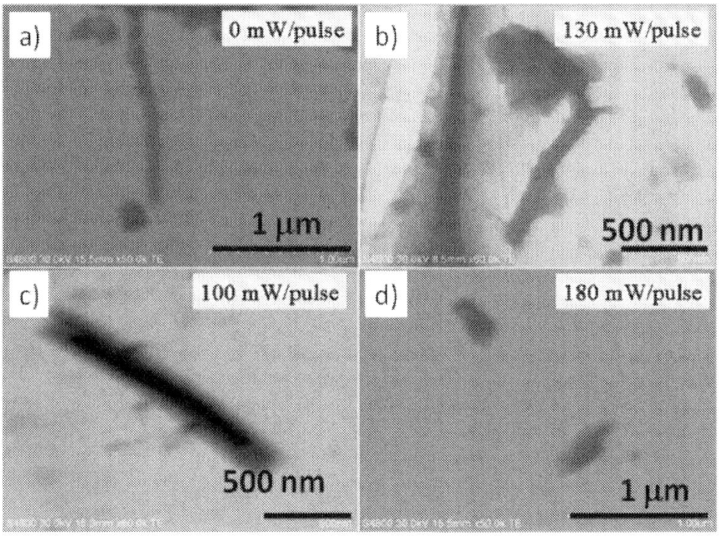

Fig. 8. (a) SEM images of Si-ncs micrograins dispersed in SOG and fragmented jointly with presence of carbon nanotubes (CNTs) at different laser fluencies ranging from 0-180 mW/pulse. With increased the laser fluence irradiation also destruction of some CNTs is observed.

4.1 Filling lipid nanotubes with freshly prepared Si-ncs by laser ablation in water

LNTs mechanical fragility and sensitivity to organic solutions do not allow applied the laser fragmentation process in organic based liquid where the filling is more efficient. Some alternative approaches that allow direct fabrication of Si-ncs in water could be applied. For instance, one approach that we explored is to fabricate Si-ncs directly by ns laser ablation when the silicon target is immersed in N-(11-cis-octadecenoyl)-β-D-glucopyranosylamine suspension. Then increase the temperature higher than 80 °C. As it is well known the glycolipid molecules in suspension around 100 °C exist as vesicle structure. These structures loose fluidity of the alkyl chains and transit to metastable helical phase via unstable lamellar structure, and finally form LNT with temperature decrease (Kamiya et al., 2005). However in our experience during the cooling down step and formation of nanotubes the Si-ncs are mostly incorporated within the lipid nanotube wall. The good dispersibility and glycolipid molecules modified Si-ncs is most likely favorable for chiral molecular packing and incorporation within the wall. However, issues whether some Si-ncs get incorporated inner wall has to be investigated in detail in near future.

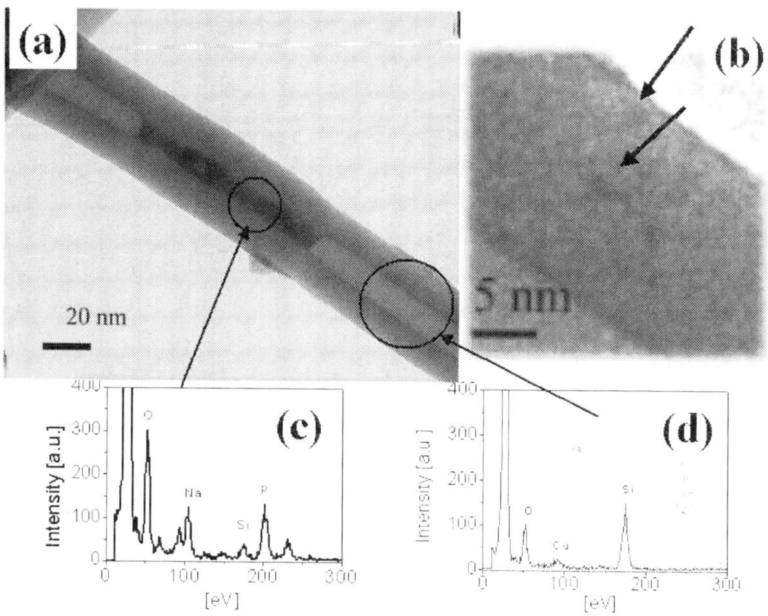

Fig. 9. (a) HR-TEM image of lipid nanotube (LNT) filled with fresh Si-ncs prepared by laser ablation of silicon target immersed in LNTs aqueous solution. (b) Corresponding detailed HR-TEM image of LNT wall. Dark spots indicated by arrows present presence of Si-ncs within the LNT wall. Images (c, d) represent corresponding EDS analysis.

Next, we prepared Si-ncs in aqueous solution by laser ablation with presence of LNTs. In order to achieve same structure of nanotubes we have removed the water from inner cavity by freeze draying step and then introduced them again in the water. The LNTs aqueous solution was putted in the glass vessel where on the bottom crystalline silicon wafer was glued. After laser ablation process some of the nanotubes were damaged however some of the nanotubes were in good shape without any significant damage. It is also observed that the presence of the LNTs in solution during ablation process inhibits Si-ncs spherical agglomeration. It is observed that same parts of the LNT contains dark areas and are possibly filled with freshly formed Si-ncs. Figure 9 (a) represents a typical HR-TEM image of

LNT after laser ablation processing with inner diameter of ~15 nm. Detailed EDS analysis showed that in this parts an increase concentration of Si is recorded (Fig. 9 (c)). However the Si-ncs were not only present in the cavity but some of them were located within the LNTs wall as well (Fig. 9 (b)). Freshly formed Si-ncs has enough energy to enter within the LNT wall when ejected from target during laser ablation process. On the other hand freshly formed Si-ncs modified with glycolipid molecules in glycolipid suspension create also reflux, which results not only introduction of Si-ncs within the wall but also within the hollow cavity.

Figure 10 presents normalized room temperature PL spectra of LNTs/Si-ncs colloidal solution (solid red symbols) and Si-ncs prepared by laser ablation in water (open blue symbols). The PL spectra of LNTs only dispersed in aqueous solution is shown for comparison and do not show PL in this part of the spectra. The PL spectra of LNTs/Si-ncs solution show clear contribution from Si-ncs. The presence of Si-ncs enhances emission band with maximum located at 420 nm. Interestingly, the PL of the LNTs/Si-ncs was visibly stronger than from Si-ncs only. This can be attributed multiple factors. Firstly, as HR-TEM images indicate low filling rate of Si-ncs low within LNTs cavity and the large number of small Si-ncs embedded in LNTs wall that facilitates the PL emission. Secondly, glycolipid molecule modified the surface of Si-ncs particles prepared in presence of LNTs. During the direct processing in LNTs solution freshly formed Si-ns surface is covered by thin layer of lipid that act as surfactant and prevent agglomeration (Mafuné et al., 2003). Then an increased PL emission also results from well separated and lipid passivated Si-ncs that prevent re-absorption and enhances PL at room temperature.

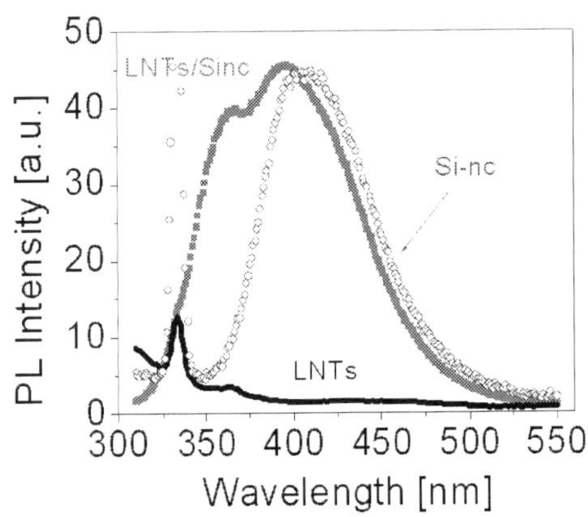

Fig. 10. Normalized room temperature photoluminecence (PL) spectra of LNTs/Si-ncs colloidal solution (full red symbols) and Si-ncs prepared by laser ablation in water only (open blue symbols). The PL spectrum of LNTs is shown for comparison (black line).

One can speculate that contrary to mechanically more stable CNTs, advantage of using laser ablation in LNTs aqueous suspension could be useful for fabrication of nanotubes composites through a simple one-step process. The modification of Si-ncs surfaces with glycolipid molecule within the lipid nanotube wall might then find important implication for luminescent glycolipid composites design. Luminescent Si-ncs particles modified with glycolipid molecules during ablation, which interact with intermediate structures of LNT such as vesicle, helical, and lamellar phases could be designed.

4.3. Filling CNTs by freshly Si-ncs prepared by laser ablation in SOG-polymer and water

Contrary to LNTs, the reflux generated during ablation process in organic based polymer or water lead to introduction of freshly produced nanocrystals in the mechanically and chemically more stable CNT cavity (Svrcek, 2008a). As mentioned above, the most important feature of the Si-ncs is the PL emission at room temperature. The silicon dioxide-based transparent polymer is proven to be an efficient liquid medium for laser processing and suitable for stabilization of Si-ncs PL emission properties (Svrcek et al., 2004). After laser processing it is observed that most of carbon nanotubes cavities are almost fully filled with luminescent Si-ncs/polymer composite (Svrcek, 2008a). Compared to the pure polymer, elemental analysis confirmed the excess of the silicon content in the CNT cavity laser processing occurred. It is observed that the ns laser processing in polymer media aside from the generation and introduction of Si-ncs in the CNT's cavity also leads at the same time to stabilization of Si-ncs PL emission at room temperature (Svrcek, 2008a).

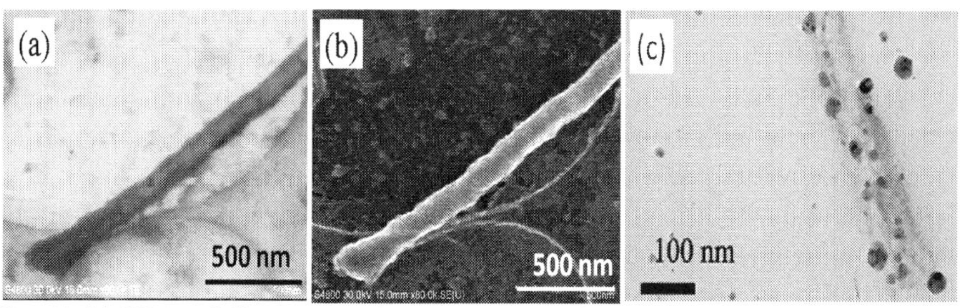

Fig. 11. (a) SEM and (b) corresponding TEM images of Si-ncs/CNT structure prepared by nanosecond laser ablation of immersed silicon wafer in a CNT/water suspension at laser fluence of 1.1 mJ/pulse. (c) Detailed HR-TEM image of CNT filled with spherically agglomerated Si-ncs particles.

On the other hand an unique wetting phenomenon for Si-ncs obtained by laser ablation of crystalline silicon target in water decrease surface tension and allow filling CNTs cavity (Svrcek, 2008a, Svrcek et al., 2009b). Figure 11 (a) and (b) shows SEM and corresponding TEM images of CNT with Si-ncs prepared in water at the laser fluence of 1.1 mJ/pulse. Contrary to the Si-ncs/LNTs the Si-ncs in CNTs/water suspension form spherically aggregated particles. As result most of them remain out of the cavity. However, some Si-ncs enter within the cavity. Detailed HR-TEM analysis was performed to confirm the presence of Si-ncs within the cavity. Figure 11 (c) shows a HR-TEM image of a filled CNT with an inner diameter of 50 nm. It is observed that spherical Si-ncs agglomerates with diameter around 25 nm are in the CNT cavity. Naturally within the cavity Si-ncs get stabilized into spherical particles, which diameter is considerably smaller as for those agglomerated out of the cavity. The Si-ncs agglomerates are found in the entire nanotube cavity far from the opened ends. Energy dispersive X-ray analysis systematically showed excess of silicon content within CNTs cavity. Corresponding electron diffraction pattern taken in the CNT cavity showed discrete spots on the circles indicate the presence of crystalline silicon in cubic phase (Svrcek, 2008a).

4.4. Induced nanotubes filling by nanosecond laser processing in liquid media

Several concurrent processes that act simultaneously are responsible for the formation of Si-ncs by nanosecond laser processing in liquid media. Particularly, superheating occurs and material ejection arises from mechanical rupture of silicon wafer or micrograins homogenously dispersed in nanotube contain liquid media (Svrcek et al., 2006d). Ejected Si-ncs particles randomly move within the liquid medium. A random walk of a particle motion in a colloid is due to the nanosecond pulsed laser irradiation (ablation) and Brownian motion. At first approximation a displacement for Brownian motion derive as follow (Watkins, 1990; Bakefi and Barret, 1987).

$$m\frac{dv}{dt} = \alpha v + F(t) \tag{1}$$

where m is the mass of the particle, v its speed, and α is the coefficient.

When the grains are fragmented, silicon target ablated or nanotube interact with pulsed laser beam the shock wave in liquid media is formed and even enhanced. As shock wave moves away a sufficiently large distance from the point of explosion it expands in the water/SOG solution and propagate mainly in the volume liquid (Svrcek et al., 2006d, Svrcek et al., 2006e), In the medium with nonlinearities as the colloids are these shock waves can develop after propagation over a characteristic distance (L) (Svrcek et al., 2006d). The L linearly depends on density and homogeneity of liquid media. Due to an increased density the ns laser fragmentation process in SOG polymer is more efficient than in water. The shock waves in colloid with higher concentration can propagate farther from the point of explosion and are stronger for filling CNTs. In case of laser processing immediately after the laser strike the target.grain a dense cloud of Si-ncs are spread within liquid medium (Svrcek et al., 2008c; Svrcek et al., 2009b]. An equation that describes well the release of energy *(E)* in explosions through Bukinghams pi theorem (Bukingham, 1915)

$$R^5 \propto \frac{Et^2}{\rho}, \tag{2}$$

where the R is the shock wave radius, t is the time and ρ is liquid medium density.

When we consider an isotropic wave propagating outward from a central point then a harmonic spherical pressure wave (i.e., having angular symmetry) is given by

$$p = \frac{A}{r}\cos(\omega t - kr + \theta) \tag{3}$$

where ω is the angular frequency, k is the wavenumber and θ is the phase. Then the radial displacement $s(r,t)$ can be given via

$$\frac{\partial^2 s}{\partial t^2} = -\frac{1}{\rho}\frac{\partial p}{\partial r} \tag{4}$$

The flux *(S)* of a spherical wave can be written by following expression

$$S = Z\left(\frac{\partial \varphi}{\partial t}\right)^2 = Z\frac{c^2\omega^2}{r^2}\cos^2(\omega t - kr + \phi) \tag{5}$$

where Z represents the wave impedance.
The propagation of a shock wave induced by a strong point explosion in a liquid media can be qualitatively evaluates as follow. The stage the wave motion at which the wave has moved away from the source of the explosion through distance comparable to scale height h is schematically sketched in Fig.12. We shall assume that we are dealing with a wave with the pressure behind the front much greater than the pressure ahead of the front (Zeldovich and Raizer, 2001). If R is the distance of the upper point of the shock wave to the explosion center, then the volume of the cavity is proportional to R^3 and the pressure can be express as follow $p \sim E/R^3$ (Svrcek, 2008). The time (t_1) required by the wave to move upward to infinity is

$$t_1 = \int_0^\infty \frac{dR}{v} \sim \frac{\rho_E^{1/2}}{E^{1/2}}\int_0^\infty E^{3/2}.\exp\left(-\frac{R}{2h}\right)dR \tag{6}$$

The physical reason for the acceleration of the shock wave is that the energy is located at the point of the explosion. The energy has the tendency to flow from the bottom and this can also accelerate the shock wave upward. This estimate appears to be for a high energy (Svrcek, 2008). As can be seen from $v \text{-} \sim (p/\rho)^{1/2}$, the wave velocity first decreases as the wave moves away from the point of explosion and has minimum at $R=3h$. The time required by the wave to move to infinity is found equal to

$$t_1 = const\left(\frac{h^5}{\rho_E E}\right)^{1/2} \tag{7}$$

The time for the wave to move from the point of the explosion to the point $R=3h$ is much smaller than t_1. Thus a shock wave produced by a strong explosion moves downward from the point of explosion through distance of about $4h$ (Svrcek, 2008). When the shock wave moves away a sufficiently large distance from the point of explosion after the shock wave emerges. The formation of these shocks during ns laser irradiation is important as they propagate through the liquid media and may cause consequently filling of CNTs nanotubes. However, in case of mechanically less strong LNTs Si-ncs most likely directly enter mainly within the soft wall where get stabilized and further cannot fill the hollow cavity.

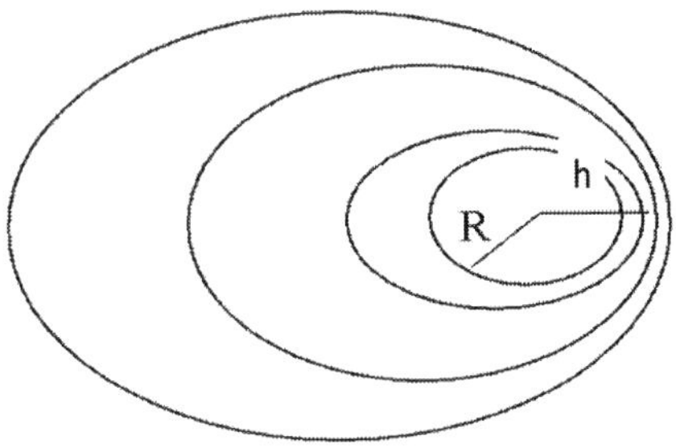

Fig. 12. Sketch of the shock front wave at successive instants.

5. Conclusions

Scalable and low cost approaches of introducing the silicon nanocrystals (Si-ncs) within carbon nanotubes (CNTs) and lipid nanotubes (LNTs) were discussed. We have shown that after opening CNTs ends and freeze-draying of the LTNs it has been possible to introduce Si-ncs inside the empted cavity of both type of nanotubes. It is demonstrated that the capillary force is strong enough to fill surfactant free Si-ncs inside nanotubes far from their opened ends. Filling of Si-ncs dispersed in ethanol and methanol polymer based solutions is possible in the case of chemically more sensitive CNTs. In the case of LNTs presence of organic solvent destroy tubular structure of the LNTs and filling cannot be achieved.

Furthermore we have demonstrated that CNTs due stronger mechanical properties compared to LNTs allow filling through the laser processing in liquid media. We argue that the nanosecond pulsed-laser processing by ablation and fragmentation in liquid medium (water, SOG) would be a reasonable solution to encapsulate Si-ncs within a CNT cavity. The generated shock waves during laser processing protect freshly formed nanocrystals and induce the entrance of freshly prepared Si-ncs into CNT opened end cavities. In the case of the soft LNTs structure the luminescent Si-ncs are mostly incorporated within the wall. It has to be noted that during the direct processing in LNTs solution freshly formed Si-ns surface is covered by thin layer of lipid that acts as surfactant, prevents agglomeration and increases PL emission from Si-ncs.

The results might have important implications to further extend the usability of Si-ncs as zero dimensional systems in colloidal solution. It is expect that those relatively simple and cheap approaches will be useful in a wide variety of applications. In case of LNTs, the direct modification of freshly formed Si-ncs surfaces with glycolipid molecule within the wall might find an important implication for luminescent glycolipid composites design. Next, direct preparation of luminescent Si-ncs in spin on glass polymer and consequent encapsulation in CNTs cavity might serves as template to align 1D Si-ncs/SOG nanocomposites. Stabilized Si-ncs in 1D alignment could be than beneficial to the stimulated emission processes, i.e. optical gain. We believe that such introduced luminescent Si-ncs nanoparticles within nanotubes will offer an attractive route for the design of a new generation of single nanoparticle-based nanodevices for biology and optoelectronics.

Acknowledgement

This work was partially supported by the research Fellowships of the Japan Society for the Promotion of Science. This study was also partially supported by Industrial Technology Research Grant Program from the New Energy and Industrial Technology Development Organization (NEDO) of Japan.

6. References

Ajayan P M, Iijima S. (1993), Capillarity-induced filling of carbon nanotubes. *Nature*, 361, pp. 333-334.

Bakefi, G. Barret, A. H., (1987), Electro*magnetic Vibrations, Waves, and Radiation (MIT* Press, Cambridge, MA,).

Bukingham, E. (1915). The principle of similitude. *Nature* 96, pp.396–397.

Canham, L. T. (1990), Silicon Quantum Wire Fabrication by Electrochemical and Chemical Dissolution of Wafer, *Appl. Phys. Lett.* 57, pp.1046-1048.

Cao, M.H. Hu, C.W. Wang, Y.H. Guo, Y. Guo, C. Wang, G, (2003), A controllable synthetic route to Cu, Cu_2O, and CuO nanotubes and nanorods, *Chem. Commun.*15, pp. 1884-1885.

Durjardin, E. Ebbessen, T. W. Hiura, H. Taginaki, K. (1994), Capillarity and Wetting of Carbon Nanotubes, *Science*, 265, pp.1850-1852.

Hirsmman, K. D. Tsybeskov, L. Duttagupta, S. P. Fauchet, P. M. (1996), Silicon-based visible light-emitting devices integrated into microelectronic circuits, *Nature* 384, pp. 338-341.

Hu, J. Odom, T.W. Lieber, C.M. (1999), Chemistry and Physics in One Dimension: Synthesis and Properties of Nanowires and Nanotubes, *Acc. Chem. Res.* 32, pp.435-445.

Iijima, S. (1991). Helical microtubules of graphitic carbon, *Nature*, 354, pp.56-58.

Kamiya, S.; Minamikawa, H.; Jung, H. J.; Yang, B.; Masuda, M.; Shimizu, T. (2005), Molecular Structure of Glucopyranosylamide Lipid and Nanotube Morphology, *Langmuir*, 21, pp.743-750.

Kanemitsu, Y.(1995), Light emission from porous silicon and related materials, *Physics Reports*, 263, pp.1-91.

Kobayashi, S. Hamasaki, N. Suzuki, M. Kimura, M. Shirai, H. Hanabusa, K. (2002), Preparation of Helical Transition-Metal Oxide Tubes Using Organogelators as Structure-Directing Agents, *J. Am. Chem. Soc.*, 124, pp. 6550-6551.

Lago R M, Tsang S C, Lu K L, et al. (1995), Filling carbon nanotubes with small palladium metal crystallites: the effect of surface acid groups. *Chem Commun*, 13, pp.1355-1356.

Mafuné, F. Kohno, J. Takeda, Y. Kondow, T. Sawabe, H. (2000), Structured and Stabilization of Silver Nanoparticles in Aqueous Solution Produced by Laser Ablation *J. Phys. Chem. B*, 104, pp.8333-8337.

Mafuné, F. Kohno, J. Takeda, Y. Kondow, T. (2003), Formation of Stable Platinum Nanoparticles by Laser Ablation in Water, *J. Phys. Chem. B*, 107. pp. 4218-4223.

Mo, M.S. Zeng, J.H. Liu, X.M. Yu, W. Zhang, S. Qian, Y. (2002), Controlled Hydrothermal Synthesis of Thin Single-Crystal Tellurium Nanobelts and Nanotubes, *Adv. Mater.* 14, pp.1658-1662.

Nath, M. Govindaraj, A. Rao, C.N.R. (2001). Simple Synthesis of MoS_2 and WS_2 Nanotubes, *Adv. Mater.* 13, pp. 283-286.

Pavesi, L. Dal Negro, L. Mazzoleni, C. et al. (2000), Optical gain in silicon nanocrystals, *Nature*, 408, pp.440-444.

Pederson, M. R. Broughton, J. Q. (1992), Nanocapillarity in fullerene tubules, *Phys. Rev. Lett.* 69, pp. 2689-2692.

Pham-Huu C, Keller N, Estournes C, et al. (2002) Synthesis of $CoFe_2O_4$ nanowire in carbon nanotubes. A new use of the confinement effect, *Chem Commun*, 17 pp.1882-1883.

Ratna, B. R. Baral-Tosh, S. Kahn, B. Schnur, J. M. Rudolph, A. S. (1992), Effect of alcohol chain length on tubule formation in 1,2-bis(10,12-tricosadiynoyl)-sn-glycero-3-phosphocholine, Chem. Phys. Lipids 63, pp. 47-53.

Shimizu, T (2002), Bottom-Up Synthesis and Structural Properties of Self-Assembled High-Axial-Ratio Nanostructures, *Macromol. Rapid Commun.* 23, pp.311-331.

Schnur, M. J. (1993) Lipid Tubules: A Paradigm for Molecularly Engineered Structures: *Science*, 262, pp.1669-1676.

Spector, M. S.;Easwaran, K. R. K.;Jyothi, G.;Selinger, J. V.;Singh, A.;Schnur, J. M. (1996), Chiral molecular self-assembly of phospholipid tubules: A circular dichroism study, *Proc Natl Acad. Sci. USA* 93, pp.12943-12946.

Švrcek, V. Slaoui, A. Muller, J.-C. (2004), Ex-situ prepared Si nanocrystals: Their elaboration and characterization in embedded silica glass, *J. Appl. Phys.* 95, pp.3158-3164.

Švrcek, V. Normand, F. L. Pham-Huu, C. Ersen, O. Joulie, S. Ledoux, M.-J. (2006a), Filling of single silicon nanocrystals within multi-walled carbon nanotubes *Appl. Phys. Lett.*, 88, pp-033112(1-3).

Švrcek, V. Ersen, O. Dintzer, T. Normand, F. L. Pham-Huu, C. Ledoux, M.-J. (2006b), Connection of silicon nanocrystals (Si-nc) with multi-walled carbon nanotubes, *Appl. Phys A*, 83, pp.153-158.

Švrcek, V. Sasaki, T. Shimizu, Y. Koshizaki, N. (2006c), Blue luminescent silicon nanocrystals prepared by ns pulsed laser ablation in water, *Appl. Phys. Lett.*, 89, pp.213113(1-3).

Švrcek, V. Sasaki, T. Shimizu, Y. Koshizaki, N. (2006d), Silicon nanocrystals formed by pulsed laser-induced fragmentation of electrochemically etched Si micrograins, *Chem. Phys. Lett.* 429, pp.483-487.

Švrcek, V. Normand, F. L. Ersen, O. Joulie, S. Pham-Huu, C. Amadou, J. Begin, D. Ledoux, M.-J. (2006e), Filling and capping multiwall carbon nanotubes with silicon nanocrystals dispersed in SiO_2-based spin on glass *J. Appl. Phys.*, 99, pp. 64306-64311.

Švrcek, V. (2008a), Fabrication of filled carbon nanotubes with fresh silicon nanocrystals produced in-situ by nanosecond pulsed laser processing in environmentally friendly solutions, *J. Phys. Chem. C* , 112, pp.13181-13185.

Švrček, V. (2008b), Nanocrystaline silicon and carbon nanotube nanocomposites prepared by pulsed laser fragmentation, *Pure and Applied Chemistry*, 80, pp. 2513-2520.

Švrcek, V. Sasaki, T. Shimizu, Y. Koshizaki, N. (2008c), Blue luminescent silicon nanocrystals prepared by ns laser ablation and stabilized in electronically compatible spin on glasses, *J. Appl. Phys.*, 103, pp.023101-023108.

Švrček, V. Mariotti, D. Kondo, M. (2009a), Ambient-stable blue luminescent silicon nanocrystals prepared by nanosecond-pulsed laser ablation in water, *Optics Express*, 17 pp. 520-527.

Švrcek, V. Sasaki, T. Katoh, R. Shimizu, T. Koshizaki, N. (2009b), Aging effect on blue luminescent silicon nanocrystals prepared by pulsed laser ablation of silicon wafer in de-ionized water, *Appl. Phys. B* **94**, pp.133-140.

Tessonnier J P, Pesant L, Ehret G, et al. (2005), Pd nanoparticles introduced inside multi-walled carbon nanotubes for selective hydrogenationof cinnamaldehyde int hydrocinnamaldehyde. *Appl Catal*, A, 288 (1-2) pp. 203-210.

Thamavaranukup N, Höppe H A, Ruiz-Gonzalez L, et al. (2004), Single-walled carbon nanotubes filled with MOH (M=K,Cs) and then washed and refilled with clusters and molecules. *Chem Commun*, 15, pp.1686-1687.

Tsang, C. S. Chen, K. Y. Harris J. F. P. Green, L. H. M. (1994) A simple chemical method of opening and filling carbon annotates, *Nature*, 372, pp. 159-162,

Ugarte, D, Chatelain A, de Heer W A. (1996) Nanocapillarity and chemistry in carbon nanotubes. *Science*, 274 pp.1897-1899.

Usui, H. Shimizu, Y. Sasaki, T. Koshizaki, N. *J.* (2005), Photoluminescence of ZnO Nanoparticles Prepared by Laser Ablation in Different Surfactant Solutions, *Phys. Chem. B*, 109, pp. 120-124.

Watkins, P. (1990), Physical *Chemistry (Oxfor*d University Press, Oxford,),

Wei, Z.X. Wan, M.X. (2003), Synthesis and characterization of self-doped poly(aniline-co-aminonaphthalene sulfonic acid) nanotubes, *J. Appl. Polym. Sci.* 87, pp.1297- 1301.

Wolkin, M. V. Jone, J. Fauchet, P.M. Allan, G. Delerue, C. (1999) Electronic States and Luminescence in Porous Silicon Quantum Dots: The Role of Oxygen, *Phys. Rev. Lett.* 82, pp.197-200.

Xia, Y. Yang, P. Sun, Y. Wu, Y. Mayers, B. Gates, B. Yin, Y. Kim, F. Yan, H. (2003), One-Dimensional Nanostructures: Synthesis, Characterization, and Applications, *Adv. Mater.* 15, pp. 353- 389.

Yang, B. Kamiya, S. Shimizu, Y. Koshizaki, N. Shimizu, T. (2004), Glycolipid Nanotube Hollow Cylinders as Substrates: Fabrication of One-Dimensional Metallic–Organic Nanocomposites and Metal Nanowires *Chem. Mater.* 16, pp.2826- 2831.

Yui, H. Shimizu, Y. Kamiya, S. Yamashita, I. Masuda, M. Ito, K. Shimizu, T. (2005), Encapsulation of Ferritin within a Hollow Cylinder of Glycolipid Nanotubes, *Chem. Lett.* 34, pp. 232-236.

Zeldovich, Y. B. Raizer, Y. P. (2001), Physics of *Shock Waves and High-Temperature Hydrodynamic Phenomena (Dover Publications*, Inc., New York, 2001).

Zhang, Q.H. Gao, L.A. Sun, J. Zheng. S, (2002), Preparation of Long TiO_2 Nanotubes from Ultrafine Rutile Nanocrystals, *Chem. Lett.* 2 pp. 226-232.

26

Microstructured Optical Fibers filled with Carbon Nanotubes: Photonic Bandgap Modification and Sensing Applications

Marco Pisco[1], Marco Consales[1], Antonello Cutolo[1], Patrizia Aversa[2], Michele Penza[2], Michele Giordano[3] and Andrea Cusano[1]
[1]*University of Sannio, Optoelectronic Division, Engineering Department, Benevento, Italy*
[2]*ENEA, C.R. Brindisi, Materials and New Technologies Unit, Brindisi, Italy*
[3]*National Research Council, Institute for Composite and Biomedical Materials, Napoli, Italy*

1. Introduction

In the recent years, a new concept is emerging in the scientific community dealing with the possibility to use optical fibers as platform to develop all-in-fiber multimaterial and multifunctional optical devices and systems (Abouraddy et al., 2007). The key feature of these new optoelectronic devices relies on the proper integration of specific materials, such as conductors, semiconductor and insulator, into the same optical fiber in order to attain advanced functionalities within a single optical fiber. A promising building block to realize these multifunctional optoelectronic devices are the Microstructured Optical Fibers (MOFs) which, being composed by a periodic distribution of micrometric air-holes running uniformly along the fiber length (Knight et al., 2003), offer an high degree of freedom in their fabrication and at the same time several opportunities of integration with specific materials. Also by manipulating the properties of the in-fiber integrated materials, the guiding features of the MOF itself can be properly changed in order to develop new tunable photonic devices (Domachuk et al. 2004; Larsen et al. 2003; Huang et al. 2004) as well as optical fiber sensors (Benabid et al. 2005; Matejec et al. 2006).
Single Walled Carbon Nanotubes (SWCNTs) constitute a very promising material for multimaterial and multifunctional photonic devices in light of their unique electrical and mechanical properties (Dresselhaus et al. 2001). Furthermore, the opto-chemical sensing properties of carbon nanotubes, deposited onto singlemode standard optical fiber (SOF) configured in buffered and not buffered reflectometric configurations, have been demonstrated to be suitable (Penza et al. 2004; Penza et al. 2005 ; Consales et al. 2006; Consales et al. 2007) to perform chemical detection of volatile organic compounds (VOCs) at room temperature.

This chapter reviews the research activities devoted to the integration of MOFs with Single Walled Carbon Nanotubes (SWCNTs) for the realization of new all-in-fiber active and passive photonic devices (Cusano et al. 2006; Pisco 2007; Pisco et al. 2008; Pisco et al. 2009).
The chapter starts with the description of the procedure to fill the MOFs with SWCNTs by Langmuir-Blodgett (LB) technique. The attention is then focused on the individuation of the main trends and correlations between the deposition process parameters and the MOF guiding properties modifications due to the SWCNTs integration. To this aim, both morphological characterizations and far field transmission measurements were performed. The activities about the integration between MOF and SWCNTs have been hence finalized and exploited for VOCs detection applications. In this framework, the sensing probes, fabricated with different process features, have been characterized by reflectance spectra measurements in order to get more information on the nanotubes filling within MOF holes. The realized sensors, employed in single wavelength reflectometric configuration, have been then exposed in a test chamber to several VOCs pulses at room temperature in order to demonstrate their capability to work for sensing applications.
The experimental results here reported demonstrate the success of the SWCNTs infiltration within the HOF holes and the capability to modify the guiding properties of HOFs by acting on deposition process features. Furthermore, the experimental measurements performed in a test chamber reveal that MOFs-SWCNTs sensing probes can be exploited to realize novel opto-chemical sensors for VOCs detection with a good sensitivity and fast response times.

2. Methodology

Several samples have been arranged by injecting SWCNTs within the MOF holes and the MOF Photonic Band-Gap (PBG) features dependence on the deposition parameters has been investigated by far field transmission measurements. The samples were composed by a piece of MOF, spliced at one end with a SOF and covered and partially filled with SWCNTs at the other termination as schematically shown in the figure 1. Before the MOF-SOF splicing, the deposition of SWCNTs was performed at atmospheric pressure by means of the LB technique described in the following.

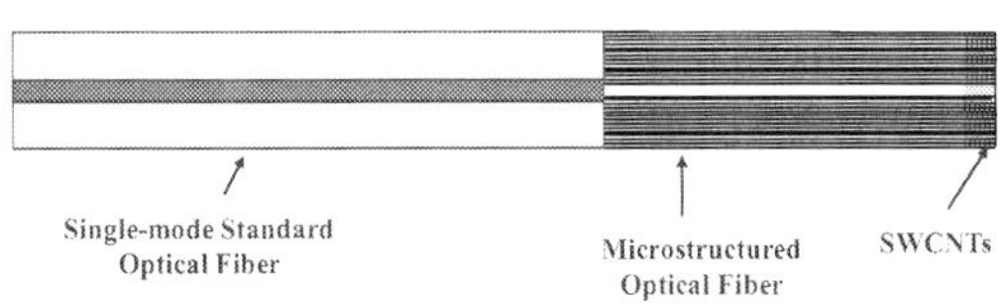

Fig. 1. Schematic view of the MOF-SWCNTs samples

2.1 The Langmuir-Blodgett Technique

The LB technique allows to deposit defect-free and molecularly ordered ultra-thin organic films of SWCNTs pristine material onto a substrate (Penza et al. 2005). This deposition technique has been yet widely used to transfer SWCNTs onto planar substrates as well as onto SOF ends (Penza et al. 2004; Penza et al. 2005) and it is here employed to integrate carbon nanotubes within the holes of MOFs. With this technique, the molecules of the films to be deposited are firstly dispersed onto the surface of a sub-phase, typically oriented with the hydrophobic part upwards and with the hydrophilic one immersed in water. Subsequently, a reduction of the surface area occupied by each molecule is performed by means of moving barriers in order to produce a solid phase of a given surface pressure in which the molecules are densely packed forming a highly ordered array. From this phase the molecules can be transferred to a properly cleaned and prepared solid substrate by its dipping through the condensed Langmuir layer.

Since the solid phase is reached only at a proper surface pressures, a continuous reduction of the moving barriers is performed when the molecules are transferred from the sub-phase to the substrate in order to keep the surface pressure constant, ensuring that the solid phase is maintained. Repeated dipping of the same substrates results in an accurate nano-sized monolayer by monolayer deposition.

In particular, for the integration within MOFs, a solution (0.2 mg/ml) of SWCNTs pristine material (purchased from Carbon Nanotechnologies Inc., Houston, USA) in chloroform has been spread onto a sub-phase constituted by deionised water (18 MΩ) with 10-4 M of $CdCl_2$. The sub-phase pH and the temperature were 6.0 and 23°C, respectively. The monolayer has been compressed with a barrier rate of 15 mm/min until the surface pressure of 45 mN/m is reached. The single layer has been deposited with a dipping rate of 3 mm/min and the transfer ratio of the monolayer from the sub-phase to the substrate surface was in the range 0.5 to 0.7.

Although the LB deposition method is a well-assessed technique, the employment of the LB technique to integrate carbon nanotubes monolayers within the MOF holes yields substantial changes in the deposition process.

Sample	Length (cm)	Monolayer
1	13.3	10
2	11.0	16
3	4.0	20
4	10.0	20
5*	6.5	20

Table 1. Samples main features (* SWCNTs deposited after the SOF-MOF splicing)

As matter of fact, the microstructuration of the substrate, constituted by the MOF, affects the deposition process in light of the micrometric dimensions of the MOF holes enabling capillarity phenomena (Nielsen et al. 2005; Zhmud et al. 2000). During the vertical MOF dipping, performed with the rate of 3 mm/min and for a depth of 1mm, it is expected that the LB suspension penetrates within the MOF holes for effect of the dipping itself and at the same time rises within the MOF holes for capillarity. It is worth to highlight that the penetration depth of the LB suspension within the MOF holes does not correspond to the penetration depth of the SWCNTs within the MOF holes. The SWCNTs deposition occurs with an high efficiency at a given suspension surface pressure while the dipping of the MOF substrate, breaking the superficial regularity of the suspension, locally changes the parameters ruling the effectiveness of the carbon nanotubes deposition on the microstructured substrate. Furthermore, the extension of the region onto the holes sides where the nanotubes adhered is limited from the quantity of carbon nanotubes "available" at the dipping time. Differently from the classic LB deposition on a planar substrate accomplished with a constant surface pressure, here, even if the overall surface pressure is held constant by the barrier movement, locally at each MOF hole, only the nanotubes sub-phase corresponding to the MOF hole affects the deposition during the dipping. In summary it is expected that the transfer of SWCNTs thin monolayers within the MOF holes by means of the LB deposition method would be the result of the dipping movement and capillarity's phenomena rising at the dipping time strongly dependent on the surface pressure. It is noteworthy that while the capillarity phenomena impose differences in the carbon nanotubes penetration depth between the core and cladding holes due to the different diameters, the dipping movement into the suspension reduces the differences, forcing the suspension penetration at the dipping time. Nevertheless, while the adhesion and penetration mechanisms are currently under investigation, the experimental results revealing the capability to deposit SWCNTs within the MOF holes and to consequently modify the MOF guiding properties are here presented.

2.2 Splicing procedure between MOF and SOF

The open end of the MOF piece has been spliced at the input end to a SOF by using an electrical arc splicing system (Fujikura FSM-50S). The fibers ends, both cleaved, have been aligned and pressed against each other using the splicer precision motors. In order to avoid the collapse of air holes during the splicing, a splicing procedure has been developed ad hoc. In particular, a series of ten arcs with short duration (100ms) and high power (80bit power) have been forced, in place of a single arc with lower power (40bit) and longer extension (800ms) usually used for SOF-SOF splicing. The splicing process was previously tested and the losses were estimated to be about 1dB by means of the comparison between the measurements of the transmitted power in butt coupling configuration and after the optical fibers fusion respectively. In addition, the bonding achieved between the two fibers, weak with respect to the bending, has been enforced by using an heating protection sleeve.

3. Experimental

The SWCNTs monolayers deposition has been accomplished on several pieces of MOF with different length on the order of few centimeters and it was performed at atmospheric pressure with the other MOF termination not connected. After the deposition, for each sample, the free termination has been then spliced at a single-mode SOF. Several samples have been realized with the LB technique by changing the monolayer number and thus the amount of carbon nanotubes able to fill the MOF holes. In the following, the analysis is focused on five samples fabricated according to the table 1. Also the influence of the external pressure at the dipping time has been investigated by depositing 20 monolayers of SWCNTs onto an MOF piece previously spliced to a SOF (sample 5).

3.1 Morphological characterization

In order to investigate the morphological characteristic of the fabricated MOF samples, a Scanning Electron Microscope (SEM) has been used. The analysis with the SEM has been performed on several samples. The retrieved images of the observed samples show a quite uniform overlay covering the fibers holes. In fig. 2 (a-b) an Atomic Force Microscope (AFM) image of a MOF before the deposition and the SEM image of the same MOF after the deposition of 10 monolayer of SWCNTs have been reported demonstrating the success of the SWCNTs deposition onto the MOFs. In figure 2 (c-d), instead, the SEM image of a bare MOF and the SEM image of the MOF after the deposition of 16 monolayer of SWCNTs (sample 2) have been reported. The SEM images, showing a particular of the MOF cladding, reveal a web-like overlay covering the MOF holes demonstrating the success of the SWCNTs deposition onto the MOFs.

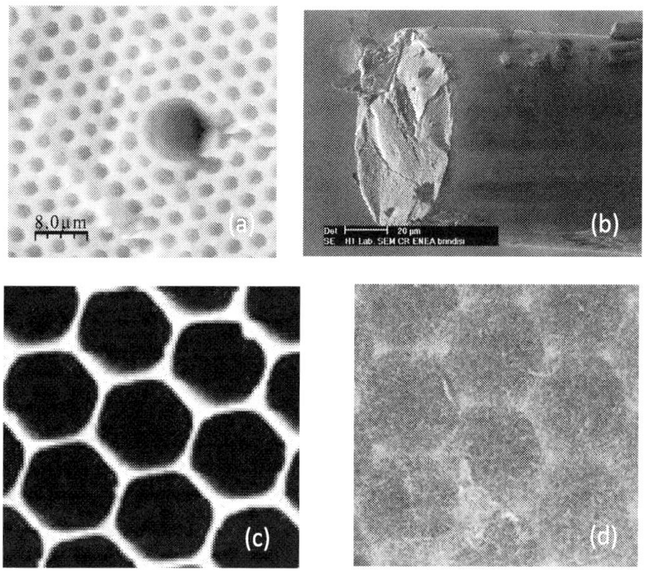

Fig. 2. AFM image of a bare MOF (a), SEM image of the MOF after the deposition of 10 monolayer of SWCNTs (b), SEM image of a bare MOF (c) and SEM image of the MOF after the deposition of 16 monolayer of SWCNTs (d). (Fig. 2 (c-d): Reprinted with permission from ref. (Pisco et al. 2009))

3.2 Far field Characterization and Bandgap Modification

In order to characterize the MOF-SWCNTs guiding properties modifications and the SWCNTs filling capability, the optical far field emerging from the samples has been collected by means of an infrared vidicon camera (Hamamatsu C2741-03) while a narrowband laser source at 1550nm lights the samples. The camera has been screen-shielded from the visible light and a proper holder for the fiber has been provided in order to guarantee the repeatability in the positioning of different samples. In figure 3 is schematically reported the experimental setup employed for the far field characterization.

The end face of the samples has been positioned in the nearby of the receiving lens. In figure 4 (a) the transmission field, revealed in far field mode, of an MOF without nanotubes is reported as reference. The far field emerging from an MOF is characterized, as known (Knight et al. 2003), by a Gaussian shape due to the MOF fundamental mode except for the presence of little peaks of circular shape corresponding to the cladding holes close to the core.
In the figure 4 (b) and 4 (c) the emerging far fields from the samples 1 and 2 realized with 10 and 16 monolayers are reported, respectively. As observable, the presence of the SWCNTs is not able to strongly modify the field distribution of the propagating mode which basically keeps the Gaussian shape. Nevertheless, while for the sample 1, a slightly higher light content can be observed in correspondence of the MOF cladding in the coated case, the same spreading of the far field is more evident for the sample 2 which exhibits a stronger increase of the field amplitude external to the corresponding fiber core.

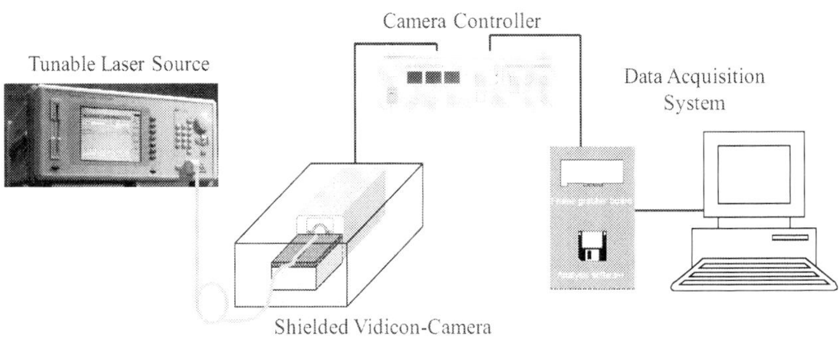

Fig. 3. Experimental setup for the Far Field characterization

The far field spreading increases with the monolayer number and thus accordingly to the higher SWCNTs content. As evident, the main effect of SWCNTs filling is a worse confinement of the fundamental mode, consistent with a refractive index contrast reduction induced by a partial filling of the core and cladding holes. It is noteworthy that the core refractive index increase (with no cladding holes filled) is expected to lead to an increase of the core confinement power, while the cladding holes refractive index increase (with core unperturbed) as well as the core and cladding holes refractive indexes simultaneous increase would lead to a diminution of the fundamental mode confinement power. Hence the field spreading observed between the samples with 10 and 16 monolayer and the bare MOF is in agreement with an increase of the cladding holes refractive index or of the core and cladding holes refractive indices. Nevertheless the exclusive cladding holes filling is not expected on the basis of the deposition technique used except for differences in the carbon nanotubes penetration between the core and cladding holes due to the different capillary diameters. With regard to the sample 3, a strong modification of the field distribution occurs and the amount of power transmitted is strongly reduced as reported in figure 4 (d). The emerging field presents a circular crown shape and the core mode is not more visible. In other words, the larger amount of SWCNTs used for sample 3 is able to induce a strong diminution of the fundamental mode power, attributable to the modification of the PBG occurring in consequence of the MOF holes filling.

Hence, the far field emerging from the sample 3, in agreement with the field exhibited from the samples 1 and 2, demonstrates the capability of the LB technique to infiltrate SWCNTs within the MOF holes. In particular, the deposition of 20 monolayers (or more) yields the functionalized MOF not more able to guide meaningfully the light. As matter of fact, the PBG modification affects particularly the field distribution of the sample 3 which was fabricated with an higher number of monolayers with respect to the samples 1 and 2.

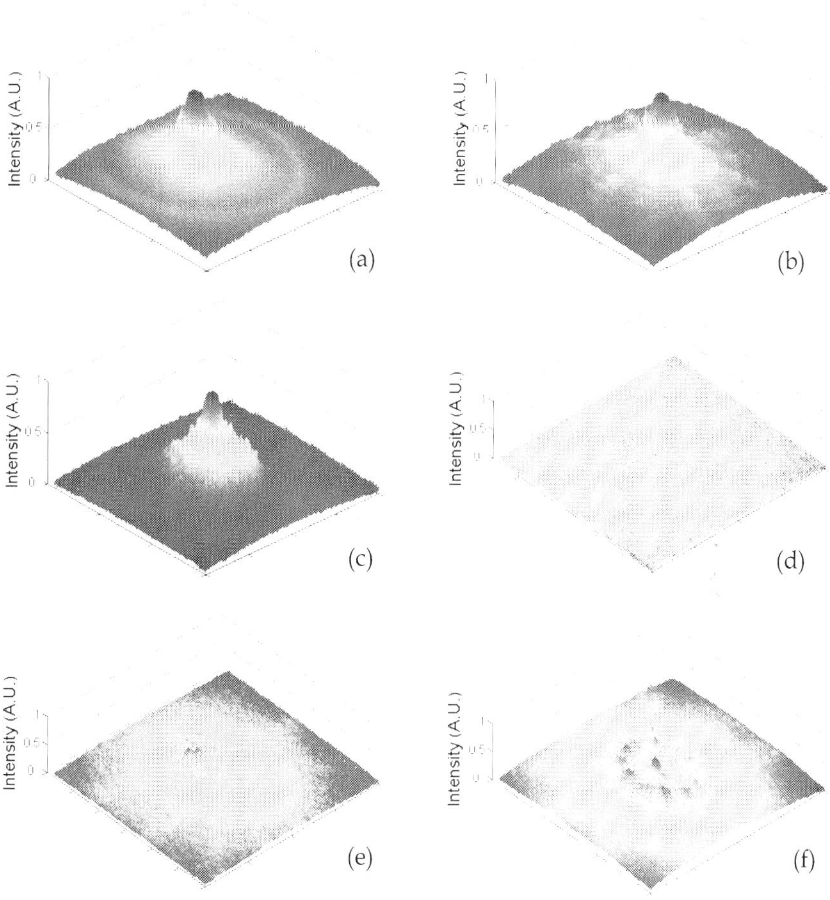

Fig. 4. Distribution of the far field of the bare MOF (a) and of the MOF samples 1-5 (b-f) respectively. (Reprinted with permission from ref. (Pisco et al. 2009))

In order to study the influence of the MOF length, the far field emerging from the sample 4 has been characterized and reported in figure 4 (e). Both samples 3 and 4 have been obtained by depositing 20 SWCNTs monolayers, but the MOF pieces are long 4 and 10cm respectively. It is observable that the far field keeps the circular crown shape and the fundamental core mode has been suppressed again. The field emerging from the sample 4, however, is slightly circularly asymmetric and the light content is slightly higher. The comparison thus reveals that the MOF length doesn't rule meaningfully the field distribution in transmission. In addition, it confirms the important role of monolayers number in the PBG modification and thus on the resulting field distribution. It is noteworthy that this result is consistent with the deposition process previously described. In fact, assuming the deposition as the resultant of dipping and capillarity mechanisms it was not expected any dependence on the fiber length. Congruently the SWCNTs deposition is strongly affected by the number of dipping and not by the capillaries lengths.

Finally, figure 4 (f) reports the far field emerging from an additional sample on which 20 monolayer of SWCNTs have been deposited after the splicing procedure. The connection at the fiber termination is responsible for a change in the air pressure within the MOF holes and thus a less efficient filling is expected. As observable, the far field emerging from the sample presents still a circular crown shape and a little peak in correspondence of the MOF core is visible. The deposition of 20 monolayer is able to strongly modify the PBG but the core mode is not completely suppressed. In this case, during the deposition stage, the LB suspension, in order to penetrate within the MOF capillaries, had to overcome the pressure offered by the air present within the MOF. The elasticity of the air thus determines a pressure variable during the MOF dipping and it represents a good exemplification of the effect of the pressure on the deposition which can be otherwise properly controlled.

In summary, the far field characterizations reveal the success of the SWCNTs deposition demonstrating the partial filling of the nanotubes within MOFs. Also, the obtained results demonstrate that the main effect of SWCNTs filling is the significant modification of the guiding properties and thus of the local PBG at the termination of the MOF. Also, by using the LB deposition method, we demonstrate how by acting on the process parameters and especially on monolayers number and pressure conditioning at the free termination, it would be possible to optimize and tailor the SWCNTs filling enabling the local control of the MOF PBG. The proper modification of the MOF PBG features, by filling (also selectively) the MOF holes with SWCNTs, can supply, in fact, to the final device advanced functionalities and can be properly exploited to develop high performances sensors based on PBG modification as well as multimaterial and multifunctional fibers.

4. Sensing Application

In this section, we report on the exploitation of the experimental results previously described about the integration between MOFs and SWCNTs in order to develop an all fiber opto-chemical sensor useful for VOCs detection.

In the recent years, the sensing properties of carbon nanotubes deposited onto singlemode SOF configured in reflectometric configurations have been widely investigated (Penza et al. 2004; Penza et al. 2005) demonstrating their capability to perform chemical detection of VOCs at room temperature.

On the basis of the previously provided demonstration about the change of the guiding properties of the MOFs due to the presence of the sensitive materials within the MOF holes, the exploitation of the integration of MOFs and SWCNTs has been successfully employed

for the development of new in-fiber opto-chemical sensors (Cusano et al. 2006; Pisco et al. 2008) and the retrieved experimental results are here summarized.
In particular, an extensive analysis of the samples 1 and 3, employed for sensing application, is reported. It is worth remembering that the former one, labelled sensor 1, is constituted by a 13.3 cm long MOF with 10 monolayer of SWCNTs, the latter, labelled sensor 3, by a 4cm long MOF and 20 SWCNTs monolayer. Furthermore, the sensors have been characterized by reflectance spectra in order to get more information on the nanotubes filling within MOF holes. Finally the aforementioned sensors, employed in single wavelength reflectometric configuration, have been exposed in a test chamber to several toluene pulses at room temperature in order to demonstrate their capability to work for sensing applications. The experimental results obtained in this framework are presented in the following and the impact of the fabrication stage on the sensing performances is also discussed.

4.1 Reflectance Characterization

Reflectance characterization of the realized probes has been carried out within the MOF bandwidth. To the aim, by means of the experimental setup sketched in figure 5, a tunable laser source and an optical spectrum analyzer have been used in synchronous mode allowing 1pm wavelength resolution in the spectral range 1520-1620nm. In figures 6 and 7, the reflected spectra are reported for the two samples. Reflectance spectra reveal several interference fringes and also their envelope offers a quite periodic behavior. In particular, in the reflectance of the sensor 1, two different harmonic contents, fast and slow, can be clearly distinguished. The observation of two different harmonic contents suggests the presence of a double cavity interferometer. From the wavelength separation of the fringes the optical path length weighted by the medium effective refractive index of each interferometer can be approximately retrieved, according to:

$$2 \cdot n_{eq} \cdot d = \frac{\lambda_1 \cdot \lambda_2}{(\lambda_2 - \lambda_1)} \quad (1)$$

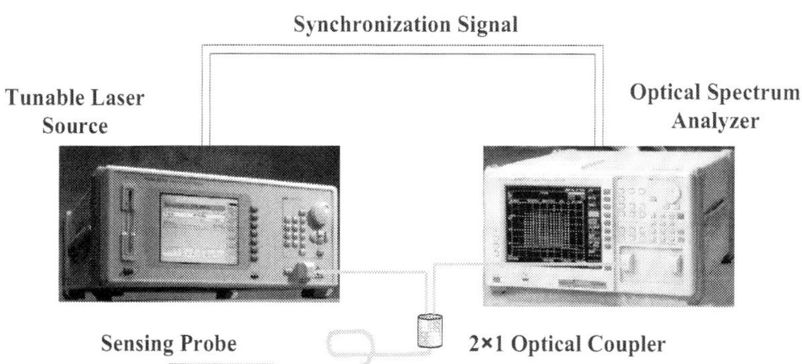

Fig. 5. Experimental setup for the spectral characterizations

where d is the length of the interferometer, n_{eq} is the medium effective refractive index of the propagating mode, λ_1 and λ_2 are the wavelengths corresponding to two adjacent maxima. Thus, by applying twice the eq. (1) at the reflectance spectrum, the features of the double interferometer, in both fast and slow regime, can be deduced. The optical path lengths weighted by the medium effective refractive index obtained for the sensor 1 are approx. 26.6cm and 40μm respectively.

The former one coincides approximately with the double length of the MOF piece, by considering the MOF effective refractive index equal to one (not filled by SWCNTs). The latter one represents the extension of the region, on micrometric scale, constituted by the deposited SWCNTs, either they are within the MOF holes or outside of the MOF end face.

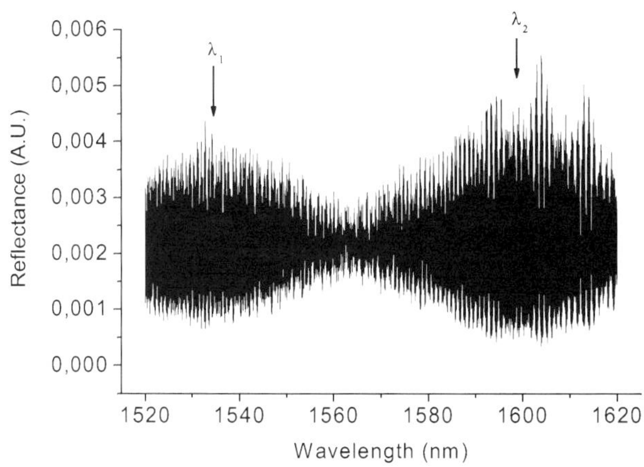

Fig. 6. Reflectance of the MOF sensor 1. The arrows indicate the wavelengths corresponding to two adjacent envelope relative maxima. (Reprinted with permission from ref. (Pisco et al. 2008))

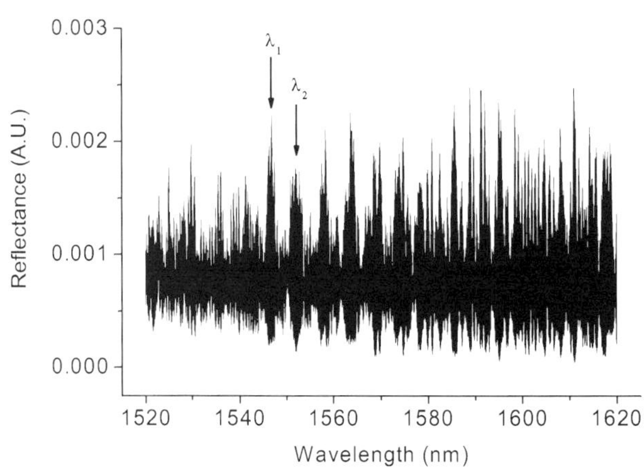

Fig. 7. Reflectance of the MOF sensor 3. The arrows indicate the wavelengths corresponding to two adjacent envelope relative maxima. (Reprinted with permission from ref. (Pisco et al. 2008))

The two cavities, hence, can be attributed to the piece of MOF and to the SWCNTs region. In turn, from the analysis of the reflectance of the sensor 3, although it is again clearly observable the fast oscillating component, the envelope of the reflectance seems to be periodic mostly in the wavelength range 1545-1575nm while it is less recognizable a periodic behavior for the envelope in the remaining spectral range.

Nevertheless, by applying also in this case twice the eq. (1), optical path lengths weighted by the medium effective refractive index of 8cm and 460μm are obtained for sensor 3. The fast oscillating behavior is still attributed to the piece of MOF, which is 4cm long, whereas the SWCNTs seem to cover a region of hundreds of microns consistently with the larger amount of SWCNTs deposited in the case of the sensor 3 (20 monolayers). Also, on the basis of previous works (Penza et al. 2005) dealing with the SWCNTs deposition onto the end face of SOFs, it results that 10 and 20monolayers of SWCNTs are not able to produce an external cavity able to produce interference fringes in the investigated wavelength range. Hence, the results obtained from the spectral characterization of both sensors reveal that the carbon nanotubes are penetrated at least in the MOF central hole and their distribution along the MOF axis, which cannot be assumed spatially uniform, extends for tens and hundred microns, respectively.

It is noteworthy that the eq. (1) is rigorously valid under the hypothesis of transparent media which in turn is not truly verified for carbon nanotubes based materials. In addition, the estimation of the equivalent cavity length by means of the interference fringes together with the trivial measurement of the length of the MOF piece before the deposition are not able to discriminate the fraction of nanotubes penetrated inside the holes from the fraction of nanotubes which composes an external overlay at MOF termination (Cusano et al. 2006). Nevertheless, an insight of the double interferometer behavior of the sensing probe is obtained and an approximated estimation of the extension of the SWCNTs region can be retrieved.

4.2 Sensor Modeling

On the basis of reflectance and far field characterizations the sensing probe reflectance of the sensor 1 can be modeled as a Fabry Perot double interferometer while the behavior of the sensor 3 cannot be trivially assumed like a simple double interferometer in agreement with the spectral and far field analysis. In eq. (2) the model of the reflectance of the sensor 1 is reported. The symbols r, n, d and A represent respectively the reflection coefficient, the complex effective refractive index, the cavity length and the absorbance (due to the coupling and splicing losses), while the subscript indices indicate the interface to which are referred as well as the subscript labels indicate the referred cavity. In particular the subscript 12 is referred to the SOF-MOF interface, 23 to the MOF-SWCNTs interface and 34 to the SWCNTs-external medium interface, while the subscript labels SOF, MOF and SWCNTs have obvious meaning. The reflectance R is ruled by several factors involved in the propagation of the lightwave through the MOF and the MOF filled with SWCNTs. In particular, any change in the SWCNTs dielectric function affects the reflectance R through the reflection coefficients r_{23} and r_{34}, the absorbance A_{23} and the propagation constant β_{SWCNTs} according to the eq. (2- 3)

$$R = \left| \frac{r_{12} + r_{23} \cdot (1 - A_{12}) \cdot e^{j \cdot \beta_{MOF}} + r_{34} \cdot (1 - A_{12}) \cdot (1 - A_{23}) \cdot e^{j(\beta_{MOF} + \beta_{SWCNTs})} + r_{12} \cdot r_{23} \cdot r_{34} \cdot e^{j \cdot \beta_{SWCNTs}}}{1 + r_{12} \cdot r_{23} \cdot e^{j \cdot \beta_{MOF}} + r_{12} \cdot r_{34} \cdot (1 - A_{23}) \cdot e^{j(\beta_{MOF} + \beta_{SWCNTs})} + r_{23} \cdot r_{34} \cdot e^{j \cdot \beta_{SWCNTs}}} \right|^2 \tag{2}$$

with

$$r_{12} = \frac{n_{SOF} - n_{MOF}}{n_{SOF} + n_{MOF}} \; ; \; r_{23} = \frac{n_{MOF} - n_{SWCNTs}}{n_{MOF} + n_{SWCNTs}} \; ; \; r_{34} = \frac{n_{SWCNTs} - n_{ext}}{n_{SWCNTs} + n_{ext}} \; ;$$
$$\beta_{MOF} = \left(\frac{4 \cdot \pi}{\lambda} \right) \cdot n_{MOF} \cdot d_{MOF} \; ; \; \beta_{SWCNTs} = \left(\frac{4 \cdot \pi}{\lambda} \right) \cdot n_{SWCNTs} \cdot d_{SWCNTs} \tag{3}$$

It is noteworthy that the sensor sensitivity to chemical induced changes in the dielectric properties of the sensitive material used to functionalize MOFs cannot be thought as derived by a simple Fabry-Perot effect in light of the PBG modifications. For instance, at the interface between the MOF and the MOF filled with SWCNTs, coupling losses due to the modal overlap between modes propagating in the sections with different guiding properties affect the reflectance R. This means that as the dielectric function of the SWCNTs is modified by chemical sorption, consequent changes are expected in the coupling coefficients as well as in the reflectance R. On the other hand, also the propagation losses within the MOF region filled with SWCNTs depend on the SWCNTs dielectric function and hence contribute to the reflectance variations. In summary, the SWCNTs effective dielectric function variations within the MOF filled with SWCNTs determine, in combination with the penetration depth of the carbon nanotubes, the sensing probe reflectance changes by affecting the propagation features of the sensing probe itself. According to the author, the key factor in the determination of the performances of this class of sensors relies in the possibility to exploit the aforementioned losses variability in order to get enhanced performances with respect to traditional SOF-SWCNTs based sensors.

4.3 Experimental Results

In order to investigate the sensing capability of the fabricated MOF sensors, they have been placed in a test chamber (Consales et al. 2006; Consales et al. 2007) and exposed to traces of VOCs. At the same time, a reflectometric system, schematically reported in figure 8, which allows reflectance measurements at single wavelength, has been employed. In order to light the sensing probes a Superluminescent Light Emitting Diode with 40nm bandwidth centered at 1550nm has been used. By means of an optical coupler the light source is split. One of the arms of the coupler is connected directly to a photodiode in order to provide a monitoring of the source power level and the other arm is connected to the sensing probes by means of an optical switch which is responsible for the time division multiplexing allowing the simultaneous interrogation of several sensors. The time-multiplexed reflected signal is collected by another photodiode. The electrical signals coming from the photodetectors are acquired by a Data Acquisition (DAQ) system and stored in a computer which provide to the time demultiplexing of the retrieved signals. The optoelectronic sensor output consists in the ratio between the reflected signal at the sensing interface and the source reference signal.
It is worth noting that since the light source used offers a bandwidth much wider than the separation of the faster interference fringes, the sensing performance exploited by each

sensor basically depends on the slow features of the reflectance spectrum and thus mainly on the SWCNTs cavity. During the measurements, the temperature within the chamber has been monitored by using a commercial thermocouple. The sensors sensitivity to temperature variations was previously characterized and used in order to properly compensate the sensors output.

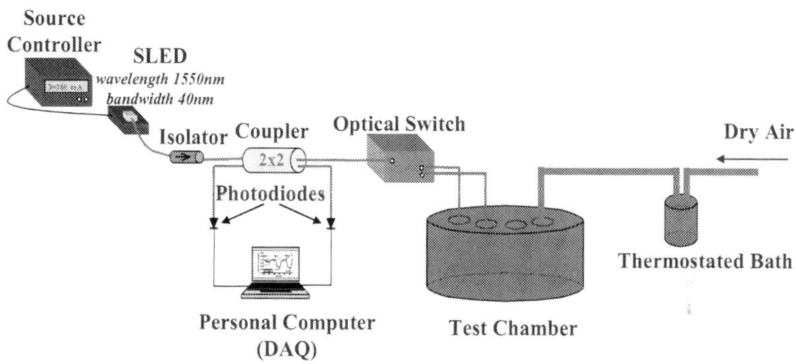

Fig. 8. Experimental setup for volatile organic compounds exposure

The optical sensors have been located in a test chamber for toluene traces exposure measurements. Dry air has been used as reference gas and carrier gas to transport the VOCs of toluene with different concentration pulses. The test cell containing the sensor had a volume of 1200ml whereas the total flow rate for each exposure has been kept constant at 1000 ml/min. The gas flow rate has been controlled by a mass flow-meter driven by a controller-unit. The VOCs vapors have been generated by the bubbling method with a thermo-stated flask containing the liquid analyte

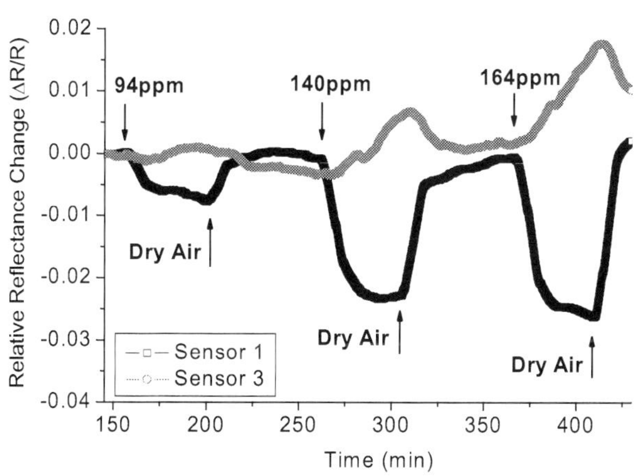

Fig. 9. Time responses of MOF sensors 1 and 3 to three toluene impulses (Reprinted with permission from ref. (Pisco et al. 2008))

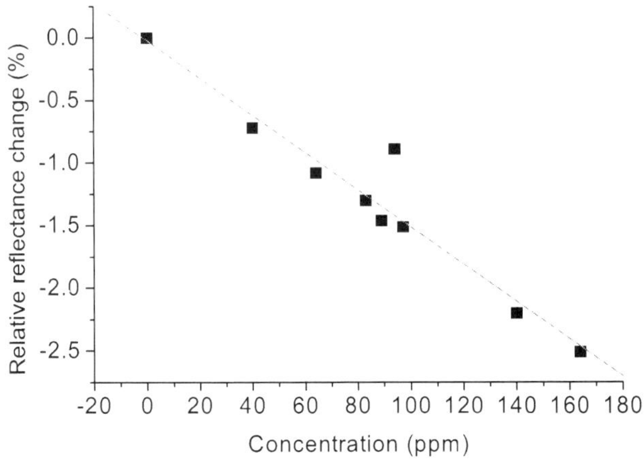

Fig. 10. Relative Reflectance Change (ΔR/R) of the sensor 1 corresponding to the steady states reached upon exposures to different concentration pulses of toluene (Reprinted with permission from ref. (Pisco et al. 2008))

Here a comparison among the performances exploited by sensors 1 and 3 is carried out by considering their relative reflectance change due to toluene exposure. To this aim, both sensors have been exposed to three toluene pulses of 45 minutes with increasing concentrations in the ppm range. Fig. 9 reports the time responses of both sensors. Upon exposures, the reflectance of the sensor 1 decreases linearly with the toluene concentration as confirmed by further exposure measurements resumed in fig. 10. Differently, sensor 3 reflectance increases upon exposure presenting a significantly lower sensitivity, opposite in sign. Moreover, sensor 3 is not able to reach a steady state within 45 minutes especially for high concentration exposures, this also limits the maximum reflectance change observed during the exposure tests. The higher response times and the lower sensitivity of the sensor 3 can be explained on the basis of the higher extension of the SWCNTs region (higher diffusion times and higher losses in the filled region). In fact, the VOCs, able to interact with the SWCNTs, lead to a change of the SWCNTs agglomerate dielectric function and the induced variations are detected as fast as short is the SWCNTs region. In addition, as demonstrated by the far field characterization, the SWCNTs penetration depth in the sensor 3 yields the functionalized MOF not more able to guide the light and thus it reduces the capability of the light to interact with the sensitive material. Further investigation is required to assess the sensitivity dependence on the SWCNTs distribution within the MOF structure and also to identify characterization features able to predict the correct functioning and the performance of the final device.

7. Conclusion

In conclusion, the research activities devoted to the integration of MOFs with SWCNTs in order to develop new in-fiber active and passive optoelectronic devices are reviewed. The infiltration of SWCNTs inside the MOF holes has been accomplished through the deposition of multiple SWCNTs monolayer by means of the LB method. The far field characterizations revealed the success of the SWCNTs deposition demonstrating the partial filling of the nanotubes within MOFs. Also, the obtained results demonstrated that the main effect of SWCNTs filling is the significant modification of the guiding properties and thus of the local PBG at the termination of the MOF. Also, by using the LB deposition method, we demonstrate how by acting on the process parameters and especially on monolayers number and pressure conditioning at the free termination, it would be possible to optimize and tailor the SWCNTs filling enabling the local control of the MOF PBG. The proper modification of the MOF PBG features, by filling (also selectively) the MOF holes with SWCNTs, can supply, in fact, to the final device advanced functionalities and can be properly exploited to develop high performances sensors based on PBG modification as well as multimaterial and multifunctional fibers. On the basis of the retrieved results, a novel opto-chemical sensors for VOCs detection has been developed and its sensing capability has been proved by exposure to VOCs traces in a proper test chamber. The presented experimental results demonstrated that the fabrication parameters affect strongly the sensors performances and that the proposed optochemical sensor is able to perform VOCs detection with a good sensitivity and fast response times.

8. References

Abouraddy A. F., Bayindir M., Benoit G., Hart S. D., Kuriki K., Orf N., Shapira O., Sorin F., Temelkuran B., Fink Y. (2007). Towards multimaterial multifunctional fibres that see, hear, sense and communicate. *Nature*, 6, 336-347

Benabid F., Couny F., Knight J. C., Birks T. A., Russel P. St J. (2005). Compact, stable and efficient all-fibre gas cells using hollow-core photonic crystal fibres. *Nature*, 434, 488-491.

Consales M., Campopiano S., Cutolo A., Penza M., Aversa P., Cassano G., Giordano M., Cusano A., (2006) .Carbon nanotubes thin films fiber optic and acoustic VOCs sensors: performances analysis. *Sensors and Actuators B*, 118 232–242.

Consales M., (2007). Carbon Nanotubes Coated Acoustic and Optical VOCs Sensors: Towards the Tailoring of the Sensing Performances. *IEEE Transactions on Nanotechnology*, 6, 601-612.

Cusano A., Pisco M., Consales M., Cutolo A., Giordano M., Penza M., Aversa P., Capodieci L., Campopiano S. (2006). Novel Opto-Chemical Sensors Based On Hollow Fibers And Single Walled Carbon Nanotubes. *IEEE Photonics Tech. Lett.*, 18, 22, 2431 - 2433

Domachuk P., Nguyen H. C., Eggleton B. J., Straub M., Gu M. (2004). Microfluidic tunable photonic band-gap device. *Applied Physics Letters*, 84, 11, 1838-1840

Dresselhaus M. S., Dresselhaus G., Avouris P. (2001). Carbon Nanotubes: Synthesis, Structure, Properties, and Applications. Topics in Applied Physics, 80, Springer, Berlin, ISBN: 978-3-540-41086-7

Huang Y., Xu Y., Yariv A. (2004). Fabrication of functional microstructured optical fibers through a selective-filling technique. *Applied Physics Letters*, 85, 22, 29, 5182-5184

Knight J. C. (2003). Photonic crystal fibres. *Nature*, 424, 847-851

Larsen T. T., Bjarklev A., Hermann D. S., Broeng J. (2003). Optical devices based on liquid crystal photonic bandgap fibres. *Optics Express*, 11, 2, 2589-2596.

Matejec V., Mrázek J., Hayer M., Kašík I., Peterka P., Kaňka J., Honzátko P., Berková D. (2006). Microstructure fibers for gas detection. *Materials Sc. & Eng. C*, 26, 2-3, 317-321

Nielsen K., Noordegraaf D., Sørensen T., Bjarklev A., Hansen T. P. (2005). Selective filling of photonic crystal fibres. *J. Opt. A: Pure Appl. Opt.*, 7, 8, L13-L20

Penza M., Cassano G., Aversa P., Antolini F., Cusano A., Cutolo A., Giordano M., Nicolais L.(2004). Alcohol detection using carbon nanotubes acoustic and optical sensors. *Applied Physics Letters*, 85, 12, 2379-2381

Penza M., Cassano G., Aversa P., Cusano A., Cutolo A., Giordano M., Nicolais L. (2005). Carbon nanotube acoustic and optical sensors for volatile organic compound detection. *Nanotechnology*, 16, 2536-2547.

Pisco M. (2007). Optoelectronic Devices based on 1-D and 2-D Photonic Bandgap Structures for Sensing and Communication Applications. PhD Thesis

Pisco M., Consales M., Cutolo A., Penza M., Aversa P., Cusano A. (2008). Hollow Fibers Integrated with Single Walled Carbon Nanotubes: Bandgap Modification and Chemical Sensing Capability. *Sensors and Actuator B: Chemical*, 129, 1, 163-170

Pisco M., Consales M., Penza M., Aversa P., Giordano M., Cutolo A., Cusano A. (2009). Photonic Bandgap Modification in Hollow Optical Fibers Integrated with Single Walled Carbon Nanotubes. *Microwave and Optical Technology Letters*, 51, 11, 2729-2732

Zhmud B. V., Tiberg F., Hallstensson K. (2000). Dynamics of Capillary Rise. *J. Colloid Interface Sci.*, 228, 263-269

Carbon Nanotubes as a New Solid Phase Extraction Sorbent for Analysis of Environmental Pollutants

Bele Constantin
University of Agricultural Sciences and Veterinary Medicine Cluj-Napoca, Romania

1. Introduction

Pretreatment is often considered to be a fundamental step in the process of successful analysis of environmental pollutants, because it helps not only to achieve low detection limits but also to clean up the sample matrix. Solid phase extraction (SPE) is an effective sample handling method and is used as an enrichment technique when low concentrations of analytes need to be determined. SPE provides higher enrichment efficiency and requires a lower volume of solvent than the traditional method of liquid-liquid extraction.In addition, SPE is simpler and easily to be automated and operated.In the procedure of SPE, the type of sorbent, its structure and its interactions with the solute play an important role in obtaining higher enrichment efficiency of analytes. Until now, several kinds of materials such as C18, Oasis HLB, bonded silica, styrenedivinyl-benzen (SDB), zeolites, carbonaceous materials have been proposed as adsorbents for SPE cartridge.

In recent years, carbon nanotubes (CNTs), a novel member in the carbon family , have attracted great attention due to its advantages that can be used for many different applications in terms of its chemical, electronic and mechanical properties as well as the unique tubular structures and large length / diameter ratio. CNTs are to be considered as a sheet of graphite that has been rolled into a tube and be classified as single - walled carbon nanotubes (SWCNTs) and multiwalled carbon nanotubes (MWCNTs) . Over the past 20 years, CNTs have been exploited in analytical and other fields such as biosensors with immobilized biomolecules, electrochemical detectors, gas sensor, catalyst supports and so on. Because CNTs surfaces have a strong interaction with other molecules, particularly with those containing benzene rings, they possess excellent adsorption ability and substitute active carbon. CNTs as SPE adsorbents has been investigated to extract organic compounds such as pesticides (carbofuran, iprobenfos, parathion-methyl, prometryn, fenitrothion etc.), polycyclic aromatic hydrocarbons, antibiotics, sulphonylurea herbicides, sulfonamides, phthalate esters, endocrine disruptors, triazines, microcystines, pyrethroids and polybrominated diphenyl ethers . In several comparative studies CNTs exhibit similar or higher adsorption capacity for environmental pollutants than silica-based sorbents or macroporous resins. CNTs can also preconcentrate volatile organic compounds.CNTs were used as SPE adsorbents for preconcentration of metal ions, such as copper, nichel, cobalt, vanadium, silver, cadmium, rare earth elements etc.

This chapter is organized in four sections including the introduction. Section 2 is devoted to the sorption properties of CNTs. Section 3 summarize the most important applications of CNTs for the enrichment of environmental pollutants.The potential factors affecting SPE and the sorption capacities of CNTs are also disscussed. The whole chapter is then concluded in Section 4.

2. Adsorption properties of carbon nanotubes

CNTs usually have a diameter in the range comprized within a tenth to tens nanometers and a length of up to centimeters. The ends of CNTs are normally capped by a fullerene-like structure. As fullerene, CNTs also exhibit limited solubility. Depending on their diameter and helicity of the graphitic sheets CNTs can be either metallic or semi-conducting (Valcarcel et al., 2008). The characteristic structures of carbon nanotubes allow a strong interaction with organic molecules via non-covalent forces, such as hydrogen bonding, π -π stacking, electrostatic forces, van der Waals forces and hydrophobic interactions (Pyrzynska et al., 2008). The presence of functionalized carbon nanotubes allows the possibility of incorporating one or more of these interactions which increase the selectivity and the stability of the system.

It was stated that CNT derivatization is required when developing special applications (e.g. retention of metals). CNTs were purified by sodium hypochlorite solutions and were employed as adsorbents to study the adsorption characteristics of zinc in water (Lu & Chiu, 2006). The properties of CNTs such as purity, structure and nature of the surface were considerably improved after purification by sodium hypochlorite which made CNTs become more hydrophilic and suitable for adsorption of Zn^{2+} from water . The adsorption of Zn^{2+} onto CNTs rises proportional to the pH increase within 1- 8 range, fluctuates very slight and reaches a maximum in the pH range of 8-11; the adsorption curve decreases at a pH of 12. The contact times to reach equilibrium are 60 min for SWCNTs and MWCNTs. The maximum adsorption capacities of Zn^{2+} calculated by the Langmuir model are 43.66 and 32.68 mg / g with SWCNT and MWCNT, respectively, at an initial Zn^{2+} concentration range of 10-80 mg/ L.

It was found that the acid treatment with a mixture of nitric acid and sulfuric acid made CNTs become more hydrophilic and suitable for adsorption of low molecular weight and relatively polar trihalomethanes (THMs) in water (Lu et al., 2005). The adsorption of THMs onto CNTs can be suitably described by both Langmuir and Freundlich models. The smallest molecule $CHCl_3$ is the most preferentially adsorbed onto CNTs, followed by $CHBrCl_2$, $CHBr_2Cl$ and then by $CHBr_3$. THMs absorption onto CNTs fluctuates very slightly in the pH range of 3 -7 but decreases with pH value when pH exceeds 7.

It was shown that carbon nanotubes can also be used as supports for adsorption materials, and the new composites have a good affinity to many metals . MWCNTs filled with Fe_2O_3 nanoparticles have been prepared and employed as adsorbent for the magnetic separation of dye contaminants (Methylene Blue and Neutral Red) in water (Qu et al., 2008). The magnetic nanoparticles have been prepared via hydrothermal reaction of shortened MWCNTs in ferric nitrate solution and subsequent calcinations. The prepared magnetic MWCNTs can be well dispersed in water and easily magnetic separated from the medium after adsorption. As compared with other adsorbents, the magnetic nanoparticles not only have high adsorption efficiency to dyes, but can also be easily manipulated by external magnetic field.

MWCNT / iron oxide magnetic composites were prepared and used for adsorptions of Ni (II) and Sr (II) (Chen et al., 2009). Scan electronic microscopy (SEM) image shows an entangled network of MWCNTs with clusters of iron oxides attached to them suggesting the formation of MWCNTs / iron oxide magnetic composites. Ni (II) adsorption on the magnetic nanoparticles is pH and ionic strength dependent and can be easily desorbed from the magnetic nanoparticles by adjusting the solution pH values.The Langmuir model fitted the adsorption isotherm data of Ni (II) better than the Freundlich model.

MnO_2 / CNTs composites were efficient for Pb (II) ion removal from aqueous solution (Wang et al., 2007b) . The optimum MnO_2 loading indicating the best performance of Mn O_2 on the Pb(II) removal is 30 wt %.The application to experimental results of the Langmuir and Freundlich models show that the Langmuir model gives a better correlation coefficient.

It was found that CNTs present a marked tendency to aggregation, which negatively affects adsorption by reducing their active surface (Valcarcel et al., 2008). In addition, when cartridges or home-made columns are employed, this tendency may increase pressure in the flow systems .

Special configurations are developed for specific applications. A complex sheet of SWCNT and polyaniline was used as a new adsorbent to remove bilirubin from plasma (Ando et al., 2009) . Bilirubin, a red- brown bile pigment, is a metabolite of heme produced from the senescent hemoglobin. If a bilirubin concentration exceeds a certain level in blood, it may cause kernicterus or liver diseases. Bilirubin CNTs adsorption capacity has been found to be much higher versus the conventional materials because of their large surface area and considerable adsorption capability for polycyclic compound molecules due to their structure similar to graphite.

A recently introduced immobilization method to link the aminoacid L-tyrosine to CNTs was described (Pacheco et al., 2009). The amount of aminoacid immobilized on CNTs surface was 3174 µmol / g. The material was tested for Co retention using a minicolumn inserted in a flow system. A 10 % (v/v) HNO_3 solution was chosen as eluent. The pH study revealed that Co binding increased at elevated pH values. The retention capacity was compared to other bivalent cations and showed the following tendency : $Cu^{2+} > Ni^{2+} > Zn^{2+} >> Co^{2+}$.

The influence of the surface functionalization on the colloidal stability of CNTs, as well as on the sorption of heavy metals was investigated (Schierz & Zanker, 2009). Uranium (VI), a chemical element of considerable public concern, was chosen as an example of a toxic heavy metal. The results indicated that acid treatment increases the amount of acidic surface groups on the CNTs. Acid treatment has an intensifying effect upon the colloidal stability of the CNTs, and on their adsorption capacity for U (VI).

The analytical potential of MWCNTs modified with a Schiff base ligand was examined for simultaneous preconcentration of Au (III) and Mn (II) in aqueous samples prior to their flame atomic adsorption spectrometric assessment (Shamspur & Mostavafi, 2009) . It was found that the sorption is quantitative in the pH range 5.0- 7.5, whereas quantitative desorption occurs instantaneously with 4.0 mL of 0.1 mol / L $Na_2S_2O_3$.

The application of the hemimicelle capped carbon nanotubes-based nanosized SPE adsorbents in environmental analysis is reported for the first time using arsenic as model target (Li et al., 2009a). The end functionalized of CNTs can introduce oxygen-containing negatively functional groups such as - COOH, - OH, or - C= O on their surface site. If cationic surfactant, such as cetyltrimethylammonium chloride (CTAC) was added to the functionalized CNTs, interactions like hydrophobic and ionic may lead to the formation of hemimicelle / admicelle aggregates on the CNTs ; this way, a new kind of adsorbents is acquired, namely hemimicelle capped CMMWCNTs. Arsenic can be quantitatively retained on the hemimicelle capped CMMWCNT at pH 5 - 6 from sample volume up to 500 mL, and subsequently eluted completely with 2 mol/ L HNO_3 in the presence of 10 mg / L CTAC.

Carbon nanotubes have also been proposed as material coatings in SPME fibers for the determination of flame retardants like polybrominated diphenyl ethers (PBDEs) in environmental and food samples (Wang et al., 2006a). The home-made fibers, which were prepared according to the method used for constructing composite electrodes, were evaluated quantitatively and compared with commercial fibers. The results demonstrated that the MWCNT coating was effective for extracting the analytes described above, and provided better enhancement factors than activated carbon and poly (5 % dibenzene- 95 % dimethylsiloxane) coatings.

3. CNTs as adsorbents in solid-phase (micro) extraction

Most CNT applications published have been developed for the extraction of water samples, which are probably the less complex samples to work with. Up to now, only few works have used CNTs (basically MWCNTs) for the extraction of environmental pollutants from matrices different than waters.

Some reprezentative examples of the use of carbon nanostructures as sorbent materials in SPE and solid phase microextraction (SPME) are given in Table 1 for an easier approach and comparison.

Carfentrazone-ethyl (a relatively novel triazolinone herbicide) residue in water was enriched by use of MWCNTs (Dong et al., 2009a). Relevant studies were developed to examine several factors affecting the recovery of the analyte, for example the pH of the water samples , sample volume, polarity and volume of eluents . It was found that MWCNT was an effective SPE adsorbent for preconcentration of carfentrazone-ethyl in water and the recovery of this herbicide from fortified water was 81.49-91.08 %.The detection limits and quantification were 0.01 and 0.03 µg / L. It was also shown that under the optimized SPE procedure, the MWCNT-packed cartridge needed only 100 mg adsorbent .

The extraction efficiency of MWCNT as a new SPE adsorbent followed by GC-ECD for the analysis of chloroacetanilide herbicides (alachlor, acetochlor, metolachlor and butalochlor) was investigated (Dong et al. 2009b). It was found that the amount of adsorbent was much less for MWCNT in comparison with the commonly used adsorbent, such as C18. As an example, in this method were used only 100 mg MWCNT, whereas for the environmental analysis routine work were applied 500 mg to 1000 mg C18 cartridges. The detection limits were situated within the range of 0.01-0.03 µg / L.

The adsorptive potential of MWCNTs was used for the extraction and clean up of eight pesticides in agricultural, ornamental and forestal soils (Asensio- Ramos et al., 2009) . Soils were first ultrasound extracted with a mixture of methanol/ acetonitrile and the evaporated

Analyte	Sample	Combined technique and detection limit (ng / mL)	Remarks	References
Carfentrazone-ethyl	Water	GC-ECD 0.01-0.03	Only 100 mg of MWCNTs as adsorbent was needed Comparison with C 18 silica	Dong et al., 2009a
Pesticides	Agricultural, ornamental, forestal soils	GC- NPD 2.97- 72.4	Low cost CNTs were used (10-15 nm o.d., 2-6 nm i.d. and 0.1-10 µm length) Satisfactory recovery values (54-91 %) were obtained	Asensio-Ramos et al., 2009
Amines	Water	GC-MS 0.005-0.016	Comparison with Lichrolut EN ,Oasis HLB, RP-C18, graphitized carbon black and fullerenes	Jurado-Sanchez et al., 2009
Pesticides	Olive oils	GC-MS 1.5-3.0	Only 30 mg of MWCNTs as adsorbent was needed	Lopez-Feria et al., 2009
Phenols	Water	HPLC-DAD 0.9-3.8	The SPME-Pt fiber coated with SWCNTs was prepared by electrophoretic deposition Linear range of 10- 300 ng / mL	Li et al., 2009b
Oxygenated ethers	Urine	GC-MS 0.003-0.01	The SWCNT fiber exhibited higher sensitivity and longer life span (over 150 times) than CAR / PDMS fiber	Rastkari et al., 2009
Atrazine, propoxur and methidathion	Reservoir waters	HPLC-UV 2-3	At flow rate higher than 5.0 mL / min the enrichment efficiencies decreased for all pesticides Linear ranges of 5-30, 3-60 and 5-40 µg / L for atrazine, methidathion and propoxur	Al- Degs et al., 2009
Chloroacetanilide herbicides	Water	GC-ECD 0.01-0.03	The recoveries were steady in the range of 200-1000 mL sample volume Linear range of 2.5-2500 µg / L	Dong et al., 2009b
Cobalt	Water	FAAS 0.05	L-tyrosine was immobilized on CNTs and used as sorbent for SPE Adsorption capacity of 37.58 µmol Co / g CNTs	Pacheco et al., 2009
Uranium	Water	ICP-MS	CNTs were modified by heating in a mixture of HNO_3 / H_2SO_4 The Langmuir model fitted the experimental data better than the Freundlich model	Schierz et al., 2009
Arsenic	Water	FI-AFS 0.002	Carboxyl modified MWCNTs with cation surfactant CTAC were used as adsorbent The recoveries remain stable when the flow rate was below 6 mL / min	Li et al., 2009a

Analyte	Sample	Combined technique and detection limit (ng / mL)	Remarks	References
Bilirubin	Plasma apheres is	VIS spectro-photometer	MWCNTs exhibit greater adsorption for bilirubin than SWCNTs	Ando et al., 2009
Organophosphate pesticides (MP)	Garlic	SWV 5	The strong affinity of MWCNTs for phosphoric group allow extracting a large amount of MP	Du et al., 2008
Linear alkylbenzene sulfonates	Water	HPLC-UV 0.02-0.03	Carboxyl modified MWCNTs were used as adsorbents Comparison with C8 and C18	Guan et al., 2008
Pesticides	Water	HPLC 0.036-0.22	MWCNTs have better ability for the extraction than C 18 silica and activated carbon	El-Sheikh et al., 2008
Diazinon	Tap water	HPLC 0.06	Preconcentration factor of 200 was achieved for 1000 mL of sample volume Linear range of 0.3-10000 ng / mL	Katsumata et al.,2008
Organophosphorus pesticides	Fruit juices	GC-NPD 1.85-7.32	Low sample pretreatment prior to the SPE procedure Only 40 mg of MWCNTs as adsorbent was needed	Ravelo-Perez et al., 2008
Aromatic amines	Water	HPLC-UV 0.04-0.13	Automated in- tube SPME using carboxylated MWCNTs Linear range of 1.02-102 µg / mL	Liu et al., 2008
Pyrethroids	Water	HPLC - UV 1.3- 4.3	The recoveries decreased except fenpropathrin when the flow rate was over 3 mL/ min	Zhou et al., 2008
Methylene Blue Neutral Red	Water	UV-VIS spectrophoto-meter	MWCNTs filled with Fe_2O_3 nanoparticles were used as adsorbent The magnetic MWCNTs have high adsorption efficiency to dyes and can be manipulated by external magnetic field	Qu et al., 2008
Non-steroidal anti-inflammatory drugs	Urine	Capillary electrophoresis-MS 1.6-2.6	Carboxylated SWCNTs (c-SWCNT) were chemically immobilized on porous glass.High sorption capacity was related with the special orientation of c-SWCNTs	Suarez et al., 2007
Pesticides	Water	GC-MS 0.01-0.03	The recoveries were constant at the flow rate in the range of 1.5- 3 mL / min Only 100 mg MWCNTs adsorbent per extraction	Wang et al., 2007a

Analyte	Sample	Combined technique and detection limit (ng / mL)	Remarks	References
Barbiturates	Pork	GC/MS/MS 0.1-0.2	MWCNTs showed better ability for the extraction of phenobarbital than C 18 Linear range 0.5-50μg / Kg	Zhao et al., 2007
Cephalosporins Sulfonamides Phenolic compounds	Water	HPLC-PDA 0.027-0.038	CNTs were much superior to C18 for the extraction of the highly polar analytesThe recoveries decreased at solution pH > 8.	Niu et al., 2007
Volatile organic compounds (VOCs)	Ambien t air	GC-FID	Comparison with Tenax TA High adsorption efficiency for collecting VOCs with low boiling points and strong volatility	Jie-Min et al., 2007
Polycyclic aromatic hydro-carbons	Water	HPLC-UV 0.005-0.058	The recoveries were steady in the range of 250-1000 mL sample volume MWCNTs have better ability for the extraction than C18	Wang et al., 2007c
Fungicides Prometryn	Water	HPLC-UV 0.003-007	Sample volume up to 1500 mL Linear range of: 0.2-80 μg / L (fungicides) 0.05-20 μg /L (prometryn)	Zhou et al., 2007
Chromium	Water	FAAS 0.90	The procedure is based on SPE of the Cr(VI)-APDC chelate on MWCNTs Adsorption capacity of Cr (VI) was 9.5 mg/ g	Tuzen & Soylak, 2007
Lead	Water	ICP	CNTs were coated with Mn oxide and used as adsorbent The Langmuir equation fitted the experimental data more closely than the Freundlich model	Wang et al., 2007b
Polybrominated diphenyl ethers	Water Milk	GC-ECD 0.0036-0.0086	MWCNTs coated fibers for SPME were compared with activated carbon and PDMS-DB coated fibers	Wang et al., 2006a
Microcystins (MCs)	Water	HPLC-DAD	The size of CNTs tube pore that is fit for molecular dimension of MCs plays a dominante role Adsorption capacity of MCs was 14.8 mg / g	Yan et al., 2006
Sulfonamides	Eggs Pork	HPLC-UV 0.004-0.010	Sample loading time up to 23 min for the flow rate of 4.5 mL / min MWCNTs gave lower detection limits, higher enrichment factors and better precisions than C18 silica	Fang et al., 2006

Analyte	Sample	Combined technique and detection limit (ng / mL)	Remarks	References
Benzodiazepines	Pork	GC-MS 2-5	Static adsorption experiments 0.2 g MWCNTs were superior to 0.5g C18 for the extraction of diazepam	Wang et al., 2006b
Atrazine and simazine	Water	HPLC-DAD 0.009-0.033	The recoveries were constant at the flow rate of 2-7 mL/ min Linear range of 0.2-100 (atrazine) and 0.02-100 ng / mL (simazine)	Zhou et al., 2006b
Organochlorine pesticides	Water	HPLC-UV 0.004-0.013	The recoveries were almost constant when the flow rate was change over the range of 2-8 mL / min for sampling loading Linear range of 0.2-60 μg / L	Zhou et al., 2006a
Sulfonylurea herbicides	Water	HPLC-DAD 5.9 - 11.2	The recoveries were constant at the flow rate in the range of 2-8 mL / min Sample volume up to 2000 mL	Zhou et al., 2006c
Zinc	Water	FAAS	The Langmuir equation is more appropriate to describe the adsorption of Zn^{2+} onto CNTs than the Freundlich model.Comparison with powdered activated carbon	Lu et al., 2006
Chlorophenols	Water	HPLC-UV 0.08-0.8	The recoveries decreased slightly with the increase of sample volume higher than 200 mL Linear range of 1-200 ng/ mL	Cai et al., 2005
Volatile organic compounds	Water	GC-FID	Comparison with Carbopack B and VOCARB 3000	Li et al., 2004
Phthalate esters	Water	HPLC-DAD 0.18-0.86	Sample volume up to 3000 mL Linear range 2-100 ng / mL	Cai et al., 2003

Table 1. Examples for adsorption of environmental pollutants on carbon nanotubes.

extract redissolved in water was passed through 100 mg MWCNT of 2-6 nm i.d. and 0,1- 10 μm length .Dichloromethane was used for the elution of analytes. In the three types of soils satisfactory recovery values (54-91 %) were registered for several pesticides (diazinon, ethoprophos, fenitrothion, malathion and phosmet).

The comparison of MWCNTs, graphitized carbon black, fullerenes, Lichrolut EN, Oasis HLB, RP- C 18 in terms of sensitivity, selectivity and reliability has been made for the retention of amine compounds including anilines, chloroanilines, N-nitrosoamines and aliphatic amines (Jurado- Sanchez et al., 2009).The analytes were retained on a SPE sorbent column and after elution, 1 μL of the extract was analysed by gas chromatography coupled with electron impact ionization . MWCNTs are adequate to retain aromatic compounds such as aromatic N-nitrosamines and despite amine aromaticity, they only interact with trichloroanilines and 2-nitroaniline, through π- π interactions, and with some

28

Hydrogen Storage Using Carbon Nanotubes

Yunjin Yao
Hefei University of Technology
China

1. Introduction

Hydrogen is the ideal candidate as an energy carrier for both mobile and stationary applications while averting adverse effects on the environment, and reducing dependence on imported oil for countries without natural resources. Hydrogen storage is clearly one of the key challenges in developing hydrogen economy. Hydrogen can be stored as (i) pressurized gas, (ii) cryogenic liquid, (iii) solid fuel as chemical or physical combination with materials, such as metal hydrides, complex hydrides and carbon materials, or produced on-board the vehicle by reforming methanol (Sakintuna, Lamari-Darkrim, & Hirscher, 2007). Compressed and/or liquid hydrogen is currently used in fuel cell powered automobiles but application issues, such as safety and the need for a hydrogen supply infrastructure, are problems, which remain to be solved. Solid-state hydrogen storage systems are expected to be simpler for the engineering design of vehicles and considerably safer than the storage of elemental hydrogen (Fig. 1).

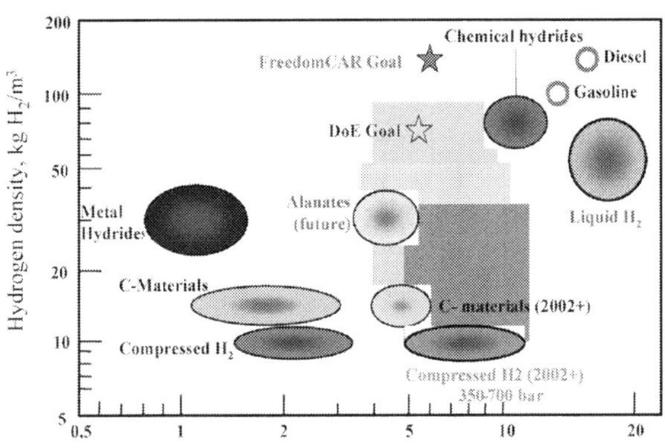

Fig. 1. Volumetric and gravimetric hydrogen density for different storage methods (R. Ströbel, J. Garche, P. T. Moseley, L. Jörissen, & G. Wolf, 2006).

The US Department of Energy (DOE) published a long-term vision for hydrogen storage applications considering economic and environmental parameters. The predicted minimum hydrogen-storage capacity should be 6.5 wt% and 65 g/L hydrogen available, at the decomposition temperature between 60 and 120 ℃ for commercial viability. It was also predicted low temperature of hydrogen desorption and low pressure of hydrogen absorption (a plateau pressure of the order of a few bars at room temperature) and nonthermal transformation between substrates and products of decomposition as reported by Schulz (Schulz et al., 1999). Furthermore, the cost of a storage medium and its toxicity properties need to be carefully considered for the realization of the set goals. The different time-dependent DOE targets are shown in Table 1.

Storage parameter	2005	2010	2015
Gravimetric capacity (specific energy)(kW h kg^{-1})	1.5	2.0	3.0
Gravimetric capacity (specific energy)(kgH$_2$ kg^{-1})	0.045	0.060	0.090
System weight (kg)	111	83	55.6
Volumetric capacity (energy density) (kW h l^{-1})	1.2	1.5	2.7
Volumetric capacity (energy density) (kgH$_2$ l^{-1})	0.036	0.045	0.081
System volume (l)	139	111	62
Storage system cost (US$ kW h^{-1})	6	4	2
System cost (US$)	1000	666	333
Refueling rate (kgH$_2$ min^{-1})	0.5	1.5	2.0
Refueling time (min)	10	3.3	2.5

Table 1. DOE targets for hydrogen on-board storage and specific data for a 5 kgH$_2$ system

Storage by absorption as chemical compounds or by adsorption on carbon materials have definite advantages from the safety perspective such that some form of conversion or energy input is required to release the hydrogen for use. A great deal of effort has been made on new hydrogen-storage systems, including metal, chemical or complex hydrides and carbon nanostructures. Carbon materials such as activated carbons, carbon nanotubes, and carbon nanofibers have been the subject of intensive research. The research on hydrogen storage in carbon materials was dominated by announcements of high storage capacities in carbon nanostructures.

However, the experimental results on hydrogen storage in carbon nanomaterials scatter over several orders of magnitude. The hydrogen-storage capacity for carbon materials is reported between 0.2 and 10 wt% . The experiments to date claiming very high values could not independently be reproduced in different laboratories. In view of today′ s knowledge although they have good reversibility properties, carbon nanostructures cannot store the amount of hydrogen required for automotive applications.

This Chapter deals with hydrogen adsorption in carbon nanotube materials over a wide range of pressure and temperature in order to establish the reliability of this phenomenon as a valuable storage technique of hydrogen. Our purpose is to give, discuss and compare the different results published comprising our works. Both experimental and theoretical investigations data concerning the hydrogen gas within nanotube material framework are given. A comparison between the different studies should enable to gain a better knowledge of hydrogen adsorption in nanotubes and then to answer the following questions: How and where did the gas adsorption occur? Moreover, we will consider the possibility of controlling tube diameters and lengths that will enable to optimize nanotube structure for a maximal hydrogen adsorption. Indeed, the possibility of controlling microscopic parameters will enable to control the main material macroscopic property: the gas adsorption. Finally, it is noticeable that both the material synthesis and purification will have to be optimized in order to enable the gas adsorption measurements. Indeed, the adsorbent material will have to be available in large amounts with high degree of purity to ensure the hydrogen storage by adsorption in carbon nanotubes

2. Hydrogen storage fundamentals

2.1. Physisorption

Physisorption is based on the van der Waals interaction (a resonant fluctuation in the charge distribution) between gases and solids. These are attractive forces, which keep graphene layers well-stacked, and are the same intermolecular forces that play an important role in the physisorption mechanism of gasses and hydrogen on carbon surfaces. Van der Walls bonds stem from intermolecular forces between atoms (and/or molecules) that result from instantaneous charge distribution in atoms and molecules when they approach each other. Under mutual interaction, an asymmetric polarization of electron charge is induced in molecules that create temporarily dipole moments, and atoms or molecules become attracted by electrostatic forces (Fig. 2).The interaction energy, also called the London Dispersion forces (E_{S-H_2}), between the substrate (S) and the hydrogen molecule (H_2), is given by

$$E_{S-H_2} \sim \frac{\alpha_{H_2}\alpha_S}{R^6} \tag{1}$$

where α is the polarizibility and R is the interaction distance. Because α_{H_2} is fixed, the only way to increase E_{S-H_2} is to use highly polarizable substrates, e.g. substrates with п-electron systems.

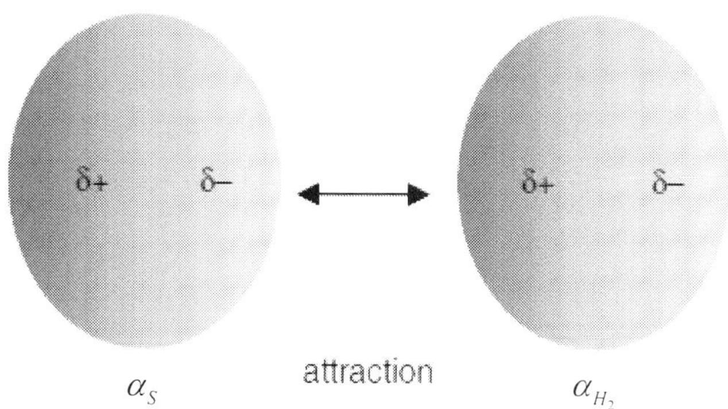

Fig. 2. Van der Walls bonding as originating from the polarizibility of electron density on hydrogen molecule and carbon atom

The equilibrium between gas attraction and gas repulsion on the surface creates an energy minimum which amounts to between 1 and 10 kJ mol^{-1}. The exact value for the hydrogen adsorption on a flat carbon surface depends on the adsorption stereometry. An average value would be about 4–5 kJ mol^{-1}. This represents a very weak interaction. Therefore, hydrogen is desorbed with increasing temperature, and very little hydrogen adsorption is observed on carbon at elevated temperatures(R. Ströbel, J. Garche, P. Moseley, L. Jörissen, & G. Wolf, 2006).

Normally only a monolayer is adsorbed above the boiling point of the adsorbent. In this case the Langmuir isotherm should apply. The maximum value of hydrogen adsorption on any particular material can be estimated as the quantity of hydrogen that can be accommodated in a monolayer. With the assumption that the structure of the adsorbed hydrogen is closed-packed face centered, the minimum surface area for the adsorption of 1 mol of hydrogen amounts to Sml (H_2) = 85.917m^2 mol^{-1}. Based on the surface area of a single graphene sheet (1315m^2 g^{-1}) the maximum value for the storage capacity of hydrogen adsorbed on graphite is calculated to be about 3 wt%. This value, however, can be reached only at very low temperatures. Based on this approximation the theoretical hydrogen storage capacity (mads) can be calculated from the specific surface of the carbon (S_{spec}): mads $\approx S_{spec} \times 2.27\times10^{-3}$ wt%; the amount of adsorbed hydrogen correlates with the specific surface of carbon. Taking this equation into account a theoretical carbon surface area of 2600m^2 g^{-1} would be necessary to meet the 2010 DOE target (6 wt%).

Fig. 3 shows the experimental relationship between different carbon surfaces and the hydrogen storage capacity under different measurement conditions.An extrapolation shows that the 2010 DOE target (6 wt %) would be reached at a carbon surface of >4000m^2 g^{-1}. This is not yet available.

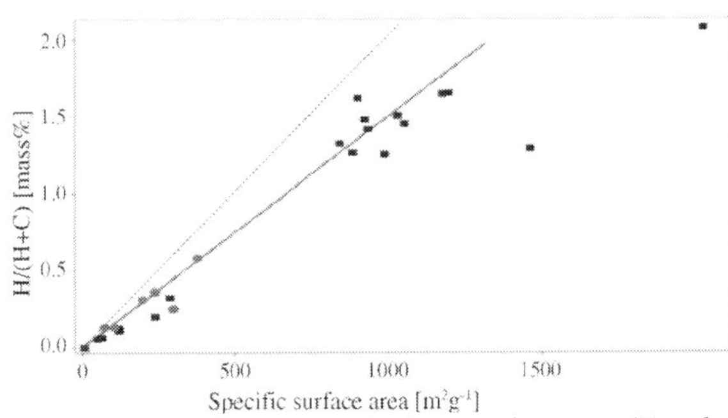

Fig. 3. Reversible amounts of hydrogen adsorbed vs. surface area (Züttel, 2003). (Circles) Carbon nanotubes, high surface area graphite (electrochemical measurement at 298 K). (Squares) Hydrogen adsorption measurement at 77K from Nijkamp et al. (Nijkamp, Raaymakers, Van Dillen, & De Jong, 2001). (Dotted line) Calculated amount of hydrogen in a monolayer at the substrate surface.

2.2. Chemisorption

If the π-bonding between carbon atoms were to be fully utilized, every carbon atom could be a site for chemisorption of one hydrogen atom. Desorption results of nanotubes treated with hydrogen under high pressure show a peak at temperatures above 400 K, which indicates chemisorption. This is ascribed to a complete layer of chemisorbed hydrogen on the outside wall of the nanotube. A first-principle calculation shows that a dissociative chemisorption of

two separate electrochemical reactions, one of which was not observed for pure palladium electrodes. Although the overall capacity for the palladium/MWNTs electrode was less than that anticipated for pure palladium, the study indicated that electrochemical hydrogen storage in MWNTs may be possible (A. Dillon & Heben, 2001).

3.3. Analysis of the results

An overview of experimental data on hydrogen adsorption in carbon nanotubes is given in the literature by Ding et al. Table 2 shows hydrogen storage values between 0.05 and 21 wt%. The high values, however, have not so far been verified independently.

The storage values are dependent on many parameters of the carbon nanotubes, including their structure, structure defects, pretreatment, purification, geometry (surface area, tube diameter, and length), arrangement of tubes in bundles and/or 'ropes', storage pressure, temperature, etc.

Hydrogen uptake varies linearly with tube diameter even though the tube volume increases as the square of the radius. This is because the uptake is proportional to the surface area (one layer adsorption), i.e. the number of carbon atoms, which increases linearly with increasing tube diameter. The tube diameter can be controlled via the synthesis conditions.

No complete agreement exists about the adsorption sites in carbon nanotubes. Theoretically the following sites exist:

- inside the tube;
- outside the tube;
- between tubes in 'bundles' and 'ropes';
- between the shells in multi-walled nanotubes.

For carbon nanotube bundles the adsorption sites are shown schematically in Fig. 6. For hydrogen storage within the tube the hydrogen either must pass through the CNTs wall or the tube must be open.

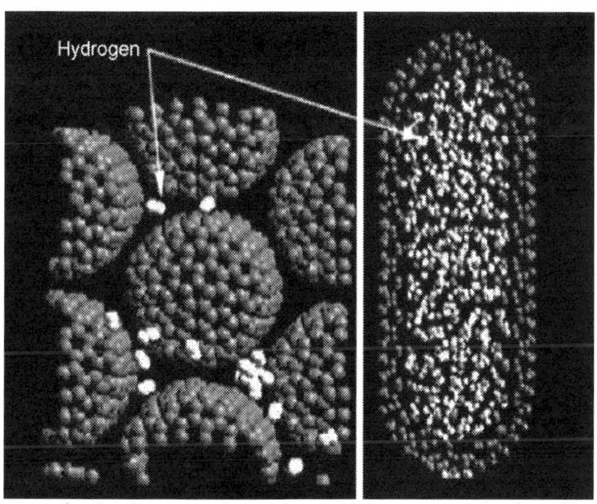

Fig. 6. Potential sites for hydrogen adsorption within a nanotube bundle; (left) interstitial sites between tubes and (right) inside the tube .

Sample	Purity (%)	T (K)	P (MPa)	H_2 (wt%)	Ref.
SWNTs	Assumed 100	133	0.04	5-10	(A. C. Dillon et al., 1997)
SWNTs	50	300	10.1	4.2	(C. Liu et al., 1999)
Aligned SWNTs	Purified	Ambient	11	4	(Liu, Yang, Tong, Cong, & Cheng, 2002)
SWNTs	High	80	7	8.25	(Y. Ye et al., 1999)
MWNTs	Purified	~300-700	Ambient	0.25	(Wu et al., 2000)
SWNTs	Purified	Ambient	4.8	1.2	(Smith Jr, Bittner, Shi, Johnson, & Bockrath, 2003)
SWNTs- Ti-6Al-4V	Purified	Ambient	0.08	1.7	(Hirscher et al., 2001)
SWNTs-Fe	Purified	Ambient	0.08	<0.005	(Hirscher et al., 2001)
Ball-milled SWNTs in Ar	<50	Ambient	0.08	<0.1	(Hirscher et al., 2002)
Ball-milled SWNTs in D2	<50	Ambient	0.9	0	(Hirscher et al., 2002)
CNTs	Purified	298-773	0.1	0	(Hirscher et al., 2002)
Li-CNTs	Purified	473-673	0.1	20	(P. Chen, X. Wu, J. Lin, & K. Tan, 1999)
K-CNTs	Purified	<313	0.1	14	(P Chen et al., 1999)
Li-CNTs (wet H_2)	Purified	473-673	0.1	12	(Yang, 2000)
Li-CNTs (dry H_2)	Purified	473-673	0.1	2.5	(Yang, 2000)
K-CNTs (wet H_2)	Purified	<313	0.1	21	(Yang, 2000)
K-CNTs (dry H_2)	Purified	<313	0.1	1.8	(Yang, 2000)
Li-CNTs	10-40	473-663	0.1	0.7-4.2	(Pinkerton et al., 2000)
SWNTs	90 vol%	298	-	0.63	(Ritschel et al., 2002)
MWNTs	Unpurified	298	-	0.5	(Ritschel et al., 2002)
CNTs	-	Ambient	Ambient	0.5	(Adu, Sumanasekera, Pradhan, Romero, & Eklund, 2001)
SWNTs	Purified	77	0.2	6	(Pradhan et al., 2002)
Aligned MWNTs	High	298	10	3	(Zhu et al., 2001)
Aligned MWNTs	High	290	10	2.4	(Cao et al., 2001)
Random MWNTs	High	298	10	0.68	(Zhu et al., 2001)
MWNTs	High untreated	300	1.0	5-7	(Y. Chen et al., 2001)
MWNTs	High acid treated	300	1.0	13.8	(Y. Chen et al., 2001)
MWNTs	High	300	7.0	0.7-0.8	. (Badzian, Badzian, Breval, & Piotrowski, 2001)
SWNTs	Unpurified	295	0.1	0.93	(Nishimiya et al., 2002)
SWNTs	Unpurified	77	0.1	2.37	(Nishimiya et al., 2002)
MER MWNTs	10-15%	298	3.6	0.03	(Tibbetts, Meisner, & Olk, 2001)
Rice SWNTs	High	298	3.6	0.05	(Tibbetts et al., 2001)
CNTs	-	298	11.35	11.26	(Chambers, Park, Baker, & Rodriguez, 1998)

Table 2. Summary of the reports of gaseous hydrogen storage capacity in carbon nanotubes

As shown in both Tables 2 extremely high hydrogen storage values (up to 21 wt% for the K-CNTs structure) have been reported. As already mentioned a worldwide effort to verify these large hydrogen storage values in various nanostructured carbons, however, has not been successful. It is surprising, therefore, that very often these values continue to be cited in the literature without any acknowledgement of their uncertainty. The reasons why confirmation of these results has not been forthcoming could include the following:

· Low purity of samples: Syntheses of carbon nanostructured samples, especially those formed shortly after the discovery of these structures, tended to produce rather impure phases. Very often only small fractions of carbon nanotubes were found in the products and the hydrogen storage capacity of those samples was extrapolated in order to predict the capacity of pure materials.

· Insufficient characterization of samples: The carbon samples are very often a mixture of single-walled nanotubes, multi-walled nanotubes, and carbon nanofibers with various diameters, lengths, chiral vectors (in the case of nanotubes), shell numbers (in the case of multi-walled tubes), open and closed structures (tubes), etc. Because of all this potential variability it would not be surprising to find a large range of hydrogen storage values.

· Contamination of samples: For example, a storage capacity of about 4.5 wt% of hydrogen was found for single-walled nanotubes at 30 kPa and 70 K. Only a few milligrams of SWNTs were used for these investigations and the tubes were opened by a high power ultrasonic procedure in nitric acid(A. C. Dillon et al., 1997). The ultrasonic horn that was used was made from a corrosion-resistant Ti−6Al−4V alloy. This alloy, however, is also a hydrogen storage material. Hirscher et al. (Hirscher et al., 2001) have shown that nearly all of the storage capacity of the sonicated samples examined in this work can be attributed to the hydrogen uptake of titanium alloy particles which have contaminated the carbon samples during sonication.

· Hydrogen contamination: Chen et al. (C Liu et al., 1999) found hydrogen storage capacities up to 20 wt% for lithium-doped multi-walled nanotubes and up to 14 wt% for lithium-doped graphite. It was shown, however , that the majority of the weight changes in this case are caused by adsorbed water, which was present as a contaminant in the hydrogen. Furthermore the Li_2O used for the carbon treatment is both hygroscopic and able to adsorb carbon dioxide. Both of these processes lead to weight gains that could be mistakenly attributed to hydrogen uptake.

· Technical problems: The relatively complicated methods of synthesis of these materials often result in the production of only a rather small mass of carbon nanotubes and/or nanofibers. There is thus considerable scope for errors during gravimetric or volumetric measurement. In addition, inaccurate cooling of the hydrogen gas after pressurization can lead to effects that simulate adsorption. Problems can also arise through leaks in measurement equipment.

4. Simulations of hydrogen adsorption in carbon nanotubes

A number of theoretical investigations have been performed. These can be classified in two different groups. On the one hand, there are the grand canonical Monte Carlo simulations, which regard the uptake as a simple physical adsorption process. Their aim is to determine the adsorption isotherms on the assumption that no chemical process takes place and that the interaction between the carbonaceous materials and the hydrogen molecules can be described via a classical empirical potential. On the other hand, there is evidence that during the uptake a chemical process does indeed take place. This is also suggested by chemical intuition. The simulation of these processes requires that the electronic states are explicitly simulated. So far only two such sets of calculations have been performed by Lee et al. (S. Lee, An, Lee, Seifert, & Frauenheim, 2001a), who studied the chemisorption process in nanotubes. calculations suggest that chemical processes do take place and that the hydrogen uptake is more complex than previously anticipated.

4.1 Classical calculations (physisorption)

Many scientific publications are devoted to the theoretical study of physical adsorption of gases on different adsorbent structures. Among these, noble gases are widely studied due to the inertness of these gases near reactive surfaces such as graphite planes and carbon nanotubes walls. These works concern both theoretical and realistic models of molecular gases in pores described by hard walls or attractive interfaces. They established that the physical adsorption of gases on a solid adsorbent can be accurately computed by numerical simulation . However, concerning carbon nanotube a few experimental works have been devoted to experimental measurements of gas adsorption in these adsorbents. The adsorption of diatomic gases such as hydrogen and nitrogen has been studied at different pressures and temperatures with the aim of investigating the physical mechanism of adsorption in carbon nanotubes. The agreement between simulation and experimental results indicates that the potential of interaction of carbon nanotubes is well described by the current potential used for planar graphite surfaces . Moreover, it is noticeable that the possibility of controlling microscopic parameters such as the nanotube diameters and lengths will enable to control material macroscopic properties such as gas adsorption. In the case of hydrogen adsorption, the storage of the gas in nanoscale-dimension material seems to be a good opportunity(FL Darkrim, Malbrunot, & Tartaglia, 2002).

Wang and Johnson have also used the Feynmann path integral in order to account for the quantum effects of hydrogen (Wang & Johnson, 1999). The results of these calculations show that the highest possible adsorption is achieved in nanofibers with large interlayer distance at 77 K. While it is not clear whether such large interplanar distances can be achieved, it is discouraging to note that the storage capacity decreases very rapidly with increasing temperature. In the calculations that involve the arrays of nanotubes, the effect of change in geometry has been studied in some detail. Change in geometry can result both from changes in internal geometry (e.g. diameter) of the nanotubes and from the the relative arrangement of the nanotubes. As usual the geometry of a single nanotube is determined by two integer numbers that describe the way in which the graphite sheets are wrapped to form the nanotubes. The two-dimensional geometrical arrangement of nanotube arrays is expressed by the van der Waals gap, g, which measures the available interstitial space. g is defined as $g = a \ D$, where a is the lattice spacing and D is the diameter of the nanotubes. The results do not show encouraging storage capacities at any temperature or pressure(Meregalli & Parrinello, 2001).

Indeed, Darkrimand Levesque (Farida Darkrim & Levesque, 2000) in recent numerical simulations have shown that adsorbent materials appropriate for hydrogen storage could be made up of carbon nanotubes constituted by nanoscale-dimension pores. Monte-Carlo numerical simulation based upon equilibrium statistical mechanics corresponds to a thermodynamic evaluation of the amount of gas adsorbed near the adsorbent surface. To compute this calculation, gas-adsorbent and gas-gas molecular interaction data are needed. Wang and Johnson (Wang & Johnson, 1999)have presented different works on hydrogen

and xenon adsorption in nanotube materials both at room temperature and at liquid nitrogen temperature.
The examples that treated the gas adsorption in such nanoscale dimension structures, seem to converge on one point: the possibility of adsorbing hydrogen at high rates for practical purposes. SWNTs present narrow pore size distribution and this characteristic should be favorable to high adsorptive material properties.
At room temperature it was established that hydrogen gas was efficiently adsorbed in carbon nanotube materials. By using Monte-Carlo simulations hydrogen adsorption was computed at moderate pressures and it was demonstrated that the amount of gas stored by adsorption in carbon nanotubes was larger than that stored by gas compression. The maximum of adsorption at 293 K was obtained with a hexagonal configuration of nanotubes distant from 0:7 nm and with diameter equal to 1:2 nm.
The differences of adsorption amounts published are mainly attributed to:

- the interaction potential models used to describe the gas-solid interaction as in the works of Wang and Johnson ;
- the sampling method used in simulations ;
- the adsorbent configurations (including the values of the tube diameters, the tube lengths and the inter-tube spacing) that permit the hydrogen gas either to move in a large adsorption volume or just to be con9ned in the interstitial pores. Moreover, due to the sterical effects that could occur within thin pores, like in nanomaterial ones, which are attributed to the tube arrangement within a row of nanotubes;
- single- or multi-walled carbon nanotubes;
- open or closed tubes which are important for the adsorption phenomenon (open tubes enable gas adsorption both inside and outside the tubes whereas closed tubes do not) and also
- the reactive surface of the adsorbent material considered. Indeed, this latter value is dependent on the distance considered between consecutive tubes the so-called inter-tube spacing. Moreover, it is important to mention this possibility seem to experimentally control this parameter. Then, considering the molecular simulation works, it is noticeable that the control of microscopic parameters during the nanotube synthesis (tube diameter and length: : :) which would enable an optimization of the material structure should then enable the control of the gas adsorption process(FL Darkrim et al., 2002).

4.2 First-principles calculations (chemisorption)

Monte Carlo calculations based on classical potentials are not well suited to study chemical processes. However, it is likely that hydrogen and carbon react chemically. In fact, the infrared experiments of Chen et al. (P Chen et al., 1999) indicate the formation of C H bonds. In order to study this kind of storage, a more sophisticated method capable of quantum-mechanically treating the bond breaking and the bond formation is required. Ab initio calculations, capable of describing chemical processes have been performed on nanotubes by Lee et al. (S. Lee & Lee, 2000)
Lee and Lee (S. Lee & Lee, 2000) have performed density-functional and density-functional-based tight binding calculations to search for hydrogen chemisorption sites and predict the maximum storage capability in single-walled carbon nanotubes (SWCNs). Their calculations are performed at zero Kelvin and show two energetically favoured configurations in which

hydrogen atoms are chemisorbed on the nanotubes. The lowest energy one (*zigzag type*) is that in which the hydrogen atoms are bonded alternately outside and inside the tube (Fig. 7c). Stable but 0.56 eV higher in energy per C−H bond is the *arch-type* arrangement, shown in Fig. 7b. These configurations are stable in spite of the deformations induced on the C−C bonds by the chemisorption process. The authors have also investigated the possibility of absorbing all the hydrogen atoms on the inside. It is, however, energetically favorable for the hydrogen atoms to recombine and form molecules, which are then physisorbed inside the nanotube, leading to physisorption. The maximum possible adsorption would be an apparently encouraging 14% in weight.

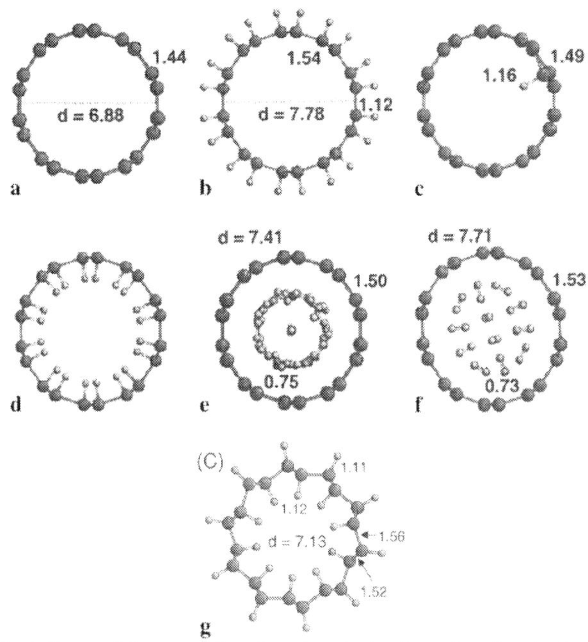

Fig. 7a–g. The calculations of Lee and Lee for chemical adsorption in singlewalled carbon nanotubes (SWCNs). a (5,5) SWCN before the storage; b covalent storage outside the tube archtype (θ = 1); c adsorption of a single hydrogen atom inside the tube; d initial and e fully relaxed geometry of adsorbed hydrogen inside the tube walls; f hydrogen adsorption inside the tube with θ = 1.2; g zigzag-type geometry

In a later publication, Lee et al. (S. Lee, An, Lee, Seifert, & Frauenheim, 2001b) addressed the issue of mechanisms for the insertion of hydrogen molecules inside the nanotubes. This is in fact a non-trivial problem even if the nanotubes are open. The very high aspect ratio makes the filling process through the open nanotubes rather inefficient. Lee et al. hypothesized a multi-step process. An hydrogen atom bonded to a carbon atom in an *arch-type* geometry pushes down the carbon atom and then flips into the C−C mid-bond; the activation barrier has an energy of 1.51 eV. After the H atom has flipped into the capillary, the C−C bond is recovered. Once the first atom has flipped in, the nearest-neighbour top site atoms can flip in more easily with the lower activation barrier of 0.74 eV, as shown in Fig. 2 II leading to a continuous flip-in process. One may also consider a zigzag flip-in process, i.e., a flip-in process for every second hydrogen (Fig. 8 III). This process has an activation barrier of 0.93 eV and will result in a zigzag-type geometry. Once the zigzag-type geometry, with a coverage of θ = 1.0, is formed, the kick-in mechanism described in Fig. 8 IV could explain hydrogen insertion in the nanotube. Repeating the kick-in process will eventually lead to the formation of molecular hydrogen inside the tube. In order to complete the storage mechanism, Lee et al. also described a similar hydrogen extraction mechanism with a low energy barrier.

Lee et al. (S. M. Lee et al., 2000) addressed the problem of storage in MWNTs. They claim that hydrogen atom chemisorption is favored between the tube walls where hydrogen storage capacity is independent of diameter, and they estimate it to be of the order of 7.7 wt%.

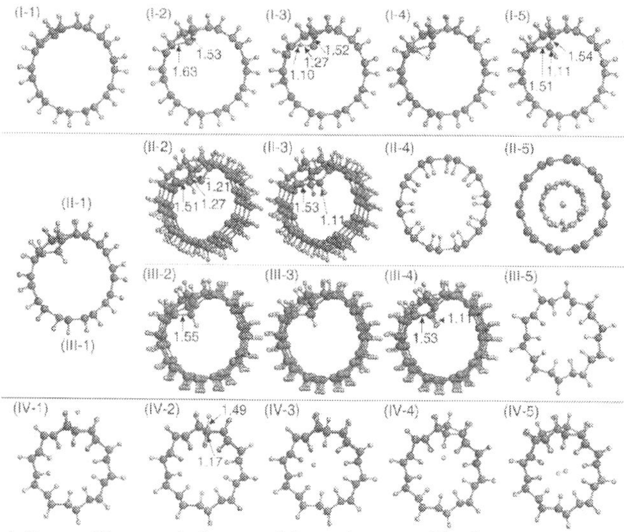

Fig. 8. The calculations of Lee et al. for possible pathways of hydrogen insertion. (I) Flip-in mechanism, (II) continuous flip-in mechanism leading to the formation of molecular hydrogen inside the tube, (III) zigzag flip-in mechanism, (IV) kick-in mechanism

Bilić et al(Bilić & Gale, 2008). modeled the energetics of hydrogen chemisorption on a (6, 6) nanotube using gradient-corrected density functional theory. The adsorption was investigated over a range of coverages, from close to the zero limit to a full monolayer. They found that the attachment of the first isolated admolecule on the tube is endothermic. However, with increasing coverage, the character of chemisorption dramatically changes to exothermic, making the dissociative addition of new H_2 molecules increasingly favorable. They explored possible routes for the formation of energetically favourable fully hydrogenated nanotubes. The mechanism termed the 1-2d exoaddition is predicted to have the potential of adsorbing between 1/3 and 1/2 ML of hydrogen on the outer face of the tube. The initial kinetics of such a process is predicted to proceed at an increasing rate with increasing degree of coverage. If a modified mechanism is considered, a combined 1-2d exo-/endoaddition, full saturation of the tube is predicted, resulting in adsorption energy of -0.618 eV per molecule and a hydrogen uptake of 7.7 wt %. The fact that such structures are not observed experimentally is attributed to the highly activated nature of the dissociative chemisorption and the less reactive character of larger, more realistic nanotubes(Meregalli & Parrinello, 2001).

5. Conclusion

The present examination of the actual state of the art of hydrogen adsorption in carbon nanotubes involves the fundamental question: is adsorption in nanotubes a reliable mode of hydrogen storage?

This work was devoted to experimental and computational studies of carbon nanotube adsorption property. These materials, if one considers some articles published in the literature, seem to be good candidates for the hydrogen storage process. However, some questions persist!! In the future, some explanations are still needed for all scientists who are interested in the hydrogen storage by adsorption in carbon nanomaterials. In particular, as concerns the surface properties of these adsorbent materials: the chemical treatment during or after the carbon nanotubes synthesis; the mechanical treatment made on the tubes in order to open their extremities and enable the gas adsorption inside the tubes; the amount of reactive nanotubes (attributed to their low purity rate after their synthesis) taken into account in the gas adsorption calculations; the presence or not of heteroatoms at their surface which can contribute to modify the electronic density at the nanotube surface and then the reactivity of the material can be reinforced (for instance it is possible to intercalate alkaline atoms between consecutive tubes); the accuracy of the volumetric and the gravimetric measurement methods should be mentioned; the definition of the hydrogen 'uptake' used by the authors in their works (excess of adsorption or total amount of the gas). Once these respective main influencing factors on the hydrogen adsorption in carbon nanomaterials are well determined, one will be able to determine, explain, and validate the various adsorption amounts published in the literature and to control their own adsorption data which are unfortunately often obtained on their own characterized materials.

6. References

Adu, C., Sumanasekera, G., Pradhan, B., Romero, H. and Eklund, P. (2001). Carbon nanotubes: a thermoelectric nano-nose. *Chemical Physics Letters,* Vol, 337 No, 1-3, pp. 31-35

Badzian, A., Badzian, T., Breval, E. and Piotrowski, A. (2001). Nanostructured, nitrogen-doped carbon materials for hydrogen storage. *Thin Solid Films,* Vol, 398, pp. 170-174

Bilic, A. and Gale, J. D. (2008). Chemisorption of molecular hydrogen on carbon nanotubes: A route to effective hydrogen storage? *Journal of Physical Chemistry C,* Vol, 112 No, 32, pp. 12568-12575,1932-7447

Bilić, A. and Gale, J. D. (2008). Chemisorption of Molecular Hydrogen on Carbon Nanotubes: A Route to Effective Hydrogen Storage? *The Journal of Physical Chemistry C,* Vol, 112 No, 32, pp. 12568-12575,1932-7447

Cao, A., Zhu, H., Zhang, X., Li, X., Ruan, D., Xu, C., Wei, B., Liang, J. and Wu, D. (2001). Hydrogen storage of dense-aligned carbon nanotubes. *Chemical Physics Letters,* Vol, 342 No, 5-6, pp. 510-514

Chambers, A., Park, C., Baker, R. and Rodriguez, N. (1998). Hydrogen storage in graphite nanofibers. *Journal of Physical Chemistry B-Condensed Phase,* Vol, 102 No, 22, pp. 4253-4256

Chen, C. and Huang, C. (2007). Hydrogen storage by KOH-modified multi-walled carbon nanotubes. *International Journal of Hydrogen Energy,* Vol, 32 No, 2, pp. 237-246

Chen, P., Wu, X., Lin, J. and Tan, K. (1999). High H2 uptake by alkali-doped carbon nanotubes under ambient pressure and moderate temperatures. *Science,* Vol, 285 No, 5424, pp. 91

Chen, P., Wu, X., Lin, J. and Tan, K. L. (1999). High H2 Uptake by Alkali-Doped Carbon Nanotubes Under Ambient Pressure and Moderate Temperatures. *Science,* Vol, 285 No, 5424, pp. 91-93

Chen, Y., Shaw, D., Bai, X., Wang, E., Lund, C., Lu, W. and Chung, D. (2001). Hydrogen storage in aligned carbon nanotubes. *Applied Physics Letters,* Vol, 78, pp. 2128

Darkrim, F. and Levesque, D. (2000). High Adsorptive Property of Opened Carbon Nanotubes at 77 K. *The Journal of Physical Chemistry B,* Vol, 104 No, 29, pp. 6773-6776,1520-6106

Darkrim, F., Malbrunot, P. and Tartaglia, G. (2002). Review of hydrogen storage by adsorption in carbon nanotubes. *International Journal of Hydrogen Energy,* Vol, 27 No, 2, pp. 193-202

Dillon, A. and Heben, M. (2001). Hydrogen storage using carbon adsorbents: past, present and future. *Applied Physics A: Materials Science & Processing,* Vol, 72 No, 2, pp. 133-142

Dillon, A. C., Jones, K. M., Bekkedahl, T. A., Kiang, C. H., Bethune, D. S. and Heben, M. J. (1997). Storage of hydrogen in single-walled carbon nanotubes. *Nature,* Vol, 386 No, 6623, pp. 377-379

Hirscher, M., Becher, M., Haluska, M., Dettlaff-Weglikowska, U., Quintel, A., Duesberg, G., Choi, Y., Downes, P., Hulman, M. and Roth, S. (2001). Hydrogen storage in sonicated carbon materials. *Applied Physics A: Materials Science & Processing,* Vol, 72 No, 2, pp. 129-132

Hirscher, M., Becher, M., Haluska, M., Quintel, A., Skakalova, V., Choi, Y., Dettlaff-Weglikowska, U., Roth, S., Stepanek, I. and Bernier, P. (2002). Hydrogen storage in carbon nanostructures. *Journal of Alloys and Compounds,* Vol, 330, pp. 654-658

Hwang, J., Gommans, H., Ugawa, A., Tashiro, H., Haggenmueller, R., Winey, K., Fischer, J., Tanner, D. and Rinzler, A. (2000). Polarized spectroscopy of aligned single-wall carbon nanotubes. *Physical Review B,* Vol, 62 No, 20, pp. 13310-13313

Kataura, H., Kumazawa, Y., Maniwa, Y., Umezu, I., Suzuki, S., Ohtsuka, Y. and Achiba, Y. (1999). Optical properties of single-wall carbon nanotubes. *Synthetic Metals,* Vol, 103 No, 1-3, pp. 2555-2558

Kubas, G. (2001). Metal–dihydrogen and -bond coordination: the consummate extension of the Dewar–Chatt–Duncanson model for metal–olefin bonding. *Journal of Organometallic Chemistry,* Vol, 635 No, 1-2, pp. 37-68

Lee, S., An, K., Lee, Y., Seifert, G. and Frauenheim, T. (2001a). A hydrogen storage mechanism in single-walled carbon nanotubes. *J. Am. Chem. Soc,* Vol, 123 No, 21, pp. 5059-5063

Lee, S., An, K., Lee, Y., Seifert, G. and Frauenheim, T. (2001b). Novel mechanism of hydrogen storage in carbon nanotubes. *JOURNAL-KOREAN PHYSICAL SOCIETY,* Vol, 38 No, 6, pp. 686-691

Lee, S. and Lee, Y. (2000). Hydrogen storage in single-walled carbon nanotubes. *Applied Physics Letters,* Vol, 76, pp. 2877

Lee, S. M., Park, K. S., Choi, Y. C., Park, Y. S., Bok, J. M., Bae, D. J., Nahm, K. S., Choi, Y. G., Yu, S. C., Kim, N.-g., Frauenheim, T. and Lee, Y. H. (2000). Hydrogen adsorption and storage in carbon nanotubes. *Synthetic Metals,* Vol, 113 No, 3, pp. 209-216,0379-6779

Liu, C., Fan, Y., Liu, M., Cong, H., Cheng, H. and Dresselhaus, M. (1999). Hydrogen storage in single-walled carbon nanotubes at room temperature. *Science,* Vol, 286 No, 5442, pp. 1127

Liu, C., Fan, Y. Y., Liu, M., Cong, H. T., Cheng, H. M. and Dresselhaus, M. S. (1999). Hydrogen Storage in Single-Walled Carbon Nanotubes at Room Temperature. *Science,* Vol, 286 No, 5442, pp. 1127-1129

Liu, C., Yang, Q., Tong, Y., Cong, H. and Cheng, H. (2002). Volumetric hydrogen storage in single-walled carbon nanotubes. *Applied Physics Letters,* Vol, 80, pp. 2389

Meregalli, V. and Parrinello, M. (2001). Review of theoretical calculations of hydrogen storage in carbon-based materials. *Applied Physics A: Materials Science & Processing,* Vol, 72 No, 2, pp. 143-146

Nützenadel, C., Züttel, A., Chartouni, D. and Schlapbach, L. (1999). Electrochemical storage of hydrogen in nanotube materials. *Electrochemical and solid-state letters,* Vol, 2, pp. 30

Nijkamp, M., Raaymakers, J., Van Dillen, A. and De Jong, K. (2001). Hydrogen storage using physisorption–materials demands. *Applied Physics A: Materials Science & Processing,* Vol, 72 No, 5, pp. 619-623

Nishimiya, N., Ishigaki, K., Takikawa, H., Ikeda, M., Hibi, Y., Sakakibara, T., Matsumoto, A. and Tsutsumi, K. (2002). Hydrogen sorption by single-walled carbon nanotubes prepared by a torch arc method. *Journal of Alloys and Compounds,* Vol, 339 No, 1-2, pp. 275-282

Pinkerton, F., Wicke, B., Olk, C., Tibbetts, G., Meisner, G., Meyer, M. and Herbst, J. (2000). Thermogravimetric measurement of hydrogen absorption in alkali-modified carbon materials. *J. Phys. Chem. B,* Vol, 104 No, 40, pp. 9460-9467

Popov, V. (2004). Carbon nanotubes: properties and application. *Materials Science & Engineering R,* Vol, 43 No, 3, pp. 61-102

Pradhan, B., Harutyunyan, A., Stojkovic, D., Grossman, J., Zhang, P., Cole, M., Crespi, V., Goto, H., Fujiwara, J. and Eklund, P. (2002). Large cryogenic storage of hydrogen in carbon nanotubes at low pressures. *Journal of Materials Research,* Vol, 17 No, 9, pp. 2209-2216

Ritschel, M., Uhlemann, M., Gutfleisch, O., Leonhardt, A., Graff, A., Täschner, C. and Fink, J. (2002). Hydrogen storage in different carbon nanostructures. *Applied Physics Letters,* Vol, 80, pp. 2985

Sakintuna, B., Lamari-Darkrim, F. and Hirscher, M. (2007). Metal hydride materials for solid hydrogen storage: a review. *International Journal of Hydrogen Energy,* Vol, 32 No, 9, pp. 1121-1140

Schulz, R., Huot, J., Liang, G., Boily, S., Lalande, G., Denis, M. and Dodelet, J. (1999). Recent developments in the applications of nanocrystalline materials to hydrogen technologies. *Materials science & engineering. A, Structural materials: properties, microstructure and processing,* Vol, 267 No, 2, pp. 240-245

Smith Jr, M., Bittner, E., Shi, W., Johnson, J. and Bockrath, B. (2003). Chemical activation of single-walled carbon nanotubes for hydrogen adsorption. *J. Phys. Chem. B,* Vol, 107 No, 16, pp. 3752-3760

Ströbel, R., Garche, J., Moseley, P., Jörissen, L. and Wolf, G. (2006). Hydrogen storage by carbon materials. *Journal of Power Sources,* Vol, 159 No, 2, pp. 781-801

Ströbel, R., Garche, J., Moseley, P. T., Jörissen, L. and Wolf, G. (2006). Hydrogen storage by carbon materials. *Journal of Power Sources,* Vol, 159 No, 2, pp. 781-801,0378-7753

Tibbetts, G., Meisner, G. and Olk, C. (2001). Hydrogen storage capacity of carbon nanotubes, filaments, and vapor-grown fibers. *Carbon,* Vol, 39 No, 15, pp. 2291-2301

Wang, Q. and Johnson, J. (1999). Molecular simulation of hydrogen adsorption in single-walled carbon nanotubes and idealized carbon slit pores. *The Journal of Chemical Physics,* Vol, 110, pp. 577

Wu, X., Chen, P., Lin, J. and Tan, K. (2000). Hydrogen uptake by carbon nanotubes. *International Journal of Hydrogen Energy,* Vol, 25 No, 3, pp. 261-265

Yang, R. (2000). Hydrogen storage by alkali-doped carbon nanotubes–revisited. *Carbon,* Vol, 38 No, 4, pp. 623-626

Ye, Y., Ahn, C., Witham, C., Fultz, B., Liu, J., Rinzler, A., Colbert, D., Smith, K. and Smalley, R. (1999). Hydrogen adsorption and cohesive energy of single-walled carbon nanotubes. *Applied Physics Letters,* Vol, 74, pp. 2307

Ye, Y., Ahn, C. C., Witham, C., Fultz, B., Liu, J., Rinzler, A. G., Colbert, D., Smith, K. A. and Smalley, R. E. (1999). Hydrogen adsorption and cohesive energy of single-walled carbon nanotubes. *Applied Physics Letters,* Vol, 74 No, 16, pp. 2307-2309

Züttel, A. (2003). Materials for hydrogen storage. *Materials today,* Vol, 6 No, 9, pp. 24-33

Zhu, H., Cao, A., Li, X., Xu, C., Mao, Z., Ruan, D., Liang, J. and Wu, D. (2001). Hydrogen adsorption in bundles of well-aligned carbon nanotubes at room temperature. *Applied Surface Science,* Vol, 178 No, 1-4, pp. 50-55

29

Carbon Nanotube Supercapacitors

Wen Lu
ADA Technologies Inc.
8100 Shaffer Parkway, Littleton, CO 80127
USA
Liming Dai
Department of Chemical Engineering
Case Western Reserve University
10900 Euclid Avenue, Cleveland, Ohio 44106
USA

1. Introduction

Supercapacitors (aka, electrochemical capacitors or ultracapacitors) are electrochemical energy storage devices that combine the high energy-storage-capability of conventional batteries with the high power-delivery-capability of conventional capacitors (Burke, 2000; Conway, 1999). Able to achieve higher power and longer cycle life than batteries, supercapacitors have been developed to provide power pulses for a wide range of applications including electric transportation technology (e.g., hybrid electric vehicles (HEVs) and plug-in HEVs), electric utility industry (e.g., emergency backup power and grid system stability improvement), consumer electronics (e.g., laptops, cell phones, pagers, and video cameras), medical electronics (e.g., portable defibrillators, drug delivery units, and neurological stimulators), and military / defense devices (e.g., communication devices, unmanned aerial vehicles, spacecraft probes, and missile systems) (Miller & Burke, 2008; Miller & Simon, 2008). In advanced electric transportation technologies, for example, supercapacitors are being developed as power assists for HEVs and plug-in HEVs, where the supercapacitor is operated to provide peak power during acceleration and hill-climbing, and it can be recharged during regenerative braking (Ehsani et al., 2005). This is extremely useful in achieving better fuel economy, decreasing harmful emissions, and reducing the reliance on petroleum sources. A recent report prepared by the US Department of Energy indicated that supercapacitors have been paid equal attention as to batteries as the future energy storage technologies (US DoE, 2007). The world market for supercapacitors has been growing steadily and rapidly. Nevertheless, in order to improve the performance of the state-of-the-art supercapacitors to satisfy the rapidly increasing performance demands for the applications mentioned above, new electrode materials having superior properties over those of the currently used activated carbon electrodes are needed. This chapter summarizes the recent research and technology in developing carbon nanotube (CNT)-based materials as a new type of electrode materials for supercapacitors.

1.1 Supercapacitors and currently used supercapacitor electrode materials

The supercapacitor concept was first described in a patent filed in 1957 by Becker, who utilized a high-surface-area carbon electrode and an aqueous H_2SO_4 electrolyte to fabricate the supercapacitors (Kötz & Carlen, 2000). In 1971, NEC (Japan) developed aqueous electrolyte-based supercapacitors under the energy company SOHIO's license for powersaving units in electronics. This activity was considered as the initial commercialization application of supercapacitors (Kötz & Carlen, 2000).

Typically, a supercapacitor consists of three essential components, namely the electrodes, the electrolyte, and the separator. The electrodes are an essential component for charge storage / delivery and play an important role in determining energy and power densities of a supercapacitor. Supercapacitors have been realized using three principal types of electrode materials, namely high-surface-area activated carbons (Frackowiak & Béguin, 2001), transition metal oxides (Zheng et al., 1995), and electroactive polymers (Rudge et al., 1994). Fig.1 illustrates the basic structure of a supercapacitor based on high-surface-area and porous carbon electrodes. In such a capacitor, energy is stored as charge separation in the double-layer (i.e., double layer capacitance) formed at the interface between the solid electrode surface and the liquid electrolyte. The ions displaced in forming the double-layers are transferred between the electrodes by diffusion through the electrolyte.

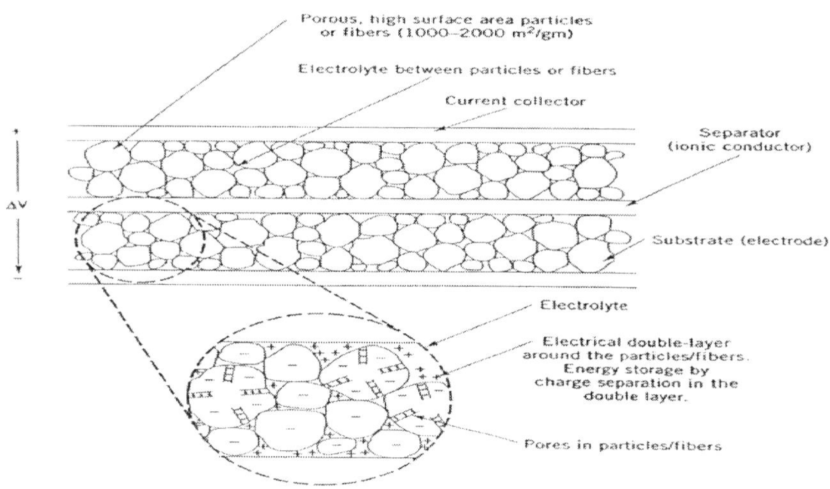

Fig. 1. Schematic of an electrical double-layer supercapacitor (Burke, 2000).

Transition metal oxides and electroactive polymers are redox materials. Faradaic reactions within the bulk of these materials provide additional charge storage, i.e., a pseudocapacitance, resulting in higher energy densities for the capacitors (i.e., pseudo-capacitors or redox capacitors). However, the application of transition metal oxides (usually noble metal oxides, e.g., RuO_2) has been limited by their high cost and that of electroactive polymers by their limited stability. High-surface-area activated carbons are still the predominant electrode materials for commercial supercapacitors.

The maximum energy (E_{max}) and maximum power (P_{max}) of a supercapacitor are given by:

$$E_{max} = (CU^2)/2 \quad (1)$$
$$P_{max} = U^2/(4R) \quad (2)$$

where C is capacitance, U is cell voltage, and R is total equivalent series resistance (ESR) of the capacitor (Conway, 1999). Overall performance of the supercapacitor is determined by the physical properties of both the electrode and the electrolyte materials. Having the advantages of relatively low cost, commercial availability, and well-established electrode production technologies, high-surface-area activated carbons (specific surface area: 1000 ~2000 m^2/g) have been the major electrode materials for commercial supercapacitors (Burke & Arulepp, 2001). Charge storage capability of these materials is usually evaluated by their capacitance, which is associated with the electrode / solution interface which varies with the type of carbon and its conditions of preparation and usually has a value of 20 ~ 50 $\mu F/cm^2$ (Conway, 1999). Taking a specific surface area of 1000 m^2/g for carbon as an example, its ideal attainable capacitance would be 200 ~ 500 F/g. However, the practically obtained values are of only a few tens of F/g. Poor accessibility of carbon surface to electrolyte (i.e., electrolyte accessibility) has been confirmed to be the most important reason for the absence of proportionality between specific capacitance and surface area of these materials.

According to the IUPAC classification, there are three classes of pore sizes: micropores (< 2 nm), mesopores (2 ~ 50 nm), and macropores (> 50 nm) (Conway, 1999). It is known that most of the surface area of activated carbon electrode materials resides in the scale defined as micropores (Frackowiak & Béguin, 2001). Pores of this scale are often poorly or non-accessible for electrolyte ions (especially for organic electrolytes) and thus are incapable of supporting an electrical double layer. In principle, the pores greater than 0.5 nm are electrochemically accessible for aqueous electrolytes. However, larger pores are required for organic electrolytes because of their larger size of solvated ions. Having wider electrochemical windows, organic electrolytes ensure higher energy density and power density, and are being employed in commercial supercapacitors. Unfortunately, these advantages have been limited by the reduced electrolyte accessibility associated with the currently used activated carbon electrode materials. Although developed surface area is preferable for the capacitance of a carbon material, increase in the surface area is usually accompanied with the decrease of the pore size. On the other hand, it is known that mesopores contribute the most to the capacitance in an electrical double-layer capacitor (Kastening & Sprinzig, 1986; Mayer et al., 1993; Tanahashi et al., 1990). Thus, an appropriate balance between the surface area and the mesoporosity of a carbon material is the key to achieve high capacitance. Currently available activated carbon materials have high surface area but unfortunately low mesoporosity, resulting in their low electrolyte accessibility and thus limited capacitance (Frackowiak & Béguin, 2001). This translates to the limited energy density of the resultant supercapacitors (Equation 1). Moreover, along with their poor electrical conductivity, the low electrolyte accessibility of activated carbons produces a high ESR (Frackowiak & Béguin, 2001) and hence a poor power density for the capacitors (Equation 2). Consequently, fabricated from these activated carbon electrodes, currently available supercapacitors possess a limited energy density (4 ~ 5 Wh/kg) and a limited power density (1 ~ 2 kW/kg) (Burke & Arulepp, 2001).

Therefore, in order to overcome the drawbacks of the presently used activated carbon electrode materials, high surface area, high electrical conductivity, high mesoporosity, and high electrolyte accessibility should be the important properties desired for an ideal electrode material. CNTs have been utilized to develop new electrode materials to meet these requirements.

1.2 Carbon nanotubes

Since their discovery by Iijima in 1991 (Iijima, 1991), carbon nanotubes (CNTs) have attracted considerable attention as a family of novel electronic and electrochemical materials. CNTs can be viewed as a graphene sheet rolled up into a nanoscale tube form to produce the so-called single-wall carbon nanotubes (SWNTs, Fig.2a). There may be additional graphene tubes around the core of a SWNT to form multi-wall carbon nanotubes (MWNTs, Fig.2b), including double wall carbon nanotubes (DWNTs) (Fig.2c) (Harris, 1999). These elongated nanotubes usually have a diameter in the range from a few angstroms to tens of nanometers and a length of several micrometers up to centimeters (allowing the fabrication of mesopore electrodes) with both ends of the tubes normally capped by fullerene-like structures containing pentagons. Morphologically, CNTs can be realized in randomly entangled and in highly aligned forms (Fig.3).

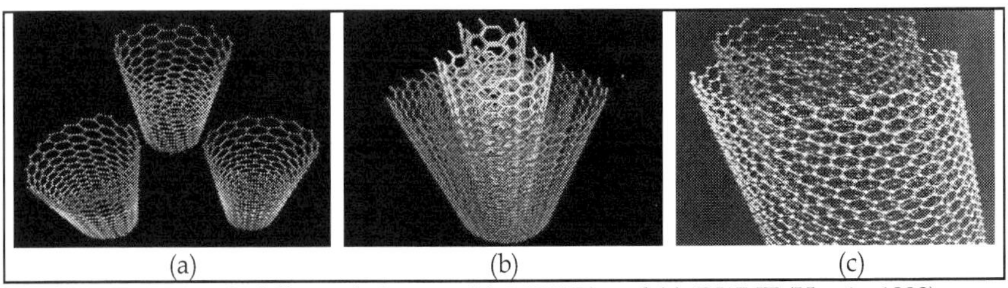

Fig. 2. (a): SWNTs with different helicities, (b): MWNT, and (c): DWNT (Harris, 1999).

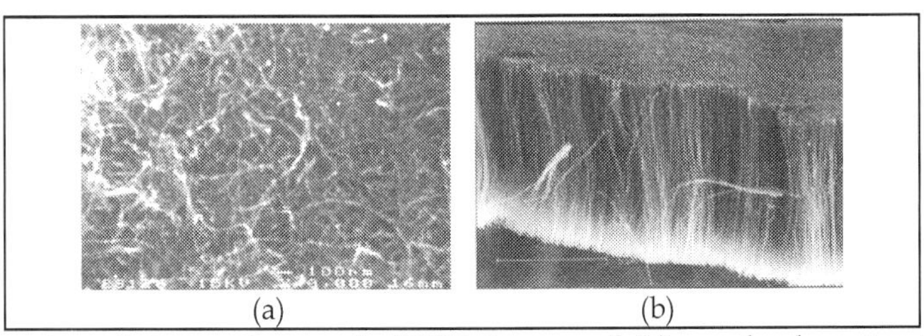

Fig. 3. Scanning electron microscopic (SEM) images of (a): randomly entangled CNTs (Niu et al., 1997), and (b): vertically aligned CNTs (Huang et al., 1999).

A wide range of approaches have been developed to synthesize CNTs including, for example, carbon arc-discharge (Ebbesen & Ajayan, 1992) and pyrolysis of hydrocarbons (Harris, 1999) for MWNTs, carbon arc-discharge (Harris, 1999), pulsed laser vaporization (Journet et al., 1997; Thess et al., 1996) and pyrolysis of hydrocarbons or carbon monoxide (Zheng et al., 2002) for SWNTs, chemical vapor deposition for aligned MWNTs (Dai, 2006;

Dai et al., 2003; Harris, 1999), and plasma-enhanced chemical vapor deposition (Qu & Dai, 2007) for aligned SWNTs. Depending on their diameter and helicity of the arrangement of carbon atoms in the walls, CNTs can exhibit semiconducting or metallic behavior (Dai, 2006; Harris, 1999), with an electrical conductivity as high as 5000 S/cm (Dresselhaus, 1996). CNTs also possess a high thermal conductivity (6000 W/mK), high thermal stability (stable up to 2800 °C in vacuum), and good mechanical properties (tensile strength 45 billion pascals) (Collins & Avouris, 2000). SWNTs have a high theoretical specific surface area of 1315 m^2/g, while that of MWNTs would be lower and is determined by the diameter of the tubes and the number of the graphene walls (Peigney et al., 2001). A reasonably high surface area (~ 400 m^2/g) has been obtained for a CNT "paper" electrode (Fig.3a) (Niu et al., 1997). These interesting properties make CNTs very attractive for a variety of potential applications, for example, sensors (Dai et al., 2002), hydrogen storage (Dillon et al., 1997), organic photovoltaic cells (Dai, 2006), energy storage (Che et al., 1998), and supercapacitors (Baughman et al., 2002).

2. Carbon nanotube supercapacitors

Specifically, owning to their novel properties of high electrical conductivity, high specific surface area, high charge transport capability, high mesoporosity, and high electrolyte accessibility, CNTs are attractive electrode materials for developing high-performance supercapacitors (Baughman et al., 2002). Like a conventional activated carbon-based supercapacitor (Fig.1), electrodes and electrolyte are two important components for a CNT supercapacitor. Research has been performed to develop different types of CNT electrode materials and combine them with various electrolytes (in different cell configurations) to improve the performance, safety, and lifetimes for supercapacitors.

2.1 Carbon nanotube electrodes for supercapacitors

Due to their relative ease of fabrication, randomly entangled CNTs are the first type of CNT materials that were studied for supercapacitor applications. Comparing with high-surface-area activated carbons, CNTs possess a moderate specific surface area. Nevertheless, higher capacitance has been demonstrated for CNTs, e.g., 102 F/g for MWNTs (Niu et al., 1997) and 180 F/g for SWNTs (An et al., 2001), in contrast to that of only tens of F/g for activated carbons. Based on the commonly realizable charge densities of 20 ~ 50 $\mu F/cm^2$ suggested for electrical double-layer capacitors (Conway, 1999), An *et al.* estimated the theoretical capacitance for their CNTs (specific surface area: 357 m^2/g) (An et al., 2001). The result of 71~178 F/g was in good agreement with the observed values (180 F/g) in the upper bound, indicating the perfect electrolyte accessibility of the CNTs. It has been widely accepted that it is their unique mesoporosity (presence and large population of mesopores) induced by the tube entanglement that makes CNTs highly accessible to the electrolyte, i.e., high electrolyte accessibility and thus high capacitance (An et al., 2001; Frackowiak & Béguin, 2001; Niu et al., 1997). This conclusion coincides with the importance of mesopores that has been pointed out previously for activated carbons, confirming that the poor mesoporosity of activated carbons is a major reason responsible for their low capacitance.
High capacitance of CNTs has been utilized to improve performance for supercapacitors. An supercapacitor fabricated from SWNT electrodes and KOH electrolyte showed a promising power density of 20 kW/kg with a maximum energy density of ~10 Wh/kg (performance

based on the mass of SWNTs) (An et al., 2001). The ESR, consisting of the resistance of the electrode itself, the resistance of the electrolyte within the porous layer of the electrode, and the contact resistance between the electrode and the current collector, of the capacitor, was very small, indicating fast delivery of the stored charge and thus a very high power density as expected (Equation 2). Excellent electrical conductivity, high mesoporosity, and high electrolyte accessibility of CNTs, resulting in a high charge transport capability, are responsible for this significance. Multiplying by a factor of 0.35 (Mastragostino et al., 2000), these active material-based performances can be roughly estimated as 7 kW/kg and ~3.5 Wh/kg for a packaged capacitor. Indeed, a power density of > 8 kW/kg (with a maximum energy density of ~ 1 Wh/kg) has been realized for a packaged supercapacitor fabricated from MWNTs electrodes and H_2SO_4 electrolyte (Niu et al., 1997). These power densities are higher than those attainable by activated carbon-based commercial capacitors (Burke & Arulepp, 2001), but the energy densities still need to improve.

It is well known that the capacitance of conventional carbon electrodes will decrease gradually with increasing discharging current density. This is because currents which are too large may block the entrances of the micropores. Therefore, the energy stored can be withdrawn only at limited discharge rates and at low frequencies or by dc techniques. In contrast, due to their high charge transport capability, fast charge injection and withdrawal are feasible for CNTs, meaning high capacitance at high frequencies (An et al., 2001; Frackowiak & Béguin, 2001; Niu et al., 1997). A frequency "knee" in the impedance spectrum (Nyquist plot) is usually used to evaluate the frequency dependence of a capacitor. At a frequency higher than the "knee", the real part of the impedance is frequency dependent and very small capacitance will be obtained. At a frequency below this value, the resistance changes weakly with frequency and the capacitor behavior tends to approach that of a pure capacitance. The knee frequency of most commercially available supercapacitors, including those specially designed for high power applications, is < 1 Hz (Miller, 1996). In contrast, a very high "knee" frequency of 100 Hz has been realized for CNT supercapacitors (Niu et al., 1997), suggesting that most of the stored energy of the capacitor is accessible at frequencies as high as 100 Hz. A larger frequency response means better power performance in demand applications.

In the recent years, aligned CNTs have also been studied for supercapacitor applications. It has been stressed that, for electrochemical applications, it is highly desirable to have aligned / patterned structures for CNTs so that their structure / property can be easily accessed and so that they can be effectively incorporated into devices (Dai et al., 2003). Recent research demonstrated that vertically aligned CNTs are advantageous over their randomly entangled counterparts for supercapacitor applications.

First, aligned CNTs are better-structured materials for supercapacitors. Unlike the irregular pore structures of randomly entangled CNTs (Fig.3a), the vertically aligned and unbundled structures and the well spacing between tubes of the aligned CNTs (Fig.3b) provide a more mesoporous and more accessible surface (Zilli et al., 2006). The aligned structures should provide improved charge storage / delivery properties as each of the constituent aligned tubes can be connected directly onto a common electrode, i.e., participation of each tube during the charge / discharge process of the capacitor. This indicates a combined charge capacity from all individual tubes of the aligned CNT electrode, and thus enhanced energy density for the capacitor. In turn, the stored energy can be delivered rapidly through each tube of the electrode, that is, excellent power density for the capacitor.

Second, aligned CNTs have enhanced effective surface area. Surface area of randomly entangled CNTs is determined by the open space between entangled fibrils (Fig.3a). Similarly, that of aligend CNTs is determined by the open space between aligned tubes. The open space between tubes of aligned CNTs is in the range of tens of nanometers (Fig.4). This provides a well-defined surface area for each of the constituent tubes to be accessible to the electrolyte ions, resulting in enhanced effective surface area for aligned CNTs. Moreover, under appropriate conditions (e.g., plasma etching (Huang & Dai, 2002), the top end-caps of aligned CNTs can be properly opened (Fig.4). The end-cap-opening then allows the electrolyte access to the inner cavity of the tubes (Frackowiak & Béguin, 2002). The well-defined alignment, in conjunction with the tip-opened structure, makes the aligned CNTs to possess a higher accessible surface area than that of randomly entangled CNTs. Therefore, the overall enhanced surface area (from both internal and external walls of tubes) translates to an increased capacitance for the aligned CNT electrode and thus an enhanced energy density for the capacitors.

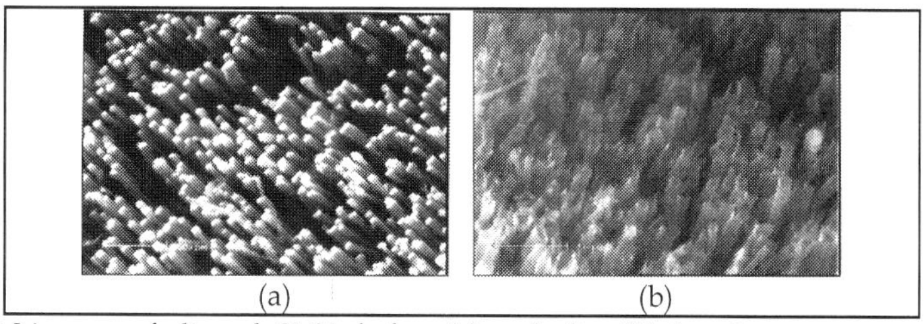

Fig. 4. SEM images of aligned CNTs before (a) and after (b) the plasma-treatment to open the top end-caps of tubes (Huang & Dai, 2002).

Recent research has demonstrated the improved rate capability for aligned CNTs over randomly entangled CNTs (Du et al., 2005; Futaba et al., 2006; Honda et al., 2007; Zhang et al., 2007; Zhang et al., 2008 (*Nano Lett*)). Also, a high capacitance has been obtained in 1 M H_2SO_4 for an aligned CNT array electrode (365 F/g) prepared by chemical vapor deposition in the template of porous anodic aluminum oxide (Chen et al., 2004) and in ionic liquid electrolytes for an aligned CNT electrode (440 F/g) prepared by a template-free chemical vapor deposition approach (Lu et al., 2009).

In addition to the pure tube electrodes as discussed above, CNTs have also been employed to fabricate electrodes in nanocomposite forms (e.g., with activated carbons, electroactive polymers, or metal oxides) to improve the performance for supercapacitors.

As mentioned earlier, comparing to activated carbons, CNTs have a relatively lower specific surface area but a higher electrical conductivity, higher charge transport capability, higher mesoporosity, and higher electrolyte accessibility. It is then desirable to combine activated carbons with CNTs to fabricate composites having the combined advantages from these two components. The resultant nanocomposites would possess an improved conductivity and an appropriate balance between the specific surface area and the mesoporosity, resulting in a higher capacitance than CNTs and a higher rate capability than activated carbons. Portet *et al.* blended activated carbons (80 wt. %), DWNTs (15%), and a polymer binder (3% carboxymethylcellulose and 2% polytetrafluoroethylene) to synthesize CNT / activated carbon composites (Portet et al., 2005). Improved capacitive behavior (lower ESR, lower capacitance loss at high frequencies, and faster charge / discharge process) has been demonstrated for supercapacitors fabricated from these composite electrodes over those from conventional activated carbon electrodes. An appropriate balance between the surface area and the mesoporosity is believed to be the key to ensure the high performance of these activated carbon- and CNT- incorporated composites.

Liu *et al.* mixed SWNTs with a polyacrylonitrile (PAN) dimethylformamide solution at the weight ratio of SWNT to PAN of 40 : 60 to fabricate a SWNT / PAN composite film electrode (Fig.5b) (Liu et al., 2003). Previously, physical or chemical activation of polyacrylonitrile or its copolymers is generally used for producing activated carbon with high specific surface area and high porosity (Addoun et al., 2002; Molina-Sabio, 1996). Liu *et al.* utilized this approach to convert their SWNT / PAN composite films into SWNT / activated carbon composite electrodes (with CO_2 activation). The specific capacitance of the resultant SWNT / activated carbon composites was significantly higher than that of a pure SWNT bucky paper electrode, which has been confirmed to be due to the higher specific surface area of the former than that of the latter. This resulted in the significantly improved power and energy densities of the SWNT / activated carbon composite electrode over those of the SWNT bucky paper electrode. For comparison, Fig.5c also plots power and energy densities of heat-treated SWNT/PAN composite film without CO_2 activation. The authors noticed that, although the specific capacitance of the heat-treated SWNT / PAN composite film without activation was similar to that of the bucky paper, the power density of the former is much higher than that of the latter.

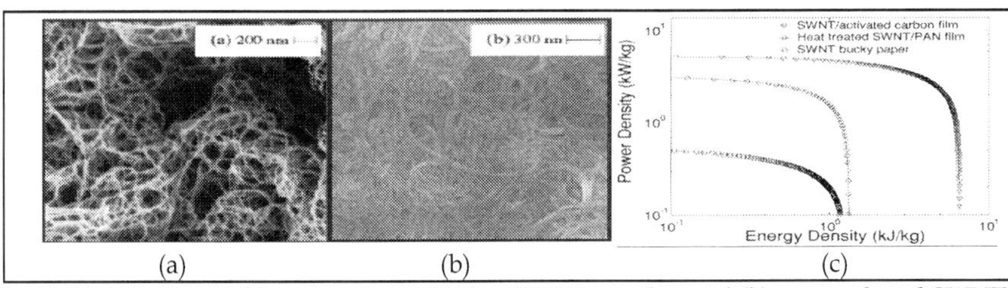

Fig. 5. SEM images of (a): as-produced HiPco SWNT powder, and (b): as-produced SWNT / PAN composite film. (c): Ragone plots for various electrodes evaluated at a discharging current of 0.01 A (Liu et al., 2003).

In order to improve energy density for commercially available activated carbon-based supercapacitors, new electrode materials with high capacitance are always desired (Frackowiak & Béguin, 2001). Able to undergo Faradaic reactions within the bulk of the materials, electroactive polymers and transition metal oxides can provide additional charge storage (i.e., pseudocapacitance) to satisfy this requirement (Burke, 2000). Furthermore, in the recent years, there is a research trend combining the high-energy-storage capability of these redox materials with the high-power-delivery capability of CNTs to fabricate new composite electrode materials with improved capacitance and rate capability.

Since their discovery in 1977 (Shirakawa et al., 1977), electroactive polymers (e.g., polyaniline, polypyrrole, polythiophene) have attracted considerable attention as a family of novel organic electrochemical materials. Able to store and release charge during their redox processes, electroactive polymers have been investigated for energy storage technologies including supercapacitors (Abrizzini et al., 1996; Rudge et al., 1994). The pseudocapacitance effects connected with the Faradaic reactions enhance the overall capacitance of electroactive polymers (up to ~ 500 F/g) (Burke, 2000). Electroactive polymers possess the advantages of low cost, light weight, and good processibility over transition metal oxides and high energy storage capability over high-surface-area activated carbons. In order to improve the electrochemical utilization (and hence capacitance) and rate capability of electroactive polymers, nanostructures have been developed in these electrode materials. Specifically, incorporating CNTs and electroactive polymers to fabricate nanocomposites is an efficient approach to fulfill this goal. The composites can be obtained both electrochemically and chemically. Combining the pseudocapacitance of electroactive polymers with the high electrical conductivity and high mesoporosity of CNTs, the resultant CNT / electroactive polymer nanocomposites possess enhanced capacitance and rate performance.

Using electropolymerization, CNT / electroactive polymer nanocomposites can be readily deposited from a monomer-containing solution onto a CNT preform or from a CNT / monomer-containing solution onto a traditional conductive substrate. Jurewicz *et al.* electrochemically coated polypyrrole onto a MWNT bucky paper electrode to fabricate CNT / polypyrrole composites (Jurewicz et al., 2001). After a 5 nm-thick polypyrrole coating, capacitance of the MWNTs increased substantially from 50 F/g to 163 F/g. The open entangled network of the nanocomposite was believed to favor the formation of a three-dimensional electrical double layer allowing a more effective contribution of the pseudofaradaic properties of polypyrrole. From an aqueous polymerization solution consisting of 0.5 M pyrrole and 0.5 M KCl, Hughes *et al.* potentiostatically grew polypyrrole onto aligned MWNTs to fabricate the nanocomposites (Hughes, et al., 2002 (*Adv. Mater.*)). The resulting polypyrrole layer was uniformly coated on the aligned nanotubes (Fig.6). With the well-aligned and conductive nanostructures of the MWNTs, the MWNT / polypyrrole composite showed a much larger capacitance (2.55 F/cm^2 vs. 0.62 F/cm^2) and higher knee frequency than a pure polypyrrole film electrode.

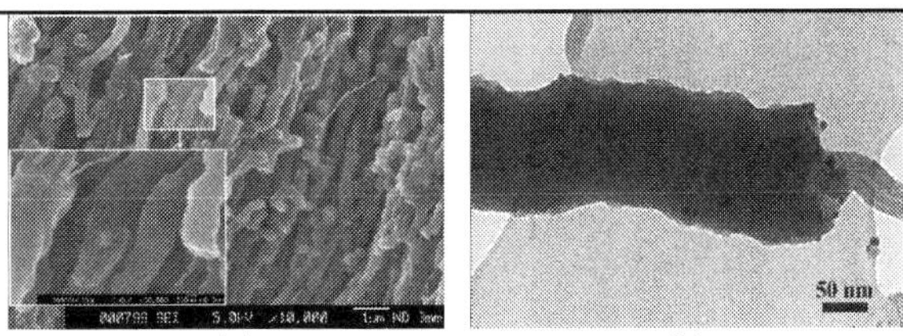

(a) (b)

Fig. 6. Polypyrrole coated aligned MWNTs (polymerization charge ~ 40 C/cm^2, nanotube length ~ 100 μm). (a): SEM image of the fractured film cross section illustrating the thickness of the polypyrrole coating relative to the uncoated MWNTs, exposed by fiber-pullout, as well as the channels available to the electrolyte between the coated nanotubes. (b): TEM image showing good interaction between the polypyrrole and MWNT (Hughes, et al., 2002 (*Adv. Mater.*)).

Onto a graphite working electrode, Hughes *et al.* electrochemically (either galvanostatically or potentiostatically) grew polypyrrole from an aqueous polymerization solution consisting of 0.5 M pyrrole monomer and 0.4 wt % of suspended oxidized MWNTs (Hughes et al., 2002

(*Chem. Mater.*)). Oxidation of MWNTs via an acid treatment attached functional groups such as hydroxyl, carbonyl, and carboxylic groups to the MWNT surface, facilitating their suspension in water. Thus, the negatively charged MWNTs acted both as a supporting electrolyte during polymerization and as a dopant in the polypyrrole deposited on the working electrode. The nanoporous three-dimensional arrangement of polypyrrole-coated MWNTs facilitated improved electron and ion transfer relative to pure polypyrrole films, resulting in a higher capacitance (192 F/g or 1.0 F/cm^2) and a higher knee frequency for the composites than pure polypyrrole films.

The CNT / electroactive polymer nanocomposites can also be prepared chemically by an oxidant. Frackowiak *et al.* deposited polypyrrole on MWNTs by chemical polymerization with $(NH_4)_2S_2O_8$ as oxidant in an acidic solution (0.1 M HCl). The electrode was prepared by blending the resultant polypyrrole-coated MWNTs with acetylene black and PVdF binder (Frackowiak et al., 2001). The capacitance obtained from nanotubes coated by polypyrrole reached 170 F/g, about twice that given by either pure nanotubes (ca. 80 F/g) or pure polypyrrole (ca. 90 F/g). The open entangled network of the nanocomposite seems to favor a better efficiency for formation of the electrical layer in polypyrrole. An *et al.* used $FeCl_3$ as oxidant to polymerize polypyrrole and deposit it on SWNTs (An et al., 2002). The electrode was then prepared by mixing the SWNT / polypyrrole composite with acetylene black and PVdF binder. Due to the uniform polypyrrole coating on the porous and conductive support of SWNTs (Fig.7), the SWNT / polypyrrole nanocomposite electrode showed a much higher capacitance (265 F/g) than pure polypyrrole and pure SWNT electrodes.

Fig. 7. SEM images of (a): as-grown SWNT, (b): pure polypyrrole, and (c): SWNT / polypyrrole composite (**A**n et al., 2002).

Noble metal oxides, such as RuO_2 and IrO_2, have been identified as the promising electrode materials for supercapacitors due to their remarkable specific capacitance, good electrical conductivity, and high chemical stability (Zheng et al., 1995). However, the high cost and scarce source of these materials retard their commercial acceptance. Recent researches have been focused on cheap transition metal oxides (e.g., vanadium oxide (V_2O_5), manganese oxide (MnO_2), and nickel oxide (Lee, J.Y. et al., 2005). Combining these metal oxides with CNTs to fabricate nanocomposites is able to improve the capacitive performance of these materials. Among them, V_2O_5 and MnO_2 are two important examples.

V_2O_5 has the advantages of non-toxicity, low cost, and high theoretical capacity (590 mAh/g) (Le et al., 1996). However, the major disadvantages of V_2O_5 are its intrinsically low electronic conductivity ($10^{-6} \sim 10^{-7}$ S/cm) (Benmoussa et al., 1995) and the slow Li^+ diffusion within the host matrix (diffusion coefficient: ~ 10^{-13} cm^2/s) (Passerini et al., 1999). One effective approach to addressing these issues is to combine V_2O_5 with CNTs to fabricate nanocomposites. The resultant high-capacity and high-rate composite electrodes have been utilized for lithium-ion (Li-ion) battery and supercapacitor applications.

SWNTs have been introduced as conductive additives into the V_2O_5 aerogels to develop V_2O_5 / SWNT nanocomposites (Sakamoto & Dunn, 2002). The SWNTs have a similar morphological character and dimensional scale as the V_2O_5 ribbons that compose the aerogel. Thus, there is an opportunity to exploit the high conductivity of SWNTs and have them provide electronic conduction without blocking electrolyte access to the active material. The V_2O_5 / SWNT composite possesses a high pore volume that ensures electrolyte access throughout the electrode, while contact between the two phases is established at the nanometer level, and occurs at multiple points along the V_2O_5 ribbons. This nanocomposite electrode shows a high specific capacity of 452 mAh/g and retains up to 65% of this capacity when the discharge rate is increased from 112 mA/g (0.2C) to 2800 mA/g (5C). Electronically conducting network developed from the SWNTs within the V_2O_5 / SWNT nanocomposite is believed to contribute to this significance.

Kim *et al.* electrochemically deposited V_2O_5 thin film onto a preformed CNT film substrate to prepare the V_2O_5 / CNT composite electrodes (Kim et al., 2006). Without the addition of a polymer binder as employed in conventional electrodes (Mandal et al. 2001), the electrochemically prepared V_2O_5 / CNT electrode has high conductivity. With a thin V_2O_5 film coating (~ 6 nm), the V_2O_5 / CNT electrode shows a very high capacity of 680 mAh/g (based on the mass of V_2O_5) at a current density of 5 A/g and retained 67% of this capacity even at a current density of as high as 100 A/g. This improved capacity is believed to be due to the combined contribution from the increased loading of V_2O_5 in the V_2O_5 / CNT electrode (due to high surface area of the CNTs), the improved electrochemical utilization of the V_2O_5 thin film (due to its very small thickness), and the double layer capacitance of the electrode (due to high surface area of the CNTs), while the high rate performance of the V_2O_5 / CNT electrode is believed to be due to the short diffusion time of Li^+ in the V_2O_5 thin film on the highly conductive CNT substrate. The porous and accessible structures of V_2O_5 at a CNT substrate (Fig.8), in contrast to the compact structures of a dense V_2O_5 film at a conventional Pt electrode, are responsible for these improvements.

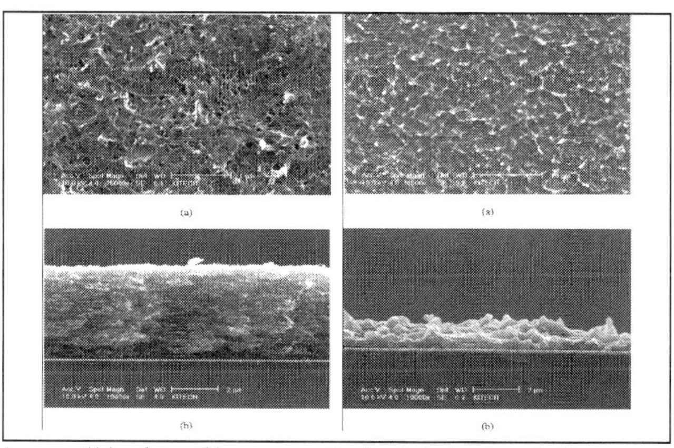

Fig. 8. SEM images ((a): plain, (b): cross-section) of electrochemically prepared V_2O_5 on a CNT electrode (LEFT) and on a Pt electrode (RIGHT) (Kim et al., 2006).

MnO_2 is also a non-toxic, low-cost, and high-capacity (theoretical: 616 mAh/g) meta oxide. In order to improve its electrochemical utilization, capacity, and rate capability for energy storage applications, MnO_2 has been combined with CNTs to fabricate nanocomposites. Different approaches have been investigated to synthesize the composites.

Lee *et al.* first deposited CNTs on a Ni substrate by electrophoresis from a 0.5 mg CNT/1 ml dimethylformamide solution, and then synthesized MnO_x onto the resultant CNTs / Ni electrode by anodic deposition in a 0.16 M manganese sulfate pentahydrate aqueous solution (Lee, C.Y. et al., 2005). In a 0.1 M Na_2SO_4 aqueous electrolyte, upon the increase in scan rate from 5 to 100 mV/s, the specific capacitance of the MnO_x / CNT nanocomposite electrode decreased slightly from 415 to 388 F/g, retaining up to 94%. After 1000 cycles of operation, this electrode maintained 79% of its original capacitance. These indicate the good electrochemical reversibility, high capacitance, and high rate of the MnO_x / CNT nanocomposites.

Using thermal decomposition of manganese nitrates, Fan *et al.* synthesized and dispersed MnO_2 onto a CNT matrix that was prepared by directly growing CNTs on a graphite disk (Fan et al., 2006). With a MnO_2 loading of 36.9 μg/cm^2, a high specific capacitance of 568 F/g was achieved for the resultant MnO_2 / CNT composite (based on MnO_2 mass) in a 1 M Na_2SO_4 aqueous electrolyte and at the charge/ discharge current density of 1 mA/cm^2. Excellent cycle stability (ca. 88% value of specific capacitance remained after 2500 charge / discharge cycles) and power characteristics have been obtained for this composite.

Upon the direct redox reaction between the CNTs and permanganate ions (MnO_4^-), the resultant MnO_2 can be in-situ coated on the CNTs. Using this method, Ma *et al.* deposited birnessite-type MnO_2 on CNTs (Fig.9) (Ma et al., 2008). In a 1 M $LiClO_4$ / propylene carbonate organic electrolyte, the resultant MnO_2 /CNT nanocomposite showed a large specific capacitance of 250 F/g even at a large current density of 1 A/g. This is equivalent to 139 mAh/g based on the total weight of the electrode material that includes the electroactive material, conducting agent, and binder. The specific capacitance of MnO_2 in the MnO_2 / CNT nanocomposite was as high as 580 F/g (320 mAh/g), indicating the excellent electrochemical utilization of the MnO_2 due to the use of highly conductive and porous CNT support. Further, the addition of CNTs as a conducting agent in the electrode considerably improved the rate capability of the MnO_2 / CNT nanocomposite.

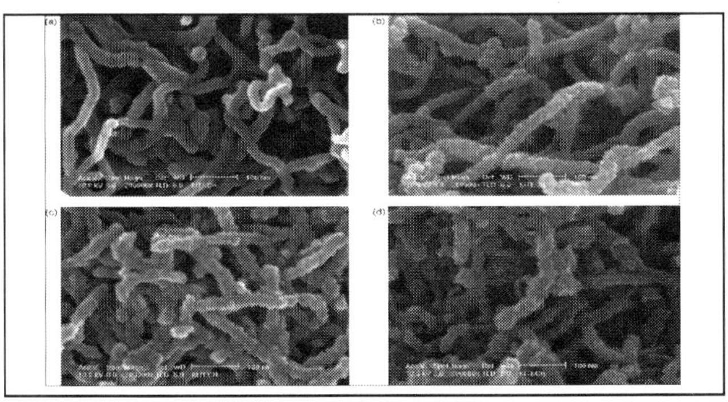

Fig. 9. SEM images of pristine CNT (a) and MnO_2/CNT nanocomposites prepared in 0.1 M $KMnO_4$ containing 1.0 g CNT at 70^{0}C under (b): pH 7, (c): pH 2.5, and (d): pH 1 of initial solution (Ma et al., 2008).

Adopting a conventional slurry procedure commonly employed in fabricating electrode materials in energy storage industry, Wang *et al.* prepared their MnO_2 / CNT nanocomposites by simply blending MnO_2 with 25 wt. % MWNTs by ultrasonic vibration in ethanol and then mixing the resultant composite thoroughly with 5 wt. % poly(tetrafluoroethylene) (PTFE) binder to make a slurry. The slurry was then coated onto an aluminum current collector to make a MnO_2 / CNT nanocomposite sheet electrode (Wang et al., 2005), which has been used for supercapacitors as discussed below.

2.2 Capacitor design for carbon nanotube supercapacitors

Along with their synthesis and characterization as discussed above, CNT nanocomposites have been utilized to fabricate supercapacitors in various configurations. Specifically, using a same electrode material for both the positive and negative electrodes (i.e., symmetric configuration, similar to that shown in Fig.1), the CNT / activated carbon composites are used to fabricate electric double layer supercapacitors and the electroactive polymer / CNT or metal oxide / CNT composites to fabricate pseudo-capacitors (aka, redox capacitors). With the improved capacitance and rate capability of the CNT / activated carbon composites, the capacitors fabricated from these composites showed enhanced energy and power densities than the conventional activated carbon-based capacitors. With the additional pseudocapacitance attainable from the electroactive polymer- or metal oxide-incorporated CNT composites, the pseudo-capacitors fabricated these materials showed a higher energy density than that of a carbon-based electric double layer capacitor. Further, introducing CNTs in these electroactive electrode materials improved the power performance for these new pseudo-capacitors.

Moreover, combining different electrode materials as the positive and negative electrodes (i.e., asymmetric configuration), the electroactive polymer / CNT and metal oxide / CNT composites have been employed to develop asymmetric supercapacitors (aka, hybrid supercapacitors) with further improved performance. Currently available commercial supercapacitors are fabricated in a symmetric configuration where two identical capacitative electrodes (usually activated carbon) are used as both the positive and negative electrodes (Fig.1). In this design, the capacitor utilizes only one half (on an area basis, F/ cm^2) or one fourth (on a mass basis, F/g) of one electrode's capacitance, resulting in a low energy density for the capacitors (Burke & Arulepp, 2001). Recently, an asymmetric design has been developed to address this issue (Naoi & Simon, 2008). Typically, an asymmetric supercapacitor is constructed by replacing one of the capacitative carbon electrodes of a symmetric capacitor with a Faradic electrode made of electroactive materials including, for example, electroactive polymers (Fabio et al., 2001), metal oxides (Bélanger et al., 2008), or battery electrode materials (Pasquier et al., 2003). Due to the infinite capacitance of the Faradic electrode comparing to that of the capacitative electrode, an asymmetric capacitor utilizes all (on an area basis) or half (on a mass basis) of the capacitative electrode's capacitance. Plus a high cell voltage deduced from the asymmetric configuration, high energy densities have been achieved for asymmetric capacitors. Ideally, such a system offers the advantages of both the capacitors (high power density and long cycle life) and the batteries (high energy density).

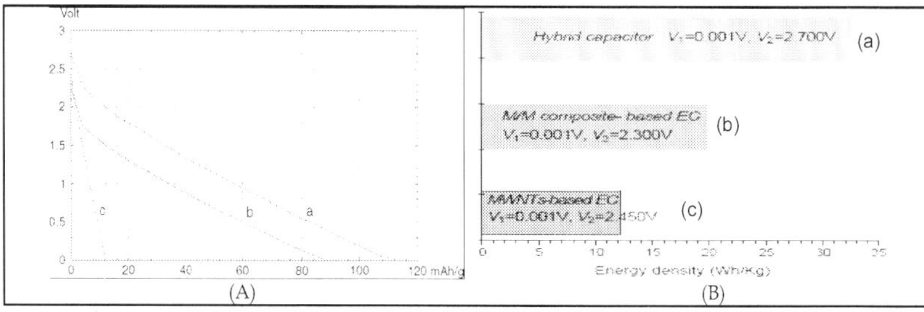

Fig. 10. Galvanostatic discharge curves (A) and energy density at 10 mA/cm^2 current density (B) of (a): MnO_2 / MWNT - MWNT asymmetric capacitor, (b): MnO_2 / MWNT - MnO_2 / MWNT symmetric capacitor, and (c): MWNT - MWNT symmetric capacitor (Wang et al., 2005).

Using their MnO_2 / CNT nanocomposites and a 1.0 M $LiClO_4$/ethylene carbonate - diethyl carbonate (1:1 in vol) electrolyte, Wang *et al.* fabricated and evaluated a MnO_2 / MWNT - MWNT asymmetric capacitor against two symmetric capacitors incorporating the same electrode materials, i.e., MnO_2 / MWNT - MnO_2 / MWNT capacitor and MWNT - MWNT capacitor (Wang et al., 2005). The asymmetric capacitor showed the highest cell voltage and highest capacity (Fig.10A). Indeed, due to its unique configuration, the asymmetric capacitor showed a maximum specific capacitance of 56 F/g, 1.3 times that of the MnO_2 / MWNT - MnO_2 / MWNT capacitor and 3.4 times that of the MWNT - MWNT capacitor. Thus, as can be expected by Equation 1, the asymmetric capacitor showed an energy density (32.91 Wh/kg) higher than that of the symmetric ones (Fig.10B). Finally, with the high rate capability deduced from the nanostructures of both the positive and negative electrodes, the nano-nano asymmetric capacitors would be able to achieve a higher power density than the current supercapacitor technology.

2.3 Electrolytes for carbon nanotube supercapacitors

The electrolyte is an essential component for charge transport between the positive and negative electrodes and plays an important role in determining the performance, safety, and lifetimes of a supercapacitor. Currently used electrolytes include aqueous, mostly H_2SO_4 and KOH, and organic, mostly propylene carbonate (PC) and acetonitrile (ACN) electrolytes.

Aqueous electrolytes have the advantage of high ionic conductivity but the disadvantage of small electrochemical window (~ 1.2 V), i.e., the potential over which the electrolyte is neither reduced nor oxidized at an electrode. Organic electrolytes are advantageous over aqueous ones mainly due to their larger electrochemical windows, resulting in a larger cell voltage (2 ~ 3 V) and thus potentially higher energy (Equation 1) and power (Equation 2) densities for the capacitor. However, some organic electrolytes suffer from serious health and safety problems as they are inherently volatile, flammable, and toxic (Xu, 2004), leading to a narrow operational temperature range and potential for explosion during outlying circumstances. In addition, electrolyte depletion is another limitation of the currently used electrolytes. Upon charge, ions of the electrolyte are transported into the double layers at the electrode/electrolyte interfaces, resulting in the decrease of salt concentration in the electrolyte (the so-called electrolyte depletion) and hence the limit of energy density of the capacitor (Zheng et al., 1997). Also, this electrolyte depletion increases the cell resistance and thus lowers the power density achievable for the capacitor. In order to overcome the limitations of currently used electrolytes, environmentally friendly electrolytes with large electrochemical windows are needed. Ionic liquids have been investigated as new electrolytes to satisfy these requirements.

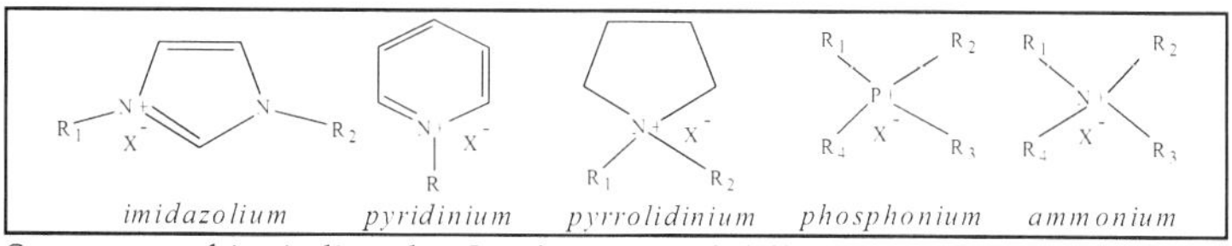

Fig. 11. Structures of ionic liquids. Combination of different cations and anions results in various ionic liquids. R = methyl, ethyl, butyl, octyl, etc. X^- = BF_4^-, PF_6^-, $CF_3SO_3^-$, $N(CF_3SO_2)_2^-$, $(C_4F_9)_3PF_3^-$,etc.

Typically, ionic liquids consist of nitrogen (or phosphorus)-containing organic cations and inorganic anions. Categorized by the cation, ionic liquids can be divided into five groups (Fig.11). Since the first ionic liquid, ethyl ammonium nitrate, was described in 1914 (Walden, 1914), ionic liquids have been used in a range of different applications such as chemical synthesis, catalysis, separation, and electrochemical devices (Freemantle, 2000). Unlike conventional solid salts, ionic liquids are salts that are fluid over a wide temperature range, including room temperature, with higher viscosities than either aqueous or organic electrolytes at room temperature. Ionic liquids are inherently conductive, making them good electrolytes. Certain unique properties of ionic liquids, including high ionic conductivity (up to 10^{-2} S/cm), large liquid phase range (-100 ~ 400 ^{0}C), wide electrochemical window (4 ~ 6 V), non-volatility, non-flammability, and non-toxicity, have made them an excellent electrolyte for various electrochemical systems including supercapacitors (Buzzeo, et al., 2004).

Given that the performance of a supercapacitor is directly proportional to the square of its cell voltage, Equations 1 & 2, the use of large-electrochemical-window ionic liquids would significantly boost the performance for supercapacitors. Compared to conventional electrolytes, ionic liquids have a unique property that they are both 100% solvents and also 100% salts. It is unnecessary to add other salts into an ionic liquid to achieve ionic conductivity. The very high ionic concentration of ionic liquids would be able to eliminate the electrolyte depletion problem as encountered with conventional electrolytes and therefore enhance the capacitor performances. Further, the superior chemical and environmental stability of ionic liquids ensures safe operation and long lifetimes for capacitors.

Conventional aqueous (An et al., 2001; Niu et al., 1997; Talapatra et al., 2006) and organic (Emmenegger et al., 2003; Futaba et al., 2006; Honda et al., 2007) electrolytes have been employed for the development of CNT supercapacitors. The aqueous electrolyte - based supercapacitors showed reasonably high power densities (> 7 kW/kg) but their energy densities (~ 4 Wh/kg) are still limited. Narrow electrochemical window of the aqueous electrolytes used, and hence small cell voltages of the capacitors, is a major reason for this drawback. Thus, the advantage of the use of organic electrolytes is mainly associated with their relatively large electrochemical windows.

Electrolytes also influence the performance of asymmetric capacitors. MnO_2 - Activated carbon is a frequently studied aqueous asymmetric capacitor system and shows a reasonably high energy density (Hong et al., 2002). Unfortunately, most of the MnO_2 - based capacitors show a limited cycle life (100 to 1000 charge/discharge cycles) (Naoi & Simon, 2008) largely due to manganese dissolution and oxygen evolution (associated with the small electrochemical window of aqueous electrolytes). Use of organic electrolytes having a larger electrochemical window can improve performance for asymmetric capacitors. For example, using a $LiBF_4$/ACN organic electrolyte, the $Li_4Ti_5O_{12}$ - Activated carbon capacitors showed a cell voltage of about 3 V and an energy density higher than 10 Wh/kg (based on package weight) (Amatucci et al., 2001). Wang *et al.* reported a cell voltage of 1.8 V and an energy density of 35 Wh/kg (based on active electrode material mass) for their aqueous $LiMn_2O_4$ - Activated carbon capacitors (Wang & Xia, 2005). When the aqueous electrolyte was replaced with an organic one, a higher cell voltage (2.5 V) and a higher energy density (56 Wh/kg) have been achieved for a similar capacitor ($LiMn_2O_4$ - MnO_2) (Ma et al., 2007).

To this end, in order to further enhance the capacitor performance, new electrolytes having even larger electrochemical windows are needed. Ionic liquids have been studied for this purpose. Initial study was performed to investigate electrochemical behavior of randomly entangled CNTs in ionic liquid electrolytes, showing a large potential window but with a limited capacitance for the CNTs (Barisci et al., 2004; Xu et al., 2006). It is likely that a facilitated access of the ionic liquid ions could not take place within the CNTs due to the mismatch between the irregular pore structures of the randomly entangled CNTs and the relatively high viscosity of the ionic liquid electrolytes (comparing to conventional aqueous and organic electrolytes). Recent research demonstrated that vertically aligned CNTs are advantageous over their randomly entangled counterparts for supercapacitor applications. Specifically, in ionic liquid electrolytes, improved rate capability has been achieved for CNTs when vertically aligned structures were employed (Zhang et al., 2008 (*J. Electrochem. Soc.*)).

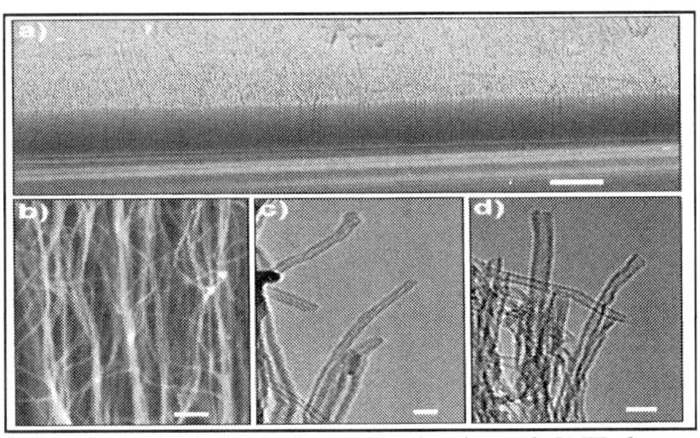

Fig. 12. (a): SEM image of a plasma-etched, vertically aligned CNT electrode (scale bar: 100 μm). (b): Higher magnification view of the electrode (scale bar: 100 nm). TEM images of the CNTs before (c) and after (d) plasma etching (scale bar: 20 nm). The multi-walled CNT array is highly aligned with the tube length of ~150 μm and an outer and inner diameter of approximately 10 and 5 nm, respectively (Lu et al., 2009).

their volumetric performance. To do this, CNT materials need to be fabricated / packed in a high-loading format. This, for example, can be fulfilled by increasing the length (e.g., super-long CNTs) (Hata et al., 2004), packing density (e.g., highly densely packed CNTs) (Futaba et al., 2006), film thickness (e.g., ultrathick CNT films) (Ci et al., 2007), and multifilaments (e.g., CNT fibers and yarns) (Zhang et al., 2004) of the CNTs. Moreover, in order to improve the shape flexibility and manufacturing integrity for CNT supercapacitors, flexible CNT electrodes have been developed, for example, by directly growing CNTs on flexible carbon layers and meta foils (Chen et al., 2008) or by embedding flexible nanoporous cellulose papers with aligned CNTs (Pushparaj et al., 2007). Fig.14 shows the examples of some of these CNT electrode materials.

3.2 Use of carbon nanotubes for high-energy and high-power energy storage technologies

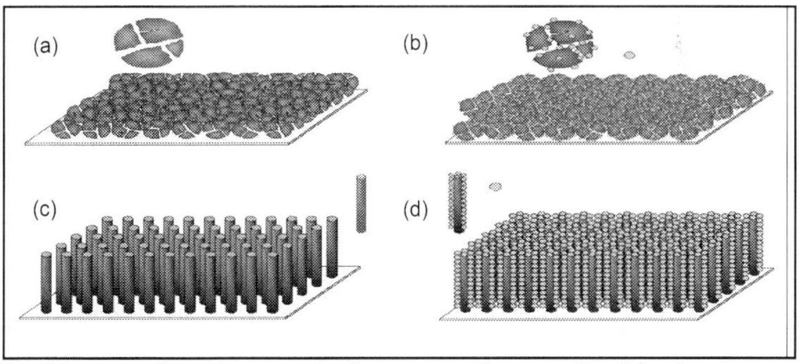

Fig. 15. Possible strategies to improve both energy and power densities for supercapacitors. a, b, Decorating activated carbon grains (a) with pseudo-capacitive materials (b). c, d, Achieving conformal deposit of pseudo-capacitive materials (d) onto highly ordered high-surface-area carbon nanotubes (c) (Simon & Gogotsi, 2008).

Achieving both high energy and high power densities for a same energy storage device has attracted considerable attention in energy storage community (Kang et al., 2006; Taberna et al., 2006). Supercapacitors have a higher power density but a lower energy density than a battery. It is then desirable to combine the high-capacity-capability of batteries with the high-rate-capability of supercapacitors to achieve a high-energy and high-power energy storage device. As discussed earlier (Section 2.1), depositing an electroactive material (electroactive polymers or metal oxides) onto the surfaces of CNTs has been investigated towards this goal. Regarding electrode architectures, vertically aligned structures of a substrate electrode have been demonstrated to be important for this requirement (Taberna et al., 2006). To this end, with their high alignment, well spacing, high effective surface area, and high electrolyte accessibility, vertically aligned CNTs have been a favorable substrate material, in conjunction with the deposition of electroactive materials, to synthesize high-capacity and high-rate electrode materials. Fig.15 illustrates the obvious advantages of vertically aligned CNTs, in contrast to traditional activated carbons, for this application. A variety of materials, e.g., electroactive polymers (Hughes et al., 2002 (*Adv. Mater.*)), manganese oxide (Zhang et al., 2008 (*Nano Lett.*)), and vanadium oxide (Fang et al., 2008), have been deposited onto aligned CNTs towards this goal. Moreover, further research will be needed to investigate the scaling-up feasibility and the electrolyte compatibility (with an appropriate electrolyte system) of these new electrode materials in order to fabricate high-energy and high-power energy storage devices with high safety and long lifetimes.

4. References

Addoun A., Dentzer J. & Ehrburger P. (2002). Porosity of carbons obtained by chemical activation: effect of the nature of the alkaline carbonates. *Carbon*; 40 (7), 1140-1143.

Abrizzini, C.; Mastragostino, M.; Meneghello, L. & Paraventi, R. (1996). Electronically conducting polymers and activated carbon: Electrode materials in supercapacitor technology. *Adv. Mater.*, 8 (4), 331-334.

Amatucci, G.G.; Badway, F.; Pasquier, A.D. & Zheng, T. (2001). An Asymmetric Hybrid Nonaqueous Energy Storage Cell. *J. Electrochem. Soc.*, 148 (8), A930-A939.

An, K.H.; Jeon, K.K.; Heo, J.K.; Lim, S.C.; Bae, D.J. & Lee, Y.H. (2002). High-Capacitance Supercapacitor Using a Nanocomposite Electrode of Single-Walled Carbon Nanotube and Polypyrrole. *J. Electrochem. Soc.*, 149 (8), A1058-A1062.

An, K.H.; Kim, W.S., Park, Y.S.; Moon, J.-M.; Bae, D.J.; Lim, S.C.; Lee, Y.S. & Lee, Y.H. (2001). Electrochemical Properties of High-Power Supercapacitors Using Single-Walled Carbon Nanotube Electrodes. *Adv. Funct. Mater.*, 11 (5). 387-392.

Barisci, N.J.; Wallace, G.G.; MacFarlane, D.R. & Baughman, R.H. (2004). Investigation of ionic liquids as electrolytes for carbon nanotube electrodes. *Electrochem. Commun.*, 6 (1), 22-27.

Baughman, R.H.; Zakhidov, A.A. & de Heer, W.A. (2002). Carbon Nanotubes--the Route Toward Applications. *Science*, 297 (5582), 787-792.

Bélanger, D.; Brousse, T. & Long, J.W. (2008). Manganese Oxides: Battery Materials Make the Leap to Electrochemical Capacitors. *Electrochem. Soc. Interf.*, 17 (1), 49-52.

Benmoussa, M.; Ibnouelghazi, E.; Bennouna, A. & Ameziane, E. L. (1995). Structural, electrical and optical properties of sputtered vanadium pentoxide thin films. *Thin Solid Films*, 265 (1-2), 22-28.

Burke, A. (2000). Ultracapacitors: why, how, and where is the technology. *J. Power Sources*, 91 (1), 37-50.

Burke A. & Arulepp M. (2001). Recent Developments in Carbon-based Electrochemcial Capacitors: Status of the Technology and Future Prospects, *Electrochemical Society Proceedings*, 2001-21, pp.576.

Buzzeo, M.C.; Evans, R.G. & Compton, R.G. (2004). Non-Haloaluminate Room-Temperature Ionic Liquids in Electrochemistry - A Review. *ChemPhysChem*, 5 (8), 1106- 1120.

Che, G.; Lakshmi, B.B.; Fisher, E.R. & Martin, C.R. (1998). Carbon nanotubule membranes for electrochemical energy storage and production. *Nature*, 393 (6683), 346-349.

Chen, J.; Minett, A.I.; Liu, Y.; Lynam, C.; Sherrell, P.; Wang, C. & Wallace, G.G. (2008). Direct Growth of Flexible Carbon Nanotube Electrodes. *Adv. Mater.*, 20 (3), 566-570.

Chen, Q.L.; Xue, K.H.; Shen, W.; Tao, F.F.; Yin, S.Y. & Xu, W. (2004). Fabrication and electrochemical properties of carbon nanotube array electrode for supercapacitors. *Electrochimica Acta*, 49 (24), 4157-4161.

Ci, L.; Manikoth, S.M.; Li, X.; Vajtai, R. & Ajayan, P.M. (2007). Ultrathick Freestanding Aligned Carbon Nanotube Films. *Adv. Mater.*, 19 (20), 3300-3303.

Collins, P.G. & Avouris, P. (2000). Nanotubes for electronics. *Sci. Am.*, Dec. 62-69.

Conway, B.E. (1999). Electrochemical Supercapacitors: Scientific Fundamentals and Technological Applications, Kluwer Academic/Plenum Publishers, New York.

Dai, L. (Ed.) (2006). *Carbon Nanotechnology: Recent Developments in Chemistry, Physics, Materials Science and Device Applications*, Elsevier: Amsterdam.

Dai, L.; Patil, A.; Gong, X.; Guo, Z.; Liu, L.; Liu, Y. & Zhu, D. (2003). Aligned Nanotubes. *ChemPhysChem*, 4 (11), 1150-1169, and references cited therein.

Dai, L.; Soundarrajan P. & Kim, T. (2002). Sensors and sensor arrays based on conjugated polymers and carbon nanotubes. *Pure Appl. Chem.*, 74 (9), 1753-1772, and references therein.

Dillon, A.C.; Jones, K.M.; Bekkedahl, T.A.; Kiang, C.H.; Bethune, D.S. & Heben, M. J. (1997). Storage of hydrogen in single-walled carbon nanotubes. *Nature*, 386 (6623), 377-379.

Dresselhaus, M. (1996). Carbon connections promise nanoelectronics. *Phys. World*, 9, 18-19.

Du, C.; Yeh, J. & Pan, N. (2005). High power density supercapacitors using locally aligned carbon nanotube electrodes. *Nanotechnology*, 16 (4), 350-353.

Ebbesen, T.W. & Ajayan, P.M. (1992). Large-scale synthesis of carbon nanotubes *Nature*, 358 (6383), 220-222.

Ehsani, M.; Gao, Y.; Gay, S.E. & Emadi, A. (2005). *Modern Electric, Hybrid Electric, and Fuel Cell Vehicles - Fundamentals, Theory, and Design*, pp. 299-332, CRC Press, Boca Raton, London, New York, Washington, D.C.

Emmenegger, Ch.; Mauron, P.; Zuüttel, A.; Nützenadel, Ch.; Schneuwly, A.; Gallay, R. & Schlapbach, L. (2000). Carbon nanotube synthesized on metallic substrates, *Applied Surface Science*, 162-163, 452-456.

Emmenegger, Ch.; Mauron, Ph.; Sudan, P.; Wenger, P.; Hermann, V.; Gallay, R. & Züttel, A. (2003). Investigation of electrochemical double-layer (ECDL) capacitors electrodes based on carbon nanotubes and activated carbon materials. *J. Power Sources*, 124 (1), 321-329.

Fabio, A.D.; Giorgi, A.; Mastragostino, M. & Soavi, F. (2001). Carbon-Poly(3-methylthiophene) Hybrid Supercapacitors. *J. Electrochem. Soc.* 148 (8), A845- A850.

Fan, Z.; Chen, J.; Wang, M.; Cui, K.; Zhou, H. & Kuang, Y. (2006). Preparation and characterization of manganese oxide/CNT composites as supercapacitive materials. *Diamond & Related Materials*, 15 (9), 1478-1483.

Fang, W.-C. (2008). Synthesis and Electrochemical Characterization of Vanadium Oxide/Carbon Nanotube Composites for Supercapacitors. *J. Phys. Chem. C*, 112 (30), 11552-11555.

Frackowiak, E. & Béguin, F. (2001). Carbon materials for the electrochemical storage of energy in capacitors. *Carbon*, 39 (6), 937-950.

Frackowiak, E. & Béguin, F. (2002). Electrochemical storage of energy in carbon nanotubes and nanostructured carbons. *Carbon*, 40 (10), 1775-1787.

Frackowiak, E.; Jurewicz, K.; Delpeux, S. & Beguin, F. (2001). Nanotubular materials for supercapacitors. *J. Power Sources*, 97-98, 822-825.

Frackowiak, E.; Metenier, K.; Bertagna, V. & Beguin, F. (2000). Supercapacitor electrodes from multiwalled carbon nanotubes. *Appl. Phys. Lett.*, 77, 2421.

Freemantle, M. (2000). *Chem. Eng. News*. 78 (20), 37.

Futaba, D.N.; Hat, K.; Yamada, T.; Hiraoka, T.; Hayamizu, Y.; Kakudate, Y.; Tanaike, O.; Hatori, H.; Yumura, M. & Iijima, S. (2006). Shape-engineerable and highly densely packed single-walled carbon nanotubes and their application as super-capacitor electrodes. *Nature Mater.*, 5 (12), 987-994.

Harris, P.F. (1999). *Carbon Nanotubes and Related Structures: New Materials for the Twenty-first Century*, Cambridge University Press: Cambridge.

Hata, K.; Futaba, D.N.; Mizuno, K.; Namai, T.; Yumura, M. & Iijima, S. (2004). Water-Assisted Highly Efficient Synthesis of Impurity-Free Single-Walled Carbon Nanotubes. *Science*, 306, 1362-1364.

Honda, Y.; Haramoto, T.; Takeshige, M.; Shiozaki, H.; Kitamura, T. & Ishikawa, M. (2007). Aligned MWCNT Sheet Electrodes Prepared by Transfer Methodology Providing High-Power Capacitor Performance. *Electrochem. Solid-State Lett.*, 10 (4), A106-A110.

Hong, M.S.; Lee, S.H. & Kim, S.W. (2002). Use of KCl Aqueous Electrolyte for 2 V Manganese Oxide/Activated Carbon Hybrid Capacitor. *Electrochem. Solid-State Lett.*, 5 (10), A227- A230.

Huang, S. & Dai, L. (2002). Plasma Etching for Purification and Controlled Opening of Aligned Carbon Nanotubes. *J. Phys. Chem. B*, 106 (14), 3543-3545.

Huang, S.; Dai, L. & Mau, A. W. H. (1999). Patterned Growth and Contact Transfer of Well-Aligned Carbon Nanotube Films. *J. Phys. Chem. B*, 103 (21), 4223-4227.

Hughes, M.; Chen, G.Z.; Shaffer, M.S.; Fray, D.J. & Windle, A.H. (2002). Electrochemical Capacitance of a Nanoporous Composite of Carbon Nanotubes and Polypyrrole. *Chem. Mater.*, 14 (4), 1610-1613.

Hughes, M.; Shaffer, M.S.P.; Renouf, A.C.; Singh, C.; Chen, G.Z.; Fray, D.J. & Windle, A.H. (2002). Electrochemical Capacitance of Nanocomposite Films Formed by Coating Aligned Arrays of Carbon Nanotubes with Polypyrrole. *Adv. Mater.*, 14 (5), 382-385.

Iijima, S. (1991). Helical microtubules of graphitic carbon. *Nature*, 354 (6348), 56-58.

Journet, C.; Maser, W.K.; Bernier, P.; Loiseau, A.; Lamy de la Chapelle, M.; Lefrant, S.; Deniard, P.; Lee, R. & Fischer, J.E. (1997). Large-scale production of single-walled carbon nanotubes by the electric-arc technique. *Nature*, 388 (6644), 756-758.

Jurewicz, K.; Delpeux, S.; Bertagna, V.; Beguin, F. & Frackowiak, E. (2001). Supercapacitors from nanotubes/polypyrrole composites. *Chem. Phys. Lett.*, 347 (1-3), 36-40.

Kang, K.; Meng, Y.S.; Bréger, J.; Grey, C.P. & Ceder, G. (2006). Electrodes with High Power and High Capacity for Rechargeable Lithium Batteries, *Science* 311 (5763), 977-980.

Kastening, B. & Sprinzig, S. (1986). Electrochemical polarization of activated carbon and graphite powder suspensions: Part II. Exchange of ions between electrolyte and pores. *J. Electroanal. Chem.*, 214 (1-2), 295-302.

Kim, Y.T.; Ito, Y.; Tadai, K.; Mitani, T.; Kim, U.-S.; Kim, H.S. & Cho, B.W. (2005). Drastic change of electric double layer capacitance by surface functionalization of carbon nanotubes. *Appl. Phys. Lett.*, 87, 234106.

Kim, I-H.; Kim, J-H.; Cho, B-W. & Kim, K-B. (2006). Synthesis and Electrochemical Characterization of Vanadium Oxide on Carbon Nanotube Film Substrate for Pseudocapacitor Applications. *J. Electrochem. Soc.*, 153 (6), A989-A996.

Kimizuka, O.; Tanaike, O.; Yamashita, J.; Hiraoka, T.; Futaba, D.N.; Hata, K.; Machida, K.; Suematsu, S.; Tamamitsu, K.; Saeki, S.; Yamada, Y. & Hatori, H. (2008). *Carbon*, 46 (14) 1999-2001.

Kötz, R. & Carlen, M. (2000). Principles and applications of electrochemical capacitors. *Electrochim. Acta*, 45 (15-16), 2483–2498.

Le, D.B.; S Passerini,.; Guo, J.; Ressler, J.; Owens, B.B. & Smyrl, W.H. (1996). High Surface Area V_2O_5 Aerogel Intercalation Electrodes. *J. Electrochem. Soc.*, 143 (7), 2099-2104.

Lee, C.Y.; Tsai, H.M.; Chuang, H.J.; Li, S.Y.; Lin, P. & Tseng, T.Y. (2005). Characteristics and Electrochemical Performance of Supercapacitors with Manganese Oxide-Carbon Nanotube Nanocomposite Electrodes. *J. Electrochem. Soc.*, 152 (4), A716-A720.

Lee, J.Y.; Liang, K.; An, K.H. & Lee, Y.H. (2005). Nickel oxide/carbon nanotubes nanocomposite for electrochemical capacitance. *Synthetic Metals*, 150 (2), 153-157.

Liu, T.; Sreekumar, T.V.; Kumar, S.; Hauge, R.H. & Smalley, R.E. (2003). SWNT/PAN composite film-based supercapacitors. *Carbon*, 41(12), 2440-2442.

Lu, W.; Qu, L.; Dai, L. & Henry, K. (2007). Superior Capacitive Performance of Aligned Carbon Nanotubes in Ionic Liquids. *ECS Trans.*, **6** (25), 257-261.

Lu, W.; Qu, L.; Henry, K. & Dai, L. (2009). High performance electrochemical capacitors from aligned carbon nanotube electrodes and ionic liquid electrolytes. *J. Power Sources*, 189 (2), 1270-1277.

Ma, S.B.; Nam, K.W.; Yoon, W.S.; Yang, X.Q.; Ahn, K.Y.; Oh, K.H. & Kim, K.B. (2007). A novel concept of hybrid capacitor based on manganese oxide materials. *Electrochem. Commun.* 9 (12), 2807-2811.

Ma, S.-B.; Nam, K.-W.; Yoon, W.-S.; Yang, X.-Q.; Ahn, K.-Y.; Oh, K.-H. & Kim, K.-B. (2008). Electrochemical properties of manganese oxide coated onto carbon nanotubes for energy-storage applications. *J. Power Sources*, 178 (1), 483-489.

Mandal, S.; Amarilla, J.M.; Ibanez, J. & Rojo, J. M. (2001). The Role of Carbon Black in $LiMn_2O_4$-Based Composites as Cathodes for Rechargeable Lithium Batteries. *J. Electrochem. Soc.*, 148 (1), A24-A29.

Mastragostino, M.; Arbizzani, C.; Paraventi R. & Zanelli, A. (2000). Polymer Selection and Cell Design for Electric-Vehicle Supercapacitors. *J. Electrochem. Soc.*, 147 (2), 407-412.

Mayer, S.T.; Pekala R.W. & Kaschmitter, J.L. (1993). The Aerocapacitor: An Electrochemical Double-Layer Energy-Storage Device. *J. Electrochem. Soc.*, 140 (2), 446-451.

Miller, J.R. (1996). ECS Symposium Proceedings, F. Delnick and M. Tomkiewicz, Eds. The Electrochemical Society, pp. 246.

Miller, J.R. & Burke, A.F. (2008). Electrochemical capacitors: Challenges and opportunities for real-world applications. *Electrochem. Soc. Interf.* 17 (1), 53–57.

Miller, J.R. & Simon, P. (2008). Electrochemical capacitors for energy management. *Science*, 321 (5889), 651–652.

Molina-Sabio, M.; Gonzalez, M.T.; Rodriguez-Reinoso, F. & Sepúlveda-Escribano, A. (1996). Effect of steam and carbon dioxide activation in the micropore size distribution of activated carbon. *Carbon*, 34 (4), 505-509.

Naoi, K. & Simon, P. (2008). New Materials and New Configurations for Advanced Electrochemical Capacitors. *Electrochem. Soc. Interf.*, 17 (1), 34-37.

Niu, C.; Sichel, E.K.; Hoch, R.; Moy, D. & Tennent, H. (1997). High power electrochemical capacitors based on carbon nanotube electrodes. *Appl. Phys. Lett.*, 70 (11), 1480-1482.

Pan, H.; Poh, C.K.; Feng, Y.P. & Lin, J. (2007). Supercapacitor Electrodes from Tubes-in-Tube Carbon Nanostructures. *Chem. Mater.*, 19 (25), 6120-6125.

Passerini, S.; Ressler, J.J.; Le, D.B.; Owens, B.B. and Smyrl, W. H. (1999). High rate electrodes of V_2O_5 aerogel. *Electrochim. Acta*, 44 (13), 2209-2217.

Pasquier, A.D.; Plitz, I.; Gural, J.; Menocal, S. & Amatucci, G. (2003). Characteristics and performance of 500 F asymmetric hybrid advanced supercapacitor prototypes. *J. Power Sources*, 113 (1), 62-71.

Peigney, A.; Laurent, Ch.; Flahaut, E.; Bacsa, R.R. & Rousset, A. (2001). Specific surface area of carbon nanotubes and bundles of carbon nanotubes. *Carbon*, 39 (4), 507-514.

Portet, C.; Taberna, P.L.; Simon, P. & Flahaut, E. (2005). Influence of carbon nanotubes addition on carbon–carbon supercapacitor performances in organic electrolyte. *J. Power Sources*, 139 (1-2), 371-378.

Pushparaj, V.L.; Shaijumon, M.M.; Kumar, A.; Murugesan, S.; Ci, L.; Vajtai, R.; Linhardt, J.; Nalamasu, O. and Ajayan, P.M. (2007). Flexible energy storage devices based on nanocomposite paper. *PNAS*, 104 (34), 13574–13577.

Qu, L. & Dai, L. (2007). Direct growth of multicomponent micropatterns of vertically-aligned single-walled carbon nanotubes interposed with their multi-walled counterparts on Al-activated iron substrates. *J. Mater. Chem.*, 17 (32), 3401-3405.

Rudge, A.; Davey, J.; Raistrick, I.; Gottesfeld S. & Ferraris, J.P. (1994). Conducting polymers as active materials in electrochemical capacitors. *J. Power Sources*, 47 (1-2), 89-107.

Sakamoto, J.S. & Dunn, B. (2002). Vanadium Oxide-Carbon Nanotube Composite Electrodes for Use in Secondary Lithium Batteries. *J. Electrochem. Soc.*, 149 (1), A26-A30.

Shirakawa, H.; Louis, E.J.; MacDiarmid, A.G.; Chiang, C.K. & Heeger, A.J. (1977). Synthesis of electrically conducting organic polymers: halogen derivatives of polyacetylene, $(CH)_x$. *J. Chem. Soc. Chem. Commun.*, 578-580.

Simon P. & Gogotsi, Y. (2008). Materials for electrochemical capacitors. *Nature materials*, 7 (11), 845-854.

Taberna, P.L.; Mitra, S.; Poizot, P.; Simon, P. & Tarascon, J.-M. (2006). High rate capabilities Fe_3O_4-based Cu nano-architectured electrodes for lithium-ion battery applications, *Nature Mater.* 5 (7), 567-573.

Talapatra, S.; Kar, S.; Pal, S.K.; Vajtai, R.; Ci, L.; Victor, P.; Shaijumon, M.M.; Kaur, S.; Nalamasu, O. & Ajayan, P.M. (2006). Direct growth of aligned carbon nanotubes on bulk metals. *Nature Nanotechnology*, 1 (2), 112-116.

Tanahashi, I.; Yoshida, A. & Nishino, A. (1990). Electrochemical Characterization of Activated Carbon-Fiber Cloth Polarizable Electrodes for Electric Double-Layer Capacitors. *J. Electrochem. Soc.*, 137 (10), 3052-3057.

Thess, A.; Lee, R.; Nikolaev, P.; Dai, H.; Petit, P.; Robert, J.; Xu, C.; Lee, Y.H.; Kim, S.G.; Rinzler, A.G.; Rinzler, D.; Colbert, T.; Scuseria, G.E.; Tomanek, D.; Fischer, J.E. & Smalley, R.E. (1996). Crystalline ropes of metallic carbon nanotubes. *Science*, 273 (5274), 483-487.

US Department of Energy, (2007). *Basic Research Needs for Electrical Energy Storage*, www.sc.doe.gov/bes/reports/abstracts.html#EES2007.

Walden, P. (1914). *Bull. Acad. Imper. Sci.* (St. Petersburg), 1800.

Wang, G.; Zhang, B.; Yu, Z. & Qu, M. (2005). Manganese oxide/MWNTs composite electrodes for supercapacitors. *Solid State Ionics*, 176 (11-12) 1169–1174.

Wang, Y.G. & Xia, Y.Y. (2005). A new concept hybrid electrochemical surpercapacitor: Carbon/$LiMn_2O_4$ aqueous system. *Electrochem. Commun.* 7 (11), 1138-1142.

Wildgoose, G.G.; Banks, C.E. & Compton, R.G. (2006). Metal Nanoparticles and Related Materials Supported on Carbon Nanotubes: Methods and Applications, *Small*, 2 (2), 182-193.

Xu, B.; Wu, F.; Chen, R.; Cao, G.; Chen, S.; Wang, G. & Yang, Y. (2006). Room temperature molten salt as electrolyte for carbon nanotube-based electric double layer capacitors. *J. Power Sources*, 158 (1), 773-778.

Xu, K. (2004). Nonaqueous Liquid Electrolytes for Lithium-Based Rechargeable Batteries. *Chem. Rev.*, 104 (10), 4303-4418.

Zhang, H.; Cao, G.; Wang, Z.; Yang, Y.; Shi, Z. & Gu, Z. (2008). Growth of Manganese Oxide Nanoflowers on Vertically-Aligned Carbon Nanotube Arrays for High-Rate Electrochemical Capacitive Energy Storage. *Nano Lett.*, 8 (9), 2664-2668.

Zhang, H.; Cao, G. & Yang, Y. (2007). Using a cut-paste method to prepare a carbon nanotube fur electrode. *Nanotechnology*, 18 (19), 195607-195610.

Zhang, H.; Cao, G.; Yang, Y. & Gu, Z. (2008). Comparison Between Electrochemical Properties of Aligned Carbon Nanotube Array and Entangled Carbon Nanotube Electrodes. *J. Electrochem. Soc.*, 155 (**2**), K19-K22.

Zhang, M.; Atkinson, R.R.; Baughman, H. (2004). Multifunctional Carbon Nanotube Yarns by Downsizing an Ancient Technology, *Science*, 306 (5700), 1358-1361.

Zheng, B.; Lu, C.; Gu, G.; Makarovski, A.; Finkelstein, G. & Liu, J. (2002). Efficient CVD growth of single-ealled carbon nanotubes on surfaces using carbon monoxide precursor. *Nano Lett.*, 2 (8), 895-898, and references cited therein.

Zheng, J.P.; Cygan, P.G. & Jow, T.R. (1995). Hydrous Ruthenium Oxide as an Electrode Material for Electrochemical Capacitors. *J. Electrochem. Soc.*, 142 (8), 2699-2703.

Zheng, J.P.; Huang, J. & Jow, T.R. (1997). The Limitations of Energy Density for Electrochemical Capacitors. *J. Electrochem. Soc.*, 144 (6), 2026-2031.

Zilli, D.; Bonelli, P.R. & Cukierman, A.L. (2006). Effect of alignment on adsorption characteristics of self-oriented multi-walled carbon nanotube arrays. *Nanotechnology*, 17 (20), 5136-5141.

30

Carbon Nanotube Membrane Solar Sails
A Challenge for Extremely Fast Space Flight

Salvatore Santoli
INT – International Nanobiological Testbed Ltd.
United Kingdom

1. Introduction

Space sailing is a concept in space propulsion technology that is based on the use of solar or other radiation for propulsion. The sunlight, or the beam of electromagnetic radiation from a laser, are reflected by the sail surface and, since photons carry momentum, their reflection changes their momentum so that a resultant force is exerted on the reflecting surface. Accordingly, spacecraft exploiting this propulsion method will carry very large, ultra-thin reflectors and will be able to transport heavy payloads both for planetary and for interstellar missions. Here the focus will be on solar sails as systems exploiting the pressure of light from the Sun for propulsion, not from a laser beamed to the sail, and their physics will be briefly analyzed from first principles involving mechanics, optics and thermodynamics. It will be then shown how the basic astrodynamic parameters and the technological and engineering requirements can be derived from first principles for interplanetary and interstellar flight purposes.

This discussion will contemplate such thumbnail sketch of the basic scientific and engineering aspects of solar sail propulsion, also known as *photonic propulsion*, just in order to allow a better understanding and appreciation of the recent technological advancements in the sail material through a proper development of carbon nanotube membranes, that has allowed the design to be realized of a solar sail featuring a cruising speed of the order of some thousandths the speed of light, i.e. a speed much higher than the speed attainable by any other propulsion methods. This is due both to the very light and reflecting sail material and to the manoeuvre of inverting the angular momentum (Vulpetti, 2002) of the sail when it has reached the perihelion according to the *Sun diving flight mode*, i.e. the mode consisting in launching the solar sail toward the Sun so that on flying by the Sun it becomes strongly accelerated to finally reach a very high cruising speed: the so-called asymptotic speed. Anyway, various other modes can be adopted, e.g. for orbiting a planet and interplanetary transportation of any payload. For detailed descriptions of solar sail structural technologies and of the astrodynamics involved in realizing some special missions, e.g. based on intra-orbit transfer of sail up to being free from planetary gravitation and ready to spacefaring, the reader is referred to the abundant specialized papers on space sailing, where also excellent treatises can be found for a systematic approach (Wright, 1993; McInnes, 1999; Vulpetti et al., 2008).

2. Solar Sails from First Principles

All the following discussion will be managing to let the reader go along the same route as I went from the basic mechanics, optics, thermodynamics and structural engineering up to the design of the latest and speediest models of solar sail whose membrane is made up of carbon nanotubes.
The problem is to design a reflector large enough to get a suitable acceleration *but* of so small a mass to be pushed and accelerated up to a significant speed by incident photons from the Sun. Accordingly, reflectance should be the highest possible with the present material technologies. But it is well known from optics that there is no such thing as a perfect reflector, some of the photons becoming absorbed by the reflector, so that the designer will have to face the problem of optimizing the combination of reflectance, density, and emittance for a correct computation of the resultant force pushing the sail, for maximizing such force as well as for keeping the sail temperature at a possible level, mainly on approaching the Sun. Indeed, thermodynamics of radiation will be thus involved. All that would not be attainable with a single material, so that at least two layers are to be planned as the sail structure: a reflecting and an emitting material. Leaving behind for the moment the problems about the nature of materials, which would also involve their structural strength, the basic approach to solar sail dynamics will be based on the following sketches depicting the mechanics, the optics and the thermodynamics of the solar sail.
In Fig. 1, I = incident light; R = reflected light; F_I = incident force; F_R = reaction force; F = the resultant; S = the sail

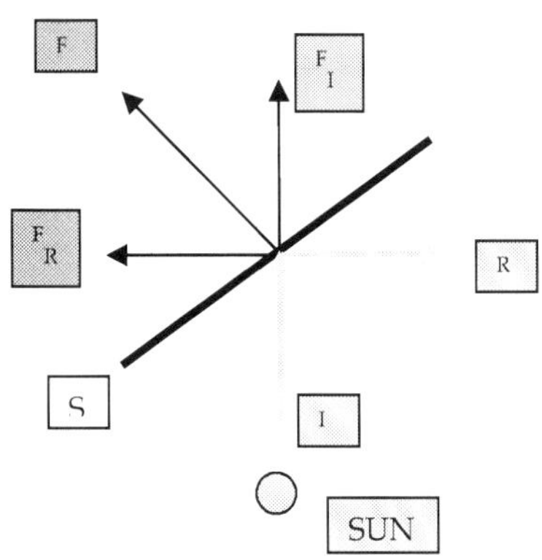

Fig. 1. Mechanics of the sail as a perfect reflector

In Fig. 2, the realistic situation for a non-ideal sail (i.e. a non perfect reflector) is depicted: the resultant will no longer be normal to the sail surface as part of incident photons will not be reflected; they will be absorbed and subsequently re-radiated as thermal radiation. The resultant will be rotated from the normal to the sail surface toward the direction of I because the force F_A from absorbed photons is quite stronger than the force F_R due to reflected photons, and the force from re-radiated photons as would be computed from thermodynamics will be normal to the surface. This will then lead to the setting forth of an optical sketch of the physical situation (Fig. 3), the latter leading then to the application of thermodynamics of radiation for evaluating the force exerted on the sail by re-radiated photons as well as the temperature of the solar sail. The situation as in Fig. 2 would also refer to the fact that stresses on the sail surface will tend to inflect the sail, which will have a strong tendency to billow so that a proper system of tie-rods and struts is to be designed. The material of the sail surface itself should be given the property of preventing any tear in the material caused by space debris or very small asteroids from running along the whole structure and to cause them to keep confined at the place of impact. It is to be considered that sail areas up to some km squared would be planned. Of course, this represents a formidable task for designers and builders to get uniform values of the optical parameters that the sail itself would embody.

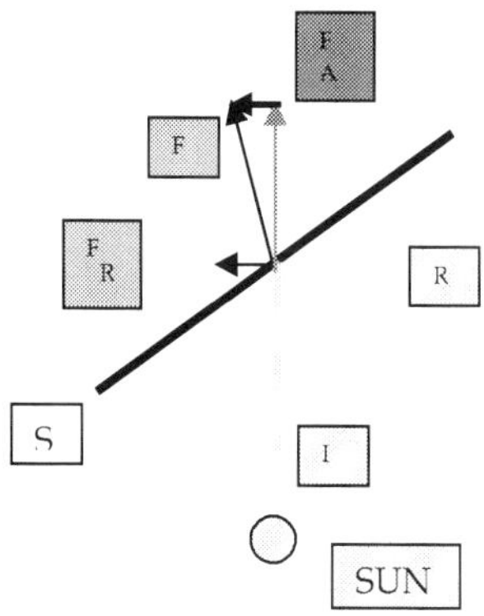

Fig. 2. Mechanics of the non-ideal sail: the resultant is shifted toward the direction of I from the normal to the surface as a result of $F_A >> F_R$ (vectors not to scale)

In the optical force model of Fig. 3, the unit vectors and the normal to the sail surface are considered because a proper calculation of forces asks for vector representations and scalar products along the unit vector **n** normal to the sail surface that defines the sail orientation, along the transverse unit vector **t** normal to **n**, along the unit vector **u** along the direction of incident photons, and along the unit vector **s** along the direction s of specularly reflected photons.

Now, with F_E ≡ force from photon emission by re-radiation; α ≡ the pitch angle of solar sail relative to the Sun line; PA ≡ radiation pressure x sail area; F_{Rs} ≡ force from fraction of incident photons specularly reflected; R ≡ fraction reflected of the incident photons; Rs ≡ fraction of reflected photons specularly reflected along **s**; F_{Rn} ≡ force along **n** from fraction of incident photons non-specularly reflected; B ≡ coefficient describing the non-Lambertian character of the surface, i.e. a surface which doesn't appear equally bright when viewed from any aspect angle, we have (cf. Fig. 3):

$$\mathbf{u} = \cos\alpha\,\mathbf{n} + \sin\alpha\,\mathbf{t}$$
$$\mathbf{s} = -\cos\alpha\,\mathbf{n} + \sin\alpha\,\mathbf{t}$$
$$\mathbf{s} = \mathbf{u} - 2\cos\alpha\,\mathbf{n}$$
$$F_A = PA\,(\cos^2\alpha\,\mathbf{n} + \cos\alpha\sin\alpha\,\mathbf{t})$$
$$F_{Rs} = -(Rs)\,PA\cos\alpha\,\mathbf{s}$$
$$F_{Rn} = BR(1-s)\,PA\cos\alpha\,\mathbf{n}$$

and for the resultant F in terms of the normal and transverse directions

$$F = PA[(Rs\cos^2\alpha + B(1-s)\,R\cos\alpha)\,\mathbf{n} - Rs\cos\alpha\sin\alpha\,\mathbf{t}\,]$$

The final force component is that due to photons which have been absorbed and then re-emitted as thermal radiation from **both the front (reflecting) and back surfaces** of the sail.

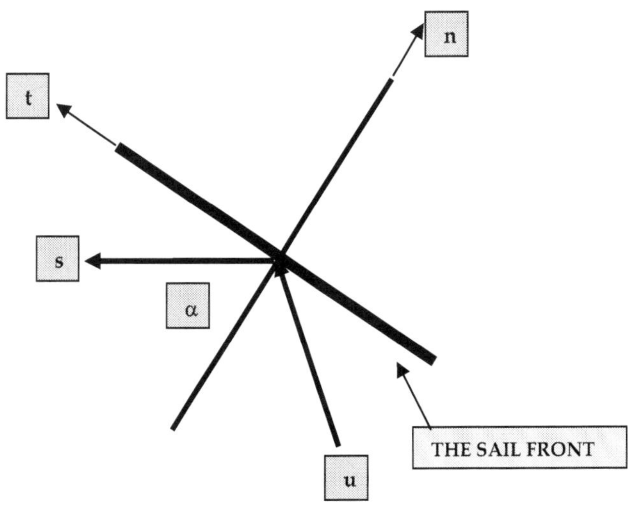

Fig. 3. Diagram of vectors for the optical model of the non-ideal solar sail

When designing the material for the solar sail film, it is important to compute the sail equilibrium temperature for two basic reasons: 1) computing the force exerted on the solar sail due to emission by re-radiation; 2) compute the sail temperature during its flight, mainly at the perihelion. Indeed, it is important to know the thermomechanical and the thermochemical behavior of the sail film during mission. For instance, in case of the so-called *light sail* made up of two-layer sails consisting of Al as the reflector and Cr as the emissive layer, as opposed to the *heavy sail* including an interposed plastic film, e.g. Kapton, it is important to estimate the whole thermal history of the double layer structure (Santoli and Scaglione, 1996) because according to the thermodynamic phase diagrams of the AlCr

system an interdiffusion process starts at a given temperature, and as a result a solid solution starts to form at the interface that goes on throughout the whole Al/Cr double layer so that all optical parameter values of the sail are changed and the whole mission fails. One of the great results stemming from the use of a carbon nanotube membrane for the sail film as a *monolayer* (Vulpetti et al.) is the perspective use of such membranes as capable of keeping all their original values of the optical parameter even after being heated up to incandescence. As to thermomechanical bevior, for instance the key point in judging about the mechanical fitness and the limitations to the performance of the Al/Cr composite double layer would be the determination of thermal stresses at the interface of that composite structure and the estimate of the possibility of occurrence of cracking, decohesion and fracture in the thin Cr film coated on the Al support (Santoli and Scaglione, 1996). *This problem would be absent from the envisaged technology of carbon nanotube monolayer films for sails that will be discussed in the following.*
The sail temperature can be calculated by the Stefan-Boltzmann law after setting forth the thermal balance of the solar sail; see Fig. 4 for the meanings of symbols and for a basic sketch stressing the vectorial nature of the problem.

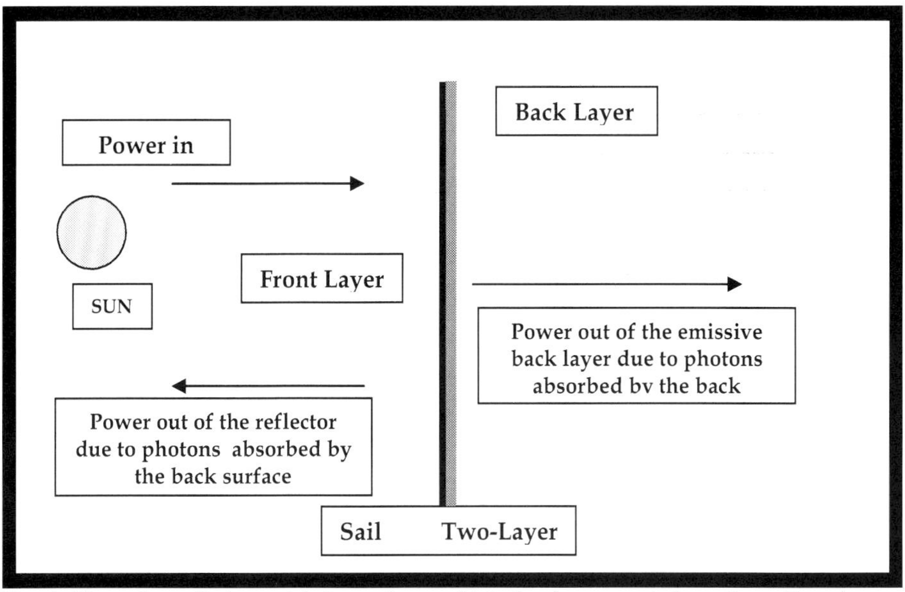

Fig. 4. The solar sail thermal balance from which the force exerted on the sail surface through emission by re-radiation and the temperature of the solar sail composite film can be calculated. This balance is of the essence for the discussion in **Section 2.1**

Now, with W as the solar flux incident on the sail, ε_f and ε_b respectively the front and back emissivities, B_b as the non-Lambertian coefficient of the back surface, σ the Stefan-Boltzmann constant, and T as the absolute temperature of the sail, we have:

Power Emitted from a Unit Area of the Sail at $T = \varepsilon_f \sigma T^4$

Force due to Emission by Re-Radiation Assuming Uniform Sail $T \equiv F_e$

$$F_e = \frac{\sigma T^4}{c}(\varepsilon_f B - \varepsilon_b B_b)\mathbf{n}$$

and, from the balance between thermal input and thermal output (cf. Fig. 4)

$$\Delta P \equiv \textit{Power in} - \textit{Power out} = (1\text{-}R)W\cos\alpha - (\varepsilon_f - \varepsilon_b)\,\sigma T^4 = 0$$

so that, taking into account that PA = W/c, both the sail equilibrium absolute temperature T and F_e can be calculated. With the aid of the vectorial optical diagram of Fig. 3, by considering all forces mentioned above it is quite trivial to get an expression of the total force exerted on the sail in terms of its normal (**n**) and transverse (**t**) components. Contrarily to the ideal sail case, for the real sail as a non perfect reflector the resultant force vector will not be in the direction normal to the sail surface due to the fact that the absorbed photons force is greater than the force due to reflected photons.
This analysis of the sail physics has been carried out by means of the particle (photon) description of the electromagnetic field, but an analysis based on the wave description by application of Maxwell equations for the field could have been carried out. This way of describing solar sail motion is important for some points in **Section 2.1**. In that section some more considerations are added for a clear understanding of solar sail physics so that some misconceptions about it can be removed. Indeed, a few remarks look like being opportune because of three reasons:

1) the objections made to the possibility of solar sailing were published in a journal enjoying a very wide audience; anyway, the Editor was asked to remove that article
2) the quite widespread erroneous description of how Crookes's radiometer works, though its correct description was given about 130 years ago by two of the most renowned scientists in the history of science: J.C. Maxwell and O. Reynolds, and
3) thermodynamics of radiation is fundamentally involved in the astrodynamics of solar sails in general, but mainly in missions contemplating ultra-light solar sails based on the emerging technology of carbon nanotube membranes, so that a clear understanding of the co-involvement of mechanical and thermodynamical laws is of the essence in designing such kinds of missions(cf. **Sections 3.4** and **4**).

2.1 Understanding the Physics of the Solar Sail

In an article published in the journal New Scientist (Parsons, 2003) some observations by Thomas Gold , Cornell University, are reported, according to which a solar sail could never fly for space missions because *"the proponents of solar sailing have forgotten about thermodynamics, the branch of physics governing heat transfer"*. His remarks, that caused a public debate, met with well reasoned disagreements (Diedrich, 2003), and the way they can be soundly refuted, are listed below:

Remark 1: when photons are reflected by a perfect mirror, they do not suffer a drop in temperature. Now, solar sails are governed by Carnot's rule for heat engines. According to that rule, there must be a degradation of energy in any machine that turns out free energy. A mirror doesn't have any degradation. According to this scenario, Carnot's rule says some energy can be extracted, so long as the object absorbing light remains cooler than the radiation itself.

Confutation: Carnot' rule applies to cyclic transformations of closed systems only. The solar sail in itself, i.e. excluding its environment made up of the whole Universe, doesn't work as a cyclic system; to describe the situation as a phenomenon in a closed system, the system would consist of the solar sail plus the whole Universe, into which the solar sail reflects back the light it receives from the Sun. **Remark 1 is neglecting the rest of the Universe**. Moreover, it would be very questionable to apply thermodynamics to the whole Universe, the more that the thermodynamics at the initial conditions (the "*Big Bang*") are not known with certainty (equilibrium? far-from-equilibrium?) (Santoli, 2009; Layzer, 1975).

Remark 2: A black sail would absorb sunlight and the momentum associated with it until it reaches thermal equilibrium, when it would stop absorbing sunlight.

Confutation: That is false. **Once an object has reached thermal equilibrium, it does not stop absorbing light!** As it has been depicted in Fig. 4, the sail would continue to absorb light after reaching thermal equilibrium, and would radiate infrared light out of the front and the back surface as a **steady state process**. Indeed, at equilibrium the temperature of the sail stops changing, and the radiated power, **as shown in the diagram**, equals the absorbed power. Should the radiated power from the front side equal the radiative power from the back side, the thrust from the radiated power would become zero, and the sail would be acted upon by the thrust from the absorbed light in this steady state. The equation for ΔP as above depicts just a steady state condition.

Remark 3: Crookes's radiometer is an example of a failed test of light pressure from reflection of photons. This device consists of four paddles attached to the arms of a rotor, inside a vacuum jar. Each paddle is silvered on one side and coated with a black absorber on the other. When placed in the sunlight, the rotor spins. If the theory of solar sailing is right, the rotor should spin with the reflecting silver surface away from the light. And it actually spins, but the other way.

Confutation: in spite of the efforts carried out about 130 years ago by J.C. Maxwell and Osborne Reynolds who were able to explain how this device, developed by the eminent physicist Sir William Crookes for measuring radiant energy of heat and light, works, very often the first erroneous explanation by Crookes himself, or other descriptions before the correct one given by J.C. Maxwell, are quoted as correct. According to Crookes's paper, light radiation pressure on the black vanes was making the rotor spin. But soon J.C. Maxwell realized that this description was wrong. Indeed, light falling on the black side should be absorbed, while light falling on the silvered side of the vanes should be reflected. As a result, there would be twice as much radiation pressure on the metal side as on the black. Accordingly, the real rotor would work the wrong way! Indeed, **the physics of Sir William Crookes's radiometer, widely known as the light mill, has nothing to do with the physics of solar sail**. It was shown (Maxwell, 1879; Reynolds, 1879) that the phenomenon is caused by the very few molecules remaining after the jar is evacuated: the radiometer work is explained by the tangential forces exerted on the **edges** of the vanes by the molecules from the warmer side of each vane which are faster than the colder molecules from the colder side. Light pressure can be correctly measured today in laboratories with proper setups. For instance, a thin fiber with vanes attached to it twists measurably when light strikes the vanes. Lasers and microwaves are used to test light pressure. And reality of light pressure was also tested in space, e.g. when the Mariner 10 mission to Mercury and Venus used solar arrays to steer itself when it ran low on propellant. The same was for the Messenger mission to Mercury: both missions demonstrated the use of solar pressure as a method of attitude control in order to save attitude control propellant. Recent discussions of the radiometer can also be found in the scientific literature (Heckenberg, 1996; Woodruff, 1968); cf. **Section 5**.

Remark 4: light cannot exert pressure because it is a scalar quantity, while Newtonian momentum is a vector.

Confutation: again this is wrong! Light is described by Maxwell equations which are vector equations. Photons are just individual packets of light waves, and their velocity is a vector given by the photon direction multiplied by the speed of light.

Remark 5: there is no degradation of energy in the reflection process on the mirror.

Confutation: this is false. Some photons are absorbed and the sail heats up to an equilibrium temperature. Moreover, photons reflected by the reflector of a solar sail undergo a Doppler shift: their wavelengths increase and energy decreases by a factor dependent on the velocity of the sail, so that energy is transferred from the **sun - photon** system to the **solar sail** system. This change of energy can be shown to be exactly equal and opposite to the energy change of the sail.

Thus, the principles of mechanics and thermodynamics are not brocken in solar sailing theory. Thermodynamics of radiation is fundamental for describing solar sail astro-dynamics, together with the laws of motion of Newtonian mechanics. This will be very clear in the astrodynamic study of missions based on carbon nanotube membrane mono-layer sails discussed in Section 4.

3. Engineering the Concept – From the Heavy Sail to the Ultralight Carbon Nanotube Membrane *Mono-Layer* Sail Technology

The following sections deal with the sailmaking aspects as spaceship- and mission-independent features and as the very core of the spaceship conception. While a very good systematic approach to what can now be called *"classical sails"* can be found in Wright's book (Wright, 1993), no source of design details can be found as to the first steps from the light all-metal sail as the objective of the **Aurora Project** (Santoli & Scaglione, 1996; Scaglione & Vulpetti, 1999; Vulpetti, 1996) toward a carbon nanotube membrane sail technology other than the very recent papers concerning this approach by myself and co-workers (Vulpetti et al., 2007; Vulpetti, 2009).

The **Aurora Project** was an Italian - British - Swiss collaboration that began in 1993 to study the general problems about an all-metal sailcraft undergoing the so-called orbital angular monentum reversal mode for fast photon solar sailing in space (Vulpetti, 1996a; 1996b). It was the very starting point and foothold for the development of the concept of the carbon nanotube membrane solar sails for even faster speeds by photonic propulsion. Accordingly, the following discussion, while illustrating some basic aspects of materials, sail shapes and support architecture engineering for a sailcraft, will reflect basicly the conceptual pathway from the heavy sails contemplated in standard books to the novel concept of the ultralight sail capable of exploiting photonic propulsion up to reaching cruise speeds whose values would be substantial fractions of the speed of light. Details concerning a design according to that novel concept are given in the case study discussed in **Section 4**.

3.1 Materials – Understanding Solar Sail Astrodynamics

In the standard design of a heavy sail, a substrate film typically provides the mechanical strength needed to carry the loads through the sail. The load will consist of the necessary structural supports for the sail not to billow too much under solar pressure - ribs, booms, masters - and of the payload - scientific instrumentation, material to be transported to other planets. Indeed, solar sails are considered as good candidates for future transportation of materials and people to other planets of the Solar System. The substrate film bears a metallic coating on each side: a reflecting and an emitter coating. An overcoating is usually put on the reflector to protect the same, e.g aluminium, from oxidation and loss of reflectance. Elimination of the substrate film can be realized in the case of thin metallic sails, thickness about 100 nm, if the sail is manufactured in space and not subjected to folding and successive unfolding, the so-called deployment, as is the case of sails manufactured on the Earth. Deployment in Earth orbit or, better, at the first Lagrange point of the Earth - Moon sytem, i.e. at the point where no gravitation is experienced, is one of the biggest problems for solar sailing. A material capable of surely surviving this manoeuvre would be preferable; the **monolayer carbon nanotube membrane would support that stress better than a self-supporting all metal sail** (Santoli & Scaglione, 1996).

In the Aurora Project (Santoli & Scaglione, 1996; Scaglione & Vulpetti, 1998) the innovative concept was developed of an all-metal sail - Al as the reflector and Cr as the emitter - bearing a thin layer of plastic material on the Cr emitter layer in order to support the sail during deployment in an Earth orbit. The thin plastic film would have been removed when in orbit by the action of UV rays from the Sun, possibly with the previous addition to the plastic film of a sensitizer to make the removal complete and rapid, in order to get a much lighter sail to be launched to orbiting the sun where, at the perihelion, the manoeuvre of inversion of the angular momentum of the sail would accelerate the sail spaceship to speeds much higher than those possible for the heavy sail based on a plastic support. Fig. 5 and 6 depict the standard heavy sail and the reflector monolayer sail, whose validity depends on the thermal conditions during mission. Fig. 7 shows the Aurora Project concept design of the Al-Cr all metal sail reinforced with the plastic layer coated on the emitter side of the sail:

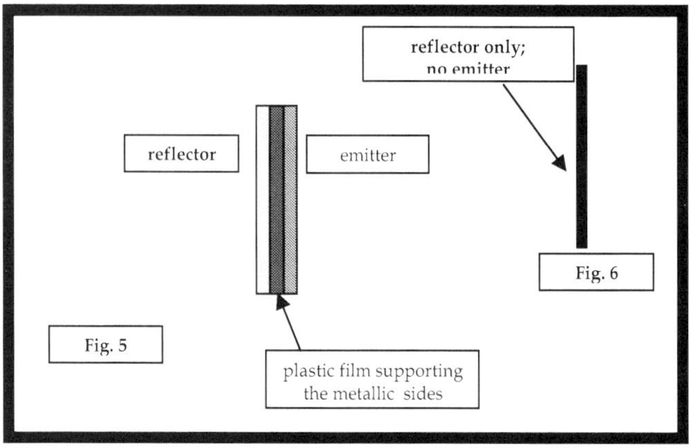

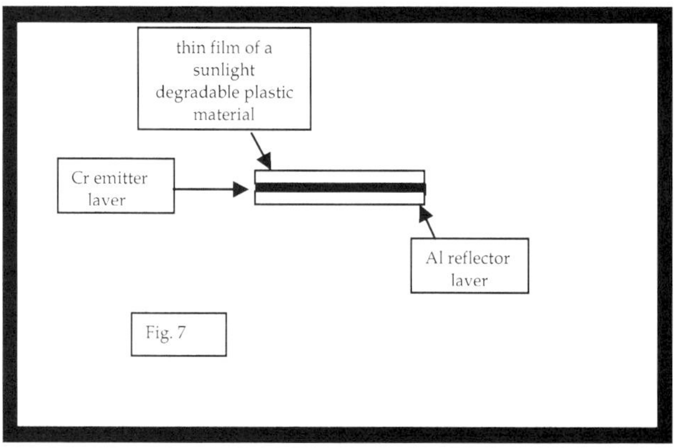

Very often the plastic film considered for the standard heavy sail with the film interposed between the reflector and the emitter was Kapton, a trademark of Du Pont, featuring good radiation and high temperature properties, capability of operating for years at 250°C, and availability at minimum thicknesses of 2 - 3μm. Such thicknesses could be reduced by chemical etching or ion bombardment etching. Kapton's density is 1.42 g/cm^2. Mylar, a trade mark of Du Pont, showed very low resistance to UV radiation, so that it would be unsuitable for long-duration missions. The idea of a supporting plastic film to overcome problems with the deployment in an Earth orbit that was coated externally on the emitter surface, not in between the reflector and emitter layer, so giving the possibility of removing the plastic film before flying, was one of the key points of the Aurora Project, mentioned above as an innovative challenging technological and scientific objective. A possible method to get the Al/Cr bilayer free from the supporting plastic was found to be the **P**lasma **C**hemical **V**aporization **M**achining (**PCVM**), with removal, after deployment, by photodegradation that would not affect the quality of the underlying Cr surface. This problem of removal of the plastic support film by photolysis without damaging the Cr layer was further investigated with positive indications (Scaglione & Vulpetti, 1998).

An important parameter for understanding the astrodynamics of a solar sailship is the lightness number λ, which is defined as the characteristic acceleration, i.e. the rate at which the sailcraft would accelerate at Earth's distance from the Sun if the sailcraft were directly facing the Sun, divided by the gravitational acceleration of the Sun at Earth's distance from the Sun. Early heavy sail spacecraft were designed so as to have lightness number values substantially less than 1, but the aim of the Aurora Project was an advanced high-performance sailcraft that might reach λ's greater than 1. Actually, lightness numbers between 0.5 and 1 could be obtained with the all-metals sail, with speeds after flybying the Sun between 60 and 80 km/s (Vulpetti, 1996; Santoli & Scaglione, 1996). But it was found (Vulpetti, 1998) that a different mathematical formalism, introducing the **lightness vector** (cf. the following section) would fully characterize the sailcraft dynamics instead of the classical equations of motion of the sail in the gravitational field; this formalism was particularly suitable to describe sailcraft motion with reversal of the orbital angular momentum and in the case of quite high lightness numbers allowing the all-metal sail to escape the solar system or to point to some distant object in the heliosphere, with velocities *somewhat* higher than the current record speed of missions beyond the planetary system, e.g. the Voyager-1 speed. This point, that leads to the astrodynamics of the extremely fast missions possible with the carbon nanotube membrane solar sails, will be discussed in a detailed way in the next section. The reader interested just in the material properties and the envisageable emerging technologies for further improvements in the performance of carbon nanotube membranes can skip this section and go directly to **Sections 3.4** and **4**.

3.2 The Astrodynamic Embodiment of the Fast Solar Sailing Concept

The introduction of very fast sailcraft technologies involves some changes in the astrodynamic mathematical formalism describing the sailcraft flight. This is of interest both for the all-metal light sail and mainly for the extremely fast carbon nanotube membrane solar sails (Vulpetti et al., 2008). For understanding the astrodynamics corresponding to any given mission and how the thermooptical parameters of the sail membrane should be designed, the introduction of the concept of sailcraft lightness vector is to be carried out and two reference frames must be used.

3.2.1 The Reference Frames for the Solar Sail Flight Description and the Lightness Vector

DEFINITIONS:

• Heliocentric Orbital Frame (**HOF**): the Cartesian reference frame with the origin in the sail center of gravity. In this reference frame, the x axis is the direction of the **Sun-to-sailcraft** position vector, the z axis is along the sailcraft orbital angular momentum, and the y axis forms with the x and z axis a counterclockwise reference frame

• Heliocentric Inertial Frame (**HIF**): the inertial Cartesian reference frame centered in the Sun

• the azimuth α is the angle that the projection of the unit vector **n** of sail orientation forms with the **HOF**

• the elevation δ is the angle that the projection of **n** forms withy the xy plane of **HOF**

Now, α and δ specify the sail orientation unit vector **n** (already used in the description of the vector optics for the sail forces) in **HOF,** with $0 \leq \alpha < 360°$ and $-90° \leq \delta \leq 90°$ and the following definitions can be given, irrespective of the material the sail is made up of:

λ_r , λ_t , λ_n are the radial, the transversal, and the normal components of the **lightness vector**, $\mathbf{L} = [\lambda_r , \lambda_t , \lambda_n]$ (cf. the vectorial optics diagram of Fig.3), and any λ_r would be mathematically represented as a component of **L** by $\lambda_r = |\mathbf{L}|$ as shown in the following.

APPLICATIONS:

The λ's depend nonlinearly on α and δ, and linearly on the thermo-optical parameters of the sail material. They are the **key parameters** for rapidly studying, once the **L** value of the sail (material and structure) and its components have been determined or set forth previously, the basic properties (circular orbiting; ability to escape through a hyperbolic orbit the Solar System toward a far point, acceleration and/or deceleration at given distances) of the sailcraft motion by computer programming of the equations for the motion of the sailcraft within the two reference frames defined above and, possibly, following the manoeuvre of reversing the angular momentum of the sailcraft (Vulpetti, 1999)

$$\alpha = 0, \delta = 0: \quad [\lambda_r , 0, 0] = |\mathbf{L}| = \lambda_r$$

$$\alpha \neq 0, \delta = 0: \quad [\lambda_r , \lambda_t, 0]$$

$$\alpha = 0, \delta \neq 0: \quad [\lambda_r , 0, \lambda_n]$$

While these are key parameters, they are not the only quantities involved in determining the thrust acceleration of sailcraft motion, whose basic equations are made up of four basic units which are *nonlinear and not all separable*: **the effective solar pressure and gravitational acceleration, sailcraft sail loading, sail thermooptical properties, and sail attitude**. The carbon nanotube membrane monolayer film solar sail concept stems just from a technology of merging the mechanical and thermooptical properties of such membranes and the astrodynamic equations for any space mission, as discussed in **Sections 3.4** and **4**.

3.3 Structural Engineering – Sail Shapes and the Supporting Architectures

Generally the sail is square-shaped or round-shaped, or in the shape of a hexagon. The sail film must be kept taut under the action of the solar light pressure. Stiff structural members must hold the sail in position. A sail can carry only tension loads since the film material has no effective resistance to compression or bending. The supporting structure must provide the stiffness that the sail itself lacks. Indeed, it is not practical to have a sail that does not deflect, nor to have a design that can distribute loads completely even across the sail. The supporting structure of the sail - booms, masts, stays, tensioners etc. - will always undergo some deflection, and the sail will then experience both large scale and local deflections, respectively resulting in billowing and wrinkling. As to the sizing of the sail, its size is designed on the basis of two main factors: the mass of the payload to be carried, and the nature of the trajectory for the mission. Stated otherwise, the analysis of the trajectory from the point of departure to the destination for the designed trip time will yield the characteristic acceleration and then the lightness number or the lightness vector that is necessary for the mission.

While the general standard design principles for the shape, size and structural members can be found in the literature (Wright, 1993), the analysis of the structural engineering and materials designed for the innovative all-metal sail of the Aurora Project (Genta and Brusa, 1996; Santoli and Scaglione, 1996) will show the preliminary stress and deformations assessment for a fast sailcraft, as the immediate precursor of the extremely fast carbon nanotube solar sail, designed to escape the Solar System for sailing toward the interstellar space at very high speed. In the case of the Aurora Project, the stressing due to light pressure is well withstood by the metallic film, made up of a 0.2 μm reflective layer and 0.01 μm emissive layer, due to its high stiffness, but the film of a plastic material is needed for overcomimg problems of storage and deployment. The sail, whose shape was square, was provided with a grid of thicker strips of Al foil for the back surface that would act as ripstops in case of punctures by micrometeoroids and as doublers to protect the thin zones which were in contact with the main structure. The sail is kept in position by a structure made of two booms along the diagonals and kept firm in position by a set of stay wires which were connected to two masts at the center of the square and perpendicular to the surface. The payload was located at the center of the spacecraft.

In operation, a sail is to be oriented to point the force in a useful direction. An attitude control system must be designed in order to execute a steering profile that will cause the sailcraft to follow a trajectory to its intended destination. The pressure is not uniform across a sail due to curvature in it. In addition to balancing the sailcraft, the control system must also control the pointing direction of the sailcraft by shifting the center of mass or the center of pressure on the sail. In the design for the Aurora Project, four ion thrusters were attached to the four corners to provide the attitude control.

3.4 From the Aurora Project to the Ultralight Carbon Nanotube Monolayer Film Concept

A real breakthrough in the field of carbon nanotube membranes as regards their application in a host of very different fields of practical commercial value but also of great importance for space technologies was the result of the study of a team of researchers. The paper with a detailed account of the techniques to obtain carbon nanotube sheets as multifunctional membranes appeared in the journal *Science* (Zhang et al., 2005). Macroscopic masses of individual carbon nanotubes assemblies could be produced at rates of 7 m/minute by cooperatively rotating carbon nanotubes in vertically oriented nanotube arrays - called usually "forests" in the nanotechnological community - such assemblies being in the shape of 5-cm-wide, meter-long transparent sheets. Such nanotube sheets were self-supporting and initially formed as highly anisotropic electronically conducting aerogels that could be densified into strong sheets that were so thin as 50 nanometers. It was very remarkable that the measured gravimetric strength of orthogonally oriented sheet arrays exceeds that of sheets of high-strength steel. Laboratory demonstrations for the microwave bonding of plastics and for making transparent, highly elastomeric electrodes, or planar sources of polarized broad-band radiation, flexible organic light-emitting diodes showed remarkable results for such carbon nanotube membranes.

The production process was quite different from the usual procedures for carbon nanotube sheets, that are made following the ancient art of paper-making, with a very slow filtration of carbon nanotubes dispersed in water followed by peeling the dried nanotubes as a layer from the filter; or according to some changes in this procedure as to the filtration rate by producing ultrathin nanotube sheets endowed with high transparency and high conductivity. Sheets with partial nanotube alignment instead of isotropic sheets were produced by the application of high magnetic fields or by mechanical rubbing of those carbon nanotubes that are vertically trapped in filter pores. According to other techniques, nanotube sheets have been fabricated form nanotube aerogels, or by Langmuir-Blodgett deposition, or by casting from oleum, or by spin coating. The strong transparent and multifunctional carbon nanotube membranes by Mei Zhang and co-workers (Zhang et al., 2005) were produced by a solid-sate process that for practical applications features its scalability for continuous high-rate production. Such membranes are highly oriented and they were produced by drawing them from a sidewall of multiwalled carbon nanotube (MWCNT) forests synthesized by catalytic chemical vapour deposition employing acetylene gas as the carbon source. The MWCNTs were about 10 nm in diameter,and the range of the investigated forest heights was form 70 to 300 nm. The membrane fabrication process was found to be robust, with no fundamental limitations as to the width and length of the membrane. The 5 cm membrane width obtained was equal to the forest width when the draw rate was 5 m/minute or lower. At constant draw rates above about 7 m/minute the membranes progressively narrowed, and MWCNT fibrils began to break at the intersection between the sheet sides and the forest.

The thickness of the MWCNT membrane increased with increasing forest height and was of about 18 μm in scanning electron microscopy (SEM) images of a membrane drawn from a 245 μm -high forest. From this thickness and the measured areal density of about 2.7 $\mu g/cm^2$ the volumetric density was calculated to be of 0.0015 g/cm^3 so that membranes produced were shown to be electronically conducting, highly anisotropic aerogels. These membranes were found to be very easily stackable, and they could support millimetre-sized liquid droplets that were 50,000 times more massive than the supporting membrane region in contact with the droplets. Such highly anisotropic membranes could be easily densified up to about 0.5 g/cm^3 and could be easily made into highly oriented membranes having thicknesses of about 50 nm. This 360-fold increase of density was obtained in a very simple way, i.e. by causing the membrane as it was produced to adhere to a planar substrate, e.g. glass, plastics, silicon, gold, copper, aluminium, steel, then vertically immersing the substrate bearing the attached MWCNT membrane in a liquid such as ethanol, along the nanotube alignment direction and then retracting the substrate from the liquid. The shrinking of the aerogel membrane down to thicknesses of about 50 nm was the result of surface tensions during ethanol evaporation. It is remarkable that the collapse of a membrane of about 20 μm down to about 50 nm without changes in lateral membrane dimensions means that the out-of-plane deviations in nanotube orientation become in-plane deviations that are noticeable in the SEM micrographs. The aerogel membranes can be effectively glued to a substrate by contacting selected regions with ethanol, and then allowing evaporation to densify the aerogel membrane. Adhesion increases because the collapse of aerogel thickness increases the contact area between the nanotubes and the substrate.

The membrane resistance in the draw direction was found to change by less than 10% upon densification by a factor of about 360, with an increase in the membrane transparency. The temperature dependence of the membrane resistivity was nearly the same for the forest-drawn densified nanotube membranes and for sheets made from the usual filtration method using the same forest-grown MWCNTs. The study of the electronic and optical properties of such membranes showed that the membrane resistance as measured in vacuum with the as-drawn MWCNT membranes has a much lower temperature dependence of electrical conductivity than does an isotropic SWCNT membrane obtained by the standard filtration process. Optical transmittance as a function of wavelength was also determined for a single MWCNT membrane before and after densification, for polarized light perpendicular to the draw direction and parallel to said direction, and for unpolarized light. The densified nanotube membranes showed high values of transparency in combination with useful electrical conductivity; this is a combination that would be needed for such applications as displays, video recorders, solar cells, and solid-state lighting. The transmittance for densified MWCNT membrane was higher than 85% for perpendicular polarization, higher than 65% for parallel polarization between 400 nm and 2μm, and higher than 85% for unpolarized radiation between 1.5 and 10 μm. These MWCNT membranes would adhere to transparencies made of poly(ethylene terephtalate) as well as to silicone rubber sheets, thereby providing transparent bilayer composites that could be bent in any direction without causing a substantial decrease in electrical conductivity, which would make them useful for a number of applications of flexible electronics.

The mechanical properties of such aerogel-like and densified MWCNT membranes showed unexpectedly high. It is remarkable to observe that the density-normalized mechanical strength could be much more accurately determined than mechanical strength, due to the fact that the membrane thickness could be less reliably measured than the ratio of the maximum force to mass-per-length in the stretch direction. Stacks of undensified membranes showed a tensile strength of between 120 and 144 MPa/(g/cm^3), and a densified stack containing 18 identically oriented membrane layers had a strength of 465 MPa/(g/cm^3), a value that decreased to 175 MPa/(g/cm^3) when neighboring layers in the stack were orthogonally oriented to make a densified biaxial structure. Such density-normalized strengths were already comparable to or greater than the value of about 160 MPa/(g/cm^3) of the strength of the Mylar and Kapton films used for ultralight spacecraft and proposed for use in solar sailing and of those for ultra-high-strength steel (about 125 MPa/(g/cm^3)) and aluminium alloy (250 MPa/(g/cm^3)) sheets.

As a nanotechnologist applied to space mission studies and belonging to the group for the Aurora Project, I was attracted by this paper in which the novel solid-state procedure for obtaining MWCNT membranes was described. Indeed, the details that I have selected and stressed above showed me clearly that such multifunctional membranes might be useful, with proper adaptation, to solve our problem of a solar sail even much lighter, and then much speedier, than the all-metal thin sail we had designed for the Aurora Project. My suggestion to the other members of the group was considered very carefully, and the idea of a preliminary technological study for the feasibility of a monolayer, ultra-light solar sail capable of exploiting the orbital angular momentum reversal manoeuvre fully (Vulpetti, 1996b) and reaching extrasolar system targets was conceived (Vulpetti, Santoli and Mocci, 2007). It was quite soon clear that speeds up to a substantial fraction the speed of light could have been obtained by properly adapting the multifunctional MWCNT membrane produced in the mentioned paper, where the membrane produced was mainly devoted to commercial applications where transparency was the basic requisite. The essentials of the preliminary investigation for such a kind of photon solar sail are presented in **Section 4** as a case study for the development of MWCNT membranes of such type for solar sail-propelled spacecraft.

4. A Case Study – Flying at 0.001•c and Beyond

The thrust acceleration equation of sailcraft motion consists of four basic blocks that are not all separable. They are: 1) effective solar pressure and solar gravitational acceleration; 2) sailcraft sail loading; 3) the sail thermo-optical quantities; 4)the sail attitude. The first block is outside our control; the second block is the only one that can cause the speed of sailcraft to make a jump, in the sense of the lightness number jump in a mission to extrasolar system targets designed on the basis of a MWCNT membrane whose reflectivity would have been tailored properly, according to the nanotechnological developments envisaged in our group (Vulpetti, Santoli and Mocci, 2007) and discussed here and in **Section 3**. Thermo-optical quantities and sailcraft attitude can give the thrust, whose vector control can be realized **if and only if** the total reflectance of the sail is positive. Putting all in all, for a leap forward as to speed of our sailcraft design according to the Aurora Project, we were confronted with the problem of finding a material of ultra-low density, sufficiently reflective to sunlight, structurally very strong, of low absorption and - a necessary condition for the telecommunication system with the basis on Earth - microwave transparent. Both the Aurora Project all-metal sail and any other metal sail or conventional semiconductor sailcraft had to be excluded. The only envisageable perspective was to consider the possibility of tailoring to such conditions the MWCNT membranes mentioned above. Their multifunctional capabilities with possibility of changes in their basic properties by changing their microscopic structures and/or their production procedures looked like pointing out a concrete way toward a material satisfying the following goals: 1) possibility of manufacturing monolayer sails on the ground; 2) possibility of safe deployment in space; 3) possibility of building a flat sail with such membrane; 4) possibility of realizing a big sail (it is quite normal for instance to build square sails with 250 m side); 5) possibility of increasing the insufficient reflectivity of the membranes as produced (Zhang et al., 2005) up to values about 70 - 80% reflectivity in a wide range of sunlight spectrum; and 6) possibility of obtaining very high lightness numbers. The astrodynamic equations for such sailcraft showed that cruising speeds could be reached of values higher than 300 km/s.

There were no big problems as to tailoring the original properties of the membranes to the requisites for employment as a sailcraft reflective material as the mechanical parameters and the capability of standing very high temperatures were concerned: the tensile strengths mentioned in **Section 3.4** were much higher than the tensile strengths of Kapton and other plastic materials that were considered for standard heavy sails, and temperatures up to incandescence of the MWCNT membranes could be reached without damage to the membrane properties. The real problem was to increase the electrical conductivity, i.e. the metallic character, of the MWCNTs in the membrane structures so as to have a sufficient value of reflectivity; moreover, the membrane should have been transparent or essentially transparent (zero reflectivity and zero absorbance) to microwaves in the band employed for telecommunication, in order to avoid complicated sail architectures, as for instance holes for letting communication signals pass.

It is known that MWCNTs can have a metallic character; however, procedures for synthesizing such carbon nanotubes produce generally mixtures of metallic and non-metallic MWCNTs. Mixtures of MWCNTs and SWCNTs are also produced. Recently (Dresselhaus, 2004; Burke, Rutherglen and Yu, 2006; Gregorczyk et al., 2006; Wang et al., 2004) it has been shown that metallic MWCNTs as random or regular arrays are so reflective that they can be used as optical nanoantennas. Indeed, such effect can be predicted theoretically; however practical experimentations (Wang et al., 2004; Gregorczyk et al., 2006) have shown that complex optical responses can be obtained from periodic and nonperiodic arrays of MWCNTs as a result of controlling the geometry and spacing of the arrays, so that it is possible to create structures that respond very strongly to specific wavelengths or bands of wavelength. Finely tuned detectors that can respond to predetermined wavelength bands ranging from the ultraviolet to the infrared region.

In our further investigation, it was found that: 1) it was possible, by controlling the arrays geometry and spacing, to build membrane microscopic structures that could respond very strongly to specific wavelengths of solar light with a high efficiency of solar reception and re-radiation. This was possible because metallic MWCNTs act as molecular one-dimensional wires endowed with high conductance leading to current densities as high as 10^9 A/mm^2 . For them, the concept of skin depth of the incoming wave of light has no practical meaning, because the wire is physically one-dimensional, i.e. electrons have no possibility of motion normal to the axis of the MWCNTs; 2) by controlling the length of the MWCNT it was possible to obtain a desirable reflectance for (useful) wavelengths of 50 nm and 20μm, while telecommunication microwaves could pass unabsorbed or reflected; 3) as the membranes are aerogels, it can be shown that their specific strength can be adjusted independently of the optical properties of the membrane.

Accordingly, any roadmap to MWCNT membrane solar sailcraft should start from the problem in Nanophotonics of obtaining massive amounts of pure metallic MWCNTs and of sizing their lengths properly for a controlled reflectance of the membranes obtained by the solid-state procedure discussed above.

5. Appendix – A Recent Accurate Procedure Demonstrating Light Pressure

It was in 2000 that the first experiment was carried out (Myrabo et al., 2000) devoted to measuring laser photonic thrust performance with real candidate sail materials. Such experiments show the validity of solar sailing theory against the objections reported above (**Section 2.1**). The experimental study was concerned with laser-beam pushed sails, which entail very different problems from solar sailing; it is anyway a good proof of principle. The test articles were 5-cm diameter laser sail discs fabricated from an ultralight carbon microtruss fabric that was sputter-coated with molybdenum on one side only, to improve its reflectivity to laser radiation of 10.6 μm, and four laser sail discs with three different areal densities (6.6 g/m^2, 27 g/m^2, and 28 g/m^2) were tested as magnetically supported pendulums with an overall length of 10 cm. Pendulum deflections for the heavier sails ranged from 2.4 to 11.4 degrees, as a function of a continuous wave laser power from 7.9 to 13.9 kW. The sails masses were 83.7, 87.3 and 88 mg each; their center of mass was at 11.5, 11.7, and 11.9 cm respectively below the magnetic bearing. Laser photon thrust ranged from 3.0 to 13.8 dynes. These values were calculated from the pendulum deflections. The vacuum chamber employed was 2.74 m long with a diameter of 2.13 m, which was evacuated to about 36 - 44 μTorr. The range of 3.3 - 6.67 N/GW showed to be feasible for sailcraft propulsion by laser beam.

6. References

Burke, P., Rutherglen, Ch. and Yu, Zh. (2006). Carbon Nanotube Antennas, *Proc. SPIE*, Vol-6328, article no. 632806.

Diedrich, B. (2003). Letter to the Editor of New Scientist : Solar Sailing Breaks the Laws of Physics, 7 July; http://www.ugcs.caltech.edu/~diedrich/solarsails/

Dresselhaus, M.S. (2004). Nanotube Antennas, Nature, Vol. 432, p. 959.

Genta, G. and Brusa, E. (1996). Project Aurora : Preliminary Structural Definition of the Spacecraft, *Proceedings of the First IAA Symposium on Realistic Near-Term Advanced Scientific Space Missions*, pp. 25-36, Levrotto & Bella, Torino, Italy

Gregorczyk, K., Kimball, B., Carlson, J.B., Pembroke, A., Kempa, K., Ren, Zh., Yelleswarapu, Ch., Kempa, Th., Benham, G., Wang, Y., Li, W., Herczynski, A., Rybczynski, J. and Rao, D.V.G.L.N. (2006). The Complex Optical Response of Arrays of Aligned, Multi-Walled Carbon Nanotubes, *Proc. SPIE*, Vol. 6321, article no.63210G

Heckenberg, N. (1996). Crookes' Radiometer and Otheoscope, *Bulletin of the Scientific Instruments Society*, Vol. 50, pp. 40-42.

Layzer, D. (1975). The Arrow of Time, *Scientific American*, Vol.233, No. 6, pp. 56-69.

Maxwell, J.C. (1879). On Stresses in Rarefied Gases Arising from Inequalities of Temperature, *Royal Society Philosophical Transactions.*

McInnes, Colin R. (1999). *Solar Sailing*, Springer - Praxis Publishing, ISBN 1-85233-102-X, Chichester, UK

Myrabo, L.N., Knowles, T.R., Bagford J.O. and Seibert, D.B. (2000). Experimental Investigation of Laser-Pushed Light Sails in a Vacuum, *Proceedings of SPIE*, Vol. 4065, pp. 521-532.

Parsons, P. (2003). Solar Sailing Breaks the Laws of Physics, *New Scientist*, issue of 4 July.

Reynolds, O. (1879). On Certain Dimensional Properties of Matter in the Gaseous State, *Royal Society Philosophical Transactions*, Part 2.

Santoli, S. (2009). Is There Any Such Thing as Self-Organization? *Keynote Address Given at the IIAS – International Institute of Advanced Studies 21st Conference*, August 3, Baden-Baden, Germany.

Santoli, S. and Scaglione, S. (1996). Project Aurora : A Preliminary Study of the Light, All-Metal Solar Sail, *Proceedings of the First IAA Symposium on Realistic Near-Term Advanced Scientific Space Missions*, pp. 37-48, Levrotto & Bella, Torino, Italy

Scaglione, S. and Vulpetti, G. (1999). The Aurora Project: Removal of Plastic Substrate to Obtain an All-Metal Solar Sail, *Acta Astronautica*, Vol. 44, Nos. 2-4, pp. 147-150.

Vulpetti, G. (1996a). The Aurora Project: Flight Design of a Technology Demonstration Mission, *Proceedings of the First IAA Symposium on Realistic Near-Term Advanced Scientific Space Missions*, June 25-27, pp. 1-16 Torino, Italy.

Vulpetti, G. (1996b). 3D High-Speed Escape Heliocentric Trajectories by All-Metallic Sail Low-Mass Sailcraft, 2nd IAA International Conference on Low-Cost Planetary Missions, paper IAA-L-0505, 16-19 April, Laurel, Maryland, USA.

Vulpetti, G. (1999). General 3D H-ReversalTrajectories for High-Speed Sailcraft, *Acta Astronautica*, Vol. 44, No. 1, pp. 67-73.

Vulpetti, G., Santoli, S. and Mocci, G. (2008). Preliminary Investigation on Carbon Nanotube Membranes for Photon Solar Sails, *JBIS*, Vol. 61, No. 8, pp. 284-289, ISSN 0007-084X

Vulpetti, G.; Johnson, L. & Matloff, G.L. (2008). *Solar Sails: A Novel Approach to Interplanetary Flight*, Springer, ISBN 978-0-387-34404-1

Vulpetti, G. (2009). Reaching Extra-Solar-System Targets via Large Post-Perihelion Lightness-Jumping Sailcraft, *Proceedings of the Sixth IAA Symposium on Realistic Near-Term Advanced Scientific Space Missions*, Aosta, Italy, July 6-9, pp. 73-78.

Wang, Y., Kempa, K., Kimball, B., Carlson, Benham, G., Li, Kempa, Th., W., Rybczynski, J. Herczynski, A. and Ren, Z.F. (2004). Receiving and Transmitting Light-Like Radio Waves : Antenna Effect in Arrays of Aligned Carbon Nanotubes, *Appl. Phys. Lett.*, Vol. 85, pp. 2607-2609.

Woodruff, A.E. (1968). The Radiometer and How It Does Not Work, *The Physics Teacher*, Vol. 6, pp. 358-363.

Wright, J. (1993). *Space Sailing*, Gordon and Breach Science Publishers, ISBN 2-88124-842-X, Reading, UK

Zhang, M., Fang, S., Zakhidov, A.A., Lee, S.B., Aliev, A.E., Williams, Ch.D., Atkinson, K.R. and Baughman, R.H. (2005). Strong, Transparent, Multifunctional, Carbon Nanotube Sheets, *Science*, Vol. 309, pp. 1215-1219.

31

Carbon Nanotube-Nanoparticle Hybrid Structures

Junhong Chen[1,2] and Ganhua Lu[1]
[1]*University of Wisconsin-Milwaukee*
U.S.A.
[2]*Tongji University*
P. R. China

1. Introduction

The discovery of carbon nanotubes (CNTs) (Iijima 1991; Iijima & Ichihashi 1993) has inspired extensive research activities unprecedented in nanoscience and nanotechnology. CNTs are very versatile due to their small diameter, high aspect ratio (Zheng et al. 2004), great structural integrity (Hinds et al. 2004; Jung et al. 2006), high electrical (Kang et al. 2007) and thermal conductivity (Berber et al. 2000), remarkable thermal (Kim et al. 2004) and chemical stability (Liu et al. 1999), and exceptional strength and resilience (Treacy et al. 1996; Saito 1998; Yu et al. 2000). Nanoparticles (NP) with sizes between 1 and 100 nm have attracted considerable interest because of their unique electronic, optical, magnetic, mechanical, and chemical properties that differ from those of both their constituent atoms/molecules and their corresponding bulk materials. CNTs decorated with NPs form a new class of hybrid nanomaterials that could potentially display not only the unique properties of NPs (Fissan et al. 2003; Scher et al. 2003) and nanotubes (Dresselhaus et al. 2001; de Heer 2004; Dresselhaus 2004), but also additional novel physical and chemical properties due to the interaction between CNTs and attached NPs. These hybrid nanomaterials have recently been shown as promising building blocks for various applications, including gas sensors (Kong et al. 2001; Sun & Wang 2007; Lu et al. 2009), fuel cells (Mu et al. 2005; Kongkanand et al. 2006; Robel et al. 2006), solar cells (Landi et al. 2005; Guldi et al. 2006; Kongkanand et al. 2007; Lee et al. 2007), Li-ion batteries (Zhang et al. 2006), hydrogen storage (Yildirim & Ciraci 2005; Anson et al. 2006), and transparent conductive electrodes (Kong et al. 2007).

This chapter begins by outlining the significance of CNT-NP hybrid structures in terms of materials advantages and potential applications. Then, we introduce the assembly of various metallic and semiconducting NPs onto the external surfaces of CNTs via the electrostatic force directed assembly (ESFDA) process that we have recently developed. Single-walled CNTs (SWCNTs), multiwalled CNTs (MWCNTs), randomly dispersed CNTs, and well-aligned CNT arrays can be decorated with a broad range of NPs (including both aerosol and colloidal NPs) through the ESFDA process for the generation of CNT-NP hybrid

---

[1] all correspondence should be addressed to jhchen@uwm.edu or 1-414-229-2615.

structures. Important factors that control the synthesis of the CNT-NP hybrid structures will be discussed for the ESFDA method. Fundamental understanding of the binding mechanism between CNTs and NPs will facilitate the rational design and use of these novel hybrid structures. The chapter also discusses the structure evolution of an example CNT-NP system under successive annealing cycles, attempting to shed light onto the interaction between the NP and the CNT. The chapter ends with a novel gas-sensing platform based on CNT-NP structures to illustrate the potential applications of such hybrid nanostructures.

2. CNT-NP: A New Class of Multifunctional Nanomaterials

Decorating CNTs with NPs has been one of the most intensively studied areas in the past decade (Georgakilas et al. 2007). These NP-loaded CNTs, regarded as a new class of multifunctional nanomaterials, are highly promising for widespread applications due to several materials advantages. Heterojunctions formed at the interface of the NP and the CNT can have an unusually minute junction area (less than 10 nm^2) and provide opportunities to tune the size, shape, and chemistry, and thus material properties (Peng et al. 2009). Electron transfer across the NP-CNT junction could be very sensitive to surrounding molecular species and photon illumination. In addition, CNTs can serve as a robust support for the dispersion and stabilization of NPs, which explains the enormous effort in the use of CNT-NP structures for catalysis.

The CNT-NP hybrid structure has been demonstrated as a new gas-sensing platform. For example, Pd-CNT systems have been used for H_2 detection with CNTs as a transducer and amplifier to convert the chemical reaction between Pd and H_2 into strong electrical signals. Kong et al. demonstrated that individual SWCNTs, as well as SWCNT networks, show enhanced H_2 sensing capability after being decorated with Pd NPs (Kong et al. 2001); the resistance of a single-SWCNT (semiconducting) device doubled upon exposure to 400 ppm H_2. Sun et al. fabricated flexible H_2 sensors on plastic substrates using necklace-like structures of Pd NPs electrochemically deposited on SWCNT networks (Sun & Wang 2007). MWCNTs decorated with discrete SnO_2 NPs showed room-temperature gas sensing capability for low concentration gases (H_2, CO, and NO_2) diluted in air (Lu et al. 2009).

In fuel cells, the catalytic activity of metal NPs for the electro-oxidation of fuel molecules strongly depends on the size and the shape of NPs (Valden et al. 1998), the type of catalyst support (Wolf & Schuth 2002), and the method of catalyst preparation (Liu et al. 1999). CNTs are considered to be strong candidates for catalyst support because of their high surface-to-volume ratio, stability, and unique mechanical properties (Treacy et al. 1996; Saito 1998; Yu et al. 2000). NPs with uniform sizes loaded onto CNTs have been proven effective for catalyzing redox reactions because of the increased surface area and the improved electrocatalytic activity (Kongkanand et al. 2006; Shao et al. 2009).

CNTs coated with CdSe and CdS NPs have been demonstrated for potential solar cell applications (Landi et al. 2005; Robel et al. 2005), in which semiconductor NPs are photoreceptors and CNTs are interconnects for conducting electrons from the photoexcitation of the semiconductor nanocrystals. The hydrogen storage capacity of Pd NPs has been shown to increase when they are supported on CNTs instead of raw carbon materials (Anson et al. 2006) due to the maximized active surface of Pd NPs on CNTs and the possible spread of H_2 from the Pd structure to CNTs (known as the hydrogen spillover effect (Conner & Falconer 1995)). Finally, it has been demonstrated both experimentally

(Kong et al. 2007) and theoretically (Khoo & Chelikowsky 2009) that the electrical conductivity of SWCNT can be improved by decorating metallic NPs, which can be attributed to a doping-induced electron depletion mechanism.

3. Synthesis of CNT-NP Hybrid Structures

Unsurprisingly, the synthesis of CNT-NP hybrid structures has attracted growing interest and a number of methods to produce CNT-NP structures have emerged in recent years. These methods can be broadly divided into two categories: wet-chemistry methods and dry routes. Most wet-chemistry methods involve the chemical functionalization of the CNT surface followed by the assembly of nanocrystals onto the CNTs via covalent (Zamudio et al. 2006), noncovalent (Li et al. 2006), or electrostatic interactions (Kim & Sigmund 2004; Xing 2004). Comprehensive reviews on chemical synthesis methods of CNT-NP hybrid structures have been carried out by Georgakilas et al. (Georgakilas et al. 2007) and Peng et al. (Peng et al. 2009). Here, we focus on a facile dry method, referred to as electrostatic force directed assembly (ESFDA), recently developed in our laboratory.

3.1 Electrostatic force-directed assembly for CNT-NP structure synthesis

The ideal technique for synthesizing a CNT-NP hybrid structure should be, in general, suitable for the deposition of NPs of a wide range of materials (metallic and semiconducting) onto CNTs with different morphologies (randomly dispersed or well-aligned). The electrostatic force directed assembly (ESFDA) is such a scalable technique, which can assemble nearly any electrically-charged aerosol NPs onto CNTs for synthesizing CNT-NP hybrid nanostructures and fabricating multifunctional devices. By taking advantage of the extremely small diameter of CNTs (Dresselhaus et al. 2001), a locally enhanced electric field in the neighborhood of CNTs can be generated, which is then used to direct the assembly of charged aerosol NPs onto the external surface of both random CNTs and vertically-aligned CNTs (or CNT arrays) (Huang et al. 1998).

A schematic diagram of the ESFDA process is shown in Fig. 1. The aerosol NPs were produced through physical vaporization of solid precursor materials using the mini-arc plasma source sustained between a tungsten cathode and a graphite anode (Chen et al. 2007). A fraction of the produced NPs was charged by the arc plasma or through thermionic emission, which made the ESFDA feasible without additional NP charging devices. These NPs were carried by room-temperature gases into a gap between two electrodes to form a stagnation flow (Q). One electrode is the grounded metal tubing that introduces the aerosol NPs; the other electrode, which is applied with a direct current (dc) voltage (V), was fabricated by dispersing CNTs onto a holey-carbon-coated copper grid. The gap distance (D) between the two electrodes was maintained with a precision-machined ceramic spacer. The nanotubes used in the experiment include both SWCNTs and MWCNTs. Surfactants have been used to disperse CNTs in DI water for more uniform coating; however, controlled experiments with pristine CNT powders and CNTs dispersed with surfactants have shown that the coating process relies on the electric field rather than the surfactants. Without the dc voltage / electric field, the NPs followed the gas flow perfectly and no particles were observed on the CNTs. With a dc voltage applied to the copper grid, the electric field near the CNT surface was significantly enhanced and both pristine and dispersed CNTs were successfully coated with NPs.

The field enhancement was facilitated by suspending CNTs on either a holey-carbon-coated copper grid or a bare copper grid. This was evidenced by significantly higher assembly efficiency with the aforementioned two grids than the carbon-coated copper grid without holes for the same operating conditions. NPs have also been observed at the edges of carbon holes and grid bars, where the electric field is highly enhanced. Two values of gap distance (0.5 and 2 mm) were used in the experiment. A lower voltage was needed to achieve the same assembly for a smaller gap. With the gap distance reduced from 2 to 0.5 mm, the required voltage decreased from 2 kV to 500 V. Based on our calculations, the maximum electric field near the surface of a 20-nm-diameter CNT reaches 2.45×10^6 V/m for V = 2 kV and D = 2 mm. The maximum electric field possible for the assembly is limited by the breakdown field of the carrier gas, e.g., about 3×10^6 V/m for dry air (Chen & Davidson 2003).

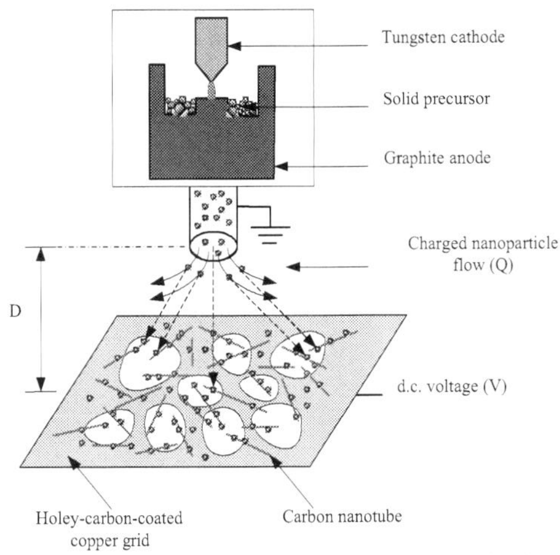

Fig. 1. Electrostatic force directed assembly (ESFDA) process. The locally enhanced electric field near CNTs results in a directed assembly of charged aerosol NPs onto CNTs. Reprinted with permission from (Chen & Lu 2006). Copyright 2006 IOP

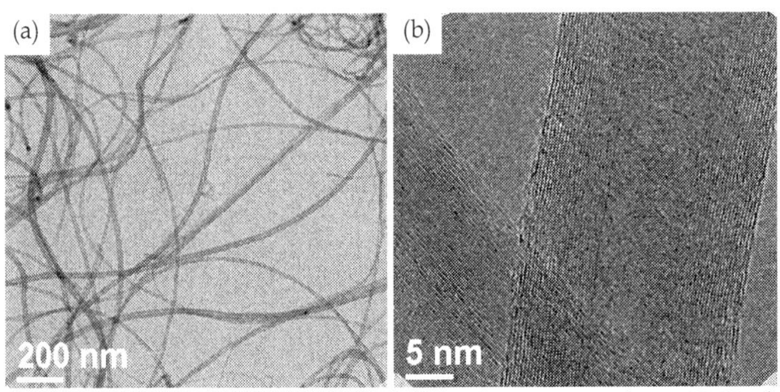

Fig. 2. Low-magnification TEM (a) and HRTEM (b) images of MWCNTs before NP assembly. Reprinted with permission from (Zhu et al. 2008). Copyright 2008 ASME

The successful assembly of NPs onto CNTs is confirmed by low-magnification transmission electron microscopy (TEM) and high-resolution TEM (HRTEM) images. As a comparison, Fig. 2 (a) shows the TEM image of MWCNTs before assembly, with a diameter of about 20-40 nm and a length of several micrometers. It is obvious that the MWCNT surface is very smooth, which can be further confirmed by the corresponding HRTEM image (Fig. 2(b)).

To demonstrate the material independence of ESFDA, NPs of semiconductor tin oxide (SnO_2) and metallic silver (Ag) were assembled onto both MWCNTs and SWCNTs. These composite structures were obtained at room temperature and atmospheric pressure with an assembly time (t) of 5.5 min. NPs have been found on both the outside wall and the tip of nanotubes. Figure 3(a) shows a representative TEM image of as-prepared MWCNTs coated with SnO_2 NPs. The ESFDA technique clearly shows the potential for efficient, large-scale decoration of CNTs with NPs. It can be seen that nonagglomerated SnO_2 NPs are quite uniformly distributed on the MWCNT surface. The hybrid structure can be more clearly seen from the HRTEM image shown in Fig. 3(b), with SnO_2 NPs less than 5 nm. Both the nonagglomerated entity and the crystallinity of NPs were preserved after the assembly. The measured lattice fringes of 0.26 and 0.33 nm correspond to the (101) and (110) crystal planes of rutile SnO_2, respectively. In principle, uncapped metal NPs tend to agglomerate even at room temperature. The nonagglomerated entity of NPs preserved during the assembly is likely attributed to the predominant unipolar charges carried by these NPs before reaching CNTs. The unipolar charges effectively prevent the NPs from agglomeration.

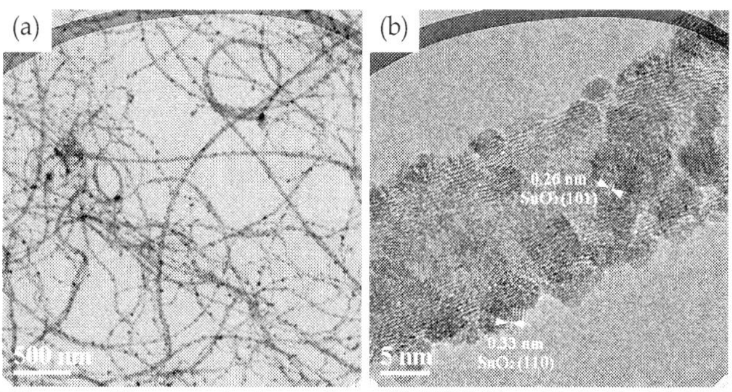

Fig. 3. (a) TEM image of large-scale decoration of MWCNTs with SnO_2. (b) HRTEM image of an MWCNT coated with SnO_2 NPs. (Operating conditions: D=2 mm, Q= 5.15 lpm, V=-2 kV, and t=5.5 min) Reprinted with permission from (Chen & Lu 2006). Copyright 2006 IOP

Similar results have been obtained for Ag NPs. Relatively uniform coverage of Ag NPs on MWCNTs is observed in Fig. 4(a). The HRTEM image of Fig. 4(b) clearly resolves the presence of Ag nanocrystals with sizes less than 10 nm. The lattice fringes of 0.23 and 0.20 nm correspond to the (111) and (200) planes of Ag crystal, respectively.

NPs of Ag and SnO_2 have also been successfully assembled onto SWCNTs using the ESFDA technique, although a significant fraction of the SWCNTs is semiconductive. Due to the strong van der Waals force between SWCNTs, small bundles instead of individual SWCNTs have been obtained after dispersion for the coating. The HRTEM images of coated SWCNT bundles are shown in Figs. 5(a) and (b) for SnO_2 and Ag, respectively. The lattice fringe of 0.33 nm shown in the HRTEM image of Fig. 5(a) corresponds to the (110) plane of rutile SnO_2, confirming the successful assembly of SnO_2 nanocrystals onto SWCNTs. Similarly, the HRTEM image of Fig. 5(b) displays the lattice fringes of 0.23 nm corresponding to the (111) plane of Ag.

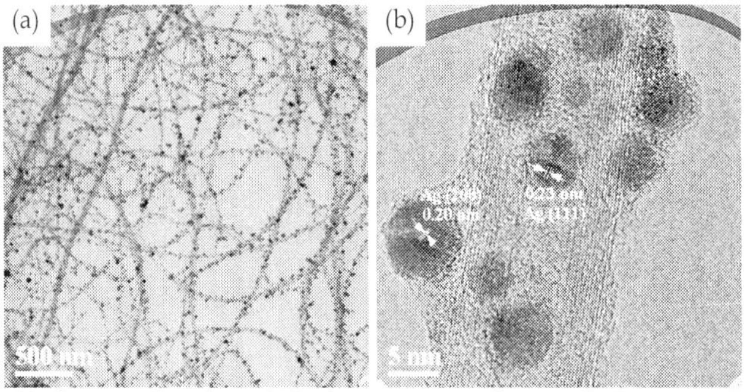

Fig. 4. MWCNTs decorated with Ag NPs (operating conditions: D=2 mm, Q=5.15 lpm, V=-2 kV, and t= 5.5 min). (a) TEM and (b) HRTEM images of Ag NP-MWCNT structure. Reprinted with permission from (Chen & Lu 2006). Copyright 2006 IOP

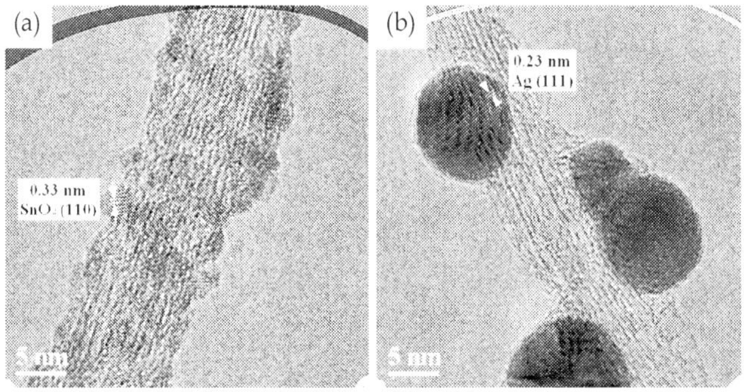

Fig. 5. HRTEM images of SWCNT bundle decorated with SnO_2 (a) and Ag (b) NPs (operating conditions: D = 2 mm, Q = 5.15 lpm, V = -2 kV, and t = 5.5 min). Reprinted with permission from (Chen & Lu 2006). Copyright 2006 IOP

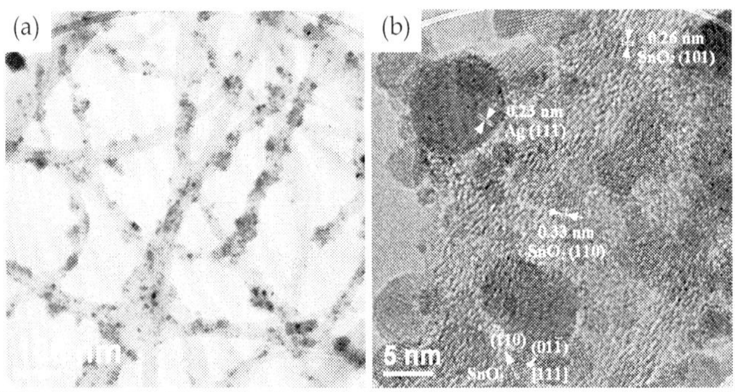

Fig. 6. MWCNTs coated with both Ag and SnO_2 NPs: (a) Low-mag. TEM image and (b) HRTEM image. Reprinted with permission from (Zhu et al. 2008). Copyright 2008 ASME

Due to the material independence of ESFDA, the range of NP materials for the assembly has been greatly expanded. As shown in Fig. 6, MWCNTs have been sequentially coated with SnO_2 and Ag nanocrystals. Figure 6(a) shows the low magnification TEM image of the double-component NP-coated MWCNTs. The uniformity of the NP coverage is not as good as that in the single-component NP-CNT sample. The HRTEM image in Fig. 6(b) provides more detailed information of the interesting hybrid structure. The lattice fringes of Ag and SnO_2 nanocrystals show simultaneously on the same MWCNT.

3.2 Size selection in the ESFDA process

The relatively broad size distribution of as-produced NPs from the mini-arc source made it possible to observe the NP size selection during the assembly. The size distribution of tin oxide NPs docking on MWCNTs (Fig. 3(a)) was measured using TEM and is shown in Fig. 7. Compared with the as-produced NPs before assembly (inset of Fig. 7, mean diameter $\overline{D_p}$ = 14.1 nm, standard deviation σ = 8.2 nm), the assembled NPs on nanotubes were much smaller ($\overline{D_p}$ = 2.9 nm) and the particle size distribution was much narrower (σ = 1.0 nm). This implies that there was an intrinsic size selection during the assembly process. The size selection feature of ESFDA is very attractive for many applications and it relaxes the requirement for the NP production process, particularly when the state-of-the-art aerosol classifiers are facing resolution challenges in the lower nanometer size range.

For the stagnation flow shown in Fig. 1, NPs migrate along the electric field lines while being carried away by the gas flow. NPs reaching CNTs are selected through their electrical mobility ($Z_p = v_p / E = (neC_c)/(3\pi\mu D_p)$) (Seinfeld & Pandis 1998), the ability of a particle to move in an electric field, where v_p is the NP velocity along electric field lines, E is the electric field, n is the number of elementary charges carried by nanoparticles, e is the elementary charge, C_c is the Cunningham slip correction factor, and μ is the flow viscosity). Only small NPs with sufficient electrical mobility can be assembled onto CNTs for a given electric field strength and flow residence time. Larger NPs with smaller electrical mobility remain in the stagnation flow and are carried away by the flow.

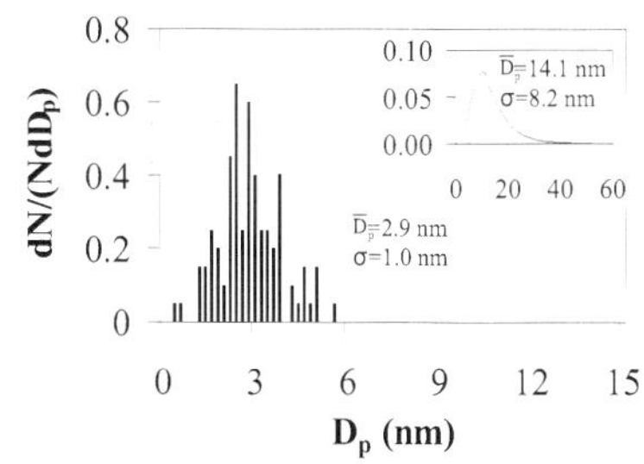

Fig. 7. SnO_2 NP size distribution before (shown in the inset) and after the assembly (operating conditions: D = 2 mm, Q = 5.15 lpm, V = -2 kV, and t = 5.5 min). Reprinted with permission from (Chen & Lu 2006). Copyright 2006 IOP

3.3 Control over the ESFDA process

The ESFDA has considerable control over the assembly process. The areal density of NPs on CNTs can be easily controlled by adjusting the assembly time for a given NP concentration (Fig. 8). As expected, the areal density increased with increasing assembly time. The entire external surface of CNTs may be coated with NPs if the assembly time is sufficiently long.

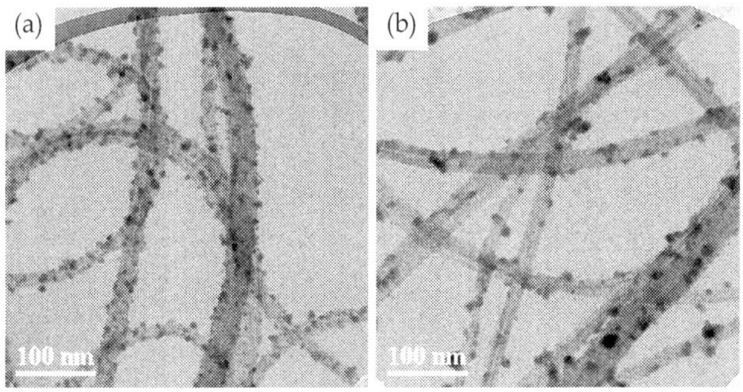

Fig. 8. Control of areal density of NPs on MWCNTs (operating conditions: D=2 mm, Q=5.15 lpm, V=-2 kV). TEM images of MWCNTs coated with SnO_2 NPs for (a) t=5.5 and (b) 2.75 min, respectively. Reprinted with permission from (Chen & Lu 2006). Copyright 2006 IOP

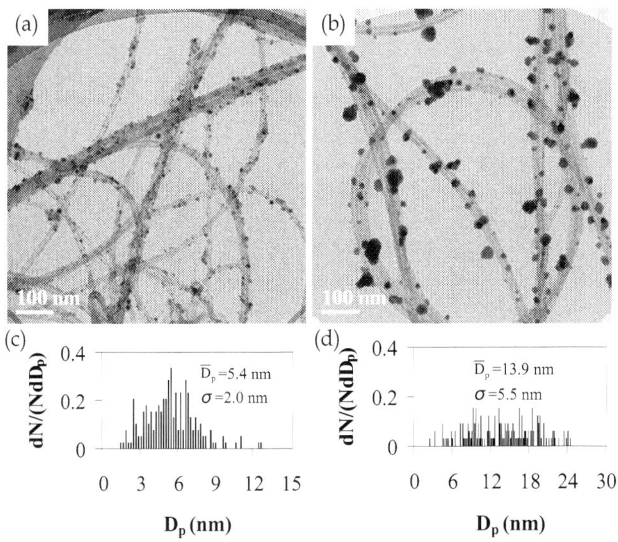

Fig. 9. Control of size distribution of Ag NPs on MWCNTs during the assembly by varying flow rate Q (operating conditions: D=2 mm, V=-2 kV, and t=5.5 min). (a) and (b), TEM images of MWCNTs coated with Ag NPs for Q=5.15 and 2.15 lpm, respectively. (c) and (d), size distribution of Ag NPs on CNTs shown in (a) and (b), respectively. Reprinted with permission from (Chen & Lu 2006). Copyright 2006 IOP

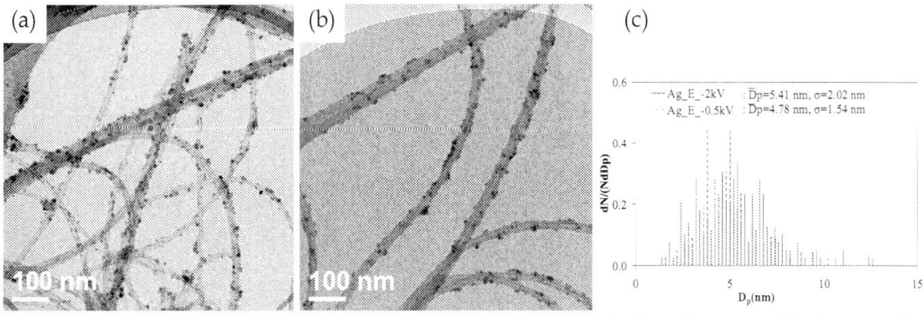

Fig. 10. Controlling the size distribution of NPs on CNTs during the assembly by varying electric field (operating conditions: D = 2 mm, Q = 5.15 lpm, and t = 5.5 min). (a) and (b), TEM images of MWCNTs coated with Ag NPs for V = -2 kV and -0.5 kV, respectively. (c) comparison of the size distributions of Ag NPs on CNTs shown in (a) and (b).

For polydisperse aerosol NPs, the size distribution of NPs on CNTs can be controlled by the aerosol flow rate and the electric field, along with the intrinsic size selection during the assembly process. With the decrease of aerosol flow rate, flow residence time increased and larger particles with smaller electrical mobility assembled onto CNTs for a fixed electric field (Fig. 9). Similar control over the particle size distribution has also been achieved by varying the electric field (Fig. 10). With the increase of electric field, larger particles with smaller electrical mobility assembled onto CNTs. The control of NP size on nanotubes is desirable since NP properties can be tuned with the particle size (Alivisatos 1996).

3.4 Assembly of NPs onto Vertically Aligned CNTs

Vertically-aligned CNTs are of great interest for practical device applications due to the potential for higher device efficiency and packing density. For instance, in fuel cells with metal catalyst nanocrystals coated on CNTs, the nature of loosely packed CNT powders could substantially decrease the effective contact area between nanocrystals and reactants, thereby decreasing the metal catalytic activity. Substituting loose CNT powders with aligned CNTs could overcome the disadvantage and increase the fuel cell efficiency.

While wet-chemistry methods work well with randomly dispersed CNTs, they are not suitable for vertically aligned CNTs—wet processing damages the aligned structure, as the upper ends of the neighboring nanotubes have been observed to bundle together and some nanotubes are caused to lie down (Ye et al. 2005). On the other hand, the ESFDA process, as a dry route, is considered suitable for producing CNT-NP structures without damaging the aligned arrangement of the CNT array. During the ESFDA process, the random CNTs must be suspended in order to maximize the electric field enhancement, which is critical to the assembly. The use of vertically-aligned CNTs represents another means to realize the field enhancement.

Figure 11 shows a schematic diagram of the process to coat a CNT array with aerosol NPs. Silver nanocrystals were carried by an inert carrier gas (Ar/N_2) and delivered into an electrode gap to form a stagnation flow. The top electrode was a grounded metal tube that introduced the aerosol nanocrystals. The bottom electrode was a Si wafer with vertically-aligned MWCNTs that had previously grown on it, and it was electrically biased with a constant voltage (-2 kV). The gap distance (D) between the two electrodes was about 2 mm.

With the bias voltage applied to the aligned MWCNTs on the Si wafer, the electric field near the MWCNT surface was significantly enhanced, especially at the tips, and the oppositely-charged Ag nanocrystals were attracted onto the surfaces of the MWCNTs.

The assembly of nanocrystals onto the CNT array is demonstrated by scanning electron microscopy (SEM) and TEM images. As a comparison, Figs. 12(a) and (b) show the SEM images of as-grown, vertically-aligned MWCNTs on an Si wafer. The CNTs were about 10 μm in length and 150 nm in diameter. The average spacing between nanotubes was about 500 nm, with an estimated areal density of 4×10^8 tubes/cm^2. A catalyst particle containing Ni caps the tip of each CNT, preventing the nanocrystals from assembling onto the interior surface of the CNTs. The external surface of the CNTs appears smooth with no particles observed. Figure 12(c) shows an MWCNT array coated with Ag NPs. The Ag NPs are distributed on the outside surfaces of the CNTs, particularly on the upper portion of the aligned MWCNTs due to the strong field enhancement effect close to the tips. Evidence of these NPs is better resolved in the higher magnification SEM image shown in Fig. 12(d).

Figure 13(a) shows a typical low-magnification TEM image of the Ag NP-MWCNTs that were transferred onto a TEM grid after being scratched off the Si wafer. Both the NP areal density and the average NP size maximize at the CNT tip, consistent with the SEM observation (Fig. 12(d)). The CNT in Fig. 13(a) bears a bamboo structure, typically from the PECVD growth process (Wen et al. 2001). The crystallinity of Ag nanocrystals on the CNTs is confirmed by lattice fringes indexed in the HRTEM image of Fig. 13(b). The measured lattice spacings of 0.20 and 0.23 nm correspond to the (200) and (111) planes of silver crystal, respectively.

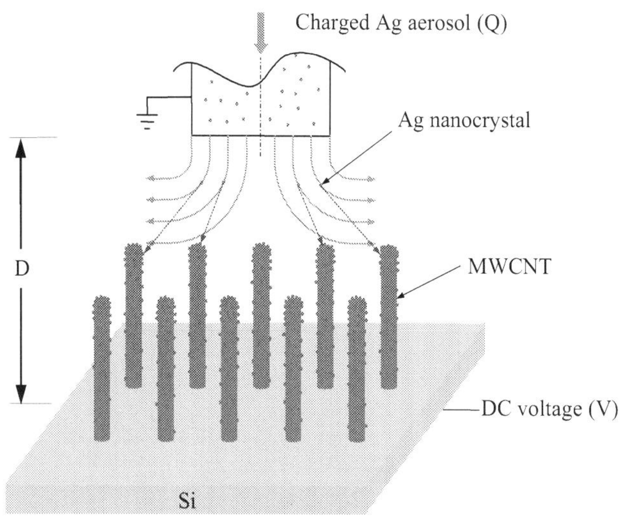

Fig. 11. Schematic diagram for the assembly of Ag nanocrystals onto vertically aligned MWCNTs. Reprinted with permission from (Lu et al. 2007). Copyright 2007 ACS

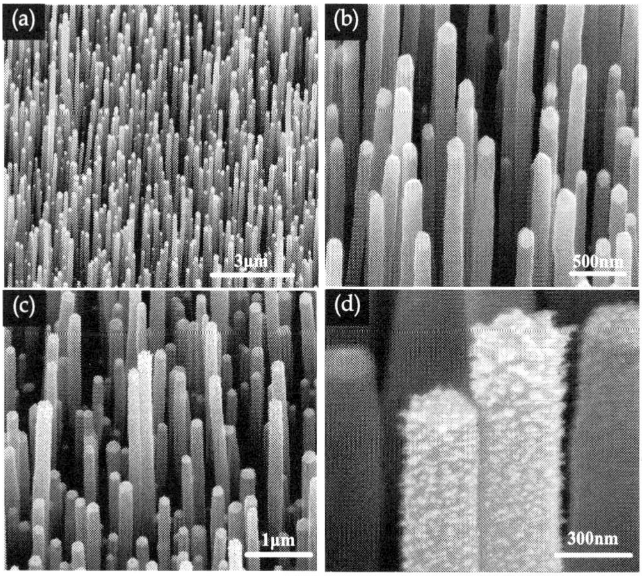

Fig. 12. SEM images of as-grown MWCNT array on a silicon substrate at (a) lower magnification and (b) higher magnification; (c) evidence of NPs coating MWCNTs at a lower magnification and (d) a higher magnification image of a region in image (c), showing NPs deposited on the external surfaces and tips of the MWCNTs. Reprinted with permission from (Lu et al. 2007). Copyright 2007 ACS

The electric field and flow residence time are two critical parameters that govern the nanocrystal assembly process. In this study, the gas flow rate (total flow residence time) was fixed and it is thought that the electric field distribution determines the final Ag NP distribution on the CNTs. There are varying CNT heights in the array shown in Fig. 12, and more NPs are distributed at the tip of taller CNTs, which is likely due to a larger field enhancement for these taller CNTs, and perhaps also due in part to shadowing effects. The CNTs in the array are close to each other and the proximity effect (field screening effect) results in field enhancement only at the tip portion of the CNTs and a larger field enhancement is expected at the tips of taller CNTs. The higher electric field can attract larger NPs with smaller electrical mobilities (for the same charge level on all NPs). The observed larger NPs near the tip of CNTs suggest the critical role of the electric field on the assembly process, since shadowing effects alone should not lead to the variation in NP size along the CNT length.

A similar variation in NP areal density and size has been observed on isolated CNTs. Figure 14(a) shows an SEM image of an isolated CNT coated with Ag NPs from the tip to the root, and with the upper portion of the tube enlarged in Fig. 14(b). Both the NP areal density and the average NP size decrease along the CNT from the tip to its root. For an isolated CNT, the proximity and shadowing effects are not present, and a variation of the NP distribution along the full tube length can only be attributed to the variation in the electric field along the tube axis. Therefore, the results clearly support the critical role of the electric field during the assembly process.

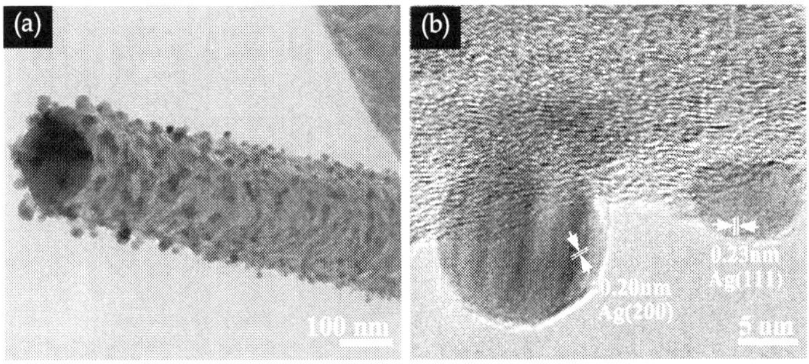

Fig. 13. (a) TEM image of a nanocrystal-coated MWCNT scratched from the Si wafer. (b) HRTEM image showing lattice fringes of Ag nanocrystals deposited on a MWCNT. Reprinted with permission from (Lu et al. 2007). Copyright 2007 ACS

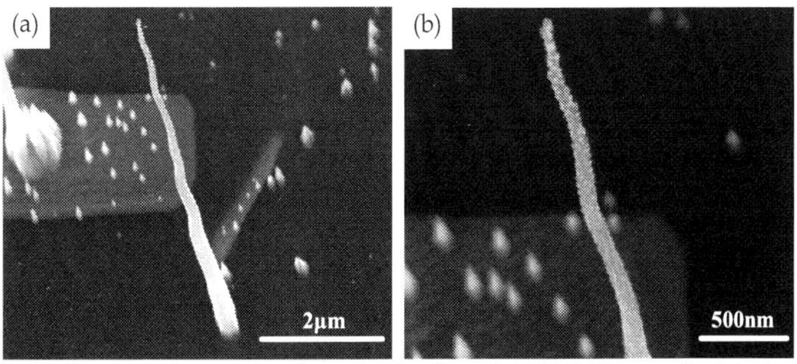

Fig. 14. (a) SEM image of an isolated vertically-aligned MWCNT coated with Ag NPs, and (b) the enlarged view of (a) showing variations of the NP size and areal density along the length of the CNT. Reprinted with permission from (Lu et al. 2007). Copyright 2007 ACS

3.5 Modeling of electric field enhancement along CNTs for two cases

The electric field distribution near the CNT surface was modeled in an attempt to understand the experimental observations. The vertical CNT was modeled as a cylinder with a hemispherical cap (Read & Bowring 2004; Podenok et al. 2006). To simplify the model, a 2-D computational domain with vertical CNT(s) against a flat plane is considered. The CNT is assumed to have the same dimensions as those used in experiments, i.e., 150 nm in diameter and 10 μm in length. The electric field distribution over the domain is computed by numerically solving the Laplace equation using "charged particle optics software" (CPO version 5.2, CPO Ltd., 2004). The field enhancement factor (γ), which is defined as the ratio of the local field to that of the uniform field that would exist between two smooth plates without the CNT(s) present, was obtained along the length of the CNT and is discussed further below.

The electric field distribution was calculated for a single isolated CNT and for a four-tube square array with the same inter-tube spacing (500 nm) as that of the experimental sample. Figure 15 shows that in both cases γ increases monotonically from the root to the tip of the CNT. The field enhancement is much weaker for the four-tube array than for the isolated

CNT due to the shielding effect, as expected. The modeling results are comparable with those published in the literature (Read & Bowring 2004; Podenok et al. 2006). Such modeling points to the possible influence of field enhancement on where nanocrystals are deposited during the ESFDA process.

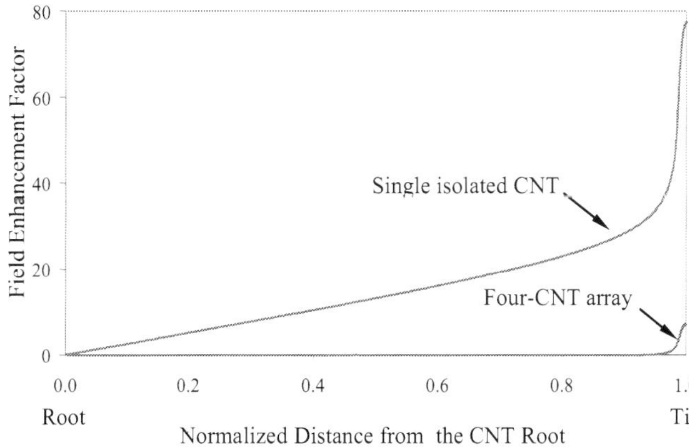

Fig. 15. The electric field enhancement factor along the length of the CNT for a single isolated, vertically-aligned CNT, and a four-tube square array with inter-tube spacing of 500 nm. The CNTs are 10 μm in length and 150 nm in diameter. The spacing between the CNT substrate and the counter electrode is 2 mm. Reprinted with permission from (Lu et al. 2007). Copyright 2007 ACS

3.6 Assembling colloidal NPs onto CNTs

Aerosol synthesis methods could be limited in producing NPs in terms of material type and control in the nanocrystal properties, and such methods typically produce NPs with a relatively broad size distribution (Fissan et al. 2003). Colloidal synthesis, on the other hand, affords considerable control over particle size and structure since the surface chemistry can be manipulated by adjusting the solution properties (Murray et al. 1993). Combining with aerosolization techniques, e.g., electrospray (Lenggoro et al. 2002), NPs in aqueous suspension can be deposited onto CNTs as well using the ESFDA technique (Mao et al. 2008).

Figure 16(a) is a schematic of the system that comprises both the electrospray process and the ESFDA process. The conductive colloidal suspension applied with a high dc voltage is extracted through a capillary submerged in the suspension due to the capillary effect and the capillary inlet/outlet pressure difference. The liquid ejected from the capillary is atomized to form charged fine droplets due to the electrohydrodynamic break-up. Then, a fraction of the aerosolized colloidal NPs is deposited on CNTs via the ESFDA process. The assembly time is typically on the order of hours and is dependent on the colloid concentration and the desired areal density of nanocrystals on CNTs.

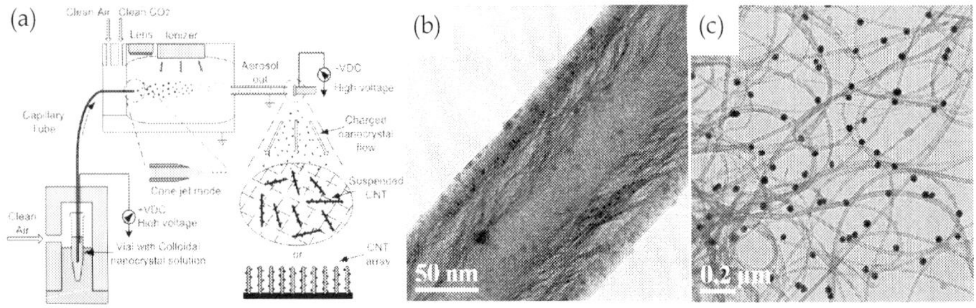

Fig. 16. (a) Schematic of the nanocrystal aerosolization by an electrospray process and the subsequent assembly of nanocrystals onto random CNTs or vertically-aligned CNTs by ESFDA. (b) and (c) TEM images of CNT(s) coated with 3 nm CdSe nanocrystals and 40 nm Au nanocrystals, respectively. Reprinted with permission from (Mao et al. 2008). Copyright 2008 IOP

Figures 16(b) is a TEM image of a single CNT scratched from a vertically-aligned CNT array coated with CdSe nanocrystals. The CdSe nanocrystals (small black dots) are distributed uniformly on the external surface of the CNT. The CdSe colloidal suspension with size-selected nanocrystals was produced by a one-pot synthesis method and the reaction temperature was accurately controlled to achieve the desired CdSe nanocrystal size (Robel et al. 2007). Figure 16(c) shows a TEM image of CNTs (20-40 nm in diameter) coated with 40 nm Au NPs. Colloidal suspensions with 40 nm Au nanocrystals in H_2O were purchased from BB International (stock #10136).

4. Interaction between CNTs and NPs

Understanding the binding mechanism between CNTs and NPs will assist the rational design and use of these novel CNT-NP hybrid structures. Attaching NPs onto CNTs can be achieved through covalent, noncovalent, or electrostatic interactions, which can influence the properties and potential applications of the CNT-NP hybrid structures. Covalent bonding between the nanocrystal and the CNT was obtained through different acid treatments to create such linking groups as carboxyl (-COOH), carbonyl (-C=O), and hydroxyl (-OH); however, the mechanical and particularly the electronic transport properties of the CNT may degrade significantly after acid treatment due to the introduction of defects (Zamudio et al. 2006). The noncovalent attachment of nanocrystals is more likely to preserve the unique properties of low-defect CNTs.

Since the ESFDA assembly is performed at room temperature in inert gas, the chance of noncovalent binding between NP and nanotube is enhanced. The behavior of the CNT-NP structure at higher-than-ambient temperatures in air is also relevant for many applications. For example, metal oxide gas sensors are typically operated at elevated temperatures (200-500°C) to maximize their sensitivity (Shimizu & Egashira 1999); therefore, it is useful to understand how particle distribution, binding, and morphology, are influenced by heating in air, following creation of the hybrid structures by the aforementioned aerosol deposition at room temperature and under inert gas. Thus, following the creation of the CNT-NP hybrid structures by the ESFDA process at room temperature and under inert gas, it is useful to investigate how particle distribution, binding, and morphology are influenced by heating in air and reveal the possible interaction between the NP and CNT.

The evolution of the hybrid structure was studied using TEM imaging after particular heating cycles were applied in air (Zhu et al. 2007). Regions on the TEM grid were identified and imaged prior to heating. The location of each region was determined by its position with respect to the central mark of the grid for a given grid orientation. After the target regions were selected, the TEM grid containing the Ag NP-CNT hybrid structures were placed into an oven with a preset temperature and held for one hour (h) at that temperature. At the end of each heating cycle, the sample was removed from the oven, quickly cooled to room temperature, and immediately imaged with the TEM. The first of the three heating cycles was performed with a preset oven temperature of 100°C, the second at 200°C, and the third at 300°C. The same sample was used for each cycle.

Figure 17(a) is a TEM image of the original sample of CNTs covered with Ag. Figure 17(b) shows the TEM image of the same Ag NP-CNT sample after being heated in air at 100, 200, and 300°C for 1 h each. Ripening evidently occurred during the heating as smaller Ag NPs move and merge with larger Ag NPs. As a result, the areal density of Ag NPs on the CNT decreases while the average diameter of Ag NPs increases.

Smaller NPs on the CNT surface may be energetically less stable than larger ones due to their relatively larger surface-to-volume ratio. As a result, the transformation of smaller particles into larger ones could be energetically favorable on the CNT surface. In addition, smaller NPs are likely to diffuse more readily than larger ones; thus, the growth of the larger particles could simply be a consequence of the smaller ones moving and colliding with the larger particles to form even larger particles. Coalescence of the merged larger particles can also be inferred based on the TEM image, as the observed particles are nearly spherical, as shown in Fig. 17(b). Although the vapor pressure of Ag NPs is known to be higher than that of bulk Ag (Blackman et al. 1968), at these relatively low temperatures and for the total exposure time of 3 h, it is unlikely that a significant fraction of Ag is lost (if any) by atomic evaporation.

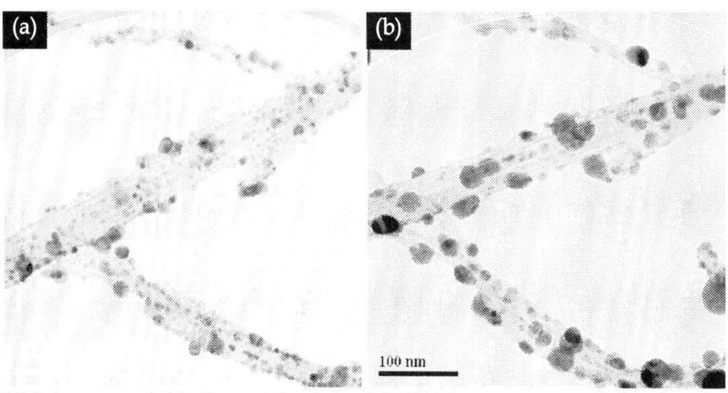

Fig. 17. TEM images of (a) the original Ag NP-CNT structure and (b) the Ag NP-CNT structure after being successively heated at 100, 200, and 300°C for 1 hr each.

The ripening of Ag NPs on CNTs can be anticipated from the studies of Ag clusters/particles located on other surfaces (Rosenfeld et al. 1998; Morgenstern et al. 1999; Meyer et al. 2007) or embedded in polymer films (Heilmann & Werner 1998). Two possible mechanisms are known to be responsible for the ripening process, namely Ostwald ripening and particle migration and coalescence (Wynblatt & Gjostein 1975; Wynblatt & Gjostein 1976). In an Ostwald ripening process, atomic species from smaller particles transport to larger particles either by surface diffusion along the substrate or by vapor phase diffusion due to the difference in the chemical potentials between smaller and larger particles. Under the particle migration and coalescence mechanism, smaller particles migrate on the substrate surface and then collide/coalesce with larger particles.

Although it is difficult to obtain a detailed mechanistic insight based solely on the change in particle size distribution before and after heating (Datye et al. 2006), additional comparison of successive TEM images has allowed us to infer the occurrence of Ag particle migration and coalescence in our experiments. Of course, we cannot exclude the possibility of Ostwald ripening, and observation using an in situ heating experiment inside a TEM chamber might further elucidate the mechanism (Liu et al. 2004). There is the possibility of competition between possible thermodesorption of the whole Ag NP from the CNT surface versus migration of the NP on the CNT surface; the rates for both are expected to increase with increasing temperature. The TEM images obtained after the three 1 h thermal treatments at 100, 200, and 300°C show that surface migration of Ag particles occurs more readily than thermodesorption of Ag NPs, even at 300°C. The more-ready surface migration thus leads to the ripening of the Ag NPs.

The migration of Ag NPs at these relatively low temperatures suggests a weak interaction between Ag NPs and CNTs. It is well known that some transition elements chemically bond to carbon (Zhang et al. 2000). This is attributed to the distribution of d electrons in such transition elements. Since Ag atoms have no d vacancy orbitals, they have a very weak affinity for bonding with carbon atoms (Durgun et al. 2003). As a result, Ag NPs most likely attach to CNTs through van der Waals forces instead of chemical bonds (Zhang & Dai 2000). Because the CNT surface is relatively smooth, we speculate that the barrier for thermally activated migration of Ag NPs is low and thus favors surface migration and ripening.

5. Gas-Sensing Platform Based on the CNT-NP Structure

SWCNTs have been used to detect gaseous species such as NO_2 and NH_3 at room temperature due to the variation in the electrical resistance of CNTs upon exposure to these gases (Kong et al. 2000). The sensing response is attributed to the *p*-type conductivity of semiconducting CNTs and the electrical charge transfer between CNTs and gaseous molecules. Bare CNTs are insensitive to some gaseous molecules such as CO and H_2 (Kong et al. 2001); in the case of H_2, it was found that the binding energy between a H_2 molecule and the outer surface of a bare CNT is so weak that the physisorption bond can easily break (Dag et al. 2005). CNT surface functionalization, e.g., with Pd nanoparticles (Kong et al. 2001) and Pt nanoparticles (Kumar & Ramaprabhu 2006), has been exploited to extend the spectrum of gases that CNT-based sensors can detect. Rutile tin oxide (SnO_2) is an *n*-type semiconducting material widely used in gas sensing applications (McAleer et al. 1987). The existing SnO_2 sensors typically operate at temperatures over 200 °C (Gopel 1996) to enhance the surface adsorption/reaction kinetics and sensitivity, which require continuous heating of the sensor and increase the sensor cost. The high-temperature operation is also undesirable in many situations, particularly in an explosive environment because high temperatures could trigger an explosion.

We have investigated the feasibility of developing a gas-sensing platform based on hybrid nanostructures consisting of discrete SnO_2 NPs uniformly distributed on the surface of MWCNTs (Fig. 18). The hybrid nanostructures were synthesized (Fig. 19a) by the ESFAD process, which assembled aerosol SnO_2 nanocrystals onto MWCNTs pre-deposited on e-beam lithographically patterned Au interdigitated electrodes. Figure 19(b) is an SEM image of the sensing device after the SnO_2 NP assembly by the ESFDA. The MWCNTs bridging a pair of neighboring electrode fingers clearly show SnO_2 nanocrystals decorated on their external surfaces.

Through the marriage between CNTs and SnO_2 NPs, the hybrid SnO_2 NP-CNT structure exhibits high sensitivity to low-concentration gases (NO_2, H_2, and CO) at room temperature. The dynamic response of the SnO_2 nanocrystal-MWCNT sensor for detecting 100 ppm NO_2 is shown in Fig. 19(c). The sensor was operated at room temperature and periodically exposed to clean air flow (2 lpm) for 10 minutes (min) to record a base value of the sensor conductance; 100 ppm NO_2 diluted in air (2 lpm) for 10 min to register a sensing signal; and clean air flow (2 lpm) again for 20 min to recover the device. Five such cycles were repeated, as shown in Fig. 19(c), and the sensing behavior appeared fairly reproducible. The sensor sensitivity, which is evaluated as the ratio of $(G_g\text{-}G_a)/G_a$, where G_a is the sensor conductance in air and G_g is that in the challenging gas, was ~ 1.8. The same SnO_2-CNT sensor also exhibited sensitivity to H_2 and CO, although bare CNTs are insensitive to these gaseous molecules. Figure 19(d) shows the room-temperature sensing performance of the SnO_2 nanocrystal-MWCNT sensor for detecting 1% and 1000 ppm H_2 and 100 ppm CO diluted in air. For CO, the sensitivity was ~ 0.11; for 1% H_2, the average sensitivity was ~ 0.17; and for 1000 ppm H_2, the average sensitivity was ~ 0.10. The H_2 sensing performance is comparable with those reported in the literature for CNTs functionalized with Pd (Kong et al. 2001) or Pt (Kumar & Ramaprabhu 2006) nanoparticles.

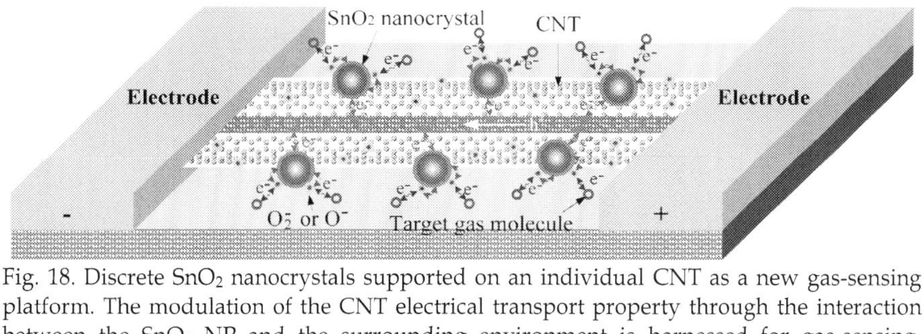

Fig. 18. Discrete SnO_2 nanocrystals supported on an individual CNT as a new gas-sensing platform. The modulation of the CNT electrical transport property through the interaction between the SnO_2 NP and the surrounding environment is harnessed for gas-sensing. Reprinted with permission from (Lu et al. 2009). Copyright 2009 WILEY

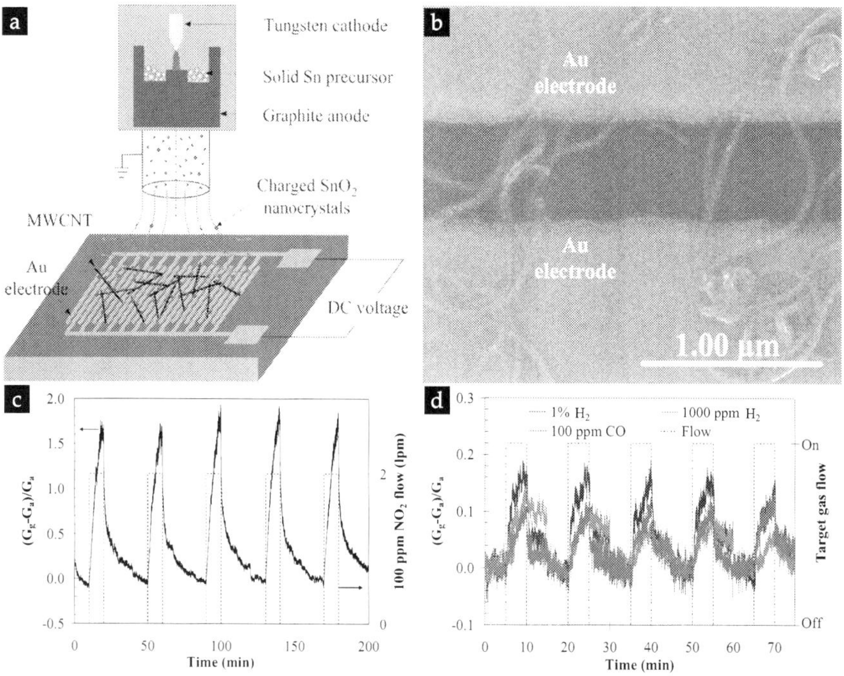

Fig. 19. (a) Assembly of charged SnO_2 NPs onto MWCNTs dispersed on an Au interdigitated electrode using ESFDA. (b) A section of the sensor after assembly of SnO_2 NPs onto MWCNTs. Dynamic response of the SnO_2 NP-MWCNT sensor for detecting room-temperature (c) 100 ppm NO_2 and (d) 1% and 1000 ppm H_2, and 100 ppm CO diluted in air. Reprinted with permission from (Lu et al. 2009). Copyright 2009 WILEY

The hybrid platform provides a radically new opportunity to engineer sensing devices with quantum-mechanical attributes by taking advantage of the electronic transfer between the nanocrystal and the CNT. The hybrid system as a sensing element is potentially superior to either of the constituent components. CNTs coated with discrete SnO_2 NPs can detect gases that are normally undetectable by bare CNTs. Attaching NPs onto CNTs leads to more active sites for the adsorption of gaseous molecules and thus a higher sensitivity than bare CNTs. The operating temperature for the hybrid sensing platform can be as low as room temperature, in contrast to the required high temperature (typically above 200 °C) for SnO_2 sensors. Moreover, the differential sensitivity between CNTs and SnO_2 NPs, the spacing between SnO_2 NPs (areal density of SnO_2 NPs on CNTs), and doping of SnO_2 NPs offer tremendous flexibility in adjusting the sensitivity of individual sensors, which can be conveniently used to engineer the selectivity of gas sensors.

6. Summary and Outlook

Our research on CNT-NP hybrid structures began as a scientific curiosity. It has advanced to a stage where the methods to produce these hybrid nanostructures are available, and much more attention has been paid to such materials because of their potential widespread applications in many areas.

The ESFDA technique can be used to efficiently coat randomly dispersed and vertically-aligned CNTs with various aerosol NPs. An intrinsic NP size selection has been observed during the assembly process. The final NP size distribution and areal density can be controlled through flow residence time/electric field or assembly time, respectively. For vertically-aligned CNTs, the electric field distribution near the CNT surface was computed using a simple model of the CNT as a cylinder with a hemispherical cap; the field enhancement factor increases from the root to the tip of the CNT for the two cases studied. The areal density and size distributions of nanocrystals observed along the CNTs in the vertical array can be rationalized by the variation of the electric field near the CNT surface; however, it is possible that shadowing effects also play a minor role on the variation in the nanocrystal areal density, but not on the nanocrystal size.

NP production is not limited to the mini-arc reactor used in our earlier study. NPs produced by other aerosol reactors or by aerosolization of colloidal NPs can also be used for the assembly, although an additional charging device may be needed in some cases. Due to the material independence of electrostatic force, the capability of ESFDA has been greatly expanded. One can imagine that NPs of multiple materials, particularly interweaving metal and semiconductor/magnetic nanocrystals, can be assembled onto CNTs. These interesting multicomponent structures will open up new opportunities in several interdisciplinary fields.

While the controlled production of the hybrid nanostructures is being perfected, a pressing area of research is to understand the synergistic interaction between the two nanocomponents, i.e., NPs and CNTs. Equally important is the understanding of the properties of hybrid nanostructures, which can be tailored for various innovative applications.

Acknowledgements

This work was financially supported by the National Science Foundation, the University of Wisconsin-Milwaukee (UWM), the UWM Research Foundation, and Science & Technology Committee of Shanghai, China. The authors would like to thank collaborators R. S. Ruoff, Z. F. Ren, M. Gajdardziska-Josifovska, and L. E. Ocola. The electrodes of sensing devices were fabricated at the Center for Nanoscale Materials at Argonne National Laboratory.

7. References

Alivisatos, A. P. (1996). Semiconductor clusters, nanocrystals, and quantum dots, *Science*, **271**, 5251, pp. 933-937.

Anson, A.; Lafuente, E.; Urriolabeitia, E.; Navarro, R.; Benito, A. M.; Maser, W. K. & Martinez, M. T. (2006). Hydrogen capacity of palladium-loaded carbon materials, *J. Phys. Chem. B*, **110**, 13, pp. 6643-6648.

Berber, S.; Kwon, Y. K. & Tomanek, D. (2000). Unusually high thermal conductivity of carbon nanotubes, *Phys. Rev. Lett.*, **84,** 20, pp. 4613-4616.

Blackman, M.; Lisgarten, N. D. & Skinner, L. M. (1968). Surface Energy and Evaporation Rate of Spherical Particles of Radii less than 500 [angst], *Nature*, **217,** 5135, pp. 1245-1246.

Chen, J. H. & Davidson, J. H. (2003). Model of the negative DC corona plasma: comparison to the positive DC corona plasma, *Plasma Chem. Plasma Process.*, **23,** 1, pp. 83-102.

Chen, J. H. & Lu, G. H. (2006). Controlled decoration of carbon nanotubes with nanoparticles, *Nanotechnol.*, **17,** 12, pp. 2891-2894.

Chen, J. H.; Lu, G. H.; Zhu, L. Y. & Flagan, R. C. (2007). A simple and versatile mini-arc plasma source for rapid nanocrystal synthesis, *J. Nanopart. Res.*, **9,** 2, pp. 203-213.

Conner, W. C. & Falconer, J. L. (1995). Spillover in Heterogeneous Catalysis, *Chem. Rev.*, **95,** 3, pp. 759-788.

Dag, S.; Ozturk, Y.; Ciraci, S. & Yildirim, T. (2005). Adsorption and dissociation of hydrogen molecules on bare and functionalized carbon nanotubes, *Phys. Rev. B*, **72,** 15, pp. 155404.

Datye, A. K.; Xu, Q.; Kharas, K. C. & McCarty, J. M. (2006). Particle size distributions in heterogeneous catalysts: What do they tell us about the sintering mechanism?, *Catal. Today*, **111,** 1-2, pp. 59-67.

de Heer, W. A. (2004). Nanotubes and the Pursuit of Applications, *Mater. Res. Soc. Bull.*, **29,** 4, pp. 281-285.

Dresselhaus, M. S.; Dresselhaus, G. & Avouris, P., Eds. (2001). *Carbon nanotubes: synthesis, structure, properties and applications*, Springer-Verlag, Berlin.

Dresselhaus, M. S. a. D., H. (2004). Carbon Nanotubes: Continued Innovations and Challenges, *Mater. Res. Soc. Bull.*, **29,** 4, pp. 237-239.

Durgun, E.; Dag, S.; Bagci, V. M. K.; Gulseren, O.; Yildirim, T. & Ciraci, S. (2003). Systematic study of adsorption of single atoms on a carbon nanotube, *Phys. Rev. B, 67, 20, pp. 201401.*

Fissan, H.; Kennedy, M. K.; Krinke, T. J. & Kruis, F. E. (2003). Nanoparticles from the gas phase as building blocks for electrical devices, *J. Nanopart. Res.*, **5,** 3-4, pp. 299-310.

Georgakilas, V.; Gournis, D.; Tzitzios, V.; Pasquato, L.; Guldi, D. M. & Prato, M. (2007). Decorating carbon nanotubes with metal or semiconductor nanoparticles, *J. Mater. Chem.*, **17,** 26, pp. 2679-2694.

Gopel, W. (1996). Ultimate limits in the miniaturization of chemical sensors, *Sensors and Actuators a-Physical*, **56,** 1-2, pp. 83-102.

Guldi, D. M.; Rahman, G. M. A.; Sgobba, V.; Kotov, N. A.; Bonifazi, D. & Prato, M. (2006). CNT-CdTe versatile donor-acceptor nanohybrids, *J. Am. Chem. Soc.*, **128,** 7, pp. 2315-2323.

Heilmann, A. & Werner, J. (1998). In situ observation of microstructural changes of embedded silver particles, *Thin Solid Films*, **317,** 1-2, pp. 21-26.

Hinds, B. J.; Chopra, N.; Rantell, T.; Andrews, R.; Gavalas, V. & Bachas, L. G. (2004). Aligned multiwalled carbon nanotube membranes, *Science*, **303,** 5654, pp. 62-65.

Huang, Z. P.; Wu, J. W.; Ren, Z. F.; Wang, J. H.; Siegal, M. P. & Provencio, P. N. (1998). Growth of highly oriented carbon nanotubes by plasma-enhanced hot filament chemical vapor deposition, *Appl. Phys. Lett.*, **73,** 26, pp. 3845-3847.

Iijima, S. (1991). Helical Microtubules of Graphitic Carbon, *Nature*, **354,** 6348, pp. 56-58.

Iijima, S. & Ichihashi, T. (1993). Single-Shell Carbon Nanotubes of 1-nm Diameter, *Nature*, **363,** 6430, pp. 603-605.

Jung, Y. J.; Kar, S.; Talapatra, S.; Soldano, C.; Viswanathan, G.; Li, X. S.; Yao, Z. L.; Ou, F. S.; Avadhanula, A.; Vajtai, R.; Curran, S.; Nalamasu, O. & Ajayan, P. M. (2006). Aligned carbon nanotube-polymer hybrid architectures for diverse flexible electronic applications, *Nano Lett.*, **6,** 3, pp. 413-418.

Kang, S. J.; Kocabas, C.; Ozel, T.; Shim, M.; Pimparkar, N.; Alam, M. A.; Rotkin, S. V. & Rogers, J. A. (2007). High-performance electronics using dense, perfectly aligned arrays of single-walled carbon nanotubes, *Nat. Nanotechnol.*, **2,** 4, pp. 230-236.

Khoo, K. H. & Chelikowsky, J. R. (2009). Electron transport across carbon nanotube junctions decorated with Au nanoparticles: Density functional calculations, *Phys. Rev. B*, **79,** 20, pp. 205422.

Kim, B. & Sigmund, W. M. (2004). Functionalized multiwall carbon nanotube/gold nanoparticle composites, *Langmuir*, **20,** 19, pp. 8239-8242.

Kim, Y. A.; Muramatsu, H.; Hayashi, T.; Endo, M.; Terrones, M. & Dresselhaus, M. S. (2004). Thermal stability and structural changes of double-walled carbon nanotubes by heat treatment, *Chem. Phys. Lett.*, **398,** 1-3, pp. 87-92.

Kong, B. S.; Jung, D. H.; Oh, S. K.; Han, C. S. & Jung, H. T. (2007). Single-walled carbon nanotube gold nanohybrids: Application in highly effective transparent and conductive films, *J. Phys. Chem. C*, **111,** 23, pp. 8377-8382.

Kong, J.; Chapline, M. G. & Dai, H. J. (2001). Functionalized carbon nanotubes for molecular hydrogen sensors, *Adv. Mater.*, **13,** 18, pp. 1384-1386.

Kong, J.; Franklin, N. R.; Zhou, C. W.; Chapline, M. G.; Peng, S.; Cho, K. J. & Dai, H. J. (2000). Nanotube molecular wires as chemical sensors, *Science*, **287,** 5453, pp. 622-625.

Kongkanand, A.; Dominguez, R. M. & Kamat, P. V. (2007). Single wall carbon nanotube scaffolds for photoelectrochemical solar cells. Capture and transport of photogenerated electrons, *Nano Lett.*, **7,** 3, pp. 676-680.

Kongkanand, A.; Vinodgopal, K.; Kuwabata, S. & Kamat, P. V. (2006). Highly dispersed Pt catalysts on single-walled carbon nanotubes and their role in methanol oxidation, *J. Phys. Chem. B*, **110,** 33, pp. 16185-16188.

Kumar, M. K. & Ramaprabhu, S. (2006). Nanostructured Pt functionlized multiwalled carbon nanotube based hydrogen sensor, *J. Phys. Chem. B*, **110,** 23, pp. 11291-11298.

Landi, B. J.; Castro, S. L.; Ruf, H. J.; Evans, C. M.; Bailey, S. G. & Raffaelle, R. P. (2005). CdSe quantum dot-single wall carbon nanotube complexes for polymeric solar cells, *Sol. Energy Mater. Sol. Cells*, **87,** 1-4, pp. 733-746.

Lee, H.; Yoon, S. W.; Kim, E. J. & Park, J. (2007). In-situ growth of copper sulfide nanocrystals on multiwalled carbon nanotubes and their application as novel solar cell and amperometric glucose sensor materials, *Nano Lett.*, **7,** 3, pp. 778-784.

Lenggoro, I. W.; Xia, B.; Okuyama, K. & de la Mora, J. F. (2002). Sizing of colloidal nanoparticles by electrospray and differential mobility analyzer methods, *Langmuir*, **18,** 12, pp. 4584-4591.

Li, L. Y.; Yang, Y.; Yang, G. L.; Chen, X. M.; Hsiao, B. S.; Chu, B.; Spanier, J. E. & Li, C. Y. (2006). Patterning polyethylene oligomers on carbon nanotubes using physical vapor deposition, *Nano Lett.*, **6,** 5, pp. 1007-1012.

Liu, H.; Kozlov, A. I.; Kozlova, A. P.; Shido, T.; Asakura, K. & Iwasawa, Y. (1999). Active oxygen species and mechanism for low-temperature CO oxidation reaction on a TiO2-supported Au catalyst prepared from Au(PPh3)(NO3) and As-precipitated titanium hydroxide, *J. Catal.*, **185**, 2, pp. 252-264.

Liu, J.; Casavant, M. J.; Cox, M.; Walters, D. A.; Boul, P.; Lu, W.; Rimberg, A. J.; Smith, K. A.; Colbert, D. T. & Smalley, R. E. (1999). Controlled deposition of individual single-walled carbon nanotubes on chemically functionalized templates, *Chem. Phys. Lett.*, **303**, 1-2, pp. 125-129.

Liu, R. J.; Crozier, P. A.; Smith, C. M.; Hucul, D. A.; Blackson, J. & Salaita, G. (2004). In situ electron microscopy studies of the sintering of palladium nanoparticles on alumina during catalyst regeneration processes, *Microsc. Microanal.*, **10**, 1, pp. 77-85.

Lu, G. H.; Ocola, L. E. & Chen, J. H. (2009). Room-Temperature Gas Sensing Based on Electron Transfer between Discrete Tin Oxide Nanocrystals and Multiwalled Carbon Nanotubes, *Adv. Mater.*, **21**, 24, pp. 2487-2491.

Lu, G. H.; Zhu, L. Y.; Wang, P. X.; Chen, J. H.; Dikin, D. A.; Ruoff, R. S.; Yu, Y. & Ren, Z. F. (2007). Electrostatic-force-directed assembly of ag nanocrystals onto vertically aligned carbon nanotubes, *J. Phys. Chem. C*, **111**, 48, pp. 17919-17922.

Mao, S.; Lu, G. H. & Chen, J. H. (2008). Coating carbon nanotubes with colloidal nanocrystals by combining an electrospray technique with directed assembly using an electrostatic field, *Nanotechnol.*, **19**, 45, pp. 455610.

McAleer, J. F.; Moseley, P. T.; Norris, J. O. W. & Williams, D. E. (1987). Tin Dioxide Gas Sensors .1. Aspects of the Surface-Chemistry Revealed by Electrical Conductance Variations, *Journal of the Chemical Society-Faraday Transactions I*, **83**, pp. 1323-1346.

Meyer, R.; Ge, Q. F.; Lockemeyer, J.; Yeates, R.; Lemanski, M.; Reinalda, D. & Neurock, M. (2007). An ab initio analysis of adsorption and diffusion of silver atoms on alumina surfaces, *Surf. Sci.*, **601**, 1, pp. 134-145.

Morgenstern, K.; Rosenfeld, G. & Comsa, G. (1999). Local correlation during Ostwald ripening of two-dimensional islands on Ag(111), *Surf. Sci.*, **441**, 2-3, pp. 289-300.

Mu, Y. Y.; Liang, H. P.; Hu, J. S.; Jiang, L. & Wan, L. J. (2005). Controllable Pt nanoparticle deposition on carbon nanotubes as an anode catalyst for direct methanol fuel cells, *J. Phys. Chem. B*, **109**, 47, pp. 22212-22216.

Murray, C. B.; Norris, D. J. & Bawendi, M. G. (1993). Synthesis and characterization of nearly monodisperse CdE (E = sulfur, selenium, tellurium) semiconductor nanocrystallites, *J. Am. Chem. Soc.*, **115**, 19, pp. 8706-8715.

Peng, X. H.; Chen, J. Y.; Misewich, J. A. & Wong, S. S. (2009). Carbon nanotube-nanocrystal heterostructures, *Chem. Soc. Rev.*, **38**, 4, pp. 1076-1098.

Podenok, S.; Sveningsson, M.; Hansen, k. & Campbell, E. E. B. (2006). Electric Field Enhancement Factors aound a Metallic, End-Capped Cylinder, *Nano*, **1**, 1, pp. 87-93.

Read, F. H. & Bowring, N. J. (2004). Field enhancement factors of one-dimensional and two-dimensional arrays of nanotubes, *Microelectron. Eng.*, **73-74**, pp. 679-685.

Robel, I.; Bunker, B. A. & Kamat, P. V. (2005). Single-walled carbon nanotube-CdS nanocomposites as light-harvesting assemblies: Photoinduced charge-transfer interactions, *Adv. Mater.*, **17**, 20, pp. 2458-2463.

Robel, I.; Girishkumar, G.; Bunker, B. A.; Kamat, P. V. & Vinodgopal, K. (2006). Structural changes and catalytic activity of platinum nanoparticles supported on C-60 and carbon nanotube films during the operation of direct methanol fuel cells, *Appl. Phys. Lett.*, **88,** 7, pp. 073113.

Robel, I.; Kuno, M. & Kamat, P. V. (2007). Size-dependent electron injection from excited CdSe quantum dots into TiO2 nanoparticles, *J. Am. Chem. Soc., 129, 14, pp. 4136-4137.*

Rosenfeld, G.; Morgenstern, K.; Beckmann, I.; Wulfhekel, W.; Laegsgaard, E.; Besenbacher, F. & Comsa, G. (1998). Stability of two-dimensional clusters on crystal surfaces: From Ostwald ripening to single-cluster decay, *Surf. Sci.*, **404,** 1-3, pp. 401-408.

Saito, R., G. Dresselhaus, and M.S. Dresselhaus, (1998). *Physical Properties of Carbon Nanotubes*, Imperial College Press., London.

Scher, E. C.; Manna, L. & Alivisatos, A. P. (2003). Shape control and applications of nanocrystals, *Philosophical Transactions of the Royal Society of London Series a-Mathematical Physical and Engineering Sciences*, **361,** 1803, pp. 241-255.

Seinfeld, J. H. & Pandis, S. N. (1998). *Atmospheric Chemistry and Physics: From Air Pollution to Climate Change,* John Wiley & Sons, Inc., New York.

Shao, Y. Y.; Liu, J.; Wang, Y. & Lin, Y. H. (2009). Novel catalyst support materials for PEM fuel cells: current status and future prospects, *J. Mater. Chem.*, **19,** 1, pp. 46-59.

Shimizu, Y. & Egashira, M. (1999). Basic aspects and challenges of semiconductor gas sensors, *Mrs Bulletin*, **24,** 6, pp. 18-24.

Sun, Y. G. & Wang, H. H. (2007). High-performance, flexible hydrogen sensors that use carbon nanotubes decorated with palladium nanoparticles, *Adv. Mater.*, **19,** 19, pp. 2818-2823.

Treacy, M. M. J.; Ebbesen, T. W. & Gibson, J. M. (1996). Exceptionally high Young's modulus observed for individual carbon nanotubes, *Nature*, **381,** 6584, pp. 678-680.

Valden, M.; Lai, X. & Goodman, D. W. (1998). Onset of catalytic activity of gold clusters on titania with the appearance of nonmetallic properties, *Science*, **281,** 5383, pp. 1647-1650.

Wen, J. G.; Huang, Z. P.; Wang, D. Z.; Chen, J. H.; Yang, S. X.; Ren, Z. F.; Wang, J. H.; Calvet, L. E.; Chen, J.; Klemic, J. F. & Reed, M. A. (2001). Growth and characterization of aligned carbon nanotubes from patterned nickel nanodots and uniform thin films, *J. Mater. Res.*, **16,** 11, pp. 3246-3253.

Wolf, A. & Schuth, F. (2002). A systematic study of the synthesis conditions for the preparation of highly active gold catalysts, *Appl. Catal., A*, **226,** 1-2, pp. 1-13.

Wynblatt, P. & Gjostein, N. A. (1975). Supported metal crystallites, *Prog. Solid State Chem.*, **9,** pp. 21-58.

Wynblatt, P. & Gjostein, N. A. (1976). Particle growth in model supported metal catalysts - I Theory, *Acta Metall. Mater.*, **24,** 12, pp. 1165-1174.

Xing, Y. C. (2004). Synthesis and electrochemical characterization of uniformly-dispersed high loading Pt nanoparticles on sonochemically-treated carbon nanotubes, *J. Phys. Chem. B*, **108,** 50, pp. 19255-19259.

Ye, X. R.; Lin, Y. H.; Wai, C. M.; Talbot, J. B. & Jin, S. H. (2005). Supercritical fluid attachment of palladium nanoparticles on aligned carbon nanotubes, *J. Nanosci. Nanotechnol.*, **5,** 6, pp. 964-969.

Yildirim, T. & Ciraci, S. (2005). Titanium-decorated carbon nanotubes as a potential high-capacity hydrogen storage medium, *Phys. Rev. Lett.*, **94,** 17, pp. 175501.

Yu, M. F.; Lourie, O.; Dyer, M. J.; Moloni, K.; Kelly, T. F. & Ruoff, R. S. (2000). Strength and breaking mechanism of multiwalled carbon nanotubes under tensile load, *Science*, **287,** 5453, pp. 637-640.

Zamudio, A.; Elias, A. L.; Rodriguez-Manzo, J. A.; Lope-Urias, F.; Rodriguez-Gattorno, G.; Lupo, F.; Ruhle, M.; Smith, D. J.; Terrones, H.; Diaz, D. & Terrones, M. (2006). Efficient anchoring of silver nanoparticles on N-doped carbon nanotubes, *Small*, **2,** 3, pp. 346-350.

Zamudio, A.; Elias, A. L.; Rodriguez-Manzo, J. A.; Lopez-Urias, F.; Rodriguez-Gattorno, G.; Lupo, F.; Ruhle, M.; Smith, D. J.; Terrones, H.; Diaz, D. & Terrones, M. (2006). Efficient anchoring of silver nanoparticles on N-doped carbon nanotubes, *Small*, **2,** 3, pp. 346-350.

Zhang, Y. & Dai, H. J. (2000). Formation of metal nanowires on suspended single-walled carbon nanotubes, *Appl. Phys. Lett.*, **77,** 19, pp. 3015-3017.

Zhang, Y.; Franklin, N. W.; Chen, R. J. & Dai, H. J. (2000). Metal coating on suspended carbon nanotubes and its implication to metal-tube interaction, *Chem. Phys. Lett.*, **331,** 1, pp. 35-41.

Zhang, Y.; Zhang, X. G.; Zhang, H. L.; Zhao, Z. G.; Li, F.; Liu, C. & Cheng, H. M. (2006). Composite anode material of silicon/graphite/carbon nanotubes for Li-ion batteries, *Electrochim. Acta*, **51,** 23, pp. 4994-5000.

Zheng, L. X.; O'Connell, M. J.; Doorn, S. K.; Liao, X. Z.; Zhao, Y. H.; Akhadov, E. A.; Hoffbauer, M. A.; Roop, B. J.; Jia, Q. X.; Dye, R. C.; Peterson, D. E.; Huang, S. M.; Liu, J. & Zhu, Y. T. (2004). Ultralong single-wall carbon nanotubes, *Nat. Mater.*, **3,** 10, pp. 673-676.

Zhu, L. Y.; Lu, G. H. & Chen, J. H. (2008). A generic approach to coat carbon nanotubes with nanoparticles for potential energy applications, *J. Heat Transfer-Transactions of the Asme*, **130,** 4, pp. 044502.

Zhu, L. Y.; Lu, G. H.; Mao, S.; Chen, J. H.; Dikin, D. A.; Chen, X. Q. & Ruoff, R. S. (2007). Ripening of silver nanoparticles on carbon nanotubes, *Nano*, **2,** 3, pp. 149-156.

32

Shock-Wave-Compaction (SWC) of Al/CNT Two Phase Systems

Noé Alba-Baena[1], Wayne Salas[2] and Lawrence E. Murr[3]
[1]*Universidad Autónoma de Ciudad Juárez, México*
[2]*Tinker Air Force Base, USA*
[3]*University of Texas at El Paso, USA*

1. Introduction

The scientific investigation and applied research on composite materials can date back to the 1940's (Schwaretz, 1997) with the advantages behind the development of metal matrix composites (MMCs) being the capability to combine phases providing a potential for tailoring material properties to meet specific and challenging requirements. Composites offer an approach for producing "designer" materials used to provide specific types of material behavior, such as their improved strength and stiffness (Ward et al, 1996), outstanding corrosion resistance (Shimizu et al, 1995), friction resistance (Akbulut et al, 1998) and wear resistance (Wang et al, 1996), high electrical and thermal conductivity (Koráb et al, 2002), and high temperature mechanical behavior (Tjong & Ma, 1997). Currently, metal matrix composites can be classified by reinforcement component into fibers (continuous or discontinuous), whiskers, particulates, or wires. Theses reinforcements have been placed in matrices of aluminum, magnesium, copper, titanium, nickel, nickel-based superalloys, and various alloys of iron. However the aluminum matrix alloy composites are those that have become an industry standard because they offer the advantage of lower cost when compared to most other MMCs. Aluminum based composites also offer the added benefits of excellent thermal conductivity, high shear strength, excellent abrasion resistance, high-temperature operation, and the ability to be formed and treated on conventional equipment (Schwaretz, 1997). Conventional metal matrix composites have been manufactured through several various processing methods. These processes can generally be categorized into four general groups: solid-state processes, liquid-state processes, deposition processes and in-situ processes. Solid-state and liquid-state processes are the two most widely used and developed methods (Everett & Arsenault, 1991). Particulate composites reinforced with micron-sized particles of various materials are perhaps the most widely utilized composites in everyday materials. Particles are typically added to enhance the matrix elastic modulus and yield strength. By scaling the particle size down to the nanometer scale, it has been shown that novel material properties can be obtained (Thostenson et al, 2001).

The nanocomposite materials, especially those using carbon nanotubes as reinforcement, have recently garnered great interest and tremendous growth from scientists and engineers in the research field (Salvetat-Delmotte & Rubio, 2002). This spurt of interest in nanocomposites stems from the unprecedented flexibility and improvements in physical properties that may be attained by using building blocks with the dimensions in the nanoscale range. It may be possible to design and create new tailored composites by using nanosize building blocks of heterogeneous dispersed phases. Thus, the materials designed from them can be multifunctional because the constituents of a nanocomposite have different structures and compositions and therefore different properties. In many cases, the interesting range may be located near the transition where properties are changing from the molecular to bulk-like, a size range in which properties can be manipulated in a positive way in order to design a material for a particular application (Oelhafen & Schuller, 2005).

It is true that nanometric sized particles help to achieve improved mechanical properties (Sahin & Acilar, 2003; Raming et al, 2004 and Lan et al, 2004) depending basically in the type of second phase nanomaterial and the dispersion strengthening due to the particle distribution and volume fraction variation. But currently much attention has been given to the possibilities of incorporating carbon nanotubes (CNTs) as the nano reinforcement in a matrix or as a second phase in two phase systems. This is not only due to the well known remarkable mechanical properties of CNTs but also in the belief that they may very well be the ultimate fiber reinforcement based on their high aspect ratio and defect-free structure. Application and innovations will take advantage of the special properties based on carbon nanotubes including electrical, mechanical, and other unique properties. The construction of composites with extraordinary properties will be related to the multifunctional materials that can be developed. Most investigators who are developing new composite materials with nanotubes work with nanotube concentrations below 10%wt due to limited availability of nanotubes. With continued developments in the synthesis and production of carbon nanotubes, new possibilities in the field of composite materials based on carbon nanotubes are emerging (Meyyappan, 2004). Although much of the present research is in the area of polymer composites, efforts in metal and ceramic matrix composites are also of interest. Studies, especially in polymers, focus on dispersion, untangling, alignment, bonding, molecular distribution, and retention of nanotube properties. Carbon nanotubes are theoretically one of the strongest and stiffest materials with a calculated tensile strength of ~200 gigaPascal and modulus of more than 1-4 teraPascal for a single walled nanotube (SWNT) (Dresselhaus et al, 2001). If the mechanical properties of SWNT could be effectively incorporated into a matrix, composites with lightweight, exceptional strength and stiffness can be achieved. It is expected that nanotube composites will be used as a replacement for existing materials where properties superior to conventional composites are achieved and to create materials for applications where composites traditionally have not been used before. Therefore, one of the most important outcomes from current nanocomposite research will be knowledge that is gained about preparing materials for the development of the nanocomposites in the future. The applications for nano-metal matrix composites (NMCs) are nanowires, lightweight structures, electronic materials for avionics, wear coatings, novel magnetic and super conducting systems and new multifunctional metals (Meyyappan, 2004 and Ajayan et al, 2003).

In order to tailor these bulk nano-based (NMCs), actual approaches have used conventional powder metallurgy with adapted techniques (as ultrasonic mixing and wetting preparation), where matrices of aluminum, copper, magnesium and silver have been used. Benefits from this approach arise based in the process work temperature that is below the melting temperature of the metals (melting temperature for Al is 660°C) (Ajayan et al, 2003). Powder metallurgy routes using powders and in some cases extrusion have produced an Al metal matrix composite, copper electrodes, and macroscopic composite wires (where extrusion was used). However, the resultant NMCs has exhibited modest improvement of its properties as compared to those characteristic of traditional metal-matrix composites (MMCs) (Salvetat-Delmotte & Rubio, 2002). Such performance can be mainly attributed to the tendency of the nanomaterials and specially the carbon nanotube (CNT) material to variously agglomerate and cluster. This leads to difficulties in the dispersion of the CNTs in the metal matrix, as well as poor wetting, or related interfacial phenomena and integrity issues. The results achieved by using high-intensity ultrasonic waves with strong micro-scale transient cavitations and acoustic streaming to successfully introduce, distribute and disperse nanoparticles into Mg alloy melts and Al matrices, thus making the production of cast high-performance nano-sized particles reinforced matrix composite promising (Rohatgi et al, 2008). In order to take full advantage of the exceptional stiffness, strength and resilience of carbon nanotubes requires a uniform dispersion of CNTs in the metal matrix. This dispersion will ensure a strong interface bonding between the CNTs and the surrounding matrix which provides effective stress transfer, as well as avoids intra-tube sliding between concentric walls within MWCNTs and intra-bundles sliding within SWCNT ropes. Efforts are in progress to overcome these difficulties. Recently, Zhong et al also studied nanocrystalline Al matrix reinforced with SWCNTs that were successfully fabricated by cold-consolidation and hot consolidation (Zhong et al, 2003). Their procedure of mixing nano-Al particles and SWNTs results in a homogeneous dispersion of SWNTs and shows that the ultrasonic energy could overcome the van der Waals force between SWNT bundles and between nano-Al particles. From this work one can infer that when used as the matrix of composites, nano-materials can be more compatible with CNTs than traditional matrices of metal, ceramic matrix, and polymer. The reported hardness of the SWNT/nano-Al composites reached a peak value of 2.89 GPa, which shows that SWNTs are a promising reinforcement for some matrices.

Conventional process seems to inhibit the diffusion of CNTs across and along the matrix grain surfaces. Sintering cannot proceed without damaging the CNTs or removing them from the liquid state matrix or if the diffusion is achieved CNTs are mostly located at grain boundaries of the matrix and are insignificant in improving material performance (Cha et al, 2005). The most important processing issue is the interfacial strength between the CNTs and the matrix. In the case of successful CNT/polymer nanocomposites, the interfacial strength between the CNTs and the polymer matrix is strong because they interact at the molecular level. In the case of CNT/metal nanocomposites, however, the interfacial strength cannot be expected to be high because the CNTs and the matrix are merely blended. In their research, Cha et al (2005) uses a novel molecular mixing approach in order to produce a metal/CNT composite rather than a two-phase system. As result, it was reported that CNTs are shown not merely lying at grain boundaries but diffusing across grain boundaries. Some work as been done on one-dimensional nanoscale composites prepared by coating the carbon nanotubes, by electroless plating, with other materials such as Co and Ni (Chen et al, 2000 and 2003). Such studies address in part the problem of interfacial adhesion between the nanotubes and the metal matrix; it is thought that high strength adhesion between nanotubes and the metal matrix can be achieved by coating the nanotubes with metallic material. Probst (2005) et al states the possibility that carbon nanotubes may not be used as a reinforcement phase without prior treatment. Probst reports that CNTs can be wetted only by liquids, or by molten metals with low surface tension at a cut-off limit lying between 100 and 200 mN/m (Jordan J. L. et al, 2001). This explains why most metals such as aluminum (surface tension of 865 mN/m), copper (1270 mN/m) or iron (1700 mN/m) are not able to wet the surface of nanotubes and achieve high interfacial adhesion. In their work, Lijie et al focuses on the interfacial reaction between CNTs and aluminum in Al/CNT composite films that were fabricated by sputtering pure Al on the surface of aligned multi-walled CNT arrays (Lijie et al, 2006). After heat treating, annealed samples show that, at various temperatures, aluminum carbide (Al_4C_3) was formed at the interface between the Al and CNT layers at defect sites and at open ends of CNTs. More recently novel approaches and fabrication processes have been reported. Bakshi et al (2008) successfully synthesized aluminum composites reinforced with CNTs by cold spraying of a blended powder. They reported an elastic modulus ranging between 40-120 GPa and attributed such variation mainly to the exhibited porosity and the agglomeration of CNTs. Laha et al (2009) used plasma spray forming (PSF) to fabricate Al-Si/MWCNTs composites measuring up to a 78% increase in the elastic modulus of the composite; also attributing such variation to the composite's porosity and the CNTs agglomerations. Lim et al (2009) also reported the production of aluminum alloy reinforced with multi-walled carbon nanotubes composite by using a friction stir processing where images and conclusions show the agglomeration of CNTs.

1.1 Two phase systems

Therefore the fabrication of two phase systems (TPS) that involves the incorporation of a second phase or reinforcer; such as particles, platelets, whiskers and fibers, to a continuous phase are differentiated primarily by the interaction between the grains (aluminum grains in this study) and the nanoparticles (carbon nanotube aggregate mixture for this study). The differences between the systems that involve more than one phase are related, in general, to the fabrication process either in the solid or liquid phases, therefore it is necessary to define a TPS as opposed to a MMC. According to Torquato (2002), a TPS is a heterogeneous material composed of domains of different materials (phases) such as composites or the same material in different states such as a polycrystal. Kennedy et al (2001) considers a two-phase system as a system where the continuous phase network (large grains) is surrounded by the second phase and the phases are interdispersed and uniformly inter-twinned with each other. This definition better expresses the concept of a TPS as used in this chapter. While a continuous phase is described for both; the difference should lie in the effectiveness of traditionally known strengthening mechanisms. The two-phase system has a different set of properties and microstructures when compared to the MMC. TPSs and MMCs can now be related to their fabrication processes: MMCs are those fabricated by liquid stage processes such as mixing the matrix grains and the particle grains by using ultrasonic dispersion, mechanical stirring and others. TPS are considered as resultant of the use of solid state processes as powder metallurgy (P/M), high temperature synthesis (SHS), mechanical mixing (MM), hot pressing (HP) and shock wave compaction (SWC) (Kennedy et al, 2001). In this case surface phenomena tend to dominate fine powders, primarily due to their high surface-to-volume ratio and in such cases van der Waals, electrostatic, and surface tension forces often have dominating effects on properties. Handling these particles is extremely difficult, and high aspect ratios (as in fine whiskers and SWCNT) further complicate the handling problem. There are cases when systems using CNTs were fabricated and reported in the literature as MMCs. In these cases the agglomeration of CNT aggregate in the nanometric regime is much larger than the molecular dimensions, so that the agglomerate may posses' properties in the macroscopic level. These MMCs may be classified as heterogeneous two phase or two component systems with heterogeneous materials ranging from dispersions with varying degrees of clustering to complex interpenetrating connected multiphases. An example of a two phase system is described by Kim et al. where the microstructure of CNT/Cu nanocomposites show that the carbon nanotubes are not homogeneously distributed in Cu matrix, but the carbon nanotubes are densely distributed in localized regions (Kim et al., 2006). The microstructure of CNT/Cu nanocomposites consists of two regions including a CNT/Cu composite region, where most CNTs are distributed, and a CNT free Cu matrix region (continuous phase). Ajayan et al (2003) consider that the relationship between particles and grains can be divided in four grain/particle types shown in Figure 1.

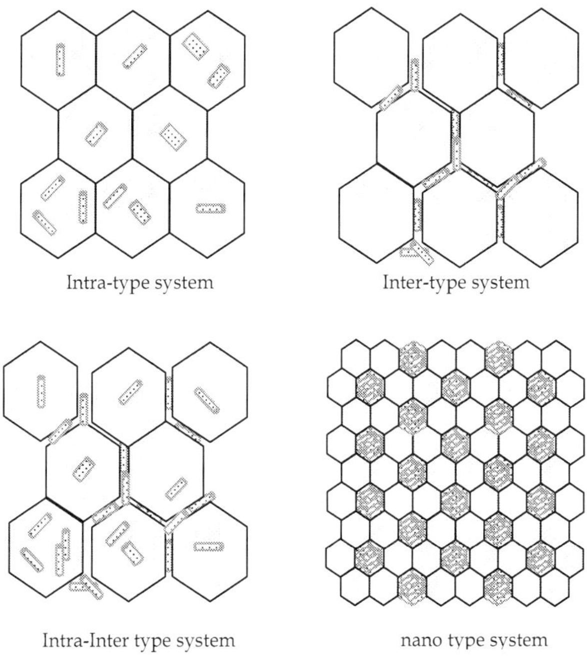

Fig. 1. Representation of different systems that can be obtained by using nanoparticles (adapted from Ajayan et al, 2003).

The first to appear is the intragranular type of interaction. This intra-type system is often addressed as the pure MMC, which results from the inclusion of the nanoparticles or CNT (for our purposes) in the grains usually during a liquid state process. For this system the grains form a matrix in which the CNTs or reinforcers are included, i. e. CNTs affect the matrix by generating internal dislocations which increase the tensile and hardness properties. The inter-type interaction shown in Figure 1 consists of the CNTs sitting along the grain boundaries of the continuous phase affecting the grain to grain bonding of the system. In this intergranular type system the grain/CNT relationship remains constant with the grains remaining without inclusions. This system is based in the volume fraction of the reinforcement with the addition of CNTs reaching a saturation point. An independent phase forms after the saturation point of the CNTs which agglomerate and form at the grain boundaries of the continuous phase. The resultant system from the interaction of the particle agglomerations and the grains can be address as a two-phase system. A combination of previously mentioned systems (inter-intra type) is also shown in Figure 1; the microstructural representation of the combination system of intragranular and intergranular usually results from the fabrication by stir casting processes which can be considered an imperfect MMC. Examples of this kind of systems have been achieved by means of traditional P/M processes where attempts to fabricate CNT/metal composites with homogeneously dispersed CNTs have been attained. Resultant systems show strong interaction of CNTs in powder form (due to Van der Waals forces) and CNTs agglomerate rather than homogenously disperse. In general, if the CNT/metal nanocomposites are manufactured by conventional processes (liquid or solid); most of the CNTs are located on the surfaces of the metal particles as agglomerates and dispersed forms. Finally, a nano/nano type system is represented in Figure 1, such system is what is hoped to be achieved in a true nano-scale two-phase system where either the nanoparticles (or CNTs) and the principal phase lie in the nanoregime. Still, the ideal metal/CNTs composite does not fit the previous descriptions and has not yet been achieved by experimental processes but the Figure 2 schematic represents a proposed case for a metal/nanotube composites. Here CNTs are shown not merely lying at grain boundaries agglomerating or immersed in the matrix grain, but nanotubes are shown diffusing across grain boundaries in an ideal fashion.

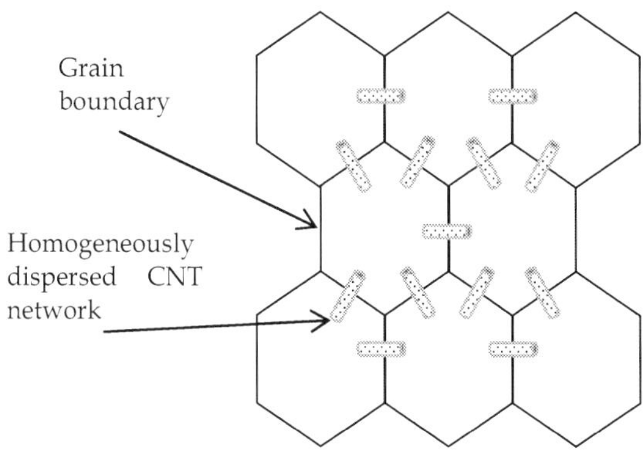

Fig. 2. Ideal microstructure of a Metal/CNT composite (Adapted from Wang et al, 1996)

2. Shockwave consolidation of powders

In an extensive report, Meyers et al (2006) reviewed and summarized the literature on the mechanical properties of nanocrystalline materials. The report defines the most important synthesis methods for powder densification (including SWC) and a number of aspects of mechanical behavior showing the potential benefits and drawbacks in the fabrication of nanocomposites. While there have been several recent examples of carbon nanotube/metal composite systems fabrication (Rao et al, 2006 and Kowbel, 2005) utilizing conventional powder metallurgy (P/M) routes, there has been no evidence of any significant improvement in properties that characterize traditional MMCs (Lan et al, 2004 and Yang et al, 2004). This is due in part to the CNT material agglomerating, and the difficulty to disperse the CNTs in the metal matrix as a consequence of poor wetting, related interfacial phenomena, or integrity issues. Also of concern is high temperature exposure required by the processing of composites by the conventional mechanical stirring, ultrasonic dispersion and powder metallurgy.

As a method to avoid the formation of intermetallic compounds and where the grain size is to be preserved, an alternative way to consolidate composites is by dynamic compaction (or SWC) (SivaKumar et al, 2001). The densification of powders by shock waves has gained revived interest as a technique for the consolidation of TPS, amorphous and nano-crystalline powders, e.g. for magnetic applications. This revised interest stems in part because TPS's are systems where microstructural and mechanical properties depend on the connectivity between a continuous-phase surrounded by an interdispersed second-phase, and also because the SWC process has a very high 'dynamic' energy rate and the energy is mainly deposited on the grain boundaries thus bulk heating is very much limited. Moreover, SWC characteristics make it even more suitable for the densification of composites of otherwise incompatible materials, such as polymeric and ceramic powders, which may lead to some distinct industrial applications (Yang et al, 2004). As an example, the full densification of metallic-ceramic mixtures of Al (30% volume fraction) and B4C has been reported, where the major drawback of this technique is the occurrence of cracks, which can be avoided when the starting materials show some plastic behavior.

For the purposes of this chapter SWC is reviewed in extend because of its possibility as an alternative for the densification and synthesis of nano-sized composites (Withers, 2005), and similarly, to consolidate nano-two-phase systems (Wang X. et al, 2004). Prummer (2001) has reviewed SWC from its origins, which now spans more than four decades. Over this time, shock-wave consolidation has been used for the densification and synthesis of ceramic compounds based on powder mixing (Dorst et al, 1997 and Jordan & Thadani, 2001), and similarly, to consolidate two-phase systems (Kennedy et al, 2001). This consolidation process has been shown to be a suitable option to produce 100% dense nanocrystalline Al_2O_3 (Torralba et al, 2003), along with a variety of ceramics and other nano-materials. SWC has also proven to be effective in the synthesis of materials (diamond for example) (Yang et al, 2004; Wang L et al, 2004; and Sherif El-Eskandarany, 1998). In addition to the different benefits of SWC introduced above, SWC is the technique lending itself best to industrial production. The costs are low therefore the process lends itself to scale up without major problems. Since the 50s many reports have been presented (Rice et al, 1958; Kimura, 1963 and Kawala et al, 1974) and recently have been published on shock consolidation of powders (Prummer & Ziegler, 1985; Thadani, 1988; Glade et al, 1995). However, publications on the explosive compaction of CNT aggregate material are limited and there is very little information available on processing of metal-matrix nanocomposites by explosive compaction. Salas et al (2007) have employed SWC as an approach to creating a two-phase monolith from mixtures of varying volume fractions of multiwalled carbon nanotube (MWCNT) aggregates with micron-size (~150 μm) aluminum powder. These two-phase systems were of special interest because the MWCNT aggregates were obtained from as-manufactured mixtures of tubes and various sizes of multi-concentric fullerenes (with diameters ranging from ~2 to 40 nm) it was not clear whether these aggregates could themselves be consolidated into a contiguous phase region, and whether this regime would be bonded, monolithically, to the consolidated aluminum particle regime.

2.1 SWC process

During SWC the densification of the powders is accomplished by the passage of a strong shock wave generated upon the impact of a flyer plate onto the green compact (target powder). Shock pressures of 3-15 GPa or higher are commonly used and a typical shock rise time is of the order of 100 ns during which the densification process is complete (Kennedy et al, 2001). The physical phenomena of such a dynamic consolidation process at the particle level are very complex and remain poorly understood.

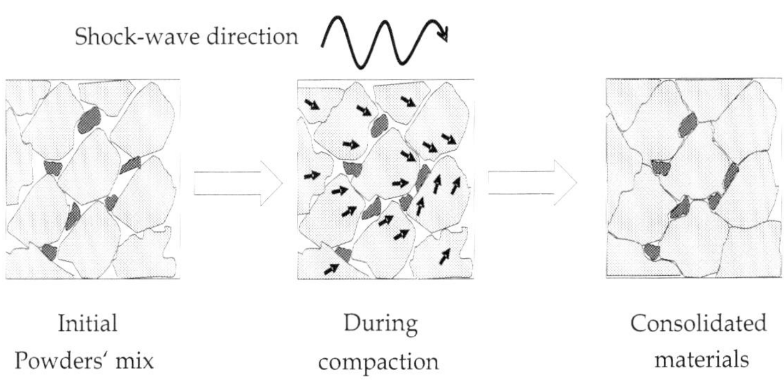

Fig. 3. Schematic view of a TPS shock-consolidation process to illustrate the shock-wave effect in the powder material grains.

However, as shown in Figure 3 and explained by Tong et al (1995), during shock-wave densification, energy is mainly deposited near the particle surfaces and the resulting heating produces softening and even melting of the particle surfaces that solidify rapidly via heat conduction into the interior of the particles before release of shock pressure. First the reduction of pores takes place through the sliding and rearrangement of particles. Then, the densification takes place by surface plastic deformation (at particle contacts) over small areas where initially free surfaces become areas of contact and the localized energy deposition at interparticle surfaces has been attributed to local plastic deformation and some frictional sliding that contributes largely to the particle-particle bonding (Tong et al, 1995) as illustrated by the single phase consolidated material images from Figure 4. Here an arrangement of tungsten rods were implosively consolidated as two dimensional grains, also one can observe the collapsed pores and the deformation of the grains (rods) after plastic sliding. Finally, 3rd image of Figure 4 illustrates the consolidation of TPS where the second phase particles are all locked up and surrounded by the solid primary phase through the densification that takes place at the same time of the pores collapse.

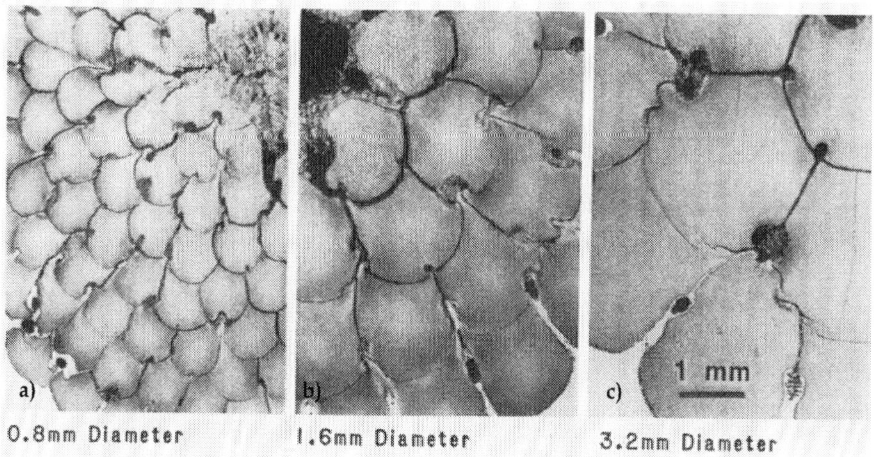

Fig. 4. Sequence showing an arrangement of tungsten rods representing a two dimensional consolidation after SWC process (from Murr, 1998).

2.2 Type of SWC methods

Murr (1998) categorizes the shock-wave consolidation processes into two major groups; the direct and indirect methods or consolidation techniques. The main difference between the two methods is the way that the pressure is applied to the sample or materials mixture. The direct methods require minimum tooling, are relatively inexpensive, and do not have geometrical limitations. The indirect methods are usually more expensive and employ fixturing, tooling, or a liquid transmitting media to achieve the same uniformity in the final product densities. The single stage or three-capsule gas-guns that were used to synthesize β-C_3N_4 (Collins et al, 2001) and intermetallics such as Ti_5Si_3 (Counihan et al, 1999) alloy and NiAl (Chen et al, 1999) and to successfully consolidate magnetic nanocomposites ($Pr_2Fe_{14}B/\alpha$-Fe) (Jin et al, 2004) are elaborate and reliable. Here the experiments are performed using guided projectiles that impact on flyer plates to incident one-dimensional shock. Shock-wave consolidation direct methods are basically two and can be exemplified with the cylindrical and plate configurations that are the most representative of these experimental setups. Plate configured experiment set ups provide higher pressures than explosives in direct contact with the material (Meyers and Wang, 1988). A well used plate configuration is the Sawaoka experimental variation. This configuration along with the Sandia calibrated shock recovery fixtures are intended to control and make reproducible the high pressure compression conditions (Murr, 1998). Plate configured experiment set ups provide higher pressures than explosives in direct contact with the material (Meyers and Wang, 1988). Here the densification of the powders is accomplished by the passage of a strong shock wave pressures generated upon the impact of a flyer onto the green compact (or packed powder) (Tong et al, 1995). Many reports on successful consolidation of different powder mixtures by using this set up can be recalled: Ti/SiCp (Tong et al, 1995), Al-SiC (SivaKumar et al, 2001), Al-Li-X alloys (Murr, 1998), carbon fiber aluminum composites (Raghundan et al, 2003), Mo-Si powder mixtures (Vandersall and Thadani, 2001) and the synthesis of diamond from fullerenes (Epanchintsev et al, 1997). Common tube configurations use single and double tube setups (Stuivinga et al, 1999). Single tube configuration is a simple design, where the green compacted powders are placed in a thin walled metal container (or pipe) encapsulated by the use of solid metal end plugs (see Figure 5) filled with the chosen explosive material and an electrical activated detonation, a fast explosive layer, is initiated at the top of the set up.

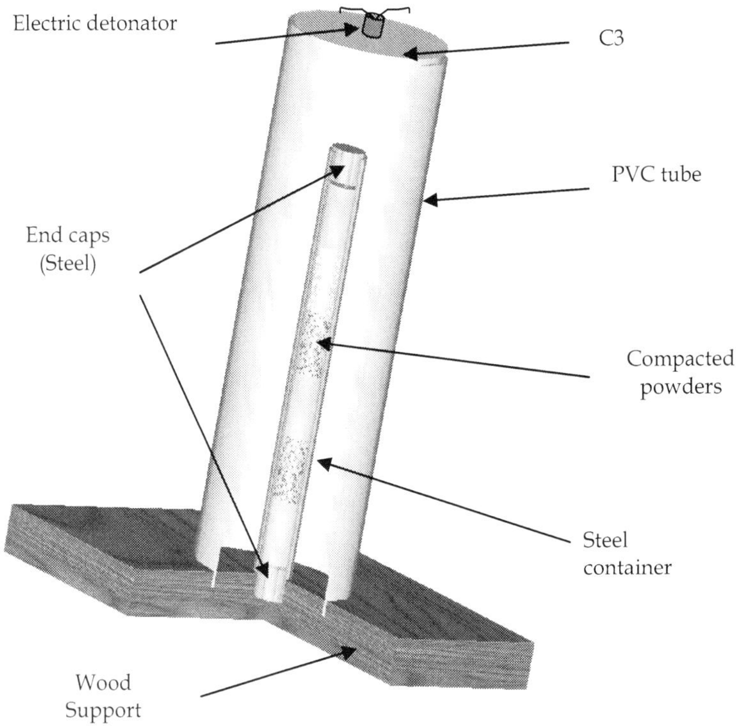

Fig. 5. Basic components of a simple cylindrical compaction set up.

The detonation wave in the explosive propagates parallel to the cylinder axis, generating oblique shock-waves to converge towards the central axis of the cylinder and if the energy applied is excessive, radial cracks and/or a mach stem can occur in the consolidation sample. Many researchers have employed explosive compaction set up, for example, to densify aluminum and Al-SiCp composite powders (SivaKumar et al, 1996), for dual phase nanocomposite systems of Y_2O_3-doped with ZrO_2 and RuO_2 (Raming et al, 2004) and more recently Al-Pb nanocomposites (Csanady et al, 2006) and nanocrystalline alumina powder (Weimar & Prummer, 2001). The double tube set up is, conceptually, analogous to the use of a flyer plate to generate high pressures in plane-wave assemblies (Meyers and Wang, 1988). The basic difference with the conventional explosive consolidation systems is that a flyer tube is placed co-axially with the container tube. The experimental setup consists of two co-axial tubes, the external one being accelerated inwards and impacting the internal tube, that contains the powder. The basic experimental set-up is similar to the described single tube set up. The explosive is placed in the cylinder, at the center of which is the assembly containing the powder (Murr, 1998). The explosive charge is detonated at the top; a detonation sheet (detasheet) booster is used to create a more uniform detonation front. This approach ensures high shock pressures while retaining a low detonation velocity which minimizes cracking and Mach stem formation. An improvement was made by Meyers and Wang (1988) proposing a variation in the central axis of the container by placing a solid rod. Substantial improvements in the quality of consolidates was obtained by using this technique, where the pressures generated in the powder are several times higher than the one for the single-tube geometry, for the same quantity and type of explosive. There has been reported significant improvement in consolidation quality of nickel-based superalloys, titanium alloys, Al-Li alloys (Kennedy et al, 2001; Kim et al, 2006). There are reports on the consolidation of superconducting YBCO (Mamalis et al, 2001), the consolidation of synthetic diamond (Deribas et al, 2001), the synthesis of intermetallic compounds such as TiAl (Prummer and Kochsier, 2001) and the densification of B_4C (Stuivinga et al, 1996) for example.

3. Al/CNT Two-Phase Systems Fabrication

The characterization of Al/2%CNT and Al/5%CNTs systems based on a previous report (Salas et al, 2007) is shown here in order to describe the fabrication of TPS by using CNTs and SWC to consolidate materials. Irregular (spherical) and small aggregates of aluminum powder, with an average primary particle size of ~150 μm, served as a base phase to which commercial aggregates of multiwalled carbon nanotubes and other assorted multi-concentric fullerenes were added in two different volume fractions (2 and 5%) for mixtures that were mechanically mixed. The CNTs percentage is decided upon from previous studies which indicate percentages of 10% and above result in a significant decrease in measured mechanical properties (Xu et al, 1999 and Feng et al, 2005) while percentages below 5% have been used in previous studies by different fabrication methods with several reported increases in mechanical properties. These mixtures, along with the pure aluminum base powder, were placed in 3.2 cm inside diameter steel tubes with one end containing a welded plug, as previously shown in Figure 5. The aluminum powder was first added to the tube and filled to accommodate a very close fitting steel mandrel which was inserted into the tube with a 4536 kg force to produce a green compact of ~70% density for a 5.08 cm aluminum test cylinder. To this initial compacted aluminum base powder the MWCNT aggregate powder/aluminum powder mixtures were added to a calculated height to produce a 5.08 cm test cylinder when compacted to ~70% density. These Al/2 and 5% MWCNT were alternated along with the pure aluminum powder base as compacted ~75%

and other two-phase powder mixtures to create a series of 6 compacted-powder sections within the steel tube. The open end was then sealed with a welded steel plug and this steel-enclosed assembly was then inserted into a wooden base and surrounded by a 15.25 cm diameter PVC tube 7.6 cm taller than the steel test cylinder, also shown in Figure 5. This PVC container was filled with ammonium nitrate-fuel oil (ANFO) and a thin sheet of detasheet with a central detonator added to the top of the ANFO-filled PVC tube to initiate the reaction. The reaction is given by

$$P=(\rho_o D^2)/2 \tag{1}$$

using an ANFO density (ρ_o) of ~1.2 g/cc and detonation velocity (D) ~3500 m/s the initial pressure was calculated (as described by Meyers and Wang, 1988) to be ~7 GPa. These arrangements or assemblies were placed on sand bags, which allowed the explosively consolidated steel tubes to be easily retrieved from the ground after detonation on the facilities of the New Mexico Tech EMRTC (Energetic Materials Research & Testing Center) installations. The recovered TPS characterization was performed on machined sections from the compacted recovery tubes. To help preserve the integrity of the microstructure of the materials a wire electric discharge machining (WEDM) was employed to cut segments for the preparation of light microscopy, SEM, (and FESEM), TEM, and hardness testing. The WEDM used is a 5 movement axis RoboFil 310 wire-electrical discharge machine (Charmilles Technologies) with a CuZn25 electrode wire in a diameter of 0.25 mm that proves to be a reliable technology for the purpose of cutting test specimens and samples. Along with the sections that contained the carbon nanotubes second phase, the consolidated aluminum was of interest as reference or control sample. Sections to be used for light metallography and SEM imaging were first ground using grit papers from 400 grit SiC down to 1200 grit SiC (using an Ecomet 6- Buehler variable speed grinder- polisher at 200 rpm), and then polished using a non-crystallizing colloidal silica suspension to a 1 micron finish, with a soap and water mixture (1:200) as the primary lubricant. For etching, a variation of Keller's reagent was used with a composition of 100 mL of water, 6 mL of nitric acid, 6 mL of hydrochloric acid, and 6 mL of hydrofluoric acid. The imaging was performed in a Reichert MEF4 A/M (Leica, Corp) light microscope and the same samples were used for SEM and field-emission SEM (FESEM, Hitachi S-4800) imaging. The TEM specimens were prepared from machined sections of the Al/MWCNT systems along with the aluminum monolith that were reduced by grinding to a thickness less than 0.5 mm, then 3 mm discs were punched from these thin-slice sections which were perpendicular to the compaction tube vertical directions (Figure 5). Electropolishing was done with a Tenupol-5 dual jet electropolisher using an electrolyte with a composition of 250 mL of nitric acid and 850 mL of methanol at -15°C. TEM imaging was performed using a Hitachi H-8000 analytical transmission electron microscope operating at 200 kV accelerating potential. An INSTRON Rockwell hardness tester, 2000 series was used to take Rockwell (E-scale) hardness reading profiles of the obtained systems, and the consolidated aluminum, for comparison. Vickers hardness measurements were also made using a Shimadzu microhardness tester employing a 25gf (0.25 N) load. Micro-tensile testing was performed using samples cut parallel to the consolidation axis and proportional to 60% of the D-638, type V ASTM standard specifications (with a gauge length of 9 mm) and an INSTRON tensile tester, (5866 series) at 0.21 mm/s. The initial materials were characterized prior to compaction and consolidation.

The primary phase (Al) has a nominal size of 100 μm with a size range (particle distribution) from the submicron to ~175, a nominal density of 2.699 g/cm^3 @ RT and EDX confirmed it was Al-1100 aluminum (Figure 6c). The powder morphology consisted of rounded and irregular granules with microdendritic structures, resulting from the vendors' powder processing, as shown in the SEM images in Figures 6a and b. Figure 7a shows a TEM image of the MWCNT aggregate mixture measuring 30 to 40 nm. The shapes were a mixture of fullerenes and naturally short MWCNTs (SLA Tubes). EDS results found the powder to be mainly carbon (Figure 7b).

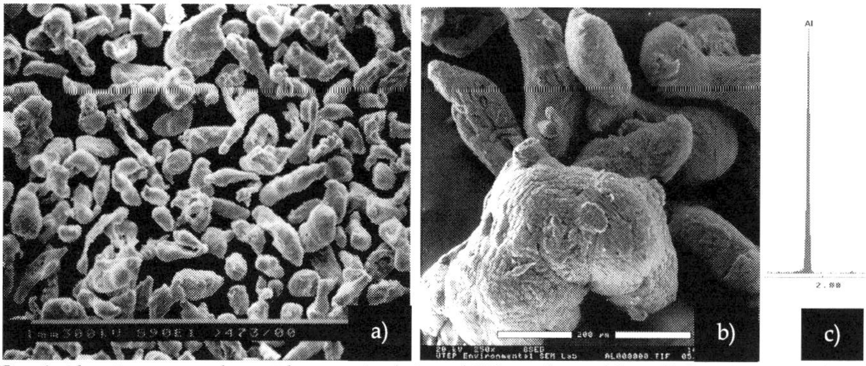

Fig. 6. Aluminum powder, with a nominal size of 150 mesh, at a) low (light microscopy) and b) high magnification (SEM) micrographs and c) EDS spectrum.

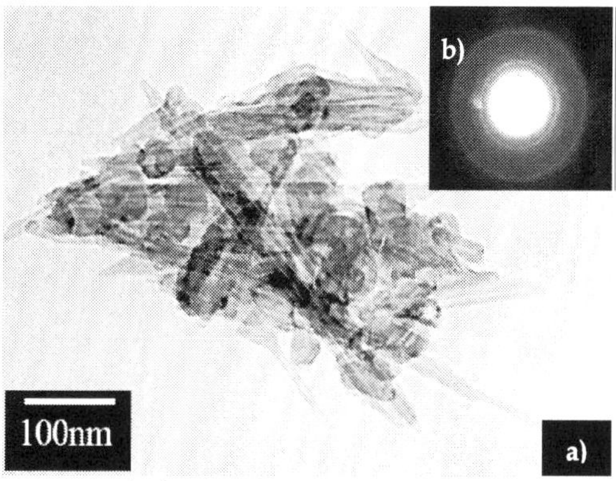

Fig. 7. Original starting material of MWCNT aggregate mixture.

Images in Figure 8 illustrate that the starting Al particles had a micro-dendritic microstructure (Figure 8a) resulting from the powder processing. Figure 8b shows for comparison a light micrograph of a section view where the micro-dendritic structure is observed to be preserved both for the shock wave consolidated (SWC) aluminum monolith and for the consolidated Al/5%MWCNT system (Figure 9b). Figure 8b also shows the high degree of compaction achieved in the aluminum single-phase during consolidation as evidenced by the absence of voids in the microstructure especially at triple grain points. This is confirmed by the achieved density of ~98% (by Archimedean method), which is consistent with a SWC process

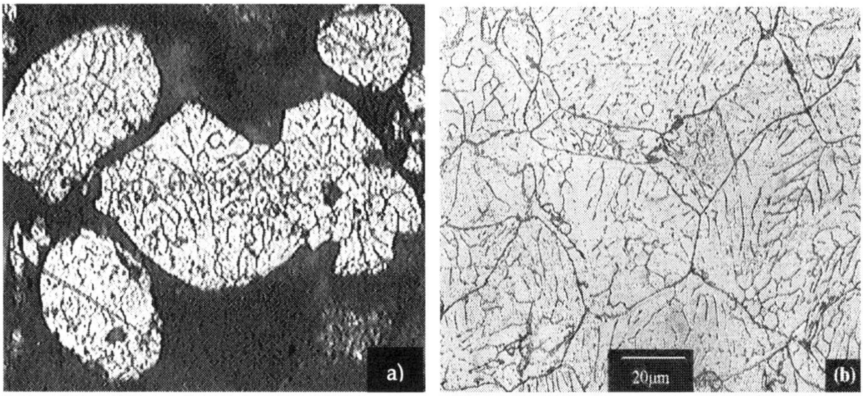

Fig. 8. (a) Light microscope image of Al-powder microstructure showing microdendritic structure and (b) light microscope image of consolidated Al phase.

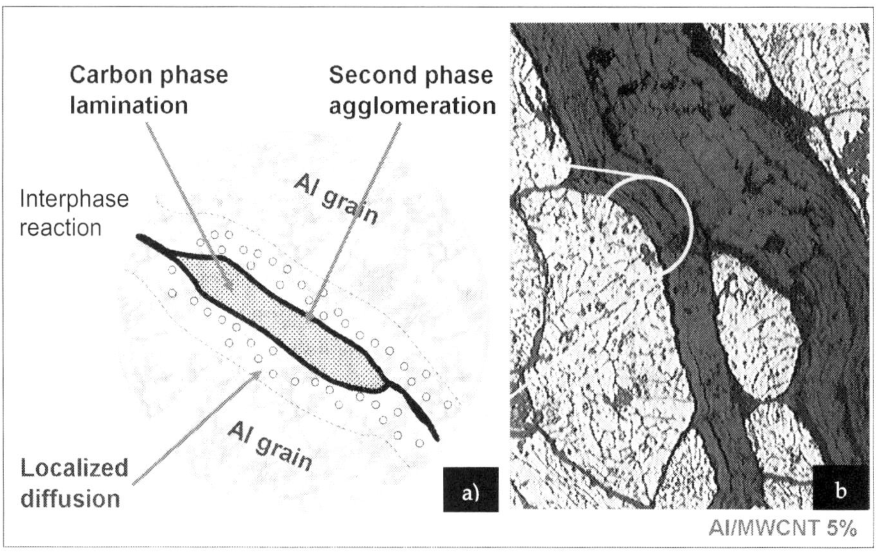

Fig. 9. a) Representation of different interactions between phases occurring during the fabrication of a metal/CNT systems and b) an image of an Al/5%MWCNT sample.

3.1 Hardness measurements

Characteristics of optimum explosive compaction strengthening include a marked increase in hardness that is mostly uniform over the cross section of the consolidated materials. As described later, the formation of agglomerates along with their distribution in the continuous phase and their size difference as compared to the continuous phase aluminum grains is expected to result in hardness variations between the consolidated systems. However the increase in hardness is mainly associated with the SWC of the aluminum continuous phase. In contrast to the initial Al powder Vickers hardness of HV 24 (HRE 22, by conversion), the explosively consolidated aluminum samples exhibit a hardness of HV 43 (HRE 40), which represents a 79% increase in hardness. The average hardness contribution from the two-phase regions (characteristic of Rockwell E scale) exhibits the ability of aluminum to shock-harden as illustrated in the TEM images of Figure 10, where it is shown the shock-induced dislocation substructures along with dynamically recrystallized regimes. There was a decrease in hardness associated with SWC of the two-phase systems as compared to the SWC of the aluminum. The Al/2%MWCNT sample had a hardness decrease to HRE 39. Increasing the volume fraction of CNT aggregates further to Al/5%MWCNT significantly lowered the hardness reading to HRE 33. This represents a decrease of ~18% from the consolidated aluminum (HRE 40). This trend in hardness reduction is consistent with Feng et al (2005), who concluded that the agglomeration of carbon nanotubes at an increasing volume fraction leads to a decrease in hardness and will eventually lead to failure under an applied load (decrease in yield strength). Feng et al (2005) also describes two weaknesses that can be found in an Al/MWCNT system: weak bonding of the agglomerate to the Al phase and also a weak bonding between the nanotube materials. It is this weak bonding that causes such a decrease in hardness. Consolidated aluminum samples showed a 0.2% offset yield for the Al of 120 MPa with a UTS of 140 MPa and an elongation of 6.6%. When comparing this data to the nominal Al-1100 data (for 1.6 mm samples) (ASM Handbook, 1990) there is an increase of ~28% in strength and a 45% reduction in the Al-base plasticity (from 110 MPa and 12% respectively). The Al/2%MWCNT aggregate TPS failed at an elongation of ~2% which is somewhat commensurate with results for two-phase Al-A356/20% fly ash (volume) where elongation was observed to be between 1 and 2% (Withers, 2005).

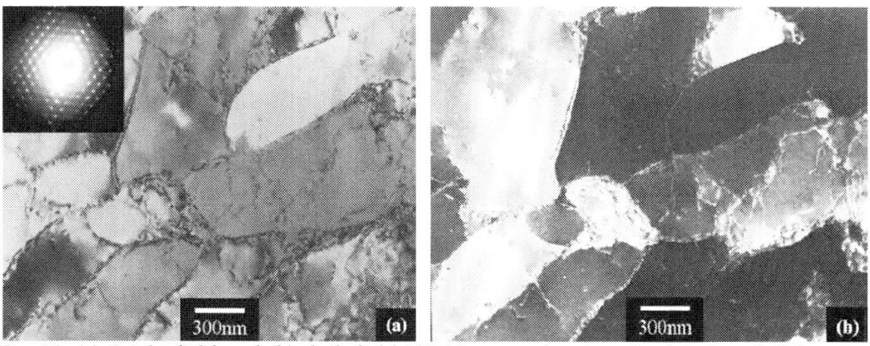

Fig. 10. (a) Bright-field and (b) dark-field TEM images showing shock-induced dislocation substructures along with dynamically recrystallized regimes.

4. CNT second phase behaviour

In addition to the achieved mechanical properties, images of the obtained samples can illustrate a set of resultant possibilities on the distribution of the second phase (refer to Figure 9a) achieved in a consolidated metal/CNT and, more specific in our case, Al/CNTs systems. As described by Salas et al (2007) the second phase can consolidate in micron-sized agglomerates (reported by Lijie et al, 2006; Laha et al, 2009; and Lim et al, 2009) and also as CNTs arranged in lamellae agglomerates (as later reported by Lim et al, 2009) or entangled clusters (Laha et al, 2009; and Lim et al, 2009). The Al-CNTs interphase can be dispersed in a connected or disrupted network (also reported by Kim et al, 2006), can have a non-reacted localized diffusion at the Al grain surface (as recently reported by Laha et al, 2009) and/or can react totally/partially at the Al-Al grain boundary forming Al_4C_3 (Lijie et al, 2006; and Alba-Baena et al, 2008).

4.1 Micron-sized agglomerates

A simplified model for the microstructure of a TPS consists of two parts, i.e. an Al/agglomerated MWCNT aggregate region and an aggregate mixture free or continuous phase region. CNTs tend to agglomerate at the primary-phase grain boundaries as illustrated in Figure 11. Figure 11a also shows large agglomerates characteristic of a high volume fraction (Al/5%MWCNT) two-phase system. The microstructure shown in Figure 11 demonstrates the high degree of compaction achieved in the aluminum phase. This is consistent with SWC process as evidenced by the absence of voids in the microstructure, especially at triple grain points. Images in Figure 11a and b reveal agglomerates located mainly at the aluminum grain boundaries and are not homogeneously distributed in the Al continuous phase but rather the carbon nanotube material is densely distributed in localized regions along the continuous phase. Figure 11b shows a higher magnification view of the agglomerations of MWCNT aggregate particles and the pore-free surface of the Al phase. Along with those observations for the aluminum monolith, i.e. low porosity and high densification, the SWC effect and plastic flow results in the accommodation of the agglomerated MWCNT aggregate mixture into the aluminum-phase grain that preserves its structure intact (shown previously by Figure 8b).

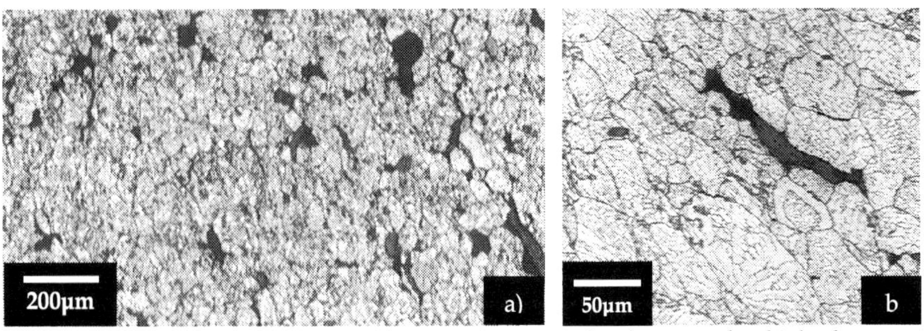

Fig. 11. a) Microstructure of an Al/5%MWCNT sample showing the high degree of compaction achieved and b) a magnified view showing the intact aluminum-grain interior but plastically deformed to accommodate the MWCNT aggregates.

4.2 Lamellae structures

CNT phase agglomerations can exhibit two types of structures: entangled clusters (described later) and laminar type arrangements (lamellae structures). This interesting laminar feature is shown in Figure 12 where the SEM view (Figure 12a) of a polished surface section (of an Al/2%MWCNT system) exhibits the laminated flow marked by the gaps separating the consolidated regions. The bonding between the MWCNT aggregate phase and the Al appears to be optimum at the arrow.

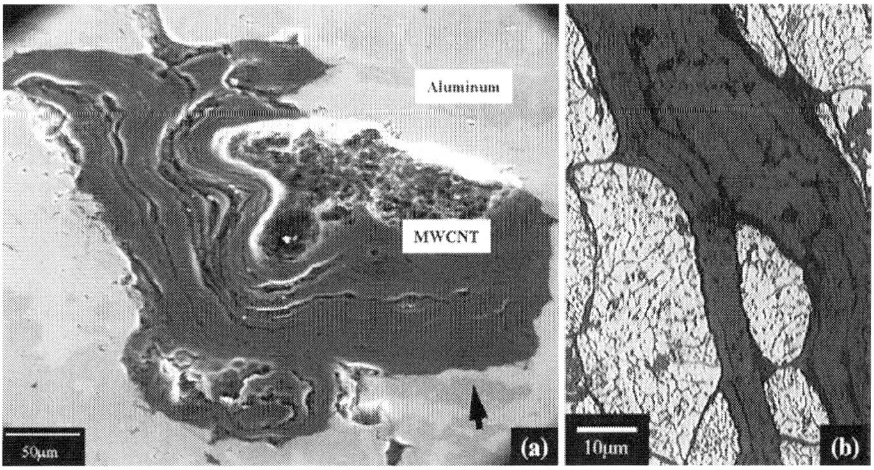

Fig. 12. a) SEM image showing a lamellae structure of a MWCNT agglomeration with bonding between phases (as indicated by arrow) and b) Light microscopy detail of the carbonaceous-laminated feature.

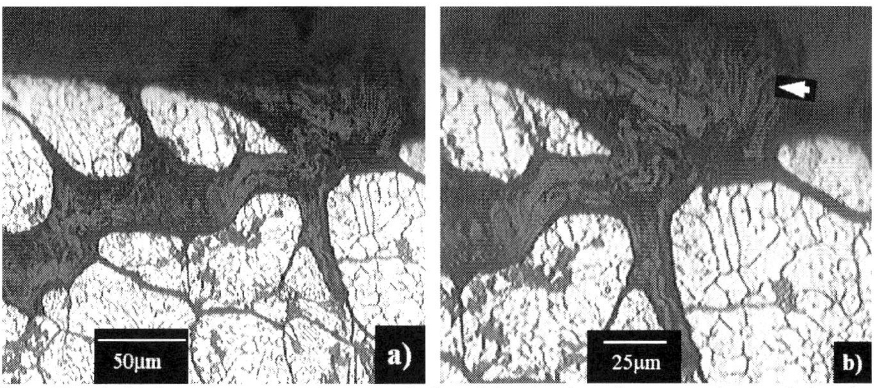

Fig. 13. a) Light microscopy image of the delaminating features observed in the carbonaceous phase on a sample edge b) arrow shows detail of the delaminating feature.

The apparent fully consolidated laminar structure, shown in Figure 12b, is resultant of the compression of such layers by the SWC process or by other highly energetic processes such as friction stir processing (Lim et al, 2009). These carbonaceous-layer regions have been reported in recent literature but not characterized. However, it is believed that such laminated regions are comprised of multi concentric tubes and fullerenes and are compacted together by the SWC process along the basal planes which experience weak Van der Waals forces. It is the attribution of these weak Van der Waals forces between layers (Figure 13) that can be a starting point to the the decohesion and delamination shown by the arrow in Figure 13b. Such delaminating features observed in the carbonaceous phase and shown in Figure 13 have similar characteistics to that observed by Peikrishvili et al (2001) during the explosive compaction of Ni and graphite powders.

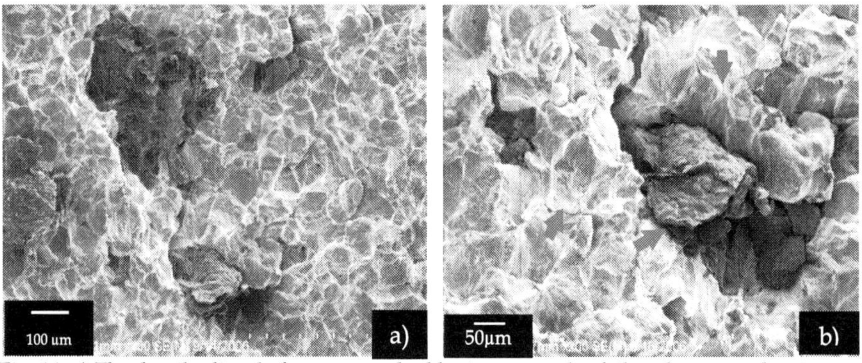

Fig. 14. a) The ductile-dimple fracture marked by intergranular failure between aluminum grains and micron-sized agglomerates of second phase and b) fracture surface that demonstrate some of the fracture modes exhibited by this TPS.

Figure 14 shows the second phase (aggregate) accommodation along the grain boundaries of the continuous aluminum phase which is observed and expected in a SWC two phase system. As known, the aggregates size variation can be the result of the mixing process and the tendency of the MWCNTs and multi-concentric fullerenes to agglomerate. Figure 14a also shows the ductile-dimple fracture characteristic of the aluminum continuous phase as noted by the intergranular failure between aluminum grains and micron-sized lamellae agglomerates that have been pulled apart (noted in Figure 14b). Figure 14b is an SEM image of a fracture surface that demonstrates some of the fracture modes exhibited by this TPS. The arrow at the top indicates the ductile behavior characteristic of the aluminum continuous phase. Intergranular particle cracking can be observed between aluminum grains in the continuous phase shown by the arrow to the left in the image. This intergranular particle failure is also exhibited between the aluminum grains and the agglomerated second phase as indicated by the arrow at the bottom of the Figure. A transgranular fracture mode is demonstrated by the region marked by the arrow to the right in the image. Here the agglomerate particle pulled apart from itself in a delayering or decohesion fashion and is pointed out by the arrow at the center of the image. Figure 15a is further detail of the layering agglomerate where the delayering mechanism after fracture has been previously described in Figure 13.

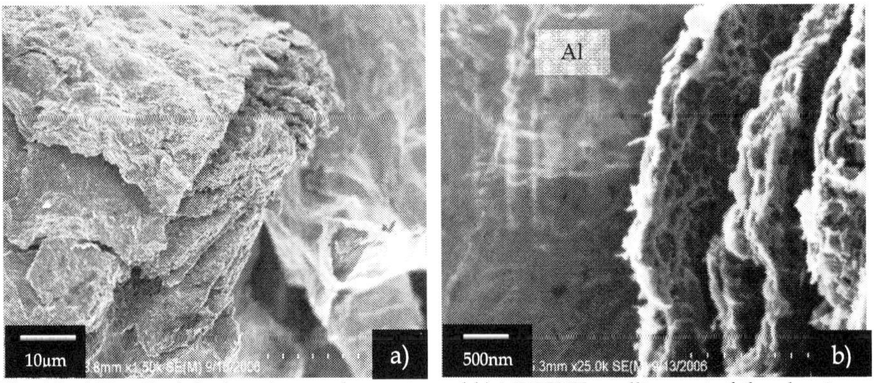

Fig. 15. a) Detail of the layering agglomerate and b) MWCNTs pulling out of the aluminum phase leaving characteristic holes.

The laminar appearance of the consolidated graphitic phase suggests that of pyrolytic graphite, especially in Figure 15b which shows details of the laminar structure and delamination of the carbonaceous arrangement. Also the Al-C interphase interaction is observed (arrow, Figure 15b) where the MWCNTs are seen emanating from the C-phase surface and MWCNTs are pulled out of the aluminum phase material and leaving characteristic holes in the surface. Finally, Figure 16 exemplifies the Al/laminar-agglomerate debonding showing a top layer of thin layered agglomerate lying on top of grain boundaries and exhibiting ductile dimple features (arrow) on the carbonaceous material which corresponds to that observed in Figure 15b.

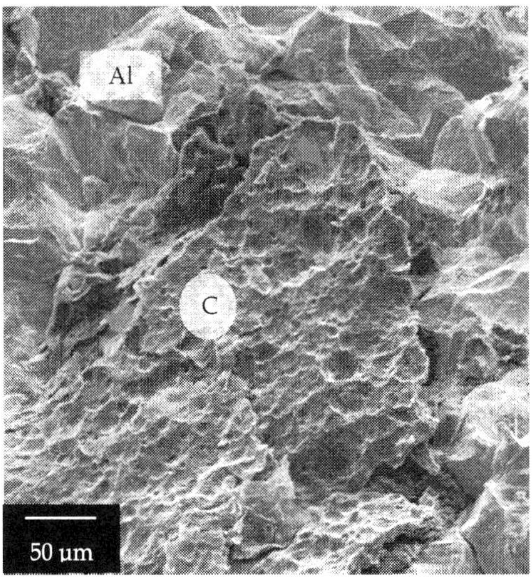

Fig. 16. Example of dimple features (arrow) of the carbonaceous material.

4.3 Entangled clusters

The agglomerate shown in Figure 17a reveals more on the nature of the entangled clustering behavior observed in between the laminar arrangements and smaller agglomerated regions. Here CNTs debond easily from the carbonaceous material. Figure 4.17b shows an enlarged view of this entangled carbonaceous material disaggregating from the Al surface (arrow). This also illustrates poor Al-C bonding that may be responsible for the reduction in elongation of this TPS's tensile samples along with the poor tensile strength.

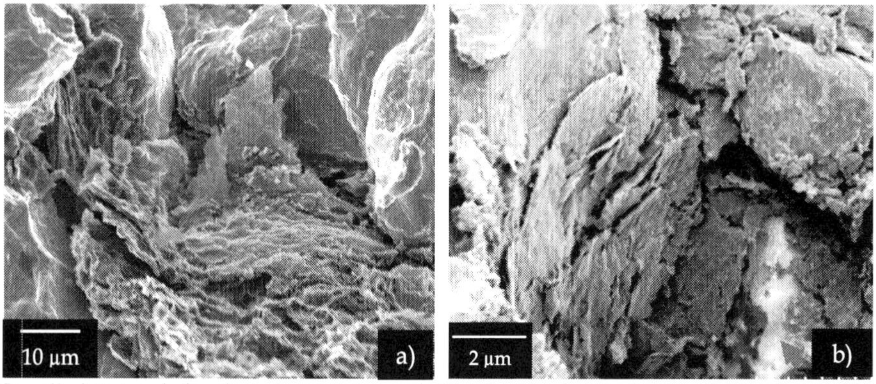

Fig. 17. a) Entangled carbon cluster surrounded by ductile-dimple fractured Al grains and b) detail view of a cluster sitting at an Al grain surface.

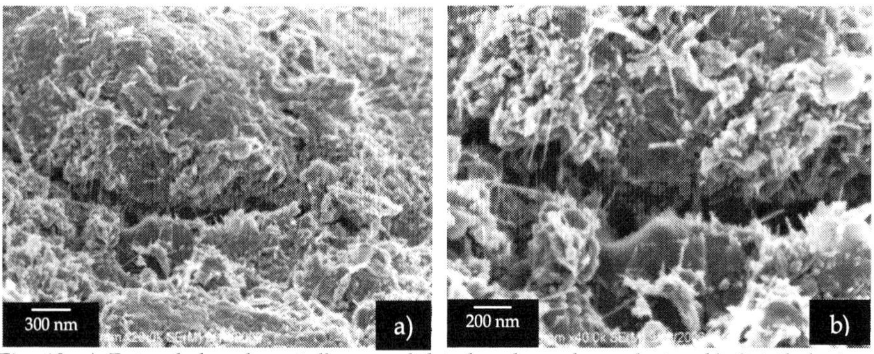

Fig. 18. a) Entangled and partially consolidated carbon phase cluster. b) Detailed view showing undamaged and randomly dispersed carbon nanotubes along the fractured surface.

An entangled cluster characterization is shown in the sequence of images shown by Figures 17-19. Figure 18a is a closer look at an entangled cluster showing different carbon phases and unconsolidated carbon nanotubes lying in between the continuous metal phase. Figure 18b is a detailed view of the cluster where the partially consolidated carbon phase is clustering carbon nanotubes with other carbonaceous phases from the original material and is a result of the SWC process. As the magnification is increased in Figure 19a the fractured graphitic phase begins to resemble features of the original MWCNT aggregate material shown in Figure 7b. The MWCNTs and individual, multiconcentric fullerenic particles and graphitic particles are clearly visible in Figure 19b as characteristic of these Al/agglomerated MWCNT aggregate clusters.

Fig. 19. Sequence showing the consolidated carbonaceous clusters along with entangled MWCNTs.

4.4 Interface diffusion

Smaller agglomerations (Figure 20) can be found in the Al-Al interphase of the TPS. These CNT agglomerations are shown as flaky unconsolidated carbonaceous particle material. Also observed are individual CNTs distributed along the Al grains interface. An interesting observation is shown in the magnified view of Figures 20a and b which are of an aluminum grain boundary that has experienced intergranular fracture. At the crevice separating these two aluminum grains we can see individual CNTs protruding out (see arrow), also there is no evidence that the nanotube had been embedded into the continuous phase as described for the inter-type of TPSs (shown previously in Figure 1).

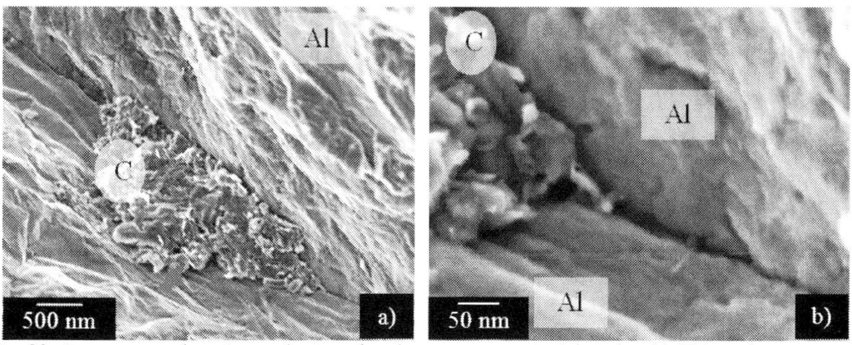

Fig. 20. A sequence showing a cluster of carbonaceous material at the Al-Al interface.

4.5 Localized diffusion

Figure 21a is a TEM micrograph for the consolidated Al/5%MWCNT TPS which shows (marked by the right arrow) that there appears to be evidence of MWCNTs embedded into the aluminum continuous phase. This embedment indicates good intra-facial bonding between the two phases that is beneficial for load transfer. These observations spur further investigation to the nature of the bonding between the MWCNTs and the aluminum continuous phase because it is an indication that there is a retention of CNTs in the continuous phase. Also in this TEM photo, aggregate material appears to lie in between the grain boundaries (left arrow) as evidence of the intergranular bonding of CNTs. This bonding also enhances the load transfer between phases and grains (refer to Figure 2). These features are also observed in Figure 21b which clearly shows remnants of etched Al grains (arrow) intermixed with the MWCNT aggregate material which is representative of a two-phase system and the agglomeration of the second phase. The edge of the thin area to the right in Figure 21b exhibits some apparent carbon/fullerene nanoparticle consolidation, but the nature of this consolidation is unclear. The SAED pattern insert is also consistent with overlapping, etched Al grains or particles and carbonaceous material.

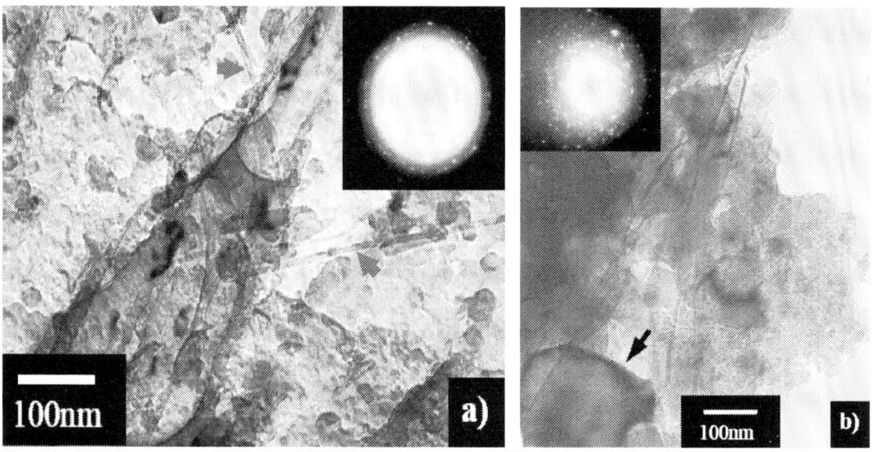

Fig. 21. TEM images of the 5%MWCNT aggregate TPS showing the CNT diffusion in and between the Al grains.

Images in Figures 22 and 23 are of a microscopic view which shows another detail in the aluminum continuous phase surface. After the SWC, the carbon diffusion at the interface is shown as having experienced partial consolidation and being accompanied by agglomerations of the carbon phase. Figure 22 illustrates the localized dispersion of CNTs along the continuous phase resultant from the SWC process; a ~1.5 μm agglomeration at the Al grain surface did not consolidate in the described laminar structure but dispersed and the CNTs that pulled out from the aluminum phase surface are shown (arrows). The fractured sample in Figure 23 illustrates how the reduction of the agglomeration sizes (below 10 μm in diameter) may improve the two-phase system's mechanical response while preserving the carbonaceous phase properties. With the use of smaller second-phase volume fractions and the reduction of the agglomeration sizes (or its elimination), improvements in the dispersion of the MWCNT phase are expected. The experimentation of these changes in the improvements in the consolidation of a TPS may lead to the consolidation of MMCs characterized by the matrix grain-size control and the MWCNT will promote the primary phase and /or TPS mechanical responses.

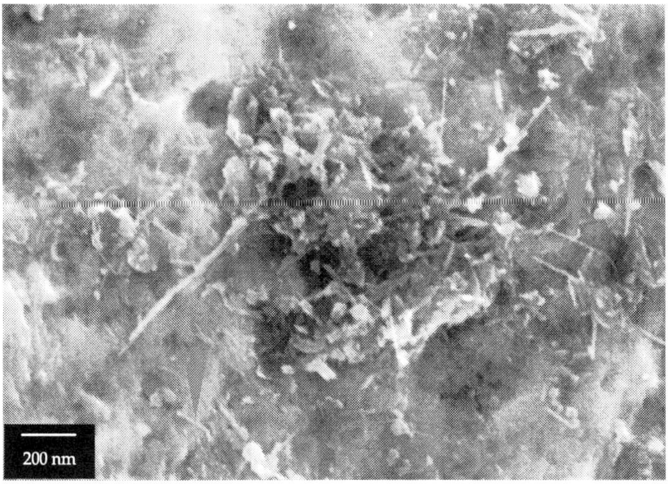

Fig. 22. Magnified image of the carbon nanotubes which can be seen to lie undamaged between grains.

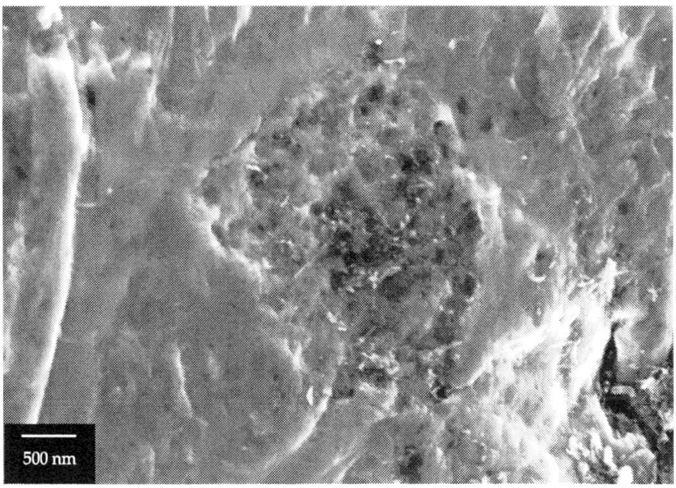

Fig. 23. An agglomerate appears to have fallen out of the continuous phase due to the inconsistency of the aggregate bonding.

4.6 Al-C consolidation

Figure 24 exemplifies the consolidation of the Al-C TPS exhibiting good bonding between the aluminum continuous phase and the second phase aggregate mixture. The carbonaceous material, indicated by the arrow on the left, has bonded to the ductile-continuous-aluminum phase. This image also shows several areas of strong bonding marked by arrows. The arrows on the right illustrates an area where the bond strength is strong enough that here failure has occurred as a result of ductile failure indicated by regions above and below the bonded carbon phase material.

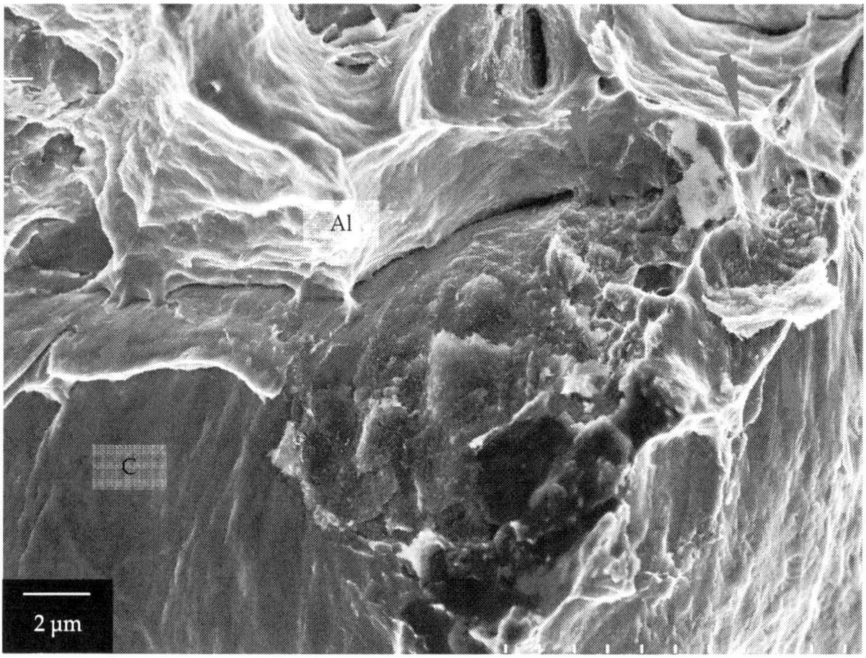

Fig. 24. Further evidence of the degree of bonding between the continuous phase and the second phase showing several areas of strong bonding marked with arrows.

4.7 Al-C Reaction

During observations on Al/MWCNT systems there was no Al-C reaction, but this can occur due to the Al plastic flow and CNT reactions through the SWC; as previously described by Alba-Baena et al (2008) for Al-SiC systems. They report nanoreaction results as hexagonal platelet-shapes (Al_3C_4) and spherical Si particles distributed at the aluminum surface. Figure 25 illustrates the observed chemical reaction products from an Al/21%SiCnp system forming a layer at the aluminum phase surface and a detail view of the Al_3C_4 platelets formed after SWC process, but the details of these reactions are still unknown.

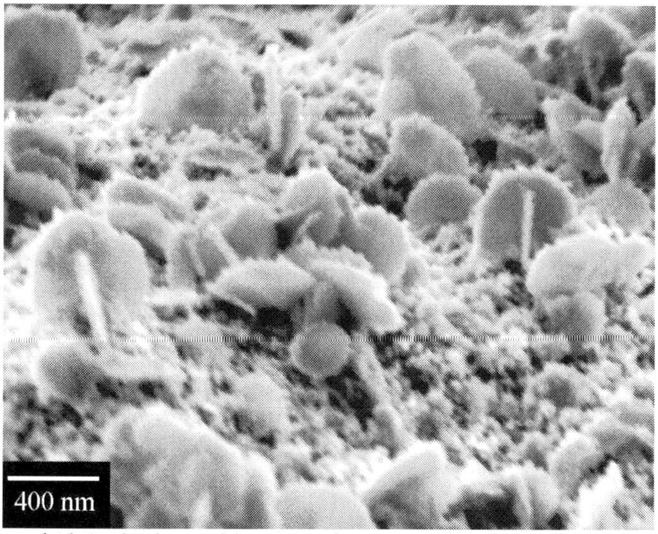

Fig. 25. Image of Al_3C_4 platelets at Al grain surface in an Al/SiC TPS (from Alba-Baena et al, 2008).

5. Conclusion

The ability in incorporating carbon nanotubes to structural and functional TPS will lead to the improvement and tailoring in the strength-to-weight ratios on materials for a wide variety of industries. This chapter has outlined further micro-nanosystems knowledge about preparing materials and using SWC for future developments. Al-MWCNT aggregate two-phase composites of 2 and 5% volume fraction were fabricated using a single tube shockwave consolidation process. This reports evidence on the achievement of aluminum-based two-phase systems describing different Al/CNTs interactions. Particularly notable is where CNT agglomerates of these two-phase systems have been shock consolidated into a contiguous phase region, and this regime can be bonded, monolithically, to the consolidated aluminum particle regime. The carbonaceous, MWCNT aggregate phase can exhibit a shock-induced, laminated flow-type, consolidation feature that appears to spread throughout the primary phase (Al) grain boundaries and consolidates as large phase regions at the aluminum triple points. At a fractured sample the MWCNT aggregate phase shown the entangled agglomerates and surface dispersion along Al grains, observed when MWCNTs pull out of the aluminum phase at fracture as well. Present descriptions and observations are in agreement with other studies that have determined that the reinforcement efficiency on the mechanical properties is dependent on dispersion, volume fraction and interfacial strength.

6. References

Ajayan, P. M.; Schadler, L. S. and Braun P. V. (2004). *Nanocomposite Science and Technology*, Wiley & sons, ISBN: 3-527-30359-6.

Akbulut, H.; Durman, M. and Yilmaz, F. (1998). *Dry wear and friction properties of δ-Al_2O_3 short fiber reinforced Al-Si (LM 13) alloy metal matrix composites*, Wear, vol. 215, No. 1-2, pp.170-179, ISSN: 0043-1648.

Alba-Baena, N. G.; Salas, W. and Murr, L. E, (2008). *Characterization of micro and nano two-phase regimes created by explosive shock-wave consolidation of powder mixtures*, Materials Characterization, vol. 59, pp. 1152-1160, ISSN: 1044-5803.

ASM Handbook, (1990). *Properties and selection: Nonferrous alloys and special purpose materials*, ASM International, pp. 3-215, ISBN 10: 0-87170-378-5.

Bakshi, S. R.; Singh, V.; Balani, K.; McCartney D. G.; Seal, S. and Agarwal, A. (2008). Carbon *nanotube reinforced aluminum composite coating via cold spraying*, Surface & Coatings Technology, vol. 202, pp. 5162–5169, ISSN: 0257-8972.

Cha, S. I.; Kim, K. T.; Arshad, S. N.; Mo, C. B. and Hong, S. H. (2005). *Extraordinary Strengthening Effect of Carbon Nanotubes in Metal-Matrix Nanocomposites Processed by Molecular-Level Mixing*, Advanced Materials, vol. 17, 11, pp. 1377-1381, ISSN: 0935-9648.

Chen, W.; Tu, J.; Wang, L.; Gan, H.; Xu, Z.; and Zhang X. (2003). *Tribological application of carbon nanotubes in a metal-based composite coating and composites*, Carbon 41, pp. 215–222, ISSN: 0008-6223.

Chen, X. Xia, J. Peng, J. Li, W.; Xie, S. (2000). *Carbon-nanotube metal-matrix composites prepared by electroless plating*, Composites Science and Technology, vol. 60, 2, pp. 301–306, ISSN: 0266-3538.

Chen, T.; Hampikian, J. M.; and Thadhani, N. N. (1999). *Synthesis and characterization of mechanically alloyed and shock-consolidated nanocrystalline NiAl intermetallic*, Acta materialia, vol. 47, 8, pp. 2567-2579, ISSN: 1359-6454.

Collins, Ch.; Thadhani, N. N,; and Iqbal, Z. (2001). *Shock-compression of C–N precursors for possible synthesis of β-C_3N_4*, Carbon, vol. 39, pp. 1175–1181, ISSN: 0008-6223.

Counihan, P. J.; Crawford, A. and Thadhani N. N. (1999). *Influence of dynamic densification on nanostructure formation in Ti_5Si_3 intermetallic alloy and its bulk properties*, Materials Science and Engineering A, vol. 267, pp. 26–35, ISSN: 0921-5093.

Csanady, A.; Sajo, I.; Labar, J. L.; Szalay, A.; Papp, K.; Balaton, G. and Kalman, E. (2006). *Al–Pb nanocomposites made by mechanical alloying and consolidation*, Current Applied Physics, vol. 6, pp. 131–134, ISSN: 1567-1739.

Deribas, A. A.; Simonov, P. A.; Filimonenko, V. N. and Shtertser A. A. (2001). *Chap. 42: Bulk samples of intermetallics, obtained by shock compaction*, Fundamental Issues and Applications of Shock-Wave and High-Strain Rate Phenomena, K.P. Staudhammer, L.E. Murr, and M. A. Meyers, (Eds.), Elsevier Science Ltd., Amsterdam, pp. 331-336, ISBN: 008043896-2.

Dresselhaus, M. S.; Dresselhaus G. and Avouris P. (2008). *Carbon nanotubes: synthesis, structure, properties and applications*, Springer, ISBN: 978-3-540-72864-1.

Drost, H.; Friedrich, M.; Mohr, R. and Gey, E. (1997). *Nanoscaled Si-C-N-composite powders with different structures by shock-wave pyrolysis of organic precursors*, Nuclear Instruments and Methods in Physics Research B, vol. 112, pp. 598-601, ISSN: 0168-583X.

Epanchintsev, O. G.; Zubchenko, A. S.; Korneyev, A. E. and Simonov, V. A. (1997). *Highly-efficient shock-wave diamond synthesis from fullerenes*, Journal of physics and chemistry of solids, vol. 58, 11, pp. 1785-1788, ISSN: 0022-3697.

Everett, R. K. and Arsenault, R. J. (1991). *Metal Matrix composites: processing and interfaces*, Academic press LTD, Elsevier Science & Technology Books, ISBN: 012341832-1.

Feng, Y.; Long Yuan, H.; Zhang, M. (2005). *Fabrication and properties of silver-matrix composites reinforced by carbon nanotubes*, Materials Characterization, vol. 55, pp. 211-218, ISSN: 1044-5803.

Glade, S. C. and Thadani, N. N. (1995). *Shock consolidation of mechanically alloyed amorphous Ti-Si powders*, Metallurgical and Material Transactions A, vol. 261, p. 2565, ISSN: 1073-5623.

Jin, Z. Q.; Chen, K. H.; Li, J.; Zeng, H.; Cheng, S. F.; Liu, J. P.; Wang, Z. L. and Thadhani, N. N. (2004). *Shock compression response of magnetic nanocomposite powders*, Acta Materialia, vol. 52, pp. 2147–2154, ISSN: 1359-6454.

Jordan, J. L. and Thadhani, N. N. (2001). *Chap. 39: Synthesis of Ti-based metal-like ternary ceramic compounds by dynamic densification and reaction synthesis* in Fundamental Issues and Applications of Shock-Wave and High-Strain Rate Phenomena, K. P. Staudhammer, L. E. Murr, and M. A. Meyers, (Eds.), Elsevier Science Ltd., Amsterdam, pp. 305-312, ISBN: 008043896-2.

Kawaka, A.; Soma, T. and Saito, S. (1974). *Structure determination of boron nitride transformed by shock compression*. Japanese Journal of Applied Physics, vol. 13, 5, p. 891, ISSN: 0021-4922.

Kennedy, G.; Keller, A.; Russell, R.; Ferranti, L.; Zhai, J.; Zhuo, M. and Thadhani. N. N. (2001). *Chap. 9: Dynamic mechanical properties of microstructurally-biased two-phase TiB_2+Al_2O_3 ceramics* in Fundamental Issues and Applications of Shock-Wave and High-Strain Rate Phenomena, K. P. Staudhammer, L. E. Murr, and M. A. Meyers, (Eds.), Elsevier Science Ltd., Amsterdam, pp. 63-70, ISBN: 008043896-2.

Kimura, Y. (1963). *Formation of zinc ferrite by explosive compression*, Japanese Journal of Applied Physics, vol. 2, p. 312, ISSN: 0021-4922.

Koráb, J.; Štefánik, P.; Kavecký, Š.; Šebo P. and Korb, G. (2002). *Thermal conductivity of unidirectional copper matrix carbon fibre composites*, Composites Part A: Applied Science and Manufacturing, vol. 33, 4, 2002, pp. 577-581, ISSN: 1359-835X.

Kowbel, W. (2005). *Hybrid Al/SiC Composite Optics for IFE Applications*, Fusion Science and Technology, vol. 47, April 2005, pp. 596-600, ISSN: 1536-1055.

Kim, K. T.; Cha, S. I.; Hong, S. H.; Hong, So. H. (2006), *Microstructures and tensile behavior of carbon nanotube reinforced Cu matrix nanocomposites*, Materials Science and Engineering A, vol. 430, pp. 27–33, ISSN: 0921-5093.

Laha, T.; Chen, Y.; Lahiri, D. and Agarwal, A. (2009). *Tensile properties of carbon nanotube reinforced aluminum nanocomposite fabricated by plasma spray forming*, *Composites Part A: Applied Science and Manufacturing*, vol. 40, 5, pp. 589-594, ISSN: 1359-835X.

Lan, J.; Yang, Y. and Li, X. (2004). *Microstructure and microhardness of SiC nanoparticles reinforced magnesium composites fabricated by ultrasonic method*, Materials Science and Engineering A, vol. 386, pp. 284–290, ISSN: 0921-5093.

Lijie, C.; Zhenyu, R.; Jin-Phillipp N. Y. and Ruhle, M. (2006). *Investigation of the interfacial reaction between multi-walled carbon nanotubes and aluminum*, Acta Materialia, vol. 54, 20, pp. 5367-5375, ISSN: 1359-6454.

Lim, D. K.; Shibayanag, T. and Gerlich, A. P. (2009). *Synthesis of multi-walled CNT reinforced aluminium alloy composite via friction stir processing,* Materials Science and Engineering A, vol. 507, pp. 194–199, ISSN: 0921-5093.

Mamalis, A. G.; Vottea, I.; Manolakos, D. E.; Salay, A. and Desgardin G. (2001). *Chap. 37: On the Numerical Simulation of Shock Compacted mMetal Sheathed High-Tc superconducting billets* in Fundamental Issues and Applications of Shock-Wave and High-Strain Rate Phenomena, K. P. Staudhammer, L. E. Murr, and M. A. Meyers, (Eds.), Elsevier Science Ltd., Amsterdam pp. 289-295, ISBN: 008043896-2.

Meyers, M. A.; Mishra, A. and Benson, D. J. (2006). *Mechanical properties of nanocrystalline materials,* Progress in Materials Science, vol. 51, pp. 427–556, ISSN: 0079-6425.

Meyers, M. A. and Wang, S. L., (1988). *An Improved Method for Shock Consolidation of Powders,* Acta Materialia, vol. 36, 4, pp. 925-936, ISSN: 1359-6454.

Meyyappan, M. (Ed.), (2004). *Carbon Nanotubes: Science and Applications,* CRC Press, ISBN 0-84932-111-5.

Murr, Lawrence E. (Ed.), (1998). *Shock Waves for Industrial Applications,* Noyes Publishing, New York, pp. 237-472.

Oelhafen, P. and Schuller, A. (2005). *Nanostructured materials for solar energy conversion,* Solar Energy, vol. 79, pp. 110–121, ISSN: 0038-092X.

Peikrishvili, A. B.; Japaridze, L. A.; Staudhammer, K. P.; Marquis, F. S.; Chikhradze, N. M.; Gobejishvili, T. G. and Bantzuri, E. G. (2001). *Chap. 32: Expplosive compaction of clad graphite powders and obtainig of coatings on their base* in Fundamental Issues and Applicationsof Shock-Wave and High-Strain Phenomena, K. P. Staudhammer, L. E. Murr, and M. A. Meyers(eds.), Elsevier Ltd, Amsterdam, p. 249, ISBN: 008043896-2.

Probst, C.; Goujon, C., Gauvin, R. and Drew, R. A. L. (2005). *Enhanced wettability by copper electroless of coating of carbon nanotubes,* Advances in Ceramic Coatings and Ceramic-Metal Systems: A Collection of Papers Presented at the 29th International Conference on Advanced Ceramics and Composites, January 23-28, Cocoa Beach, Florida, Ceramic Engineering and Science Proceedings, vol. 26, 3, pp. 263-270, ISBN: 978-1-57498-233-6.

Prummer, R and Kochsier, D. (2001) *Chap. 44 Bulk samples of intermetallics, obtained by explosive compaction* in Fundamental Issues and Applications of Shock-Wave and High-Strain Rate Phenomena, K. P. Staudhammer, L. E. Murr, and M. A. Meyers, (Eds.), Elsevier Science Ltd., Amsterdam, 2001, pp. 345-349, ISBN: 008043896-2.

Prummer, R. A. and Ziegler, G. (1985). *Structure and annealing behavior of explosively compacted aluminum powders,* Powder Metallurgy International, vol. 9, 1, p. 1035, ISSN: 0048-5012.

Prummer R. (2001), *Chap. 30: State or the art of explosive compaction* in Staudhammer K.P., Murr L. E., Meyers M. A., Editors. Fundamental Issues and Applications of Shock-Wave and High-Strain Rate Phenomena. Amsterdam: Elsevier Science Ltd.; pp. 235 -244, ISBN: 008043896-2.

Raghukandan, K.; Hokamoto, K.; Lee, J. S.; Chiba, A. and Pai, B. C., (2003). *An investigation on underwater shock consolidated carbon fiber reinforced Al composites,* Journal of Materials Processing Technology, vol. 134, pp. 329-337, ISSN: 0924-0136.

Raming, T. P.; van Zyl, W. E.; Carton, E. P. and Verweij, H. (2004). *Sintering, sinterforging and explosive compaction to densify the dual phase nanocomposite system Y_2O_3- doped ZrO_2 and RuO_2,* Ceramics International, vol. 30, pp. 629–634, ISSN: 0272-8842.

Rao, B. S.; Hemambar, Ch.; Pathak, A. V.; Patel, K. J.; Rödel, J. and Jayaram, V. (2006). *Al/SiC carriers for microwave integrated circuits by a new technique of pressureless infiltration,* IEEE Transactions on Electronics Packaging Manufacturing, January 2006, vol. 29, 1, pp. 58-63, ISSN:1521-334X.

Rice, M. H.; Mcqueen, R. G. and Walsh, J. M. (1958). *Compression of solids by strong shock waves* in: F. Sectz, D. Turnball (Eds.), Solid State Physics, No. VI F, Academic Press, New York.

Rohatgi, P. K.; Schultz B. and Ferguson J. B., (2008), *Metal Matrix Nanocomposites for Structural Applications,* University of Wisconsin-Milwaukee Center for Composites, Milwaukee, WI.

Sahin Y. and Acilar M., (2003). *Production and properties of SiCp-reinforced aluminium alloy composites,* Composites: Part A, v. 34, pp. 709–718, Composites Part A: Applied Science and Manufacturing, vol. 34, 8, pp. 709-718, ISSN: 1359-835X.

Salas, W.; Alba-Baena, N. G. and Murr, L. E., (2007). *Explosive Shock-Wave Consolidation of Aluminum Powder/Carbon Nanotube Aggregate Mixtures: Optical and Electron Metallography,* Journal Metallurgical and Materials Transactions A, vol. 38, 12, ISSN: 1073-5623.

Salvetat-Delmotte, J. and Rubio, A. (2002). *Mechanical properties of carbon nanotubes: a fiber digest for beginners,* Carbon, vol. 40, 10, pp. 1729–1734, ISSN: 0008-6223.

Schwaretz, M. M. (1997). Composite Materials Handbook, vol. 2, Prentice Hall, ISBN: 0133000397.

Sherif El-Eskandarany, M. (1998). *Mechanical solid state mixing for synthesizing of SiC/Al nanocomposites,* Journal of Alloys and Compounds, vol. 279, pp. 263–271, ISSN: 0925-8388.

Yoshiaki, Sh.; Toshiyasu, N. and Iwao, M. (1995). *Corrosion resistance of Al-based metal matrix composites,* Materials Science and Engineering A, vol. 198, 1-2, pp.113-118, ISSN: 0921-5093.

SivaKumar, K.; Soloman, R.P.; Balakrishna, B. T. and Hokamoto, K. (2001). *Studies on Shock consolidated 2124 Al-40 vol.% SiCp composites,* Journal of Materials Processing Technology, vol. 115, pp. 396-401, ISSN: 0924-0136.

SivaKumar, K.; Hokamoto, K.; Nakano, S. and Fijita, M. (2001). *Chap. 40 Fabrication of 2124 Al-SiC metal matrix composites by one dimensional underwater shock* consolidation in Fundamental Issues and Applications of Shock-Wave and High-Strain Rate Phenomena, K. P. Staudhammer, L.E. Murr, and M. A. Meyers, (Eds.), Elsevier Science Ltd., Amsterdam, pp. 313-320, ISBN: 008043896-2.

Sivakumar, K.; Balakrishna, B. T. and Ramakrishnan, P. (1996). *Dynamic consolidation of aluminium and A1-20 V/o SiCp composite powders,* Journal of Materials Processing Technology, vol. 62, pp. 191-198, ISSN: 0924-0136.

Stuivinga M.; Verbeek H. J. and Carton E. P. (1999). *The Double Explosive Layer Cylindrical Compaction Method,* Journal of Materials Processing Technology, vol. 85, pp. 115–120, ISSN: 0924-0136.

Thadani, N. N. (1988). *Shock compression processing of powders,* Advanced Materials and Manufacturing Processes, vol. 3, 4, pp. 493–549, ISSN: 1042-6914.

Thostenson, E. T.; Ren, Z. and Chou, T. W. (2001). *Advances in the science and technology of carbon nanotubes and their composites: a review.* Composites Science and Technology, vol. 61, pp. 1899–1912, ISSN: 0266-3538.

Tjong S. C. and Ma, Z. Y. (1997). *The high-temperature creep behaviour of aluminium-matrix composites reinforced with SiC, Al_2O_3 and TiB_2 particles"*, Composites Science and Technology, vol. 57, 6, pp. 697-702, ISSN: 0266-3538.

Tong, W.; Ravichandrant, G.; Christman, T. and Vreeland Jr, T. (1995). *Processing SiC-particulate reinforced titanium-based metal matrix composites by shock wave consolidation*, Acta Metallurgica et Materialia, vol. 43, 1, pp. 235-250, ISSN: 1359-6454.

Torralba, J. M.; da Costa, C. E. and Velasco, F. (2003). *P/M aluminum matrix composites: an overview*, Journal of Materials Processing Technology, vol. 133, pp. 203–206, ISSN: 0924-0136.

Torquato, S., (2002). *Random Heterogeneous Materials: Microstructure and Macroscopic Properties*, Springer, ISBN: 0-387-95167-9.

Vandersall, K. S. and Thadhani, N. N. (2001). *Chap. 34 Investigation of shock-induced chemical reactions in Mo-Si powder mixtures* in Fundamental Issues and Applications of Shock-Wave and High-Strain Rate Phenomena, K.P. Staudhammer, L.E. Murr, and M. A. Meyers, (Eds.), Elsevier Science Ltd., Amsterdam, pp. 267-274, ISBN: 008043896-2.

Wang, L.; Jiang, W. and Chen, L. (2004). *Fabrication and characterization of nano-SiC particles reinforced TiC/SiC nano composites*, Materials Letters, vol. 58, pp. 1401–1404, ISSN: 0167-577X.

Wang, H. Z.; Wu, S. Q. and Tjong, S. C. (1996). *Mechanical and wear behavior of an Al/Si alloy metal-matrix composite reinforced with aluminosilicate fiber*, Composites Science and Technology, vol. 56, 11, pp. 1261-1270, ISSN: 0266-3538.

Wang, X.; Wu, G.; Sun, D.; Qin, Ch. and Tian, Y. (2004). *Micro-yield property of sub-micron Al_2O_3 particle reinforced 2024 aluminum matrix composite*, Materials Letters, vol. 58, pp. 333– 336, ISSN: 0167-577X.

Ward, P. J.; Atkinson, H. V.; Anderson, P. R. G.; Elias, L. G.; Garcia, B.; Kahlen L. and Rodriguez-Ibabe, J.M. (1996). *Semi-solid processing of novel MMCs based on hypereutectic aluminium-silicon alloys*, Acta Materialia, vol. 44, 5, pp. 1717-1727, ISSN: 1359-6454.

Weimar, P. and Pruemmer, R. (2001). *Chap. 36 Explosive compaction of nanocrystalline alumina powder* in Fundamental Issues and Applications of Shock-Wave and High-Strain Rate Phenomena, K.P. Staudhammer, L.E. Murr, and M. A. Meyers, (Eds.), Elsevier Science, Ltd., Amsterdam, pp. 283-287, ISBN: 008043896-2.

Withers, G. (2005). *ULTALITE Aluminum Composites*, Advanced Materials & Processes, September, pp. 44-48, ISSN: 0882-7958.

Xu, C. L.; Wei, B. Q.; Ma, R. Z.; Liang, J.; Ma, X. K. and Wu D. H. (1999). *Fabrication of aluminum-carbon nanotube composites and their electrical properties*, Carbon, vol. 37, 5, pp. 855-858, ISSN: 0008-6223.

Yang, Y.; Lan, J. and Li, X. (2004). *Study on bulk aluminum matrix nano-composite fabricated by ultrasonic dispersion of nano-sized SiC particles in molten aluminum alloy*, Materials Science and Engineering A, vol. 380, pp. 378–383, ISSN: 0921-5093.

Zhong, R.; Cong, H. and Hou, P. (2004). *Fabrication of nano-Al based composites reinforced by single-walled carbon nanotubes*, Carbon, vol. 41, 4, pp. 848–851, ISSN: 0008-6223.

33

Superconductivity in carbon nanotubes

Junji Haruyama[1-3]

[1]*Aoyama Gakuin University, 5-10-1 Fuchinobe, Sagamihara, Kanagawa 229-8558, Japan*

[2]*Institute for Solid State Physics, University of Tokyo, Kashiwanoha 5-1-5, Kashiwa, Chiba 277-8581, Japan*

[3]*Japan Science and Technology Agency-CREST, 4-1-8 Hon-machi, Kawaguchi, Saitama 332-0012, Japan*

Abstract

Carbon nanotubes (CNTs), which are one-dimensional (1D) molecular conductors, have attracted considerable attention from viewpoints of the novel structures, electronic states, quantum phenomena, and also applications to new opt and electronic devices. However, mostly none reported on superconductivity (SC) and its correlation with 1D electron phenomena. In the present study, I report on SC with the world-highest transition temperature (T_c) of 12K (1) in arrays of multi-walled CNTs (MWNTs) with entirely end-bonded electrode junction and (2) in highly uniform thin films of boron-doped single-walled CNTs (*B*-SWNTs). In the former one, I reveal correlation of SC with Tomonaga-Luttinger liquid (TLL) state, which is a collective phenomenon arising from repulsive Coulomb interaction between electrons confined in 1D ballistic charge transport regimes. In the latter one, I clarify correlation of SC with van Hove singularities (VHS) in electronic density of states (DOS) in the SWNT. Based on this thin film, I show fabrication of paper-like thin films consisting of pseudo two-dimensional network of weakly coupled *B*-SWNTs (the so-called Buckypaper) and show an enhancement of the onset T_c up to 19 K by applying only a small pressure. In section 4, I discuss about possibility for realization of quantum spin entangler and quantum bit utilizing a hybrid system of CNTs/superconductor, which is associated with the section 1. Because CNTs have strong spin coherence and possibly strong spin entanglement in spin singlet (Cooper pair), it must allow realization of quantum spin entangler and bit without decoherence problem in near future. The present observations will shed light to feasibility of CNTs as a 1D superconductor and also research of 1D electron correlation. They promise application to novel quantum devices.

1. Superconductivity in multi-walled carbon nanotubes: Interplay of Tomonaga -Luttinger liquid and superconductive phase

It is well known that the phonon- mediated attractive electron-electron interaction leads to a superconductive instability in 2D and 3D conductors. In particular, high-Tc SC favors 2D space such as cupper-oxide high-T_c superconductors. Here, how does SC like 1D space? In CNTs, which is one of typical 1D conductors, a variety of 1D quantum phenomena have been reported. However, intrinsic SC has been reported in the CNTs only by a few groups [1-3]. Nevertheless, the SC in a CNT is attracting considerable attention for the following reasons: (1) The curvature resulting from a small diameter (<< ~1 nm) can bring high a T_c due to a strong coupling of σ-π band electrons, which arises from formation of sp^3 hybrid orbitals and σ-π mix band, with radial breathing phonon mode. (2) The perfect alignment of the Fermi level (E_F) to a van Hove singularity (VHS) can also lead to high T_c due to the appearance of an extremely large electronic density of states (DOS). (3) It can provide an insight into the 1D electron correlation. Thus, study of SC in CNTs can shed lights to feasibility of 1D conductors as superconductors.

In contrast, many quantum phenomena and electronic states tend to obstruct appearance of SC in 1D space. For instance, Coulomb repulsion in TLLs, energy gaps arising from Pierls transition, very small electronic DOS in energy regimes out of VHSs, and phase fluctuation can such obstructions. In particularly, TLL is one of typical 1D quantum phenomena. The TLL provides the relevant theoretical description [4–6], representing a collective electronic state (e.g., showing spin-charge separation) that arises from the repulsive Coulomb interaction between electrons confined in a 1D ballistic conductance regime. Many previous works have reported TLL features in CNT systems. It is known that the tunneling density of states $\nu(E)$ for the TLLs decreases following a power-law dependence on energy, $\nu(E) \propto E^{\alpha}$, where the power α is given by different expressions according to the different tunneling regimes [4, 5, 12]. Most previous works on CNTs have been dealing with the regime of strong Coulomb interaction.

Here, SC is a phenomenon arising from attractive Coulomb interaction via phonon, which is in contradiction to the Coulomb repulsion leading to the TLL state. This leads naturally to the question of how the SC can develop in the CNTs, with the consequent transition from the TLL [7–11, 14]. The interplay between the TLL state and intrinsic SC has been only reported in our MWNTs with a significant number N of shells participating in the conduction [2]. The results were qualitatively consistent with low-energy theories of CNTs [8], which had already predicted the breakdown of the TLL at low temperatures in nanotube bundles with a strong intertube electrostatic coupling between the different 1D conduction channels.

In the MWNTs, however, the interplay of the SC with the TLL and the value of T_c as high as 12 K still had to be quantitatively understood, as the samples were not intentionally doped and the number of 1D conduction channels was unknown. On the other hand, the high T_c value was in approximate agreement with T_c's = ~10K of CaC_6 and boron(B)-doped diamond. Indeed, low-energy theories of MWNTs have suggested the importance of having a large number of 1D conduction channels for the development of a SC instability [10]. Therefore, it is crucial to clarify the correlation between the appearance of the SC phase and carrier doping in MWNTs. B-doping has been actually successfully achieved following known methods [15, 16].

In this section, I report on SC in array of the MWNTs and reveals correlation with TLL state. I find an abrupt resistance drop due to SC at a T_c as high as 12K in the entirely end-bonded MWNTs. In contrast, the partially end-bonded MWNTs show very small resistance drop at low temperatures. It reveals that a gradual transition from TLL behavior to a SC regime can start only from low eV/kT values at low temperatures, but the development of the SC enhances the regime to high eV/kT region. Phase diagram obtained from low-energy theory reveals that such a transition is actually possible. Nuclear Magnetic Resonance (NMR) measurements ascertain the presence of substitutional boron doping for SC in the MWNTs.

At first, I report on the possibility of unintentional *B*-doping in our MWNTs, which were synthesized by chemical vapor deposition using Fe/Co catalyst and methanol gas in the nanopores of the alumina template [2, 19]. Boron has been intentionally used only for the enhancement of the chemical reaction of the Fe/Co catalyst [2, 19]. Figure 1 shows results of nuclear magnetic resonance (NMR) measurements of our MWNTs with three different boron concentrations (N_B) in the catalyst. Evident peaks can be observed at 0 ~ 5 ppm in individual samples. The peak position implies the presence of chemical bonds of *B-C* and, hence, substitutional *B* doping in the carbon network of the MWNTs. Moreover, at least five-times higher intensity of NMR signal peak for the *B-C* was detected in the *B*-MWNTs with N_B = ~2 at.%, compared with other *B*-MWNTs with different N_B values. Very importantly, we find that only the *B*-MWNTs with N_B = ~2 at.% showed SC with high T_c (Fig. 3(a)) and also present signature of SC (Fig. 3(b)) with the highest reproducibility. This result implies that the most suitable condition for substitutional *B* doping into the MWNTs, which leads to appearance of the SC, is ~2 at.%. This is consistent with our recent discovery [20] and a theory [21]. The small *B* concentration allows the CNTs remaining a 1D ballistic charge transport regime.

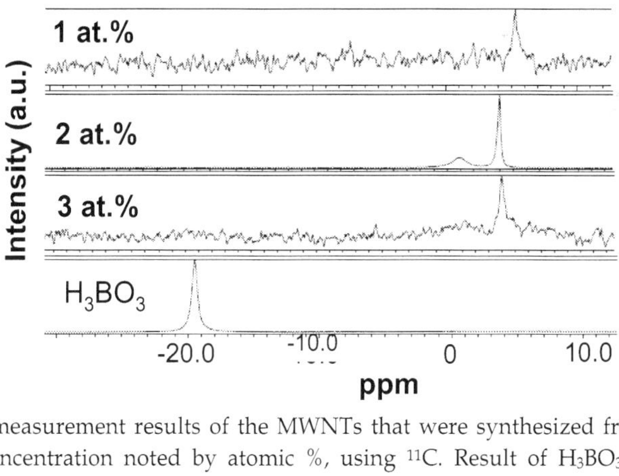

Fig. 1. NMR measurement results of the MWNTs that were synthesized from Fe/Co with different *B* concentration noted by atomic %, using ^{11}C. Result of H_3BO_3 is a reference. H_3BO_3 was mixed in $FeSO_4$ and $CoSO_4$, and then Fe/Co catalyst including *B* was electrochemically deposited into the bottom ends of nanopores of alumina template [5, 19].

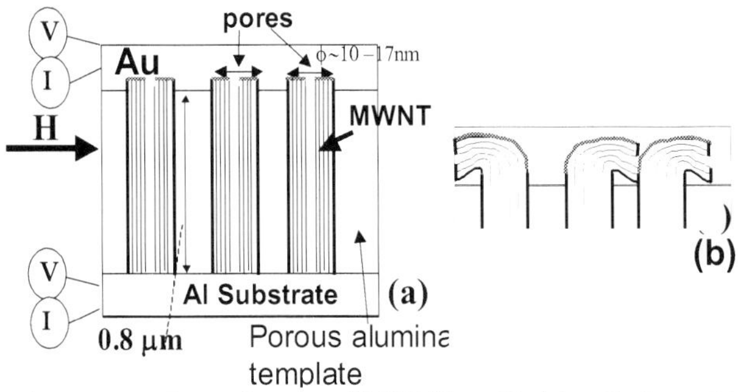

Fig. 2. Schematic cross sections of array of MWNTs with (a) entirely end-bonded and (b) partially end-bonded junctions between Au electrode and top ends of MWNTs. The red lines indicate the interfaces between the MWNTs and the electrodes.

Figure 2 shows the schematic cross sections of our samples, which include arrays of *B*-MWNTs with (a) entirely end-bonded (N = 9) and (b) partially end-bonded ($1 < N < 9$) junctions to Au electrodes [2]. They were fabricated by entirely and partially cutting the excess top portions of MWNTs, which were grown up from the nanopores and accumulated on alumina membrane, respectively. No cutting resulted in bulk junction, which has electrical contact to only the outermost shell (i.e., N = 1).

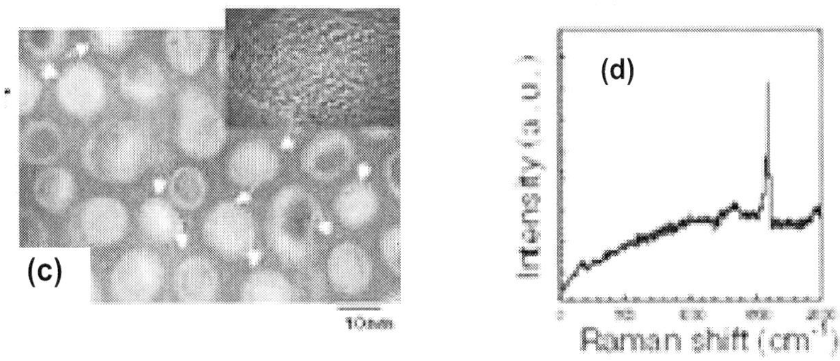

Fig. 2. (c) Plane TEM image of an array of MWNTs. Inset: High resolution TEM image of one MWNT. (d) Resonant Raman spectrum of the MWNT. The significantly sharp peak indicates G band.

Figure 2(c) shows a planer TEM image of the sample [2]. One can confirm presence of MWNTs in most pores. Figure 2(d) shows measurement result of resonant Raman spectrum. Peak height of G band is significantly greater than that of D band. This strongly indicates high quality of the MWNTs.

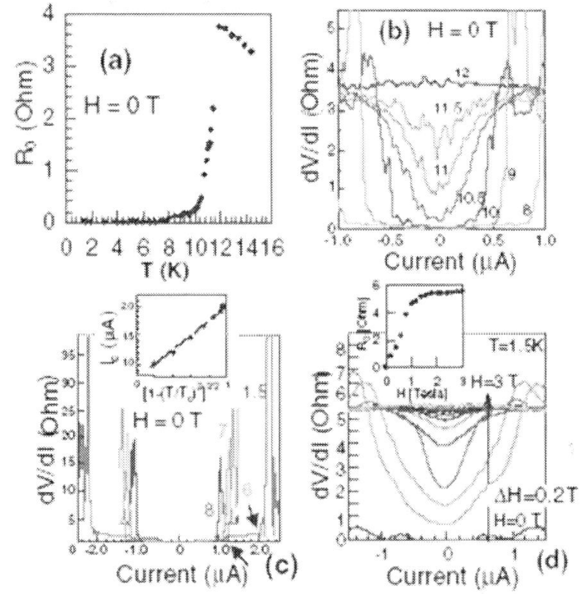

Fig. 3. (a) Superconductivity with T_c=12K in the sample with junction type (a) (b) Differential resistance vs. Temperature (Kelvin) for (a) (T>8K). (c) that for T<8K Inset : Critical current vs. Normalized temperature for GL theory (d) Differential resistance vs. magnetic fields applied perpendicular to tube axis

Figure 3 shows superconducting behaviors of the entirely end-bonded MWNTs as shown in Fig. 2(a). In Fig. 3(a), zero-bias resistance (R_0) abruptly starts to drop from T_c = 12K and attains to zero-ohm at T = 7.8K, when residual resistance, which arises from contact resistance, was subtracted. Figure 3(b) shows the differential resistance dips for various temperatures > 8K. As temperature decreases, the dip monotonically deepens and mostly attains to zero-ohm at the center point at T = 8K. This is consistent with Fig. 3(a). Figure 3(c) shows differential resistance at T < 8K. When current increases at each temperature, zero differential resistance dip suddenly disappears at critical current (I_c). The inset shows relationship of I_c vs. temperature normalized for GL equation. It exhibits a linear relationship and turns out that Figs.3(a) - (c) are attributed to SC. Figure 3(d) shows differential resistance dips as a function of magnetic field. The dip rapidly disappears as the field increases.

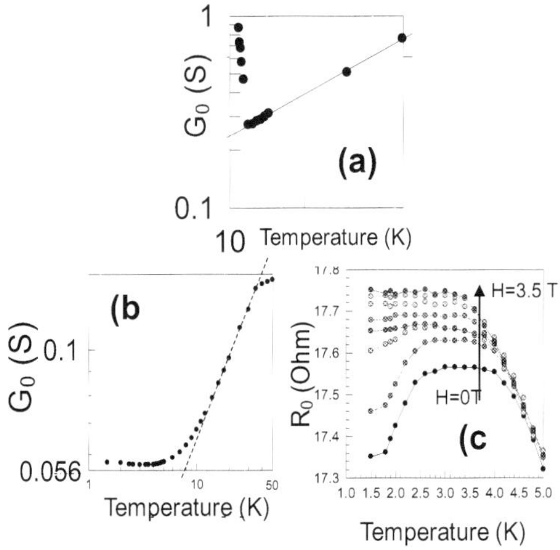

Fig. 4. Doubly logarithmic scales of zero-bias conductance G_0 and temperature for (a) entire and (b) partial end junctions. (c) Magnetic field dependence of zero-bias resistance R_0 as a function of the temperature, corresponding to (b).

Figure 4 shows the doubly logarithmic scales of zero-bias conductance G_0 and temperature for the two different junctions shown in Fig. 2. The observed power law behaviors in the different types of junctions ($G_0 \propto T^{\alpha}$; α = 0.7 and 0.8 in (a) and (b) respectively) are in excellent agreement with the previous reports of TLL behavior in CNTs [2, 5]. As shown in Fig. 4(a), the abrupt G_0 increase (i.e., SC regime) appears overcoming the TLL behavior (i.e., the power law at T > 12K) at T_c = 12 K in the entirely end-bonded *B*-MWNTs. The origin of this effect was in the large screening of the Coulomb interaction, arising from the electrostatic coupling between a large number of 1D conduction channels due to boron doping and also the large *N* value originating from entire end-bonding of the *B*-MWNTs [2, 8].

In contrast, in the partially end-bonded *B*-MWNTs, only a slight and gradual G_0 increase at T < 2.5K (i.e., a signature of SC correlations) can be observed as shown in Fig. 4 (b). As we implied in ref. [5], the appearance of SC and the correlations of power laws with G_0 increase were strongly correlated with the values of *N* as shown in Fig. 3. Because we could confirm that the abrupt G_0 increase in Fig. 4(a) was attributed to SC as mentioned above [2], it can be concluded that the small G_0 increase observed in Fig. 4(b) is also attributed to SC. The gradual G_0 increase corresponds to the slight and gradual resistance drop shown in Fig. 4(c). With an increase in the magnetic field, the drop in R_0 disappears rapidly. This behavior actually agrees with that of conventional superconductors.

In Fig. 4(b), the power law is observable up to T = ~40 K for α = ~0.7, while it gradually starts to deviate below T = ~12 K. At T < ~6 K, it saturates completely. Subsequently, a small increase in G_0 due to the SC correlations appears at T < ~2.5 K. However, the trend towards the SC transition is not completed above T = 1.5 K in this case. This is consistent with the fact that the SC phase requires the coherent transport of Cooper pairs along the directions transverse to the CNTs, which cannot be established when only part of the shells are electrically active and the number of conductance channels is smaller as in the partially end-bonded *B*-MWNTs with the small *N* values [2, 8]. This also implies that the Coulomb interaction cannot be sufficiently screened to allow for the development of SC correlations. In fact, *B*-MWNTs with $N = 1$ exhibited no G_0 increase, consequent with the absence of screening of the Coulomb interaction.

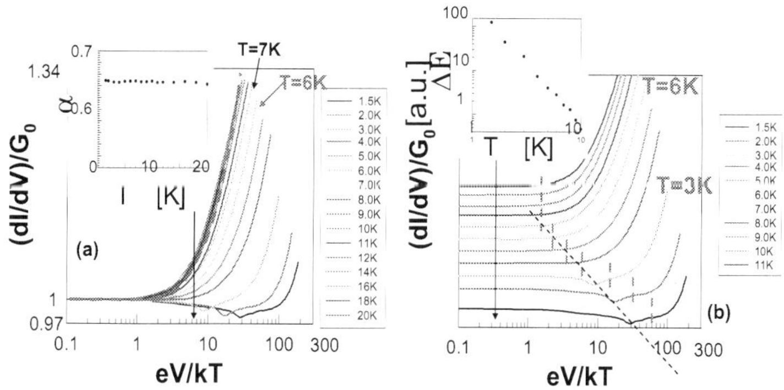

Fig. 5. **(a)** Doubly logarithmic scales on normalized differential conductance vs. eV/kT, corresponding to TLL formula in $eV >> kT$ regime [13]. The red dotted line gives the best fitting result by eq.(1). **Inset**: power value α vs. temperature estimated from main panel. **(b)** Transform of $(dI/dV)/G_0$ in Fig. 5(a) to arbitrary units. The long black-dotted line connects the right edges of the conductance peaks. **Inset**: ΔE_c vs. temperature on doubly logarithmic scales. $\Delta E_c = E_c\,(T) - E_c\,(T = 7\text{ K})$, where $E_c(T)$ is the boundary point between the saturation and power law regions at each temperature, as shown by the short red-dotted lines in the main panel.

On the other hand, besides the dependence on temperature, the power law behavior of the observables as a function of the bias voltage V provides absolute evidence of the TLL behavior [4, 5, 13]. Thus, it is crucial to investigate the regime $eV >> kT$ to discuss in depth the transition from TLL behavior to the SC phase. In this sense, Fig. 5(a) exhibits doubly logarithmic plots of the normalized differential conductance vs. eV/kT, which is one of the typical dependences to check the TLL behavior for $eV >> kT$ [13]. It shows that the entire data collapse suitably onto a single universal curve at $T \geq \sim 7$ K. This agrees with the results reported in previous investigations of TLL behavior [4, 5, 13]. As shown by the red dotted line, this universal curve can be fitted by the following result for a TLL [13]

$$\frac{G(V,T)}{G(V=0,T)} = \cosh\left(\gamma\frac{eV}{2kT}\right)\frac{1}{\left|\Gamma\left(\frac{1+\alpha}{2}\right)\right|^2}\times\left|\Gamma\left(\frac{1+\alpha}{2}+\gamma\frac{ieV}{2\pi kT}\right)\right|^2 \tag{1}$$

where V is the applied voltage; $\Gamma(x)$, the Gamma function; k, the Boltzmann constant; and γ, the inverse number of the measured junctions M weighted by their resistance [13]. For M junctions in series, the value of γ should be within $\gamma = 1/M$. The best fit gives the parameter γ = ~0.7, which is mostly in good agreement with $\gamma = 1/M = 0.5$ for $M = 2$ of our individual MWNTs with two end junctions. Therefore, we conclude that the universal curve can be actually attributed to TLL behavior at T ≥ ~7 K [2].

However, we importantly find appearance of a deviation from the universal curve in the low temperature regime (e.g., at T ≤ 6 K) in Fig. 5. As shown by red lines in Fig. 5(b), this deviation originates from a monotonic increase in the width of the low-eV/kT saturation region (i.e., eV/kT < ~2 at T = 6 K) in the universal curve at T ≤ ~6 K. The inset of Fig. 5(b) shows ΔE_c (width of the saturation region) as a function of temperature in a doubly logarithmic scale. Indeed, we can confirm that ΔE_c actually increases with a decrease in temperature, following power law dependence.

Quite remarkably, it should be noted that the value of the critical temperature of 6 K for the deviation of the universal curve is in excellent agreement with that of the saturation region in Fig. 4(b). This agreement strongly suggests that the deviation at T ≤ ~6 K in Fig. 4 for eV >> kT is associated with the appearance of the SC regime. Figure 5(b) supports this argument, because it suggests that the increase in the width of the saturation region is caused by the appearance (insertion) of the conductance peak due to emergence of the SC regime. In particular, this is evident at T ≤ 3 K, where G_0 is seen to increase in Fig. 4(b) [22]. This implies that the emergence of the SC regime at low eV/kT leads to the deviation from the TLL states [17].

In the conventional theory of a TLL, the width of the low-eV/kT saturation region (< ~1) reflects the regime of $eV/kT < \Delta E = hv_F/L$, where ΔE is the energy spacing of the quantized electronic orbital formed in the CNTs in the ballistic charge transport regime [18]. This means that the noise from thermal fluctuations smears the effects of the bias voltage (i.e., power laws) out in the low-eV/kT regime (< ~1). In contrast, one should notice that insertion of the conductance peak in Fig. 5(b) enhances this thermal smearing region to high-eV/kT regime (~100) at T = 1.5 K. This is very anomalous and, hence, strongly implies a possibility that appearance of the SC regime introduces superconducting gap Δ_{SC} in the system, resulting in $\Delta E + \Delta_{SC}$.

Moreover, the inset of Fig. 5(a) shows the values ofα, which were estimated from the main panel of Fig. 5(a), as a function of temperature. It is very important to note that the α value is constant (~0.67) in all the temperature range, despite the emergence of conductance peak due to the SC regime at low eV/kT regime below T = 6K. This implies that the transition from the TLL behavior to the SC regime starts to occur only in the low eV regime, when the values of eV are drastically lowered under constant low kT values.

Here, in Fig. 6, we show that such a transition is actually possible on the light of the low-energy theories of doped MWNTs [10]. The two regimes are represented in the phase diagram of Fig. 6, which has been obtained following the approach of Ref. [10]. We observe that a minimum N value per shell (10) in a MWNT is needed for the onset of the SC regime, and that subsequent peaks arise in the critical temperature as the Fermi level(ε_F) crosses the top of a new subband (i.e., a Van Hove singularity). This result is in good agreement with the present experimental results, in which the N_B value is below 2 at.% with the small N values leading to shift of ε_F in ~0.3 eV [20, 21] and corresponding T_c of 3 ~ 12K.

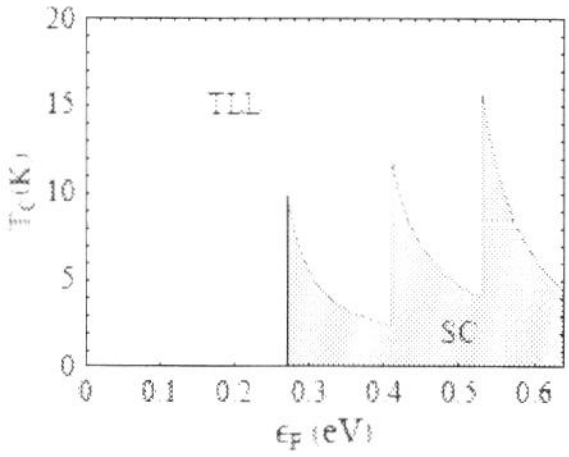

Fig. 6. Phase diagram of doped MWNTs (incorporating the average over different geometries of the shells) in terms of the doping level, represented by the shift in the Fermi energy ε_F with respect to the charge neutrality point, and the temperature].

The characterization of doping in the MWNTs provides actually the clue to account for the SC regime. This appears as a consequence of the large screening of the Coulomb interaction, which arises in doped MWNTs, as in the case of CNT ropes [8], from the electrostatic coupling of a large number of 1D conduction channels. The effect is such that the long-range part of the Coulomb potential is largely suppressed, placing the system on the verge of a pairing instability. This takes place as the intertube Cooper-pair tunneling is progressively enhanced at low energies [10]. In this respect, the Cooper pairs do not find the obstruction that single electrons have to tunnel between the incommensurate CNT lattices of a MWNT. When intertube coherence can be established at some low temperature T_c under a low *eV* value, the system undergoes the transition from the TLL to the SC regime as shown in Fig. 6. Although further quantitative investigation is expected, the present phenomena must shed light on understanding of interplay of SC with 1D electron correlations in *B*-CNTs.

2. Superconductivity in thin films of boron-doped carbon nanotubes

New carbon-based superconductors - calcium- intercalated graphite (C_6Ca) and highly boron (*B*)- doped diamond [23-25] - have been discovered recently and have attracted attention. It is well known that the small mass of carbon can promote high T_c in Bardeen-Cooper-Schrieffer (BCS)-type SC and novel behaviors of SC can also be expected. As mentioned in section 1, the SC in a CNT is attracting significant attention [26-30] for the following reasons: (1) The curvature resulting from a small diameter can bring high T_c due to a strong electron-phonon coupling [31]. (2) The alignment of the E_F to a VHS can also lead to high T_c due to the presence of an extremely large DOS [31]. (3) It can provide an insight into the 1D electron correlation [29, 32-36].

Here, it is also a well-known fact that effective carrier doping is crucial for realizing high-T_c SC in any material. In fact, SCs in C_6Ca and *B*-diamond have been associated with it. Carrier doping into CNTs has also been studied in many previous works [37-39]. Ref.[37] have reported a systematic study of SWNT bundles using the pulsed laser vaporization technique, in which the *B* doping concentration (N_B) is controlled by the amount of elemental *B* mixed into a Co/Ni catalyst impregnated targets. Evidence for substitutional *B* doping and enhanced electronic DOS at the E_F has been reported previously [37-39].

Nevertheless, experimental studies on the SC in carrier-doped CNTs are lacking. Although only I reported on SC in the *B*-MWNTs as explained in section 1, the boron doping was accidental and could not be in a controlled manner. This could be due to (1) difficulties associated with substitutional doping in CNTs (with diameter as small as ~1 nm) without compromising structural integrity of the hexagonal carbon network, (2) since the conventional size of Cooper pairs and the path of the Meissner shielding current cannot be smaller than 1 nm, SC cannot be realized in an isolated single CNT [30], (3) in individual CNTs, strong 1D electron correlation phenomena tend to destroy Cooper pairs and the alignment of the E_F to a VHS requires highly optimized doping rate. From these viewpoints, the assembled *B*-doped SWNTs with low *B* concentration reported in this section are crucial for the observation of Meissner effcet [30, *33*, *46*]. Theoretical studies also suggests that SC can be realized in *B*-SWNTs due to small-dopant-induced (≤ 0.4 at. %) tuning of the E_F with the VHS present in individual (10, 0) SWNTs [40] and also in carrier-doped MWNTs [36].

In this section, I report on SC in the thin films of *B*-SWNTs and clarify correlation with VHS in electronic DOS in 1D conductor. I report on the Meissner effect and show that only SWNT films consisting of low boron concentration leads to evident Meissner effect with T_c = 12 K. I also show that a highly homogeneous ensemble of the SWNTs is crucial for realizing the Meissner effect. The first-principles electronic-structure study of the *B*-SWNT reveals strongly correlation of the boron concentration for Meissner effect with VHSs.

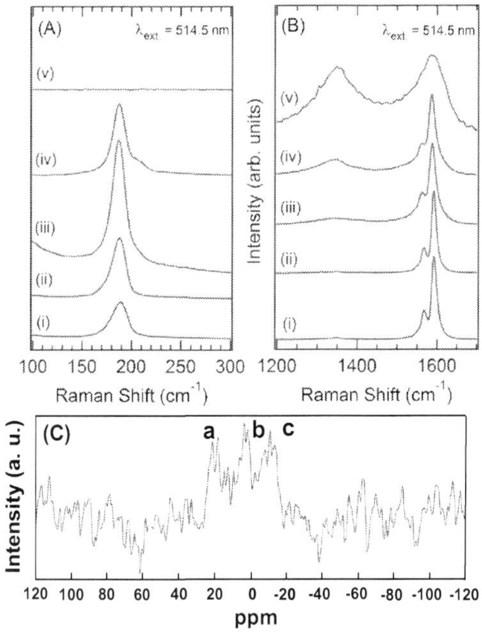

Fig. 7. (a)(b) Room temperature Raman spectra of semiconducting *B*-SWNT bundles synthesized from targets containing boron concentrations (N_B) of ~ (i) 0 (ii) 1.5 (iii)2.0 (iv) 3.0 (v) 4.5 atomic %. An excitation wavelength λ_{ex} = 514.5 nm is used to probe the *B*-induced changes in the Raman spectra. The *B*-induced changes in (a) the *RBM* (present around 200 cm^{-1}) and (b) the *D* band (present around 1300–1350 cm^{-1}) are shown. (c) Result of NMR measurement of *B*-SWNTs synthesized with N_B = ~1.5 at. % target. Three peaks noted as a, b, and c are evidently observed.

Figure 7 (a) and (b) show the room-temperature Raman spectra of semiconducting SWNT bundles prepared from targets containing *B* in the range $0 < N_B < 4.5$ at. % (described above and in Ref. [37, *46*]). The present catalytically *B*-doping method allows *B* doping in low concentration in the SWNTs, avoiding destruction of carbon networks. The data set shown in Fig. 7 clearly reveals the following: (1) In Fig. 7(a), the intensity of the radial breathing mode (*RBM*) varies nonlinearly with the N_B and exhibits the highest values around $N_B = 1.5$ and 2 at. %. These intensity changes result from a change in the resonance conditions induced by the presence of *B* in the SWNT lattice, which shifts the transition energies between the VHSs in the electronic DOS. (2) No SWNTs are present in the products prepared from targets with N_B in excess of 4 at. %, since the *RBM* is absent in their spectra in Fig. 7(a). This decline is strongly correlated with the well-known saturation limit of ~4 at. % for N_B in graphite at 1200 K. (3) In Fig. 7(b), a systematic increase in the intensity of the disorder-induced band (*D* band) with increasing N_B is evident implying that a change in the degree of ordering in the hexagonal lattice of the SWNTs results due to the incorporation of *B* in the lattice. Similar results are observed in the bundles of metallic SWNTs.

Figure 7(c) shows the result of nuclear magnetic resonance measurement (JNM-ECX400) of *B*-SWNTs synthesized target with $N_B = 1.5$ at. %. It evidently exhibits presence of three peaks (a, b, and c). The peaks noted as *b* and *c* originate from chemical bonds between *B* and carbon, in which the peak heights are highly sensitive to *B* concentration in the target. In contrast, peak *a* originates from *B*-oxide bond. The thermoelectric power (TEP) measurements also provide complementary evidence for the *B*-induced changes in the electronic properties since temperature dependent TEP studies revealed *p*-type characteristics [37]. These results collectively provide strong evidence for substitutional doping of *B* in the SWNTs.

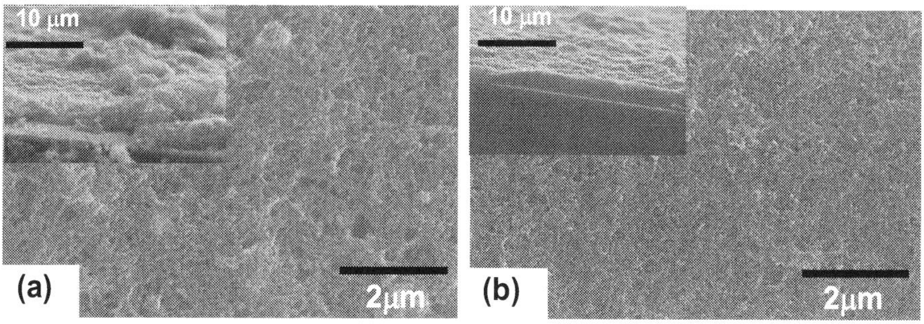

Fig. 8. Scanning electron microscope (SEM) images of thin films consisting of assembled *B*-SWNTs prepared on Si substrates. (a) Film fabricated without spin coating and (b) film fabricated by spin coating (500 rpm). Insets: Cross-sectional SEM images of individual samples observed at a 30° tilt.

The magnetization measurements of the thin films consisting of *B*-SWNTs [40] were performed using a superconducting quantum interference device (Quantum Design, MPMS) [24]. Figure 9 shows normalized magnetization ($M_N = M(T) - M(T{=}40K)$) as a function of the temperature (*T*) (M_N–*T* relationship) in field-cooled (FC) and zero-field -cooled (ZFC) regimes in the highly uniform film with $N_B \sim 1.5$ at. % as shown in Fig. 8(b) . Evident drop in

M_N is observed below T_c = 12 K in the ZFC regime, and below T = ~8 K in the FC regime. Interestingly, this value of T_c = 12 K exactly agrees with the T_c value for an abrupt resistance drop observed in the array of entirely end-bonded MWNTs [29]. The small M_N drop observable at T_c > 12 K in the ZFC regime is due to diamagnetism of graphite structure of bundle of SWNTs and not associate with Meissner effect [41]. It is found that the magnitude of the drops observed in the M_N–T relationship becomes considerable at magnetic field (H) < ~1400 Oe as H values increase. In contrast, it decreases for H > ~1400 Oe, and at H = ~3500 Oe, the magnitude becomes almost zero.

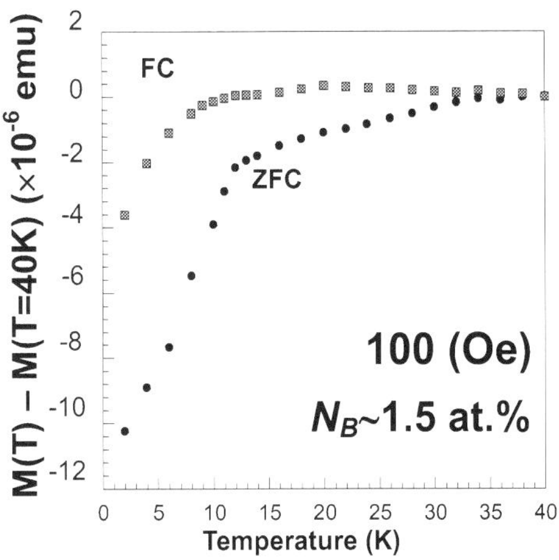

Fig. 9. Normalized magnetization as a function of temperature at magnetic fields (H) of 100 Oe in FC and ZFC regimes in thin films of assembled B-SWNTs, which are synthesized from a target with N_B = ~1.5 at. %. The films are prepared on a Si substrate by spin coating at 500 rpm.

Figure 10 shows the values of M_N as a function of H (M_N–H relationship) for various T in the ZFC regime in the sample shown in Fig. 9. At each temperature, M_N values decrease at H < ~1400 Oe, while they increase at H > ~1400 Oe. The magnitude of the drops decreases monotonically as T increases. These behaviors observed in the shown M_N–T and M_N–H relationships are qualitatively in good agreement with the Meissner effect in type-II superconductors.

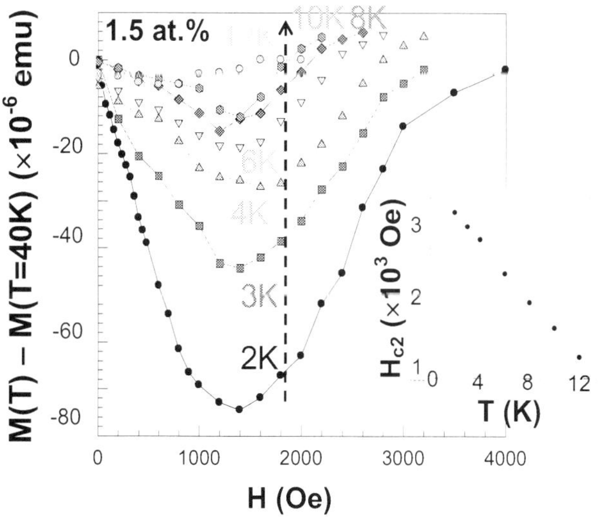

Fig. 10. Normalized magnetization vs. magnetic field for various temperatures in the sample shown in Fig. 9. **Inset:** Upper critical field (H_{c2}) vs. temperature relationship estimated from main panel. We determine H_{c2} at the *H* value for which $M_N = 0$ at each temperature by extrapolating the linear slope of the M_N-*H* relationship for $H > \sim 1500$ Oe in main panel.

In order to confirm the Meissnser effect, we focus on the temperature dependence of the upper critical magnetic fields (H_{C2}) in the M_N–*H* relationships shown in Fig. 10. The inset of Fig. 10 shows the result of the relationship between the estimated H_{c2} and *T*. The values of H_{c2} decrease linearly as the temperature increases. This result is also qualitatively in good agreement with the results for type-II superconductors. Residual H_{c2} around T_c = 12K is attributed to diamagnetism of graphite as mentioned above.
From a quantitative viewpoint, the values of H_{c2} (T = 0) = ~1700 Oe can be estimated from the inset of Fig. 10 and the relationship

$$H_{c2}(T = 0) = -0.69(dH_{c2}/dT|_{Tc})\, T_c \tag{2}$$

$$dH_{c2}/dT|_{Tc} = -200 \text{ and } T_c = 12\text{ K} \tag{3}$$

The Ginzburg-Landau (G-L) superconductive coherence length

$$\xi = [\Phi_0/2\pi H_{c2}(T = 0)]^{1/2} \tag{4}$$

where $\Phi_0 = h/2e$ is the quantum magnetic flux, can be estimated as ξ = ~ 17 nm from the H_{c2} (T = 0) = ~1700 Oe. Further, the penetration length of the magnetic field

$$\lambda = (m^*/\mu n_s e^2)^{1/2} \quad (5)$$

is estimated to be of the order of ~100 nm. This λ value is significantly larger than ξ = ~ 17 nm. This result apparently supports the fact that the present films of *B*-SWNTs are type-II superconductors. Moreover, the value of ξ = ~ 17 nm is in good agreement with ξ = ~11 nm in the arrays of MWNTs [8], ξ_{ab} = 13 nm in C_6Ca [23, 24], and ξ = 10 nm in *B*-doped diamond [25]. Therefore, we conclude that the M_N–T drops shown in Fig. 3 and the corresponding M_N–H relationship shown in Fig. 10 are attributed to the Meissner effect observed in type-II superconductors. The gradual and unsaturated M_N drops in the Meissner effect imply inhomogeneous *B* doping similar to the cases of C_6Ca and *B*-diamond at early stage.

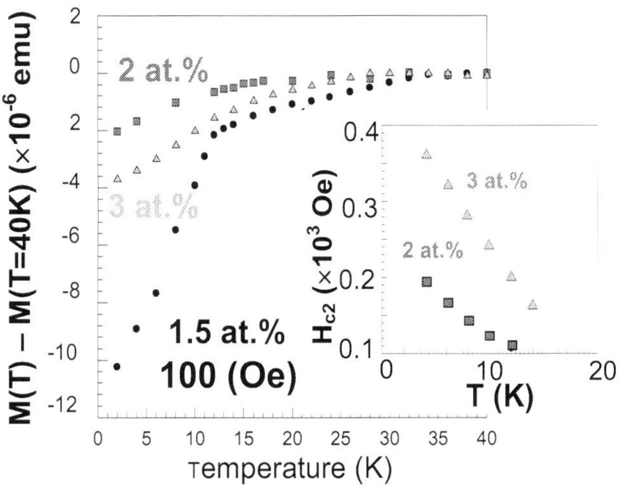

Fig. 11. Correlation of normalized magnetization drops with N_B in targets. Contribution of graphite structure on magnetization is not subtracted.
Inset: H_{c2} vs. temperature relationships estimated from M_N-H relationship in the thin films consisting of high N_B -value SWNTs (2 and 3 at. %).

Here, we find that appearance of the Meissner effect is extremely sensitive to degree of uniformity of the thin *B*-SWNT films, which can be controlled by the conditions used in the spin coating process [40]. Figure 8 shows the SEM top view and cross-sectional images (observed at a tilt of 30 degree) of the thin films consisting of assembled *B*-SWNTs fabricated without spin coating (a) and by spin coating at 500 rpm (b). The difference in the uniformities of the thin films shown in Figs. 8(a) and (b) is very evident. Further, in the case of spin coating at 1000 rpm, the extremely large number of rotations results in poor uniformity, similar to Fig 8(a). All 10 samples fabricated by the spin coating at 500 rpm exhibit the evident Meissner effect, as shown in Figs. 9 and 10. In contrast, in the samples fabricated without spin coating and by spin coating at 1000 rpm, only one sample out of 20 exhibits the Meissner effect. This result suggests that the Meissner shielding current or a superconducting vortex is not confined to individual SWNTs but exists across assembled

B-SWNTs in the thin films. Indeed, ref.[42] has predicted that Cooper pairs have a conducting probability across assembled SWNTs (with different chilarity) higher than that in individual electrons This suggestion is also consistent with ξ = ~17 nm mentioned above, because ξ corresponds to the diameter of Cooper pair and the diameter of individual SWNT is ~1 nm at most.

Figure 11 shows the correlation of the M_N–T and M_N–H behaviors with the N_B values in the catalyst in the ZFC regime. It clearly indicates strong sensitivity to N_B; i.e., (1) the M_N drops become considerably smaller and ambiguous [41] in the samples with N_B values of 2 at.% and 3 at.% as compared to those with N_B = 1.5 at.% and (2) the M_N drop in the sample with N_B = 2 at.% is slightly more than that with N_B = 3 at.%. The M_N–H relationships similar to that shown in Fig. 10 were found in the samples with N_B = 2 and 3 at.%. However, the H_{c2} values of the order of 100 Oe as shown in the inset of Fig. 11 are considerably smaller than those of the order of 1000 Oe shown in Fig. 10. The estimated values of H_{c2} (T = 0) = ~280 Oe and ξ = ~7 nm for the N_B = 3 at.% sample are also considerably smaller than the corresponding values in the N_B = 1.5 at.% sample (H_{c2} (T = 0) = ~1700 Oe and ξ = ~17 nm). These results imply that low *B* concentration yields stronger SC behavior. This is mostly consistent with the intensity change in the *RBM* as shown in Fig. 7, which exhibited the highest intensity around 1.5 and 2 at. %.

This result is also qualitatively in good agreement with the first-principles electronic-structure study of the *B*-SWNT [43]. From the electronic band structure and DOS of (10,0) SWNT with N_B values of 0.83 at.% to 2.50 at.%, it can be predicted that the occurrence of SC should be sensitive to the correlation of the position of the Fermi level with that of the VHS, and that even lower N_B value should be preferable for realizing SC as shown in inset of Fig. 12. Main panel of Fig. 12 shows the electronic structure of BC_{159} which corresponds to the N_B value as low as 0.625 at.% using the local-density approximation in the framework of the density functional theory [44]. It clarifies that the Fermi-level DOS of this system is considerably larger than those of higher N_B -value systems [45].

A resistance drop has not yet been detected owing to very high resistance (>> ~MΩ), because of (1) difficulty in establishment of good ohmic-contact between individual SWNTs and metal electrodes, (2) impossibility in having just a single *B*-SWNT aligned within the 1 μm electrode spacing since the maximum length of most SWNTs is ~1 μm, and (3) difficulty in the entire end-bonding of the *B*-SWNTs [29] since the *B*-SWNTs are present inhomogeneously in the film.

The present results assure that further optimized *B* doping into CNTs, and forming highly uniform ensemble of the *B*-SWNTs could lead to considerably high T_c (e.g., up to 30 ~ 40 K as in alkali-metal-doped fullerenes and MgB_2). Homogeneously assembled *B*-CNTs, which can provide weakly interacted CNTs (quasi-1D property) so as to maintain both the 1D properties (e.g., contribution of a VHS and strong curvature in one SWNT) and the 3D property (e.g., Meissner shielding-current path across assembled SWNTs), are promising as a novel structure which is expected to open doors to the fields of carbon-based SC.

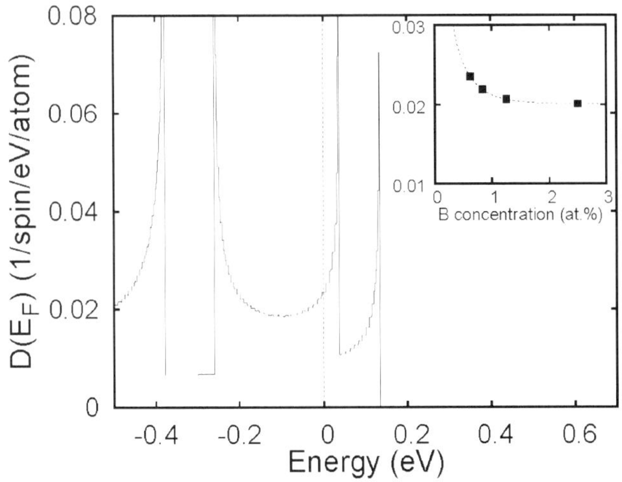

Fig. 12. Electronic density of states of individual *B*-doped semiconducting SWNT, BC_{159} obtained by using the local-density approximation in the framework of the density-functional theory. Plane-wave basis with the cutoff energy of 50 Ry is used. The host CNT is (10, 0) SWNT (i.e., the so-called zigzag nanotube) with the N_B value as low as 0.625 at.%. Energy is measured from the E_F.
Inset: Fermi-level density of states as a function of N_B.

3. Pressure-induced superconductivity at 19 K in boron-doped Buckypapers

As mentioned in section 2, the BCS-type superconductors are expected to exhibit a T_c, because of the high phonon frequency and Debye temperature of carbon atoms with low mass. Recently developed novel carbon-based superconductors (e.g., calcium-intercalated graphite and highly *B*-doped diamond [47–49]) have attracted significant attention from this viewpoint. However, the highest T_c obtained is still now lower than 15 K.

On the other hand, with CNTs, one expects much higher T_c, because (1) the curvature resulting from a small tube diameter (e.g., << 1 nm) can yield a strong coupling between the σ-π electrons and RBM [50, 51] and (2) the alignment of the E_F to a VHS in the electronic DOS (EDOS) can lead to an extremely high electronic density of states [51, 52]. In the sections 1 and 2, I reported SC in the entirely MWNTs [53–55] and in the uniform thin films consisting of *B*--SWNTs [56]. The highest T_c observed in these CNTs, however, is still now 12 K.

In section 2 [56], I showed synthesis of *B*-SWNTs by mixing elemental *B* powder into catalyst in order to intentionally generate charge carriers for SC and assembling them to thin films. We revealed that (1) *B* concentration (N_B) as low as 1.5 atomic % (at.%) and (2) assembling into highly uniform thin films led to Meissner effect with the onset T_c of 12K. I argued that this Meissner effect was realized by (1) the better alignment of E_F to a VHS in the EDOS of the *B*-SWNT [52,56] and (2) providing of a loop current path for Meissner diamagnetism across multiple *B*-SWNTs by the abovementioned two factors, respectively

[54,56]. In these viewpoints, assembling the *B*-SWNTs to film structure is crucial for realizing SC.

Recently, thin films consisting of SWNTs are attracting considerable attention for application to electrically-conductive transparent films and CNT -field effect transistors (FETs) [57]. Thus, thin films of *B*-SWNTs are also highly expected as superconducting transparent films. Here, in study of the above-mentioned previous *B*-SWNT thin films, I had proposed the following measures for increasing T_c: (1) employing lower N_B values ($N_B << 1$ at.%); (2) using thinner *B*-SWNTs (e.g., diameter << 1 nm); (3) forming a dense assembly of thin films; and (4) applying pressure to the films.

In this section, I show fabrication of paper-like thin films consisting of pseudo two-dimensional network of weakly coupled *B*-SWNTs (the so-called Buckypaper) by sufficiently dissolving as-grown ropes of SWNTs and densely assembling them on silicon substrate. I show T_c of 8K under absent pressure in the Buckypaper and find enhancement of the onset T_c up to 19 K by applying only a small pressure. An increase in the phonon frequency of the radial breathing phonon mode under applied pressure is also found and its correlation with the pressure-induced T_c is discussed.

B-SWNTs were synthesized by the pulsed laser vaporization technique [56], in which N_B was controlled by the amount of elemental *B* mixed with the Co/Ni-catalyst-impregnated targets. Substitutional *B* doping was confirmed by NMR (JNM-ECX400) and Raman spectra measurements. Conventional thin film samples of *B*-SWNTs and the present *B*-Buckypaper were prepared by solubilizing the *B*-SWNTs in dichloroethane solution with densities of 0.5–1 mg/mL and 3–5 mg/mL by centrifugation at 5000 rpm for 1 day and 10000 rpm for 2 days (Tomy; low-speed centrifuge) and ultrasonification (As One, US cleaner) for 2 days and 5 days, respectively. Then, the solutions were spin coated at 500 rpm on a Si substrate.

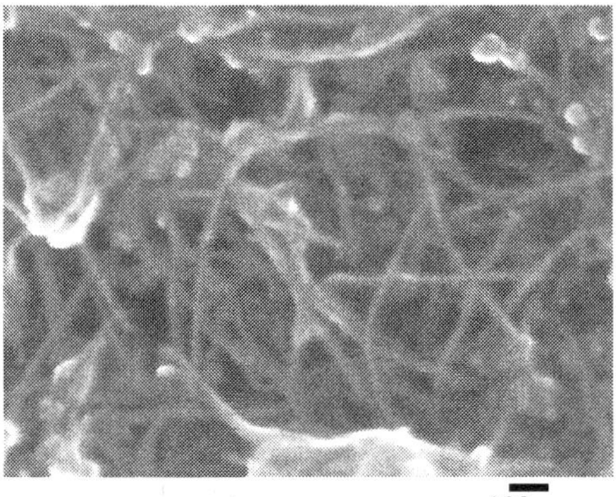

Fig. 13. (a) Top-view FESEM images of the conventional *B*-SWNT thin film fabricated by the conditions following our previous method. It consists of ropes of SWNTs.

Figure 13(a) shows the top-view field-emission scanning electron microscope (FESEM) images of the *B*-SWNTs thin films prepared by a previously reported method [56]. Although we tried to fabricate uniform thin films with dissolving ropes by the previous conditions, Fig. 13(a) reveals that the film still consists of thin ropes of as-grown *B*-SWNTs with ~10 nm diameter.

On the other hand, Fig. 13(b) shows the FESEM image of the present novel thin film consisting of *B*-SWNTs assembled densely after very strong centrifuges and ultrasonication of the SWNT solution. This figure clearly shows the absence of thick *B*-SWNT ropes (bundles) in the film and indicates that the film is sufficiently solubilized into ensemble of individual *B*-SWNTs. The fabrication conditions are harsher than those used for obtaining the film shown in Fig. 13(a). The film shown in Fig. 13(b) resembles a pseudo two-dimensional network of *B*-SWNTs deposited in the form of paper fibers. Such a SWNT network is known as a Buckypaper [57].

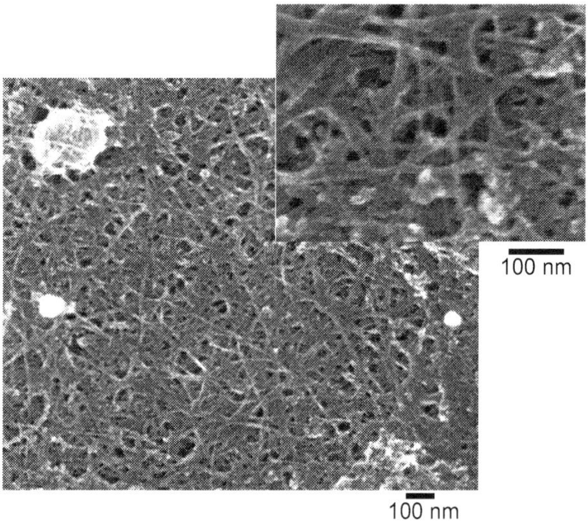

Fig. 13. (b) Top-view FESEM images of the present novel *B*-SWNT thin film (*B*-Buckypaper) fabricated using SWNTs sufficiently dissolved by strong ultrasonication. Ropes are unobservable.

We carried out magnetization measurements on the samples using a superconducting quantum interference device (Quantum Design, MPMS) and by embedding the samples in an oil-filled cell for applying pressure. Fig. 14 shows the normalized magnetization ($M_N = M(T) - M(T = 30\ K)$) of the sample shown in Fig. 13(a) as a function of temperature (T) and pressure (P) ($N_B \sim 1.5$ at.%) at a magnetic field (H) of 100 Oe in the zero-field-cooled (ZFC) regime. We could observe an apparent T_c at ~8 K and P = 0. We have confirmed that diamagnetism observed below T_c is attributed to Meissner diamagnetism for type-II SC on the basis of the magnetic field dependence of M_N (M_N–H relationship) at each temperature following our previous report [56], while the diamagnetism above T_c that appears gradually from high temperatures is due to the graphitic structure of the *B*-SWNT ropes, as also explained in ref.[56]. Figure 14 indicates that the T_c value does not change even at P value as high as 3000 MPa. As P increases further, graphitic diamagnetism becomes well pronounced. Then, T_c corresponding to Meissner diamagnetism becomes no longer apparent. This tendency is also consistent with Fig. 13(a), which indicated presence of many ropes of *B*-SWNTs that produce strong graphitic diamagnetism.

Meissner diamagnetism and graphite-originate diamagnetism were identified from measurements of M_N-H characteristics following [56] and section 2. In the Meissner diamagnetism, magnitude of diamagnetism at each temperature increases with an increase in H, while it decrease above a critical H value of ~100 Oe (lower H_c; H_{c1}) and becomes zero at a H value of ~1000 Oe (upper H_c; H_{c2}). We could confirm mostly linear relationship in the H_{c2} vs T and estimate superconducting coherence length of ~10nm from the slope value, which is within a reasonable value compared with other carbon-based superconductors. Moreover, hysteresis in the M_N-H curves could be observed on an increase and decrease in H. In contrast, in the graphite-originate diamagnetism, magnitude of the diamagnetism also increases with an increase in H. However, it does not decrease even at very high H values, e.g., as large as ~10000 Oe. Moreover, this diamagnetism appeared from high temperatures (even from T=~50K), very gradually. They have very different from those in Meissner effect.
Figure 15 shows the M_N-T relationships for the sample shown in Fig. 13(b) (B-doped Buckypaper (B-Buckypaper) with N_B ~ 1.5 at.%) at $P \leq 20$ MPa (Fig. 15(a)) and $P \geq 20$ MPa (Fig. 15(b)) at H = 100 Oe in the ZFC regime. At P = 0, T_c is evident at T = ~8 K in Figs. 15(a) and (c). Meissner diamagnetism at $T_c \leq 8$ K and graphitic diamagnetism at $T_c \geq 8$ K could be identified from the M_N–H relationships, as in the case of Fig. 14.

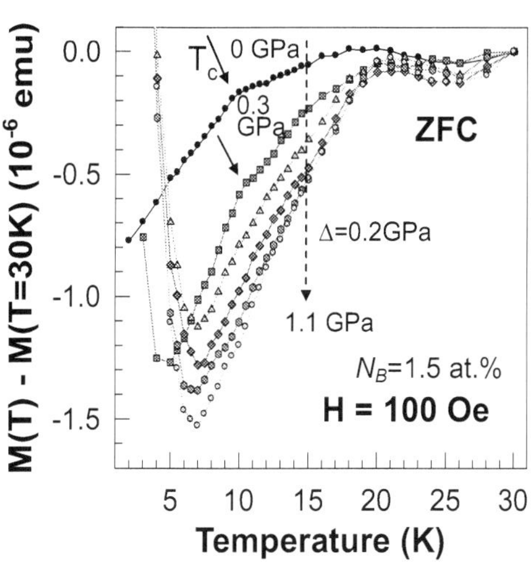

Fig. 14. Pressure dependence of normalized magnetization vs. temperature relationships for sample shown in Fig. 13(a) (conventional B-SWNT film) in ZFC regime.

In contradiction to Fig. 14, as P increases, T_c shifts to higher temperatures over two times (i.e., T_c is 8K at 0 Pa, 12 K at 10 MPa, and 19 K at 20 MPa). This behavior of T_c is well apparent in Fig. 15(c), which was identical to Fig. 15(a) but has Y-axis in arbitrary unit. In contrast, as can be seen in Fig. 15(b), as P exceeds 20 MPa, M_N increases almost over the entire temperature region, and the value of T_c becomes unclear (i.e., possibly T_c = 10 K at 30 MPa and 8 K at 40 MPa). The overall T_c-P relationship is shown in Fig. 3(d). These behaviors are very different from those shown in Fig. 14 and suggest that the pressure dependence of the Meissner effect is highly sensitive to the structure of the B-SWNT assembly.

Moreover, Fig. 15 shows that graphitic diamagnetism appearing above T_c does not become stronger with an increase in P; in contradiction to that shown in Fig. 14. This behavior is consistent with the absence of the *B*-SWNT ropes in the *B*-Buckypaper as shown in Fig. 13(b). Furthermore, the M_N–T behavior observed with increasing P in Fig. 15(a)(b) is analogous to that observed in type-II SC with increasing H. This indicates a possibility of SC regions in the *B* -Buckypaper being compressed and becoming thin under applied pressure, because the aforementioned behavior can be observed when magnetic flux partially penetrates the superconductor as H is increased.

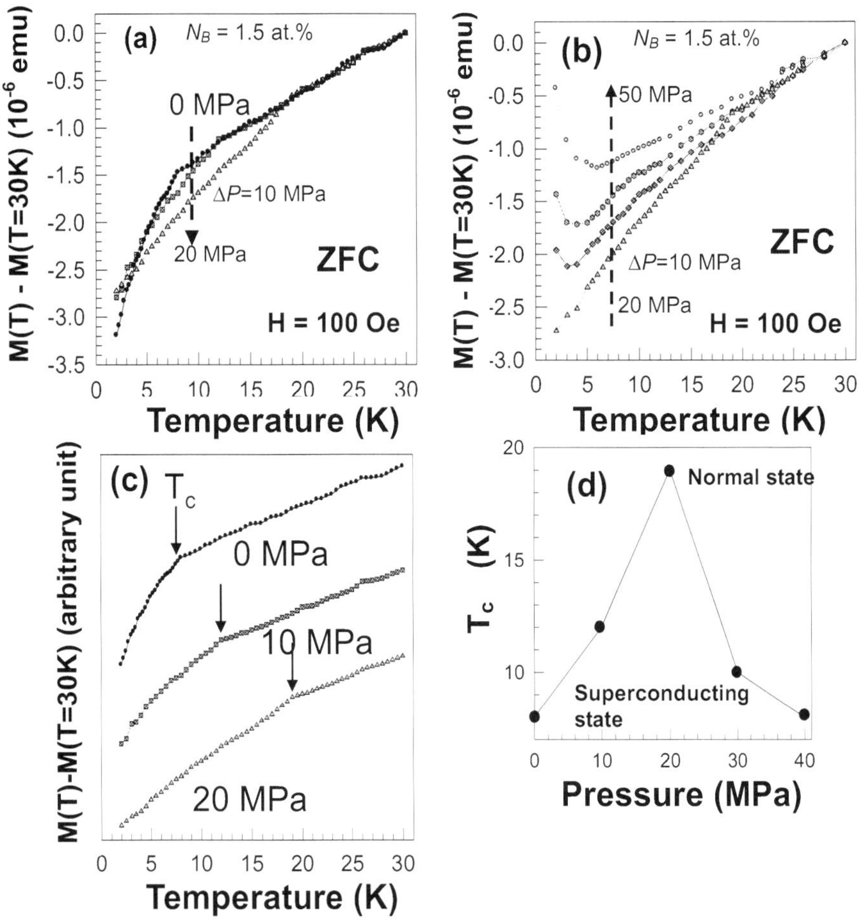

Fig. 15. (a), (b) Pressure dependence of normalized magnetization vs. temperature relationships of sample shown in Fig. 13(b) (*B*-Buckypaper) at P ≤ 20MPa (a) and P ≥ 20MPa in zero-field cooled regime. (c) Identical to (a) but with Y-axis in arbitrary unit. (d) Pressure dependence of T_c.

Here, we discuss the following two causes for the difference in the behavior of T_c with an increase in *P* (as shown in Figs. 14 and 15); (1) Increase in phonon frequency and (2) Change in charge transfer in the *B-C* bonds. From the viewpoint (1), it is known that application of pressure to the SWNT ropes (bundles) compresses the triangular lattice of an SWNT rope and thus increase the phonon frequency. This in turn can cause an upshift in the frequency of the RBM and the G band in the Raman spectra [58, 59]. If a strong electron-phonon coupling exists when the frequency of the phonons increases under the application of pressure, T_c should also increase in the manner shown in Fig. 15.

Figure 16 shows the pressure dependence of the Raman spectra for the sample shown in Fig. 13. Figures 16(a) and 16(b) correspond to the conventional film shown in Fig. 13(a), and Figs. 16 (c) and 16(d) correspond to the *B*-Buckypaper shown in Fig. 13(b). The pressure dependences of the peak positions observed in Figs. 16(a)-(d) are shown in Figs. 16(e) and 16(f). From Figs. 16(e) and 16(f), it is clear that at $P = 0$, the peak positions observed for the *B*-Buckypaper are lower than those observed for the conventional film in both the RBM (Figs. 16(a) and 16(c)) and the G-band mode (Figs. 16(b) and 16(d)). This trend is consistent with the observation in Fig. 13, because the phonon frequency increases by intertubule van der Waals coupling of the SWNTs in a rope structure as shown in Fig. 13(a). This is also strong evidence for the sufficient dissolving of ropes in the *B* -Buckypaper, as shown in Fig. 13(b).

With an increase in *P*, all the peak positions show a monotonic upshift for ratios of ~1 cm^{-1}/100 MPa and ~2 cm^{-1}/100 MPa in the RBM (Fig. 16(e)) and ~0.7 cm^{-1}/100 MPa and ~1 cm^{-1}/100 MPa in the G-band mode (Fig. 16(f)) for the conventional film and the *B*-Buckypaper, respectively. These observations imply that the increase ratios for the *B*-Buckypaper are larger than those for conventional film in both modes. In particular, the RBM peak position for the *B*-Buckypaper shows the highest upshift ratio, which is twice the upshift ratio observed for the conventional film. Here, although the mean diameter of the present *B*-SWNTs is ~1 nm, the *B*-Buckypaper can include SWNTs with diameters smaller than 1nm (e.g., ~0.6nm). It is known that σ-π electrons resulting from the sp^3 hybrid orbitals in such thin SWNTs yield strong coupling with the RBM for formation of the loop current for Meissner diamagnetism [51]. Thus, the significant upshift in the RBM phonon frequency of *B*-Buckypaper shown in Fig. 16(e) can be origin for the pressure-induced T_c in Fig. 15. Thus, the increase in T_c seen in Fig. 15 can be mainly attributed to the increase in the RBM phonon frequency in the regions assembled by such thin *B*-SWNTs.

Also in the SWNT ropes shown in Fig. 13(a), pressure-induced phonon frequency could be observed in the RBM (Fig. 16(e)). However, the small increase ratio implies that the increase in phonon frequency is more significant in the *B*-Buckypaper. This is because the RBM is not restricted by van der Waals coupling in the individual SWNTs of the sufficiently solubilized *B*-Buckypaper and the phonon frequency can be easily induced under the application of pressure.

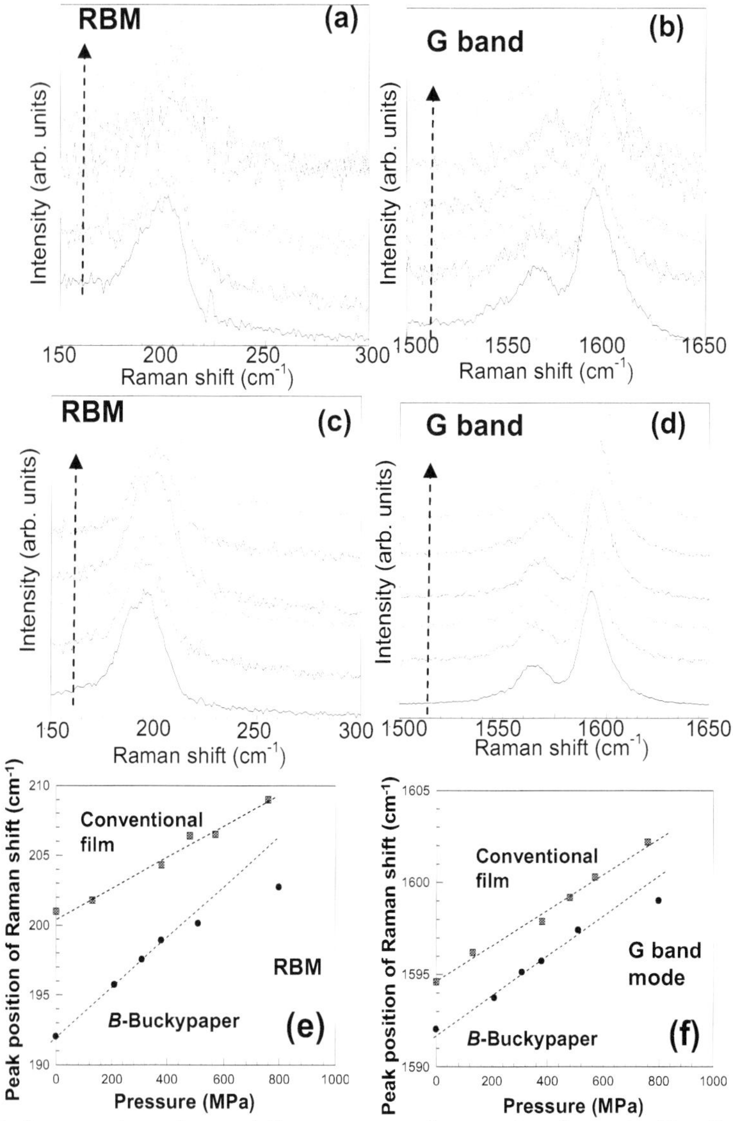

Fig. 16. Pressure dependence of Raman spectra for samples shown in Fig. 13; (a), (b) conventional film and (c), (d) *B*-Buckypaper. Spectra were measured by an irradiating Ar laser of 488 nm wavelength at room temperature. *P* values are 0, 130, 380, 480, 570, 760 MPa for (a) and (b), and 0, 210, 310, 380, 510, 800 MPa for (c) and (d) from the bottom to the top curves.

Peak positions vs. pressure relationships in (e) RBM and (f) G-band mode. The peak positions are determined by data fitting based on Gaussian-Lorentzian curves in (a)-(d).

On the other hand, the following difficulties may arise for the abovementioned interpretation: (1) the applied pressure in Fig. 15(a) is as small as 20 MPa, which corresponds to an increase of 0.4 cm^{-1} in the phonon frequency; (2) in the case of the sample shown in Fig. 13(a), T_c does not change in spite of the increase in the phonon frequency. However, these difficulties can be understood by difference in the electron -phonon coupling between the thin SWNTs and the ropes of SWNTs. With respect to difficulty (1), it has been predicted [51] that in very thin SWNTs, T_c increases drastically even for a small increase in the phonon frequency owing to the strong electron -phonon coupling (the strong superconducting coupling strength λ_{SC}). With respect to difficulty (2), it can be said that electron-phonon coupling is weak in the case of the sample shown in Fig. 13(a). This is due to reduction in the electron-phonon interactions in the rope structure, as it is known that intertube coupling in the rope structure weakens the electron-phonon coupling [51].

Here, from the standpoint (2) for the different T_c behavior with an increase in *P*, changes in charge transfer caused by the *B-C* bond modulation under the application of pressure can result in two effects for the increase in T_c.

One is the improved alignment of E_F to a VHS by a decrease in the charge transfer. According to [56], with a decrease in N_B (< 1 at %) and charge transfer, alignment of E_F to a VHS is improved, resulting in a drastic increase in EDOS and in T_c. When the *B-C* bonds become weak and charge transfer decreases under the application of pressure, this phenomenon can appear in the present system. In fact, in *B*-doped diamond, T_c decreases under high pressures [49], because of the decrease in charge transfer in the *B-C* chemical bonds. Thus, a similar phenomenon can occur in the present *B*-Buckypaper, although its crystal structure is different from that of *B*-doped diamond. In our present experiment, a decrease in charge transfer and consequent shift of the E_F position are small, because *P* is small. However, an increase in the EDOS is extremely sensitive to the alignment of the E_F to the VHS [52, 56] and can induce T_c drastically.

Moreover, *B-C* bonds can also be located inside a rope structure in the film shown in Fig. 13(a), where they are not significantly affected by the applied pressure. In contrast, the *B-C* bonds in individual SWNTs are easily affected by the pressure, as in the case of the well-solubilized *B*-Buckypaper shown in Fig. 13(b). Consequently, the modulation in the *B-C* bonds can be the second cause for the increase in T_c more or less. However, showing evidence for the modulation of *B-C* bonds is indispensable.

The other is hole doping by an increase in the *B-C* bonds and an increase in phonon frequency renormalized by hole-phonon interactions [60, 61]. However, this is not the present case, because the charge transfer in the *B*-C bonds is induced by applied pressure based on the above-mentioned discussion.

Further investigation of the causes for the pressure dependence of T_c is necessary. However, the successful fabrication of the *B*-Buckypaper studied herein shows the feasibility of producing superconducting materials from light-mass atoms such carbon and boron. Moreover, the present results strongly imply a possibility for obtaining higher T_c above 20K in CNT ensembles by optimization of boron doping and film structures.

4. Possibility of quantum computation by utilizing carbon nanotubes

-Cooper pair splitting by Tomonaga-Luttinger liquid-

Challenges for quantum computation (quantum bit) have been carried out in many material systems, such as superconductor- and semiconductor-quantum dots with single carrier charging effect, flux-controlled superconductor rings, Bose-Einstein condensed atoms and molecules, and NMR. No studies of quantum computation, however, have been reported in CNTs. It is well known that CNTs have strong electron spin coherence due to the unique electronic states and, hence, electron spins can be effectively transported through CNTs without losing the phase over a long distance and in a long time [62]. This characteristic gives a good chance for realizing quantum computation.

One of realistic ways for such challenges using CNTs is creation of quantum spin entangler (QSE) utilizing Cooper pairs injected from s-wave superconductor. Indeed, two theoretical studies have predicted possibilities of QSE by proposing a hybrid system of a superconductor and two CNTs within TLL states. They predicted separation of Cooper pairs into individual spins by strong repulsive Coulomb interaction of TLLs in CNTs [63][64]. As explained in section 1, TLL is non-Fermi liquid states consisting of a repulsive Coulomb interaction between electrons confined in 1D conductors and has been frequently reported in CNTs by observing power laws in relationships of conductance (G) versus energies (E) (i.e., $G \propto E^{\alpha}$) [65][66]. Strong repulsive electron-electron interaction in TLLs of CNTs can efficiently separate spin singlet of Cooper pairs, which are injected from s-wave superconductors, into individual spins and allows injection of the separated spins into different CNTs. Ref.[64] predicted differences in formulas for power laws between the cases of tunneling of separated Cooper pairs and non-separated Cooper pairs from a superconductor electrode to CNTs. Ref.[63] predicted that efficiency of the splitting of Cooper pairs strongly depends on tube spacing, dimension of superconductor, and strength of TLLs, etc.. Moreover, one experimental study reported on presence of possibly strong spin entanglement in spin singlet of Cooper pairs in CNTs by observing co-existence of Kondo singlet in Kondo effect and spin singlet in Cooper pairs in CNT quantum dots [67].

However, there are no any other experimental studies, which revealed behaviors of coherence and entanglement of Cooper pairs in CNTs, because it was difficult to efficiently inject Cooper pairs into CNTs owing to high interface resistances of CNTs/metal electrode junctions. In contrast, I successfully realized high-transparent interfaces of Nb/MWNTs by utilizing end-bonded MWNTs synthesized in nanopores of alumina templates and found efficient injection of Cooper pairs (i.e., proximity-induced superconductivity) into CNTs [68]. In addition, I reported on possibilities of strong spin entanglement in Cooper pairs and separation of them into individual spins by TLLs [75].

Here, in this section, I report on (1) Evidence of spin separation in Cooper pairs by the TLLs and injection into different CNTs and (2) Possibility of molecular spin entangler by utilizing them. For the 1st term, I report that injected Cooper pairs are separated by the TLLs following theories and individual spins exist in different CNTs. Magnetic-field dependence in these spins is for the first time investigated. For the 2nd term, I propose a hybrid system of one superconductive CNT and two Aharonov-Bohm (AB) ring CNTs within TLLs based on the 1st and 2nd terms. I explain that this system can be a good candidate for molecular spin quantum bit and entangler without decoherence problem.

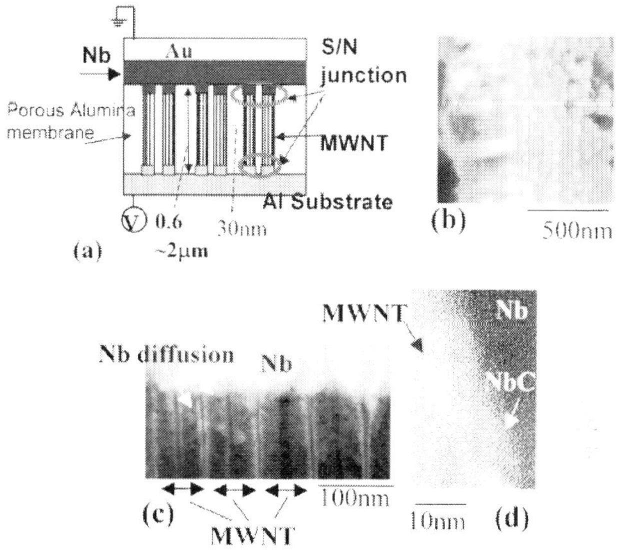

Fig. 17. (a) Schematic cross sections of an array of end-bonded Au/Nb/MWNT/Al junctions, synthesized into the nanopores of alumina membranes by chemical vapor deposition of methanol gas, each with a thickness of 1μm for Au/Nb. T_c of our Nb was about 8.1 - 9K and H_c was about 1500 G. The mean outer and inner diameters are 100 and 60 nm, respectively, with a shell thickness as large as 20 nm. (b) SEM overview image of a MWNT array, standing in the alu,ima membrane, implying high quality and open top portion of MWNTs. (c) High-angle annular dark-field image of a cross sectional TEM (CSTEM) around Nb/MWNT interface array annealed by the optimal conditions, implying a high diffusion of Nb atoms. No diffusion of Nb into lower ends of the MWNT was also reconfirmed by EDX analyses. **(d)** High resolution CSTEM image in (b).

Figure 17(a) shows the schematic cross-sectional view of the sample structures (i.e., array of Nb/multi-walled CNs (MWNTs)/Al junctions), which were prepared in array of nanopores of alumina templates. That's similar to the system in which I reported on intrinsic SC in section 1. However, in the present case, electrode metal is superconductor (Nb) and the MWNT was synthesized by acetylene gas. I have reported on a variety of quantum phenomena in CNTs and metal nano-wires in this system [69]. This structure allows us to end-bond the MWNTs by evaporating metal electrode on the top ends of MWNTs (see Figs.17 (b) and (c)), as mentioned in section 1. This end-bonding has significantly enhanced interface transparency of metal electrode/CNTs junctions. I have successfully reported on proximity-induced superconductivity in this structure.

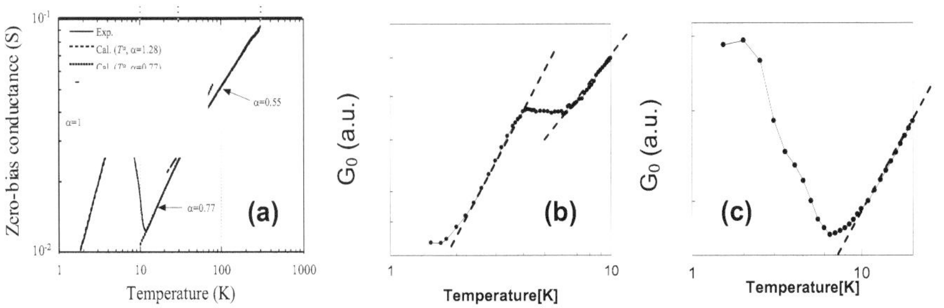

Fig. 18. Temperature dependence of the zero-bias conductance G_0 on a doubly logarithmic scale in NbN/MWNTs/Al junctions for different tube spacing with the 600 nm-length MWNTs within a ballistic charge transport regime and TLL state. Tube spacing in arrays are (a) 20 nm, (b) 30nm, and (c) 40nm. α in (a) is power value of power law behaviors in each temperature region.

Figure 18(a) shows the doubly logarithmic scales of zero-bias conductance versus temperature relationship in the sample with L_{tube} = 600 nm and 20-nm tube spacing in a MWNT array. When the length of MWNTs for Fig. 18(a) was reduced below mean fee path ~ 700nm like this case, the MWNTs took a transition to a ballistic charge transport regime and showed power law behavior at temperatures > T_c~10K. Such power laws in the relationships of conductance vs. energies have been evidence for TLLs in both MWNTs [65] and SWNTs [66], with the following various powers α.

$$\alpha^{bulk} = (g^{-1}+g-2)/8 \quad (6)$$

for tunneling from electrode to the bulk of CNTs and

$$\alpha^{end} = (g^{-1}-1)/4 \quad (7)$$

for tunneling from electrode to the ends of the CNTs

The observed correlation exponent g, the Luttinger parameter, is dictated by the strength of the electron-electron interaction, with ranges $g < 1$, $g > 1$ reflecting respectively its repulsive or attractive character. It determines α, as low as ~ 0.2 stressed the presence of a strong repulsive Coulomb interaction in CNTs.

Because the observed power exponent α = ~ 0.8 is in good agreement with α reported for Au-end junction [62], which just corresponds to our junction structure, this implies presence of TLLs. Correlation exponent g can be estimated to be g = ~ 0.25 from the formula for (7) [65]. In contrast, this behavior disappears and conductance suddenly increases around T_c and, then, it starts to decrease again at T < 7K, following power law with the α ~ 1.28 larger than α~ 0.8 mentioned above (Fig. 18(a)). This result stresses separation of Cooper pairs, injected from Nb, into individual spins by TLLs of the MWNTs in accordance with the theory in [64].

Ref.[64] predicted that a tunneling probability Γ_{AB} of Cooper pairs from one superconductor to ends of two CNTs within TLLs is given by

$$\Gamma_{AB} = \sim (eV/h)(kT/\varepsilon)^{[1/g - 1]/2} \tag{8}$$

when Cooper pairs are separated into two spins, which tunnel into different two CNTs. We can obtain $\alpha = \sim 1.5$ by inserting $g = \sim 0.25$ into this formula. This $\alpha = \sim 1.5$ is approximately in agreement with the observed $\alpha = \sim 1.28$ in Fig. 18(a) and, hence, suggests possible splitting of Cooper pair by TLLs. In order to reconfirm this argument, the results, which were measured in the samples with tube spacing of 30 nm and 40 nm, are shown in Fig. 18 (b) and (c), respectively. In Fig. 18(c), the conductance decrease, which appeared at T < 7K in Fig. 18(a), entirely disappeared and only a conductance increase (i.e., proximity-induced superconductivity) is observable down to T = 1.5K. In contrast, a conductance decrease following power laws appears at T < 3 K, which is lower than T = 7K in (a), in Fig. 18(b). These results are qualitatively in good agreement with the prediction by theory [63] as follows.

The theory [63] predicted inequality

$$A(2\mu/\Delta)^{2\gamma} < (1/k_F\delta r)^{d-1} \tag{9}$$

as the efficiency of Cooper pair splitting by TLLs in the system similar to ref.[64], where d and Δ are the dimension of superconductor and superconducting gap, respectively, μ is the applied bias voltage between the superconductor and the leads, δr is the tube spacing, and γ is the exponent for tunneling into the bulk of a single TLLs, which is proportional to *g*. As this inequality is easily satisfied, the efficiency of Cooper pair splitting increases. This inequality implies that increase in tube spacing δr results in a lower probability in Cooper pair splitting. Because emergence of only PIS in Fig. 18(c) for the largest δr means no splitting of Cooper pairs, this agrees with the theory [63]. Shift of critical temperature for power law from T = 7K in Fig. 18(a) to T = 3K in Fig. 18(b) also supports this argument. Consequently, the results in Fig. 3 are qualitatively in good agreement with this theory [63].

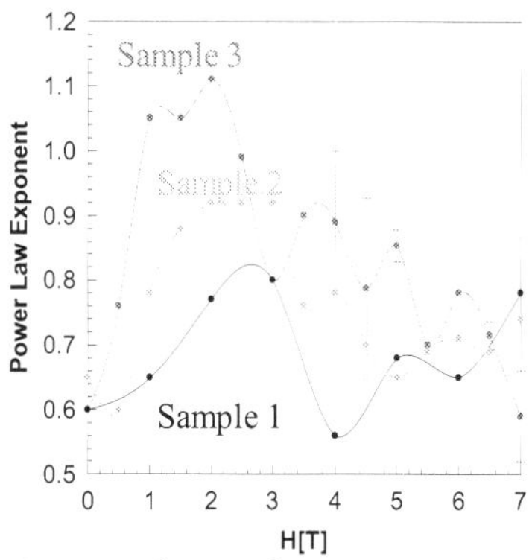

Fig. 19. The value of power α in power laws at low temperatures in Fig. 18(a) as a function of magnetic field for different three samples.

Figure 19 shows the values of powers α in the low temperature region in Fig. 18(a) as a function of magnetic fields (H). Power laws still survive, even if H = 7T is applied. The value of α is sensitive to H. We find that the value of α increase as H starts to increase at H < ~ 3T, while it monotonically decreases at H > ~ 3T with exhibiting small oscillative behaviors. Ref.[71] predicted that the value of α monotonically decreases as H increases in TLLs within a short range Coulomb interaction, because spins with opposite momenta localize at opposite ends of a CNT due to the applied H and, hence, a repulsive Coulomb interaction between the electrons is reduced and the value of g increases, resulting in a decrease in α. This theory is qualitatively in good agreement with the decrease in α in the region of H > 3T in Fig. 19. Hence, this can be evidence for TLLs. Because the tube length of about 600 nm is large enough for a short-range interaction and Coulomb interaction is easily reduced by the edge localization, this agreement is appropriate.

In contrast, the increase in α at H < 3T cannot be explained by this theory. The difference between conventional TLLs described by theory [71] and our system is at least that separated Cooper pairs are injected into the ends of MWNTs in our case. Hence, this increase in α at low fields can be attributed to this difference. We propose one interpretation for this result as follows. When separated Cooper pairs are injected as individual spins into different MWNTs at H = 0 T, momentum of spins are highly aligned in comparison to those in single spins injected from normal conductor electrodes. When two spins separated from Cooper pair are still entangled state, this alignment is much stronger. Moreover, MWNTs have spin coherence. Hence, when low fields are applied to such highly aligned spins, electron-electron interaction may be enhanced, resulting in the increase in α as shown in Fig. 19. As the fields increase further and electron start to localize, this situation is destroyed and the value of α decreases. Therefore, this can be evidence for entangled spins existing in different MWNTs.

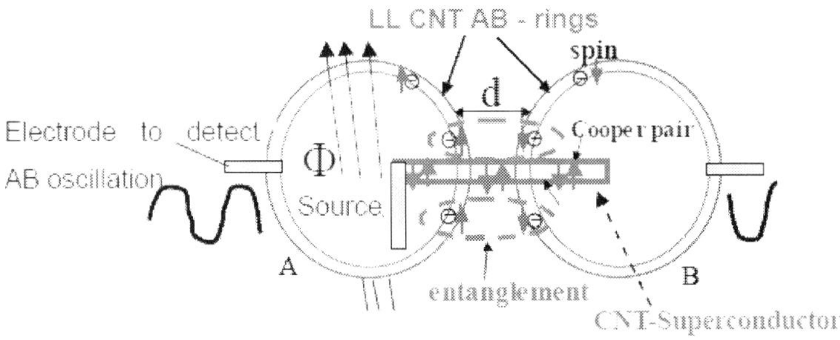

Fig. 20. An example of molecular quantum spin entangler; A hybrid system of one superconductive CNTs and two Aharonov-Bohm (AB) ring CNTs within TLLs. Cooper pairs, which are injected from the SC-CNT to AB CNT-ring, are separated to individual spins by repulsive Coulomb interaction of TLL state and injected to different rings. Strong spin coherence and entanglement of Cooper pairs in CNTs might realize spin quantum bit with dissolving decoherence problem.

Here, Fig. 20 shows one of examples of molecular QSE, based on a hybrid system of one superconductive CNT and two normal conductive CNTs within TLL states discussed above. Normal conductive CNTs form AB rings in this system. Cooper pairs are efficiently separated to individual spins by TLLs in the AB-ring CNTs, because superconductor CNT is within d = 1 regime and following the inequality (9) [63], efficiency of Cooper pair splitting

can be enhanced by reduction of dimension of superconductor (d). The separated spins are injected into the different CNT-AB rings with opposite spin momenta maintaining memory of spin singlet. As shown in Fig. 19, electron-electron interaction is enhanced at low fields. Hence, when a low magnetic field is applied so as to penetrate only into one AB ring, AB oscillation can be very sensitive to the applied fields. If the two spins existing in the different AB rings are still within entangled state, spin existing in the other AB ring should show opposite-phase AB oscillation even without applying magnetic fields. Therefore, one can investigate spin entanglement by simultaneously observing phase of these two AB oscillations. If Cooper pairs in CNTs actually have a strong entanglement [67][75], separated spins retain a strongly entangled state crossing the different CNTs. This results in realization of good molecular QSE. One can explore this by increasing AB ring spacing.
Therefore, CNTs can be good candidates for realizing molecular quantum computation with strong spin coherence and entanglement.

Acknowledgements

The article described here was the results from many collaborators. J.H. sincerely acknowledges H.Shinohara, H. Fukuyama, J. Akimitsu, T. Ando, S.Maruyama, S. Saito, R.Saito, S.Tarucha, Y.Iye, N. Nagaosa, H.Takayanagi, M.Tsukada, Y.Hasegawa, T.Sugai, S. Bandow, M.Dresselhaus, A. M. Rao, J. Gonzalez, H. Bouchiat, C. Schoenenberger, D. Loss, R. Egger, A. Bachtold, P.E.Lindelof, J.-P. Leburton, J.Reppert, P.Recher, T.Koretsine, T.Nishio, H.Sano, Y.Awano, H.Nihey S.Okazaki, Y. Muranaka, O. Kamo, T.Nakanishi, T.Akazaki, W.Izumida, H.Sugiura, M.Tachibana, and my students and stuff.

5. References

1. M.Kociak, et al., Phys.Rev.Lett. 86, 2416 (2001)
2. I.Takesue, J.Haruyama et al., Phys.Rev.Lett.96, 057001 (2006). Power laws have been also discussed from viewpoints of Coulomb blockade coupled with its external electromagnetic environment (i.e., MWNT acts as a *LC* transmission line connected to tunnel junctions) [4] and also one-dimensional localization. However, the present case does not correspond to these models, because the calculation result of our MWNT as LC transmission line does not agree with the observed power laws and also the present interplay of power laws with SC cannot be understood by these models.
3. Z.K.Tang, et al., Science 292, 2462 (2001)
4. A.Bachtold, et al., Phys.Rev.Lett. 87, 166801 (2001)
5. M.Bockrath, et al., Nature 397, 598 (1999); Z.Yao, et al., Nature 402, 273 (1999)
6. R.Egger, Phys.Rev.Lett. 83, 5547 (1999)
7. I.Takesue, J.Haruyama, H.Takayanagi et al., Physica E 24, 32 (2004); P.Recher et al, Phys.Rev.B 65, 165327 (2002); C.Bena et al., Phys.Rev.Lett.89, 037901 (2002)
8. J.Gonzalez, Phys.Rev.Lett. 88, 076403 (2002); A.Sedeki, et al., Phys.Rev.B 65, 140515(R) (2002); J.Gonzalez, Phys.Rev.Lett. 87, 136401 (2001)
9. D.Loss and T.Martin, Phys.Rev.B 50, 12160 (1994-II)
10. E.Perfetto and J.Gonzalez, Phys.Rev.B 74, 201403(R) (2006)

11. R.Egger and A.O.Gogolin, Phys.Rev.Lett. 79, 5082 (1997)
12. C.Kane, L.Balents et al., Phys.Rev.Lett. 79, 5086 (1997)
13. M.Monteverde, G.Garbarino et al., Phys.Rev.Lett. 97, 176401 (2006)
14. A.De Martino and R.Egger, Phys.Rev.B 67, 235418 (2003), Phys.Rev.B 70, 014508 (2004)
15. K.McGuire, et al., Carbon 43, 219 (2005)
16. K.Liu, et al., Phys.Rev. B 63, 161404R (2001)
17. Experimental observation of TLL states and its confirmation by theory requires ν(E) in CNTs [1], while SC transition may not favor a tunnel junction. However, Ref [12] predicted that even attractive Coulomb interactions could lead to power law behaviors along with perfect transmission of Cooper pairs via tunnel junction. Thus, the present result is consistent with this theory.
18. v_F (= hk_F/m^*) and L are the Fermi velocity and the tube length of the 1D conductor, respectively. The previous studies on TLL states in CNTs reported that the saturation regimes fall onto a universal curve in any temperature ranges.
19. N. Murata, J.Haruyama, S.Maruyama, H.Shinohara et al., Phys. Rev. B 71, 081744 (2007)
20. N. Murata, J.Haruyama, J.Reppert, A.Rao et al., Phys.Rev.Lett. 101, 027002 (2008).
21. T. Koretsune, S. Saito, Phys. Rev. B 77, 165417 (2008)
22. As the critical temperature of 6 K for the appearance of the deviation is in excellent agreement with that from Fig. 4(b), the SC regime potentially exists even at 3 K $< T \leq$ 6 K, though the conductance peak is not seen there due to the obstruction by the TLL regime.
23. T. E. Weller et al., *Nature Physics* 1, 39 (2005).
24. N. Emery et al., *Phys. Rev. Lett.* 95, 87003 (2005).
25. E. A. Ekimov et al., *Nature* 428, 542 (2004).
26. M. Kociak et al., *Phys. Rev. Lett.* 86, 2416 (2001).
27. Z. K. Tang et al., *Science* 292, 2462 (2001).
28. M. Ferrier et al., *Phys. Rev. B* 73, 094520 (2006).
29. I. Takesue, J.Haruyama H.Shinohara et al., *Phys. Rev. Lett.* 96, 057001 (2006). "The T_c = 12K" can be a magic number. The common T_c=12K implies presence of common origins in ensemble of *B*-CNTs (e.g., the best alignment of E_F to a VHS under this N_B value). In contrast, because C_6Ca has a different structure, the agreement of T_c may be occasional.
30. N. Murata, J.Haruyama, S.Maruyama, H.Shinohara et al., *Phys. Rev. B* 71, 081744 (2007).
31. R. Barnett et al., *Phys. Rev. B* 71, 035429 (2005).
32. M. Matsudaira, J.Haruyama., H.Shinohara et al., *Physica E* 40/7, 2299-2304 (2008)
33. M. Matsudaira, J.Haruyama., et al., *Phys. Rev. Lett.* (2008) in submission.
34. I.Takesue, et al., *Physica E* 24, 32 (2004); P. Recher, D. Loss, *Phys. Rev. B* 65, 165327 (2002); C. Bena et al., *Phys. Rev. Lett.* 89, 037901 (2002).
35. J. Gonzalez, *Phys. Rev. Lett.* 88, 076403 (2002).
36. E. Perfetto, J. Gonzalez, *Phys. Rev. B* 74, 201403(R) (2006).
37. K. McGuire et al., *Carbon* 43, 219 (2005).
38. K. Liu et al., *Phys. Rev. B* 63, 161404R (2001).
39. S. Bandow, S. Numao, S. Iijima, *J. Phys. Chem. C* 111, 11763 (2007).
40. http://www.ee.aoyama.ac.jp/Labs/j-haru-www/

41. Graphite diamagnetism in ensemble of SWNTs very gradually appears in *M-T* curves from high temperatures (e.g., > 50K) even in undoped samples and the amplitude does not decrease in *M- H* features even at high magnetic fields (e.g., > 2000 Oe). They are very different from those in Meissner effect observed here. However, because the component is very small in the samples shown here and different by respective samples, we could not entirely delete it in this study.

42. J.Gonzalez, Phys.Rev.Lett. 88, 076403 (2002)
43. T. Koretsune, S. Saito, *Phys. Rev. B* 77, 165417 (2008)
44. http://www.pwscf.org/

45. Because we have measured SWNTs with $N_B \geq 1.5$ at.%. in the present experiment, $N_B = 1.5$ at.% is the lowest value and, hence, it is qualitatively consistent with the theory. For quantitative agreement, the present theory should take into account influence of "ensemble" of B-SWNTs. Here, if the SWNTs are undoped, the difference in chirality (i.e., semiconducting /metallic behaviors) of the respective SWNTs is important for producing SC. In contrast, in carrier-doped SWNTs, SC is no longer sensitive to them, because one can freely control the position of E_F.

46. N.Murata, J.Haruyama, M.Matsudaira, J. Reppert, A.Rao, T.Koretsune, S.Saito, Y.Yagi, Phys.Rev.Lett. 101, 027002 (2008)
47. T. E. Weller et al., *Nature Physics* 1, 39 (2005).
48. N. Emery et al., *Phys. Rev. Lett.* 95, 87003 (2005).
49. E. A. Ekimov et al., *Nature* 428, 542 (2004).
50. M. Kociak et al., *Phys. Rev. Lett.* 86, 2416 (2001).
51. R. Barnett, E. Demler et al., *Phys. Rev. B* 71, 035429 (2005).
52. T. Koretsune, S. Saito, *Phys. Rev. B* 77, 165417 (2008)
53. I. Takesue, J. Haruyama H. Shinohara, S. Maruyama et al., *Phys. Rev. Lett.* 96, 057001 (2006).
54. N. Murata, J. Haruyama, S. Maruyama, H. Shinohara et al., *Phys. Rev. B* 71, 081744 (2007).
55. M. Matsudaira, J. Haruyama, S. Maruyama, H. Shinohara et al., *Physica E* 40/7, 2299 (2008).
56. N. Murata, J. Haruyama, J. Reppert, A.M. Rao, T. Koretsune, S. Saito, et al., *Phys. Rev. Lett.* 101, 027002 (2008).
57. J.A. Fagan, M.L. Becker, J. Chun, E.K. Hobbie, *Advanced Materials* 20, 1609 (2008).
58. U.D. Venkateswaran, A.M. Rao, R.E. Smalley, P.C. Eklund at al., *Phys. Rev. B* 59, 10928 (1999-II)
59. U.D. Venkateswaran, A.M. Rao, P.C. Eklund et al., *Phys. Stat. Sol.* (b) 223, 225 (2001)
60. J.C. Tsang, M. Freitag, V. Perebeinos, J. Liu, Ph. Avouris, *Nature Nanotechnologies* 11, 321 (2007).
61. A.M. Rao, P.C. Eklund, S. Bandow, A. Thess, R.E. Smalley, *Nature* 388, 257 (1997).
62. L.Langer, V.Bayot, J.-P.Issi, et al., Phys.Rev.Lett 76, 479 (1996)
63. P.Recher and D. Loss, Phys. Rev. B 65, 165327 (2002);
64. C.Bena et al., Phys.Rev.Lett.89, 037901 (2002)
65. A.Bachtold, C.Shonenberger, et al., Phys.Rev.Lett. 87, 166801 (2001); M.R.Buitelaar, et al., Phys.Rev. Lett. 88, 156801 (2002); A.Bachtold, et al., Nature 397, 673 (1999);
66. M.Bockrath, et al., Nature 397, 598 (1999); H.Ishii, H.Kataura, et al., Nature 426, 540 (2003)
67. M.R.Buitelaar, T.Nussbaurner, C.Shonenberger, Phys.Rev. Lett. 89, 256801 (2002)

68. J.Haruyama, et al., Phys.Rev.B 68, 165420 (2003); Appl.Phys.Lett. 84, 4714 (2004); Microelectronics Journal 34, 537 (2003); Physica C 408, 85 (2004)
69. J. Haruyama, et al., Phys. Rev. B 65, 33402 (2002); Phys.Rev.B, 073406 (2001); Appl.Phys.Lett. 81, 3031 (2002); Appl.Phys.Lett. 77, 2891 (2000)
70. A. Yu. Kasumov, H. Bouchiat, et al., Science 284, 1508 (1999); Phys.Rev.B 68, 214521 (2003)
71. S.Bellucci and P.Onorato, Cond-mat 0504170
72. M. Kociak, H. Bouchiat, et al., Phys. Rev. Lett. 86, 2416 (2001)
73. M.Tinkam, *Introduction to Superconductivity* (McGraw-Hill, New York 1996)
74. J.Gonzalez, Phys.Rev.Lett. 88, 076403 (2002); A.Sedeki, et al., Phys.Rev.B 65, 140515(R) (2002); J.Gonzalez, Phys.Rev.Lett. 87, 136401 (2001)
75. J.Haruyama, et al., Physica Stat. Sol. (b) 242(2), 265 (2005)

34

Nucleic acid interaction and interfaces with single-walled carbon nanotubes

Dovbeshko Galina[1], Fesenko Olena[1], Gnatyk Olena[1], Shtogun Yaroslav[2], Woods Lilia[2], Bertarione Serena[3], Damin Alessandro[3], Scarano Domenica[3] and Adriano Zecchina[3]

[1]*Department of Physics of Biological Systems, Institute of Physics, National Academy of Sciences of Ukraine, Prospect Nauki 46, Kiev, Ukraine 03680.*

[2]*Department of Physics, University of South Florida, Tampa, Florida 33620.*

[3]*Department of Chemistry IFM, Centre of Excellence of Nanostructured Interfaces and Surfaces and INSTM Centro di Riferimento, University of Torino, Italy.*

The discovery of the carbon nanotubes has opened up a new field in biomedical research. Indeed, the recognition of carbon nanotubes by DNA, DNA assisted separation of carbon nanotubes, DNA immobilization on carbon nanotubes surfaces have been demonstrated. The knowledge of carbon nanotubes-DNA interaction is of fundamental importance for using carbon nanotubes/biomechanical complexes. In this study experimental and theoretical study of the DNA-interface coupling is performed to achieve better understanding of the properties of many carbon nanotubes - based biosystems, as well as novel phenomena caused by the interaction of carbon nanotubes with biomolecules. This paper is focused on the study of interactions between single walled carbon nanotube (SWCNT) surfaces with DNA, poly-A, and individual nucleotides, to clear up the conformation changes in these complexes. New spectroscopic technique based on the effect of enhancement of infrared (IR) absorption by rough metal surface (SEIRA) together with surface enhanced Raman spectroscopy (SERS) and atomic force microscopy (AFM) for registration of structural changes in carbon nanotubes and DNA/carbon nanotubes complexes were applied. SEIRA spectroscopy data was compared with *ab initio* quantum-chemical calculations performed for thymine and adenine adsorbed on carbon nanotubes. A possible model of interaction between nucleic acid bases, double and single nucleic acid strands and the carbon nanotube surfaces is derived from the experiments and calculations.

1. Introduction

The discovery of carbon nanotubes in 1991 (Iijima, 1991), and afterwards, the production of bulk quantities (Dresselhaus, 1996) have paved the way to exploration of the physical and chemical properties of these quasi-one dimensional graphite structures. Carbon nanotubes have been envisioned in a variety of applications such as gas and battery devices, as efficient

energy storage elements, sensors and electromechanical systems, biocompatible agents and recognition elements, and more (Dresselhaus et al., 1996), (Dresselhaus et al., 2000).
In regard to the above mentioned carbon nanotubes application, one of the most promising and interesting properties is the ability of the carbon nanotubes to interact with biological molecules providing to new properties of the carbon nanotubes/biomolecule complex.
The discovery of carbon nanotubes has opened up a new field in biomedical research as well. Indeed, functional bio-nanosystems, which are formed by functionalizing inorganic nanostructures with biochemical molecules, utilize unique properties of biochemical species recognition, DNA recognition, drug delivery systems, DNA transfection and more. The use of carbon nanotubes in this regard is of particular importance. The functionalization of carbon nanotubes with proteins, nucleic acids, DNA and RNA has been pioneered. More specifically, the DNA recognition of carbon nanotubes (Williams et al., 2002), DNA assisted separation of carbon nanotubes (Zheng et al., 2003 (a)), DNA immobilization on CNT surfaces (Guo et al., 1998) have been demonstrated.
In connection to using carbon nanotubes/biomechanical complexes, the knowledge of carbon nanotubes/DNA interaction is of fundamental importance. Today several models on DNA interaction with carbon nanotubes are known. The most popular model is a DNA wrapping around carbon nanotubes proposed by Smalley (O'Connell et al., 2001). In (Zheng et al., 2003 (a)) a molecular modeling have suggested that single stranded DNA (ssDNA) can bind to carbon nanotubes through π-stacking via bases. The authors of (Zheng et al., 2003 (a)) have demonstrated the poly-T wrapping around carbon nanotubes and concluded about many ways in which short ssDNA strands can interact with nanotube surface, namely, wrapping with right-handed turns, wrapping with left-handed turns, (absorb) on the surface with a linearly extended structure. In other paper (Zheng et al., 2003 (b)) the authors concluded that ssDNA can adopt many different modes of binding to carbon nanotubes with little difference in binding free energy. These modes include helical wrapping with different pitches and forming a fever allowed conformation than ssDNA.
However, due to complexity, full comprehension of the mechanism of interaction of carbon nanotubes and nucleic acids (NA) has not been established up to date. That is why we did the experimental measurements and theoretical modeling of the DNA/interface couplings to achieve better understanding of the properties of the carbon nanotubes-NA, as well as the induced novel phenomena. It will also help us to establish new concepts of controlling and tuning the performance of such systems and facilitate the design and optimization of DNA/carbon nanotubes based functional devices. *The goal* of our paper was to study the interaction of single walled carbon nanotubes with the double stranded DNA dsDNA), Poly-A in double and single strands, thymine and adenine with modern sensitive methods - surface enhanced infrared absorption (SEIRA) and surface enhanced Raman scattering (SERS), as well atomic force microscopy (AFM) and *ab initio* calculation.

2. Methods and Materials

2.1 SEIRA and SERS spectroscopy

The enhancement of optical process by a factor $10^2 \ldots 10^6$ near rough surface of a metal is already known for optical transitions in adsorbed molecules and the processes which do not depend on the presence of the molecules on the metal surface. These processes are SERS, SEIRA, metal enhanced fluorescence, second harmonics generation. For the first time an enhancement of IR signal from molecules chemisorbed on Au and Ag surface by factor up to 10^3 has been registered by Harstein and colleagues in 1980 (Harstain et al., 1980). Only in the beginning of the nineties the scientists came back to study of the enhanced IR effect, mainly due to very low limit of the molecules that could be detected by IR and its some advantages (non-destructivity, higher signal-noise ratio, less distorted spectra) in comparison with SERS. In 1991 the effect was named as SEIRA by Osawa and Ikeda (Osawa & Ikeda, 1991). The effect can not be explained in a simple way, it includes several mechanisms, such as: 1) increase of electromagnetic field near the rough metal surface or metal film, 2) increase of dipole transition moment of adsorbed molecules; and was theoretically described by V. Kosobukin (Kosobukin, 1983; Kosobukin, 1985).

The advantages of SEIRA compared to other spectroscopic methods show that SEIRA is desirable when biological complexes are studied. This is dictated by the fact that when one works with biological complexes, usually very small amounts with desirable properties are available. SEIRA provides a mechanism for enhancing the biological substances' spectroscopic properties. In addition, biological molecules are with very complicated structures and many properties from different constituents of the biological compound entangle and mix. SEIRA provides an opportunity to monitor specific parts of the biological compound and its conformations. Therefore, the surface enhancement methods has very attractive features which will be useful when DNA/carbon nanotubes conformations are described. SEIRA spectroscopy was applied to study a conformation of DNA, Poly-A, at the single-walled carbon nanotubes (SWCNT) interface for an ensemble of SWCNT/biological molecule with of the use of a special SEIRA support (Dovbeshko, 2004 (a); Dovbeshko, 2004 (b)). The SEIRA application for the study of nanotubes gives a possibility to increase a sensitivity of IR spectroscopy by 3-10 times and enhances the absorbance of biological molecules adsorbed on the nanotube surface located on the rough gold surface (Fig. 1).

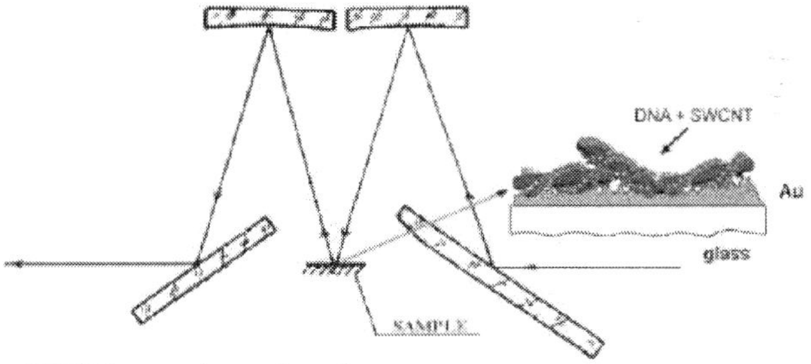

Fig. 1. Scheme of SEIRA experiment in reflectance geometry

A drop from the transparent part of the solution was then taken for SERS, AFM, and SEIRA studies. The characterization of individual nanotubes wrapped by Poly-A, as well as the Poly-A conformation on the separated nanotubes, was probed and enhanced with a conventional Klarite SERS support. The factor of enhancement in SERS experiments on the Klarite support was equal to 10^4-10^6 for different vibrations in comparison with a conventional Al support. The spot of the field for SERS and AFM study was taken as 1x1 µm. SERS spectroscopy give us a possibility to register a signal from 1-2 nanotubes and Poly-A adsorbed on the nanotubes.

2.2 FTIR and Raman spectra evaluation

A Bruker IFS-66 instrument with a reflection attachment (the incidence angle of 16.5°) was used for the registration of IR spectra in the reflectance mode in the region of 400-4000cm^{-1}. Spectra of DNA, Poly-A, adenine, thymine and their complexes were collected in reflectance mode. Evaluation of the spectra has been done with Opus 5.5. The position of the bands has been estimated with the use of the method of second derivative and/or standard method. The intensity of the bands was normalized to the band with maximal intensity in each spectrum, namely, to OH-NH-CH stretching vibration in the region 3340-3400 cm^{-1}. The bands have been assigned to the certain molecular groups in DNA, poly-A, according to references: (Taboury et al., 1985; Taillandier et al., 1985; Tajmir-Riahi et al., 1995; Schrader, 1995; Parker, 1983; Litvinov, 1991; Dovbeshko, 2008, Dovbeshko, 2009). Reproducibility of the frequency in the IR spectra was equaled to ±0.5 cm^{-1} and for absorption ±0.0005 a.u.). Earlier we studied the validity of the use of the effect of enhancement of infrared absorption by metal surface for nucleic acids, lipids, amino acids and albumin (Dovbeshko et al., 2002; Dovbeshko et al., 2004, Dovbeshko et al., 2006 (a); Dovbeshko et al., 2006 (b)) deposited onto Au substrate in the reflectance experiments. The results show that when using the SEIRA method for the study of IR absorption of nucleic acids (and for other molecules), no deformation of contour of the absorbance bands is observed as compared with the spectrum on neutral CaF_2. Some frequency shifts are present, but for the most part of the bands they are small (1-2 cm^{-1}).

In our experimental conditions we registered the DNA with 60% humidity and poly-A with 65% humidity. The microscope experiment was done using a Via Raman Microscope (Renishaw) instrument with laser excitation at 785 nm with 0.3-60 mW power.

2.3 AFM imaging

The microphotograph of the Au substrate for SEIRA, Klarite support for SERS as well as Poly-A adsorbed on nanotubes placed on Klarite support was obtained by atomic force microscopy (AFM). We used the tapping mode under AFM imaging of the gold surface of home-made support with a commercial Nanoscope IIIa (Digital Instrument, Santa Barbara, CA). Tapping force mode scans were performed using commercially available AFM tips (silicon nitride). The scanning frequency was approximately 1 Hz in all experiments.

AFM microscopy with a Park Scientific Instrument Auto Probe LS was used to analyze the nanotube-Poly-A interface. We have obtained images on a SWCNT in the contact mode with a silicon nitride ultra sharp tip (all images are obtained in air).

2.4 Au home-made support for SEIRA and Klarite support for SERS

Au home-made support for SEIRA was produced in the Institute of Physics of NASU (Kiev, Ukraine) (Dovbeshko et al., 2004 (b)). Gold thin layer was obtained by vacuum deposition of 99.999% pure Au upon glass supports (TF-1 glass, 20x20 mm) via an intermediate adhesive Cr layer. Before Au deposition, glass surface was cleaned by NH_4OH:H_2O_2:H_2O and HCl:H_2O_2:H_2O solution subsequently, both 1:2:2 by volume concentration during 5 minutes at boiling temperature. Then it was rinsed in bidistilled water and dried in a flow of pure nitrogen. The gold was evaporated from molybdenum heater and deposited at a rate of 1.0-1.5 nm s^{-1} on room temperature substrate. The thickness of gold surface was within 300-350 Å in the different experiments. The Cr interlayer does not exceed 10-20 Å. The gold surface just after deposition looks like hydrophobic surface with wetting angle close to 80° and random roughness about 50 Å (Fig. 2).

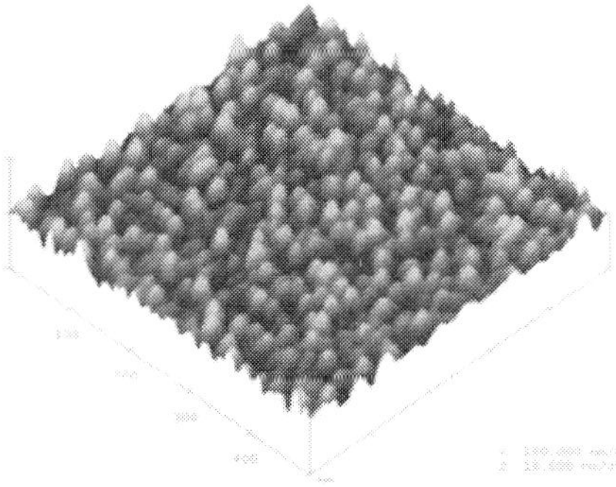

Fig. 2. AFM image of gold surface of home made support used in SEIRA

Fig. 3. AFM Image of gold surface of Klarite support used in SERS

We used standard Klarite support for SERS produced by Mesophotonics (Fig.3).

2.5 Principal component analysis

Principal component analysis was applied to SEIRA spectra of DNA. As one principal component was chosen a relative intensity of phosphate asymmetrical band (at 1240 cm^{-1}) to intensity of maximum in the 3400 - 2300 cm^{-1} region assigned to OH stretching vibration. According to (Shie, 1977), this component characterizes the number of water molecules per 1 nucleotide. The second principal component was ratio of intensity at 1712 cm^{-1} and 1700 cm^{-1} that is a characteristics of conformation state of DNA relating to A, B or Z-form. Earlier we have estimated the contribution of above-mentioned components in principal components and found that they have shown the preferential contribution (Dovbeshko et al, 2004 c). The amount of the molecules of water per nucleotides in DNA was estimated according to a formula:

$$W_w^{\alpha} = \frac{N_{H_2O}}{N_{\text{nucleotide}}} = \frac{\mu^{\alpha}_{H_2O}}{\mu_{\text{nucleotide}}} = \frac{D^{\alpha}_{3400}}{D^{97\%\text{relative humidity}}_{1228^{-1}_{cm}}} \cdot \frac{\varepsilon_{1228cm^{-1}}}{\varepsilon_{3420cm^{-1}}}; \quad (1)$$

$\varepsilon_{3420cm^{-1}}$ = 101 ± 7 $L \cdot M^{-1} \cdot cm^{-1}$ (value obtained by averaging the data from different authors) $\varepsilon_{1228cm^{-1}}$ = 530 ± 9 $L \cdot M^{-1} \cdot cm^{-1}$. Then we obtained:

$$W_w^{\alpha} = 5,25 \frac{D_{water}}{D_{\text{phosphate}}} \quad (2)$$

The error of calculation of amount of molecules of water was calculated as a square root from the sum of squares of standard errors averaged by all multipliers in a formula 1. Standard averaged error is determined as: $\sqrt{\frac{\sigma^2}{n}}$, where σ is standard deviation, n-amount of experimental points. Standard deviation for $\varepsilon_{3420cm-1}$ equaled to 6,9%, for $\varepsilon_{1228cm-1}$

cm^{-1} =1,7%. Then standard averaged error for $\varepsilon_{3420cm-1}$ cm^{-1} equaled to 3,2 %, for $\varepsilon_{1228cm-1}$ cm^{-1} =0,8%. An error for D is determined by systematic instrumental errors and is 0,1%. Then a total error of calculation of amount of molecules of water equaled to 3,3%.

2.6 Method of the dispersion of SWCNT

SWCNTs have been synthesized by arc discharge between two graphite electrodes in a He atmosphere at the Moscow Institute of Physics (Obraztsova et al., 1999). The nanotube content in the raw material was about 30-40 wt%. The contaminants are amorphous carbon, metallic catalyst particles, soot, etc. The material has been certified by high resolution transmission electron microscopy and Raman scattering (Dovbeshko et al., 2003; Dovbeshko et al., 2008,). The length of SWCNTs was 1-2 µm, and the diameters are 0.9-1.6 nm with preferential fraction of 1.4-1.5 nm. The probe of SWCNTs was mixed with DNA and/or Poly A aqueous solution (0.25-0.5mg/ml) (0.1-1 weight part of nanotubes to 1 weight part of Poly-A) by an ultrasonic mixer Sonorex TK52 (60 W, 35 kHz) during 60 min (Fig. 1) with a subsequent centrifugation under 30 000 turn/min during 30-60 min. Na-salt of calf thymus DNA was purchased from Servo; Polyadenilic acid potassium salt (polyA), Adenine (Ade) was purchased from Fluka (Germany), Thymine (Thy) was purchased from Sigma and used without additional purification. Aqueous solution of DNA (0.25 mg DNA/ 0.5 ml H_2O) was heated up to 100 °C, then 0.5 mg SWCNT was added. Compounds of the solution were mixed, stirred and then precipitated onto quartz substrate covered with gold layer. The concentration of adenine and thymine solutions in SEIRA experiment was 1 mg/ml. In the experiment first we isolated the nanotubes with aqueous (alcoholic) solution of Thy (Ade) (1:1 weight ratio) by ultra-sound mixing (UZDN-A, Sumy, Ukraine) for 1 hour. Then we have deposited a drop of the solution of Ade (Thy) with carbon nanotubes on gold and CaF_2 substrate. The output of nanotubes during the process was 2-8%, the concentration of poly-A in different experiments was 10^{-3}-10^{-4} M, and the concentration of nanotubes was 10^{-4}-10^{-5}M as estimated according to (Dai, 2005). This procedure separates nanotubes from a bundle for their further individual characterization and usage (Fig. 4). After the treatment, the inhomogeneous solution contained 3 parts of different densities. The upper and bottom parts contained a black dust with nanotubes and contaminations such as soot, metallic particles, etc., while the mid transparent part (Fig. 4) contained a solution with isolated nanotubes.

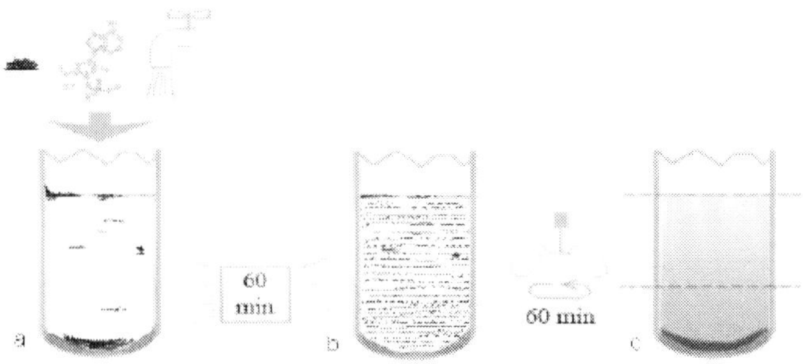

Fig. 4. Scheme of the nanotube-Poly-A solution: (a) before mixing in an ultrasonic mixer, (b) bottom and (c) top part of the solution after centrifugation

The pH of the solution was measured with a pH-150 M (Belorussia, Gomel) instrument with 0.01 accuracy.

2.7 Method of Calculation

The method for this work will be based on Density Functional Theory (DFT) (Hohenberg & Kohn, 1964)] electronic structure calculations. The calculations were done utilizing the Vienna Ab Initio Simulation Package (VASP). This is state of the art Projection Augmented Wave (PAW) simulation methods which use exact valence wave functions improving the reliability of the potential (Kresse & Joubert, 1999). The code is available and supported at the USF computing facility.

This code uses a plane wave basis and a periodic supercell method with Vanderbilt pseudopotentials or a projector-augmented wave method to account for the electron core. The adsorption on the metallic SWCNT (6,6) with length of 12,25 Å, and the adsorption of thymine on the semiconducting (8,0) SWCNT with length of 17,035 Å, were considered with nonspin-polarized LDA (local density approximation) for the exchange-correlation functional. DFT is known to be suitable for the description of short ranged strong forces and to give an imperfect description for long range interactions as the local density approximation (LDA) tends to overestimate and the generalized gradient approximation (GGA) for the exchange correlation energy tends to underestimate the binding (Shtogun et al., 2007), (Woods et al., 2007). DFT gives a better description of these systems than empirical models or classical molecular dynamics models and it has proven to be a powerful tool in the adsorption description of carbon nanotubes (Girifalco & Hodak, 2002).

3. Result and Discussion

3.1. Characterization of carbon nanotubes

Carbon nanotubes are extended cylindrical structures with a diameter from one to a few ten of nanometers and length of a few centimeters (Iijima, 1991; Dresselhaus et al., 2000). They consist of one or more hexagonal graphene planes rolled up in a tube with a hemispheric head. Characterization of the carbon-based materials could be done with Raman spectra being a fingerprint of it.

The main feature of the Raman spectra of graphite structure is, so called G mode (1589 cm^{-1}) according to the theory characterizes the tangential vibrations of carbon atoms of graphene layer of E_{2g}, E_{1g}, A_{1g} symmetry in the Γ point of Brillouin zone and correlates with the ordering of crystal lattice of graphite structure. Frequency shift in high frequency region means reduction of the areas of coherent scattering of the materials under study (Saito & Dresselhaus, 1998). The second characteristic modes of graphite-like materials, so called D-mode, with A_{1g} symmetry that characterizes the disorder of lattice of graphene layer. It refers to breathing vibrations of rings of graphene layer in the K point of Brillouin zone (Saito & Dresselhaus, 1998) and has been registered at 1291 cm^{-1} for nanotubes (Fig.5). The second order mode of this vibration (so called D_1) well registered at 2568 cm^{-1} in SWCNT and they have the intensity which usually exceeds the intensity of second-order vibrations (Fig.5). The last fact could be the evidence that nanotubes consist of similar structures that manifest a strong electron-phonon interaction and strong dispersion dependence of D-mode. Other carbon-like materials don't have such peculiarity.

SWCNT have a low frequency mode of significant intensity that is assigned to the breathing vibration of a whole tube (Saito & Dresselhaus, 1998).

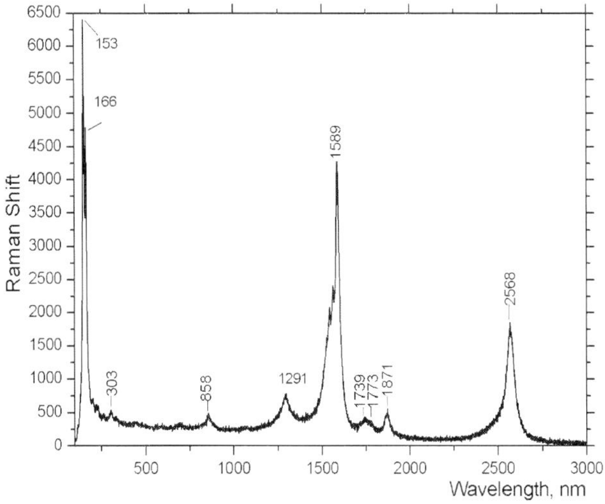

Fig. 5. Raman spectra of initial fraction of nanotubes

The wavenumbers, halfwidth and relative intensity of above mentioned bands could be used for characterization of type of allotropic forms of carbon, the nanotubes and their diameters.

3.2 Characterization of complex SWCNT/Poly A

Visualization of the separate nanotubes with poly-A was done with AFM (Fig. 6). According to AFM, size of nanotubes after separation was 4.3 nm in vertical axis and 10 times more- 40.31 nm in horizontal axis (Dovbeshko et al. , 2008). It is known that probe method, AFM included, does not give real object image, it is the product of probe only. Due the fact that curvature of probe is much more than size (diameter) of the object under study, we should expect that lateral size of the object is much more than its real size. In the same time the height of the object keeps real size as the nanotube is single object and method is valid for height estimation without error. So, shows that height of nanotube is equaled to 4 nm, and its lateral size 40 nm estimated with big error (Dovbeshko et al., 2008). According to this data we could suppose that 1 or 2 nanotubes have been separated from the bundle and registered by AFM (Fig.6).

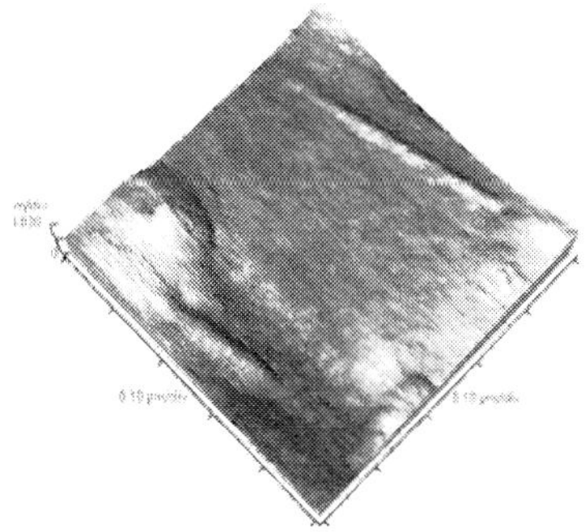

Fig. 6. AFM image SWCNT

Breathing mode in RS spectra is sensitive to diameter of nanotubes and density of states and depends on frequency of excited light. Linear dependence between reverse diameter and position of breathing mode is expressed by the formula:

$$\nu\,(cm^{-1}) = \frac{223.75}{D_1\,(nm)} \tag{3}$$

(Chen et al., 1998) that obtained from *ab initio* calculation. In the most cases nanotubes in the bundle are studied by scientists, so Van der Waals interaction between nanotubes in the bundle are realized and determined by the diameter of nanotubes and number of nanotubes in the bundle, that is why additional constant are appeared in the formula or coefficient in the term of fraction are changed. Different formulas are applied by different authors for calculation of diameter of nanotubes. We estimated a diameter of nanotubes in initial fraction of nanotubes as well in solution, accordingly, with the formula in the following form:

$$\nu\,(cm^{-1}) = \frac{234}{D_2\,(nm)} + 10 \tag{4}$$

(Bokova et al., 2003) and/or

$$\nu\,(cm^{-1}) = \frac{248}{D_6\,(nm)} \tag{5}$$

(Jorio et al., 2001) that is in the good agreement with data of electron microscopy.

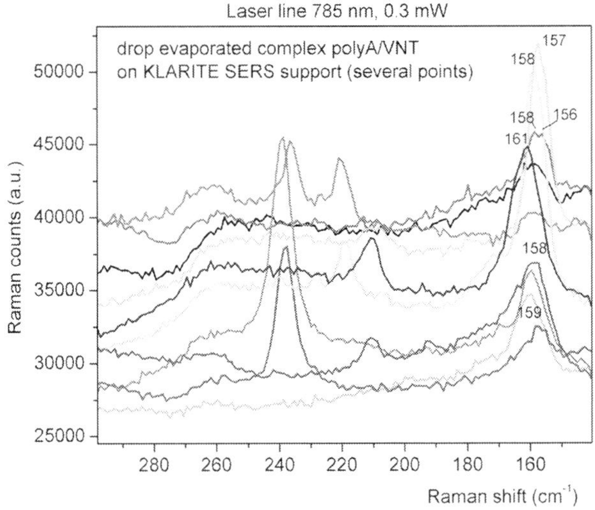

Fig. 7. SERS of isolated nanotubes in the region of «breathing modes» in the different points of the support

Fit peaks for breathing modes was done with Peak Fit Program. The nanotubes have a number of "breathing modes" with the frequency: 136, 148, 156, 162, 169, 177, 219, 233, 264 cm^{-1} (Dovbeshko et al., 2006(c)). Diameter of isolated single wall carbon nanotubes wrapped by poly-A has been estimated by formula (3) and (5) that is used for estimation of diameter of isolated nanotubes (Dovbeshko et al. , 2008). For isolated nanotubes by poly-A (Fig.7) the frequencies of breathing modes are 157, 158, 159, 161, 221, 236, 238 cm^{-1} and diameter of this nanotubes are in the range 0,94-1,58 nm with preferential fraction of diameter about 1 nm (Dovbeshko et al. , 2008).

3.3 DNA conformation under SWCNT interactions

DNA is the basic building block of the cell. Hereditary information is encoded in the chemical structure of DNA and reproduced in all living organisms. In Fig. 8 we show dsDNA although, other forms exist, such as ssDNA, B-DNA, A-DNA, Z-DNA, etc. with various stretching and twists (Saenger, 1988; Strick et al., 2003). The common structure to all of them is special combination of bases, sugar and phosphorous groups. Describing adsorption of different complexes on CNT surfaces in general, is a complex process, which stems from different adsorption mechanisms possible (Bruch et al., 1997). The adsorption can occur through a chemical bond between the CNT and the compound, which is a chemisorption process and it is a result from strong short range forces. The adsorption can also be physisorption where no chemical bond is formed and it is due to long range dispersive forces such electrostatic, Van der Waals, dipolar forces. The last type of adsorption will be discussed here. No functionalization nanotubes will be done in the experiment. In the case of DNA-SWCNT complex, we obtained the changes of the vibrational modes in comparison with reference DNA according to the marker bands of the conformational state of DNA (Schrader, 1995; Liquier et al., 1991). It is known (Schrader, 1995; Liquier 1991), that marker bands of DNA in B-helix conformation are the following: 1715-1710 cm^{-1} (base pairing), 1228-1220 cm^{-1} (phosphate asymmetric band), 1088-1085 cm^{-1} (phosphate symmetric band), 890 and 835 cm^{-1} ($C_{2'\text{-}endo/anti}$ conformation). A-helix of DNA has characteristic features at 1705-1700 cm^{-1} (base pairing), 1240-1230 cm^{-1} (phosphate asymmetric band), 1092-1090 cm^{-1} (phosphate symmetric band), 890,878,860 and 805 cm^{-1} ($C_{3'\text{-}endo/anti}$ conformation). It is known (Schrader, 1995), that marker bands of DNA in Z conformation are the following: 1700-1692 cm^{-1} (base pairing), 1434 cm^{-1} (A-T pairs), 1215 cm^{-1} (phosphate asymmetric band), and 925-929 cm^{-1} (C-O-C deoxyribose), 867-864 and 838-834 cm^{-1} ($C_{3'\text{-}endo}$ conformation).

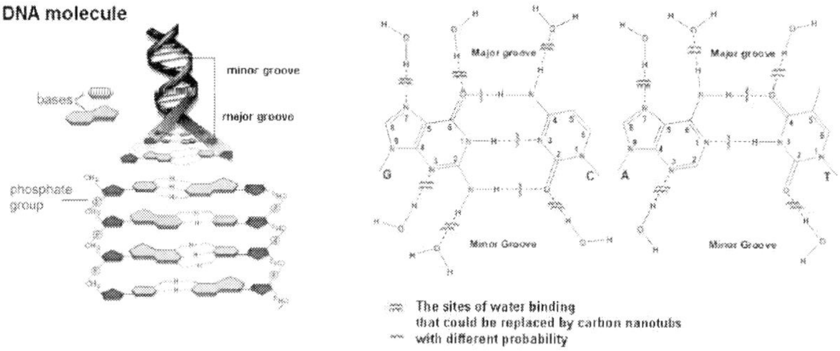

Fig. 8. DNA structure, minor and major grooves with water molecules

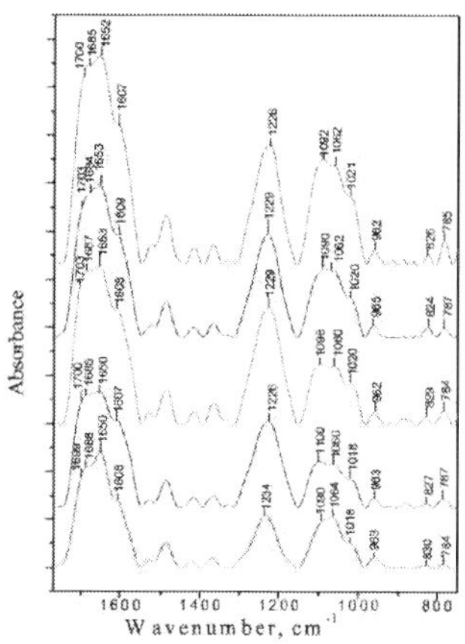

Fig. 9. SEIRA spectra of different parts of the DNA-SWCNT sample in comparison with the reference DNA (bottom spectra).

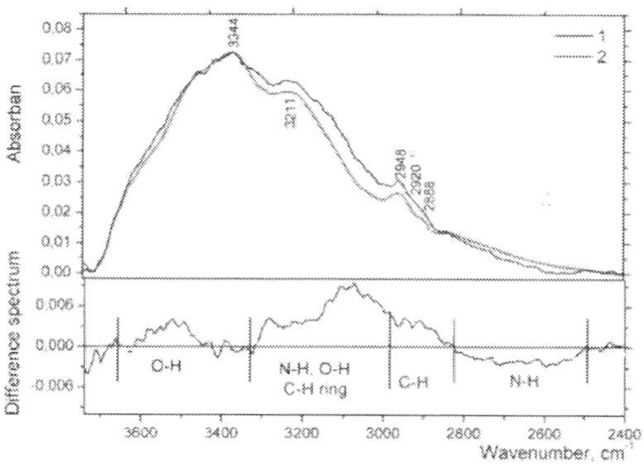

Fig. 10. SEIRA spectra of reference DNA (curve 1), DNA with carbon nanotubes (curve 2) on the gold substrate in the region 2400–3800 cm^{-1} and its difference spectrum. Both spectra was normalized to intensity of the maximum of the band in the 2400-3700 cm^{-1} region.

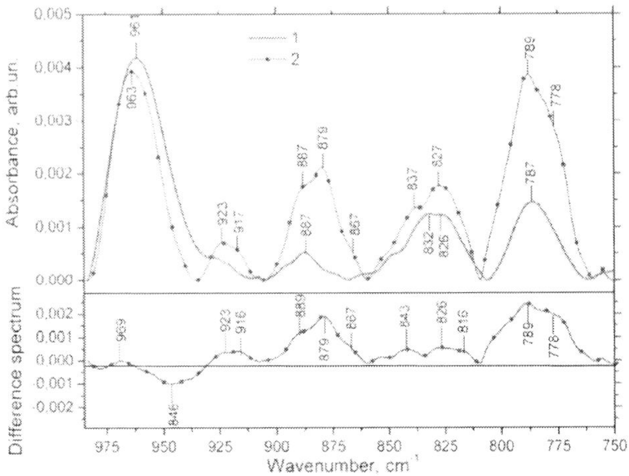

Fig. 11. SEIRA spectra of reference DNA (curve 1), DNA with carbon nanotubes (curve 2) on the gold substrate in the region 750-1025 cm-1. Difference spectrum - DNA with carbon nanotubes-reference DNA

As it could be seen from Fig.9, the spectra of different DNA-SWCNT parts of the same sample changed drastically in comparison with the reference DNA (bottom spectra). The phosphate I band shifts in high frequency region from 1090 cm^{-1} in reference DNA to 1100, 1096 (non-A- or B-form or Z-form), 1092 cm^{-1} (A-form) in the DNA-SWCNT, the phosphate II band has low-frequency shift from 1234 cm^{-1} in reference DNA to 1226, 1229 cm^{-1} (B-form) in the DNA-SWCNT. The most contribution in phosphate asymmetrical band in DNA (1220-1240 cm^{-1}) is assigned to PO_2^- vibration. The phosphate symmetrical band (1085-1092 cm^{-1}) is more complex and deoxyribose (C-O, C-C) gives a great contribution and influences the shape of this band. However, no known forms of DNA with the position of phosphate I at 1100 cm^{-1}. It says about drastic curvature of DNA backbone as well about restoring of helix structure in DNA (frequency moves from 1064 to 1060 cm^{-1}). So, the changes in phosphate backbone along with sugar moieties could be induced by SWCNT and explain the position of phosphate symmetrical band that is not A- or B-form in the case of DNA- SWCNT. In the case of DNA-SWCNT complex, the essential increase of the absorption of C=O molecular groups at 1712, 1700, 1685 cm^{-1} and C=N at 1608 cm^{-1} was registered. The base bands show the presence of a high-frequency shoulder at 1700 cm^{-1} (A-form) - 1712 cm^{-1} (B-form) (in comparison with 1699 cm^{-1} in reference DNA. Abovementioned data indicates the strengthening of interaction between the DNA with SWCNT and restoring the helix form of DNA. The integral intensity of the base band has been increased by factor of 6.24 for the DNA-SWCNT complex. Some of the features shows that in the case of DNA-SWCNT complex the B-helical structures DNA are restored. By the way, we registered an increase of absorbance in 1600-1700 cm^{-1} in comparison with reference DNA that could indicate to changes in base-pairing as well specific interaction with carbon nanotubes. This could be connected with enhancement of SEIRA signal of DNA under their adsorption on carbon nanotube and/or decreasing of stacked basis, etc. The increase of the bands in the sugar region (900-750 cm^{-1}) (Fig. 11) and shift of sugar vibration at 830 cm^{-1} (reference DNA) to 827,

829, 824, 826 cm^{-1} in DNA-SWCNT is an evidence of some changes in conformations of sugar and bases in DNA under SWCNT interaction. Registration of the band at 925 cm^{-1} (deoxiribose in Z-conformation), 784 cm^{-1} (syn-form of dG- marker of Z -conformation) indicates to the existence of DNA in Z-form. That is why we could suppose that SWCNT causes to B-A and B-Z transitions in DNA sugar-phosphate backbone.

Some transformation of H-bonds in the region of OH-, NH- and CH-vibrations was observed (Fig.10). The increase (about 50 cm^{-1}) of the halfwidth of H-bonded O-H and N-H stretching vibration has been registered for DNA-SWCNT complex. Thus, for reference DNA the halfwidth of the OH-NH-CH band is equal to 478 cm^{-1}, for DNA-nanotubes complex this parameter is equal to 527 cm^{-1}. This halfwidth increase in DNA-SWCNT is due to increase of shoulder in the 2950-3200 cm^{-1} region. This indicates the formation of strongly H-bonded NH and OH molecular groups (Badger & Bauer, 1937; Semenov et al., 1994; Maleev et al., 2003) in DNA with SWCNT and could be seen from the difference spectra in Fig. 10.

We registered a drastic decrease of water molecules in DNA-SWCNT. Number of water molecules per nucleotide for reference DNA is equaled to 17.5, for DNA-SWCNT - 10.5, 7.9, 8.9, 7.6 for different points. So, we suppose that nanotube replaces the water molecules from their cites under interaction with DNA Seems, that the cites of DNA interaction with CNT is bases due to their hydrophobic properties. The nanotubes could be inserted in major and/or minor grooves of DNA (Fig.8). The authors of (Li et al., 2006) consider that minor groove of G-C has preference for interaction with carbon nanotubes than A-T minor groove. The interaction of nanotube with G-C pairs is stronger than those for A-T. We think that this point is connected with appearance of additional cites of interaction in the case of G-C-SWCNT binding in comparison with A-T-SWCNT binding. Indeed, in DNA-SWCNT complex a number of water molecules bound with DNA became less 7-9 molecules that is in coincidence with a total number of the sites occupied of water molecules in DNA bases in the major and minor grooves., see Fig.8.

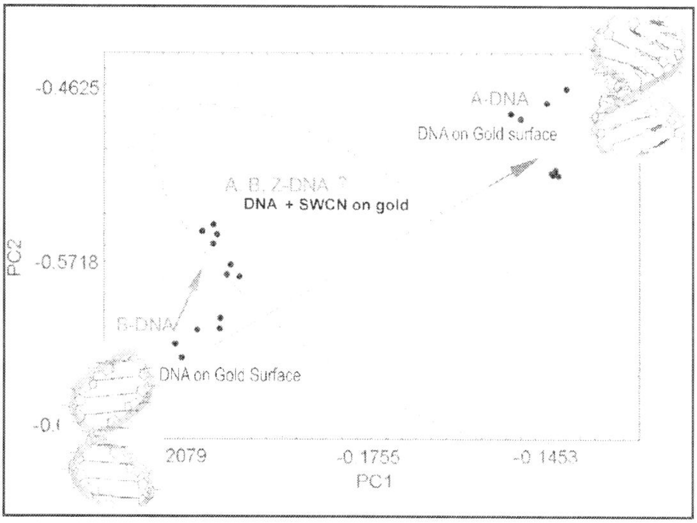

Fig. 12. Principal component analysis of DNA

So, according to spectroscopic data and principal component analysis, in the SWCNT-DNA/complex the main marker bands correspond to different conformations of DNA- B, A, Z-forms and un-known form (Fig.12).

3.4 Conformation of poly-A adsorbed on the surface of carbon nanotubes

Poly-A has 2 ordered forms: under neutral and alkaline pH, poly-A shows single strand, a transition in double strand takes place lower pH 5,25 (Scovell, 1978). Under pH 5,25 and lower adenine in poly-A is protonated on N_1 cites. Formation of H-bonding pair with other adenine with participation of N_7 of one adenine and N_{10} another molecules as well as an aggregation of helixes with formation of double strand poly-A occurs. The structure is kept in aqueous solution, under cold temperature and under fiber spinning. Stability of helix form of poly-A is going due to dispersive, vertical stacking interaction for both ordered forms as well as to formation of intramolecular interstrand and intrastrand H-bonds for double strand of poly-A (Zenger, 1987).

Conformation form of poly-A is determined by pH of solution, under pH 5.25-5.75 the poly - A has single strand form. Under protonation from pH 5.27 to 4.62 the poly-A has transition from single strand form to double strand form (Scovell, 1978) and marker bands of poly-A indicate to it.

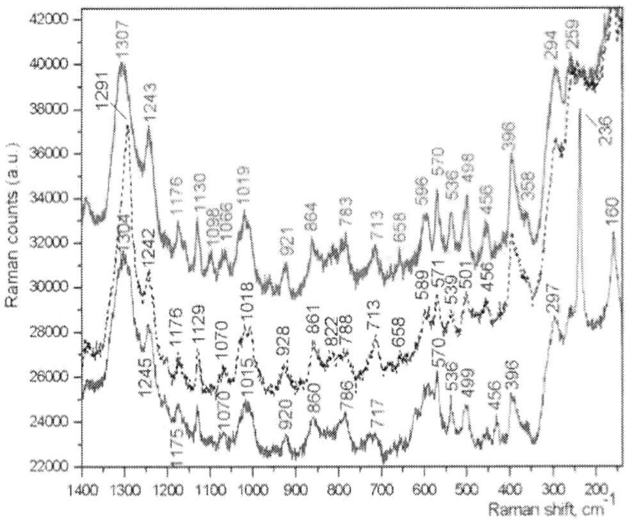

Fig. 13. SERS of poly-A (under pH=5.27) adsorbed at carbon nanotube surface in the region of 1400-400 cm^{-1} in 3 different points

SERS (Fig. 13) and SEIRA (Fig. 14, 15) data was applied for characterization of conformation changes of poly-A at carbon nanotubes surfaces. As a result of interaction between poly-A and nanotubes we have derived the poly-A in single strand conformation (pH=5.27) (Scovell, 1978; Zenger, 1987). With SERS spectroscopy we registered numerous breaks and deformations of poly-A in single strand adsorbed on nanotubes, however the poly-A keeps their initial form. E.g. sugar conformation in 3 different points has marker bands at 786, 788, 783 cm^{-1} that indicate to backbone of poly-A in helix right hand form. We could registered the signal of poly-A adsorbed on isolated nanotubes (1-3) wrapped by poly-A of concentration of 10^{-3}-10^{-4} M due to usage of Klarite substrate and enhancement of signal about 10^4-10^5 in comparison with Al substrate. We got the selection of nanotubes by poly-A preferentially with diameter size close to 1 nm, meanwhile the diameter of the nanotubes in the bundle was preferentially about 1.4 nm.

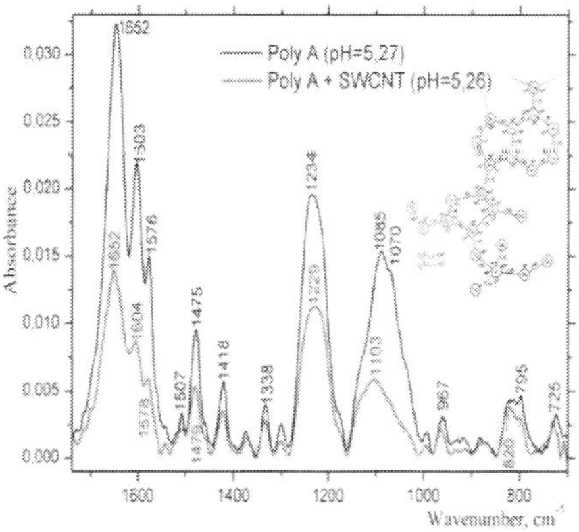

Fig. 14. SEIRA spectra of poly-A and poly-A/SWCNT in neutral solution in the region of 1700-800 cm^{-1}

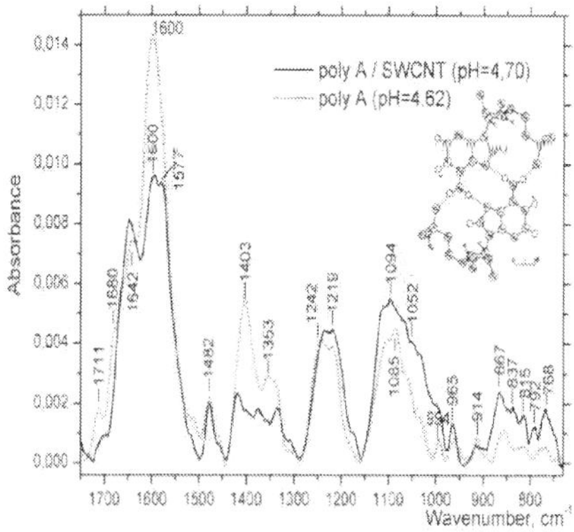

Fig. 15. SEIRA spectra of poly-A and poly-A/SWCNT in acid solution in the region of 1700-800 cm^{-1}

According to SEIRA data we have not observed changes in spectra of poly-A on nanotubes for pH 5.26 in the region of bases 1700-1500 cm^{-1} (Fig.14). However, we have observed strong changes in phosphate bands: symmetrical phosphate shits from 1085 cm^{-1} to 1103 cm^{-1} and asymmetrical phosphate shifts from 1234 cm^{-1} to 1229 cm^{-1} after poly-A adsorption on nanotubes as well as increase of sugar vibration about 4 times. We suppose that the characteristic cites of interaction of poly-A with nanotubes are adenine that induces the curvature of sugar-phosphate backbone.

Double strand poly-A interaction with nanotubes lead to increase of pH of initial solution of poly-A from 4.62 to 4.7 (Fig.15). Spectroscopic features of double strand poly-A/SWCNT complex are similar to DNA/SWCNT. The addition of nanotubes to poly-A prepared with pH 6.47 lead to decrease of pH complex SWCNT/poly-A to 6.37, conformation of poly-A adsorbed at nanotube surface keep single strand form with minor changes.

In poly-A/ SWCNT complex the most drastic changes are observed in the region of OH band (Dovbeshko et al.,2008). A decreasing of water molecules has been calculated in 2 times. We suppose that in poly-A/SWCNT complex, a formation the possible sites of interaction is similar to those that under process of dehydration. In is known that in Poly-A an adenine bounds 4 H_2O molecules, phosphate- 4, ribose -1 (Semenov et al., 1994; Maleev et al., 2003). The sites of H_2O in poly-A, seems, are replaced by nanotubes as in DNA.

Thus, the poly-A under SWCNT interaction originating from the data obtained for the ensemble of the molecules keeps the initial conformation with numerous change, disordering and breaks of backbone derived from the data obtained for separate poly-A molecules. Double strand poly-A in SWCNT complex, seems, could adopt different forms similar to DNA.

3.5 Modeling of DNA Base Interaction with Carbon Nanotubes

It is well-known from calculated data (Shtogun et al., 2007; Meng et al., 2007) that a process of adsorption of nucleobases on carbon nanotubes occurs by π-π stacking between carbon nanotube surface and bases resulting in almost flat-laying thymine (or adenine). We are going to compare the calculation data with experimental data obtained with vibration spectroscopy for Thy and Ade adsorbed on SWCNT.

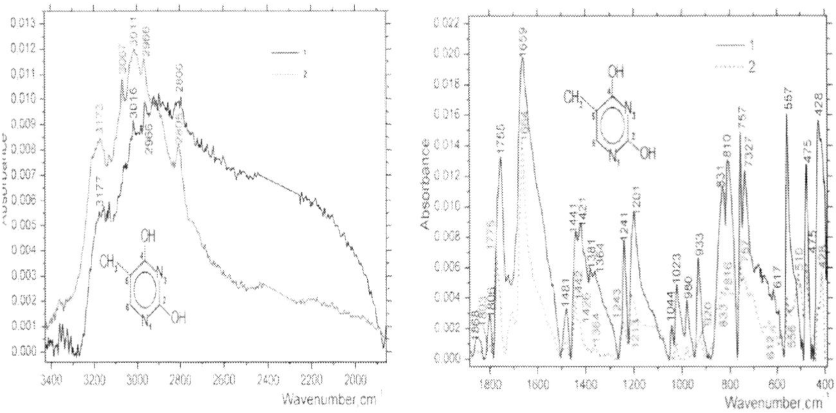

Fig. 16. SEIRA spectra: 1- Thy-SWCNT on gold and 2- Thy on gold

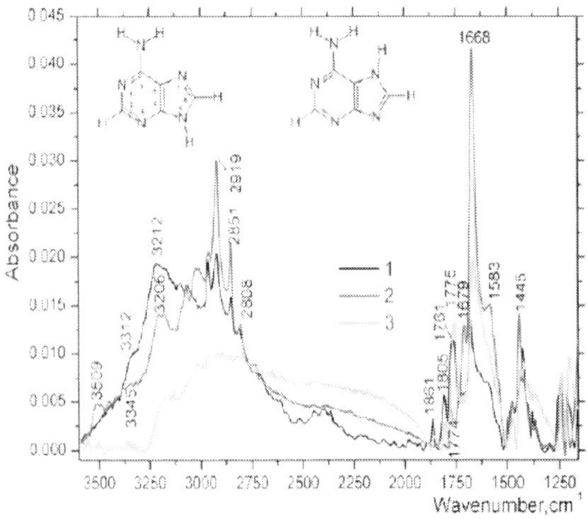

Fig. 17. FTIR spectra: 1 - Thy-SWCNT on CaF_2 and 2 – Thy on CaF_2 , 3 - Thy-SWCNT on gold

Nanotube-Thy interaction resulted to the following spectroscopic features:
1) Widening of H-bonded NH vibrations in the range of 3500-2000 cm^{-1} (Fig.16) and enhance of the intensity of the low-frequency shoulder.
2) Missing of Thy band at 3509 cm^{-1} assigned to N_1H under absorption of SWCNT for both Au and CaF_2 substrates (Fig.17) (Dovbeshko et al., 2009). The last point could be connected with protonation of the Thy ring. We suppose that possible point of Thy interaction with SWCNT is N_1 position.
3) Enhancement of a intensity of the band of NH for the Thy-SWCNT on CaF_2 substrate and a shifting from 3345 (Thy) to 3312 cm^{-1} (Thy-SWCNT) (Dovbeshko et al., 2009). One more NH vibration (seems N_3) shows an increase of the intensity and shift from 3206 (Thy) to 3212 cm^{-1} (Thy-SWCNT).
5) In Thy-SWCNT at Au and CaF_2 we could registered C=O vibrations such as 1861,1805 that could be assigned to C_2=O and C_4=O, respectively, for isolated Thy molecules. We supposed that it connected with adsorption of separated molecules of Thy on carbon nanotubes resulted to film formations due the π-π stacking interaction.
6) Thy-SWCNT on CaF_2 (Fig.17) shows decrease of the intensity of C_4=O vibration and a shift from 1668 to 1679 cm^{-1}, indicating the breaking of some of the inter- or intra-molecular H-bonds with participation of C_4=O.
In Thy-SWCNT (Fig. 17) the C_2=O vibration shows the enhancement of the intensity and a shift is in the low-frequency region from 1774 to 1761 cm^{-1} (Thy-SWCNT on CaF_2) and to 1755 cm^{-1} (Thy-SWCNT on Au) that indicates the formation of new intermolecular H-bonds with C_2=O participation.
8) The bands at 1023, 831, 557, 475, 428 assigned to CH_3 and N_1C_2, C_2=O, C_4=O, N_3H, N_3C_4, N_3C_2=O SWCNT for Thy-SWCNT on gold show a 3 to 5 times enhancement of their intensity (Fig. 16). It could be connected with enhancement of local electrical fields in the system SWCNT-gold. In general, one more reason of the enhancement of intensity of vibrations of Thy adsorbed at nanotubes in low frequency region is connected with crystalline packing of Thy molecules. For Ade-SWCNT on gold (Fig.18) the enhancement of intensity of the numerous vibration NH_2, C=C in the region of stretching and deformation bands in the 3400-1100 cm^{-1}, and decrease of many vibrations lower 1100 cm^{-1} up to 400 cm^{-1} were observed also in contrast to Thy adsorption on nanotubes. This fact demonstrates a difference in behavior of the intensity of the bands of Thy in comparison with Ade molecules and, from our point of view, could be connected with different formation of layers by Thy and Ade molecules. Ade, seems, forms the isolated islands and does not cover the nanotube by monolayer or multilayer films as it does Thy by preference.

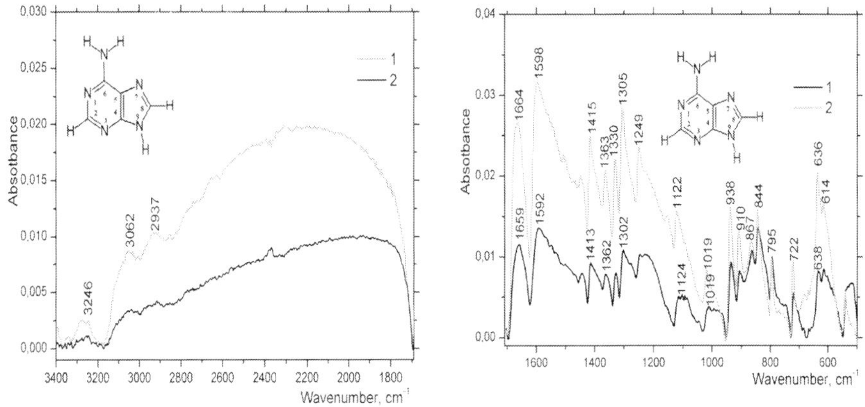

Fig. 18. FTIR spectra on Au: 1-Ade in H_2O, 2- Ade-SWCNT in H_2O

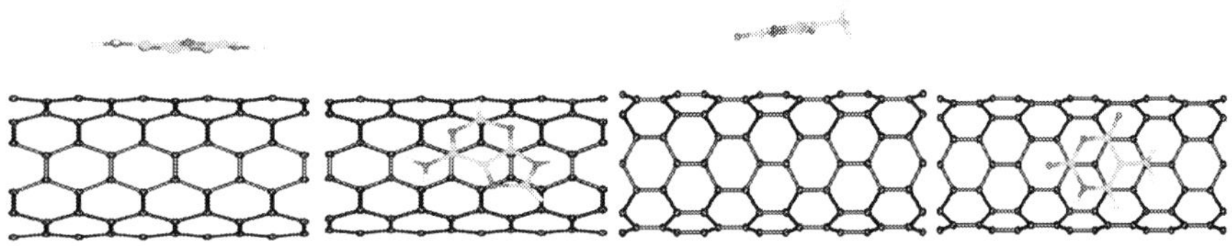
Fig. 19. Possible orientation of a adenine and thymine molecule on surface of SWCNT.

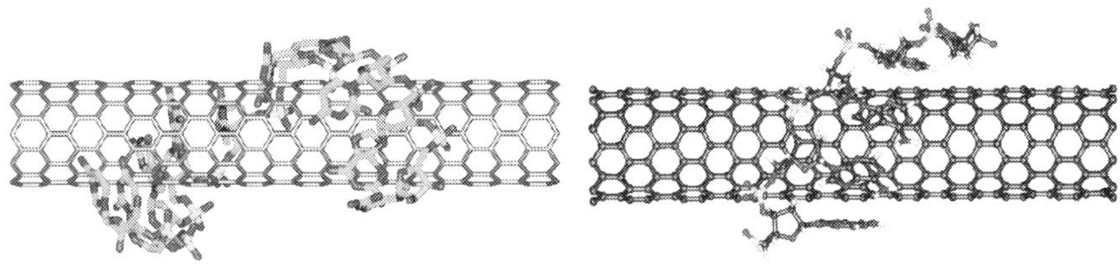
DNA a Poly-A b

Fig. 20.Visualization of possible SWCNT interaction with ss-DNA (a) and Poly-A (b)

Modeling and calculation of the process of the adsorption of thymine on carbon nanotubes have shown the numerous cites of its adsorption on the carbon nanotube surface with weak energy of interaction (Shtogun et al., 2007; Meng et al., 2007; Gowtham et al., 2007). On the basis of the spectral data and the Badger-Bauer rule ($\Delta H = 0{,}3(\Delta \nu - 40)^{1/2}$, where ΔH – enthalpy of H-bond formation) (Sokolov, 1981) we have estimated the energy of H-bond formation and break under Thy and Ade interaction with carbon nanotubes. We suppose that break of the H-bonds is due to nanotube interaction with bases, so this energy was accepted by us as an marker for base-nanotube interaction. Data for Ade and Thy in gas phase was taken from (Nowak et al., 1996; Aamouche et al., 1997), respectively correspondingly and in our calculation stretching vibration was taken into account.

Sample	Thy on nanotubes		Thy on graph sheet	Ade on nanotubes		Ade on graph sheet
	Calc. (Shtogun , 2007)	Exp. (our date)	Calc. (Meng, 2007)	Calc. (Meng, 2007)	Exp. (our date)	Calc. (Gowtham , 2007)
Energy, kcal/ mol	5,70-7,29	0,03-3,13	8,16	6,48-7,47	0,09-0,94	10,6
Distance, Å	3,1-3,31	-	3,14	3,05-3,25	-	3,1

Table 1. Energy of interaction of the bases with nanotubes and graphene sheet

Table 1 shows that energy of interaction estimated from SEIRA spectra is less than calculated one. The main reason of non-coincidence of the experimental and calculated data is the fact that more complicated system occurs in the experiment than in calculation. In the calculation we did not take into account an aqueous solution, IR substrate, defects in nanotubes moreover we suppose that covering of nanotube by Thy or Ade is a single molecule as not in the experiment (Fig 19).
Thus, nucleic acid could interact with nanotubes forming the stable complexes by wrapping (Fig. 20), attachment or insertion of NA into SWCNT surface. In the DNA/ SWCNT complex, DNA could adopt left, right helix and un-known forms. The bases are the cites of interaction of NA with SWCNT. The π-π stacking is the main type of interaction in the systems under study that is coincide with data obtained by other scientists (Zheng et al., 2003a ; Zheng et al., 2003b; Gladchenko et al., 2006).

4. Conclusion

1. SERS could be used for characterization of carbon nanotubes and their complexes. SEIRA spectra could effectively indicate the conformation state of nucleic acids in the complex with carbon nanotubes. The analysis of vibrational modes of DNA with SWCNT has shown that the structural changes in DNA occur. They can be interpreted as B-A, B-Z transition and stabilization of DNA structure in some DNA fragments. Un-known conformation forms of DNA induced by SWCNT could be supposed also.
2. The model of DNA wrapping around SWCNT on the basis of replacing of the water molecules by SWCNT is proposed for the process of carbon nanotube interaction with DNA.
3. We developed a technology of selection of nanotubes with diameter of 1 nm by poly-A, meanwhile the bundle has nanotubes of 1.4 nm mostly.
4. We registered a Raman signal from separated nanotubes wrapped by poly-A in single strand as well numerous marker band of poly-A indicated the presence of poly-A on the surface of nanotubes in their initial form. SEIRA data shows the conformation form of poly-A/nanotube in assemble is practically non-changeable in comparison with poly-A.
5. With SERS spectroscopy we registered numerous breaks and deformations of poly-A in single strand adsorbed on nanotubes, however the poly-A keeps their initial form. The characteristic cite of interaction of poly-A with nanotubes is adenine. We could registered the

signal of poly-A adsorbed on nanotubes with poly-A concentration of 10^{-3}-10^{-4} M due to usage of Klarite substrate and enhancement of signal about 10^4 -10^5 in comparison with Al substrate. Nanotubes induces the breaks into double strand in poly-A, redistribution of the H-bond net, break of a number of H-bonds.
6. Ade adsorbed on carbon nanotubes forms mostly islands and Thy forms monolayer or multilayer films by π-π stacking interaction with tilted bases to plane of carbon nanotube.

Acknowledgments

We thank the Ukrainian program Nanostructured Systems, Nanomaterials, Nanotechnology ("Bionanosystems" N0107U008449, 2007-2009) and Centre of Excellence of Nanostructured Interfaces and Surfaces and INSTM "Centro di Riferimento", Italy for financial support. L.M. Woods acknowledges partial support from the American Chemical Society Petroleum Research Fund and from the Center for Integrated Functional Materials (CIFM) through Grant No. USAMRMC-07355004.

5. References

Aamouche, A., Ghomi, M., Coulombeau, C., Grajcar, L., Baron, M., Jobic, H., Berthier, G. (1997). Neutron Inelastic Scattering, Optical Spectroscopies, and Scaled Quantum Mechanical Force Fields for Analyzing the Vibrational Dynamics of Pyrimidine Nucleic Acid Bases. 2. Thymine J. *Phys. Chem. A*, Vol. 101, No. 10, pp. 1808-1817, ISSN: 1089-5639.

Badger, R., Bauer, S. (1937). Spectroscopic Studies of the Hydrogen Bond. II. The Shift of the O-H Vibrational Frequency in the Formation of the Hydrogen Bond *J.Chem. Phys*, Vol. 5, pp. 839-851, ISSN 0021-9606

Bokova, S., Kononov, V., Obratscova, E., Osadchiy, A. (2003). Laser-induced effects in Raman spectra of single-wall carbon nanotubes. *Quantum electronics*, Vol. 33, pp. 645-650, ISSN 0368-7147.

Bruch, L., Cole, M. & Zaremba, E. (1997). *Physical Adsorption: forces and phenomena*, Dover Publications, **ISBN:** 9780486457673.

Chen, J., Hamon, A., Hu, H., Chen, Y., Rao, A. (1998). Solution properties of single-walled carbon nanotubes Science., 282, 95-97, ISSN 0193-4511.

Chopra, S., McGuire, K., Gothard, N., Rao, A. M. & Pham, A. (2003). Selective Gas Detection Using a Carbon Nanotube Sensor, *Appl. Phys. Lett.*, Vol. 83, pp.2280-2282. ISSN: 0003-6951.

Dovbeshko, G., Chegel, V., Gridina, N., Repnytska, O., Shirshov, Y., Tryndiak, V., Todor, I., Solyanik, G. (2002). Surface Enhanced IR Absorption of Nucleic Acids from Tumor Cells: FTIR Reflectance Study. *Biospectroscopy*, Vol. 67, pp. 470-486, ISSN 10754261.

Dovbeshko, G., Repnytska, O., Obraztsova, E., Shtogun, Y. (2003). DNA interaction with single-walled carbon nanotubes: a SEIRA study. *Chem. Phys. Letters*, Vol. 372, pp. 432-437, ISSN 0009-2614.

a Dovbeshko G., Fesenko O., Shirshov Yu., Chegel V. (2004). The enhancement of optical processes near rough surface of metals. *Semiconductor, Quantum Electronics and Optoelectronics*. Vol. 7, No 4. pp.411-424, ISSN 1560-8034.

b Dovbeshko G., Shirshov Yu., Chegel V., Fesenko O. (2004). Experimental and calculated enhancement factor in the SEIRA method. *SPIE*. Vol. 5507. pp. 386-395, ISSN 0277-786X.

c Dovbeshko G., Repnytska O., Pererva T., Miruta A., Kosenkov D. (2004). Vibrational spectroscopy and principal component analysis for conformational study of virus nucleic acids. Proceedings of SPIE, Vol. 5507. pp.309-316, ISSN 0277-786X.

a Dovbeshko G., Fesenko O., Chegel V., Shirshov Yu., Kosenkov D., Nazarova A. (2006). Effect of nanostructured gold surface on the SEIRA spectra of nucleic acid, albumin, α-Glycine and Guanine. *Asian Chemistry Letter*, Vol. 10, № 1-2, pp. 33-48, ISSN 1001-8417.

b Dovbeshko G., Fesenko O., Nazarova A. (2006). Effect of nanostructured metal surface on SEIRA spectra of albumin and nucleic acids. *Journal of Physical Studies*. Vol. 10, No 2. pp. 127-134, ISSN 1027-4642.

c Dovbeshko, G., Obrastzova, E., Fesenko, O., Yakovkyn, K. (2006). Reactivity single walled carbon nanotubes under interaction with biological molecules - DNA and proteins. *Sensor Electronics and Microsystem Technologies.*, Vol. 1, N 36, ISSN 1815-7459.

Dovbeshko G., Fesenko O., Yakovkin K., Bertrione S., A.Damin, Scarano D., Zecchina A., Obraztsova E.. (2008). The poly-A interaction and interfaces with carbon nanotubes. *Molecular Crystals Liquid Crystals*. Vol 496, No. 345331. pp. 170-185, ISSN 1542-1406.

Dovbeshko, G., Woods, L., Shtogun, Ya., Fesenko, O. (2009). Modeling of DNA base interactions with carbon nanotubes: ab initio calculations and SEIRA data. *AIP C Pr.* Vol. 1148, pp. 416-420, ISSN 0732-183X.

Dresselhaus, M., Dresselhaus, G. & Eklund, P. (1996). *Science ofFullerenes and Carbon Nanotubes*, 965 p., Elsevier Science Ltd, ISBN 0-12-221820-5, New York, U.S.A.

Dresselhaus, M., Dresselhaus, G. & Avouris, P. (2000). *Carbon Nanotubes: Synthesis, Structure, Properties, and Applications*, Springer, Berlin, ISBN 3540410864.

Довбешко, Г., Фесенко, О., Образцова, Е., Алахвердиев, К., Кайя, А. (2009). Конформационный анализ нуклеиновых кислот и белков при их адсорбции на углеродные одностенные нанотрубки. *Журнал структурной химии*. Т. 50, No 5. pp.991-999, ISSN 0136-7463.

Girifalco, L. & Hodak, M. (2002). Van der Waals Binding Energies in Graphitic Structures, *Phys. Rev. B*. Vol. 65, pp. 125404, ISSN 1098-0121.

Gladchenko, G., Karachevtsev, M., Leontiev, V., Valeev, V., Glamazda, A., Plokhotnichenko A. & Stepanian S. (2006). Interaction of fragmented double-stranded DNA with carbon nanotubes in aqueous solution. *Molecular Physics*, Vol. 104, pp. 3193-3201, ISSN 0026-8976.

Gowtham, S., Scheicher, R., Ahuja, R., Pandey, R., Karna, Sh. (2007). Physisorption of nucleobases on graphene: Density-functional calculations *Physical Review B*, Vol. 76, pp. 33401-33404, ISSN: 1098-0121.

Guo, Z., Sadler, P.J., Tsang, S.C. (1998). Immobilization and Visualization of DNA and Proteins on Carbon Nanotubes, *Adv. Mat.* Vol. 10, No9, pp.701- 703, ISSN 0935-9648

Harstain, A., Kirtley, J.R. and Tsang, J.C., (1980), Infrared absorption of the molecular monolayers with thin metal overlayers, *Phys. Rev. Lett.*, Vol. 45, pp. 201-209, ISSN 0031-9007.

Hohenberg, P. & Kohn, W. (1964). Inhomogeneous Electron Gas, *Phys. Rev.* Vol. 136, pp. B864-871, ISSN 1050-2947.

Iijima, S. (1991). Helical microtubules of graphitic carbon. *Nature (London)*, Vol.,354, No 6348, pp. 56-58, ISSN 0028-0836.

Jorio, A., Satio, R., Hafner, (2001). Structural (n, m) Determination of Isolated Single-Wall Carbon Nanotubes by Resonant Raman Scattering. J. *Phys. Rev. Lett.*, 86,1118, ISSN: 0031-9007.

Kam, N.W.S., O'Connell, M., Wisdom, J., Dai, H. (2005). Carbon Nanotubes as Multifunctional Biological Transporters and Near-infrared Agents for Selective Cancer Cell Destruction. *PNAS.*, Vol. 102, pp. 11600-11605, ISSN 1091-6490

Kosobukin, V.A. (1983). Effect of enhancement of external electric field near metal surface and its manifestation in spectroscopy. *Surface. Physics, chemistry, mechanics,* Vol. 12, pp. 5-20. ISSN 02073528

Kosobukin, V.A. (1985). Collective modes in enhancement of external electric field on the surface of the metals. *Reparts of Academy of Sciences of USSR, Physical edition,* Vol. 49, No. 6, pp. 1111-1120, ISSN 0-7803-8142-4.

Kresse, G. & Joubert, J. (1999) From Ultrasoft Pseudopotentials to the Projector Augmented Plane Wave Method. *Phys. Rev. B,* Vol. 59, pp. 1758-1775. URL for the ABINIT project www.abinit.org, ISSN 1550-235X.

Li, Xi, Peng, Y. & Qu, X. (2006). Carnon nanotubes selective destabilization of duplex and triplex DNA and inclucing B-A transition in solution. *Nucleic Acids Research,* Vol.34, No13, pp.3670-3676, ISSN 0305-1048.

Litvinov, G. (1991). Infrared vibration spectra of plant viruses and its molecular compounds. *Biopolimers and cells,* Vol. 7, No. 6, pp. 32-47, ISSN 0092-8674.

Liquier, J., Akhebat, A., Taillandier, E., Ceolin, F., Huynh Dinh, T. Igolen, J. (1991). Characterization by FTIR spectroscopy of the oligoribonucleotide duplexes r(A-U)6 and r(A-U)8. *Spectrochimica Acta A,* Vol. 47, pp. 177-186, ISSN: 1386-1425.

Maleev V. Ya., Semenov M. A., Kruglova E. B., Bolbukh T. V., Gasan A. I., Bereznyak E. G., Shestopalova A. V. (2003) Spectroscopic and calorimetric study of DNA interaction with a new series of actinocin derivatives. J. *Mol. Struct.* Vol. 645, pp. 145-148, ISSN: 0166-1280.

Meng, S., Maragakis, P., Papaloukas, C., Kaxiras, E. (2007). DNA Nucleoside Interaction and Identification with Carbon Nanotubes *Nano Lett.* Vol. 7, pp. 45, ISSN: 1530-6984

Moghaddam, M., Taylor, S., Gao, M., Huang, S., Dai, L. & McCall, M. (2004). Highly Efficient Binding of DNA on the Sidewalls and Tips of Carbon Nanotubes Using Photochemistry, *Nano Lett.* Vol. 4, No 1, pp. 89-93, ISSN 1530-6984.

Nowak, M., Lapinski, L., Kwiatkowski, J., Leszczynski, J. (1996). Molecular Structure and Infrared Spectra of Adenine. Experimental Matrix Isolation and Density Functional Theory Study of Adenine 15N Isotopomers J. *Phys. Chem.,* Vol. 100, No. 9, pp. 3527-3534. ISSN: 0022-3654.

Obraztsova, E., Bonard, J.-M., Kuznetsov, V., Zaikovskii, V., Pimenov, S., Pozarov, A., Terekhov, S., Konov, V., Obraztsov, A., Volkov, A. (1999). Structural measurements for single-wall carbon nanotubes by Raman scattering technique. *Nanostruct. Mater.* Vol. 12, pp. 567, ISSN 0965-9773.

O'Connell, M., Boul, P., Ericson, L., Huffman, C., Wang, Y., Haroz, E., Kuper, C., Tour, J., Ausman, K.D., Smalley, R.E. (2001). Polymer-wrapped single wall carbon nanotubes. *Chem. Phys. Lett.* Vol. 342, pp. 265 - 271, ISSN 0009-2614

Osawa, M. & Ikeda, M. (1991). Surface-enhanced infrared absorption of p-Nitrobenzoic acid deposited on silver island films: contributions of electromagnetic and chemical mechanisms. J. *Phys. Chem.,* Vol. 95, pp.9914-9919, ISSN: 1089-5639.

Parker, F. (1983). *zBiochemical application of Raman and resonance Raman spectroscopies,* New-York, London, Acad. Press, ISBN-10: 0306412063.

Saenger, W. (1988). *Principles of Nucleic Acid structure,* Springer-Verlab, ISBN 0123695074, New York Berlin Heidelberg Tokyo.

Saito, R., Dresselhaus, M., Dresselhaus, G. (1998). Physical Properties of Carbon Nanotubes; Imperial College Press: Cambridge.

Schrader, B. (1995), *Infrared and Raman Spectroscopy,* Weinheim-New York-Basel-Cambridge-Tokio, p.765, ISBN 3-527-26446-9

Scovell, W. (1978). *Biopolymers,* Vol. 17, pp. 969, ISSN: 0301-0104

Semenov, M., Bol'buh, T., Krasnitskaya, A. & Maleev, V. (1994). Hydration and structural state of liver DNA from rats exposed within the Chernobyl NPP zone. *Radiation Biology. Radioekology,* Vol. 34, No (3), pp.328-335, ISSN 0739-1102

Shie, M. (1977). The study of conformations of free and intrafagic DNA with inrared absorption spectra in the region of sugarphosphate backbone vibrations. *Proc. Japanese-United States Congress of Pharmaceutical Sciences.* Honolulu (Hawaii). Present at the Pushchino, Moscow, Institute of Biophysics, AN USSR, pp. 22.

Shtogun, Ya., Woods, L., Dovbeshko, G. (2007). Adsorption of adenine and thymine and their radicals on single-wall carbon nanotubes. J. *Phys. Chem. C,* Vol. 111, 18174-18181, ISSN 1932-7447.

Sokolov, N. D. (1981). *Handbook. H-bonding,* Nauka, Moskva.

Strick, T., Dessinges, M., Charvin, G. Dekkar, N., Allenmand, J. (2003). Stretching of macromolecules and proteins, *Rep. Prog. Phys.* Vol. 66, No. 1, ISSN 0034-4885.

Taboury, J.A., Liquier, J., Taillandier, E. (1985). Characterization of DNA structure by infrared spectroscopy: double helical forms of poly(dG-dC)poly(dG-dC),poly(dD8G-dC)poly(dD8-dC), and poly(dG-dm5C)poly(dG-dm5C). *Can.J.Chem.,* Vol. 63, pp. 1909-1904, ISSN 0008-4042

Taillandier, E., Liquier, J. & Taboury, J.A. Infrared spectral studies of DNA conformation. (1985). *Advances in Infrared and Raman Spectroscopy,* Clarc, R. J. H., Hester, R. E. and Heyden, W. (Eds), pp. 65-114, ISBN-10: 0471262153.

Tajmir-Riahi, H.A., Neault, J.F. and Naoui, M. (1995). Does DNA acid fixation produce left-handed Z structure? *FEBS Letters,* Vol. 370, pp. 105-108, ISSN 0014-5793.

Williams, K., Veenhuizen, P.T.M., Torre, B., Eritija, R. & Dekker, C (2002). Nanotechnology - Carbon nanotubes with DNA recognition, *Nature,* Vol. 420, No 6917, pp. 761, ISSN 0028-0836.

Woods, L.M., Badescu, S.C., Reinecke, T.L. (2007). *Adsorption of Simple Benzene Derivatives on Carbon Nanotubes,* J. Phys. Rev. B, Vol. 75, pp. 155415, ISSN 1098-0121.

Zenger, V. (1987). *The principals of structural organizing of nucleic acids.* Wiley: Mir, ISBN-10:0387907610 Moscow.

a Zheng, M., Jagota, A., Semke, E.D., Dinner, B.A., McLean, R.S., Lustig, S.R., Richardson, R.E. & Tassi N.G. (2003). DNA-assisted dispersion and separation of carbon nanotubes. *Nature matherials.* Vol.2, pp.338-342, ISSN 1476-1122

b Zheng, M., Jagota, A., Strano, M.S., Barone, P., Chou, S.G., Diner, B.A., Dresselhaus, M.S., McLean, R.S., Onoa, G.B., Samsonidze, G.G., Semke, E.D., Usrey, M. & Walls, D.J. (2003). Structure based carbon nanotube sorting by sequence-dependent DNA assembly, *Science.* Vol. 302, No 5650, pp. 1545-1548, ISSN 0193-4511.

35

DNA-Wrapped Carbon Nanotubes: From Synthesis to Applications

Germarie Sánchez-Pomales, Coral Pagán-Miranda, Lenibel Santiago-Rodríguez and Carlos R. Cabrera
Department of Chemistry, Institute for Functional Nanomaterials and Center for Advanced Nanoscale Materials, University of Puerto Rico, Río Piedras Campus San Juan, Puerto Rico

1. Introduction

Carbon nanotubes (CNTs) have been the focal point of many studies since their discovery two decades ago (Iijima, 1991). They are allotropes of carbon with cylindrical shape and sp^2 hybridization, which are formed by benzene-type hexagonal rings of carbon atoms. CNTs can be classified as either single-walled (SWNTs) or multi-walled carbon nanotubes (MWNTs) depending on their structure. They possess a very high aspect ratio (i.e. length/diameter), with diameters in the range of nanometers and lengths that can reach microns. Additionally, CNTs can be divided according to their conductivity, into metallic or semiconducting tubes.

Nanotubes possess outstanding structural, mechanical and electronic properties due to the unique combination of their dimension, structure and topology. For example, CNTs are extremely strong, can be highly conducting, and exhibit high thermal and chemical stability (Mintmire et al., 1993; Mintmire & White, 1998). Also, they are elastic and have many times the tensile strength of steel (Service, 1998). Due to these unique properties, CNTs have been proposed for many applications, ranging from scanning probes (Dai et al., 1996) and hydrogen storage (Dillon et al., 1997), to biosensors (Nguyen et al., 2002).

Several methods have been commonly used to synthesize CNTs, including arc discharge, chemical vapor deposition and laser ablation, among others (Ando et al., 2004). After the growth process, CNTs contain impurities such as metal catalyst nanoparticles, amorphous carbon, carbon nanoparticles, and fullerenes, which must be removed. Typical CNTs purification techniques include ultrasonication (Shelimov, et al., 1998), microfiltration (Bandow et al., 1997), chromatographic techniques (Nigoyi et al., 2001), microwave heating (Harutyunyan et al., 2002), gas-phase oxidation (Ebbesen et al., 1994), and acid oxidation (Rinzler et al., 1998). Nevertheless, their purification is not a trivial task, and either improvement on the existing techniques or development of new purification methodologies are required in order to facilitate the use of CNTs and their integration into nanodevices and composite materials.

Another main challenge to overcome is the need for the development of new functionalization chemistries that can increase the solubility of CNTs without altering their properties. Carbon nanotubes are extremely difficult to manage due to their low solubility in both aqueous and organic solvents, which restricts the extent of their applications. To increase the solubility of CNTs opens many possibilities, such as simplifying the purification process and making the nanotubes more manageable, which consequently facilitates their use as standard chemical reagents.

One of the first and more commonly used approaches to increase the solubility of carbon nanotubes is to use strong acids to attack and shorten the tubes (Liu et al., 1998). Even though this method is effective in increasing the CNTs solubility, it has the disadvantage of getting rid of one of the attractive properties of the nanotubes: their high aspect ratio. Since then, different approaches using chemical functionalization have been probed in order to increase the solubility and manageability of CNTs without altering their desirable properties.

2. Non-covalent functionalization of carbon nanotubes

Chemical functionalization has been the most successful method to solubilize CNTs thus far. Functionalization of CNTs offers many advantages such as: improving the solubility of the nanotubes, transforming them into more manageable materials, and combining the unique properties of CNTs with those of other materials.

Chemical functionalization of CNTs can be achieved by either non-covalent or covalent interactions. Contrary to covalent functionalization methods, which disrupt the extended π-networks on the CNTs surfaces and can modify their mechanical and electronic properties, non-covalent functionalization preserves the desired properties of CNTs, at the same time as it improves their solubility. The non-covalent functionalization of CNTs has been performed with molecules as diverse as polymers, biomolecules, surfactants, and polyaromatic compounds (Tasis et al., 2006). The main driving forces for the non-covalent interaction between the CNTs and these molecules are van der Waals and π–stacking forces.

A wide range of polymers have been used in the formation of supramolecular complexes of CNTs including, among others, polyvinyl pyrrolidone, polystyrene sulfonate, poly(1-vinyl pyrrolidone-co-vinyl acetate), dextran, dextran sulfate, poly(1-vinyl pyrrolidone-coacrylic acid), poly(1-vynil pyrrolidone-co-dimethylaminoethyl methacrylate), bovine serum albumin, starch and deoxyribonucleic acid (DNA) (O'Connell et al, 2001; Star et al., 2002; Zheng et al., 2003a).

The formation of supramolecular complexes between DNA and CNTs has drawn much attention in the past years, since these hybrids can take advantage of the unique properties of the nanotubes and the remarkable recognition capabilities of DNA. DNA-wrapped CNTs exist as a well-defined chemical entity in aqueous solution due to the strong non-covalent interaction between the DNA and the CNTs (Zheng, 2007). Compared to other polymers used, DNA offers the advantage of defined length and sequence, high dispersion efficiency (i.e. up to 4 mg/mL) and well-developed chemistries for further functionalization of the DNA-CNT hybrid through either covalent or non-covalent functionalization (Zheng et al., 2003a).

In the next sections of this chapter, we will describe the different methods used to synthesize DNA-wrapped CNTs, as well as the main techniques employed to characterize these complexes in both aqueous solutions and solid substrates. In addition, the main applications of the DNA-CNT complexes will be discussed.

3. DNA-wrapped carbon nanotubes

3.1 Synthesis

The first reports on the synthesis of DNA-wrapped CNTs were published by several groups in 2003 (Zheng et al., 2003a; Dovbeshko et al., 2003a; Dovbeshko et al., 2003b; Nakashima et al., 2003; Zheng et al., 2003b). The preparation of the DNA-wrapped CNTs complexes is a simple and straightforward procedure, which consists on adding the CNTs to an aqueous solution of DNA, and then mixing the components by stirring and/or sonicating in an ice-water bath. Wrapping of the CNTs by DNA (Fig. 1) occurs when the aromatic hydrophobic DNA bases interact with the sidewall of the CNT via π-stacking, whereas the hydrophilic sugar-phosphate backbone of the DNA strand is left exposed to interact with the aqueous solvent (Zheng et al., 2003a). Changes in either the type of DNA or CNTs used are common variations to the synthesis methodology. For example, Dovbeshko and coworkers synthesized the DNA-CNT complexes by using denatured calf thymus DNA and SWNTs produced by the arc discharge method (Dovbeshko et al., 2003a; Dovbeshko et al., 2003b). Nakashima's group, in contrast, used purified HiPCO (High-Pressure CO conversion) SWNTs and a solution of DNA from salmon testes (Nakashima et al., 2003). On the other hand, Zheng *et al.* reported the first time use of short synthetic oligonucleotide sequences for the preparation of DNA-wrapped CNTs (Zheng et al., 2003a; Zheng et al., 2003b). In addition to short synthetic DNA strands (Zheng et al., 2003a) and denatured genomic DNA(Dovbeshko et al., 2003a; Dovbeshko et al., 2003b; Gigliotti et al., 2006), CNTs have been dispersed and functionalized with short fragmented DNA (Gladchenko et al., 2006; Xu et al., 2007), DNA synthesized by asymmetric polymerase chain reaction (PCR) (Liang et al., 2007), and single-stranded DNA (ssDNA) obtained by rolling circle amplification (Zhao et al., 2006). Some reports have also shown that long denatured fragments of lambda DNA (Gigliotti et al., 2006) and double-stranded DNA (dsDNA) produced by symmetric PCR (Liang et al., 2007) lacked the capability to efficiently wrap around and disperse the CNTs.

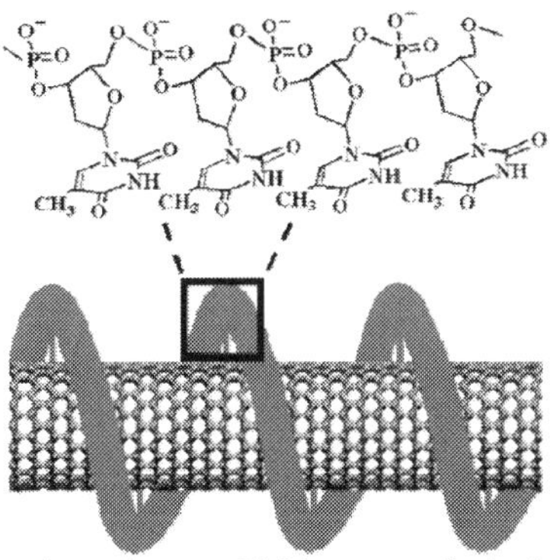

Fig. 1. Graphical representation of an ss-DNA-wrapped single-walled carbon nanotube. Reprinted with permission from (Ma et al., 2006a). Copyright (2006) American Chemical Society.

Many studies have been performed in order to optimize the dispersion efficiency of CNTs in DNA. Variations in parameters such as DNA sequence, DNA length, sonication time, type of CNTs, and solvent conditions, have been reported. For example, the first report on the effect of DNA sequence and length on the dispersion efficiency showed that among 60-mer homopolymers, poly(T) resulted in the best yield, whereas among the four different lengths of poly(T) studied (15, 21, 30 and 60), T_{30} displayed the highest dispersion efficiency (Zheng et al., 2003a). Additionally, $d(GT)_n$ (with n = 10 to 45) DNA sequences have been shown to have both a high dispersion yield and the highest efficiency in facilitating CNTs separation by diameter and type with anion exchange chromatography (Zheng et al., 2003b). A subsequent report demonstrated that shorter DNA sequences unexpectedly displayed excellent dispersion yields, and that a mixture of the complementary oligonucleotides $d(GT)_3$:$d(AC)_3$ showed the highest efficiency among all the sequences studied (Vogel et al., 2007). Additionally, a recent study showed that the compactness and selectivity of DNA wrapping around SWNTs are dependent on the length of the DNA sequence, and that by increasing the strand length, both the degree of compactness of the wrapping around the nanotube and the diameter selectivity decrease (Yang et al., 2009).

A study on the effects of sonication time and CNT type on the dispersion efficiency of CNTs by DNA showed that sonication times higher than 90 minutes were required with $poly(T)_{30}$ in order to obtain well-dispersed individual DNA-wrapped CNTs (Lahiji et al., 2007). Increasing the sonication time to 120 minutes increased the dispersion efficiency, but decreased the size of the CNTs, suggesting that they break during long sonication periods. Moreover, CNTs synthesized by electric-arc (EA) and HiPCO processes were compared and the results demonstrated the presence of particulates or impurities in the EA sample that were not observed in the HiPCO CNTs. These observations confirmed that the purity of as-received HiPCO CNTs is higher than that of EA CNTs. As-received and chemically oxidized CNTs were also compared after DNA wrapping. The oxidized CNTs had a shorter length and showed an increase in the number of DNA molecules on the CNT surface.

Variations in solvents and DNA:CNT ratios have also been studied. For instance, it has been reported that the dispersion efficiency can be increased by changing the solvent from normal water (H_2O) to deuterated water (D_2O) (Yang et al., 2007b). Additionally, the results of this study showed that CNTs with small diameters are preferentially dissolved in D_2O, whereas larger diameters CNTs are easily dissolved in H_2O. On the other hand, Taeger and collaborators reported that the best dispersion efficiency is achieved at DNA:CNT ratios of 1:1 or 1:2 (Taeger et al., 2005). Furthermore, a ratio of 1:2 is preferred, since it minimizes the amount of unbound DNA.

In contrast to the typical approach for creating DNA-wrapped CNTs, which requires sonication, milling has also been used to synthesize the DNA-CNT complexes. Nepal and coworkers reported the synthesis of DNA-wrapped CNTs based on a solid-state mechanochemical reaction (Nepal et al., 2005). Their "one-pot synthetic method" consisted on adding CNTs and DNA to a ball-milling apparatus, and milling the mixture for 30 minutes to 1 hour. Aqueous solutions of DNA-wrapped CNTs with high stability (> 6 months), and a uniform length distribution were obtained. Additionally, a high-speed vibration milling technique was used to solubilize SWNTs with nucleotides (Ikeda et al., 2006). The dispersion efficiency was dependent on the type of base used and the number of phosphate groups.

Finally, DNA-wrapped CNTs have been used as a template to create more complex conjugates. DNA-CNTs have been functionalized with platinum nanoparticles (Ostojic, et al., 2008), quantum dots (Campbell et al., 2008), cytochrome c (Heering et al., 2006), biotin-avidin-biotin conjugated antimouse IgG (Cao et al., 2009), gold nanoparticles (Li et al., 2007; Chen et al., 2007), a photosensitizer (Zhu et al., 2008), and other DNA-wrapped nanotubes (Li et al., 2007; Chen et al., 2007), and they have also been incorporated into a polymeric film (Ma et al., 2006a; Ma et al., 2006b). Hersam's group, for example, used cisplatin and potassium tetrachloroplatinate as DNA cross-linkers and subsequently reduced the bound platinum salts to functionalize the DNA-CNT complexes with platinum nanoparticles (Ostojic, et al., 2008). This methodology provided strong binding between the Pt nanoparticles and the DNA without perturbing the desirable properties of the nanotubes. The prepared Pt/DNA-CNT complexes might find use in applications as diverse as fuel cells and biosensors.

Thiol-functionalized DNA strands have also been used to attach gold nanoparticles and quantum dots to DNA-CNT conjugates. Rigid arrays of gold nanoparticles aligned on the surface of CNTs were obtained by Han and coworkers through the bonding between the thiol group in the modified DNA strands wrapped around the CNTs, and the gold nanoparticles (Han et al., 2007). DNA-wrapped CNTs were also functionalized with CdSe/ZnS quantum dots by using thiol-modified DNA wrapped around the nanotubes (Campbell et al., 2008). The binding of the quantum dots to the DNA-CNT complexes allowed the elucidation of the location of the DNA strands on the nanotube surface by atomic force microscopy. Conversely, two groups functionalized DNA-wrapped CNTs with gold nanoparticles and with other DNA-wrapped CNTs (Fig. 2) though hybridization between complementary DNA sequences (Li et al., 2007; Chen et al., 2007).

On the other hand, a composite between poly-(anilineboronic acid) and DNA-wrapped CNTs was prepared by Ma *et al.* by in situ polymerization of 3-aminophenylboronic acid) in the presence of the DNA-CNT complexes. The resultant composites showed enhanced stability, conductivity and redox chemistry.

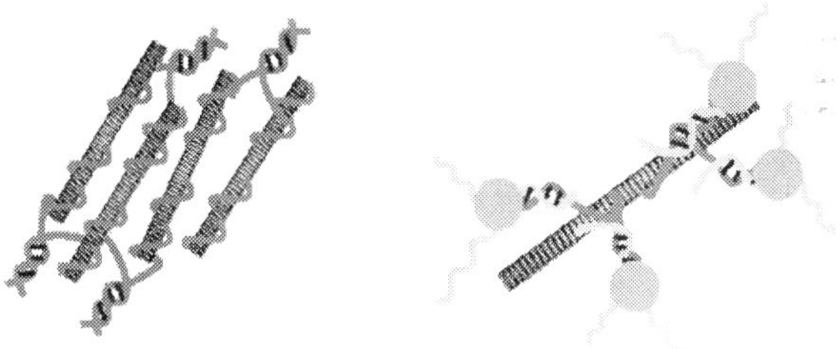

Fig. 2. Schematic representation of the complexes formed between several DNA-wrapped CNTs modified with complementary DNA sequences (left), and between a DNA-wrapped CNT and Au nanoparticles modified with complementary DNA strands (right). Reprinted with permission from (Chen et al., 2007). Copyright (2007) American Chemical Society.

3.2 Main properties and characterization

DNA-wrapped CNTs have been studied by numerous techniques, including, among others, atomic force microscopy (AFM) (Zheng et al., 2003a; Nakashima et al., 2003; Zheng et al., 2003b; Taeger et al., 2004; Takahashi et al., 2006), transmission electron microscopy (TEM) (Nakashima et al., 2003; Zhao et al., 2006; Rajendra, et al., 2004; Malik et al., 2007; Ikeda et al., 2006; Shoda et al., 2009), Raman spectroscopy (Dovbeshko et al., 2003a; Dovbeshko et al.,

2003b; Zheng et al., 2003b; Shoda et al., 2009; Chou et al., 2004a; Kawamoto et al., 2006a; Kawamoto et al., 2006b; Yang et al., 2007a), Fourier transform infrared spectroscopy (FTIR) (Dovbeshko et al., 2003a; Dovbeshko et al., 2003b), absorption spectroscopy (Zheng et al. 2003a; Nakashima et al., 2003; Zheng et al., 2003b; Malik et al., 2007; Meng et al., 2007; Nair et al., 2006), scanning electron microscopy (SEM) (Sánchez-Pomales et al., 2007a; Nepal et al., 2005; Li et al., 2007; Hu et al., 2005; Karachevtsev et al., 2006a), fluorescence spectroscopy (Nepal et al., 2005; Strano et al., 2004; Hobbie et al., 2005; Luo et al., 2006; Fagan et al., 2007b), scanning tunneling microscopy (STM) (Iijima et al., 2005; Yarotski, et al., 2009), circular dichroism (Rajendra et al., 2004; Rajendra & Rodger, 2005; Dukovic et al., 2006), small angle neutron scattering (Hobbie et al., 2005), optical microscopy (Badaire et al., 2005; Li et al., 2009), photoluminescence spectroscopy (Luo et al., 2006; Chou et al., 2005; Torrens et al., 2006), energy dispersive spectroscopy (EDS) (Sánchez-Pomales et al., 2007a; Hu et al., 2005), capillary electrophoresis (Khripin et al., 2009b) and electrochemistry (Hu et al., 2005; Li et al., 2009; Heering et al., 2006; Napier et al., 2005).

Specifically, the structural properties of the DNA-wrapped CNTs have been studied by AFM, Raman spectroscopy and photoluminescence spectroscopy, among others. For example, Qian and coworkers were able to differentiate between DNA-wrapped and unwrapped CNT segments by using high-resolution tip-enhanced near-field microscopy to monitor the photoluminescence energy shift along the nanotube surface (Qian et al., 2008). In another study it was reported that SWNTs induce destabilization and a B-A transition on DNA, and that these processes depend on the DNA sequence (Li et al., 2006).

Atomic force microscopy and photoluminescence spectroscopy results showed that diminishing the nanotube concentration could decrease the mean bundle diameter in DNA-SWNTs dispersions (Cathcart et al., 2008). The results demonstrated that once a dispersion of DNA-SWNTs is obtained, spontaneous debundling occurs upon dilution, which means that dispersions with a higher content of individually dispersed nanotubes can be obtained without the need for additional sonication.

Toita and coworkers prepared DNA-SWNTs dispersions with nanotubes synthesized by two methods: HiPCO and arc discharge (Toita et al., 2008). Their AFM study suggests that the DNA-wrapping mechanism for SWNTs prepared by arc discharge may be different from HiPCO SWNTs due to differences in nanotube diameters. A dependence on the diameter was also reported in a study of DNA-wrapped CoMoCAT SWNTs (Chou et al., 2004). Additionally, a study on the time dependence of DNA wrapping in SWNTs dispersions was performed with photoluminescence (Cathcart et al., 2008). Photoluminescence intensity increased over time, and this observation was correlated to the formation of a monolayer coating of DNA on the nanotube walls, as evidenced by high-resolution TEM and circular dichroism.

A systematic resonance Raman study of DNA-SWNT complexes elucidated that the intensity of the ratio between the D-band and the G-band (I_D/I_G) depends on the nanotube length (Chou et al. 2007b). Also, an increase in the intensity is observed for several features of the intermediate frequency mode region of the Raman spectra of DNA-SWNTs, as the nanotube length decreased below the wavelength of light (Chou et al. 2007a). Raman spectroscopy was also used to show a strong dependence of the G-band of the spectra of the DNA-SWNTs on the relative humidity (Kawamoto et al., 2006b). On the other hand, the relationship between the optical properties of DNA-SWNT complexes and their length was studied and it was shown that as the nanotube length increases, the intensity for the absorption, near-infrared fluorescence and Raman scattering features of the samples are enhanced (Fagan et al. 2007a). Alternatively, a study by Kawamoto and coworkers demonstrated that the asymmetric profile of the Breit-Weigner-Fano (BWF) Raman line disappears when HiPco SWNTs are individually dispersed due to DNA wrapping around the SWNTs (Kawamoto et al., 2006c), which suggests that the asymmetric feature of the BWF line can be attributed to bundling effects in SWNTs dispersions. Another study by this group revealed that the asymmetric feature of the BWF Raman line reappears when the DNA-SWNTs dispersion is dried in air, and demonstrated the reversibility of the changes in the BWF line by switching between the air-dried and hydrated states of the DNA-SWNTs dispersions (Kawamoto et al., 2006a).

Several groups have also studied the stability of the DNA-CNT complexes. For instance, it has been demonstrated that DNA-SWNTs prepared with double-stranded DNA are very stable after one month of storage, even when free double-stranded DNA was removed from the dispersion (Noguchi et al., 2008). The stability of different dispersions of DNA-SWNTs was also tested by heating the dispersions and following the flocculation process via UV-visible spectroscopy (Vogel et al., 2007a; Vogel et al., 2007b). The thermal denaturation experiments demonstrated that the stability of the DNA-SWNT complexes increases with the length of the oligonucleotide used, since longer DNA chains can engage in a larger number of multivalent interactions. On the other hand, a study on the effect of UV irradiation on DNA-wrapped CNTs showed that nanotubes wrapped with $(GT)_{15}$ are more stable under UV irradiation than CNTs wrapped with $(T)_{15}$ (Yoon et al., 2008).

Studies on the electrical properties of the DNA-wrapped CNTs have also been reported in the past few years. Hembram and Rao studied the electrical properties of DNA/MWNTs composites by modifying alumina substrates with dispersions of DNA/MWNTs with equal DNA concentration, but varying CNT composition, and by analyzing them by electrochemical impedance spectroscopy (EIS) (Hembran & Rao, 2009). The electrochemical results revealed that the charge transfer resistance of the system decreases (i.e. higher conductivity) with increases in CNT concentrations. This enhancement in conductivity was attributed to an increase in the conducting path due to the increase in the amount of CNTs, as evidenced by SEM. Alternatively, differences in the conductivity type of the nanotubes were reported after functionalization with DNA. Fantini and coworkers performed Raman and optical absorption measurements to demonstrate that an enrichment of semiconducting versus metallic nanotubes occurs when the nanotubes are dispersed with DNA instead of with sodium dodecyl sulfate (Fantini et al., 2007).

An SWNT field effect transistor (FET) with a DNA-wrapped SWNTs, was fabricated in order to study the effect of DNA wrapping on the electronic transport characteristics of SWNTs (Hwang et al., 2008). This report showed that after DNA wrapping, the conduction type of SWNTs was changed from p-type to n-type, and that the charge transport occurred through both the DNA and SWNT in the fabricated FET. Subsequently, the transition between a metallic SWNT to a semiconductor due to DNA wrapping around the nanotube, was demonstrated by FET-type measurements (Cha et al., 2009).

UV-visible spectroscopy has been used to monitor the electron transfer resistance between small molecule redox agents and DNA-wrapped nanotubes and to determine that there is a facile electron transfer between the redox reagents studied and the semiconducting CNTs (Zheng & Diner, 2004). For example, $KMnO_4$ and K_2IrCl_6 can oxidize the nanotubes, whereas $NaBH_4$ and Na_2SO_4 can reduce them. Furthermore, Xu and colleagues showed that the redox chemistry of DNA-wrapped SWNTs with H_2O_2 depends on whether or not the suspension was purified by chromatography, since the purified DNA-SWNTs suspensions appeared to be less sensitive to hydrogen peroxide (Xu et al., 2008).

Electrochemical methods were employed to report the first-time electrocatalytic oxidation of DNA-wrapped CNTs in solution in the presence of $Ru(bpy)_3^{2+}$ (Napier et al., 2005). Conversely, absorption spectroscopy was used to study the donor-acceptor complex formed by Ag^+ and DNA-CNTs, and the results demonstrated the capability of the DNA-CNT complexes to catalyze the photosynthesis of Ag nanoparticles (Zheng & Rostovtsev, 2006).

Additional properties, ranging from pH dependence to cytotoxicity, have been studied by spectroscopic and microscopic techniques. Kelley and coworkers reported that the optical transitions of DNA-wrapped semiconducting nanotubes are dependent on pH (Kelly et al., 2005). Variations in pH have been used to control the aggregation of DNA-SWNTs (Han et al., 2008). At low pH values, the DNA-SWNTs dispersion aggregated, whereas at higher pH values, the aggregates disappeared and a clear dispersion was obtained.

Polarized photoluminescence measurements on DNA-SWNT complexes demonstrated that differences in magnetism exist among SWNTs chiralities (Torrens et al., 2007), whereas optical absorption spectroscopy served to report that anisotropic hypochromicity is observed for the optical transitions of DNA bases interacting with SWNTs (Hughes et al., 2007). On the other hand, the cytotoxicity of SWNTs suspended in different surfactants was studied by phase contrast light microscopy and absorbance spectroscopy, and the measurements demonstrated that while SWNTs dispersed with sodium dodecyl sulfate and sodium dodecylbenzene sulfonate were toxic to human astrocytoma cells, DNA-wrapped SWNTs did not affect the proliferation and viability of the cells (Dong et al., 2008).

3.3 Theoretical and simulation experiments

Theoretical and simulation experiments have been crucial to the understanding of the non-covalent interaction between DNA and CNTs, and to the elucidation of the structure of DNA-wrapped CNTs. The first theoretical report on DNA-wrapped CNTs was published in 2003 (Zheng et al., 2003a). Their simulations showed that short ssDNA strands (i.e. poly (T)) can bind to the surface of a (10,0) carbon nanotube in several allowed configurations, including helical wrapping with right- and left-handed turns, or surface adsorption with a linearly extended structure. Furthermore, the thermodynamics of binding of DNA to CNTs were estimated by comparing the minimized energy of the DNA-CNT complex with that of the DNA strand and the individual CNT. Their calculations demonstrated that the binding energy of ssDNA onto the surface of CNTs could compete effectively with the tendency of CNTs to aggregate and form bundles in order to maximize the interaction with each other.

Molecular dynamic simulations were subsequently performed on both DNA adsorption on the nanotubes walls and encapsulation inside the CNTs (Gao & Kong, 2004). It was determined that due to van der Waals forces between the DNA and CNTs, the DNA strands could be encapsulated inside or wrapped around the CNTs. The effect of nanotube size, temperature, DNA sequence and CNT-end group on the functionalization process was also studied. Conversely, Ranjan and coworkers performed density functional theory calculations of poly (T) and poly (CT) with CNTs, to show that no covalent interaction exists between the DNA and the CNTs, but longe-range van der Waals interactions are the main force (Ranjan et al., 2005).

Computer simulations were used by Gladchenko and coworkers to study the binding between short fragmented double stranded DNA and CNTs (Gladchenko et al., 2006). The results revealed that approximately 60% of the bases are out of the stacking with the nanotubes. Their model suggests that the denatured parts of the double stranded DNA wrap around the CNTs, whereas no stacking interaction was found between the double-stranded portion of the DNA and the CNT.

Manohar and coworkers used scaling arguments and molecular dynamics to study the factors that contribute to the formation of the DNA-CNT hybrids (Manohar et al., 2007). The results suggest that a fraction of the bases unstuck from the CNT, and that at low ionic strength, the main forces contributing to the free energy are the adhesion between the bases and the CNT and electrostatic repulsion due to the charged phosphate groups in the DNA.

The differences in adsorption of a 12-base pairs DNA fragment on uncharged, positively- and negatively-charged SWNTs was also studied (Zhao & Johnson, 2007). The simulations demonstrated that DNA can bind to the surface of uncharged or positively charged SWNTs, whereas DNA does not adsorb on negatively charged SWNTs. Furthermore, the results showed that the hydrophobic end groups of DNA are attracted to the hydrophobic SWNT surface, while the hydrophilic backbone does not bind to the nanotube.

Atomistic molecular dynamic simulations on DNA and CNTs were used to determine that the presence of SWNTs induces a conformational change in DNA in order to enable the wrapping of DNA around SWNTs with right- and left-handed helices, achiral loops and disordered, kinked structures (Johnson et al., 2008). These results also demonstrated that purines exhibit the strongest binding towards nanotubes, and that this binding occurs via the π-π stacking interaction between DNA bases and the nanotubes sidewalls. On the other hand, Stepanian and coworkers used both *ab initio* MP2 (second order Møller-Plesset perturbation theory) and DFT (density functional theory) methods to study the stacking interaction of cytosine and to determine that the interaction energy of the DNA-CNT complex formed is < 42.4 kJ/mol (Stepanian et al., 2008).

Shtogun and coworkers used DFT to study the interactions between DNA bases and their radicals with the nanotube surface (Shtogun et al., 2007). Their studies suggest that both DNA bases and their radicals can be easily adsorbed on nanotube surfaces due to the non-covalent interaction between the delocalized π orbitals from the DNA and the nanotubes. DFT was also used to study the adsorption of nucleic bases on nanotubes of small diameter (Gowtham et al., 2008). The calculated binding energies followed the hierarchy G>A>T>C>U and the molecular polarizability dominated the interaction between the bases and the nanotubes. Alternatively, Das and coworkers used the *ab initio* Hartree-Fock method to study the binding of DNA bases to CNTs with and without considering the solvation effects of water (Das et al., 2008). By including solvation energy, the relative binding energy preference changes from G>A>T>C to G>T>A>C, which was in good agreement to their experimental results. The same order of binding preference in gas-phase was reported by DFT in another study (Wang et al., 2008). Nevertheless, it was also found that for aqueous solutions, the binding preference varies from A>G>T>C for (10,0) SWNTs to G>A>T>C for (5,5) SWNTs, showing that cross-stacking has a weak dependence on the nanotube diameter.

Even though theoretical studies have given us a better understanding of the type of forces and interactions involved on DNA-wrapping around CNTs, there are still many unanswered questions, as well as some discrepancies on the data reported. Additional studies that can consider environmental conditions such as solvent effects and ionic strength could provide a clearer picture of the interactions between DNA and CNTs. Furthermore, it is very difficult to obtain data that can be universally applied to all types of DNA-CNT complexes, because of the many possible permutations due to the different DNA sequences, DNA lengths, and types of CNTs (e.g. metallic, semiconducting, chiral, etc.) available.

3.3 Assembly on solid substrates

The development of a straightforward methodology for the ordered assembly of CNTs on solid substrates is a very desirable goal, because it would facilitate the use of CNTs as sensors, and at the same time, it would make easier the study of CNTs by surface techniques such as XPS, SEM, and ellipsometry, among others. Individually dispersed DNA-wrapped CNTs have been assembled and characterized on solid substrates, and have also been proposed for biosensing applications. These DNA-CNT complexes have been deposited as a film on substrates such as glassy carbon (Li et al., 2007), glass (Hu et al., 2005), and gold (Zangmeister et al., 2007), aligned on a SiO_2 surface with a gold boundary (Mclean et al., 2006), and as self-assembled monolayers (SAMs) on gold electrodes (Sánchez-Pomales et al, 2005; Sánchez-Pomales et al, 2006; Sánchez-Pomales et al, 2007b; Sánchez-Pomales & Cabrera, 2007).

Hu and coworkers prepared an electrode with DNA-wrapped CNTs (Fig. 3) by depositing the DNA-CNTs complexes onto a glass substrate (Hu et al., 2005). The individually dispersed DNA-CNTs adhered tightly to the substrate and formed a uniform, stable film, which was subsequently used for the electrochemical detection of dopamine. SEM analysis of the electrode confirmed that the DNA-CNTs were well dispersed in the film, and had an average length of 600 nm. FT-IR and EDS were also used to verify the presence of the DNA-wrapped CNTs on the glass surface. Conversely, electrochemical analysis revealed that the DNA-CNTs modified electrode possessed a wide and flat potential window and fast electron transfer for the $Fe^{3+/2+}$ redox couple, which are advantages for electrodes used in typical electrochemical detection applications. On a subsequent report, the properties of the DNA-CNTs film were compared versus those of an electrode prepared with oxidized MWNTs (Hu et al., 2007). This study showed that even though the oxidized MWNTs exhibited a wider potential window with lower background current in KCl, higher peak currents and smaller difference of peak potentials in $Fe(CN)_6^{3-}$/ $Fe(CN)_6^{4-}$, the film made with the DNA-CNTs performed better in the detection of low concentration of dopamine in the presence of excess ascorbic acid.

Zangmeister and coworkers investigated the adsorption properties of DNA-wrapped CNTs on bare gold and on self-assembled monolayers of neutral, positively and negatively charged species (Zangmeister et al., 2007). X-ray photoelectron spectroscopy (XPS), reflection absorption infrared spectroscopy (RAIRS) and Raman spectroscopy results demonstrated that the DNA-CNT complexes adsorb preferentially to positively charged amine-terminated SAMs or to bare gold, in comparison to neutral methyl-terminated or negatively charged carboxylate terminated SAMs. Their results suggest that electrostatic interactions between the negatively charged phosphate groups of the DNA and the charge of the species on the Au surface influence the adsorption of the DNA-wrapped CNTs.

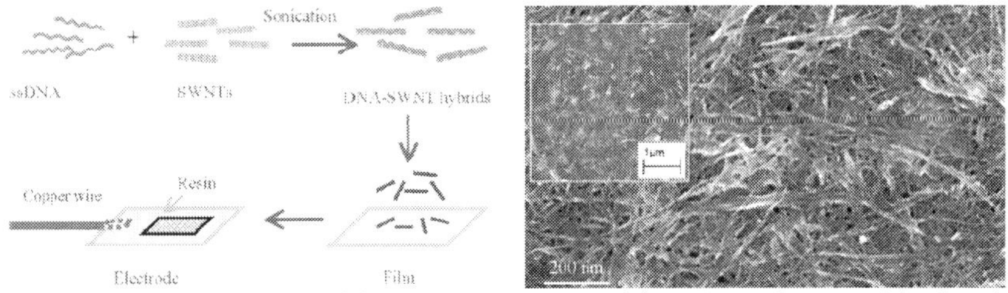

Fig. 3. Schematic showing the fabrication process used by Hu *et al.* to modify the glass substrate with DNA-wrapped SWNTs (left) and SEM image showing the topography of the modified electrode (right). Reprinted with permission from (Hu et al., 2005). Copyright (2005) American Chemical Society.

Glassy carbon electrodes were modified with either DNA partly-wrapped SWNTs (DNA-p-SWNTs) or DNA fully-wrapped SWNTs (DNA-f-SWNTs) by H. H. Li and coworkers (Li et al., 2007). The DNA-p-SWNTs were separated from DNA-f-SWNTs and from free DNA by differential centrifugation. The modified electrodes were characterized by SEM and cyclic voltammetry in $Fe^{3+/2+}$, and the results showed that the DNA-p-SWNTs form a uniform and porous film, which displays better electrochemical properties that the DNA-f-SWNTs.

DNA-wrapped SWNTs have also been deposited on silicon wafers and mica substrates (Hopkins et al, 2007). Alignment and organization of the DNA-CNT complexes was achieved by using an inkjet printer for the deposition of the DNA-SWNTs, and characterization of the modified substrates was performed by SEM and AFM.

Mclean and coworkers studied the deposition of dilute solutions of DNA-wrapped CNTs on SiO_2 and they showed that the adsorption of the DNA-CNT complexes on SiO_2 is dependent on pH, and at pH values >5 (pKa=5), the DNA-CNTs do not adsorb on the substrate, whereas at pH values <5, DNA-CNTs were adsorbed with random orientations (Mclean et al., 2006). However, if the SiO_2 substrate was coated with a hydrophobic layer, adsorption of the DNA-CNTs occurred over a broad pH range, and the complexes were aligned parallel to each other. The results showed that the thickness of the hydrophobic layer influenced the alignment of the DNA-CNTs, and for chains with C_2 and C_8, alignment was observed, whereas the adsorption was random for C_{18}. Additionally, they studied the factors that affect the DNA-CNT complexes adsorption rate and they reported that longer incubation times, lower pH and higher ionic strength increase the adsorption rate onto the modified SiO_2 substrate. Deposition experiments with patterned metal electrodes on the SiO_2 surface were also performed, and it was found that the metal boundary tends to impose a local orientation on the CNTs.

Alignment of DNA-CNT complexes was also observed by Gigliotti and coworkers (Gigliotti et al., 2006). Their AFM results demonstrated that high density arrays of aligned DNA-CNT complexes, produced by long genomic ss-DNA, could be formed at the edges of air-drying droplets on aminopropyltrietoxysilane (APTES)-terminated silicon substrates. Alternatively, deposition and meniscus alignment of DNA-CNT complexes was performed on a silicon wafer coated with an alkyl-silane monolayer, and the effects of pH, ionic strength and time on the density of the deposited DNA-CNTs were studied (Khripin et al., 2009a).

Electrochemical oxidative polymerization of ethylenedioxythiophene (EDOT) was used by Bae and coworkers to deposit complexes of ds-DNA-CNTs on indium tin oxide (ITO) electrodes (Bae et al., 2004). Their results revealed that the DNA-CNT complexes possessed a fibrous morphology, and that the ability of DNA to bind an intercalator (i.e ethidium bromide) was retained.

Alternating current dielectrophoresis has also been used to deposit DNA-SWNTs across Au electrodes (Tallin et al., 2004). The results showed that debris-free individual SWNTs were bridging the electrodes when DNA-SWNTs complexes were used, whereas when SWNTs dispersed with a surfactant were deposited, carbon debris was attached to the SWNTs, and that there was a tendency towards nanotube aggregation into bundles. On the other hand, a study by Karachevtsev *et al.* revealed that after a film of DNA-SWNTs dried over a Si substrate, the SWNTs aggregated, as evidenced by an increase in the Raman G^- peak intensity, and a broadening of the G^+ peak (Karachevtsev et al, 2006b).

Fibers of DNA/SWNTs/poly(ethylene oxide) with an average diameter of 150 nm were electrospun on Pt-coated glass slides, as confirmed by Raman spectroscopy and SEM (Liu et al., 2008). The fibers showed improved electroactive behavior and demonstrated their capability as an immobilization matrix for a glucose oxidase biosensor.

Electrochemical deposition was used to immobilize DNA-CNT composites on glassy carbon electrodes (Li et al, 2008). The modified electrodes exhibited good stability, and the DNA-CNT composites showed good electrocatalytic activity towards the detection of biomolecules such as dopamine, uric acid and ascorbic acid.

DNA-SWNTs were also deposited over an n-octadecyl mercaptan monolayer on Au, and the modified electrode was studied by cyclic voltammetry and electrochemical impedance spectroscopy (Zheng et al., 2009). Furthermore, the release of the DNA-SWNTs from the solid/liquid interface was controlled by applying either a positive (0.90 V vs. Ag|AgCl) or a negative (-1.40 V vs. Ag|AgCl) potential.

DNA-SWNT complexes were immobilized on gold via the formation of mixed self-assembled monolayers (Fig. 4), and our group was the first to report that the DNA-SWNTs were attached in a perpendicular fashion to the gold electrodes (Sánchez-Pomales et al., 2005; Sánchez-Pomales et al., 2006). Such unusual attachment geometry was, up to that time, obtained only by direct growth of the CNTs on a surface by chemical deposition or by post-synthesis attachment of oxidized/activated CNTs by covalent means on previously formed self-assembled monolayers. The developed methodology represents a straightforward alternative to the assembly of aligned CNTs on solid surfaces, and it offers the advantage of allowing the tailoring of the surface by controlling simple parameters such as concentration and modification time, among others. The efficiency of the immobilization process was determined by microscopic, electrochemical, and spectroscopic techniques. Mixed self-assembled monolayers, containing aggregates of DNA-SWNT hybrids with diverse surface coverage were obtained (Sánchez-Pomales & Cabrera 2007). Our group was able to tailor the surface coverage by varying the initial concentration of DNA-SWNTs or by using reductive electrochemical desorption (Sánchez-Pomales et al., 2006; Sánchez-Pomales & Cabrera 2007). The results revealed that the use of DNA-SWNT hybrids to form SAMs on gold provides advantages over traditional DNA SAMs (Sánchez-Pomales et al., 2007b). Additionally, vertically aligned SAMs of DNA-SWNTs on gold, prepared by Viswanathan and coworkers, were subsequently coated with polyaniline, modified with acetylcholinesterase and used as a biosensor for the detection of pesticides such as methyl parathion and chlorpyrifos (Viswanathan et al., 2009).

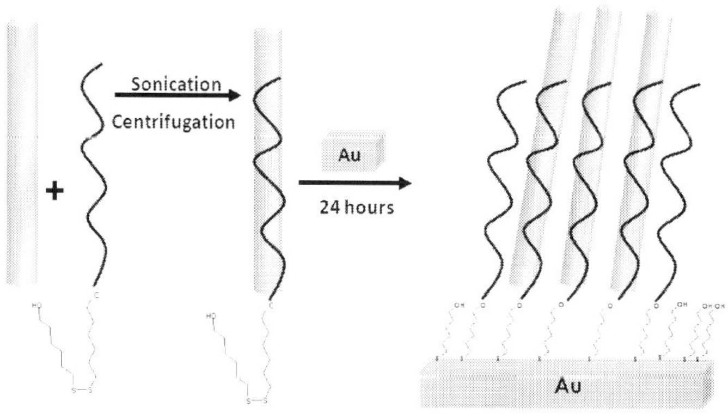

Fig. 4. Schematic representaiton of the functionalization and DNA-mediated attachment of carbon nanotubes on gold. (The figure does not necessarily represent the correct ratio between the species chemisorbed on the gold surface, and the drawings are not to scale.)

3.4 Applications

Since the first reports on the synthesis of DNA-wrapped CNTs, the amount of scientific articles on the application of these hybrids has increased exponentially. The DNA-CNT supramolecular complexes have been proposed for applications ranging from biological transporters and biosensors (Sánchez-Pomales et al., 2009) to fibers for artificial muscles (Shin et al., 2008) and bioelectrodes for fuel cells (Lee et al., 2010; Sánchez-Pomales et al., unpublished b). Furthermore, the use of DNA-wrapped CNTs has helped in the purification and separation of CNTs by techniques as diverse as electrophoresis, size-exclusion chromatography and ultracentrifugation, among others. The main findings of these reports will be discussed in the next sections.

3.4.1 Chemical biosensors

DNA-wrapped CNTs have been used as biosensors for the detection of diverse species, including glucose (Karachevtsev et al., 2007; Xu et al., 2007), peroxide (Liang et al., 2007; Xu et al., 2007), dopamine (Ma et al., 2006; Hu et al., 2005), pesticides (Viswanathan, et al., 2009), vapors (Johnson et al., 2006), proteins (Wu et al., 2009), and ions (Heller et al., 2006; Jin et al., 2007), among others. Hu and coworkers reported the first application of DNA-wrapped SWNTs for biosensing (Hu et al., 2005). DNA-wrapped CNTs deposited over glass substrates were used as an electrochemical biosensor for the detection of dopamine. Their results showed that the prepared dopamine biosensor possesses excellent selectivity and sensitivity, even in the presence of interferences such as ascorbic acid. Additionally, the reported detection limit was low, demonstrating that the DNA-SWNTs-modified electrode can be used as an analytical dopamine biosensor. Dopamine was also detected by He's group by using DNA-wrapped SWNTs in a self-doped polyaniline nanocomposite (Ma et al., 2006; Ali et al., 2007; Cheung et al., 2009). The biosensing capability of the DNA-SWNT/polyaniline composite was tested, and concentrations as low as 1 nM of dopamine were detected, with a linear range from 1 nM to 10 nM.

Non-covalent functionalization of DNA-wrapped SWNTs with glucose oxidase (GOx) was performed with the purpose of using DNA as an interface between the enzyme and the nanotubes, so that the enzyme retains its native structure and activity (Karachevtsev et al., 2007). Micromolar concentrations of glucose were detected by monitoring the changes in the near infrared (NIR) luminescence intensity of the GOx-DNA-SWNTs complexes upon addition of glucose in the presence of potassium ferricyanide. The emission of the SWNTs was quenched by potassium ferricyanide, and it was subsequently restored after adding glucose. The possible application of DNA-wrapped SWNTs for glucose and peroxide biosensors was studied by Zhao's group (Xu et al., 2007; Tu et al., 2007). DNA SWNT hybrids were allowed to interact with glucose, in the presence of glucose oxidase and peroxide, and decreases in the intensity of the S_{11} band of the absorption spectra of the hybrids were recorded. Peroxide was detected by electrochemical means with a DNA-wrapped SWNTs biosensor (Fig. 5) in the presence of hemoglobin (Hb) (Liang et al., 2007). Hb/ssDNA-SWNTs films were analyzed by cyclic voltammetry in buffer in the presence of H_2O_2 and it was shown that the peroxidase activity of Hb was preserved in the films. These results show that DNA-SWNT hybrids may be successfully used as peroxide and glucose analytical biosensors, and might find applications in the chemical, clinical and biological areas, among others (Xu et al., 2007).

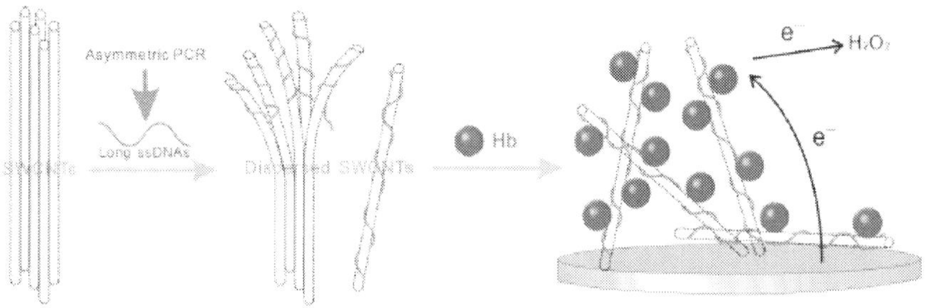

Fig. 5. Schematic showing the peroxide biosensor prepared with DNA-SWNTs complexes. Reprinted with permission from (Liang et al., 2007). Copyright (2007) by MDPI.

DNA-CNT complexes have also been used for the detection of gas odors, including methanol, propionic acid, trimethylamine (TMA), dinitrotoluene (DNT) and dimethyl methylphosphonate (DMMP) (Johnson et al., 2006; Staii et al., 2005; Poonam & Deo, 2008). A field effect transistor (FET) device based on a DNA-CNT complex was used to detect five different odors that a bare CNT FET device was not able to detect. The device's response was enhanced by adding DNA, due to an increase in the binding affinity of the DNA layer towards the analytes. The device showed good sensor response (which was specific to the base sequence used), low recovery times and excellent reproducibility for repeated measurements with a single device and among different devices.

Reports by several groups have elucidated the potential of DNA-SWNT hybrids for ion detection. For example, Strano's group studied the decrease in the NIR emission energy of the DNA-wrapped nanotubes produced when divalent metal cations, which stabilize the Z form of DNA, induced a conformation change of the DNA adsorbed on the nanotube surface (Heller et al., 2006; Jin et al., 2007). A shift in band gap was observed in the emission and absorption spectra, and an ion sensitivity of $Hg^{2+} > Co^{2+} > Ca^{2+} > Mg^{2+}$ was reported. Furthermore, mercuric ions were detected in whole blood, black ink, and cells and tissues, which suggest that the DNA-wrapped SWNTs can be used as novel probes in biological media. Conversely, Gao and coworkers detected trace levels of mercuric ions in aqueous solutions containing DNA-wrapped SWNTs, by monitoring the decrease in the induced circular dichroism signal of the DNA-SWNTs produced by the increase of DNA pitch that occurs when the mercuric ions bind to DNA (Gao et al., 2008). Additionally, tin and arsenic ions have been detected by electrochemical means by using electrodes modified with the DNA-nanotube complexes (Liu & Wei., 2008; Ferancová et al., 2007).

3.4.2 DNA hybridization biosensors

Several groups have studied the hybridization event between a DNA-wrapped CNT and a free complementary DNA strand. Nevertheless, there are discrepancies between the different groups on whether hybridization occurs on the nanotube surface or whether the nanotube is displaced once the DNA strands hybridize.

Cheng and Zhang reported that delamination of DNA and precipitation of the nanotube occur when a complementary DNA sequence, free in solution, is allowed to interact with a DNA-wrapped CNT complex, as evidenced by a visual test, UV-visible spectroscopy, and gel electrophoresis (Chen & Zhang, 2006). Tan's group has also provided evidence that shows that the hybridization event occurs in solution rather than on the nanotube surface (Yang et al., 2008a; Yang et al., 2008b). For their studies, they have used DNA probes labeled with dyes to demonstrate that the fluorescence of the dye is quenched in the DNA-SWNT complex due to the proximity of the dye to the nanotube surface, but the fluorescence is subsequently restored when a complementary DNA strand is added to the system. Fluorescence anisotropy data and fluorescence intensity results obtained after the system was dialyzed were used to support the hypothesis that DNA hybridization does not occur on the nanotube surface. Label-free DNA strands (both linear and hairpin structures) have been successfully used by the same group to detect hybridization between DNA-SWNT complexes and the complementary DNA strands in the presence of a solution ethidium bromide, a DNA-intercataling dye (Liu et al., 2009).

The opposite theory has been supported by Stano's, Baik's and Cabrera's groups. Strano's group reported that hybridization between a target DNA sequence and the complementary probe sequence that was wrapped around a SWNT occurs on the nanotube surface (Jeng et al., 2006). The hybridization event was detected by using the optical modulation of the NIR fluorescence of the SWNTs. An energy shift in the fluorescent peak of DNA-SWNTs that occurred upon addition of the complementary DNA strand, and not when a non-complementary sequence was used, was attributed to a denser surface coverage of DNA in the SWNT surface after hybridization. Forster resonance energy transfer (FRET) between the fluorescent labeled DNA strands was used to corroborate hybridization at the SWNTs surface, and by using the optical modulation of the SWNTs NIR fluorescence, a detection limit of target DNA in the nanomolar range was obtained. On a subsequent report, Strano's group compared the kinetics of DNA hybridization at DNA-SWNTs complexes versus free DNA, and the data showed that the kinetics of DNA hybridization at DNA-SWNT complexes is slower than the kinetics in free DNA solution, with both entropic and activation energy barriers higher for DNA-SWNTs in comparison to free DNA (Jeng et al., 2007). A two-step model, which includes a first step where free, target DNA strands are adsorbed to the DNA-SWNTs probes, followed by a second slower hybridization step was proposed. Balik's group, on the other hand, studied the hybridization event at the nanotube surface by monitoring the changes in the absorption spectra of DNA-wrapped SWNTs (Cao et al., 2008). Furthermore, Cabrera's group provided microscopy, electrochemical and spectroscopic evidence that demonstrated that the hybridization event between SAMs of the complementary DNA sequence and DNA-CNT complexes occurs on the surface of the DNA-CNT, and that therefore, the nanotubes can be deposited on gold surfaces previously modified with a DNA SAMs (Sánchez-Pomales et al., unpublished a).

Instead of performing hybridization on the surface of DNA-SWNTs complexes, Huang and coworkers used a DNA complex containing both a single-stranded and a double-stranded portion to functionalize SWNTs, and they were able to detect a particular DNA sequence within a complex genome (Hwang et al., 2006). The single-stranded part of the DNA was preferentially bound to the SWNTs, whereas the double-stranded (ds) portion was freely exposed on the exterior. Denaturation of the ds portion was achieved by heating, and the now exposed ss-probe sequence was allowed to hybridize with target genomic, as evidenced by a southern blotting employing Raman spectroscopy.

Field effect transistors (FETs) made with DNA-wrapped CNTs have been used to detect specific nucleic acid sequences (Onoa et al., 2006; Tang et al., 2006). The functionalization of CNTs with DNA facilitates the placement of highly purified nanotubes on the devices, and the DNA strands can be subsequently removed from the nanotube by thermal denaturation, if desired (Onoa et al., 2006). Nevertheless, these reports showed that the presence of DNA on the nanotube surface does not alter significantly the carrier transport properties of CNTs.

3.4.3 Biological transporters

Studies by different groups have established the capabilities of DNA-wrapped CNTs as cellular transporters, due to the higher biostability of these complexes in comparison with free DNA strands. Wu and coworkers reported that DNA probes bound to SWNTs are protected from nuclease digestion and from interference from DNA binding proteins (Wu et al., 2008). These protected DNA probes, which target mRNA inside the cells, possess increased intracellular biostability and enhanced self-delivery capabilities, providing significant advantages over free DNA probes.

Dai's group, on the other hand, demonstrated the potential of DNA-wrapped CNTs as intracellular transporters for DNA, and showed that DNA-wrapped CNTs are capable of transporting oligonucleotides into living cells via an energy-dependent endocytosis mechanism (Kam & Dai, 2005a; Kam et al., 2007). In another study by the same group, it was revealed that short NIR laser pulses can be used to release DNA cargoes from SWNTs, and that the released oligonucleotides can translocate into the cell nucleus (Kam et al., 2005b). Alternatively, continuous NIR will cause extensive local heating of the SWNTs and will induce cell death. Selective cancer cells destruction was reported for SWNTs functionalized with a folate moiety, demonstrating their capabilities for drug delivery and cancer therapy.

3.4.4 Purification and separation

Dispersion of carbon nanotubes by DNA also has the benefit of purifying the nanotubes (Taeger et al., 2004; Sánchez-Pomales et al., 2007a; Tu & Zheng, 2008) and of opening the door for separation applications as previously reviewed (Tu and Zheng, 2008).

Chromatographic techniques have been used to separate DNA-wrapped CNTs by conductivity type, diameter, chirality and length. Zheng's group published the first report on the chromatographic separation of DNA-CNT complexes in 2003 (Zheng et al., 2003). Ion exchange chromatography was used to separate DNA-CNT complexes into fractions of different conductivity and length and even though the separation of the nanotubes into metallic and semiconducting tubes was convoluted by variations in length, this report served as the basis for additional efforts on the separation of nanotubes by chromatographic means (Strano et al., 2004; Zheng, 2004; Lustig et al., 2006). A subsequent paper by the same group revealed that, by using ion-exchange chromatography on the DNA-CNTs complexes, earlier fractions could be enriched in smaller diameter and metallic CNTs, whereas late fractions were enriched in larger diameter semiconducting CNTs (Zheng et al., 2003). Furthermore, size exclusion chromatography has been successfully used to separate DNA-wrapped CNTs by length (Huang et al., 2005; Bauer et al., 2007; Bauer et al., 2008), and a combination of size exclusion chromatography followed by ion exchange chromatography was used to separate the complexes by diameter, length, conductivity and chirality (Zhang et al., 2008) and to obtain fractions enriched in a single chirality SWNTs (Zheng & Semke, 2007). Additional separation of the nanotubes by length, diameter and conductivity has been achieved by several techniques, including flow-field flow fractionation (Chun et al., 2008), ultracentrifugation in aqueous density gradients (Arnold, et al. 2005), dielectrophoresis (Sickert et al., 2005), and agarose gel electrophoresis (Vetcher et al., 2006).

4. Summary and future prospects

Chemical functionalization of CNTs is currently one of the most efficient methods for increasing their solubility, for facilitating their use as standard chemical reagents and for combining their unique properties with those of other materials with interesting characteristics. Chemical functionalization of CNTs by DNA has drawn much attention in the past years, since DNA-CNT hybrids can take advantage of the unique properties of the nanotubes and the outstanding recognition capabilities of DNA. Particularly, non-covalent functionalization of CNTs by DNA wrapping is a relatively simple procedure that produces a distinct chemical entity in aqueous solution due to the strong non-covalent interaction between the DNA strands and the CNTs. Non-covalent functionalization of CNTs by DNA increases the solubility of the tubes in aqueous solution, and has the benefits of purifying the nanotubes, facilitating separation applications, and allowing further functionalization of CNTs without altering their unique properties.

DNA-CNT complexes are more manageable than as-received CNTs, and their assembly into different solid substrates has been reported. Additionally, the DNA strands have offered additional recognition capabilities to the system, and have therefore allowed the use of DNA-CNT complexes as sensors for biomolecules, ions and gases, as a biosensor for DNA hybridization, and as biological transporters. The optimization of the current methods available for the controllable integration of the DNA-CNT hybrids into solid substrates is a desirable goal, since it provides opportunities for additional applications of these complexes as sensors. For example, the modification of CNTs with DNA aptamers (Chen et al., 2009) and molecular beacons, and their application as ion, protein or DNA sensors are interesting potential areas of research. Furthermore, additional studies, both theoretical and experimental are needed in order to solve the current discrepancy on whether DNA hybridization occurs on solution or on the tube surface. A clearer understanding of the interactions involved in the hybridization event with DNA-CNTs might allow the development of improved DNA hybridization sensors.

On the other hand, further studies of these DNA-CNT complexes must be performed in order to fully control the properties of the complexes and to exploit all of their possible uses and applications. Theoretical studies and simulations have provided us with a better understanding of the type of forces and interactions involved on DNA-wrapping around CNTs, but there are still numerous unanswered questions. Additional theoretical studies that can consider the effects of variations in DNA sequence and length, CNT type, and solvent conditions, among others, can enhance our knowledge of all the interactions involved during the functionalization process, and would allow us to finely tailor and control the DNA-CNT complexes depending on our needs.

Acknowledgments

This work was supported in part by the Center of Advanced Nanoscale Materials, NASA-URC Grant numbers NCC3-1034 and NNX08BA48A. GSP and LSR acknowledge support from a NSF-EPSCoR Institute for Functional Nanomaterials Fellowship, Grant # OIA-0701525, and a Puerto Rico Industrial Development Company (PRIDCO) Scholarship. CPM was supported by NIH-MARC U* Program, Grant # 5T34GM07821-27.

5. References

Ali, S. R.; Ma, Y.; Parajuli, R. R.; Balogun, Y.; Lai, W.Y.-C.; He, H. (2007). A Nonoxidative Sensor Based on a Self-Doped Polyaniline/Carbon Nanotube Composite for Sensitive and Selective Detection of the Neurotransmitter Dopamine. *Anal. Chem.*, 79, 2583-2587.

Ando, Y.; Zhao, X.; Sugai, T.; Kumar, M. (2004). Growing Carbon Nanotubes. *Mater. Today*, 7, 22-29.

Arnold, M. S.; Stupp, S. I.; Hersam, M. C. (2005). Enrichment of Single-Walled Carbon Nanotubes by Diameter in Density Gradients. *Nano Lett.* 5, 713-718.

Badaire, S.; Zakri, C.; Maugey, M.; Derré, A.; Barisci, J. N.; Wallace, G.; Poulin, P. (2005). Liquid crystals of DNA-stabilized carbon nanotubes. *Adv. Mater.*, 17, 1673-1676.

Bae, A.-H.; Hatano, T.; Nakashima, N.; Murakami, H.; Shinkai, S. (2004). Electrochemical Fabrication of Single-Walled Carbon Nanotubes-DNA Complexes by Poly(Ethylenedioxythiophene) and Photocurrent Generation by Excitation of an Intercalated Chromophore. *Organic Biomol. Chem.*, 2, 1139-1144.

Bandow, S.; Rao, A. M.; Williams, K. A.; Thess, A.; Smalley, R. E.; Eklund, P. C. (1997). Purification of Single-Wall Carbon Nanotubes by Microfiltration. *J. Phys. Chem. B*, 101, 8839-8842.

Bauer, B. J.; Becker, M. L.; Bajpal, V.; Fagan, J. A.; Hobbie, E. K.; Migler, K.; Guttman, C. M.; Blair, W. R. (2007). Measurement of Single-Wall Nanotube Dispersion by Size Exclusion Chromatography. *J. Phys. Chem. C* , 111, 17914-17918.

Cao, C.; Kim, J. H.; Yoon, D.; Hwang, E.-S.; Kim, Y.-J.; Baik, S. (2008). Optical Detection of DNA Hybridization Using Absorption Spectra of Single-Walled Carbon Nanotubes. *Mater. Chem. Phys.*, 112, 738-741.

Cao, C.; Kim, J. H.; Kwon, Y.-J.; Kim, Y.-J.; Hwang, E.-S.; Baik, S. (2009). An Immunoassay using Biotinylated Single-Walled Carbon Nanotubes as Raman Biomarkers. *Analyst*, 134, 1294-1296.

Cathcart, H.; Quinn, S.; Nicolosi, V.; Kelly, J. M.; Blau, W. J.; Coleman, J. N. (2007). Spontaneous Debundling of Single-Walled Carbon Nanotubes in DNA-Based Dispersions. *J. Phys. Chem. C*, 111, 66-74.

Cathcart, H.; Nicolosi, V.; Hughes, J. M.; Blau, W. J.; Kelly, J. M.; Quinn, S. J.; Coleman, J. N. (2008). Ordered DNA Wrapping Switches on Luminescence in Single-Walled Nanotube Dispersions. *J. Am. Chem. Soc.*, 130, 12734-12744.

Cha, M.; Jung, S.; Cha, M.-H.; Kim, G. & Ihn, J. (2009). Reversible Metal-Semiconductor Transition of ssDNA-Decorated Single-Walled Carbon Nanotubes. *Nano Lett.*, 9, 1345-1349.

Chen, H.; Yu, C.; Jiang, C., Zhang, S., Liu, B. & Kong, J. (2009). A Novel Near-Infrared Protein Assay Based on the Dissolution and Aggregation of Aptamer-Wrapped Single-Walled Carbon Nanotubes. *Chem. Commun.*, 5006-5008.

Chen, R. J. & Zhang, Y. (2006). Controlled Precipitation of Solubilized Carbon Nanotubes by Delamination of DNA. *J. Phys. Chem. B*, 110, 54-57.

Chen, Y.; Liu, H.; Ye, T.; Kim, J.; Mao, C. (2007). DNA-Directed Assembly of Single-Wall Carbon Nanotubes. *J. Am. Chem. Soc.*, 129, 8696-8697.

Cheung, W.; Chiu, P. L.; Parajuli, R. R.; Ma, Y.; Ali, S. R. & He, H. (2009). Fabrication of High Performance Conducting Polymer Nanocomposites for Biosensors and Flexible Electronics: Summary of the Multiple Roles of DNA Dispersed and Functionalized Single Walled Carbon Nanotubes. *J. Mater. Chem.*, 19, 6465-6480.

Chou, S. G.; Ribeiro, H. B.; Barros, E. B.; Santos, A. P.; Nezich, D.; Samsonidze, G. G.; Fantini, C.; Pimenta, M. A.; Jorio, A.; Filho, F. P.; Dresselhaus, M. S.; Dresselhaus, G.; Saito, R.; Zheng, M.; Onoa, G. B.; Semke, E. D.; Swan, A. K.; Ünlü, M. S.; Goldberg, B. B. (2004). Optical Characterization of DNA-Wrapped Carbon Nanotube Hybrids. *Chem. Phys. Lett.*, 397, 296-301.

Chou, S. G.; Plentz, F.; Jiang, J.; Saito, R.; Nezich, D.; Ribeiro, H. B.; Jorio, A.; Pimenta, M. A.; Samsonidze, Ge. G.; Santos, A. P.; Zheng, M.; Onoa, G. B.; Semke, E. D.; Dresselhaus, G.; Dresselhaus, M. S. (2005). Phonon-Assisted Excitonic Recombination Channels Observed in DNA-Wrapped Carbon Nanotubes Using Photoluminescence Spectroscopy. *Phys. Rev. Lett.*, 94, 127402.

Chou, S. G.; Son, H.; Zheng, M.; Saito, R.; Jorio, A.; Dresselhaus, G.; Dresselhaus, M. S. (2007a). Finite Length Effects in DNA-Wrapped Carbon Nanotubes. *Chem. Phys. Lett.*, 443, 328-332.

Chou, S. G.; Son, H.; Kong, J.; Jorio, A.; Saito, R.; Zheng, M.; Dresselhaus, G.; Dresselhaus, M. S. (2007b). Length Characterization of DMA-Wrapped Carbon Nanotubes Using Raman Spectroscopy. *Appl. Phys. Lett.*, 90, 131109.

Chun, J.; Fagan, J. A.; Hobbie, E. K.; Bauer, B. J. (2008). Size Separation of Single-Wall Carbon Nanotubes by Flow-Field Flow Fractionation. *Anal. Chem.*, 80, 2514-2523.

Dai, H.; Hafner, J. H.; Rinzler, A. G.; Colbert, D. T.; Smalley, R. E. (1996). Nanotubes as nanoprobes in scanning probe microscopy. *Nature*, 384, 147-150.

Das, A.; Sood, A. K.; Maiti, P. K.; Das, M.; Varadarajan, R.; Rao, C. N. R. (2008). Binding of Nucleobases with Single-Walled Carbon Nanotubes: Theory and Experiment. *Chem. Phys. Lett.*, 453, 266-273.

Dillon, A. C.; Jones, K. J.; Bekkedahl, T. A.; Kiang, C. H.; Bethune, D. S.; Heben, M. J. (1997). Storage of Hydrogen in Single-Walled Carbon Nanotubes. *Nature*, 386, 377-379.

Dong, L.; Joseph, K. L.; Witkowski, C. M.; Craig, M. M. (2008). Cytotoxicity of Single-Walled Carbon Nanotubes Suspended in Various Surfactants. *Nanotechnology*, 19, 255702.

Dovbeshko, G. I.; Repnytska, O. P.; Obraztsova, E. D.; Shtogun, Y. V. (2003a). DNA Interaction with Single-Walled Carbon Nanotubes: A SEIRA Study. *Chem. Phys. Lett.*, 372, 432-437.

Dovbeshko, G. I.; Repnytska, O. P.; Obraztsova, E. D.; Shtogun, Y. V.; Andreev, E. O. (2003b). Study of DNA Interaction With Carbon Nanotubes. *Semicond. Phys. Quantum. Electron. & Optoelectron.*, 6, 105-110.

Dukovic, G.; Balaz, M.; Doak, P.; Berova, N. D.; Zheng, M.; Mclean, R. S.; Brus, L. E. (2006). Racemic Single-Walled Carbon Nanotubes Exhibit Circular Dichroism when Wrapped with DNA. *J. Am. Chem. Soc.*, 128, 9004-9005.

Ebbesen, T.; Ajatan, A.; Hiura, H.; Tanigaki, K. (1994). Purification of Carbon Nanotubes. *Nature*, 367, 519.

Fagan, J. A.; Bajpai, V.; Bauer, B. J.; Hobbie, E. K. (2007a). Anisotropic Polarizability of Isolated Semiconducting Single-Wall Carbon Nanotubes in Alternating Electric Fields. *Appl. Phys. Lett.*, 91, 213105.

Fagan, J. A.; Simpson, J. R.; Bauer, B. J.; Lacerda, S. H. D. P.; Becker, M. L.; Chun, J.; Migler, K. B.; Walker, A. R. H.; Hobbie, E. K. (2007b). Length-Dependent Optical Effects in Single-Wall Carbon Nanotubes. *J. Am. Chem. Soc.*, 129, 10607-10612.

Fantini, C.; Jorio, A.; Santos, A. P.; Peressinotto, V. S. T.; Pimenta, M. A. (2007). Characterization of DNA-Wrapped Carbon Nanotubes by Resonance Raman and Optical Absorption Spectroscopies. *Chem. Phys. Lett.*, 439, 138-142.

Ferancová, A.; Adamovski, M.; Gründler, P.; Zima, J.; Barek, J.; Mattusch, J.; Wennrich, R.; Labud, J. (2007). Interaction of Tin(II) and Arsenic(III) with DNA at the Nanostructure Film Modified Electrodes. *Bioelectrochemistry*, 71, 33-37.

Gao, H. & Kong, Y. (2004). Simulation of DNA-nanotube interactions. *Annu. Rev. Mater. Res.*, 34, 123-150.

Gao, X.; Xing, G.; Yang, Y.; Shi, X.; Liu, R.; Chu, W.; Jing, L.; Zhao, F.; Ye, C.; Yuan, H.; Fang, X.; Wang, C.; Zhao, Y. (2008). Detection of Trace Hg2+ Via Induced Circular Dichroism of DNA Wrapped Around Single-Walled Carbon Nanotubes. *J. Am. Chem. Soc.*, 130, 9190-9191.

Gigliotti, B.; Sakizzie, B.; Bethune, D. S.; Shelby, R. M.; Cha, J. N. (2006). Sequence-Independent Helical Wrapping of Single-Walled Carbon Nanotubes by Long Genomic DNA. *Nano Lett.*, 6, 159-164.

Gladchenko, G. O.; Karachevtsev, M. V.; Leontiev, V. S.; Valeev, V. A.; Glamazda, A. Y.; Plokhotnichenko, A. M.; Stepanian, S. G. (2006). Interaction of Fragmented Double-Stranded DNA with Carbon Nanotubes in Aqueous Solution. *Mol. Phys.*, 104, 3193.

Gowtham, S.; Scheicher, R. H.; Pandey, R.; Karna, S. P.; Ahuja, R. (2008). First-Principles Study of Physisorption of Nucleic Acid Bases on Small-Diameter Carbon Nanotubes. *Nanotechnology*, 19, 125701.

Han, X.; Li, Y.; Deng, Z. (2007). DNA-Wrapped Single-Walled Carbon Nanotubes as Rigid Templates for Assembling Linear Gold Nanoparticle Arrays. *Adv. Mater.*, 19, 1518-1522.

Han, X.; Li, Y.; Wu, S.; Deng, Z. (2008). General Strategy Toward pH-Controlled Aggregation - Dispersion of Gold Nanoparticles and Single-Walled Carbon Nanotubes. *Small*, 4, 326-329.

Harutyunyan, A. R.; Pradhan, B. K.; Chang, J.; Chen, G.; Eklund, P. C. (2002). Purification of Single-Wall Carbon Nanotubes by Selective Microwave Heating of Catalyst Particles. *J. Phys. Chem. B*, 106, 8671-8675.

Heering, H. A.; Williams, K. A.; de Vries, S.; Dekker, C. (2006). Specific Vectorial Immobilization of Oligonucleotide-Modified Yeast Cytochrome c on Carbon Nanotubes. *ChemPhysChem*, 7, 1705-1709.

Heller, D. A.; Jeng, E. S.; Yeung, T.-K.; Martinez, B. M.; Moll, A. E.; Gastala, J. B.; Strano, M. S. (2006). Optical Detection of DNA Conformational Polymorphism on Single-Walled Carbon Nanotubes. *Science*, 311, 508 511.

Hembram, K. P. S. S. & Rao, G. M. (2009). Studies on CNTs/DNA composite. *Mater. Sci. Eng. C*, 29, 1093.

Hobbie, E. K.; Bauer, B. J.; Stephens, J.; Becker, M. L.; McGuiggan, P.; Hudson, S. D.; Wang, H. (2005). Colloidal Particles Coated and Stabilized by DNA-Wrapped Carbon Nanotubes. *Langmuir*, 21, 10284-10287.

Hopkins, A. R.; Kruk, N. A.; Lipeles, R. A. (2007). Macroscopic Alignment of Single-Walled Carbon Nanotubes (SWNTs). *Surf. Coat. Technol.*, 202, 1282-1286.

Hu, C.; Zhang, Y.; Bao, G.; Zhang, Y.; Liu, M.; Wang, Z. L. (2005). Functionalized Single-Walled Carbon Nanotubes for Electrochemical Detection. *J. Phys. Chem B*, 109, 20072-20076.

Hu, C. G.; Feng, B.; Xi, Y.; Zhang, Z. W.; Wang, N. (2007). Modification of Carbon Nanotubes and their Electrochemical Detection. *Diam. Relat. Mater.*, 16, 1988-1991.

Huang, X.; Mclean, R. S.; Zheng, M. (2005). High-Resolution Length Sorting and Purification of DNA-Wrapped Carbon Nanotubes by Size-Exclusion Chromatography. *Anal. Chem.*, 77, 6225-6228.

Hughes, M. E.; Brandlin, E.; Golovchenko, J. A. (2007). Optical Absorption of DNA-Carbon Nanotube Structures. *Nano Lett.* , 7, 1191-1194.

Hwang, E.-S.; Cao, C.; Hong, S.; Jung, H.-J.; Cha, C.-Y.; Choi, J.-B.; Kim, Y.-J.; Baik, S. (2006). The DNA Hybridization Assay Using Single-Walled Carbon Nanotubes as Ultrasensitive, Long-Term Optical Labels. *Nanotechnology*, 17, 3442-3445.

Hwang, J. S.; Kim, H. T.; Kim, H. K.; Son, M. H.; Oh, J. H.; Hwang, S. W.; Ahn, D. (2008). Electronic Transport Characteristics of a Single Wall Carbon Nanotube Field Effect Transistor Wrapped with Deoxyribonucleic Acid Molecules. *J. Phys. Conf. Ser.*, 109, 012015.

Iijima, M.; Watabe, T.; Ishii, S.; Koshio, A.; Yamaguchi, T.; Bandow, S.; Iijima, S.; Suzuki, K.; Maruyama, Y. (2005). Fabrication and STM-Characterization of Novel Hybrid Materials of DNA/carbon Nanotube. *Chem. Phys. Lett.*, 414, 520-524.

Iijima, S. (1991). Helical Microtubules of Graphitic Carbon. *Nature*, 354, 56-58.

Ikeda, A.; Hamano, T.; Hayashi, K.; Kikuchi, J.-I. (2006). Water-Solubilization of Nucleotides-Coated Single-Walled Carbon Nanotubes using a High-Speed Vibration Milling Technique. *Org. Lett.*, 8, 1153-1156.

Jeng, E. S.; Moll, A. E.; Roy, A. C.; Gastala, J. B.; Strano, M. S. (2006). Detection of DNA Hybridization Using the Near-Infrared Band-Gap Fluorescence of Single-Walled Carbon Nanotubes. *Nano Lett.*, 6, 371-375.

Jeng, E. S.; Barone, P. W.; Nelson, J. D.; Strano, M. S. (2007) Hybridization Kinetics and Thermodynamics of DNA Adsorbed to Individually Dispersed Single-Walled Carbon Nanotubes. *Small*, 3, 1602-1609.

Jin, H.; Jeng, E. S.; Heller, D. A.; Jena, P. V.; Kirmse, R.; Langowski, J.; Strano, M. S. (2007). Divalent Ion and Thermally Induced DNA Conformational Polymorphism on Single-Walled Carbon Nanotubes. *Macromolecules*, 40, 6731-6739.

Johnson, A. T. C.; Staii, C.; Chen, M.; Khamis, S.; Johnson, R.; Klein, M. L.; Gelperin, A. (2006). DNA-Decorated Carbon Nanotubes for Chemical Sensing. *Phys. Stat. Sol. (b)*, 13, 3252-3256.

Johnson, R. R.; Johnson, A. T. C.; Klein, M. L. (2008). Probing the Structure of DNA-Carbon Nanotube Hybrids with Molecular Dynamics. *Nano Lett.*, 8, 69-75.

Kam, N. W. S.; Dai, H. (2005). Carbon Nanotubes as Intracellular Protein Transporters: Generality and Biological Functionality. *J. Am. Chem. Soc.*, 127, 6021-6026.

Kam, N. W. S.; Liu, Z.; Dai, H. (2005a). Carbon Nanotubes as Intracellular Transporters for Proteins and DNA: An Investigation of the Uptake Mechanism and Pathway. *Angew. Chem. Int. Ed.*, 45, 577-581.

Kam, N. W. S.; O'Connell, M.; Wisdom, J. A.; Dai, H. (2005b). Carbon Nanotubes as Multifunctional Biological Transporters and Near-infrared Agents for Selective Cancer Cell Destruction. *Proc. Nat. Acad. Sci. U.S.A.*, 102, 11600-11605.

Karachevtsev, V. A.; Glamazada, A. Yu.; Dettlaff-Weglikowska, U.; Leontiev, V. S.; Mateichenko, P. V.; Roth, S.; Rao, A. M. (2006a). Spectroscopic and SEM Studies of SWNTs: Polymer Solutions and Films. *Carbon*, 44, 1292-1297.

Karachevtsev, V. A.; Glamazada, A. Yu.; Leontiev, V. S.; Mateichenko, P. V.; Dettlaff-Weglikowska, U. (2006b). SWNTs with DNA in Aqueous Solution and Film. *AIP Conf. Proc.*, 786, 257-261.

Karachevtsev, V. A.; Glamazda, A. Y.; Leontiev, V. S.; Lytvyn, O.S.; Dettlaff-Weglikowska, U. (2007). Glucose Sensing Based on NIR Fluorescence of DNA-Wrapped Single-Walled Carbon Nanotubes. *Chem. Phys. Lett.*, 435, 104-108.

Kawamoto, H.; Uchida, T.; Kojima, K.; Tachibana, M. (2006a). G Band Raman Features of DNA-Wrapped Single-Wall Carbon Nanotubes in Aqueous Solution and Air. *Chem Phys. Lett.*, 432, 172-176.

Kawamoto, H.; Uchida, T.; Kojima, K.; Tachibana, M. (2006b). Raman Study of DNA-Wrapped Single-Wall Carbon Nanotube Hybrids Under Various Humidity Conditions. *Chem. Phys. Lett.*, 431, 118-120.

Kawamoto, H.; Uchida, T.; Tachibana, M. (2006c). The Feature of the Breit-Wigner-Fano Raman Line in DNA-Wrapped Single-Wall Carbon Nanotubes. *J. Appl. Phys.*, 99, 094309.

Kelley, K.; Pehrsson, P. E.; Ericson, L. M.; Zhao, W. (2005). Optical pH Response of DNA Wrapped HiPco Carbon Nanotubes. *J. Nanosci. Nanotechnol.*, 5, 1041-1044.

Khripin, C. Y.; Zheng, M.; Jagota, A. (2009a). Deposition and Meniscus Alignment of DNA-CNT on a Substrate. *J. Colloid Interface Sci.*, 330, 255-265.

Khripin, C.Y.; Manohar, S.; Zheng, M. & Jagota, A. (2009b). Measurement of electrostatic properties of DNA-carbon nanotube hybrids by capillary electrophoresis. *J. Phys. Chem. C*, 113, 13616-13621.

Lahiji, R. R.; Dolash, D. D.; Bergstrom, D. E.; Reifenberger, R. (2007). Oligodeoxyribonucleotide Association with Single-Walled Carbon Nanotubes Studied by SPM. *Small*, 3, 1912-1920.

Lee, J. Y.; Shin, H. Y.; Kang, S. W.; Park, C.; Kim, S. W. (2010) Use of Bioelectrode Containing DNA-Wrapped Single-Walled Carbon Nanotubes for Enzyme-Based Biofuel Cell, *J. Power Sources*, 195, 750-755.

Li, H.-H.; Zhang, Z.-L.; Shen, X.-C.; Zhang, M.-X.; Pang, D.-W. (2007). Glassy Carbon Electrodes Modified with DNA-Partly-Wrapped Single-Walled Carbon Nanotubes. *Electrochem. Commun.*, 9, 2729-2733.

Li, X.; Peng, Y.; Qu, X. (2006). Carbon Nanotubes Selective Destabilization of Duplex and Triplex DNA and Inducing B-A Transition in Solution. *Nucleic Acids Res.*, 34, 3670-3676.

Li, Y.; Han, X.; Deng, Z. (2007). Grafting Single-Walled Carbon Nanotubes with Highly Hybridizable DNA Sequences: Potential Building Blocks for DNA-Programmed Material Assembly. *Angew. Chem. Int. Ed.*, 46, 7481-7484.

Li, Y.; Wang, P.; Li, F.; Huang, X.; Wang, L.; Lin, X. (2008). Covalent Immobilization of Single-Walled Carbon Nanotubes and Single-Stranded Deoxyribonucleic Acid Nanocomposites on Glassy Carbon Electrode: Preparation, Characterization, and Applications. *Talanta*, 77, 833-838.

Li, Z.; Wu, Z. & Li, K. (2009). The high dispersion of DNA-multiwalled carbon nanotubes and their properties. *Anal. Biochem.*, 387, 267-270.

Liang, Z.; Lao, R.; Wang, J.; Liu, Y.; Wang, L.; Huang, Q.; Song, S.; Li, G.; Fan, C. (2007). Solubilization of Single-Walled Carbon Nanotubes with Single-Stranded DNA Generated from Asymmetric PCR. *Int. J. Mol. Sci.*, 8, 705-713.

Liu, J.; Rinzler, A. G.; Dai, H.; Hafner, J. H.; Bradley, R. K.; Boul, P. J.; Lu, A.; Iverson, T.; Shelimov, K.; Huffman, C. B.; Rodriguez-Macias, F.; Shon, Y. S.; Lee, T. R.; Colbert, D. T.; Smalley, R. E. (1998). Fullerene Pipers. *Science*, 280, 1253-1256.

Liu, Y.; Chen, J.; Anh, N. T.; Too, C. O.; Misoska, V.; Wallace, G. G. (2008). Nanofiber Mats from DNA, SWNTs, and Poly(ethylene oxide) and their Application in Glucose Biosensors . *J. Electrochem. Soc.*, 155, K100-K103.

Liu, Y. & Wei, W. (2008). Layer-by-Layer Assembled DNA Functionalized Single-Walled Carbon Nanotube Hybrids for Arsenic(III) Detection. *Electrochem. Commun.*, 10, 872-875.

Liu, Y.; Wang, Y.; Jin, J.; Wang, H.; Yang, R.; Tan, W. (2009). Fluorescent Assay of DNA Hybridization with Label-Free Molecular Switch: Reducing Background-Signal and Improving Specificity by Using Carbon Nanotubes. *Chem. Commun.*, 665-667.

Luo, Z.; Pfefferle, L. D.; Haller, G. L.; Papadimitrakopoulos, F. (2006). Abundance Evaluation of Single-Walled Carbon Nanotubes by Fluorescence and Absorption Spectroscopy. *J. Am. Chem. Soc.*, 128, 15511-15516.

Lustig, S. R.; Jagota, A.; Khripin, C.; Zheng, M. (2006). Structure-Based Carbon Nanotube Separations by Ion-Surface Interactions. *Mater. Res. Symp. Proc.*, 923, 1-6.

Ma, Y.; Ali, S. R.; Dodoo, A. S.; He, H. (2006a). Enhanced Sensitivity for Biosensors: Multiple Functions of DNA-Wrapped Single-Walled Carbon Nanotubes in Self-Doped Polyaniline Nanocomposites. *J. Phys. Chem. B*, 110, 16359-16365.

Ma, Y.; Ali, S. R.; Wang, L.; Chiu, P. L.; Mendelsohn, R.; He, H. (2006b). In Situ Fabrication of a Water-Soluble, Self-Doped Polyaniline Nanocomposite: The Unique Role of DNA Functionalized Single-Walled Carbon Nanotubes. *J. Am. Chem. Soc.*, 128, 12064-12065.

Malik, S.; Vogel, S.; Rösner, H.; Arnold, K.; Hennrich, F.; Köhler, A.; Richert, C.; Kappes, M. M. (2007). Physical Chemical Characterization of DNA-SWNT Suspensions and Associated Composites. *Compos. Sci. and Technol.*, 67, 916-921.

Manohar, S.; Tang, T.; Jagota, A. (2007). Structure of Homopolymer DNA-CNT Hybrids. *J. Phys. Chem. C*, 111, 17835-17845.

Mclean, R. S.; Huang, X.; Khripin, C.; Jagota, A.; Zheng, M. (2006). Controlled Two-Dimensional Pattern of Spontaneously Aligned Carbon Nanotubes. *Nano Lett.*, 6, 55-60.

Meng, S.; Wang, W. L.; Maragakis, P.; Kaxiras, E. (2007). Determination of DNA-Base Orientation on Carbon Nanotubes through Directional Optical Absorbance. *Nano Lett.*, 7, 2312-2316.

Mintmire, J. W.; Robertson, D. H.; White, C. T. (1993). Properties of Fullerene Nanotubules. *J. Phys. Chem. Sol.*, 54, 1835-1840.

Nair, N.; Usrey, M. L.; Kim, W.; Braatz, R. D.; Strano, M. S. (2006). Estimation of the (n,m) Concentration Distribution of Single-Walled Carbon Nanotubes from Photoabsorption Spectra. *Anal. Chem.*, 78, 7689-7696.

Nakashima, N.; Okuzono, S.; Murakami, H.; Nakai, T.; Yoshikawa, K. (2003). DNA Dissolves Single-Walled Carbon Nanotubes in Water. *Chem. Lett.*, 5, 456-457.

Napier, M. E.; Hull, D. O.; Thorp, H. H. (2005). Electrocatalytic Oxidation of DNA-Wrapped Carbon Nanotubes. *J. Am. Chem. Soc.*, 127, 11952-11953.

Nepal, D.; Sohn, J.; Aicher, W. K.; Lee, S.; Geckeler, K. E. (2005). Supramolecular Conjugates of Carbon Nanotubes and DNA by a Solid-State Reaction. *Biomacromolecules*, 6, 2919-2922.

Nguyen, C. V.; Delzeit, L.; Cassell, A. M.; Li, J.; Han, J.; Meyyappan, M. (2002). Preparation of Nucleic Acid Functionalized Carbon Nanotube Arrays. *Nano Lett.*, 2, 1079-1081.

Niyogi, S.; Hu, H.; Hamon, M. A.; Bhowmik, P.; Zhao, B.; Rozenzhak, S. M.; Chen, J.; Itkis, M. E.; Meier, M. S.; Haddon, R. C. (2001). Chromatographic Purification of Soluble Single-Walled Carbon Nanotubes (s-SWNTs) [1]. *J. Am. Chem. Soc.*, 123, 733-734.

Noguchi, Y.; Fujigaya, T.; Niidome, Y.; Nakashima, N. (2008). Single-Walled Carbon nanotubes/DNA Hybrids in Water are Highly Stable. *Chem. Phys. Lett.*, 455, 249-251.

O'Connell, M. J.; Boul, P.; Ericson, L. M.; Huffman, C.; Wang, Y.; Haroz, E.; Kuper, C.; Tour, J.; Ausman, K. D.; Smalley, R. E. (2001). Reversible Water-Solublization of Single-Walled Carbon Nanotubes by Polymer Wrapping. *Chem. Phys. Lett.*, 342, 265-271.

Onoa, B.; Zheng, M.; Dresselhaus, M. S.; Diner, B. A. (2006). Carbon Nanotubes and Nucleic Acids: Tools and Targets. *Phys. Stat. Sol. (a)*, 203, 1124-1131.

Ostojic, G. N.; Ireland, J. R.; Hersam, M. C. (2008). Noncovalent Functionalization of DNA-Wrapped Single-Walled Carbon Nanotubes with Platinum-Based DNA Cross-Linkers. *Langmuir*, 24, 9784-9789.

Poonam, P. & Deo, N. (2008). Current Correlation Functions for Chemical Sensors Based on DNA Decorated Carbon Nanotube. *Sens. Actuators B*, 135, 327-335.

Qian, H.; Araujo, P. T.; Georgi, C.; Gokus, T.; Hartmann, N.; Green, A. A.; Jorio, A.; Hersam, M. C.; Novotny, L.; Hartschuh, A. (2008). Visualizing the Local Optical Response of Semiconducting Carbon Nanotubes to DNA-Wrapping. *Nano Lett.*, 8, 2706-2711.

Rajendra, J.; Baxendale, M.; Rap, L. G. D.; Rodger, A. (2004). Flow Linear Dichroism to Probe Binding of Aromatic Molecules and DNA to Single-Walled Carbon Nanotubes. *J. Am. Chem. Soc.*, 126, 11182-11188.

Rajendra, J. & Rodger, A. (2005). The Binding of Single-Stranded DNA and PNA to Single-Walled Carbon Nanotubes Probed by Flow Linear Dichroism. *Chem. Eur. J.*, 11, 4841-4847.

Ranjan, N.; Seifert, G.; Heine, T. (2005). Wrapping Carbon Nanotubes with DNA: A Theoretical Study. *AIP Conf. Proc.*, 786, 448-451.

Rinzler, A.; Liu, J.; Dai, H.; Nikolaev, P.; Huffman, C.; Rodriguez-Macias, F. J.; Boul, P.; Lu, A.; Heymann, D.; Colbert, D. T.; Lee, R. S.; Fischer, J.; Rao, A.; Eklund, P. C.; Smalley, R. E. (1998). Large-Scale Purification of Single-Wall Carbon Nanotubes: Process, Product, and Characterization. *Appl. Phys. A*, 67, 29-37.

Sánchez-Pomales, G.; Morales-Negrón, Y.; Cabrera, C. R. (2005). Study of Self-Assembled Monolayers of DNA and DNA-Carbon Nanotube Hybrids. *Rev. Adv. Mater. Sci.*, 10, 261-265.

Sánchez-Pomales, G.; Rivera-Vélez, N. E.; Cabrera, C. R. (2006). DNA-Wrapped Carbon Nanotubes Assembled on Gold Substrates. *ECS Trans.*, 3, 21-29.

Sánchez-Pomales, G.; Santiago-Rodríguez, L.; Rivera-Vélez, N. E.; Cabrera, C. R. (2007a). Characterization of the DNA-Assisted Purification of Single-Walled Carbon Nanotubes. *Phys. Stat. Sol. (a)*, 204, 1791-1796.

Sánchez-Pomales, G.; Rivera-Vélez, N. E.; Cabrera, C. R. (2007b). DNA-Mediated Self-Assembly of Carbon Nanotubes on Gold. *J. of Phys.: Conf. Ser.*, 61, 1017-1021.

Sánchez-Pomales, G. & Cabrera, C. R. (2007). Vertical Attachment of DNA-CNT Hybrids on Gold. *J. Electroanal. Chem.*, 606, 47-54.

Sánchez-Pomales, G.; Santiago-Rodríguez, L.; Cabrera, C. R. (2009). DNA-Functionalized Carbon Nanotubes for Biosensing Applications. *J. Nanosci. Nanotechnol.*, 9, 2175-2188.

Sánchez-Pomales, G.; Pagán-Miranda, C.; Santiago-Rodríguez, L.; Cabrera, C. R. "Hybridization-Mediated Deposition of Carbon Nanotubes on DNA Self-Assembled Monolayers on Gold", Unpublished a.

Sánchez-Pomales, G.; Wang, D.; Wang, S.; Cabrera, C. R.; Jiang, S. P. Pt-Ru Decorated DNA-Carbon Nanotube Complexes as Catalysts for the Electrochemical Oxidation of Methanol. Unpublished b.

Service, R. F. (1998). Superstrong Nanotubes Show They are Smart, Too. *Science*, 281, 940-942.

Shelimov, K. B.; Esenaliev, R. O.; Rinzler, A. G.; Huffman, C. B.; Smalley, R. E. (1998). Purification of Single-Wall Carbon Nanotubes by Ultrasonically Assisted Filtration. *Chem. Phys. Lett.*, 282, 429-434.

Shin, S. R.; Lee, C. K.; So, I.; Jeon, J.-H.; Kang, T. M.; Kee, C.; Kim, S. I.; Spinks, G. M.; Wallace, G. G.; Kim, S. J. (2008) DNA-Wrapped Single-Walled Carbon Nanotubes Hybrid Fibers for Supercapacitors and Artificial Muscles. *Adv. Mater.*, 20, 466-470.

Shoda, M.; Bandow, S.; Maruyama, Y. & Iijima, S. J. (2009). Probing Interaction Between ssDNA and Carbon Nanotubes by Raman Scattering and Electron Microscopy. *Phys. Chem. C*, 113, 6033-6036.

Shtogun, Y. V.; Woods, L. M.; Dovbeshko, G. I. (2007). Adsorption of Adenine and Thymine and their Radicals on Single-Wall Carbon Nanotubes. *J. Phys. Chem. C*, 111, 18174-18181.

Sickert, D.; Taeger, S.; Neumann, A.; Jost, O.; Eckstein, G.; Mertig, M.; Pompe, W. (2005). Separation and Assembly of DNA-dispersed Carbon Nanotubes by Dielectrophoresis. *AIP Conf. Proc.*, 786, 271-274.

Staii, C.; Johnson Jr., A. T.,; Chen, M.; Gelperin, A. (2005). DNA-Decorated Carbon Nanotubes for Chemical Sensing. *Nano Lett.*, 5, 1774-1778.

Star, A.; Steuerman, D. W.; Heath, J. R.; Stoddart, J. F. (2002). Starched Carbon Nanotubes. *Angew. Chem. Int. Ed.*, 41, 2508-2512.

Stepanian, S. G.; Karachevtsev, M. V.; Glamazada, A. Yu.; Karachevtsev, V. A.; Adamowicz, L. (2008). Stacking Interaction of Cytosine With Carbon Nanotubes: MP2, DFT and Raman Spectroscopy Study. *Chem. Phys. Lett.*, 459, 153-158.

Strano, M. S.; Zheng, M.; Jagota, A.; Onoa, G. B.; Heller, D. A.; Barone, P. W.; Usrey, M. L. (2004). Understanding the Nature of the DNA-Assisted Separation of Single-Walled Carbon Nanotubes Using Fluorescence and Raman Spectroscopy. *Nano Lett.*, 4, 543-550.

Taeger, S.; Jost, O.; Pompe, W.; Mertig, M. (2004). Purification and Dispersion of Carbon Nanotubes by Sidewall Functionalization with Single-Stranded DNA. *AIP Conf. Proc.*, 723, 185-188.

Taeger, S.; Xuang, L. Y.; Günther, K.; Mertig, M. (2005). Noncovalent Sidewall Functionalization of Carbon Nanotubes by Biomolecules: Single-stranded DNA and Hydrophobin. *AIP Conf. Proc*, 262-265.

Takahashi, H.; Namao, S.; Bandow, S.; Iijima, S. (2006). AFM Imaging of Wrapped Multiwall Carbon Nanotube in DNA. *Chem. Phys. Lett.*, 418, 535-539.

Talin, A. A.; Dentinger, P. M.; Jones, F. E.; Pathak, S.; Hunter, L.; Léonard, F.; Morales, A. M. (2004). Assembly and Electrical Characterization of DMA-Wrapped Carbon Nanotube Devices. *J. Vac. Sci. Technol. B*, 22, 3107-3111.

Tang, X.; Bansaruntip, S.; Nakayama, N.; Yenilmez, E.; Chang, Y.-I.; Wang, Q. (2006). Carbon Nanotube DNA Sensor and Sensing Mechanism. *Nano Lett.*, 6, 1632-1636.

Tasis, D.; Tagmatarchis, N.; Bianco, A.; Prato, M. (2006). Chemistry of Carbon Nanotubes. *Chem. Rev.*, 106, 1105-1136.

Toita, S.; Kang, D.; Kobayashi, K.; Kawamoto, H.; Kojima, K.; Tachibana, M. (2008). Atomic Force Microscopic Study on DNA-Wrapping for Different Diameter Single-Wall Carbon Nanotubes. *Diam. Relat. Mater.*, 17, 1389-1393.

Torrens, O. N.; Milkie, D. E.; Zheng, M.; Kikkawa, J. M. (2006). Photoluminescence from Intertube Carrier Migration in Single-Walled Carbon Nanotube Bundles. *Nano Lett.*, 6, 2864-2867.

Torrens, O. N.; Milkie, D. E.; Ban, H. Y.; Zheng, M.; Onoa, G. B.; Gierke, T. D.; Kikkawa, J. M. (2007). Measurement of Chiral-Dependent Magnetic Anisotropy in Carbon Nanotubes. *J. Am. Chem. Soc.*, 129, 252-253.

Tu, X.; Pehrsson, P. E.; Zhao, W. (2007). Redox Reaction of DNA-Encased HiPco Carbon Nanotubes with Hydrogen Peroxide: A Near Infrared Optical Sensitivity and Kinetics Study. *J. Phys. Chem. C*, 111, 17227-17231.

Tu, X. & Zheng, M. (2008). A DNA-Based Approach to the Carbon Nanotube Sorting Problem. *Nano Res.*, 1, 185-194.

Tu, X.; Manohar, S.; Jagota, A.; Zheng, M. (2009). DNA Sequence Motifs for Structure-Specific Recognition and Separation of Carbon Nanotubes. *Nature*, 460, 250-253.

Vetcher, A. A.; Srinivasan, S.; Vetcher, I. A.; Abramov, S. M.; Kozlov, M.; Baughman, R. H.; Levene, S. D. (2006). Fractionation of SWNT/nucleic Acid Complexes by Agarose Gel Electrophoresis. *Nanotechnology*, 17, 4263-4269.

Viswanathan, S.; Radecka, H.; Radecki, J. (2009). Electrochemical Biosensor for Pesticides Based on Acetylcholinesterase Immobilized on Polyaniline Deposited on Vertically Assembled Carbon Nanotubes Wrapped with ssDNA. *Biosens. Bioelectron.*, 24, 2772-2777.

Vogel, S. R.; Kappes, M. M.; Hennrich, F.; Richert, C. (2007). An Unexpected New Optimum in the Structure Space of DNA Solubilizing Single-Walled Carbon Nanotubes. *Chem. Eur. J.*, 13, 1815-1820.

Vogel, S. R.; Müller, K.; Plutowski, U.; Kappes, M. M.; Richert, C. (2007). DNA-Carbon Nanotube Interactions and Nanostructuring Based on DNA. *Phys. Stat. Sol. (b)*, 244, 4026-4029.

Wang, Y. (2008). Theoretical Evidence for the Stronger Ability of Thymine to Disperse SWCNT than Cytosine and Adenine: Self-Stacking of DNA Bases vs Their Cross-Stacking with SWCNT. *J. Phys. Chem. C*, 112, 14297-14305.

Wu, Y.; Phillips, J. A.; Liu, H.; Yang, R.; Tan, W. (2008). Carbon Nanotubes Protect DNA Strands During Cellular Delivery. *ACS Nano*, 2, 2023-2028.

Wu, Z.; Zhen, Z.; Jiang, J. -H.; Shen, G. -L. & Yu, R. -Q. (2009). Terminal Protection of Small-Molecule-Linked DNA for Sensitive Electrochemical Detection of Protein Binding via Selective Carbon Nanotube Assembly. *J. Am. Chem. Soc.*, 131, 12325-12332.

Xu, Y.; Pehrsson, P. E.; Chen, L.; Zhang, R.; Zhao, W. (2007). Double-Stranded DNA Single-Walled Carbon Nanotube Hybrids for Optical Hydrogen Peroxide and Glucose Sensing. *J. Phys. Chem. C*, 111, 8638-8643.

Xu, Y.; Pehrsson, P. E.; Chen, L.; Zhao, W. (2008). Controllable Redox Reaction of Chemically Purified DNA-Single Walled Carbon Nanotube Hybrids with Hydrogen Peroxide. *J. Am. Chem. Soc.*, 130, 10054-10055.

Yang, Q.-H.; Gale, N.; Oton, C. J.; Li, F.; Vaughan, A.; Saito, R.; Nandhakumar, I. S.; Tang, Z.-Y.; Cheng, H.-M.; Brown, T.; Loh, W. H. (2007a). Controllable Redox Reaction of Chemically Purified DNA-Single Walled Carbon Nanotube Hybrids with Hydrogen Peroxide. *Nanotechnology*, 18, 405706.

Yang, Q. -H.; Gale, N.; Oton, C. J.; Li, H.; Nandhakumar, I. S.; Tang, Z.-Y.; Brown, T.; Loh, W. H. (2007b). Deuterated Water as Super Solvent for Short Carbon Nanotubes Wrapped by DNA. *Carbon*, 45, 2701-2703.

Yang, Q. -H.; Wang, Q.; Gale, N.; Oton, C. J.; Cui, L.; Nandhakumar, I. S.; Zhu, Z.; Tang, Z.; Brown, T. & Loh, W. H. (2009). Loosening the DNA wrapping around single-walled carbon nanotubes by increasing the strand length. *Nanotechnology*, 20, 195603.

Yang, R.; Jin, J.; Chen. Y.; Shao, N.; Kang, H.; Xiao, Z.; Tang, Z.; Wu, Y.; Zhu, Z.; Tan, W. (2008a). Carbon Nanotube-Quenched Fluorescent Oligonucleotides: Probes That Fluoresce Upon Hybridization. *J. Am. Chem. Soc.*, 130, 8351-8358.

Yang, R.; Tang, Z.; Yan, J.; Kang, H.; Kim, Y.; Zhu, Z.; Tan, W. (2008b). Noncovalent Assembly of Carbon Nanotubes and Single-Stranded DNA: An Effective Sensing Platform for Probing Biomolecular Interactions. *Anal. Chem.*, 80, 7408-7413.

Yarotski, D. A.; Kilina, S. V.; Talin, A. A.; Tretiak, S.; Prezhdo, O. V.; Balatsky, A. V.; Taylor, A. J. (2009). Scanning Tunneling Microscopy of DNA-Wrapped Carbon Nanotubes. *Nano Lett.*, 9, 12-17.

Yoon, D.; Cao, C.; Choi, J.-B.; Kim, Y.-J. & Baik, S. (2008). Optical Characterization of DNA-Wrapped Single-Walled Carbon Nanotubes Irradiated with Ultraviolet Light. *J. Nanosci. Nanotech.* 8, 5135-5138.

Zangmeister, R. A.; Maslar, J. E.; Opdahl, A.; Tarlov, M. J. (2007). Adsorption Behavior of DNA-Wrapped Carbon Nanotubes on Self-Assembled Monolayer Surfaces. *Langmuir*, 23, 6252-6256.

Zhang, L.; Zaric, S.; Tu, X.; Wang, X.; Zhao, W.; Dai, H. (2008). Assessment of Chemically Separated Carbon Nanotubes for Nanoelectronics. *J. Am. Chem. Soc.*, 130, 2686-2691.

Zhao, W.; Gao, Y.; Brook, M. A.; Li, Y. (2006). Wrapping Single-Walled Carbon Nanotubes with Long Single-Stranded DNA Molecules Produced by Rolling Circle Amplification. *Chem. Commun.*, 3582-3584.

Zhao, X. & Johnson, J. K. (2007). Simulation of Adsorption of DNA on Carbon Nanotubes. *J. Am. Chem. Soc.*, 129, 10438-10445.

Zheng, D.; Li, X.; Ye, J. (2009). Adsorption and Release Behavior of Bare and DNA-Wrapped-Carbon Nanotubes on Self-Assembled Monolayer Surface. *Bioelectrochemistry* 2009, 74, 240-245.

Zheng, M.; Jagota, A.; Semke, E. D.; Diner, B. A.; McClean, R. S.; Lustig, S. R.; Richardson, R. E.; Tassi, N. G. (2003a). DNA-Assisted Dispersion and Separation of Carbon Nanotubes. *Nat. Mater.*, 2, 338-342.

Zheng, M.; Jagota, A.; Strano, M. S.; Santos, A. P.; Barone, P.; Chou, S. G.; Diner, B. A.; Dresselhaus, M. S.; Mclean, R.S.; Onoa, G. B.; Samsonidze, G. G.; Semke, E. D.; Usrey, M.; Walls, D. J. (2003b). Structure-Based Carbon Nanotube Sorting by Sequence-Dependent DNA Assembly. *Science*, 302, 1545-1548.

Zheng, M. (2004). Manipulating Carbon Nanotubes with Nucleic Acids. *AIP Conf. Proc.*, 723, 75-80.

Zheng, M. & Diner, B. A. (2004). Solution Redox Chemistry of Carbon Nanotubes. *J. Am. Chem. Soc.*, 126, 15490-15494.

Zheng, M. & Rostovtsev, V. V. (2006). Photoinduced Charge Transfer Mediated by DNA-Wrapped Carbon Nanotubes. *J. Am. Chem. Soc.*, 128, 7702-7703.

Zheng, M. (2007). Redox Chemistry and Electrochemistry of DNA-Wrapped Carbon Nanotubes. *ECS Trans.*, 6, 29-32.

Zheng, M. & Semke, E. D. (2007). Enrichment of Single Chirality Carbon Nanotubes. *J. Am. Chem. Soc.*, 129, 6084-6085.

Zhu, Z.; Tang, Z.; Phillips, J. A.; Yang, R.; Wang, H., Tan, W. (2008). Regulation of Singlet Oxygen Generation Using Single-Walled Carbon Nanotubes. *J. Am. Chem. Soc.*, 130, 10856-10857.

36

The Unlikely Surfactant: DNA as a Ligand for Single-Walled Carbon Nanotubes

Katharina Müller [a] and Clemens Richert [a,b]
[a]*Institute for Organic Chemistry and Center for Functional Nanostructures (CFN), University of Karlsruhe (TH) and KIT*
[b]*Institute for Organic Chemistry, University of Stuttgart*
Germany

1. Introduction

Single-walled carbon nanotubes (SWCNTs) were first described almost 20 years ago (Iijima, 1991). Their remarkable properties like mechanical strength, low density, stiffness, chemical and thermal stability, as well as their extraordinary electric and heat conductivity have made them candidates for applications in several areas of nanotechnology. Some of these applications call for pristine carbon nanotubes of a specific chirality. This is particularly true for applications in nanoelectronics. Most production processes generate complex mixtures of metallic and semiconducting SWCNTs, however, making purification a critical step in the overall process. For purification to be successful, several obstacles have to be overcome. One such obstacle is breaking up the very strong interactions between individual carbon nanotubes that lead to bundling. Due to their geometrical structure with a huge length-to-diameter ratio, nanotubes are almost one-dimensional molecules. These cylindrical molecules, which may be regarded as rolled-up graphene sheets, have very hydrophobic properties and engage in extensive van-der-Waals interactions through their π electron-rich surfaces, most probably strengthened by solvophobic effects in polar solvents, such as water. Intermolecular forces sum up over very long distances due to the shape complementarity of the binding partners. Solutions or suspensions of single-walled carbon nanotubes in conventional solvents are therefore metastable, even when the solvents are lipophilic (Liu et al., 1998; Chen et al., 1998; Liu et al., 1999; Bahr et al., 2001; Furtado et al., 2004). To generate suspensions at all, strong mechanical forces, usually in the form of ultrasonication, have to be employed. But, upon time, re-bundling of nanotubes sets in after sonication, again decreasing the interface between tubes and solvent molecules.

To overcome the poor solubility, the surface of nanotubes may be modified chemically, resulting in covalently attached functionalities or side chains. Being *all*-sp^2 carbon frameworks, whose covalent modification introduces sp^3 centers, this necessarily disrupts the electronic structure of the nanotubes (Jung et al., 2004; Cosnier & Holzinger, 2008). Thus, their most characteristic property is often affected in a hard-to-predict fashion with undesirable consequences for any resulting nanostructured device. To obtain pristine, monodisperse nanotubes in solution, bundles have to be broken up and rebundling has to be prevented. Experimental work of the last decade has shown that ligands that shield the surface of individual nanotubes immediately after break-up of bundles can indeed lead to suspensions of surprising stability in aqueous solution. Usually, these ligands are amphiphilic molecules with a hydrophobic part and a polar or charged "head group". Amphiphilic molecules with such properties are commonly surfactants. In water, the hydrophobic moiety can associate with the hydrophobic nanotube surface, while the polar part forms an outer shell, well solvated by the surrounding medium. Particularly if a charged head group is involved, the nanotubes thus encapsulated, repulse each other, preventing re-bundling.

2. Solubilization of SWCNTs

2.1 Detergents as Surfactants

Initially, detergents like pyrene salts (Nakashima et al., 2002) and tensides, such as sodium dodecyl sulfate (SDS), sodium dodecyl benzene sulfonate (SDBS), or sodium cholate (SC) were used to generate stable nanotube suspensions (Ausman et al., 2000; Chen et al., 2001; O'Connell et al., 2002; Islam et al., 2003). The molecular structure of some of these is shown in Figure 1.

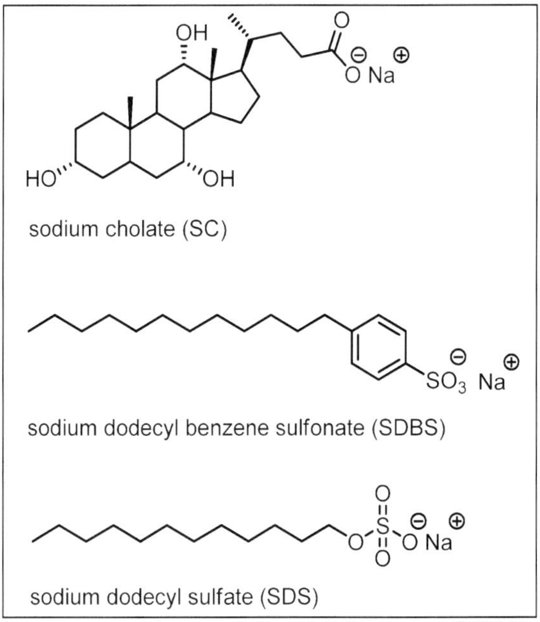

Fig. 1. Structure of most common detergents for solubilization of SWCNTs.

A drawback of these small molecule surfactants is that a large excess of detergent molecules is needed to keep SWCNTs in solution (compare Figure 2), most probably because of the fast off-rate for the complexes with the nanotubes. Removal of the excess surfactant leads to rebundling of the nanotubes. The free surfactants affect many protocols, though, complicating practical applications and the characterization of the nanotube suspensions.

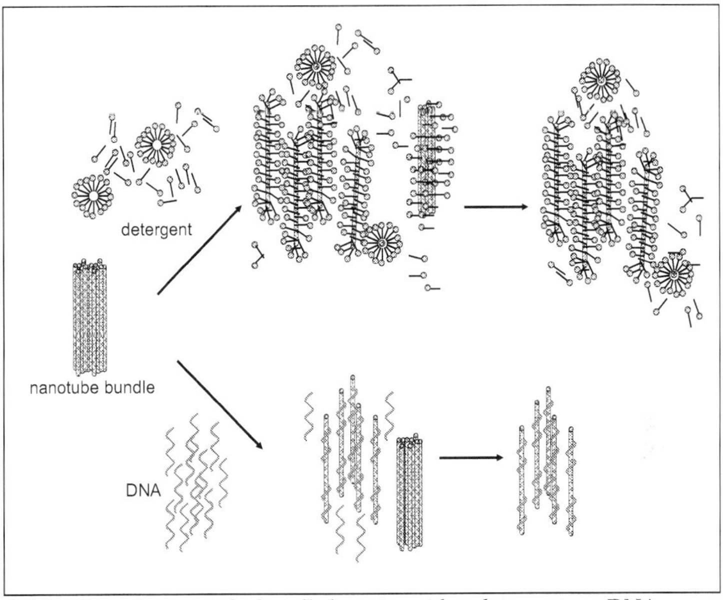

Fig. 2. Solubilization of a nanotube bundle by using either detergents or DNA.

2.2 Oligonucleotides as Solubilizing Ligands

Overcoming the problem of fast off-rates necessitate the use of a solubilizing ligand that engages in cooperative binding interactions with SWCNTs. Macromolecules may engage in multivalent interactions, resulting in large binding energies for their complexes. These macromolecules may be either synthetic polymers or biomacromolecules. Nanotube suspensions have been obtained with macromolecules like gum arabic (Bandyopadhyaya et al., 2002), amylomaize (Star et al., 2002; Kim et al., 2003), cyclodextrines (Chen et al., 2001; Dodziuk et al., 2003), peptides (Dieckmann et al., 2003; Wang et al., 2003), polymers (O'Connell et al., 2001), or fluorene-based polymers (Chen et al., 2007 a; Nish et al., 2007) as "surfactants". Among the biomacromolecules used to solubilize carbon nanotubes, DNA sticks out as the one producing particularly high yields of solubilized tubes (Figure 2). So, even though its natural role as carrier of genetic information does not suggest that DNA may act as a surfactant, synthetic DNA strands have proven particularly effective in solubilizing single-walled carbon nanotubes and aiding their purification.

Oligodeoxynucleotides - short strands of DNA - may also be attached covalently to SWCNTs (Baker et al., 2002; Dwyer et al., 2002; Hazani et al., 2003; Hazani et al., 2004), but it is the non-covalent complexes (Zheng et al., 2003 a; Nakashima et al., 2003) with nanotubes that we and others find particularly fascinating. They have many favorable properties while retaining the uninterrupted structure of the tubes. For example, very high thermal stability was observed for aqueous suspensions of SWCNTs produced with DNA strands, with half life times of flocculation in the range of many hours at 90 °C (Vogel et al., 2007 a; Vogel et al., 2007 b). Further, suspensions can be obtained with DNA that remain stable for at least one year upon storage at room temperature. Simulations have suggested that close molecular contacts between DNA and SWCNTs mainly involve the aromatic nucleobases of the oligonucleotides and the surface of the nanotubes (Gowtham et al., 2008). For long single-stranded DNA a helical wrapping around the nanotubes is predicted (Enyashin et al., 2007). This leaves the negatively charged backbone of the oligonucleotides exposed to the solvent (water). The repulsion between the negative charges of the phosphodiesters then efficiently prevents re-bundling. In contrast to small molecule surfactants as solubilizers, excess DNA can be removed from the suspensions of SWCNTs, e.g. by filtration (Kam et al., 2005; Vogel et al., 2007 a) , without flocculation of the tubes. Due to their shape, nanotubes cannot pass through the channels inside membrane of the filter, whereas the more globular, unbound DNA molecules are enriched in the filtrate of the centrifugation. After filtration, DNA-SWCNT complexes retained on the filter can be resuspended.

Synthetic DNA has the advantage of being available in monodisperse form with any given sequence of the four canonical nucleobases (A/C/G/T) at moderate cost. So, once DNA was established as a good solubilizing agent for SWCNTs, several groups set out to find the optimal DNA sequence and strand length to solubilize a maximum amount of nanotubes. The first group to systematically vary sequence and length of the oligonucleotide solubilizer was that of Tassi, Walls and colleagues at DuPont (Zheng et al., 2003 a; Zheng et al., 2003 b). Their work suggested that sequences of poly d(T) and alternating sequences of deoxyguanosine and thymidine of a length of 20 to 90 nucleotides ($d(GT)_{10-45}$) were best suited for generating concentrated suspensions suitable for purification. Another group questioned these conclusions (Gigliotti et al., 2006). They solubilized nanotubes with single-stranded DNA of 100 nucleotides average length and found that shorter stands, consisting of approx. 50 nucleotides, and made up of mixed sequences, yielded lower amounts of solubilized SWCNTs. It was then reported that SWCNTs can be solubilized by very long single-stranded DNA (> 1000 nucleotides) produced by amplification (Zhao et al., 2006) and asymmetric polymerase chain reaction, PCR (Liang et al., 2007). Others had found independently that double-stranded DNA can be used to obtain very concentrated suspensions (Barisci et al., 2004), but it is not clear whether denaturation of the duplexes occurred during the solubilization protocol, making single-stranded DNA the relevant species in solution. Studies with single nucleotides, like 2'-deoxyadenosine-5'-diphosphate and 2'-deoxyguanosine-5'-monophosphate, as surfactants also produced fairly concentrated nanotube suspension (Ikeda et al., 2006), but it was not studied in detail what the kinetic stability of these suspensions is. As expected, mononucleotides with the smaller and less lipophilic pyrimidine nucleobases uracil and cytosine proved to be poorer surfactants for solubilization of SWCNTs than their purine counterparts.

The solubilization of SWCNTs with modified, thiol-terminated oligonucleotides was reported in 2007 (Han et al., 2007). The group described an assembly of gold nanoparticles and the DNA-SWCNT complexes, in which the sulfur of the former thiol groups of the oligonucleotides binds to the nanoparticle surface. Atomic force microscopy (AFM) showed nanoparticles in the immediate vicinity of nanotubes. This is an early example of hybrid complexes in which the (modified) DNA acts as the "molecular glue", holding together complexes with SWCNTs.

Because of the diverse set of results obtained by different researchers, our own group performed a detailed study on the length and sequence dependence of the solubilizing power of synthetic DNA used as surfactants for HiPco SWCNTs (Vogel et al., 2007a; Vogel et al., 2007b). Oligonucleotides with sequences of $d(AC)_n$ and $d(GT)_n$, with n = 2, 3, 5, 10, 20, and 40, as well as mixtures of equal length strands of both sequences were employed. A maximum amount of HiPco tubes was solubilized with a mixture of the two hexamers $d(AC)_3$ and $d(GT)_3$. Surprisingly, longer oligonucleotides yielded less concentrated suspensions. Loh and coworkers also reported that longer oligonucleotides interact less well with nanotubes and show looser wrapping around the tubes, as confirmed by AFM, Raman and photoluminescence measurements (Yang et al., 2009). Work by Kim and colleagues demonstrated enrichment of (6,5)-nanotubes when using genomic, salmon DNA (Kim et al., 2008), which may or may not be the result of sequence-specific interactions, as genomic DNA contains a range of different sequence motifs. As detailed below, our own work on DNA-SWCNT complexes now employs sequences with a high propensity to fold, leading to stable intramolecular structures (Müller et al., 2009).

Studies of the kind mentioned above benefit from the fact that the nanotube content of suspensions can be determined directly from UV-Vis/NIR absorption spectra, using inexpensive instrumentation. The absorption spectra also provide a good first impression of the composition of the nanotube mixture. Figure 3 shows such spectra of samples prepared in our lab from HiPco SWCNTs (black line) or CoMoCAT tubes (red line). The advantage of CoMoCAT tubes as starting material is its richness in certain chiralities, particularly (6,5)-tubes that are formed selectively during the production process (Tan & Resasco, 2005). A characteristic absorption peak of CoMoCAT nanotubes appears at a wavelength of 989 nm (see Figure 3) and its intensity may be taken as a measure of solubilized nanotubes. Since bundles often produce a broad background signal, the absorption value should be corrected for the background, measured e.g. at 1090 nm. More detailed information on the composition of nanotube samples may be obtained from photoluminescence (PL) maps (O'Connell et al., 2002).

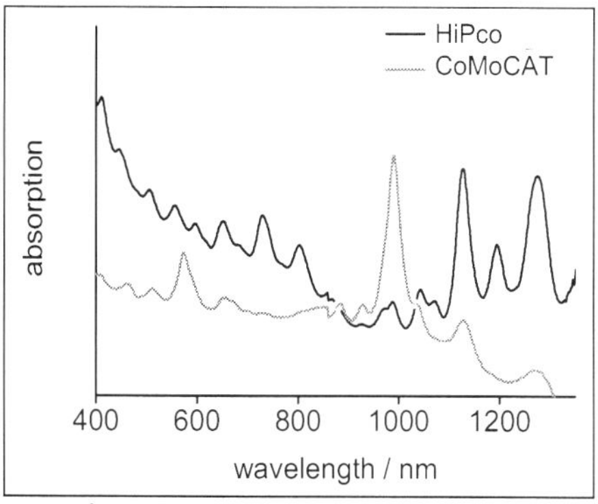

Fig. 3. Absorption spectra of HiPco and CoMoCAT SWCNTs solubilized with DNA as surfactant. Both prominent peaks in the CoMoCAT spectrum can be assigned to the E_{11} (989 nm) and E_{22} (575 nm) transitions of (6,5)-nanotubes, respectively. The HiPco sample (black line) contains many different chiralities, and the spectrum is a superposition of absorption spectra of different (n,m)-nanotubes.

3. Purification and Separation of SWCNTs

3.1 General Methods

While solubilizing nanotubes in aqueous media may be a mature technology, separating the mixtures of different tube diameters, chiralities and lengths into samples of uniform structures certainly is not. For more demanding applications, e.g. in nanoelectronics, such a separation is highly desirable, if not necessary. There are some techniques producing nanotubes within a small diameter range that, together with improved separation techniques, may yield spectrally uniform SWCNTs of a high purity. Enrichment of either semiconducting or metallic SWCNTs by chemical conversion of nanotubes has also been reported. Other approaches, relying on non-covalent complexes, allow for enrichment of nanotube species without covalent alteration of their structure. Dielectrophoresis (Krupke et al., 2003; Lee et al., 2005) and density gradient ultracentrifugation (Arnold et al., 2005; Arnold et al., 2006) have been used in several studies to enrich metallic nanotubes or nanotubes with a specific diameter/chirality. Separation by density gradient ultracentrifugation has led to semiconducting tubes enriched in (6,5)- and (7,5)-tubes (Arnold et al., 2006; Crochet et al., 2007). A different approach is based on extraction after polymer wrapping SWCNTs (Shigeta et al., 2006). Finally, fluorene-based polymers have a high preference to extract individual species of SWCNTs (Chen et al., 2007 a; Nish et al., 2007). Selectivity for specific (n,m)-species apparently relies on the stability of specific SWCNT-polymer complexes in solution. "Chirality" is used loosely, though, as nanotube suspensions are usually optically inactive, containing equal amounts of left- and right-handed tubes. But, using diporphyrins (chiral "nano-tweezers"), Osuka and coworkers enriched nanotubes of a defined handedness to obtain optically active SWCNT samples (Peng et al., 2007).

3.2 DNA-assisted Separation and Purification

Interestingly, DNA has been reported not only to produce stable nanotube suspensions, but to also to facilitate nanotube separation (Zheng et al., 2003 a; Strano et al., 2004). It was reported that complexes of certain DNA sequences and SWCNTs, when submitted to ion-exchange chromatography, separate into individual fractions enriched in semiconducting or metallic tubes (Zheng et al., 2003 a; Zheng et al., 2003 b). In addition, size-exclusion chromatography of DNA-solubilized nanotubes has led to samples of nanotubes with a narrow length distribution (Huang et al., 2005; Bauer et al., 2007; Bauer et al., 2008). The researchers described a separation of DNA-coated SWCNTs by size, using flow-field fractionation (Chun et al., 2008). Dai and coworkers reported separation of HiPco-grown nanotubes solubilized with DNA by length, diameter, and chirality with different chromatography methods (Zhang et al., 2008). They combined size-exclusion chromatography and ion exchange chromatography and identified the chiral vector of the nanotubes by photoluminescence excitation spectroscopy, absorption spectroscopy, and electrical transport measurements. One of the most recent publications described certain DNA sequences, identified from a DNA "library", that wrap preferentially around semiconducting SWCNTs of one specific chirality (Tu et al., 2009). The authors write that they have separated more than 10 types of nanotubes of different chiralities from HiPco material by ion-exchange chromatography. The fractions were characterized mainly by optical absorption spectroscopy. The authors also presented structures of hydrogen bonded antiparallel non self-complementary dodecamer strands on a nanotube that they obtained by modeling. They hypothesize that each of these DNA barrel structures is formed on a particular (n,m)-nanotube. It will be interesting to see whether direct experimental evidence for the base pairs proposed can be obtained.

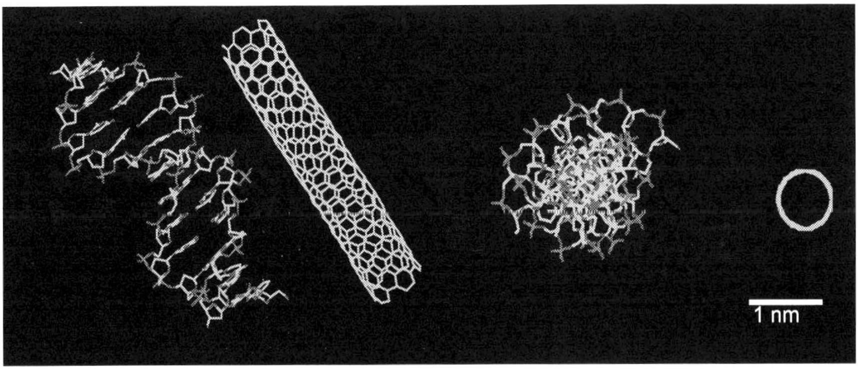

Fig. 4. The shape of double stranded DNA and (6,5)-nanotubes, as seen perpendicular to the helix axes (left-hand side) or along the helix axes ("top view", right-hand side). Structures are drawn to scale to visualize shape complementarities (or lack thereof).

Coordinates of the nanotube were generated with TubeGen 3.3 (http://turin.nss.udel.edu/research/tubegenonline.html by Frey and Doren, University of Delaware, Newark DE, 2005) and visualized using VMD (Humphrey et al., 1996).

Fundamentally, there is no simple shape complementarity between base-paired, canonical duplex DNA and carbon nanotubes. This can be seen from an inspection of Figure 4. The DNA double helix does not have a central hole that could accommodate even a small (6,5)-nanotube, and neither of the two grooves is sufficiently large or hydrophobic to bind to tubes.

4. Assemblies

4.1 Overview

Oligonucleotides are attractive nanoscale building blocks because they are addressable sequence-specifically. It is therefore an interesting question to what extent complexation of oligonucleotides with SWCNTs blocks the ability of the DNA to form duplexes with complementary strands. If the adsorption on the tube surface forces the DNA to adopt a conformation that is incompatible with Watson-Crick duplexes, the programmable recognition capability of the DNA will be lost. Further, if multivalent interactions or unspecific interactions dominate, the strength of the interactions will be difficult to program into the DNA sequences and difficult to control on the experimental level. Currently there is no general answer to these questions, but some encouraging results have been obtained. For example, hybridization between strands of partly DNA-wrapped SWCNTs has been reported to induce the formation of aggregates (Figure 5a). Furthermore, hybrid structures of SWCNTs and gold nanoparticles (AuNP) have been generated, based on hybridization (see Figure 5b) (Chen et al., 2007 b). The new materials were analyzed by gel electrophoresis, dynamic light scattering (DLS), optical absorption spectroscopy, and atomic force microscopy (AFM). In the case of nanotube assemblies (Figure 5a), the authors measured a decrease in absorbance. For the hybrid materials (Figure 5b), AFM images showed evidence for SWCNT-AuNP complexes. In either case, the DNA-SWCNT complexes were generated by sonicating the tubes with SDS as surfactant and a probe sonicator, followed by ultracentrifugation. Afterwards single-stranded DNA was added at room temperature without any further sonication. Under these conditions, the DNA strands may not displace more than a small fraction of the SDS on the tubes. Being only loosely bound to the surfactant-wrapped tubes, much of the base pairing capability of the DNA may be retained.

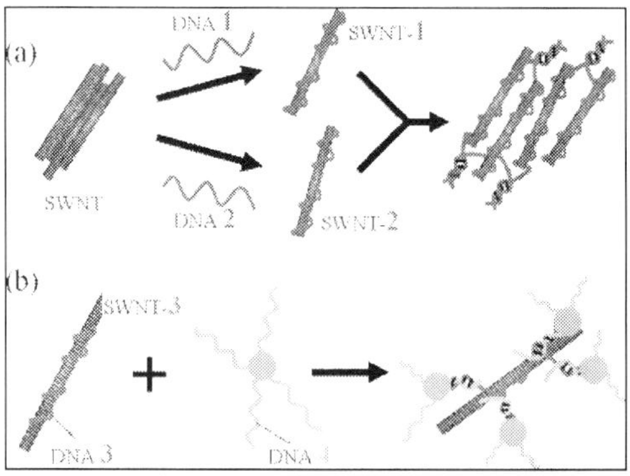

Fig. 5. DNA hybridization controls aggregation of SWNTs (a) and formation of SWNT-gold nanoparticle hybrid structures (b). Reprinted with permission from Chen et al., 2007 b. Copyright 2007 American Chemical Society.

In a related study, controlled aggregation of nanotubes mediated by hybridizing DNA strands (Li et al., 2007) was induced by two different strategies (Figure 6). The first, dubbed "tail strategy" involves a hybridizable tail of the nanotube-wrapping oligonucleotide that forms a duplex with a so-called splint strand. The second strategy, referred to as "tail-free strategy" involves direct hybridization of two tube-wrapping single strands. The former strategy gave much faster hybridization kinetics than the latter strategy. The authors state that in the latter case the DNA sequences have a strong tendency to wrap around the SWCNTs, leading to a dramatically decrease in base pairing ability. For this study, DNA-SWCNT complexes were obtained by sonication for 30 min with a probe sonicator under ice cooling, followed by ultracentrifugation for 1 h. In the "tail strategy" the mediating splint strands were added after the solubilization step to pre-formed DNA-SWCNT suspensions. Again characterization relied on gel electrophoreses and AFM.

So, there are ways to make tube-bound DNA strands addressable to complementary sequences. In our experience, it is difficult to prevent overhang DNA from adsorbing to the remaining free tube surfaces, though, making designs that require unbound portions of partially bound strands dependant on exquisite control over kinetic parameters. Further, once DNA-induced complex formation has occurred, the size of the aggregates can make characterization difficult.

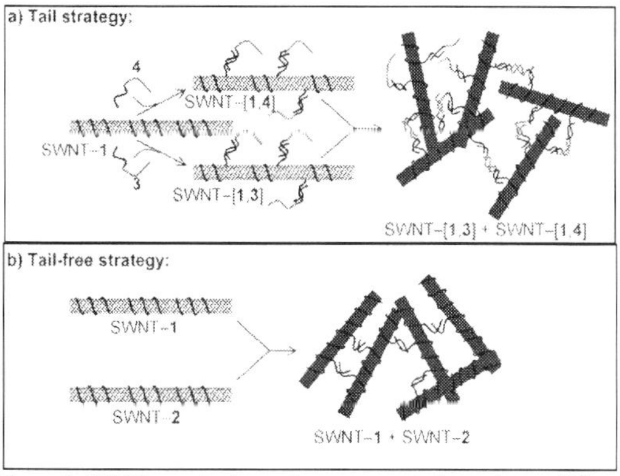

Fig. 6. Self-assembly of DNA-coated carbon nanotubes into 3D aggregates through hybridization: Tail strategy (a) and tail-free strategy (b). Reproduced with permission from Li et al. 2007, Copyright Wiley-VCH Publishers.

In yet another study by Tan and coworkers, DNA adsorbed on nanotubes was shown to engage in base pairing with other DNA strands. In one of their publications, this group chose a molecular beacon (MB) DNA motif (Yang et al., 2008 a). An MB is a single-stranded oligonucleotide with self-complementary ends linked by a sequence of non-complementary nucleotides in the middle. If the ends form an intramolecular duplex, the resulting structure has the shape of a hairpin. If both ends of the hairpin are labeled, one with a fluorophore and the other with a quencher, breaking up the hairpin separates fluorophore and quencher, with concomitant increase in fluorescence intensity. The term "molecular beacon" is used in analogy to the beacons that lead mariners approaching the land. Molecular beacons are interesting tools for studying the often enigmatic interactions between DNA and SWCNTs. In the study mentioned above, complexes of SWCNTs and MBs were treated with an oligonucleotide complementary to the MB, leading to a structural change from hairpin to an extended, intermolecular duplex (Figure 7). Hybridization led to the expected change in fluorescence intensity. Apparently, hybridized beacons are unable to hold on to nanotubes.

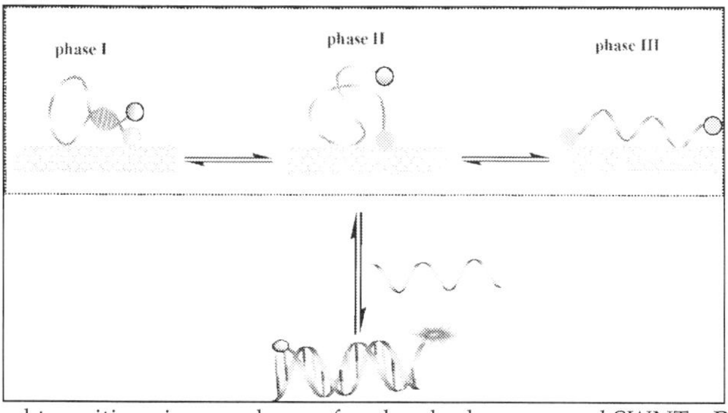

Fig. 7. Structural transitions in complexes of molecular beacons and SWNTs. Reprinted with permission from Yang et al., 2008 b. Copyright 2008 American Chemical Society.

In another study, the same authors described a new class of fluorescing sensors for biomolecular interactions, based on assemblies of SWCNTs and single-stranded DNA (Yang et al., 2008 b). Again, changes in fluorescence intensity were measured. When the fluorophore is in the vicinity to the nanotube, its fluorescence is quenched by the tube. Upon hybridization with a complementary strand, the fluorescence is recovered because the fluorophore-bearing duplex is released from the nanotube surface. The SWCNT-DNA complexes were prepared as follows. First, nanotubes were solubilized in dimethylformamide (DMF) by sonication with a probe sonicator for 5 h, followed by the addition of oligonucleotides and incubation for 5-15 min. Finally the hybridization partner was added, and after a few hours, the mixture was submitted to ultracentrifugation.

So, all hybridization experiments with non-covalent DNA-nanotubes complexes have one important detail in common: Successful hybridization is favored by a protocol in which the DNA is not present during sonication with the probe sonicator. Instead, adsorption of the DNA to be addressed on the SWCNTs is usually induced *after* nanotube suspensions have been generated. This may explain why we found DNA adsorbed during the sonication step to be unaddressable (Vogel et al., 2007 a), whereas groups did not.

4.2 Outlook: Towards DNA-Mediated Nanostructuring

How may the programmable molecular recognition properties of DNA be harnessed to steer DNA-coated SWCNTs into designed three-dimensional architectures, based on predictable base pairing interactions? While results from recent studies are promising, significant challenges remain. Many of the complexes used are metastable. Keeping the SWCNT-DNA complexes within their local minimum in the energy landscape is difficult. Small changes in the chemical environment can trigger decomposition of these complexes. If sufficient kinetic energy is applied, as during sonication with a probe sonicator, DNA will fully denature and then adsorb in a conformation that makes it unavailable for base pairing. In these unproductive complexes, oligonucleotides are probably fully extended and fully wrapped around the nanotubes, leaving no unbound overhang that can be addressed with a complementary sequence. Transmission electron microscopy images seem to confirm this view (Malik et al., 2007). Even if a metastable state is established, with sufficiently high activation barriers preventing re-bundling of the tubes or conformational changes in the DNA, binding to specific positions in very long nanotubes is not yet feasible. Even if it was, the sheer strength of the multivalent interactions between nanotubes and surfaces or other nanoscale objects will continue to make it difficult to place an individual SWCNT at a precise location in a designed, DNA-driven structure. Single-walled carbon nanotubes are not only molecules with unusual electronic properties; they are also very large molecules with a massive intrinsic binding energy!

Ideally, one would want to develop a receptor-like DNA folding motif that binds to a specific nanotube chirality and retains its ability to engage in base pairing. Extensive searches in DNA sequence space have led to some oligonucleotides with surprising selectivity (Tu et al., 2009). Still, these oligonucleotides have not been shown to be available for conventional intermolecular base pairing. We believe that a combination of DNA strands and suitable surfactants may lead to complexes that do both bind and remain addressable. If free nanotube surface areas between DNA binding sites are covered with surfactant, adsorption of addressable overhangs of DNA strands may be kept from adsorbing. Further, assembling tight and specific coatings from a suitable combination of small molecules and

oligonucleotides should be simpler and less expensive than wrapping long nanotubes in a single polynucleotide of a length inaccessible by conventional DNA synthesis. Addressability of the DNA may then be achieved by employing higher order structures, such as triplexes, quadruplexes, or I-motifs (Nonin et al., 1997; Zhao et al., 2009).
So, to get addressable nanotubes, we assume that it is useful to identify a combination of surfactants and structured DNA. We are currently engaging in a project with this goal, using a two-step protocol. First, the SWCNTs are dispersed with surfactants only, without DNA. Then, folded (pre-structured) DNA is added under conditions under which no more than partial wrapping of the tubes is expected. To obtain nanotubes suitable for generating nanostructured electronic devices, we employ tubes of the greatest sample homogeneity possible. Currently, we apply a literature-known density gradient ultracentrifugation protocol to further enrich CoMoCAT SWCNTs in (6,5)-tubes. This protocol employs a gradient of iodixanol and a buffer containing two weight percent of a mixture of sodium cholate and sodium dodecyl sulfate (Arnold et al., 2006; Crochet et al., 2007). Since nanotubes of different diameters have different densities, ultracentrifugation yields bands containing nanotubes of the same diameter. We then expose the pre-purified and cholate-wrapped tubes to oligonucleotides folded to a loop designed to bind the nanotubes.
To demonstrate that the DNA bound to the SWCNTs is indeed an addressable binding motif, we use hybridization assays with fluorophore-labeled complementary strands. Control experiments with labeled strands of a different sequence are critically important to demonstrate that binding is not the result of unspecific interactions that do not rely on base pairing. Binding equilibria can be studied by using filters that retain the nanotubes, but allow unbound DNA to pass. In either sample (with complementary and non-complementary DNA) the tubes retained on the filter are resuspended and the resulting solutions studied by fluorescence spectroscopy. Similar experiments can be performed with quantum dots (QDs) as nanoscale "cargo", bound to DNA-coated SWCNTs (Müller et al., 2009). Centrifugation separates bound from unbound species, allowing one to determine binding constants and to establish sequence-specific base pairing.

5. Conclusion

In this chapter, oligonucleotide-SWCNTs complexes are discussed, emphasizing that DNA has surfactant properties that make it useful for solubilizing carbon nanotubes. Stable, largely monodisperse suspensions can be obtained after sonication and separation processes, such as density gradient ultracentrifugation. In ideal cases, complexes are obtained that are predominantly of a single nanotube diameter and chiral index and that engage in sequence-specific base pairing interactions. Based on models of unperturbed duplex DNA, it appears as if there was no shape complementarity between DNA duplexes and single-walled carbon nanotubes. So, when the DNA adsorbs on the nanotube surface, it can be expected to be distorted so much that it loses its ability to form Watson-Crick duplexes where T pairs with A and G pairs with C. DNA-mediated nanostructuring involving SWCNT-DNA complexes may be achieved by generating complexes where part of the DNA strands remains unbound. Usually, such complexes are metastable structures, since full adsorption of the DNA would lead to a thermodynamically more stable structure. Unpublished results from our laboratories shows that such kinetically stable, addressable DNA can be generated from stably folded DNA motifs. Engaging them in sequence-specific base pairing interactions constitutes an important step towards nanoconstruction with carbon nanotubes. Since DNA is one of the most powerful materials for soft matter nanostructuring (Park et al., 2008; Nykypanchuk et al., 2008; Meng et al., 2009), and consequently, DNA nanostructing is one of the fastest growing fields of present-day science, (Niemeyer, 1999; Seeman, 2003; Gothelf and LaBean, 2005; Lu et al., 2008; Shin et al., 2009), we believe that harnessing the complexes the DNA and SWCNTs will lead to the discovery of fascinating new functional nanomaterials.

Acknowledgment

The authors are grateful to J. Crochet and T. Hertel for an introduction to SWCNT purification by density gradient ultracentrifugation, and to M. M. Kappes, S. Malik and S. Vogel for discussions. K. Müller was a recipient of a graduate fellowship of the state of Baden-Württemberg. This work was supported by the DFG-funded Center for Functional Nanostructures (CFN, subproject C 4.3) at the University of Karlsruhe.

6. References

Arnold, M. S.; Stupp, S. I. & Hersam, M. C. (2005). Enrichment of Single-Walled Carbon Nanotubes by Diameter in Density Gradients. *Nano Letters,* Vol. 5, No. 4, 713-718, ISSN 1530-6984

Arnold, M. S.; Green, A. A.; Hulvat, J. F.; Stupp, S. I. & Hersam, M. C. (2006). Sorting carbon nanotubes by electronic structure using density differentiation. *Nature Nanotechnology,* Vol. 1, No. 1, 60-65, ISSN 1748-3387

Ausman, K. D.; Piner, R.; Lourie, O. & Ruoff, R. S. (2000). Organic Solvent Dispersions of Single-Walled Carbon Nanotubes: Toward Solutions of Pristine Nanotubes. *Journal of Physical Chemistry B,* Vol. 104, No. 38, 8911-8915, ISSN 1089-5647

Bahr, J. L.; Mickelson, E. T.; Bronikowski, M. J.; Smalley, R. E. & Tour, J. M. (2001). Dissolution of small diameter single-wall carbon nanotubes in organic solvents? *Chemical Communications,* No. 2, 193-194, ISSN 1359-7345

Baker, S. E.; Cai, W.; Lasseter, T. L.; Weidkamp, K. P. & Hamers, R. J. (2002). Covalently Bonded Adducts of Deoxyribonucleic Acid (DNA) Oligonucleotides with Single-Wall Carbon Nanotubes: Synthesis and Hybridization. *Nano Letters,* Vol. 2, No. 12, 1413-1417, ISSN 1530-6984

Bandyopadhyaya, R.; Nativ-Roth, E.; Regev, O. & Yerushalmi-Rozen, R. (2002). Stabilization of Individual Carbon Nanotubes in Aqueous Solutions. *Nano Letters,* Vol. 2, No. 1, 25-28, ISSN 1530-6984

Barisci, J. N.; Tahhan, M.; Wallace, G. G.; Badaier, S.; Vaugien, T.; Maugey, M. & Poulin, P. (2004). Properties of Carbon Nanotube Fibers Spun from DNA-Stabilized Dispersions. *Advanced Functional Materials,* Vol. 14, No. 2, 133-138, ISSN 1616-301X

Bauer, B. J.; Becker, M. L.; Bajpai, V.; Fagan, J. A.; Hobbie, E. K.; Migler, K.; Guttman, C. M. & Blair, W. R. (2007). Measurements of Single-Wall Nanotube Dispersion by Size Exclusion Chromatography. *Journal of Physical Chemistry C,* Vol. 111, No. 48, 17914-17918, ISSN 1932-7447

Bauer, B. J.; Fagan, J. A.; Hobbie, E. K.; Chun, J. & Bajpai, V. (2008). Chromatographic Fractionation of SWNT/DNA Dispersions with On-Line Multi-Angle Light Scattering. *Journal of Physical Chemistry C,* Vol. 112, No. 6, 1842-1850, ISSN 1932-7447

Chen, F.; Wang, B.; Chen, Y. & Li, L.-J. (2007 a). Toward the Extraction of Single Species of Single-Walled Carbon Nanotubes Using Fluorene-Based Polymers. *Nano Letters,* Vol. 7, No. 10, 3013-3017, ISSN 1530-6984

Chen, J.; Hamon, M. A.; Hu, H.; Chen, Y.; Rao, A. M.; Eklund, P. C. & Haddon, R. C. (1998). Solution Properties of Single-Walled Carbon Nanotubes. *Science,* Vol. 282, No. 5386, 95-98, ISSN 0036-8075

Chen, J.; Dyer, M. J. & Yu, M.-F. (2001). Cyclodextrin-mediated soft cutting of single-walled carbon nanotubes. *Journal of the American Chemical Society,* Vol. 123, No. 25, 6201-6202, ISSN 0002-7863

Chen, Y., Liu, H., Ye, T., Kim, J. & Mao, C. (2007 b). DNA-Directed Assembly of Single-Wall Carbon Nanotubes. *Journal of the American Chemical Society,* Vol. 129, No. 28, 8696-8697, ISSN 0002-7863

Chun, J.; Fagan, J. A.; Hobbie, E. K. & Bauer, B. J. (2008). Size Separation of Single-Wall Carbon Nanotubes by Flow-Field Flow Fractionation. *Analytical Chemistry,* Vol. 80, No. 7, 2514-2523, ISSN 0003-2700

Cosnier, S. & Holzinger, M. (2008). Design of carbon nanotube-polymer frameworks by electropolymerization of SWCNT-pyrrole derivatives. *Electrochimica Acta,* Vol. 53, No. 11, 3948-3954, ISSN 0013-4686

Crochet, J.; Clemens, M. & Hertel, T. (2007). Quantum Yield Heterogeneities of Aqueous Single-Wall Carbon Nanotube Suspensions. *Journal of the American Chemical Society,* Vol. 129, No. 26, 8058-8059, ISSN 0002-7863

Dieckmann, G. R.; Dalton, A. B.; Johnson, P. A.; Razal, J.; Chen, J.; Giordano, G. M.; Muñoz, E.; Musselman, I. H.; Baughman, R. H. & Draper, R. K. (2003). Controlled Assembly of Carbon Nanotubes by Designed Amphiphilic Peptide Helices. *Journal of the American Chemical Society,* Vol. 125, No. 7, 1770-1777, ISSN 0002-7863

Dodziuk, H.; Ejchart, A.; Anczewski, W.; Ueda, H.; Krinichnaya, E.; Dolgonos, G. & Kutner, W. (2003). Water solubilization, determination of the number of different types of single-wall carbon nanotubes and their partial separation with respect to diameters by complexation with η-cyclodextrin. *Chemical Communications,* No. 8, 986-987, ISSN 1359-7345

Dwyer, C.; Guthold, M.; Falvo, M.; Washburn, S.; Superfine, R. & Erie, D. (2002). DNA-functionalized single-walled carbon nanotubes. *Nanotechnology,* Vol. 13, No. 5, 601-604, ISSN 0957-4484

Enyashin, A. N.; Gemming, S. & Seifert, G. (2007). DNA-wrapped carbon nanotubes. *Nanotechnology,* Vol. 18, No. 24, 1-10, ISSN 0957-4484

Furtado, C. A.; Kim, U. J.; Gutierrez, H. R.; Pan, L.; Dickey, E. C. & Eklund, P. C. (2004). Debundling and Dissolution of Single-Walled Carbon Nanotubes in Amide Solvents. *Journal of the American Chemical Society,* Vol. 126, No. 19, 6095-6105, ISSN 0002-7863

Gigliotti, B.; Sakizzie, B.; Bethune, D. S.; Shelby, R. M. & Cha, J. N. (2006). Sequence-Independend Helical Wrapping of Single-Walled Carbon Nanotubes by Long Genomic DNA. *Nano Letters,* Vol. 6, No. 2, 159-164, ISSN 1530-6984

Gothelf, K. V. & LaBean, T. H. (2005). DNA-programmed assembly of nanostructures. *Organic & Biomolecular Chemistry,* Vol. 3, No. 22, 4023-4037, ISSN 1477-0520

Gowtham, S.; Scheicher, R. H.; Pandey, R.; Karna, S. P. & Ahuja, R. (2008). First-principles study of physisorption of nucleic acid bases on small-diameter carbon nanotubes. *Nanotechnology,* Vol. 19, No. 12, 1-6, ISSN 0957-4484

Han, X.; Li, Y. & Deng, Z. (2007). DNA-Wrapped Single-Walled Carbon Nanotubes as Rigid Templates for Assembling Linear Gold Nanoparticle Arrays. *Advanced Materials,* Vol. 19, No. 11, 1518-1522, ISSN 0935-9648

Hazani, M.; Naaman, R.; Hennrich, F. & Kappes, M. M. (2003). Confocal Fluorescence Imaging of DNA-Functionalized Carbon Nanotubes. *Nano Letters,* Vol. 3, No. 2, 153-155, ISSN 1530-6984

Hazani, M.; Hennrich, F.; Kappes, M. M.; Naaman, R.; Peled, D.; Sidorov, V. & Shvarts, D. (2004). DNA-mediated self-assembly of carbon nanotube-based electronic devices. *Chemical Physics Letters,* Vol. 391, No. 4-6, 389-392, ISSN 0009-2614

Huang, X.; McLean, R. S. & Zheng, M. (2005). High-Resolution Length Sorting and Purification of DNA-Wrapped Carbon Nanotubes by Size-Exclusion Chromatography. *Analytical Chemistry,* Vol. 77, No. 19, 6225-6228, ISSN 0003-2700

Humphrey, W.; Dalke, A. & Schulten, K. (1996). VMD - Visual Molecular Dynamics. *Journal of Molecular Graphics,* Vol. 14, No. 1, 33-38, ISSN 0263-7855

Iijima, S. (1991). Helical microtubules of graphitic carbon. *Nature,* Vol. 354, No. 6348, 56-58, ISSN 0028-0836

Ikeda, A.; Hamano, T.; Hayashi, K. & Kikuchi, J. (2006). Water-Solubilization of Nucleotides-Coated Single-Walled Carbon Nanotubes Using High-Speed Vibration Milling Technique. *Organic Letters,* Vol. 8, No. 6, 1153-1156, ISSN 1523-7060

Islam, M. F.; Rojas, E.; Bergey, D. M.; Johnson, A. T. & Yodh, A. G. (2003). High Weight Fraction Surfactant Solubilization of Single-Wall Carbon Nanotubes in Water. *Nano Letters,* Vol. 3, No. 2, 269-273, ISSN 1530-6984

Jung, D.-H.; Ko, Y. K.; & Jung, H.-T. (2004). Aggregation behavior of chemically attached poly(ethylene glycol) to single-walled carbon nanotubes (SWNTs) ropes. *Materials Science & Engineering,* Vol. 24, No. 1-2, 117-121, ISSN 0928-4931

Kam, N. W. S.; Liu, Z. & Dai, H. (2005). Functionalization of carbon nanotubes via cleavable disulfide bonds for efficient intracellular delivery of siRNA and potent gene silencing. *Journal of the American Chemical Society,* Vol. 127, No. 36, 12492-12493, ISSN 0002-7863

Kim, O.-K.; Je, J.; Baldwin, J. W.; Kooi, S.; Pehrsson, P. E. & Buckley, L. J. (2003). Solubilization of Single-Wall Carbon Nanotubes by Supramolecular Encapsulation of Helical Amylose. *Journal of the American Chemical Society,* Vol. 125, No. 15, 4426-4427, ISSN 0002-7863

Kim, S. N.; Kuang, Z.; Grote, J. G.; Farmer, B. L. & Naik, R. R. (2008). Enrichment of (6,5) Single Wall Carbon Nanotubes Using Genomic DNA. *Nano Letters,* Vol. 8, No. 12, 4415-4420, ISSN 1530-6984

Krupke, R.; Hennrich, F.; v. Lohneysen, H. & Kappes, M. M. (2003). Separation of Metallic from Semiconducting Single-Walled Carbon Nanotubes. *Science,* Vol. 301, No. 5631, 344-347, ISSN 0036-8075

Lee, D. S.; Kim, D. W.; Kim, H. S.; Lee, S. W.; Jhang, S. H.; Park, Y. W. & Campbell, E. E. B. (2005). Extraction of semiconducting CNTs by repeated dielectrophoretic filtering. *Applied Physics A: Materials Science & Processing,* Vol. 80, No. 1, 5-8, ISSN 0947-8396

Li, Y.; Han, X. & Deng, Z. (2007). Grafting Single-Walled Carbon Nanotubes with Highly Hybridizable DNA Sequences: Potential Building Blocks for DNA-Programmed Material assembly. *Angewandte Chemie International Edition,* Vol. 46, 7481-7484, ISSN 1433-7851

Liang, Z.; Lao, R.; Wang, J.; Liu, Y.; Wang, L.; Huang, Q.; Song, S.; Li, G. & Fan, C. (2007). Solubilization of Single-Walled Carbon Nanotubes with Single-Stranded DNA Generated from Asymmetric PCR. *International Journal of Molecular Science,* Vol. 8, No. 7, 705-713, ISSN 1422-0067

Liu, J.; Rinzler, A. G.; Dai, H.; Hafner, J. H.; Bradley, R. K.; Boul, P. J.; Lu, A.; Iverson, T.; Shelimov, K.; Huffman, C. B.; Rodriguez-Macias, F.; Shon, Y.-S.; Lee, T. R.; Colbert, D. T. & Smalley, R. E. (1998). Fullerene Pipes. *Science,* Vol. 280, No. 5367, 1253-1256, ISSN 0036-8075

Liu, J.; Casavant, M. J.; Cox, M.; Walters, D. A.; Boul, P.; Lu, W.; Rimberg, A. J.; Smith, K. A.; Colbert, D. T. & Smalley, R. E. (1999). Controlled deposition of individual single-walled carbon nanotubes on chemically functionalized templates. *Chemical Physics Letters,* Vol. 303, No. 1-2, 125-129, ISSN 0009-2614

Lu, H.; Schöps, O.; Woggon, U. & Niemeyer, C. M. (2008). Self-assembled donor comprising quantum dots and fluorescent proteins for long-range fluorescence resonance energy transfer. *Journal of the American Chemical Society,* Vol. 130, No. 14, 4815-4827, ISSN 0002-7863

Malik, S.; Vogel, S.; Rösner, H.; Arnold, K.; Hennrich, F.; Köhler, A.-K.; Richert, C. & Kappes, M. M. (2007). Physical chemical characterization of DNA-SWNT suspensions and associated composites. *Composites Science and Technology,* Vol. 67, No. 5, 916-921, ISSN 0266-3538

Müller, K.; Malik, S. & Richert, C. (2009). Sequence-Specifically Addressable Hairpin DNA-SWCNT Complexes for Nanoconstruction. *ACS Nano,* submitted, ISSN 1936-0851

Meng, M.; Ahlborn, C.; Bauer, M.; Plietzsch, O.; Soomro, S. A.; Singh, A.; Muller, T.; Wenzel, W.; Bräse, S. & Richert, C. (2009). Two base pair duplexes suffice to build a novel material. *ChemBioChem,* Vol. 10, No. 8, 1335-1339, ISSN 1936-0851

Nakashima, N.; Tomonari, Y. & Murakami, H. (2002). Water-soluble Single-Walled Carbon Nanotubes via Noncovalent Sidewall-Functionalization with a Pyrene-Carrying Ammonium Ion. *Chemistry Letters,* No. 6, 638-639, ISSN 0366-7022

Nakashima, N.; Okuzono, S.; Murakami, H.; Nakai, T. & Yoshikawa, K. (2003). DNA Dissolves Single-walled Carbon Nanotubes in Water. *Chemistry Letters,* Vol. 32, No. 5, 456-457, ISSN 0366-7022

Niemeyer, C. (1999). Progress in "engineering up" nanotechnology devices utilizing DNA as a construction material. *Applied Physics A: Materials Science & Processing,* Vol. 68, No. 2, 119-124, ISSN 0947-8396

Nish, A.; Hwanh, J.-Y.; Doig, J. & Nicholas, R. J. (2007). Highly selective dispersion of single-walled carbon nanotubes using aromatic polymers. *Nature Nanotechnology,* Vol. 2, 640-646, ISSN 1748-3387

Nonin, S.; Phan, A. T. & Leroy, J.-L. (1997). Solution structure and base pair opening kinetics of the i-motif dimer of d(5mCCTTTACC): a noncanonical structure with possible roles in chromosome stability. *Structure,* Vol. 5, No. 9, 1231-1246, ISSN 0969-2126

Nykypanchuk, D.; Maye, M. M.; van der Lelie, D. & Gang, O. (2008). DNA-guided crystallization of colloidal nanoparticles. *Nature,* Vol. 451, No. 7178, 549-552, ISSN 0028-0836

O'Connell, M. J.; Boul, P.; Ericson, L. M.; Huffman, C.; Wang, Y.; Haroz, E.; Kuper, C.; Tour, J.; Ausman, K. D. & Smalley, R. E. (2001). Reversible water-solubilization of single-walled carbon nanotubes by polymer wrapping. *Chemical Physics Letters,* Vol. 342, No. 3-4, 265-271, ISSN 0009-2614

O'Connell, M. J.; Bachilo, S. M.; Huffman, C. B.; Moore, V. C.; Strano, M. S.; Haroz, E. H.; Rialon, K. L.; Boul, P. J.; Noon, W. H.; Kittrell, C.; Ma, J.; Hauge, R. H.; Weisman, R. B. & Smalley, R. E. (2002). Band Gap Fluorescence from Individual Single-Walled Carbon Nanotubes. *Science,* Vol. 297, No. 5581, 593-596, ISSN 0036-8075

Park, S. Y.; Lytton-Jean, A. K. R.; Lee, B.; Weigand, S.; Schatz, G. C. & Mirkin, C. A. (2008). DNA-programmable nanoparticle crystallization. *Nature,* Vol. 451, No. 7178, 553-556, ISSN 0028-0836

Peng, X.; Komatsu, N.; Bhattacharya, S.; Shimawaki, T.; Aonuma, S.; Kimura, T. & Osuka, A. (2007). Optically active single-walled carbon nanotubes. *Nature Nanotechnology,* Vol. 2, No. 6, 361-365, ISSN 1748-3387

Seeman, N. C. (2003). DNA in a material world. *Nature,* Vol. 421, No. 6921, 427-431, ISSN 0028-0836

Shigeta, M.; Komatsu, M. & Nakashima, N. (2006). Individual solubilization of single-walled carbon nanotubes using totally aromatic polyimide. *Chemical Physics Letters,* Vol. 418, No. 1-3, 115-118, ISSN 0009-2614

Shin, S. R.; Lee, C. K.; Lee, S. H.; Kim, S. I.; Spinks, G. M.; Wallace, G. G.; So, I.; Jeong, J.-H.; Kang, T. M. & Kim, S. (2009). Switchable redox activity by proton fuelled DNA nano-machines. *Chemical Communications,* No. 10, 1240-1242, ISSN 1359-7345

Star, A.; Steuerman, D. W.; Heath, J. R. & Stoddart, J. F. (2002). Starched Carbon Nanotubes. *Angewandte Chemie International Edition,* Vol. 41, No. 14, 2508-2512, ISSN 1433-7851

Strano, M. S.; Zheng, M.; Jagota, A.; Onoa, G. B.; Heller, D. A.; Barone, P. W. & Usrey, M. L. (2004). Understanding the Nature of the DNA-Assisted Separation of Single-Walled Carbon Nanotubes Using Fluorescence and Raman Spectroscopy. *Nano Letters,* Vol. 4, No. 4, 543-550, ISSN 1530-6984

Tan, Y. & Resasco, D. E. (2005). Dispersion of single-walled carbon nanotubes of narrow diameter distribution. *Journal of Physical Chemistry B,* Vol. 109, No. 30, 14454-14460, ISSN 1520-6106

Tu, X.; Manohar, S.; Jagota, A. & Zheng, M. (2009). DNA sequence motifs for structure-specific recognition and separation of carbon nanotubes. *Nature,* Vol. 460, No. 7252, 250-253, ISSN 0028-0836

Vogel, S. R.; Kappes, M. M.; Hennrich, F. & Richert, C. (2007 a). An unexpected New Optimum in the Structure Space of DNA Solubilizing Single-Walled Carbon Nanotubes. *Chemistry - A European Journal,* Vol. 13, No. 6, 1815-1820, ISSN 0947-6539

Vogel, S. R.; Müller, K.; Plutowski, U.; Kappes, M. M. & Richert, C. (2007 b). DNA-carbon nanotube interactions and nanostructuring based on DNA. *Physica Status Solidi B,* Vol. 244, No. 11, 4026-4029, ISSN 0370-1972

Wang, S.; Humphreys, E. S.; Chung, S.-Y.; Delduco, D. F.; Lustig, S. R.; Wang, H.; Parker, K. N.; Rizzo, N. W.; Subramoney, S., Chiang, Y.-M. & Jagota, A. (2003). Peptides with selective affinity for carbon nanotubes. *Nature Materials,* Vol. 2, No. 3, 196-200, ISSN 1476-1122

Yang, Q.-H; Wang, Q.; Gale, N.; Oton, C. J.; Cui, L.; Nandhakumar, I. S.; Zhu, Z.; Tang, Z.; Brown, T. & Loh, W. H. (2009). Loosening the DNA wrapping around single-walled carbon nanotubes by increasing the strand length. *Nanotechnology,* Vol. 20, No. 19, 1-5, ISSN 0957-4484

Yang, R.; Jin, J.; Chen, Y.; Shao, N.; Kang, H.; Xiao, Z.; Tang, Z.; Wu, Y.; Zhu, Z. & Tan, W. (2008 a). Carbon Nanotube-Quenched Fluorescent Oligonucleotides: Probes that Fluoresce upon Hybridization. *Journal of the American Chemical Society,* Vol. 130, No. 26, 8351-8358, ISSN 0002-7863

Yang, R.; Tang, Z.; Yan, J.; Kang, H.; Kim, Y.; Zhu, Z.; & Tan, W. (2008 b). Noncovalent Assembly of Carbon Nanotubes and Single-Stranded DNA: An Effective Sensing Platform for Probing Biomolecular Interactions. *Analytical Chemistry,* Vol. 80, No. 19, 7408-7413, ISSN 0003-2700

Zhang, L.; Zaric, S.; Tu, X.; Wang X.; Zhao, W. & Dai, H. (2008). Assessment of Chemically Separated Carbon Nanotubes for Nanoelectronics. *Journal of the American Chemical Society,* Vol. 130, No. 8, 2686-2691, ISSN 0002-7863

Zhao, W.; Gao, Y.; Brook, M. A. & Li, Y. (2006). Wrapping single-walled carbon nanotubes with long single-stranded DNA molecules produced by rolling circle amplification. *Chemical Communication,* No. 34, 3582-3584, ISSN 1359-7345

Zhao, C.; Song, Y.; Ren, J. & Qu, X. (2009). A DNA nanomachine induced by single-walled carbon nanotubes on gold surface. *Biomaterials,* Vol. 30, No. 9, 1739-1745, ISSN 0142-9612

Zheng, M.; Yagota, A.; Semke, E. D.; Diner, B. A.; McLean, R. S.; Lustig, S. R.; Richardson R. E. & Tassi, N. G. (2003 a). DNA-assisted dispersion and separation of carbon nanotubes. *Nature Materials,* Vol. 2, No. 5, 338-342, ISSN 1476-1122

Zheng, M.; Yagota, A.; Strano, M. S.; Santos, A. P.; Barone, P.; Chou, S. G.; Diner, B. A.; Dresselhaus, M. S.; McLean, R. S.; Onoa, G. B.; Samsonidze G. G.; Semke, E. D.; Usrey, M. & Walls, D. J. (2003 b). Structure-based carbon nanotube sorting by sequence-dependent DNA Assembly. *Science,* Vol. 302, No. 5650, 1545-1548, ISSN 0036-8075